McGraw-Hill
Yearbook of
Science &
Technology

1987

McGraw-Hill
Yearbook of
Science &
Technology

1987

COMPREHENSIVE COVERAGE OF
RECENT EVENTS AND RESEARCH AS
COMPILED BY THE STAFF OF THE
McGRAW-HILL ENCYCLOPEDIA OF
SCIENCE AND TECHNOLOGY

McGRAW-HILL BOOK COMPANY

New York   St. Louis   San Francisco
Auckland   Bogotá   Hamburg   Johannesburg
London   Madrid   Mexico   Milan   Montreal
New Delhi   Panama   Paris   São Paulo
Singapore   Sydney   Tokyo   Toronto

The photograph facing the title page shows part of a
quasicrystal diffraction pattern. (From D. Shechtman and
I. A. Blech, The microstructure of rapidly solidified Al₆Mn,
Met. Trans., 16A:1005-1012, 1985)

Q1
M13
1987

1234567890 DODO 8932109876

The Library of Congress has cataloged this serial
publication as follows:

McGraw-Hill yearbook of science and technology.
1962–   . New York, McGraw-Hill Book Co.

v. illus. 26 cm.
Vols. for 1962–   compiled by the staff of the
McGraw-Hill encyclopedia of science and
technology.

1. Science—Yearbooks. 2. Technology—
Yearbooks. 1. McGraw-Hill encyclopedia of
science and technology.
Q1.M13   505.8   62-12028

ISBN 0-07-046182-1
ISSN 0076-2016

TABLE OF CONTENTS

FEATURE ARTICLES

Consulting Editors

Consulting Editors (cont.)

Contributors

A list of contributors, their affiliations, and the articles they wrote will be found on pages 503–508.

The 1987 *McGraw-Hill Yearbook of Science and Technology*, continuing in the tradition of its predecessors, presents the outstanding recent achievements in science and technology. Thus it serves as an annual review and also as a supplement to the *McGraw-Hill Encyclopedia of Science and Technology*, updating the basic information in the fifth edition (1982) of the Encyclopedia.

The Yearbook contains articles reporting on those topics that were judged by the consulting editors and the editorial staff as being among the most significant recent developments. Each article is written by one or more authorities who are actively pursuing research or are specialists on the subject being discussed.

The Yearbook is organized in two independent sections. The first section includes 7 feature articles, providing comprehensive, expanded coverage of subjects that have broad current interest and possible future significance. The second section comprises 165 alphabetically arranged articles.

The *McGraw-Hill Yearbook of Science and Technology* provides librarians, students, teachers, the scientific community, and the general public with information needed to keep pace with scientific and technological progress throughout the world. The Yearbook has long served this need through the ideas and efforts of the consulting editors and the contributions of eminent international specialists.

SYBIL P. PARKER
EDITOR IN CHIEF

McGraw-Hill
Yearbook of
Science &
Technology

1987

CONTINENTAL DRILLING

ROBERT D. HATCHER, JR.

Robert D. Hatcher, Jr., is professor of geology at the University of South Carolina and editor of the "Geological Society of America Bulletin." He received his doctor's degree at the University of Tennessee in 1965. His research interests include structural geology, tectonics of mountain chains, crustal structure, and regional geophysics. He has published more than 70 papers on structural geology and tectonics of the Appalachians, and a physical geology text.

Observations in the earth sciences have largely been two-dimensional. Most unequivocal knowledge of earth materials and structures is derived from the Earth's surface or the sea floor. Vertical rock exposures are limited to the seacoasts, where cliffs rise to several hundred feet, and exposures in mountains, where 3300 to 10,000 ft (1000 to 3000 m) of vertical relief may exist. Geologic structures and rock units that plunge or are inclined into the Earth may also be projected to depths of several miles. Indirect geophysical methods of observation of the Earth's crust and mantle have provided important data which add immensely to knowledge of the Earth's interior. They include the techniques of seismic reflection and refraction; measurements of terrestrial gravity, magnetic fields, and electrical conductivity; and the recently developed technique of tomographic imaging of the Earth's interior. Still, the best first-hand means of understanding the Earth's crust at any depth is by tunneling or by drilling. The deepest mines at the present time range only about 2 to 2.5 mi (3 to 4 km) in depth. Continental crust averages more than 18 mi (30 km) in thickness, while oceanic crust is only about 3 mi (5 km) thick, beneath an average of 2.5 mi (4 km) of water (Fig. 1).

The drill has been used as a tool for both exploration and recovery of valuable minerals and hydrocarbons for more than a century. Core drilling is an accepted exploration tool in the mining and petroleum industries (Fig. 2). It has been used to calibrate geophysical methods and to test extrapolations to depth-of-surface geologic data and structure. Drilling is also the primary means by which liquid or gaseous hydrocarbons are recovered from the Earth.

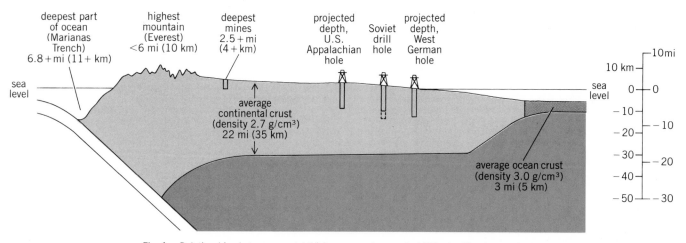

Fig. 1. Relationships between crustal thicknesses and present abilities to directly sample the deeper parts of the crust.

Any technique which produces interpretable information has its limits. The use of the drill to recover rock material from depths up to several miles provides an opportunity for interpretation of the structure and composition of the Earth's crust at greater depths than are possible through surface observation.

Scientists interested in studying the Earth's crust

Fig. 2. Shallow wireline coring rig.

and obtaining fresh samples of crustal materials have in the past largely relied upon samples gathered for industrial or engineering studies. These have been termed holes of opportunity and have provided a valuable source of information about inaccessible crustal rocks covered by alluvium, glacial deposits, or younger rocks.

In the Soviet Union a hole has been drilled and cored for scientific purposes to a depth of approximately 7 mi (12 km). The initial goal was to begin drilling in very old lower crustal rocks exposed at the surface to study their composition and origin. During the 15 years of drilling, researchers have discovered that water exists at much greater depths in the Earth than was ever anticipated. They have also produced important developments in deep drilling technology, particularly in the area of core recovery using downhole fluid-driven turbine drilling motors.

The West German government has recently committed funds to drill and core a 9-mi (15-km) hole into the deep crust during the next 5 to 10 years. Several other countries, including Japan, France, Belgium, and the United States, are attempting to begin continental scientific drilling programs which will provide information on various aspects of the structure and evolution of continental crust.

Continental Scientific Drilling Program. The need for a coordinated Continental Scientific Drilling Program became evident shortly after the establishment of the Deep Sea Drilling Project in 1968. Most of the holes drilled on the continents have been located at sites most likely to contain economic deposits of either hydrocarbons or minerals. Often these do not coincide with the places having the potential of yielding the greatest amount of information concerning the Earth's crust and its history. As a consequence, the necessity of a more broadly based Continental Scientific Drilling Program became evident early on. As a result, a workshop on continental drilling for scientific purposes was convened by the U.S. Geodynamics Committee for the

Fig. 3. Core which has just been removed from a 20-ft (6.5-m) wireline core barrel.

National Academy of Sciences near Los Alamos, New Mexico, in 1978. It resulted in the formation of the Continental Scientific Drilling Committee of the National Academy.

Once organized, the Continental Scientific Drilling Committee was structured around several focal areas or major objectives. These included basement structures and deep continental basins, thermal regimes, mineral resources, earthquakes and active faults, as well as the more technologically oriented drilling, logging, and instrumentation areas, and the curation of sample materials.

Ultimately, proposals for target problems were generated by the various topical subcommittees of the Continental Scientific Drilling Committee. In addition, geothermal regimes, hydrothermal systems, and the nature of cooling magmas were already being investigated by drilling because of the fundamental interest in both these areas by the Department of Energy (DOE). This long-term interest by DOE will continue to be very active and should generate a great deal of useful information for earth scientists.

Proposals were submitted by the various topical panels to the Continental Scientific Drilling Committee and were also solicited from the scientific community. As a result, a number of excellent possibilities for scientific study using drilling (mostly core drilling) were submitted. These include the drilling of 3-billion-year-old lower crustal rocks exposed in the Minnesota River valley and the billion-year-old rocks in the Adirondacks in New York; rift structures in the Mississippi embayment in Arkansas and the Midcontinent Rift in Kansas; the area

in Central Texas which had been the subject of a recent Consortium for Continental Reflection Profiling (COCORP) seismic reflection profile; detachment faults in the Basin and Range Province in Utah or Nevada; and active faults on the west coast, with the San Andreas Fault the prime target. Another suggestion involved drilling several mineral deposits to help understand the nature and chemical variations within large altered plutons which influence the zonation of elements in a mineral deposit. Other projects receiving considerable interest and support were continued drilling of Long Valley in California, and the Rio Grande rift area in New Mexico, as well as studies of the hydrothermal alteration and the nature of fluids and brines in the Salton Sea geothermal area.

The two projects which received the highest priorities initially by the Continental Scientific Drilling Committee as possible targets for drilling include: (1) the mineral deposit at Creede, Colorado, involving investigation of the subsurface nature and zonation of different components of the deposit through a carefully designed drilling program with one or two core holes up to 2 to 3 mi (3 to 5 km) deep; and (2) the core hole in the southern Appalachians, which would involve testing the hypothesis of large-scale thin-skinned thrusting of crystalline rocks by drilling and coring a hole through the Blue Ridge–Piedmont thrust sheet into the untransported rocks beneath. This would require a hole on the order of 4.8 to 6 mi (8 to 10 km) deep with present-day technology.

Criteria for drilling and site selection. No formal criteria have been suggested for the selection of either problems or drilling sites, but several objectives should be addressed as possible criteria for problem and site selection.

The most obvious criterion is that drilling must be necessary for the solution of the problem. If the

Fig. 4. Core from the Appalachian Piedmont. The light-colored rock is granite, the darker bands are amphibole gneiss. Knife is 4 in. (10 cm) long.

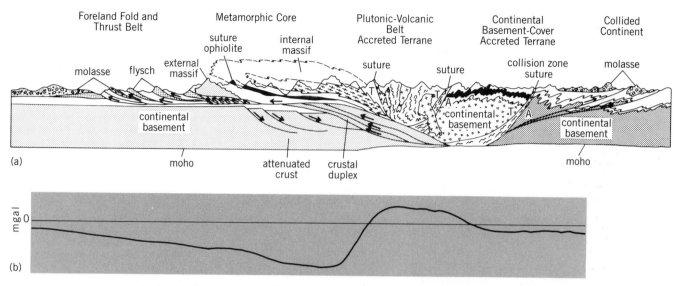

Fig. 5.   Idealized mountain chain. (*a*) Model showing properties possessed by most ancient and modern mountain chains; none possesses all of the properties. The area enclosed by broken lines is the projection of the structure above the present erosion surface. T = toward, A = away. (*b*) An ideal gravity profile across the chain.

problem can be solved effectively by other means, then there is no need to expend a great deal of effort and money to undertake a drilling project to solve it. This is even more acute where problems involving deep drilling are addressed.

A second criterion is that the target geometry should be one that will be readily hit by the drill. There is no point in drilling a hole to any depth if the target has a shape which is not likely to be encountered by drilling.

The objective of any major project undertaken as part of a Continental Scientific Drilling Program project should address fundamental principles, particularly if the hole to be drilled is deep and involves the expenditure of very large sums of money. Examples are basic problems in tectonics, problems related to fault mechanics, the chemistry of hydrothermal mineral deposits, or the mechanisms by which a magma crystallizes. Regional and local problems would also be addressed.

It would also be advisable to include intermediate-depth objectives constituting spinoff from actual drilling and completion of the deeper hole. Such objectives are not necessary for reaching ultimate target depth and would not justify initiation of the project. However, in the event that drilling encounters problems preventing the completion of the deeper-hole objective, they would provide important scientific results and would ensure a degree of success of the project.

The kinds of objectives in any drilling project might ultimately be based on the projected depth of the hole and the target to be drilled. If it is a very shallow hole and the cost relatively slight, or if it involves low-cost deepening of an existing hole of opportunity, then more local or regional problems in earth science might be considered to be a worthwhile undertaking. However, if the hole is to be deep to ultradeep, 3 to 6 mi (5 to 10 km) or more, then fundamental principles should be addressed in both the ultimate target and the intermediate objectives. Projects requiring ultradeep drill holes should probably be multidisciplinary.

Another aspect of problem and site selection involves the availability of existing data; there must be an adequate data base to define the problem and to locate a suitable site. Many areas already contain this data base.

Drilling technology. Technology is available for drilling and coring to depths of 2.5 to 3 mi (4 to 5 km). Conventional wireline drilling technology is believed to be capable of achieving depths of 3 to 5 mi (6 to 8 km), or even 6 mi (10 km), without having to be modified. However, the amount of trip time needed to retrieve a wireline core barrel (Fig. 3) becomes very large once depths of 3 to 3.5 mi (5 to 6 km) are reached. Actually, wireline core holes to depths of 3 mi (5 km) were drilled in South Africa in the late 1940s and early 1950s. Also, the mining industry in South Africa routinely drills and recovers core to 2 to 2.5 mi (3 to 4 km) by using wireline technology.

It appears that it will be possible to adapt wireline drilling technology to drilling all but the deepest core holes. The Soviet ultradeep hole uses a technique which involves a downhole turbine motor rotated by pressure on the drilling fluid. The core is collected in a revolving cylindrical assembly at the

Fig. 6.   Map showing the tectonic subdivisions of the southern Appalachians, and an enlargement showing the site study area for locating an ultradeep core hole, as well as the location of the cross section A-A' in Fig. 8. The question marks indicate uncertainty of the projected boundaries.

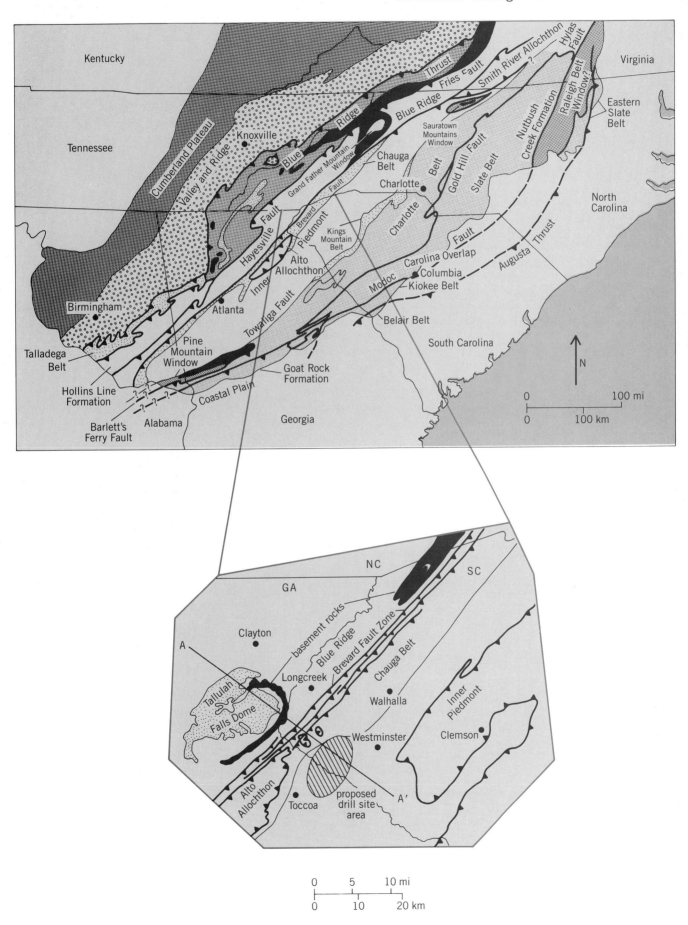

bottom and retrieved intermittently as this cylindrical device becomes filled with core.

If it were desirable only to drill and not to core a hole, drilling rigs and drilling technology presently available in the United States could be used to drill a hole 9 to 12 mi (15 to 20 km) deep. Such rigs and supporting equipment are designed to drill to great depths and to complete hydrocarbon exploration or development goals. However, drilling of ultradeep holes for multidisciplinary research will require recovery of as much sample as possible. Consequently, coring technology must be developed and modified so that it becomes possible not only to drill but also to core drill deep holes routinely to the desired depth. This technology could be readily developed within a few years prior to drilling an ultradeep hole in the United States.

Ultradeep program in the Soviet Union. Soviet scientists have begun a program of ultradeep drilling and have completed one core hole on the Kola Penninsula in Arctic Russia to a depth of more than 7 mi (12 km), with an objective depth of close to 9 mi (15 km). This in itself is quite an accomplishment. They are presently well ahead of the western countries in the development of coring technology and use of downhole motors for drilling to great depths. This hole also involves recovery of other samples and permits different kinds of downhole measurements. The objectives of this hole involve investigations of deep crust in an area where the rocks are very old and have been subjected to high-grade metamorphism allowing problems of lower crustal geology and geophysics to be addressed.

Possible targets in the United States. Several workshops held over the last several years have yielded a number of possible drilling targets within the United States. Many address major problems in tectonics, petrology, geophysics, and other disciplines of the earth sciences. The Continental Scientific Drilling Committee at its Fall 1983 meeting in Albuquerque, New Mexico, voted unanimously to select the ultradeep drill hole in the southern Appalachians as its first major undertaking as part of a continental scientific program. Other targets such as at Creede, Colorado, have also been identified as high-priority targets, along with the site at Cajon Pass, California, near the San Andreas Fault, the Salton Sea and Long Valley, California, and Valles Caldera in New Mexico.

A workshop held in May 1984 in Tarrytown, New

Fig. 7. Cross section through part of the southern Appalachians showing the large horizontal fault beneath the Foreland, Blue Ridge, Inner Piedmont, and Avalon Terrane. Note that the fault gets deeper toward the southeast. The state map shows location of the fault. CP = Cumberland Plateau; V & R = Valley and Ridge; CHB = Chauga Belt; CB = Charlotte Belt; CSB = Carolina Slate Belt; BRF = Blue Ridge Fault; MS = Murphy Syncline; HF = Hayesville Fault; BF = Brevard Fault; TF = Towaliga Fault; MF = Modoc Fault; AF = Augusta Fault.

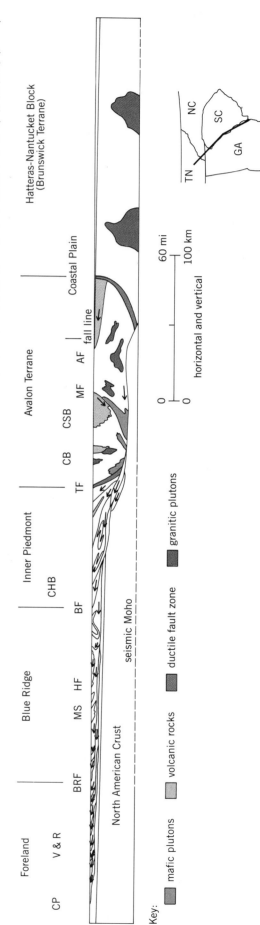

York, considered these and other projects as well as the needs for deep drilling in the United States and elsewhere. Its principal goal was to assemble an international group of scientists to discuss continental drilling. Projects described at this meeting ranged over both active and passive tectonic processes, the evolution of mountain chains, the generation of ore deposits and hydrothermal systems, and studies of the oldest rocks in the continents.

Many properties of rocks which can be measured in place require contact of the measuring device with rock materials in a drill hole. Drilling is required to make such measurements whether they are in either an active or passive setting. Several appeals have been made for drilling to measure various physical properties of materials in drill holes, because these properties change appreciably when measurements are made on core.

Many proposed studies emphasize the necessity for recovery of core (Fig. 4). Coring provides the opportunity for collection of a continuous suite of samples which should record subtle changes in physical properties, structure, and composition that take place within a body of rock. Core also provides opportunities for description of the structural, stratigraphic, and petrologic changes which may not be observable on the surface even though the rock body may be well exposed. This is considered sufficient argument to undertake drilling of shallow-to-moderate depths to sample a number of rock masses which may be only partially accessible on the surface. Commercial studies and holes drilled for hydrocarbon exploration into sedimentary basins frequently have not collected core, and as a consequence critical data are lacking.

Southern Appalachians. Many potential deep and shallow targets exist for scientific drilling. A long list of potential targets could be easily compiled that would be worthy of drilling over the next several decades. The Continental Scientific Drilling Committee of the National Academy of Sciences recommended that the hole in the southern Appalachians be drilled as the first in a national program of continental scientific drilling. In 1984 the Office of Science and Technology Policy approved a site selection study in the southern Appalachians.

The Appalachian ultradeep core hole represents an attempt to address several fundamental problems in the construction of mountain chains and the evolution of continental crust. The process of mountain building is a corollary to the overall process of plate motion, generation and destruction of oceanic crust, and the general increase through time in the total amount of continental crust. J. Tuzo Wilson described this as a cyclic process beginning with the rifting and spreading apart of two continents, and the opening of a major ocean and its closing again by subduction and ultimately by continental collision.

It is also now recognized that prior to continent-continent collision, smaller masses of lithosphere, suspect terranes, may be accreted to a continental

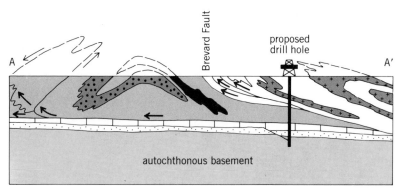

Fig. 8. Cross section A-A' showing a location of the proposed 6-mi (10-km) Appalachian ultradeep core hole. The broken lines indicate the structure above the present erosion surface. The arrows indicate related motion on faults.

margin. These are masses of rocks which are totally exotic with respect to those of the adjacent continent to which they are attached; they may be derived from offshore, another continent, another part of the same continent, or an island arc.

Mountain chains, such as the Alps, Appalachians, and Himalayas, are products of a complete Wilson cycle. However, some mountain chains, including the Andes and United States Cordillera, have never experienced continent-continent collision and still have most of the characteristics of collisional chains. Continental margin, syn- and post-orogenic sedimentation, arc volcanism, plutonism, metamorphism, and deformation are all part of the process of mountain building. Areally large, thin sheets of crystalline rocks have been thrust as much as several hundred miles over the old continental margin along which the mountain chain was being built. These are dominant structures in a number of mountain chains. Their existence has been demonstrated in the Alps, Scandinavian and British Caledonides, Himalayas, North American Cordillera, and the Appalachians. Figure 5 shows a model of an ideal mountain chain containing the features commonly observed in orogenic belts ranging from the Precambrian up through those presently being built. The Appalachian ultradeep core hole would confront several major problems in the mountain building process, but its primary focus is to drill through and test the existence of the large crystalline thrust sheet.

Certain criteria were established by which this problem was selected and the site study undertaken. The project would be undertaken in three phases: (1) a site study, probably taking 1½ to 2 years to complete; (2) an engineering phase during which the core hole would be designed and a shallow experimental hole drilled in the area to determine the engineering properties of the rocks to be encountered in the deep hole; and (3) the actual drilling phase, where a deep hole would be cored and the study completed. A fourth phase would involve long-term study in the ultradeep hole, perhaps for several decades, to measure various kinds of physical property changes which might occur in the rocks, including variations in elastic or time-dependent elasticoviscous strain and thermal flux which might occur after the hole is drilled. It may also be possible to conduct long-term seismic monitoring deep inside the crust.

The southern Appalachian hole meets such criteria for problem and site selection quite well (Fig. 6). It involves a problem which would address fundamental principles of both mountain building and continental evolution primarily by testing the hypothesis that a very large horizontal thrust fault exists beneath the Blue Ridge and Piedmont of the southern Appalachians, as indicated by both surface geologic studies and seismic reflection data (Fig. 7).

Similar structures exist in Scandinavia, the Alps, Great Britain, and the western United States, but many have been dismembered by erosion or subsequent tectonic activity. This particular site area was selected because the hypothesis can be tested in an intact thrust sheet by using largely existing technology, and drilling would take place in a rather benign environment of low temperatures in the absence of caustic fluids or gases which would damage drilling equipment. The site study area is located at the widest part of the large thrust sheet whose leading edge outcrops in Tennessee, northern Georgia, and southwest Virginia (Fig. 6) and, if the thrust is present, would be penetrated at a depth which is accessible by existing technology. The target geometry of a hor-

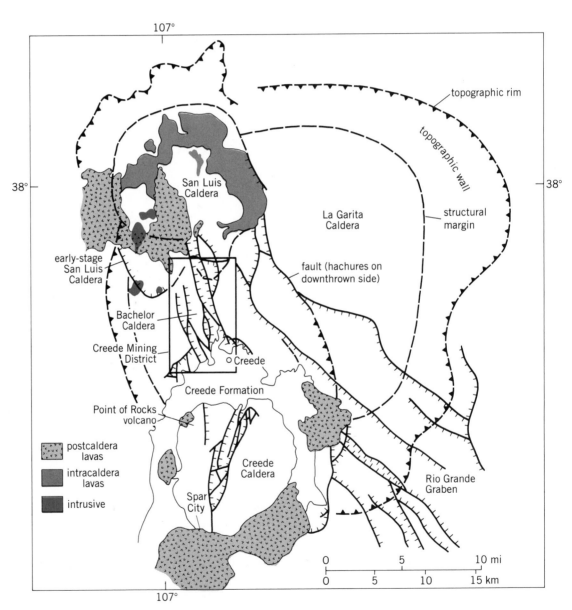

Fig. 9.   Generalized geology of the Creede and San Luis calderas in relation to remnants of the Bachelor and La Garita calderas. Control is moderate to good at solid boundaries and conjectural at broken-line boundaries. Areas denoted by patterns are specific geologic or topographic units. (*After T. A. Steven and P. W. Lipman, Geologic map of the Spar City Quadrangle, Mineral County, Colorado, USGS Map GQ-1052*)

izontal surface is an ideal target (Fig. 8). The project would also address regional and local problems as spinoff from the primary objectives. Intermediate objectives include measurement of stresses at depths in the crust where they have not been measured before, studies of fault rocks, heat flow and heat production at great depths, and opportunities to conduct geochronological, vertical gravity, magnetic, and seismic studies. Other fundamental problems to be considered will be the relation of metamorphic assemblages which outcrop on the surface to those which would be encountered in the same rocks at depths several miles below the Earth's surface, and how the pressure-temperature conditions of formation of these rocks compare. The site study area in northwestern South Carolina, northeastern Georgia, and nearby North Carolina has already been intensively studied on the surface for at least 18 years; a very large high-quality data base of the surface geology along with several diverse kinds of geophysical data already exists for this location. Except for New England, there is no data base of similar quality and quantity anywhere else in the Appalachians.

There is an advantage in drilling this hole in the southern Appalachians rather than in another country where smaller or more demonstrable examples of large thrust sheets exist which are likewise well studied. In the southern Appalachians, however, there is a very large, intact crystalline thrust sheet available for study by drilling. This particular sheet has not been dismembered tectonically or by erosion, so it is probably the best structure of its kind in North America or even in the entire world to address the problem of large-scale overthrusting of thin crystalline sheets of rock.

One concern is whether, if the hole is drilled and no sedimentary rocks occur beneath the thrust fault (with crystalline rocks against crystalline rocks) in the drill hole, it will be possible to determine that a large fault has actually been crossed. In fact, it can be determined that the fault had been encountered for several reasons: (1) The character of the rocks, even though they are crystalline, should exhibit significant differences in composition and texture on either side of the fault. (2) The rocks within the fault zone should exhibit greater deformation, whereas the rocks beneath the fault zone should be deformed less and have a character totally discordant to the deformation within the transported sheet. (3) Rocks above the fault should contain structures which are recognizable on the surface in the Blue Ridge and in the Piedmont of the Appalachians.

If no fault is encountered at all, two possibilities for interpretation exist: the fault was not drilled, and it is deeper than anticipated from all projections of surface geologic and geophysical studies; or there is no fault at all and the hypothesis would be disproved.

Creede Mining District, Colorado. The Creede Mining District, located in southwestern Colorado,

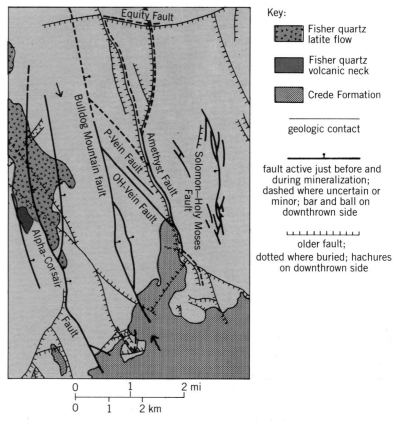

Fig. 10. Detailed geologic map showing the locations of major mineralized faults; the location of the cross section shown in Fig. 11 is indicated by the arrows. (*After T. A. Steven and G. P. Eaton, Environment of ore deposition in the Creede mining district, Colorado: I, Geologic, hydrologic and geophysical setting, Econ. Geol., 70:1023–1037, 1975*)

is of interest as a site for shallow drilling because of the opportunity to investigate the chemical and petrologic interactions between intrusive and extrusive igneous rocks, hydrothermal fluids, and the emplacement of a complex hydrothermal silver-lead-zinc-copper ore deposit (Fig. 9). The ore deposits appear to be localized along several mineralized fault zones (Fig. 10). Ore emplacement was accompanied by extensive hydrothermal alteration of the wall rocks. The relationships between vertical zonation in the veins and a speculative buried pluton that could contain copper-molybdenum mineralization as a porphyry deposit at depth would be investigated by drilling (Fig. 11). This proposed project offers the opportunity for scientific investigation of a hydrothermal system which has an extensive data base of readily available surface and industry-derived shallow borehole data. A number of geochemical and isotopic studies have also been conducted here.

Drilling would be conducted in two phases. Phase I consists of shallow coring of two holes to approximately 0.6 mi (1 km) to test the hypothesis that the sediments filling an ancient lake constituted the reservoir for most of the Creede ore fluids, as well as to provide an isotopic, mineralogic, petrologic, and

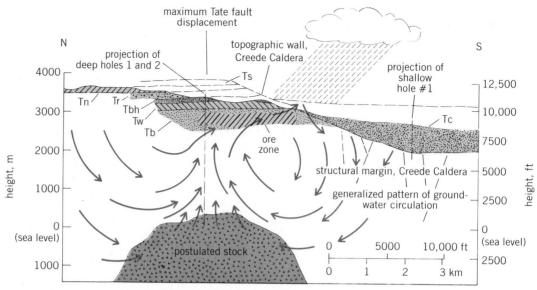

Fig. 11. Idealized north-south section through the Creede mining district. Existing rocks are patterned; rocks eroded away and restored by extrapolation, open. Tc, Creede Formation; Ts, Snowshoe Mountain Tuff; Tn, Nelson Mountain Tuff; Tr, Rat Creek Tuff; Tbh, andesite of Bristol Head; Tw, Wason Park Tuff; Tb, Bachelor Mountain Member, Carpenter Ridge Tuff. Coarse stipple in the Rat Creek Tuff and at the top of the Bachelor Mountain Member indicates soft, relatively impermeable tuff. (After T. A. Steven and G. P. Eaton, *Environment of ore deposition in the Creede mining district, Colorado: I, Geologic, hydrologic and geophysical setting, Econ. Geol., 70:1023–1037, 1975*).

hydrologic data base upon which the evolution of pore fluids and their incorporation into the Creede hydrothermal system can be systematically modeled. Phase II of the drilling program would consist of a deep, continuously cored hole 2 to 3 mi (3 to 5 km) deep into the roots of the Creede hydrothermal system. This hole is intended to establish continuity between the well-studied shallow part of the complex and the deeper heat transfer region. It should also provide material for study of the mineralogy, rock alteration, petrology, and structure at depth, and define boundary conditions for modeling the environment of the entire deposit.

Site investigations will be undertaken to locate the best places for the drillholes and provide the necessary information for interpretation of the new drill data in the context of the existing data base. A number of postdrilling scientific studies are also planned.

The Creede Mining District Project represents an opportunity to attempt to solve a complex problem related to the evolution of magmatic systems and the origin of hydrothermal ore deposits. The study would be multidisciplinary and address one or more fundamental problems in the earth sciences.

Cajon Pass, California. The Cajon Pass Project (Figs. 12–13) proposal involves deepening an existing 1.2-mi (2-km) borehole to 3 mi (5 km). The main problem to be addressed is to attempt to resolve a paradox regarding the level of shear stress along the San Andreas Fault. Lack of understanding of whether the average shear stresses that resist plate motion are on the order of 100 or 1000 atm (10 or 100 megapascals) limits understanding of the forces that move lithospheric plates, deformation at plate or transform boundaries, and the mechanics of earthquake generation. Measurements of heat flow near the San Andreas Fault suggest that there is no large component of frictional heat generated along the fault, and also that the upper limit for shear stress on the fault does not exceed 200 atm (20 MPa). Near-surface stress measurements, which are

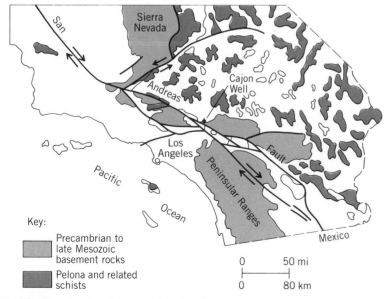

Key:

Precambrian to late Mesozoic basement rocks

Pelona and related schists

Fig. 12. Generalized geologic map of southern California showing the location of the Cajon Pass well relative to major faults in the region. Arrows represent movement along the fault.

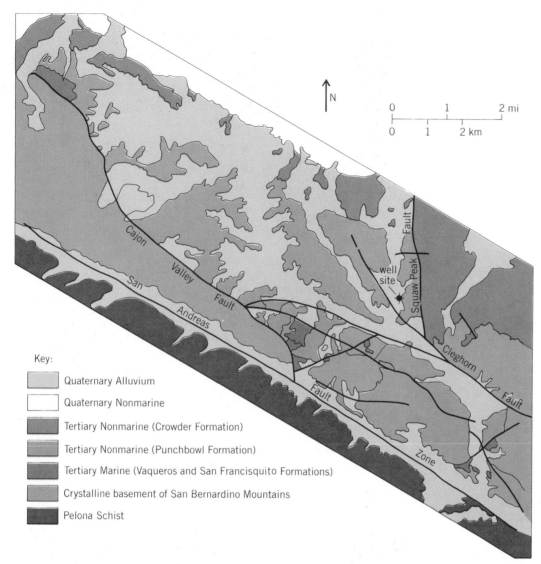

Key:

- Quaternary Alluvium
- Quaternary Nonmarine
- Tertiary Nonmarine (Crowder Formation)
- Tertiary Nonmarine (Punchbowl Formation)
- Tertiary Marine (Vaqueros and San Francisquito Formations)
- Crystalline basement of San Bernardino Mountains
- Pelona Schist

Fig. 13. Geologic map in the vicinity of the Cajon Pass well. (*After M. D. Woodburne and D. J. Golz, Stratigraphy of* *the Punchbowl Valley, Southern California, University of California Publications in Geological Science, 92:73, 1972*)

consistent with laboratory measurements, suggest that shearing stresses should increase to values greater than this at depths of 1–2 mi (2–3 km), but it is difficult to extrapolate near-surface data to these depths. Consequently, it would be very useful to be able to make stress measurements at depths of 3 mi (5 km) and to extrapolate surface-derived knowledge to these depths. Measurements of thermal gradients, hydrologic studies, and geochemical and isotopic analyses of core will help shed light upon the question of the nature of pore fluid migration near the San Andreas Fault. All of these studies have a direct bearing upon the problems of predicting earthquakes on large faults.

A wealth of supportive geologic and geophysical data already exists near the Cajon Pass site. Stress measurements have been made in the upper part of the hole; and the rock type in the hole, massive granodiorite, is well suited to study and should provide excellent drilling and measuring conditions.

Project Upper Crust. Knowledge of the Precambrian crust in the midcontinent region of the United States is derived from exposures in domal uplifts, such as the St. Francois Mountains in Missouri, the Arbuckle and Wichita Mountains in Oklahoma, the Black Hills (Wisconsin) dome, and the Rocky Mountains; from cuttings or small sections of core; from holes of opportunity sited for other purposes; and from magnetic, gravity, and seismic reflection data. Despite this random selection of sparse data, a number of important gains have been made in knowledge of the Precambrian of this region (Fig. 14). There is a southward decrease in the age of crustal rocks here, suggesting that there was a significant period of formation of new crust about 1.8 billion years ago followed by formation of an exten-

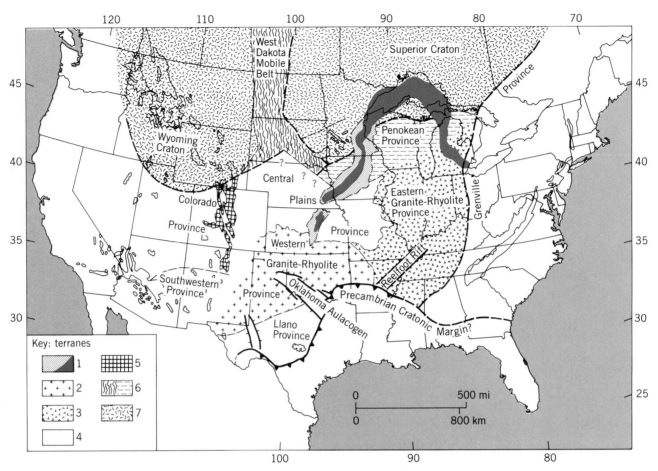

Fig. 14.   Generalized geologic map of the Precambrian basement of the continental interior showing major geologic provinces. Broken lines and question marks indicate approximate boundaries. Terrane 1 = Midcontinent Rift System; stippled = flanking sedimentary basins, black = main rift basin complex of volcanic, plutonic, and sedimentary rocks. Terrane 2 = Western Granite-Rhyolite Province. Terrane 3 = Eastern Granite-Rhyolite Province. Terrane 4 = 1.6–118 $\times 10^9$ year terranes of Colorado, Southwestern, and Central Plains provinces (undivided). Terrane 5 = exposed Precambrian bedrock of Colorado Province. Terrane 6 = principally 1.8–2.0 $\times 10^9$ year old orogenic belts of the Penokean Province (horizontal dashes) and the Western Dakotas Mobile Belt (vertical pattern). Terrane 7 = undivided rocks of the Archean cratons. (*W. R. Van Schmus, University of Kansas*)

sive belt of younger volcanic rocks about 1.3 billion years ago.

This project proposes to site several shallow holes in order to obtain core samples in critical areas and to continue the established program of isotopic dating and petrologic study of samples from these holes and other holes of opportunity. This should, over the long term, provide greater characterization of the United States midcontinent and contribute to the fundamental concepts of the evolution of continental crust.

Future of the Continental Scientific Drilling Program. The Continental Scientific Drilling Program is presently in its infancy. It has been established through a grant from the National Science Foundation to a nonprofit management group called Direct Observation and Sampling of the Earth's Continental Crust (DOSECC). This group will coordinate drilling contracts, receive proposals for drilling projects, and set priorities for scientific drilling throughout the United States. It will also be involved with the U.S. Geological Survey and the Department of Energy in drilling activities. Already DOSECC has held two workshops to discuss continental drilling and to solicit proposals for drilling projects.

The ultimate success or failure of this program will be dependent upon funding. DOSECC's goal is to achieve a funding level of at least $20 million per year. This would allow several shallow holes to be drilled per year, to conduct one or more site investigations, and to be drilling one ultradeep hole whose total cost would be on the order of several tens of millions of dollars.

[ROBERT D. HATCHER, JR.]

Bibliography: Committee on Ocean Drilling, *Options for Scientific Ocean Drilling,* 1982; Continental Scientific Drilling Committee, *Continental Scientific Drilling Program,* 1979; Continental Scientific Drilling Committee, *Mineral Resources: Research Objectives for Continental Scientific Drilling,* 1984;

Continental Scientific Drilling Committee, *Priorities for a National Program of Continental Drilling for Scientific Purposes*, 1984; Y. A. Kozlovsky, The world's deepest well, *Sci. Amer.*, 251(6):98–105, 1984; C. Petit, Tectonics on the Archean earth, *Mosaic*, 15(6):38–48, 1984; J. T. Wilson, Did the Atlantic close and re-open?, *Nature*, 211:676–681, 1966.

TAILORED MATERIALS

PETER H. ROSE

As president of Nova Associates, a division of the Eaton Corporation, Peter H. Rose is actively engaged in designing equipment for ion-beam-associated processes. Earlier, working with Robert J. Van de Graaff at the High Voltage Engineering Corporation, he helped develop the tandem accelerator. Holder of a doctor's degree in electron-nuclear scattering from the University of London, Rose has published numerous papers on atomic and accelerator physics.

All of the familiar materials are made by chemical reactions which have occurred either naturally during the cooling of the Earth or artificially as the materials were created or processed by humans. The science of chemistry lies behind these materials, and temperature and pressure are the energy sources used to bring about the chemical reactions. This science is well developed and appears to be able to create an almost infinite number of different compounds from the 90 naturally occurring atomic species. Limitations, however, do exist and can be explained most simply by the thermodynamics arguments, which imply that some reactions can be prevented from going in the desired direction at attainable combinations of temperature and pressure. There are also processing limitations which may occur during fabrication which inhibit the use of heat during the later stages of manufacturing.

The shape and size of a fabricated part usually determine its function, but frequently its performance is governed by the behavior of its surface, either untreated or coated with some film to prevent corrosion or wear. Many desirable coatings cannot be employed because of the high temperatures needed to apply them or because of the change in dimensions caused by their additional thickness. Because of the importance of such films, new techniques for the modification of the surface of a material without affecting the core structure have become of increasing interest in the past 20 years. Irradiation by ion beams, energetic electrons, or laser beams is being used to change the constitution and morphology of surfaces.

ION IMPLANTATION

Among these processes, ion implantation is perhaps the most interesting because the introduction of energetic ions into a material can create all the effects associated with electrons and electromagnetic radiation, and can cause the displacement of the target atoms during the stopping process. The ion also remains in the material, changing the chemical and atomic composition in ways that are not achievable by less energetic techniques. Ion implantation is a nonequilibrium method for introducing atoms into solids, and the method allows any ion to be implanted into almost any solid object. Reactions due to diffusion, precipitation, segregation, and solid solubility can usually be avoided. The effects of ions depend very much upon the sort of material into which they are accelerated. For example, the microstructure of the target is important, and the modifications depend on whether the material is amorphous, polycrystalline, or single-crystal.

The violent passage of ions through the target breaks apart the atoms among which it passes, so the strength of the bonds between the atoms influences the extent of the disorder left behind. These effects are superimposed on materials which themselves show an infinite variability. For example, glasses are amorphous and have strong ionic bonding, metals have complicated bonding modes and are polycrystalline, and crystalline materials like diamond and silicon have weak covalent bonding. Only some of the many changes produced are useful, but all are of scientific interest and contribute to knowledge.

The drawbacks of implantation are that a vacuum is necessary for ion beam processing and that ions move in straight lines, making the treatment of surfaces of complicated objects difficult. Although the vacuum is unavoidable, it does have the advantage of allowing the ions to be used without difficulty in combination with other vacuum film-deposition techniques. As a result, a combination of processes is available to tailor materials in ways which cannot be duplicated by other methods.

Range of applications. The low mass transfer rates of ion beams, measured in micrograms per second, have always been a conceptual difficulty when considering their use for practical material modifications. The fortunate discovery that tiny amounts of implanted material would tailor semicon-

Table 1. Material properties modified by ion implantation or ion beam mixing

Electrical-optical	Chemical	Mechanical
Creation of insulating or conductive layers*	Sensitivity to etch*	Elimination of surface porosity
Magnetic	Corrosion	Friction
Doping semiconductors*	Oxidation	Wear
Photoconductivity*	Catalysis	Hardness
Resistivity-conductivity*	Adhesion	Fatigue
Refractive index	Biocompatibility	Toughness
Reflectivity		Ductility
Dielectric constant		
Superconductivity		
Color		

*Primarily concerns semiconductors.

ductors better than any other process has led to a sizable ion implanter industry. Some 2000 machines are in use for semiconductor production, and there are a sizable number of specialized implanters which have been designed for other uses such as research or the implantation of metals. As new applications are discovered where the value added justifies the use of implantation, additional commercial uses will be found, and ions of many species are likely to find application in industry for material modification.

Except for semiconductor applications where ion implantation is already an accepted process step, the known examples of surface modification are scattered over many properties (Table 1) without any one of them having achieved significant acceptance, except in a limited way for metals. It is not surprising that applications outside the semiconductor industry are developing slowly. The technique seems exotic to potential users, and the equipment is complicated, difficult to maintain, and expensive.

Physical processes. The energy ranges of the three basic ion-material interactions—implantation, sputtering, and ion beam deposition (Fig. 1)—broadly overlap. Sputtering, or removal of material by ion bombardment, is the mechanism that restricts ion beam deposition to low energies and affects ion implantation to some extent up to very high energies. Depending on the energy range, each of these mechanisms can be used to change or create new materials. Unlike the other processess, sputtering removes rather than adds material and can be used to change surface texture.

Stopping mechanisms of implanted ions. When an energetic ion enters a solid, it loses energy by two mechanisms: loss of energy to the electron clouds surrounding a target atom; and loss of energy due to elastic collisions with the nuclei of the target atoms. Fortunately, at the energies usually considered for implantation, no inelastic nuclear collisions occur, and the only radiation observed is the low-energy photons from the decay of excited electrons from the atoms involved in the collisions. The rate of energy loss increases as the ion slows down, reaching a peak near the end of its range. The ion-electron collisions do not deflect the ion very much because the

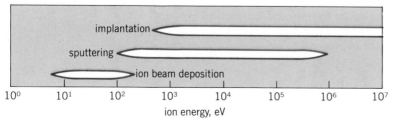

Fig. 1. Energy span of the three basic ion material interactions.

electrons are light and only bound loosely to their atoms, but the elastic nuclear collisions introduce straggling so the ions come to rest in a somewhat random manner. Ions come to rest in an approximately gaussain distribution (Fig. 2) and the sideways straggling is small, only becoming important when the implanted ions are used to implant submicrometer-sized structures, as is now increasingly the case in semiconductor manufacturing.

Damage. The ions enter the solid at velocities of the order of 10^7 ft/s (10^8–10^9 cm/s) and come to rest in about 10^{-13} s, leaving behind them an intense trail of damage. The track is enlarged by collision cascades recoiling into the surrounding medium. The collisional energy left behind appears as heat, and the track experiences temperatures as high as 10^8 K for a very short time before cooling. This intense but very localized disturbance, if repeated often enough by introducing a very high concentration of ions, eventually destroys all the original surface structure. The rate of energy loss or damage is highest just before the ion comes to rest, and as a result the damage reaches a maximum before the peak in the ion stopping distribution. As might be expected from the severity of the structural changes, damage plays a large role in determining the properties of the implanted material.

Range calculation. The depth of penetration depends upon the energy and mass of the implanted ions and on the target materials. Progress with theoretical modeling has made it possible to calculate ranges, and graphs and tables are available from which ranges can be estimated, sometimes with an accuracy of a few percent. Unfortunately, changes in local composition or crystalline effects such as channeling impair the accuracy of these calculations. For many applications, knowing the range with any accuracy is not so important, but for semiconductors it is essential to be able to predict both the damage and dose profiles within a few percent. The range of boron and arsenic in amorphous silicon is shown in Fig. 3, which illustrates the shallowness of the region that can be modified by implantation. It is even shallower for heavier target materials.

Limitations. The two most important limitations to material modification are the shallow depths of penetration of the incident ions, and sputtering which removes surface layers and restricts the maximum surface concentration. For practical accelerating energies, the range of the ions is limited at the most to a few micrometers, which is very thin compared with conventional coatings. That the effects are sometimes claimed to persist to greater than 10 times that depth is probably due to the occurrence of temperatures which lead to radiation-enhanced diffusion of the implanted material.

Sputtering. The erosion of a surface by sputtering results from collisions between the incident ion and the target atom or chemical interaction between the ion and the target. The former involves the transfer of kinetic energy to the target atoms, which can eventually result in some of them being ejected

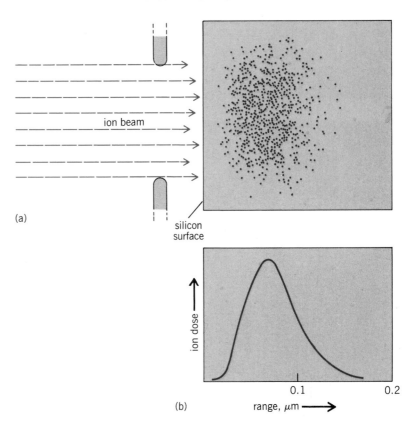

Fig. 2.   Behavior of collimated beam of 100-keV arsenic ions entering amorphous silicon. (*a*) Distribution of resting points of ions, showing the range distribution and lateral straggling. (*b*) Graph of range distribution.

through the surface of the target. Chemical reactions between the incident ions and the target can form volatile molecules, although in the case of energetic implanted ions the molecules created by chemical reactions would occur in the body of the material and be trapped. The maximum surface concentration of the implanted ion depends inversely on the sputtering coefficient and sometimes can limit the implanted ion population to only a few percent of base material. Sputtering depends strongly on the choice of target material and the energy and mass of the incident ion (Fig. 4).

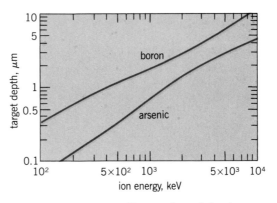

Fig. 3.   Calculated range of boron and arsenic ions in amorphous silicon.

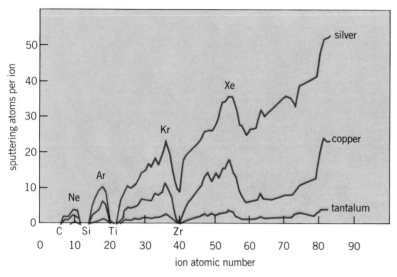

Fig. 4. Sputtering yields for various ions impacting at normal incidence on silver, copper, and tantalum at an energy of 45 keV. (*After O. Almen and G. Bruce, High energy sputtering, in L. Preuss, ed., Transactions of the 8th National Vacuum Symposium, Pergamon Press, 1962*)

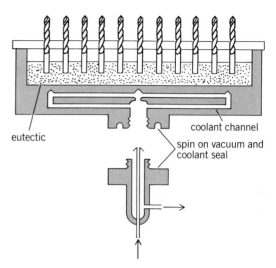

Fig. 6. Method of cooling metal objects by clamping them in a cooled tray containing a low-temperature eutectic. Arrows indicate coolant flow into and out of system.

Rates also depend very strongly on the angle of incidence, the surface morphology, and possibly the temperature of the target. Because of the sensitivity of sputtering to numerous variables, the most careful consideration must be given to each situation if the desired implantation profile is to be achieved. The problem can be clearly illustrated by considering the retained dose on the walls of a cylinder held in front of a parallel beam (Fig. 5).

Target heating. Where an implant has been discovered that shows promise as a processing application, another limitation immediately appears as the process is scaled up. When the beam intensity is increased to improve the rate of production and to make the process economic, the increased beam power causes target heating. At high temperatures the implant may fail, or bulk properties of the target, like toughness, may be spoiled. It is frequently necessary to cool the target to avoid these problems. One possible method is shown in Fig. 6, where twist drills are cooled by being inserted into a vessel containing a low-melting-point eutectic metal which is frozen around the drill stems to hold them in place and provide a heat sink during implantation.

Combination with other processes. When a high surface concentration of implanted atoms is needed, lower implant energies are used, and sputtering can severely limit the amount of new material in the surface. A way of overcoming this problem is to bombard sputter- or evaporation-deposited films with heavy ions having sufficient energy to pass through the film-substrate interface. Either the ions and the film material can be deposited simultaneously, which is called ion-beam-enhanced deposition (IBED), or the film can be bombarded with the ions after deposition, which is called ion beam mixing (IBM; Fig. 7). Key features of ion beam mixing include ion energies of about 100–400 keV; formation of nonequilibrium structures; competition between mixing and sputtering that may limit the process; and the possibility of enhanced diffusion due to high temperatures. Features of ion-beam-enhanced deposition include ion energies of about 10 eV–100 keV; formation of equilibrium structures; the availability of added energy for film growth processes; and the absence of any limit to film thickness.

There is much to be learned about the details of ion beam mixing; the role of the thermal spike accompanying the implanted atom is important but so are many other phenomena such as nuclear recoils, and most of the results reported cannot be explained in detail. One of the main features of ion beam mixing is the improved adhesion of the film to the sub-

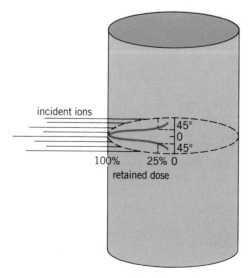

Fig. 5. The retained dose implanted around a cylinder by an ion beam normal to the cylinder axis. (*After A. B. Wittkower and J. K. Hirvonen, Some practical aspects of ion implantation for wear reduction, in J. F. Ziegler and R. L. Brown, eds., Proceedings of the 5th International Conference on Equipment and Techniques, North Holland, 1984*)

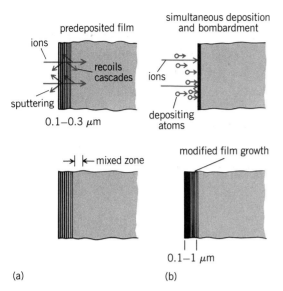

predeposited film

simultaneous deposition
and bombardment

ions

recoils
cascades

ions

sputtering

depositing
atoms

0.1–0.3 μm

→| |← mixed zone

modified film growth

0.1–1 μm

(a)

(b)

Fig. 7. Comparison of (a) ion beam mixing and (b) ion-beam-enhanced deposition. In each part of the figure, the surface during bombardment is shown above and the surface after bombardment is shown below.

Fig. 8. Ion implanter used for processing silicon wafers. (*Eaton Corp.*)

strate, which is accompanied by the interchange of film and target material. The ion that mixes can be either an inert species like argon or one which when incorporated into the material behaves in an active way. This leads to a whole range of materials, some with conventional properties but others in nonequilibrium phases. At high implant energies of approximately 100 keV or above, the bombardments at high doses tend ultimately to lead to amorphous mixtures and metastable phases, but as the energy is reduced to a few hundred electronvolts, structural growth as well as disorder reactions seem to be possible. In particular, it appears that thin crystalline layers can be grown at low temperatures on many different substrates by using low-energy ion bombardment, a result of great importance for the future development of three-dimensional semiconductors. Some combinations of ions and films that are under investigation are shown in Table 2.

Equipment. An ion implanter must have a low-pressure gas plasma to generate ions, electric and magnetic fields to control them, complicated mechanical contrivances to move the targets in and out of the vacuum system, and a computer-based control system. The machine shown in Fig. 8 is representative of implantation systems and contains the components shown schematically in Fig. 9. The machine can be divided into two parts: an ion gun (Fig.

9a) and the target chamber region (Fig. 9b or c).

The ions are generated in an ion source, usually a chamber in which a low-pressure plasma is generated. The ion source must generate a plasma that preferentially produces the ions that are required, and at the same time allows them to be pulled out or extracted from the source in a well-collimated beam. How well this is done determines the parameters (and cost) of the rest of the implanter; of the many sources available, only the source that was developed by Harwell is generally accepted by the industry.

Following the ion source, there is an analyzing magnet which separates the ions by mass to produce a clean ion beam; ions of higher or lower mass are bent less or more by the magnetic field and do not pass through the resolving aperture. If higher energies are required, the ions can be accelerated again after analysis before they enter the target chamber. Figure 9b and c shows two variations, one designed to implant silicon wafers and the other an experimental chamber for implanting metallic objects of different shapes. The latter includes a film-deposition system so that investigations of ion beam mixing

Table 2. Combinations of ions and films being investigated by ion-assisted techniques

Application	Ion beam	Evaporant species	Mode
Corrosion reduction	Ar^+, P^+	Chromium	Ion beam mixing
Adherent metal coatings	Ar^+	Aluminum, copper, gold	Ion beam mixing
Adherent hard coatings (TiN)	N^+	Titanium	Ion beam mixing, ion-beam-enhanced diffusion
Hard film growth (cubic BN)	N^+	Boron	Ion-beam-enhanced diffusion
Superconducting films	P^+, S^+	Molybdenum	Ion beam mixing

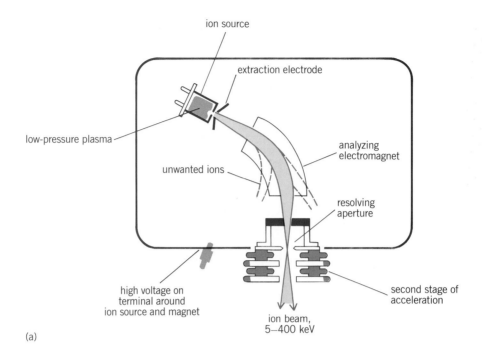

ion source

extraction electrode

low-pressure plasma

unwanted ions

analyzing
electromagnet

resolving
aperture

high voltage on
terminal around
ion source and magnet

second stage of
acceleration

ion beam,
5–400 keV

(a)

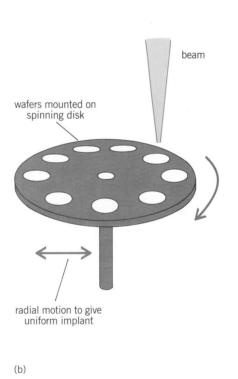

beam

wafers mounted on
spinning disk

radial motion to give
uniform implant

(b)

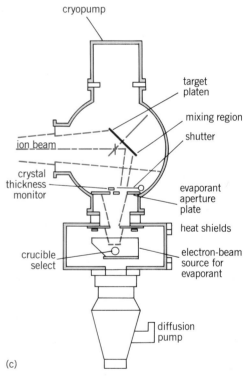

cryopump

target
platen

mixing region

shutter

ion beam

crystal
thickness
monitor

evaporant
aperture
plate

heat shields

crucible
select

electron-beam
source for
evaporant

diffusion
pump

(c)

Fig. 9. Components of ion implanter. (*a*) Ion gun. (*b*) Target chamber of the type used to process wafers. (*c*) Experimental target chamber with an electron beam evaporation system for use in beam mixing. This target chamber can also be used for implanting small production batches.

or ion-beam-enhanced deposition can be conducted.

Efforts are focused on developing machines with high reliability and throughput to make them more competitive. Figure 10 shows a laboratory-prototype very high-current (100–200-mA) implanter for making the buried silicon dioxide layers of the sort described below. This machine is already producing beam powers an order of magnitude higher than any accelerator designed for ion implantation. To cover the expanding spectrum of energy and beam current, new variations of machine are continually being introduced, the chief driving force being the new process requirements of the semiconductor industry. In addition to powerful equipment for the generation of buried oxide or nitride layers, production-rated megavolt energy systems for very deep penetration are due to be available in the very near future.

A special class of implanters, not usually associated with material modification, in which beams focused to micrometer or submicrometer diameters are employed for semiconductor mask making or repair, is also under development. Experimental machines for projecting fine patterns also exist. The interest in these machines presently lies in the realm of pattern generation or other semiconductor-related applications. Future uses may include specially patterned surfaces made to cause catalytic effects on chemical or genetic materials, but these types of applications have not been tried yet.

EXAMPLES OF TAILORED MATERIALS

Extensive research has been done on the effect of ion beams on the properties shown in Table 1. But it is only for semiconductors that enough knowledge has been gained to use the technique in an engineering sense, and even for these materials there is much to be understood, especially in the treatments to remove the radiation damage. The importance of semiconductors justified intensive research which resulted in silicon becoming the best characterized and purest material in existence. The reproducibility of silicon has made it possible to take advantage of some of the principal features of implantation, such as the flexibility made available by the choice of implant species and the precision given by choosing the energy and the dose. Other materials are seldom well enough characterized to obtain completely reproducible results unless they are an unusually pure material or better still a single crystal, as is the silicon wafer. However, many fascinating and important effects are observed with nonsemiconductor materials, and while it is not yet certain that commercial realization of ion beam modification of these materials will be economic, this research may ultimately revolutionize how materials are processed from the viewpoints of conservation and of obtaining substantially improved performance.

Optical materials. Indices of refraction, reflection, and optical absorption are all very susceptible to the changes caused by ion beam interactions. Many lenses, mirrors, filters, and other optical elements are in use in severe environmental conditions. Implantation or improved films made by ion-assisted processes can increase hardness and corrosion resistance and are known to eliminate pinhole formation in films. These process characteristics alone are of great importance, especially if they can be coupled to the other ways of modifying the materials made possible by ion implantation. For example, mutilayer filters could be made inside a material by a series of deep implantations of different ions with the advantage of more precise control and possibly more rugged optical elements.

There is no doubt that the ion-assisted techniques already essential in integrated optics will play an

Fig. 10. Prototype very high-current oxygen implanter for forming buried silicon layers. (*Eaton Corp.*)

expanding role in the future for the fabrication of optical mixers, waveguides, and other microoptical devices. These circuits are an optical equivalent to analogous electronic devices and have become very important in communication, as exemplified by the optical fibers used increasingly for signal transmission because of the large number of signals they can carry.

Ceramic materials. The highly purified fine-grain ceramics are among the most exciting new materials. Their bonding structure is particularly susceptible to near-surface modification by ion beams, and ion-induced changes in their chemical and mechanical properties have been studied. Several applications of these materials are particularly likely to benefit from ion implantation (Fig. 11) either because of the induced damage or the presence of the implanted material. For example, in situations demanding the high strength and the high temperature resistance of these composites, the brittle nature of ceramic can be a problem and may be lessened by introducing a strong compressive surface stress by ion implantation. Ceramic appears to be an excellent prosthetic material, and there is a possibility that healing and bonding to tissue could be promoted by implanting selected ions.

Metal modification. This subfield of ion beam modification has received attention second only to semiconductors, but while many exciting material modifications have been made and much new understanding has been gained, industry has been slow to accept the advantages of the implant process. This

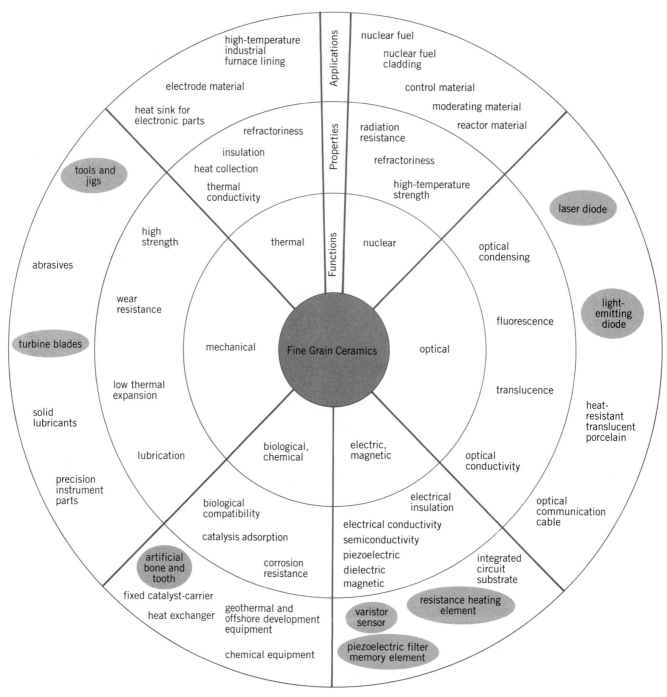

Fig. 11. Examples of the properties and application of fine-grain ceramics. Applications in which ion implantation could play a role are highlighted in color.

is partly because of the conservative nature of the metals industry and the presence of competing processes such as plasma nitriding and chemical vapor deposition, which also provide improved corrosion resistance, wear, and hardness. These processes require high temperatures and, unlike implantation, can cause distortions and dimensional changes. Implantation offers the complete range of elements to modify surface properties without the limitations imposed by chemistry, and with the advantages of low

processing temperatures, and the usually beneficial effects of the "damage" caused by the implanted ion.

The items in Table 3 are representative of those being implanted with 100-keV nitrogen ions from the Zymet implanter, the only machine designed specifically for metals. The major benefit of a nitrogen implant is believed to be the formation of a hardened, wear-resistant nitride layer. The low processing temperatures and unchanged dimensions are

Table 3. Improvements in water lifetime obtained by implantation of nitrogen ions

Application	Material	Lifetime increase, additional benefits
Scoring die for aluminum beverage can lids	D2 tool steel	3X
Forming die for aluminum beverage can bottoms	D2 steel	Lowered wear, markedly lowered material pickup
Wire guides	Hard Cr plate	3X without significant wear
Finishing rolls for copper rod	H-13 steel	Negligible wear after 3X normal lifetime; improved surface finish of product
Paper slitters	1.6% Cr, 1% C steel	2X
Synthetic rubber slitters	WC-6% Co	12X
Punches for acetate sheet	Cr-plated steel	Improved product
Taps for phenolic resin	M2 high-speed steel	Up to 5X
Thread-cutting dies	M2 high-speed steel	5X
Tool inserts	4% Ni, 1% Cr steel	Reduced tool corrosion by 3X
Forming tools	12% Cr, 2% C steel	Greatly reduced adhesive wear
Fuel injectors and metering	Tool steel	100X in engine tests
Cam followers	Steel	Improved lifetime
Plastic cutting	Diamond tools	2–4X lifetime
Tinsel rolling	Chrome steel roller	Highly successful, prevents pickup
Hip joint prostheses	Ti/6A1/4V	100X in laboratory tests
Dental drills	WC-Co	2–7X, significantly lower cutting force required, lower patient discomfort
Precision punches for electronic parts	WC-15% Co	Greater than 2X
Punch and die sets for sheet steel laminations	WC-Co	4–6X, improvement remains after resharpening
Dies for copper rod	WC-6% Co	5X throughput, improved surface finish
Dies for steel wire	WC-6% Co	3X
Deep drawing dies	WC-6% Co	Improved lifetime, markedly reduced material pickup
Sheet steel chopper blades	WC-Co	Greater than 3X, reduced chipping
Injection-molding nozzle, molds, screws, gate pads for glass and mineral, filled plastics	Tool steels and hard Cr plated steels	4–6X
Thermally nitrided steel molds	Tool steels	Combination better than either process alone
Profile hot die for plastic extrusion	P-20 tool steel	4X

important in all applications shown in Table 3. Because of the hard wear conditions associated with cutting tools, other applications with lighter load conditions, such as those found in ball races, should offer more certain advantages. Reductions in lubricated wear and improved corrosion resistance in bearings are being investigated, and so far the results are promising. Nitrogen implantation of many additional materials has been widely studied, but the results are often difficult to verify under industrial conditions. The demand for the applications of Table 3 is sufficient to support a number of service centers.

Modified semiconductor implanters are often used for experiments with other ion species. Boron implants often provide 5- to 10-fold improvement in hardness and wear, and implantations with chromium, yttrium, and tantalum ions give improved corrosion resistance.

Corrosion is a very serious industrial problem, but ion modification will play a role only where the expense of implantation is justified. This may occur where alloying alone cannot produce all the necessary properties, such as when ion implantation provides corrosion resistance to a very hard material in which the temper would be destroyed by a high-temperature process. Metals which do not have the usual corrosion-resistant oxide coating or have a coating that is too thin and can be easily penetrated by abrasion can be protected by ion implantation. The ability to make new alloys from materials not normally soluble in each other, like tantalum in iron, creates unique corrosion resistance not obtainable by other means.

Amorphous metals or metallic glasses have become of great interest because of the lack of long-range order which gives the material great hardness. These materials are usually obtained by rapid quenching from the liquid or gaseous state, but it appears that ion implantation can closely duplicate these properties when ions of the same species as the base material are implanted at saturation doses. If ions of different species are employed, new metallic glasses can be tailored quite outside the solubility limits possible with the usual quenching process. Furthermore, these materials can be formed at far lower temperatures than conventional methods which start with a liquid or gaseous phase. The amorphization of the surface layer by implantation is caused by the cumulative effect of the different features associated with the passage of fast ions in the target, such as the quenching of the material in the thermal spikes along the track of the incident particle, the accompanying radiation damage, and massive compressive stresses introduced by the presence of the implanted atom.

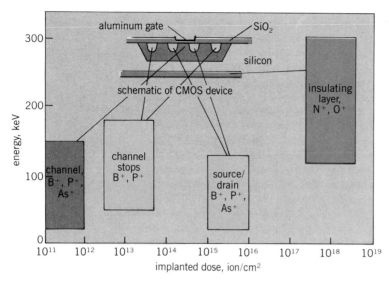

Fig. 12. Range of dose and energy required to process a typical CMOS device. The ions employed are p-type boron (B⁺) and n-type phosphorus (P⁺) and arsenic (As⁺). Oxygen ions (O⁺) and nitrogen ions (N⁺) are used at very high doses for buried insulating-layer formation.

Semiconductor materials. Tailoring semiconductors by ion implantation has largely replaced the diffusion technique, in which atoms of the dopant species surrounding the semiconductor material in a gaseous form are diffused into the silicon. This technique is slow, needs high temperatures, and is difficult to control. In contrast, implantation gives uniform and measurable doses at low temperatures and allows the doping profiles to be tailored to give the best performance by varying the energy. Although it is necessary to anneal the implant damage after an implant, the temperatures are lower, a common property of all ion modification techniques.

Many acres (hectares) of semiconductor silicon are processed each year by ion implantation. Ions of p-type or n-type at different energies and doses are embedded into silicon to make devices of all types. Figure 12 shows the large range of different applications of implantation for processing complementary metal oxide semiconductor (CMOS) circuits. The dose range is impressive, from 10^{11} ions to 10^{16} ions/cm² or five orders of magnitude; this corresponds to volume concentrations of 10^{-5} percent at the low end to 10^{-2} percent at the highest dose levels. It is very easy to vary the dose and the energy of an implanter, a flexibility of great importance to the manufacturer or circuit designer.

Some of the more novel applications to semiconductor materials will be discussed. The examples are selected on the basis that implantation may be the only way that these materials can be made.

Multilayer structures. Ions can be implanted deep into a semiconductor to produce layers of implanted atoms, and recently it has been found that buried layers of oxygen can be used to eliminate the effects produced by radiation in circuits used for satellite communication or in military systems that require

radiation hardening. When radiation passes through the comparatively thick base of silicon on which the devices are built, charges are generated which can drift back to the surface where the circuits are located. If this random charge is sufficiently large, it can cause an error. This is particularly a problem in CMOS memory circuits because they work by storing small quantities of charge, and as devices are scaled to smaller sizes the amount of charge becomes correspondingly smaller and the device becomes even more vulnerable to the radiation-induced charges. The provision of an insulating layer between the device region and the base material would greatly reduce the effect of these drifting charges. This is done at present by growing thin layers of high-quality crystalline silicon on insulating sapphire crystal. Sapphire (crystalline aluminum oxide) luckily has an atom-to-atom or lattice spacing very close to that of crystalline silicon, which makes it possible to grow good-quality silicon crystal on top of it, though because the match is not perfect there are limits to the quality of the silicon film that can be formed. In addition, the aluminum in the sapphire can diffuse into the silicon, causing device degradation.

As an alternative to this method, the new material created by ion implantation offers the promise of being less expensive and producing a higher-quality crystalline silicon layer. In the proposed method, a beam of 100–200-keV oxygen ions is implanted into silicon wafers at such large doses that a layer of silicon dioxide is formed. The implanted oxygen first reacts with the silicon, forming silicon monoxide (SiO), and then as more oxygen is implanted, silicon dioxide (SiO_2) is formed, creating an insulating barrier between the crystalline surface and the base material.

Unlike most semiconductor implantations which use small doses of ions to modify materials, this is an example of enough ions being injected to transform one material into another. Two oxygen ions must be implanted for every silicon atom in the stopping region. A good insulating layer can also be formed by injecting nitrogen ions to form silicon nitride (Si_3N_4), which has the advantage, other things being equal, that fewer nitrogen atoms are needed to form the nitride. The value of these processes is that the layer of silicon above the oxide or nitride can retain its crystallinity in spite of the intense damage caused by the transiting ions. To make this happen, it is necessary to keep the wafer at a high temperature during implantation so that annealing of the damage can take place; otherwise, the upper layer would become completely amorphous, and there would be no way of restoring the surface crystallinity afterward so that devices could be built on it.

Another way of hardening devices has been suggested which offers the possibility that it can perhaps be used after the devices have been fabricated. The proposal is to implant a buried conducting grid which traps the unwanted charge before it reaches

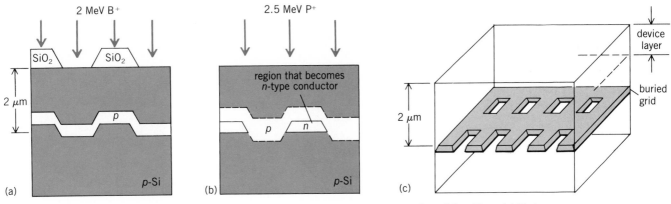

Fig. 13. Proposed process for implanting a conducting grid below the device region on the surface of the silicon. (*a*) First step. (*b*) Second step. (*c*) Final product.

the circuit, rather than using an insulating layer (Fig. 13). This is again an example of material modification rather than material transformation. Boron and phosphorus of the same range, but different energies because of their different energy loss rates, are implanted 2 micrometers deep, well below the device region. A silicon dioxide grid pattern created by lithographic techniques, shown on the surface of the wafer in Fig. 13*a*, decreases the penetration of the boron into the wafer where it obstructs the ion beam, creating regions where the *n*-type doping of the phosphorus is not canceled by the larger boron doses, and leaving an *n*-type conducting grid.

Strained layer superlattices. Superlattices are an important new class of semiconductor materials formed by layering alternating thin crystalline films of material with different atomic spacings. To maintain overall crystallinity in layers that are sandwiched between other layers of smaller atomic spacing, the superlattice must undergo a strong compressive strain which, repeated over many layers, results in a strained lattice with electronic properties different from the original materials. These superlattices are usually grown by metallorganic chemical vapor deposition. To make devices from them, they have to be doped just like other semiconductor materials, and implantation would appear to be the only way

that this can be done in these unique substances.

Other materials. The field of ion beam modification is developing rapidly, and only a few of the new materials that can be created have been discussed. There are examples of materials that can be transformed into superconductors, such as palladium bombarded by helium, and of amorphous coatings changed by ion bombardment to materials like cubic boron nitride and diamondlike carbon. Whether these ion-beam-tailored materials will gain general industrial acceptance is still an open question, but then few would have expected 10 years ago that ion beam modification would become the indispensable tool in semiconductor production that it is today.

[PETER H. ROSE]

Bibliography: Division of Material Sciences, U.S. Department of Energy, Panel report on coatings and surface modifications, *Mater. Sci. Eng.*, 70:1–89, 1985; C. M. Preece and J. K. Hirvonen (eds.), *Ion Implantation Metallurgy*, Metallurgical Society of AIME, 1980; H. Ryssel and H. Glawischnig (eds.), *Ion Implantation Techniques*, 1982; J. S. Williams and J. M. Poate (eds.), *Ion Implantation and Beam Processing*, 1984; J. F. Ziegler and R. L. Brown (eds.), *Proceedings of the 5th International Conference on Ion Implantation Equipment and Techniques*, 1985.

UNDERGROUND BIOLOGY

DAVID C. COLEMAN

David C. Coleman is research professor of entomology and ecology at the University of Georgia. His research interests are soil ecology, nutrient dynamics and organic matter turnover, and natural and managed agroecosystems. Coleman is on the editorial boards of three journals and has written over 100 scientific papers. He is a member of the NRC Committee for the Study of the Role of Alternative Farming Methods in Modern Production Agriculture (1985–1986).

THEODORE V. ST. JOHN

Theodore V. St. John is research ecologist at the Laboratory of Biomedical and Environmental Biology, University of California at Los Angeles. He received the Ph.D. degree in physiological plant ecology from the University of California, Irvine, in 1976. He has worked as a research scientist at the National Institute for Amazon Research (INPA), Manaus, Brazil, and the Natural Resource Ecology Laboratory in Fort Collins, Colorado. Author of about 30 technical publications, he is currently studying the quantitative role of mycorrhizal fungi in nature and their use in the restoration and management of damaged ecosystems.

Since the mid-1970s, ecologists have gained a new understanding of the atmosphere, plants and animals in the environment, and the soil. A major finding has been the role played by underground biological processes in the economy of nature. Because soil is a complex mixture of varied constituents, providing important physical, chemical, and biological properties, soil ecologists have used a holistic approach, considering physics, chemistry, and biology of belowground systems.

Studies of underground biology and ecology will do much to increase understanding of basic and applied aspects of humans' interactions with soils. Wise use of natural resources requires knowledge of interactions of biota, such as roots, microorganisms, and soil animals. Many important and immediate problems, such as how to preserve the critical organic matter in agricultural soils, how to reestablish vegetation after strip mining for coal or oil shale, and how to manage natural areas (such as national parks) can only be resolved with a better understanding of underground biology. The soil-related processes, so critical in these and other important problems, are the topic of this article.

Definition. Underground biology is concerned with the array of plants, animals, and microorganisms that occur at or below the soil surface. The organisms are arranged in certain spatial and temporal relationships to each other, and the physical and chemical milieu in which the organisms exist governs their activities. Because there are so many species (estimated at more than 1000 species under a person's footprint in a pine forest floor, for example), scientists often consider various functional roles played by the organisms in the soil ecosystem. Often the functions of the organisms, or the kinds of biological processes they mediate, are of more interest than the identity or life history of any individual kind of organism.

ENERGY AND NUTRIENT ELEMENTS

The energy that sustains all biological activity on Earth is captured from sunlight by photosynthesis of green plants. Two consequences of this simple but profoundly important observation summarize the essential aspects of soil ecology: (1) organic matter, remains of dead plant parts, is ultimately derived from photosynthesis and provides the energy that drives all biological processes in soil; (2) nutrient elements, without which photosynthesis cannot proceed, are the great contribution of soil organisms to all other forms of life.

Organic matter as energy source. The majority of organic matter in soil consists of plant remains in various stages of decay, and the remains of soil organisms that grew and reproduced as they consumed the live or decaying plant parts. Plant material reaches the soil from aboveground as dead leaves, stems, flowers, fruits, and woody tissues. Probably of greater significance through most of the soil volume, however, is the addition of dead roots, along with particles and chemical substances liberated by live roots during their normal growth and function. Only a small number of the fine absorbing roots growing in soil survive to thicken and become major support or extension roots. Most roots live only a few weeks, absorbing nutrients from a local portion of the soil, giving rise to branches if conditions are favorable, then dying at the end of the growing season or when the supply of nutrients in the vicinity has been exhausted. The remains of these short-lived rootlets provide much of the annual input of organic matter in most soils. Roots contribute organic materials during their active phase as well. Cell division takes place near the tip of a growing root, and a steady progression of new cells become root tissues, expand, and push the root apex forward into new soil. Just forward of the zone of cell division is the root cap, which is continuously abraded away by the soil, sloughing its cells and the associated mucilages and gels into the surrounding environment (Fig. 1). A variety of soluble sugars, amino acids, and other compounds are exuded, creating a rich medium for microbial growth in the immediate vicinity of the root. This region, the rhizosphere, is 2–4 mm (1/10 in.) thick. A large fraction of the total photosynthetic production finds its way into the soil through these pathways—perhaps as much as 80% in such ecosystems as Colorado's shortgrass steppe (Table 1).

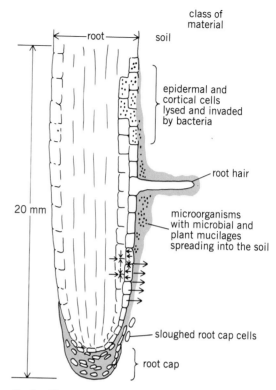

Fig. 1.   Diagram of a root cap, zone of elongation, and root hairs, which are extensions of epithelial cells. 20 mm = 0.8 in. *(After A. D. Rovira, R. C. Foster, and J. K. Martin, Origin, nature and nomenclature of the organic materials in the rhizosphere, in J. L. Harley and R. Scott Russell, eds., The Soil-Root Interface, pp. 1–4, Academic Press, 1979)*

Nutrient elements. Nutrient elements are the chemical elements needed by plants for metabolic activities, including photosynthesis. There is a long list of elements known to be required by plants, most of these only in small quantities. A shorter list includes the macronutrients—the elements used in large amounts, such as nitrogen, phosphorus, calcium, magnesium, and potassium. Iron and sulfur are sometimes included with the macronutrients since, like the others, they may sometimes be available in insufficient amounts to support vigorous plant growth.

Nutrient elements reach the soil through several pathways. The most important pathway is through the return of plant residues to the soil. Some of the stock of nutrient elements present in living tissue is translocated to other parts of the plant. This is the

Table 1. Distribution of organic matter production, g/(m²) (year), in four natural ecosystems and one agroecosystem*

Production	Deciduous forest (Tennessee)	Lightly grazed shortgrass prairie (Colorado)	Cool desert, *Atriplex* shrub complex (Utah)	Arctic tundra (Barrow, Alaska)	Irrigated alfalfa field (Minnesota)
Aboveground	700 (370)†	118	120	150	920
Belowground	900	576	270	100	560
Belowground/total	0.56 (0.71)‡	0.83	0.69	0.40	0.38

*Modified from D. C. Coleman et al., Energy flow and partitioning in selected man-managed and natural ecosystems, *Agro-Ecosystems*, 3:45–54, 1976.
†Parentheses indicate increments other than woody tissue.

case with nitrogen: as a leaf dies and turns brown, the green chlorophyll is broken down and moved to parts of the plant that continue to metabolize. Other nutrients, along with water and some organic compounds, also are moved out of the leaf. Other plant parts have similar processes; in most cases, dead plant tissues contain smaller amounts of nutrients than live tissues. Even annual plants, which die at the end of the growing season, are often poorer in nutrients than when alive because large amounts of nutrients in the tissues were translocated into the seeds as they formed. Even so, much of the original stock of nutrients remains in the dead tissue and separates from the plant as the leaf is shed. Thus return of plant parts is an important source of both soil nutrients and soil organic matter.

Even though substantial amounts of nutrient elements are returned to the soil in plant detritus, most of the existing stock was originally derived from the soil's parent material. In the original rock the nutrients are usually present in chemical forms that are unavailable to plants, but they are released steadily by moisture and chemical activity of the soil microorganisms (weathering).

Nitrogen is unusual among the nutrient elements in that it is supplied primarily from the atmosphere. Molecular nitrogen, which makes up 78% of the atmosphere, is inert and of no use to plants. Fortunately, however, several processes can transform nitrogen to ammonium, nitrate, and certain other forms that can be absorbed by plants. Where thunderstorms are frequent, chemical reactions in the presence of lightning transform significant amounts of molecular nitrogen to available forms; this nitrogen reaches the soil in rain water. Biological nitrogen fixation, the transformation of molecular nitrogen by specialized microorganisms, is an important source of nitrogen in favorable circumstances, and will be discussed below in some detail.

The nutrient elements in plant residues reach the soil in a variety of chemical forms. Some, like amino acids, can be absorbed directly by at least some kinds of plants. Many metabolic intermediates (which are present in plant residues) contain high concentrations of nutrients and are also readily available in unmodified or only slightly modified form. Such compounds are important sources of energy for microorganisms, however, and may be taken up by microorganisms before they can diffuse to plant roots. The lartest fraction of the nutrient stock in plant residues is linked chemically with resistant organic materials, or is adsorbed in ionic form in cell walls, and only becomes available to plant roots through microbial action. Frequently, these nutrients are incorporated into microbial biomass as part of the transition to plant-available form; the nature of the processes that complete the cycle are the subject of much of the rest of this article.

Heterogeneity of soil resources. Plant residues are deposited in discrete particles, most of which, on a microbiological scale, are large. This packaged deposition has the important consequence that both organic energy compounds and nutrient elements are distributed unevenly through the soil. The heterogeneity of the soil is a fundamental quality at every size scale, to which both plants and microorganisms respond.

BIOLOGY OF THE SOIL

The roots and other belowground organs (rhizomes, tubers, and so on) of higher plants are the most important of the soil organisms. They are the main source of carbon compounds that provide energy to other organisms, and they remove large amounts of nutrient elements from the soil.

While soil organisms are all "powered" by organic material, mostly from plants, the pathways by which photosynthetic energy is delivered are diverse. Some kinds of soil organisms derive energy directly from living roots. These can be classified as parasites and pathogens, which extract a living to the detriment of the plant, and symbionts, which provide beneficial effects to the plant that more than compensate for the photosynthetic energy they require.

The decomposers are species that operate on dead plant parts, at early or late stages of degradation. The decomposers are preyed upon and parasitized by a variety of other organisms, many of which, in turn, have their own predators.

Pathogenic organisms. A wide range of soil organisms cause diseases of plants. Fungi invade root tissues, often killing cells or destroying cell walls in the process of obtaining nourishment. Some kinds of fungi attack seedlings at the soil surface, causing damping-off diseases. Bacteria may damage the structure of roots; some kinds kill the plants by preventing uptake of water. Nematodes (roundworms) may insert their stylets, in the manner of an aphid, into a root and withdraw its fluid components. A variety of other groups also include pathogenic organisms, but most soil species are not able to harm plants. Pathogens tend to be specialized, often infecting only one or a few species of plant. Plants possess several kinds of defenses against pathogens, and most soil organisms that might otherwise invade roots are successfully turned away by the plant's defenses.

Symbiotic organisms. Those organisms that are not detrimental to the plant are called symbiotic. Especially important are mycorrhizae and nitrogen fixers.

Mycorrhizae. Among the soil fungi are a group of beneficial mycorrhizal fungi. When the threadlike filaments, or hyphae, of this fungus encounter the growing plant root, they penetrate the cortical cells and establish metabolic connections with the plant. The fungus extracts energy compounds (such as reduced carbon, sugars, and amino acids) from the host, and transfers nutrients back through the hyphae and into the host plant. This symbiotic exchange (or mutualism) is so beneficial to the plant, even though it requires additional energy (when compared with a nonmycorrhizal plant), that mycor-

Fig. 2. Scanning electron micrograph of *(a)* nonmycorrhizal pine root and *(b)* pine mycorrhiza colonized with ectomycorrhizal fungus. *(From J. G. Mexal, C. P. P. Reid, and E. J. Burke, Scanning electron microscopy of lodgepole pine roots, Bot. Gaz., 140:318–323, 1979)*

rhizal plants can sometimes grow as much as 30 times more rapidly than comparable nonmycorrhizal plants. Further, mycorrhizal plants may be more resistant to periods of drought and better able to repel disease organisms than nonmycorrhizal plants.

Energy costs of mycorrhizal symbionts are significant, but not overly large. Investigators working with soybeans with and without mycorrhizae found that overall photosynthesis rates for plants with mycorrhizae were 4–12% higher than nonmycorrhizal plants of equivalent dry mass, and similar phenological stage. The mycorrhizal plants took up significantly more inorganic phosphorus than the nonmycorrhizal plants.

Phylogenetically, certain of the mycorrhizae, the vesicular-arbuscular mycorrhizae (VAM), arose early in the history of land flora. Vesicular-arbuscular mycorrhizae, the so-called endomycorrhizae, are characterized by vesicles (oil-containing storage structures) and arbuscules (branching structures) inside the cortical cells of roots of several families of plants. They appeared contemporaneously with the earliest land plants, in the early-to-mid Devonian (some 400 million years ago). Presently, nearly 80% of all land plants form endomycorrhizae or VAM types.

Several families of trees and shrubs, particularly conifers, form a morphologically different association, the ectomycorrhizae (Fig. 2). Ectomycorrhizae are characterized by an elaborate structure formed outside the root, the Hartig net, which has long strands, or hyphae, extending several meters away from the root-fungus (mycorrhizal) structure (Fig. 3b). This association arose in the late Mesozoic era (about 100 million years ago), at a later time than the vesicular-arbuscular mycorrhizae.

Nitrogen-fixing microorganisms. Within certain groups of plants, including the legume family (which includes beans and alfalfa), there is another important group of symbionts. Bacteria of the genus *Rhizobium* invade the root tissue of the legume and form a nodule, a swollen structure on the root surface that houses an active population of bacteria. Nitrogen fixation takes place in the root nodules under reduced O_2, or microaerophilic, conditions. By using photosynthetic energy from the plant, and a specialized array of enzymes, nitrogenases, the *Rhizobium* bacteria can incorporate (fix) molecular nitrogen (breaking its triple covalent bonds) into organic compounds, which are then mineralized and made available to the plant. The energy demand is considerable, with 12 molecules of adenosinetriphosphate (ATP) required per molecule of N_2 fixed. This biological source of nitrogen frees legumes from a dependence on soil-borne sources, and allows them to grow vigorously where many other kinds of plants could not thrive without fertilizers.

Certain other groups of plants have symbiotic partners of a different bacterial genus, *Frankia*. These Actinomycetes (filamentous bacteria) are able to fix nitrogen and, like *Rhizobium*, occupy nodules on the root systems of their host plants. Alder trees usually harbor *Frankia* symbionts, which fix quantities of nitrogen quite comparable to those fixed by the legume-*Rhizobium* symbiosis. The amounts of nitrogen fixed per unit ground surface are quite similar for either *Rhizobia*- or *Frankia*-nodulated plants.

Decomposer organisms. The organisms that decompose organic plant residues to carbon dioxide and water, extracting energy in the process, are the primary decomposers. Most decomposers operate by releasing enzymes into their surroundings; the enzymes chemically cleave large molecules into smaller, more easily absorbed subunits. Fungi, bacteria, actinomycetes, and some protozoa function in this way. Small soil animals, including many arthropods (mites, and insects of the order Collembola), chew organic particles into finer fragments, exposing a greater surface area to enzymatic activity. Larger fauna, such as earthworms and termites, also are active in fragmentation. While fragmentation, or comminution, is in itself a physical rather than chemical process, it hastens degradation by other organisms, and the animals that cause comminution are an important part of the decomposition process.

Grazing and predatory organisms. The fauna that comminute organic materials are, in most cases, not directly utilizing the organic residues for energy. They are in fact ingesting, or "grazing" on, bacteria, fungi, and other primary decomposers. They are part of a much more extensive group of animals that have been studied by various specialists, but seldom by applied soil scientists. In addition to arthropods, the soil animals include enchytraeids (related to earthworms) and a vast array of nematodes. Some of the latter feed on bacteria, some on fungi, others on plant roots, and yet others, true predators, feed on other soil animals. Protozoa include a wide range of grazers and predators: amebas and ciliates are among the forms known to feed extensively on primary decomposers. These animals are in turn preyed upon by other large animals, such as nematodes and small arthropods, collectively known as the mesofauna.

CRITICAL BIOLOGICAL PROCESSES IN SOIL

The soil organisms that process mineral nutrients release nutrient elements as a by-product of energy acquisition for growth and reproduction. Because of the linkage between energy flow and nutrient release, the utilization of organic, or carbon, compounds by the various soil organisms is of primary importance in soil biology.

Energy flow. The dead roots, sloughed cells, and soluble organic products left in the soil by growing plants become the focus of intense and immediate biological activity. Microorganisms of many types are present in a dormant state throughout the soil; when fresh plant remains are deposited in their vicinity, the microorganisms quickly become active. They colonize the organic material and reproduce their own kind, populating the surface of the material and quickly using up the most readily available, "labile" or "soft" fractions of the material. If moisture and temperature conditions are favorable for microbial growth, there is intense competition for the labile materials. The organisms that win the competition are often those that can reproduce and colonize the material most rapidly. By simply being in place before other species can become established, the fast colonizers make it more difficult for the slower kinds to succeed. As the labile materials are exhausted, however, the fast-growing species can no longer grow and reproduce. They reenter dormancy, and slower-growing but more specialized kinds of organisms replace them. These later waves of microbial colonizers have specialized enzymatic capabilities. They are able to break down and capture energy from more refractory organic components (that is, those with high molecular weights or resistant chemical bonds).

Sugars, proteins, amino acids, and other soluble constituents of protoplasm are representative of the labile organic fractions. Cellulose and related compounds, the primary building materials of living plant tissues, are more difficult and are utilized by specialized organisms. Some plant products are very difficult to decompose such as lignin, which provides part of the cell walls in woody tissues. Also among the most resistant are some compounds that appear to defend the living plant from attack by parasites and disease organisms. Tannins, which can be abundant in many kinds of living plants, form almost irreversible chemical linkages with proteins and certain other molecules, protecting both the tannin and the protein from microbial attack. This quality of the tannins is well illustrated by their application in the leather industry. Animal hide, which consists largely of protein, is in itself readily available for microbial degradation. Once linked chemically with plant tannins, however, leather can endure many years without noticeable deterioration. Plant remains contain varying amounts of these most resistant substances, the degradation of which can be accomplished only by slow-growing, highly specialized organisms. When refractory plant products and microbial metabolites combine, they form very recalcitrant compounds, called humic acids. These may have a residence time in soil of thousands of years.

Even when the original plant materials consist primarily of easily available products, as may be true of most grass and crop roots, a portion of the original material is transformed into resistant substances. This can be brought about through the formation of degradation products by the soil microbiota. Some of the degradation products are residues of the microorganisms that used up the labile fractions. They may consist of compounds that resemble, either chemically or functionally, the most persistent plant components. Like plant compounds, some microbial products can bind proteins or other large molecules into complexes that have greater resistance to degradation than any of the parent substances. Other means of rendering organic materials unavailable include chemical binding with clay and various kinds of physical shielding.

Organic material that has been degraded to a quite inert state, and that no longer bears any resemblance to its original source material, is classified as humus. Highly degraded organic substances coat fine mineral particles, particularly in the upper layers of soil, and contribute important physical and chemical properties to the soil even though they no longer contribute significant amounts of energy of soil organisms.

Microbial immobilization. The relationship between energy flow and nutrient release is a complex one. Relatively undegraded plant residues are high in energy content and relatively low in nutrient elements. The primary decomposers require the same range of nutrient elements needed by plants, and tend to incorporate the nutrients from plant residues into their own tissues. This microbial immobilization predominates when energy sources for microorganisms in the soil are relatively more abundant than nutrient elements. The incorporation of wheat straw into agricultural soil provides a ready illustration of microbial immobilization. Plants growing in that soil may suffer deficiencies of nitrogen, and sometimes of other elements, until the readily available carbon components of the wheat straw have been used up and the nutrients previously immobilized by microorganisms are made available.

Mineralization and nutrient transformations. Since the primary decomposers tend to immobilize nutrient elements, rather than release them, as they degrade organic material, subsequent mineralization must occur, or there would be no subsequent plant growth. Some events that kill microbial populations, such as drought and freezing temperatures, may cause lysis of cells and release of nutrients. These stresses are normal parts of the natural environment, however, and when drought or cold temperatures come about gradually, the organisms are able to go into dormant resting states, and lose little of their nutrient supply to the environment. Under normal field conditions, it is probably predation on the primary decomposers that accounts for most nutrient release.

The mesofauna, like the primary decomposers upon which they feed, require both energy and nutrient elements. The concentrations of nutrient elements in the tissues of both groups are approximately equal. In metabolizing their prey, the predators must use a portion of the energy materials (often half or more) for respiration; that is, the organic compounds are lost as carbon dioxide, rather than incorporated into the mesofaunal biomass. The predators therefore use only half the nutrient content, corresponding to the amount of organic material remaining after respiration, to build their own biomass. The other half is excreted as waste products, in forms available for uptake by plants or primary decomposers.

Specialized soil organisms carry out various transformations of mineral nutrients, which influence the

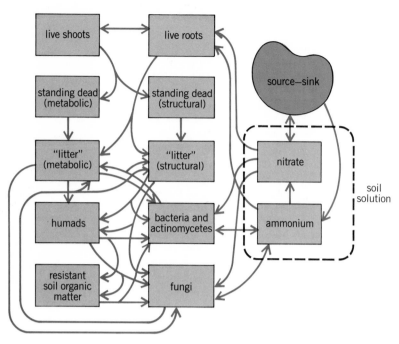

Fig. 3. Compartment diagram for nitrogen flow in soil. The majority of nitrogen flow passes into, and out of, the primary decomposers (bacteria, actinomycetes, and fungi). *(After W. B. McGill and C. V. Cole, Comparative aspects of cycling of organic C, N, S, and P through soil organic matter, Geoderma, 26:267–286, 1981)*

through metabolic action of the bacteria and fungi, they are quickly taken up and retained by the decomposers themselves. Where energy sources are abundant, as in the region behind a growing root tip, the expanding microbial population can absorb (immobilize) large amounts of nitrogen, phosphorus, and other elements which might otherwise have been available to the roots. Microorganisms (bacteria, actinomycetes, and fungi) that break down organic matter are eaten, in turn, by microbial predators, including nematodes, protozoa, and microarthropods, such as mites and collembolid insects. The predators use the energy gained from their prey, and a portion of the nutrients, but excrete much of the nutrient material as waste products such as ammonium, phosphate, and sulfate ions. The waste products are available for uptake by the roots, which completes the decomposition-mineralization cycle.

Utilization of soil volume. As the root system extends into the soil, it does not grow and branch equally in all parts of the soil volume. Most fine roots stop growing, absorb nutrients for a time, and die, to be replaced by others in other parts of the root system. Some roots, however, chance upon particularly rich microsites in the soil. A rich site may be a fertilizer band in an agricultural field, or it may be a buried deposit of decomposing organic material. Physiological processes within the plant cause the roots to proliferate in the vicinity of the favorable nutrient sources. Such root growth at favorable sites proceeds at the expense of growth elsewhere in the plant. The mature root system becomes concentrated in areas that furnish the greatest return in nutrients for the lowest energetic investment in new growth. On a microscopic scale, the same kind of

availability of the nutrient elements to plant roots. Ammonification, the transformation of ammonium to nitrate, is an example. Ammonium ion is usually the first ionic form of nitrogen released either by degradation of organic nitrogen or by nitrogen fixation. Ammonium ion can be taken up by roots but is toxic in very high concentrations. Nitrate is taken up more readily by most plants.

Nitrogen fixation. Nitrogen fixation proceeds in the presence of favorable soil conditions and when nitrogen-fixing plant-microorganism symbionts occupy the site. Nitrogen fixation can be suppressed by the presence of sufficient nitrogen already in the soil, by a lack of phosphorus or other elements, or by highly acid soil.

Uptake of mineral nutrients. Photosynthesis and its concomitant metabolic functions require several elements in addition to carbon and hydrogen. Nitrogen, phosphorus, calcium, magnesium, and potassium are all required in considerable quantities, and a variety of other elements are needed in lesser amounts. Nitrogen and phosphorus are particularly likely to be in short supply in the soil. The nutrient elements are absorbed by roots in water-soluble, ionic forms, most of which are adsorbed to mineral and organic particles in soil and diffuse rather slowly. For this reason, the supply of ions near the root surface quickly becomes exhausted. The growing root tip moves into relatively rich, unexploited soil, takes up the ions in its vicinity, and leaves a depleted cylinder of soil surrounding the root.

Bacteria, fungi, and other soil organisms require the same array of nutrient elements required by higher plants. As these elements are liberated

Table 2. Two-year cumulative C and N flows to microorganisms and litter*

	C		N	
	g·m^{-2}	%	g·m^{-2}	%
Flow to microorganisms				
Structural to bacteria	35.9	3.8	0.81	1.7
Structural to fungi	771.8	82.2	17.3	35.8
Metabolic to bacteria	31.6	3.4	8.73	18.1
Metabolic to fungi	69.8	7.4	18.9	39.2
Humads to bacteria	1.52	0.2	0.12	0.3
Humads to fungi	12.1	1.3	0.95	2.0
Resistant to bacteria	1.22	0.1	0.12	0.3
Resistant to fungi	14.7	1.6	1.39	2.9
Litter production				
Shoot death	105.9	12.3	4.68	10.0
Root death	234.8	27.2	6.92	14.7
Bacterial death	32.1	3.7	2.96	6.3
Fungal death	489.2	56.8	32.5	69.0

*After W. B. McGill et al., Phoenix, a model of the dynamics of carbon and nitrogen in grassland soils, in F. E. Clark and T. Rosswall (eds.), Terrestrial nitrogen cycles, *Ecol. Bull.* (Stockholm), 33:49–115, 1981.

process takes place with the external hyphae of symbiotic, mycorrhizal fungi. Fragments of soil organic matter, the decomposing remains of insects and other soil animals, and particles of fertilizer all provide locally rich microsites that are vigorously exploited by the fungal hyphae, while other organically poor portions of the soil go largely uncolonized.

Nutrient cycling. The varied and complex flows of nitrogen are depicted in Fig. 3. There is an impressive web of movement from litter, into and out of the primary decomposers (bacteria, actinomycetes, and fungi), and from them into the inorganic soil solution (nitrate and ammonium). The inorganic nitrogen then flows to growing microorganisms and to live roots, into shoots, and the process is repeated.

It is possible to estimate flows of nutrients in this system, which is of intermediate-level complexity. In comparing flows of both carbon and nitrogen over two calendar years (Table 2), it is apparent why the microorganisms are considered the "primary actors" in decomposition processes. The bacterial and fungal biomass, around 3 g · m^{-2} dry weight nitrogen, has annual flows into and out of it of many times that static, or "standing crop," amount.

The soil microorganisms and fauna also play a major role regulating flows of labile organic and inorganic phosphorus into and out of the soil solution (Fig. 4). The total nutrient flow is rather complex, as shown by flows into and out of stable organic phosphorus, and conversion to occluded forms (from solution phosphorus) in more weathered soils. The predominant flows involve the labile inorganic and organic phosphorus pools, however.

Because of its chemical similarity to phosphorus, sulfur behaves in a rather similar fashion to phosphorus in soil systems. As noted with phosphorus, the major role of microorganisms and associated micro- and mesofauna play a key part in mineralization of organic sulfur (stable, clay-protected, and labile organic sulfur in Fig. 5). This pathway in certain ecosystems (such as aerated grassland surface soils) accounts for up to 95% of the total sulfur mineralized throughout the year.

As with phosphorus and nitrogen, principal labile sulfur flows operate from soil solution → roots → shoots → and animal residues → various organic forms of sulfur, and then, via microbial-faunal activity into sulfate solution.

While there are numerous similarities, there are also some interesting differences in the cycling of carbon, nitrogen, sulfur, and phosphorus. Some North American soil scientists have proposed a dichotomous scheme, in which nitrogen and part of the soil sulfur are stabilized as a result of direct association with soil carbon (N—C and carbon-bonded sulfur). These entities are mineralized as a result of carbon oxidation, so-called biological mineralization, to provide energy. Organic sulfur and phosphorus exist as esters (C—O—S and

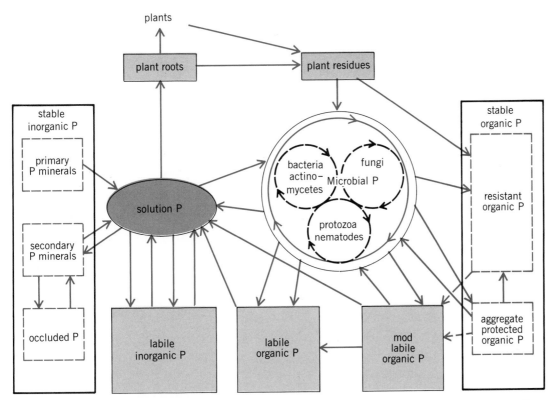

Fig. 4.   Flow chart of inorganic and organic forms of phosphorus in soil. Microbial and faunal interactions, via feeding and excretion, affect flows of inorganic and organic labile forms. *(After J. W. B. Stewart and R. B. McKercher, Phosphorus cycle, in R. G. Burns and J. H. Slater, eds., Experimental Microbial Ecology, pp. 221–238, Blackwells, 1982)*

C—O—P), and are stabilized by adsorption and precipitation with soil components. The ester-bonded sulfur and phosphorus are mineralized by enzymes in response to the need for these specific elements. The latter process, termed biochemical mineralization, operates mostly outside the cell via extracellular enzymes and is controlled by the need for the element released.

In summary, although some of the general pathways of nitrogen, phosphorus, and sulfur cycling are similar, the nature of the biology and chemistry involved differs, and appears to go across a gradient from all biological to all (or most) biochemical mineralization. The gradient thus goes from nitrogen to sulfur to phosphorus with sulfur (since it shares both types of mineralization) occupying the intermediate position.

ACTIVITY IN CONTRASTING SOILS

The physical and chemical properties of different soils result in different soil biological processes. The type of vegetation present both depends on and strongly influences soil biology.

Native vegetation soils. In stark contrast to a rich garden or agricultural soil is the soil of the great coniferous forests that cover vast tracts of North America and northern Eurasia. These soils are comparatively impoverished in mineral nutrients. Biologically important elements enter the soil from weathering and atmospheric inputs at a very slow rate, and the vegetation that covers these soils has evolved mechanisms that parsimoniously conserve the existing nutrient capital. Organic litter from the aboveground vegetation forms a layer of decaying matter on the forest floor. Mycorrhizal roots and fungal hyphae penetrate this material, tying it

together into a mat of organic matter, roots, and fungal hyphae. The decomposition and mineralization processes never occur very far from a living root or hypha, and ions freed by these processes are quickly taken up and reused by a plant.

The distinct layering of the soil and the organic material at its surface is characteristic of coniferous forests. The native plants are species that can grow with a limited intake of nutrients, and they often possess ectomycorrhizae, which appear to be advantageous in especially poor soils. Forests with surface organic layers, mats of roots at the surface, ectomycorrhizae, and numerous forest floor arthropods (both large and small) are also found in other parts of the world in the most impoverished soils. Large areas of the Amazon basin and tropical southeast Asia support forests that are of this type, even though the tropical tree species there are evolutionarily very distant from the conifers of the northern forest.

In soils under grasslands, plant litter at the surface of the soil is rapidly incorporated into the soil. The fine grass roots are produced rapidly, resulting in the development of an organic-rich upper layer of soil. The high content of organic matter lends valuable water and mineral-holding properties to such soils, and they are among the most fertile of all soils. Most such soils have been converted to agricultural use because of their high native fertility. Under intensive agriculture, however, the organic content tends to decrease.

Agricultural soils. There are some intriguing problems in agricultural land management which arise as a result of the desire to continually boost crop productivity. For example, in the vast wheat-growing areas of the Great Plains, particularly those

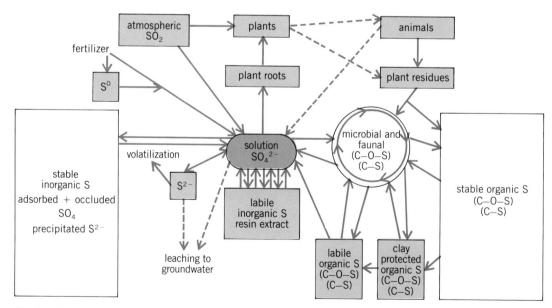

Fig. 5. Flows of principal forms of inorganic and organic sulfur in soil. As with phosphorus and nitrogen, microbial and faunal interactions influence flows of labile inorganic and organic sulfur. *(After J. W. B. Stewart, Interrelation of carbon, nitrogen, sulfur and phosphorus cycles during decomposition processes in soil, in M. J. Klug and C. A. Reddy, eds., Current Perspectives in Microbial Ecology, American Society for Microbiology, 1984)*

Table 3. Management effects on carbon and nitrogen losses in northern United States Great Plains*

Property	Depth, cm	Loss from virgin grassland, %	
		Stubble mulch	Conventional till
Carbon	0–15	27	38
	15–30	7	14
Nitrogen	0–15	23	33
	15–30	5	10

*After D. C. Coleman, C. V. Cole, and E. T. Elliott, Decomposition and nutrient cycling in agro-ecosystems, in R. Lowrance, B. R. Stinner, and G. J. House (eds.), *Agricultural Ecosystems: Unifying Concepts*, Wiley Interscience, 1984.

with annual precipitation of less than 20 in. (50 cm), there is a need to store soil water to enhance plant growth. Summer fallowing, such as by leaving alternate strips of land fallow every other year, is often used to store water for subsequent crop growth. In the process of maintaining summer fallow, the fallowed areas are tilled or weeded (with mechanical weeders or chemicals) several times per year. The end result, over a few decades, is that water is conserved, but at the expense of diminishing content of organic matter (Table 3).

Alternatives to the usual tillage practices offer some hope of reestablishing more natural and fertile soils. Among the alternatives are continuous cultivation and growth of crops every year (ensuring more frequent organic matter inputs), or another technique, which is gradually gaining acceptance, zero tillage. This technique is not without drawbacks, as certain insect pests and rodents may thrive in the surface crop residues. Application of specific insecticides or crop rotations may alleviate this undesirable side effect.

A variety of management techniques, such as reduced cultivation (minimal tillage, or stubble mulching) or no-till, have been developed to con-

Table 4. Management and contents of organic matter in soils*

Soil and location	Treatment	Organic matter, %
Chernozem Leningrad Oblast, Soviet Union	Virgin soil Old arable	4.33 4.00
Silt soil Lincolnshire, England	Grassland 100 years Arable 25 years	7.58 2.16
Alfisol Nigeria	Nontilled Tilled	4.52 3.38
Red-brown earth Australia	Pasture 30 years Wheat fallow rotation	5.30 2.08
Alfisol Nigeria	Cover crop Weed fallow	3.14 2.74

*After J. M. Tisdall and J. M. Oades, Organic matter and water-stable aggregates in soils, *J. Soil Sci.*, 33:141–163, 1982.

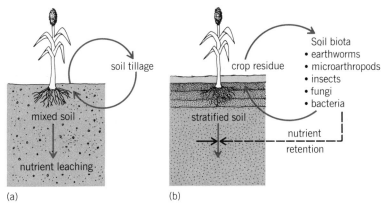

Fig. 6. Soil profiles in *(a)* conventional versus *(b)* no-tillage agroecosystems. After several years of minimal or no cultivation, there is a greater array of soil biota located near the soil surface. *(After G. J. House et al., Nitrogen cycling in conventional and no-tillage agroecosystems on the Southern Piedmont: An ecosystem analysis, J. Soil Water Conserv., 39:194–200, 1984)*

serve fossil fuel costs for plowing, disking, and other activities. An additional advantage is that minimum tillage over the course of a 25-year experimental investigation reduced losses of organic matter by about one-half, when compared with the older-style conventional tillage (Table 3).

In no-till systems, weeds are controlled by herbicides, and the layering and structure of the soil is allowed to develop as under native vegetation (Fig. 6). This permits the development of a more diverse array of soil fauna.

The historical development of no-till agriculture illustrates an intersecting, or melding, of short-term economic needs and those necessary for greater long-term agroecosystem stability. The tendency for cultivation to cause significant losses of organic matter is a worldwide problem. In some cases, well over half of the organic matter is lost due to cultivation, in either high- or low-rainfall regions of the world (Table 4).

It is hoped that no-till agriculture will permit high-yield farming to continue without the long-term soil damage (loss of organic carbon and nitrogen) that often attends conventional tillage.

[DAVID C. COLEMAN; THEODORE V. ST. JOHN]

Bibliography: R. V. Anderson, D. C. Coleman, and C. V. Cole, Effects of saprotrophic grazing on net mineralization, in F. E. Clark and T. Rosswall (eds.), Terrestrial Nitrogen Cycles, *Ecol. Bull.* (Stockholm), 33:201–216, 1981; D. C. Coleman, C. V. Cole, and E. T. Elliott, Decomposition, organic matter turnover, and nutrient dynamics in agroecosystems, in R. Lowrance, B. R. Stinner, and G. J. House (eds.), *Agricultural Ecosystems: Unifying Concepts*, 1984; D. C. Coleman, C. P. P. Reid, and C. V. Cole, Biological strategies of nutrient cycling in soil systems, *Adv. Ecol. Res.*, 13:1–55, 1983; T. V. St. John, and D. C. Coleman, The role of mycorrhizae in plant ecology, *Can. J. Bot.*, 61:1005–1014, 1983.

OFFICE AUTOMATION

PAUL WALLICH

Associate editor of the IEEE "Spectrum," Paul Wallich covers automation, robotics, computer-aided design, computer architecture, artificial intelligence, and advanced software. He is also involved in "Spectrum" coverage of government controls on technical information. Earlier, he was a writer for "SciQuest," where he worked on articles with topics ranging from high-energy physics to electronic games. He holds a B.A. in physics from Yale University.

Although the term office automation has been widely used for several years, its meaning has never been clearly defined. It has been identified with word processing, computer networks, and number crunching, and depending on the office being automated it may refer to any or all of these areas. Perhaps this uncertainty has contributed to the slow pace at which automation of offices has progressed beyond primitive beginnings such as word processing or data entry.

One problem with defining office automation is that offices themselves come in all shapes and sizes, from one-person cubbyholes combining sales, finance, engineering, and strategic planning, to enormous departments which may employ hundreds of people, all working on a single task. Expecting to automate all of them in the same way would be like expecting to use an automotive welding robot in a semiconductor assembly plant.

However, in general, an office may be defined as a place where workers process information rather than materials. Stacks of paper may flow into an office and out again, but the important thing is what is on each individual sheet; a single sentence may sometimes be of more value than the contents of an entire filing cabinet. Automating the office thus means giving office workers tools and systems to manipulate information, much as automating the factory means giving workers tools and systems to cut, weld, or stamp raw materials into a finished product.

And just as the most sophisticated pneumatic screwdrivers, robot arms, or enormous stamping presses in a factory are of little value without the assembly line to bind them together, so word-processing programs, database managers, spreadsheets, and other office tools do not constitute office automation without an overarching framework to let workers pass information back and forth efficiently, using the right tool for each stage in the information-handling process.

Almost any kind of computer system can be used for office automation, provided that it gives knowledge workers sufficient power to accomplish their tasks; however, most available systems are based either on individual processors at each workstation or

Fig. 1. Vectra personal computer, featuring modular design for application-specific configurations, IBM PC/AT compatibility with higher performance, and high-resolution text and graphics. (*Hewlett-Packard Co.*)

Fig. 2. 1165 workstation in operation. (*Xerox Corp.*)

on terminals connected to a host computer. Systems based on a single central processing unit can handle certain large jobs more efficiently than distributed-processing systems, but they may slow down excessively when too many people are working on complex tasks at the same time. Individual processors, such as personal computers (Fig. 1) or workstations (Fig. 2), run at the same speed no matter how heavily loaded the system as a whole may be.

Office automation based on networks of personal computers (Figs. 3 and 4) appears to be a significant trend, especially as the cost of connecting each personal computer to a network declines and network standards are more widely promulgated. Many companies offer networks based on the Ethernet (IEEE 802.3) standard or other, lower-cost protocols.

Data sharing. The chief reason, of course, for the network is to allow people to share data electronically, rather than having to print out reports at one location and then type them back into the computer somewhere else. Most networks have one or more file servers which act as central repositories for data, while individual workstations may also contain their own hard disks or floppy-disk drives to store information needed by only one person.

A system based on a central processor makes it easy for workers to share files, since all of them are using the same computing facilities, and data does not need to be transferred from machine to machine. Such a system often speeds up access to data, since files do not have to be transferred through a network, but can just be read in the central processor's memory. A 50-page document could take several seconds to transmit along a crowded Ethernet, and up to half a minute on the slower connections used by some vendors.

Sharing files can raise problems in certain applications, particularly databases that must be queried and updated by several people. Most systems allow only one person to access a database at a time, and some have been known to destroy files when two people attempt to input information simultaneously. Office procedures may have to be adjusted to cope with such idiosyncrasies.

Software. Much of the software that an office automation system runs is relatively simple: word processing and similar programs. Simple programs that are also included in true office automation are electronic mail and time management programs.

Time management programs. Many personal computer–based systems have no sense of time. That is, they may be able to tell the user the time and date if asked, but since they cannot perform several tasks concurrently, they cannot easily take a particular action at a given time. A simple example of this would be to ring an alarm to warn a worker of a meeting or an appointment; more sophisticated applications might include running a particular data collection program once a month to generate progress reports required by management.

Electronic mail. Another important component of the office automation system is electronic mail (Fig. 5): the ability to send messages from one worker to another (or to a group), edit them, include text from other files in them, and so on. An office can hardly be said to be automated when a memo can be edited and printed electronically but then must be copied and delivered by hand.

In a properly automated office, or in an entire automated corporation, electronic mail can be the preferred method of communication. It arrives at its destination within minutes but stays unread until its recipient is ready for it, thus combining the speed of a telephone call with the lack of interruption and permanence of a letter.

Communication. Electronic mail is, of course, most useful for communicating among far-flung groups of people, be they sales representatives on a trip or divisions of a company located in different time zones. But office automation should include other links to the world in addition to the immediate office or department. For example, workers should be able to search corporate or third-party databases for information that they need; this requires communications software and modems.

In fact, office automation functions effectively only when the people using an automated office system can communicate quickly and easily with people and automated systems in many other parts of a company. Depending on the kind of product or service an organization provides, this communication may take many forms.

For example, in a manufacturing company, the offices of the marketing department should be able to communicate quickly with the production divisions, so that they know current production rates and can give their customers accurate delivery dates, price quotes, and other information. Simultaneously, the production line should be able to query databases in marketing to find the latest information on order rates, product preferences, customer comments, and the like. And the engineering department should have access to customer service records so that it can determine what problems in a product can be fixed by small changes in the design.

A typical example of such a process occurs in most software companies, where customer-support representatives enter bug reports and requests for new features into a database which is then used by the development group to perfect and enhance the program. The development people in turn enter into the same database the comments on software updates and ways to work around bugs, so that the customer-support people will be able to provide the most up-to-date information and aid.

There are many ways that such interactions can be structured, depending on the kinds of office and factory automation systems in place; all data could be kept in a single central database, or appropriate summaries could be sent out via electronic mail. In a true office automation system, people are not the only entities that can send electronic mail; a report-writing program can also be directed to send its output to an appropriate list of people, for example.

Specialized requirements. Certain business sectors call for particular ways of implementing office

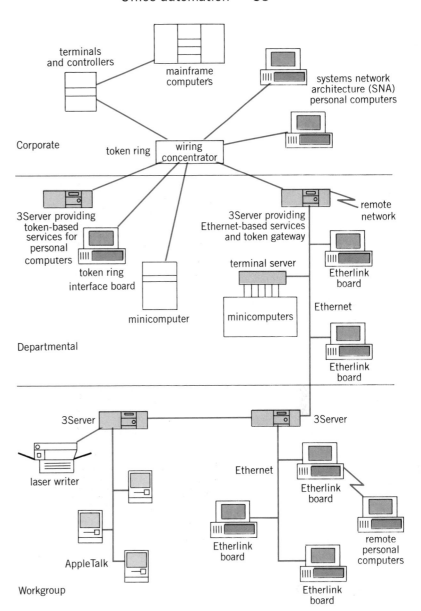

Fig. 3.   Typical corporate personal computer network. (*3Com Corp.*)

automation because of their demands for specialized kinds of information. Banking and financial services, for example, require a large current database of stock prices, interest rates, and the like, and also a complete file of all customer accounts and transactions. Every financial institution's central data-processing department is responsible for maintaining and updating such files. But now the office automation system that brokers, loan officers, and other financial workers use for more mundane tasks, such as writing reports or making projections, must also be integrated into the mainframe-based management information systems if these workers are to function most effectively.

For example, a loan officer at a bank should be able to extract pertinent data on problem loans, analyze them by using a spreadsheet program, and incorporate the resulting statistics into a report sent to the appropriate people by electronic mail. This re-

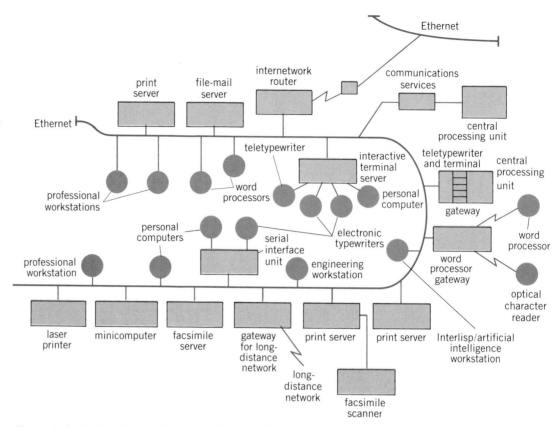

Fig. 4.   Typical portion of personal computer network in a single department of a corporation. (*Xerox Corp.*)

quires databases and spreadsheets that use the same format, or programs for converting between the two, a word processor that can accept the output of the spreadsheet, and an electronic mail program that can incorporate text files in its messages. In addition, it is useful to have graphics programs that draw charts from information contained in spreadsheets (Fig. 6). All of these requirements are relatively easy to meet, but few of them are met by most existing software.

Databases. Many businesses already have large, centralized databases, and office automation planning must work with existing resources. This requirement, however, can be a source of departmental friction as well as technical difficulties. Even after the problems of converting from one data format to another, standardizing file transfer protocols, and the like have been ironed out, policy questions will still remain. Among them are: who will maintain the centralized database; who will be responsible for adding information to it; how will data downloaded from the central database to an individual workstation be kept up to date; and who will be responsible for the security of information in the central database and in downloaded versions of the database?

Differences from data processing. While a company's data-processing department might seem the logical center of power for computer decisions, there are also valid reasons why a data-processing department should not be entrusted with decisions about what kinds of machines to put on people's desks.

Factors that end users value highly, such as ease of use and interactivity, may be less valued by technically adept people, and factors that appear technically sound, such as computational throughput and gross cost per station, may be less important in an office environment.

One primary difference between data processing and office automation is provided by measures of machine utilization and loading. Data-processing departments typically use their mainframes as close to 100% of capacity as possible, so as to maximize cost effectiveness. If a huge processor sat idle most of the day, it would be considered wasteful. If a workstation sits idle most of the day, it may still be considered cost-effective if it generates significant savings when used. Similarly, if certain tasks take hours or more to execute on a mainframe, the situation may still be acceptable so long as total throughput is adequate. To require workers at a desk to desist from useful work because their workstation is busy, on the other hand, is generally not acceptable. This underscores the dichotomy: while the object in data-processing use of computers is to achieve maximum utilization of computing resources, the object in office automation uses of computers is to achieve maximum utilization of human resources.

Making computers serve people, then, requires significantly more computing power than might be thought at first glance: if people worked steadily, there would be no problem, but the computer must

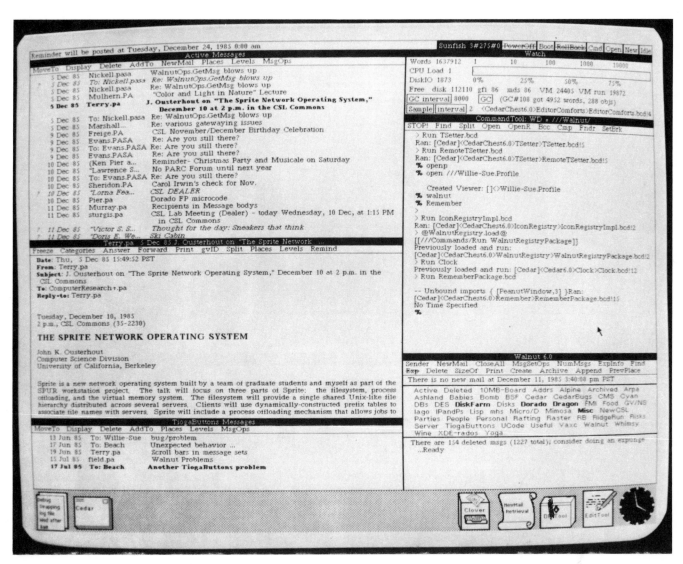

Fig. 5. Typical display of Walnut electronic mail system, entering a message into its reminder system. (*Xerox Corp.; Sun Microsystems*)

alternate between periods of inaction, when the user is thinking of what task to assign next, and frenetic activity, when the user has decided what to do and wants the results immediately.

Limits on computing power. If users do any kind of significant computing in addition to typical document-preparation tasks, then the demand for computing power goes up again, occasionally to levels that cannot be met by any personal computer now offered. In an engineering office, a user can ask the desktop machine to calculate the stresses in a skyscraper under wind loading, or the transient reponse of a 10,000-transistor integrated circuit, but even in more mundane areas it takes only a short time to specify analyses on a microcomputer that would be beyond the capabilities of even a mainframe. This is especially true when the microcomputer is attempting to deal with a significant portion of a mainframe database. For example, a detailed financial model of a division and its sales growth might occupy a spreadsheet only a few dozen columns wide by a few

hundred rows deep, but a request to check it against the last 2 or 3 years of monthly data could take hours or more, which would be impracticable. In the last analysis, cost or availability limits the computing power available to end users of office systems, but ways can generally be found to take advantage of whatever power is available.

PROBLEMS

Obviously, office automation provides many benefits, such as increased efficiency, the ability to react faster to changes in business conditions or policies, and a tendency to better, higher-quality work due both to machine checking and to the ease of revision. However, it has disadvantages as well.

Non-office automation systems. One of the most pressing problems is not with office automation as a concept, or with systems that help implement an automated office, but with the fact that the term office automation has become a "buzzword" that vendors have seized upon to differentiate their products.

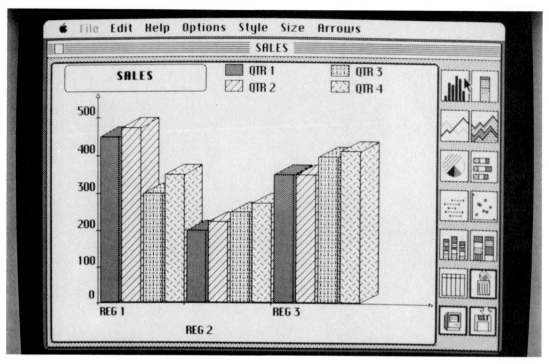

Fig. 6. Typical display of Ensemble integrated software system with charts drawn from information in spreadsheets. The system has 10 standard charts which can be customized, and a built-in word processor. (*Hayden Software Co.*)

Many of the systems being sold as office automation are nothing of the kind: word-processing systems, collections of stand-alone personal computers, or central processors with a few terminals and inadequate software.

Word processing. The selling of word processing as office automation is perhaps the most common deception. While many office workers spend much of their time producing various reports and texts, a system that does only word processing is flawed because it does not address the processes that must go on in the office to produce the text.

Without electronic mail and database access, for example, the information required to make good decisions may be lacking. Without spreadsheet software and other analytical tools, office workers must resort to calculators and pencils to produce the numbers which they can then type up in automatically tabulated formats. Without real-time functions, such as alarms or appointment reminders, someone in the supposedly automated office will still have to call people into meetings, inform them of telephone calls, and so forth.

Stand-alone computers. If word-processing is not enough for automated offices, stand-alone personal computers might seem to be adequate. They can run whatever software is purchased for them. Any individual who needs to access a database can buy a modem and communication software, and even rudimentary real-time capabilities can be grafted onto some personal computers. Some of the desk-accessory programs, for example, permit the personal computer users to set alarms at specific times and dates.

But without communication between personal computers, there is no office automation. Without a network, people can exchange information only by shuttling floppy disks from machine to machine, and true central repositories of information, such as customer or vendor lists, are impossible. With stand-alone personal computers, information is tied to the person who originally generated it, completely defeating the attempt to automate information flow within an office. If workers start to share information by making copies of a single database or report, then each of them will soon have a slightly different version, and collating them will become virtually impossible.

Even though stand-alone personal computer software packages can perform almost any desired function, it is often difficult to make them work together and to transfer information from a database into a spreadsheet or to incorporate it in a report. While so-called integrated software packages go part of the way toward solving these problems, they are usually only good at one or two tasks, and often do not address the problems of sharing information at all.

Partial automation. Partial automation of an office can be as bad as no automation at all; in some cases it can be worse. If only a few people have computer equipment in an office, strange bottlenecks may be created. This is not true when functional areas, such as word processing or order entry, are automated, but that is also not office automation.

An obvious example of such a bottleneck arises if an unautomated person in an automated department wants to send a memo to some colleagues. With a workstation, the memo could be distributed electronically, but instead it must be composed by hand, entered into the system for editing and filing, printed out, and distributed by hand.

In many automated offices, traditional secretarial functions are performed by the people who create information. There is no need to draft a report, have someone type it, make written corrections, and so on, when it is possible to come out with a completed document in the same time it would have taken for the first draft alone. This frees secretaries' time, allowing them to become more like administrative assistants, but only if all the people in an office have equal access to automated tools; experience shows that a single unautomated person in an office can generate more work than a single secretary can handle.

If only lower-level employees in an office use workstations while managers abstain, then imbalances can also result, since the managers will not have as rapid access to information as their subordinates. If electronic mail is used to exchange information, reach a consensus, and make decisions, then managers who do not read and send mail may be left out of the process entirely. In this fashion, an automated office can run away from the unautomated manager. One way to put a stop to such an out-of-control situation is to restrict the use of electronic mail and other automated aids, so that nonautomated workers can keep up, but that solution obviates most of the benefits of office automation.

Managers can also lose touch if they are not proficient in using computer-based tools. If some managers must ask for help in formatting a memo or retrieving information from a database, they may lose face, but will eventually get the job done. On the other hand, if a manager must rely on a subordinate to prepare projections that will determine the choice between competing policy alternatives, that manager may be abdicating his or her role as decision maker, not realizing that the data can be manipulated to show whatever an interested party desires. The deception need not be deliberate; a "natural" choice of assumptions in building a model will lead to the "obvious" result with unimpeachable logic, and the power of the computer can be invoked to defend the decision.

Design of procedures. Once management has made the decision to install an office automation system, the first question is what will be automated. One choice is to build a perfect electronic copy of the existing office systems and procedures. Unfortunately, this will likely perpetuate idiosyncrasies whose utililty no one has examined in years, if at all, and lead to ludicrous anachronisms. For example, the fact that three different copies of paper invoices were always filed in three different locations for easy access does not mean that electronic invoices should be similarly replicated. Once the invoices are stored, they are only a few keystrokes away from any terminal.

Need for flexibility. Another choice is to imagine an idealized flow of paper and information, and then to build a system that rigidly enforces it. That pro-

cedure, too, creates problems: people make mistakes or are forced to work with incomplete information. For example, if a single part number is missing on an electronic order form, that should not prevent the rest of it from being processed, just as a worker might scrawl "hold this item" on a paper form. So-called "perfect" systems often overlook the reasons behind the way things are really done.

In one case, software developers designing an order-processing module proposed to eliminate the job of a person on the loading dock who logged all outgoing shipments, reasoning that the computer could generate the log automatically. Only after the beginnings of a disaster did they realize the purpose of the loading-dock log was to ensure that someone read all the waybills for outgoing shipments. The job was handled by an experienced employee who could spot errors in addresses, inconsistent orders, and other mistakes. Thus, what appeared to be a rote copying job was actually quality control.

The sensible choice is to implement a flexible system, one that can adapt to the needs of its users as they find the best ways to use it. It is all very well to have a memo go to a dozen people at the touch of a button, but not if changing the recipients requires a systems analyst. It is better to sacrifice some execution speed to gain flexibility.

Changes in structures and roles. Redesigning office procedures to make effective use of automated tools may call into question traditional structures and roles. As noted above, many people who had previously eschewed typing will be called upon to do it, and many of those whose primary task was typing may find their roles extended to all sorts of facilitation. As office work is redefined, so are the structures that go with it, with results that have yet to be fully understood.

Productivity control. In offices where information is generated with production-line regularity, automation brings the potential for added control and drudgery. Keying data from printed forms into a database is one example of such make-work. Unfortunately, since this kind of work can also be quantified more easily than more knowledge-intensive tasks, it is more susceptible to automation, and also to automated control: some software packages include provisions for counting forms processed, keystrokes typed, or other simpleminded measures of productivity. Workers who labor under such systems generally agree that they make an unpleasant job even more unpleasant, and they may retaliate by rapid-fire keying-in of gibberish to raise productivity figures.

Information control. Even in the more conventional office, many issues of control are opened by office automation. Electronic mail, for example, is a very good thing for communication, but there is the issue of whether restrictions should be placed on who can send what mail to whom. The executive officer of a company, for example, might not enjoy logging on and receiving messages from disgruntled employees. Middle managers, on the other hand, might be disconcerted by the speed with which they can be bombarded by directives from the top.

There is also the issue of who should read electronic mail. Several cases have been made public where personal electronic correspondence fell into the wrong hands, but some degree of flexibility is required if people are to communicate quickly and simply with each other.

The issue of who should read electronic mail is simply a subset of the issue of who should read whose files. In a paper office, documents are kept in file cabinets or desk drawers, both of which can be locked. People generally do not rummage through each other's desks, but the corresponding operation is much easier and less apparently invasive in the automated office.

If people are not allowed to read each other's files (a prohibition it would be senseless to make absolute), then the question of what people should be allowed to keep in their files may also arise. In some cases, employees have been disciplined for keeping address lists or other personal information in office computers. If the personal computer is the analog of the desk, then this is equivalent to disciplining someone for keeping a personal address book in the top drawer, something that few employers would make policy. It is not immediately clear where to draw the lines, and office workers probably have to be relied on to exercise their discretion until such time as a more comprehensive set of computer ethics is developed. A good model might be the Unix system's default protection scheme, where people are allowed to read each other's files unless declared private, and to execute each other's programs, but not to alter or delete the property of others.

Personal use. The Unix system also sets another interesting precedent: it comes with games. The implication here is that a certain amount of personal use of office computer systems is probably in the best interests of all concerned.

Transition. The transition to the automated office should be given at least as much consideration as its design. Among the problems to be considered in implementing an office automation system are system design, training, service, and security.

Design process. The process of designing the procedures for an automated office may be as important as the procedures themselves. Generally, the people who are doing the work have the best idea of how it is done or how it should be done. Their knowledge should be tapped for political as well as technical reasons, since even the most technically perfect system requires that people use it to be effective.

Training. Training goes hand in hand with the design process as a critical factor in implementing a system that people are expected to use. Courses taught either in house or by vendors must take into account schedules, needs, and personalities. Lecturing on the intricacies of a footnoting package to people who will mostly be writing memos is unnecessary, for example. Another problem, especially in vendor training, is the tendency to treat everyone in an office, from secretaries to executives, as if they had equal comprehension levels or were completely absorptive. The ideal training program for office automation should emphasize teaching each person how to do enough to accomplish the tasks that need doing, and leaving more advanced features for later training or to be acquired on the job.

Service. Once a system is installed, of course, service is paramount: the system cannot be used if it is not working. While a large collection of sophisticated electronic components is likely to have a few that fail, the demands on office equipment are not that great, and it should not fail often. If it does, it must be repaired quickly. If equipment is purchased from a single vendor, that vendor must be responsible for service; if a system is put together from components, a service contract from an independent organization might be in order. Although these observations are truisms, they are all too often neglected.

Security. One issue that may not seem important at first glance is security: the protection of information in an automated office system from viewing, alteration, or deletion by unauthorized persons inside or outside the office. Access from outside is easily prevented by eliminating dial-up phone lines, although at some cost in convenience to those who would work at home or on the road. Within the office, it must be possible to protect certain documents, or even certain areas of storage, with access codes; otherwise, sensitive information, such as salary records, must be kept manually. It may also be desirable for individual employees to be able to protect parts of their personal files such as reports that have not yet been put in final form, or memos that are to go only to a single party.

CURRENT STATUS

True office automation is still at an early stage. Many vendors offer glorified word-processing sys-

Advantages and disadvantages of office automation methods

Method	Advantages	Disadvantages
Central computer	High throughput Central file storage Good electronic mail	Can be slowed by computer-intensive tasks Terminals may be slow Software may be unfriendly
Stand-alone personal computers	Cheap Easily expandable Wide range of software Can be implemented quickly	Limited power Cannot share information No electronic mail
Networked personal computers (more powerful)	Files widely accessible Electronic mail High throughput	Limited software Few vendors

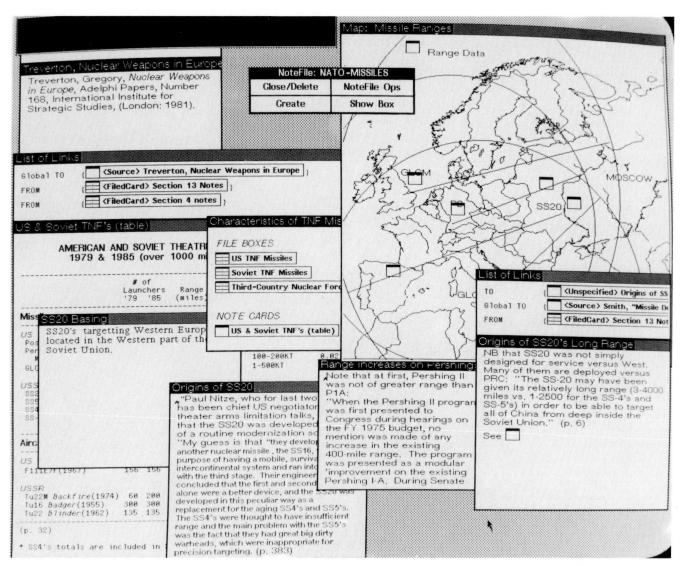

Fig. 7. Typical display of Note Cards system for authoring, information structuring, and idea processing that runs on artificial intelligence workstations and provides an integrated set of tools for manipulating idea-sized units of text, images, sketches, maps, and other electronic information stored in a data base. (*Xerox Corp.*)

tems as office automation, but they lack such facilities as electronic mail, adequate security or communications capabilities, or multiple applications that work well with each other. Some vendors offer office automation systems that are more comprehensive, but these are often based on a central computer and can thus experience performance problems. In addition, since they are based on traditional techniques of interacting with minicomputers, such products may not be very user-friendly. They do, however, offer good functionality.

A much more potentially usable system is one based on powerful and sometimes very expensive personal computers using networks, graphics screens, and "mouse" pointing devices (Fig. 7). As yet, these systems are hampered by lack of a full selection of software; an office can be antomated with them only in very particular ways. On the other hand, as they become more widely applicable and cheaper, such systems may offer much more power for each user than systems based on central computers. The advantages and disadvantages of various office antomation methods are summarized in the table.

In short, current office automation offerings have a long way to go, and it is unclear just what specific form the office systems of the future will take. Clearly, it is not necessary to equip order-entry clerks with bit-mapped screens capable of displaying multiple fonts, backed by a computing engine with the power of a present-day minicomputer. On the other hand, forcing a middle manager to communicate and make decisions by using only a text terminal and a microprocessor developed 8 or 10 years ago is also foolish. But even though a systematic approach to office automation has yet to appear and be validated, offices are being automated by those who inhabit them and are willing to spend money to make them run faster.

[PAUL WALLICH]

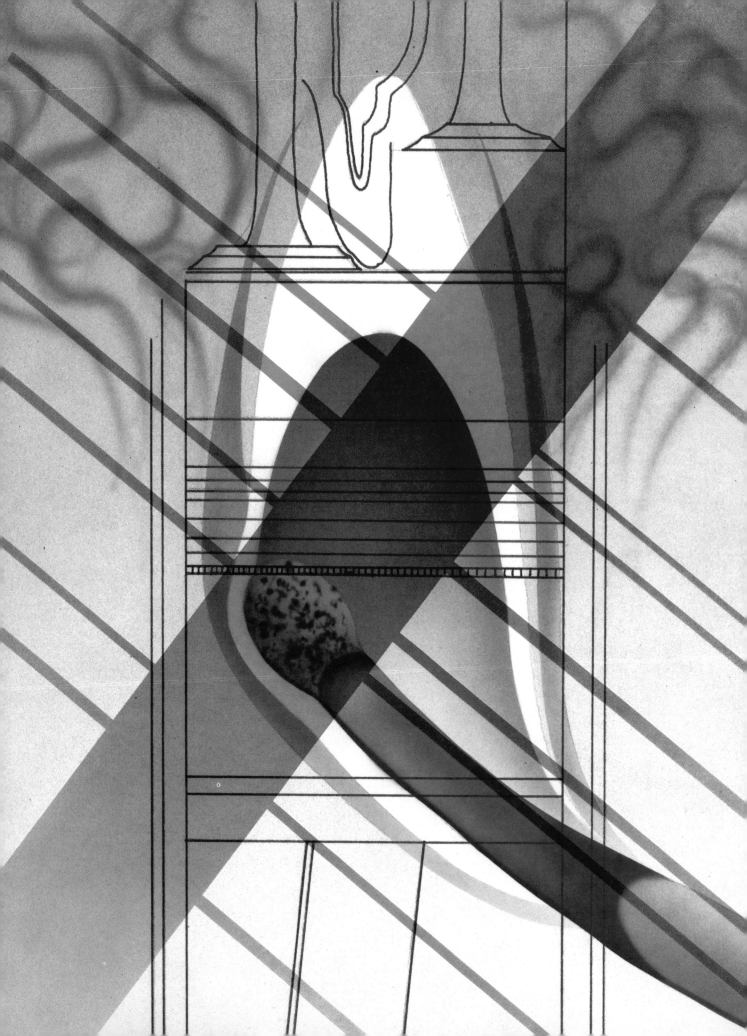

COMBUSTION CHEMISTRY AND PHYSICS

DAN HARTLEY

Dan Hartley is the director of Combustion and Applied Research at Sandia National Laboratories in Livermore, California. He was a pioneer in the development and application of laser-based diagnostics to the study of turbulent-fluid-mechanic processes. In 1974 he conceived the national Combustion Research Facility, a scientific complex dedicated to the advanced methods in combustion science, of which he is also director. Hartley is the United States technical representative to the Combustion Research Group of the International Energy Agency, and was involved in initiating the group in 1976. He serves on several national review committees and is currently a member of the board of directors of The Combustion Institute (Pittsburgh).

Humans have dealt with fire for over a half million years. Since the paleolithic period, they have enjoyed the energy and comforts that combustion provides. Today combustion is used to heat homes, generate electricity, run cars, drive factories, and cook food. Great benefits have been extracted from combustion with little effort.

Lately, however, the price that is actually being paid for this luxury has become evident. The supply of fuel for combustion systems is a limited one, and undesirable combustion by-products are threatening the health of the world. Combustion systems must now be designed, controlled, and refined with a complete understanding of the consequences to fuel supplies and to the environment. This is a formidable task of extreme complexity. Fortunately, during the last few years it has been possible to attack these problems with the help of advanced technologies: new measurement methods based on lasers and new tools using highly specialized computers.

THE SITUATION TODAY

During World War II, significant advances in combustion were motivated by aircraft propulsion needs and artillery and rocket propellant uses. Following the war, combustion research in the United States received systematic government support from the Department of Defense. As a result, more combustion research was accomplished between the 1950s and the 1980s than during the preceding three centuries.

The 1950s saw the introduction of air-breathing jet aircraft for military purposes, while rocket propulsion technology led to the development of spacecraft. Combustion had been harnessed in ways that moved society forward rapidly, but in the 1970s the very real limits of the environment were realized. Thus in the early 1970s major problems in air pollution were encountered, and combustion scientists were challenged with learning how pollutants formed and how they could be controlled. Major advances were made, especially considering the limited research tools available at the time. Then, in the late 1970s another crisis developed: dwindling energy supplies. Combustion engineers were confronted with the dilemma of designing efficient combustion systems that were also clean-burning. Unfortunately, many methods to increase efficiency also increase pollutants.

And so the 1980s began by posing the challenge of finding the more difficult answers to the design of better combustion systems. The 1990s will bring demands for increased, cleaner, and more efficient coal utilization. The next century will certainly introduce a plethora of fuels and fuel mixes as the struggle to match limited fuel supplies to growing needs continues. All of this will continue to force the combustion researcher to dig deeper, to seek fine details, to predict the consequences of new combustion designs, and to find the optimum solutions to combustion problems.

Progress in virtually all areas of science has moved in parallel with the development of better experimental tools and analytical methods. For combustion research, the introduction of mass spectrometers a few decades ago advanced the field. The ability to measure even the chemically stable species in a reacting flow provided new insight into the gross behavior of flames. However, the clean-air crisis created the need for new analytical methods for exhaust-gas measurement. Chemiluminescence detectors for nitric oxide and infrared absorption detectors for carbon monoxide were developed, while simultaneously a host of spectroscopic information about these pollutant molecules and other important chemical intermediates emerged. Although these tools were developed primarily to validate combustion-product limits for environmental control, they quickly were adopted for general use in flame studies. By the time of the energy crisis, the combustion community had already become familiar with a whole family of spectroscopic techniques based on the use of the special properties of laser light. The simultaneous introduction of high-power lasers covering broad spectral ranges and availability of large-scale computers then presented a synergistic opportunity to tackle the important issues of combustion research in a new, very detailed fashion.

It was in this era that the U.S. Department of Energy created the Combustion Research Facility. These engineers, chemists, and physicists can unite their skills in developing and applying new tools to critical problems in combustion science. Similar facilities and programs have since been developed in France, Germany, the United Kingdom, and Japan. In the United States, combustion research has grown rapidly and is also carried out at many universities and in most comergy-related industries. The diagnostics, engine, and coal research programs at the

Combustion Research Facility will be used to illustrate the nature of that effort.

DIAGNOSTICS RESEARCH

To develop combustion systems that can operate both efficiently and cleanly, it is necessary to understand in detail the multiple interactions among the physical and chemical processes that occur in a flame. Detailed descriptions of fluid motion, turbulence, chemistry, and their effects on each other must be provided to create predictive design models—the computer programs that attempt to simulate the combustion process and predict overall performance. The ability to provide the required level of detail with adequate space and time resolution has been made possible since the mid-1970s by advanced laser-optical diagnostics. The development of this technique has permitted new insight, on a microscopic scale, into the chemistry and physics of combustion processes. As computer models become more complex with hundreds of chemical reactions, laser measurement techniques become more exacting in terms of the precision, specificity, and resolution which they must provide.

Laser diagnostics, which have allowed researchers to overcome difficult problems associated with physical access to harsh experimental environments, have themselves been rapidly improved. Laser diagnostic techniques generally rely on analysis of light scattered either from particles (Mie scattering) or from gas molecules (Rayleigh, Raman, and fluorescence scattering). This weakly scattered light can easily be masked by a flame's highly luminous back-

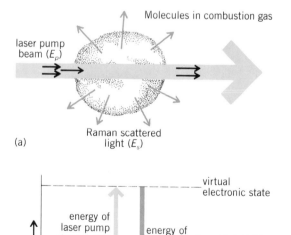

Fig. 1.   Raman effect. (a) Laser pump beam interacting with molecules to generate Raman-scattered light. (b) Energy-level diagram of molecular processes. (*After New light on combustion research, Sandia Technol., pp. 2–17, May 1984*)

ground. However, newly developed laser-probing methods, such as coherent anti-Stokes Raman spectroscopy (CARS), allow measurements to be made in these harsh environments. With such laser-based methods, fast (less than 6 nanoseconds) and spatially accurate (100-micrometer-resolution) measurements that had previously been impossible are made in research flames, internal combustion engines, combustors, and jet engines. Accurate determinations of temperatures, species concentrations, turbulent flow structure, and trace species can now be made in flames whose properties preclude use of conventional diagnostic methods.

Results obtained with these new techniques verify or modify the computer-based analytical models that improve understanding of important combustion processes. Some of these methods have been adopted by industry, in particular the automobile industry, where they are being applied to research and development activities.

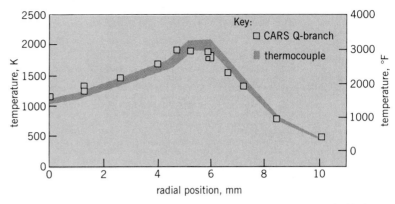

Fig. 3. Temperatures obtained from interpretation of CARS data compared with thermocouple measurements in a sooting, highly luminous diffusion flame. 1 mm = 0.04 in. (*After New light on combustion research, Sandia Technol., pp. 2–17, May 1984*)

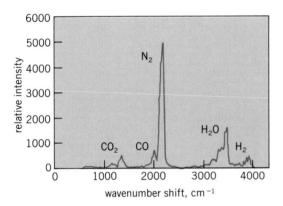

Fig. 2. Spontaneous Raman spectrum of the exhaust of a propane-fueled internal combustion engine, showing individual spectra of the species present. (*After New light on combustion research, Sandia Technol., pp. 2–17, May 1984*)

Spontaneous Raman spectroscopy. Spontaneous Raman spectroscopy is a long-established diagnostic technique in the study of molecular spectra and structure. The Raman effect can be thought of as an interaction of a molecule with a laser photon (Fig. 1a). Most of the time, the molecule prefers to radiate at the same energy E_P as the incident laser photon (a phenomenon called Rayleigh scattering). However, about once in every thousand times, the molecule (Fig. 1b) instead radiates light at a frequency (energy E_S) different from the incident radiation. This difference E_R is the energy gained by the molecule and corresponds to energy levels characteristic of, and specific to, the molecule; it is thus the basis for analytical measurements. The process can be thought of as an inelastic collision of the laser photon with the molecule.

In the most common form of Raman spectroscopy, a laser beam is focused into the medium to be studied (in combustion research, a flame or combustion environment), and the Raman-scattered light is collected at right angles to the beam. Typically, spon-

taneous Raman spectroscopy is used to characterize species whose concentrations are greater than 1000 parts per million (ppm). The sample volume is about $0.1 \times 0.1 \times 1$ mm, which is determined primarily by the optical system collecting the scattered light. An optical spectrometer provides spectral analysis of the light emerging from the combustion medium.

Spontaneous Raman spectroscopy is straightforward and relatively simple to perform, and has been refined to the extent that useful data have been obtained from environments as severe as those in a working internal combustion engine (Fig. 2). However, it has the disadvantage that the Raman effect is extremely weak. Furthermore, the light that is scattered is distributed uniformly over all angles, so that much less than the total scattered Raman light can be collected by the optical detection system. As a result, there are many situations in which useful data can be acquired only by integrating the signal for a long time.

CARS. In recent years, CARS has received considerable attention because it has many of the positive attributes of spontaneous Raman scattering,

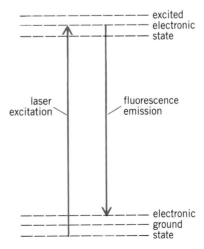

Fig. 4. Energy-level diagram illustrating the molecular processes involved in laser-induced fluorescence (LIF). Both ground and excited electronic states have vibrational and rotational sublevels, as shown. (*After New light on combustion research, Sandia Technol., pp. 2–17, May 1984*)

with the potential to avoid most of the problems discussed above.

CARS signals result from the interaction of three laser beams with the combustion-gas molecules: Two "pump" beams from one laser source, having the same photon energy (or wavelength), are focused on the point of interest in the combustion region for measurement purposes. A third, "probe" laser beam whose photon energy, less the photon energy of the pump beam, corresponds to a Raman transition energy of molecule, is also focused to the same point. Properly oriented with respect to the polarizations of the beams, these normally high-intensity laser fields interact with the molecule being probed in such a fashion that an additional "signal" beam is created. This CARS signal beam contains spectral information that identifies the species being probed and yields its temperature. Furthermore, the emitted radiation, being coherent, is highly collimated in space so that all of the emitted light can be collected with little background interference. Finally, large enhancements in signal strength are possible in comparison with spontaneous Raman scattering.

Because of these properties, CARS methods can provide Raman-like information from hostile environments that would prevent use of spontaneous Raman methods. An example is shown in Fig. 3, where temperatures obtained from the interpretation of CARS data from sooting, highly luminous portions of a diffusion flame are compared with detailed thermocouple measurements. Spontaneous Raman measurements in this flame are impossible because of intense background radiation.

However, CARS has drawbacks. Because of the nonlinear nature of the effect, the amplitude of the measured signal depends in complicated ways upon laser intensity, species concentration, and other experimental parameters. This behavior makes data analysis more involved than for linear spectroscopies. Also, CARS experiments are complex, involving at least two lasers, with considerable optics and electronics equipment required.

Laser-induced fluorescence. Another widely used laser diagnostic technique is based on the laser-induced fluorescence (LIF) effect. The physical principle underlying this approach is quite simple (Fig. 4): a laser excites a molecule from one level (or set of sublevels) to a higher electronic energy state. This mechanism is quite different from Raman scattering: in laser-induced fluorescence, the

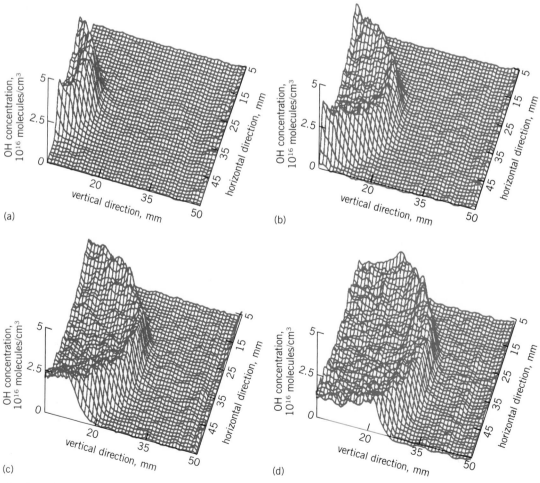

Fig. 5.   Single laser pulse hydroxyl (OH) concentration maps (a–d) of a methane flame propagating in a constant-volume combustion chamber at different times after ignition. (*After Combustion Institute, 20th Symposium (International) on Combustion, 1984*)

molecule is excited to, and remains in, a particular state for some time; then the molecule emits a new photon of radiation, with a frequency corresponding precisely to the difference in energy between the excited state and the lower electronic level to which the molecule returns, and hence characteristic of the particular molecule under study. When the molecule absorbs and radiates between the same two levels, the process is called resonance fluorescence.

One of the more advanced contributions of laser-induced fluorescence to flame diagnostics is its application to planar, two-dimensional measurements of a slice through a flame. The probe laser beam is expanded to a sheet of light by using cylindrical lenses; laser-induced fluorescence from this sheet of light is collected by a high-speed digital imaging system. Figure 5 shows an example of a two-dimensional, time-dependent display of hydroxyl (OH) concentrations emanating from an ignition source in a methane-air mixture. The hydroxyl molecules have also been used as a thermometer by performing laser-induced measurements at two different laser frequencies, each corresponding to a different initial vibrational-rotational level in the electronic ground state. By analyzing the data and determining the relative populations of the two levels, which is a function of temperature, a two-dimensional temperature distribution for the flame can be formed. This two-dimensional laser-induced fluorescence capability can greatly improve understanding of unsteady and turbulent flames.

Another variation of laser-induced fluorescence has been applied to kinetics experiments that measure the rate at which chemical reactions occur un-

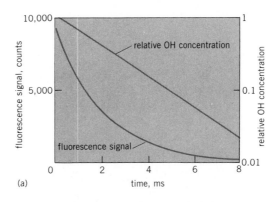

(a)

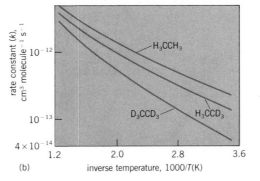

(b)

Fig. 7. Typical measurements from laser-induced fluorescence experiments. (*a*) Decay of hydroxyl (OH) concentration after hydroxyl radical is created by photolysis laser. (*b*) Rate constant data deduced from these measurements for the reactions of hydroxyl radical and three isotopes of ethane.

der carefully controlled conditions. These data are vital inputs to the chemical mechanisms and models that are being developed to explain combustion. In operation (Fig. 6), a photolysis pulse from an excimer laser creates free radicals within a controlled gas flow in a glass reactor, and fluorescence induced by a picosecond dye laser characterizes the decay rate of the chemical species under study. In the setup shown in Fig. 6, the beam from the dye laser is focused into a temperature-tuned, frequency-doubling crystal to convert the wavelength of the visible laser beam into the ultraviolet range required for the transition of the hydroxyl (OH) radical under study. Figure 7a shows typical measurements of the decay of the hydroxyl concentration after creation of the radical by photolysis. The quality of these data surpasses previously achievable results by one or two orders of magnitude. From measurements such as these, rate-constant data for the reaction as a function of temperature can be deduced (Fig. 7b).

Newer methods. The techniques discussed above have been well developed since the mid-1970s and are finding their way into numerous combustion laboratories around the world, including several industrial laboratories. For example, CARS systems are used in industry to monitor vapor concentrations over molten steel baths, measure diesel engine conditions, and monitor jet engine exhausts. Yet the quest for better sensitivity continues, and a number of new methods are being developed. Among them

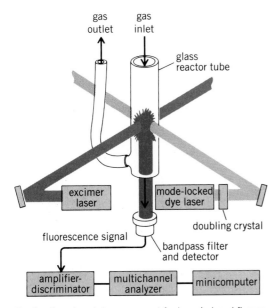

Fig. 6. Experimental arrangement for laser-induced fluorescence measurement on reactions of hydroxyl (OH) radical with ethylene (C_2H_4). Pulses from argon-ion laser-pumped dye laser have energy of 3–6 nanojoules each, duration of 8–10 picoseconds, and repetition rate of 246 MHz. (*After New light on combustion research, Sandia Technol., pp. 2–17, May 1984*)

are Raman gain-loss spectroscopy, optogalvanics, optical Stark effect spectroscopy, and a variety of adaptations of Raman, Mie scattering, and fluorescence for multidimensional imaging. The coupling of three factors—detailed understanding of the molecular physics contributing to spectroscopic behavior, specially designed lasers, and very sophisticated, computer-based detection systems—potentially provides the opportunity to measure any desired property of a combustion system with remarkable accuracy.

ENGINE COMBUSTION RESEARCH

Before trade-offs between performance, fuel economy, and emissions can be made to optimize the performance of internal combustion engines, the governing physical and chemical processes inside operating engines need to be understood. One goal of combustion research is to provide this scientific understanding and thereby lead to improved efficiency in present and future engine designs, as well as help in the redesign of engines to make use of alternative fuels. Although engines have been studied and refined for most of this century, only since the mid-1970s have engine designers been able to use this detailed information.

Goals. The critical issues in stratified-charge engines, which are designed to provide a fuel-rich cloud near the ignition and a fuel-lean condition elsewhere, rest with understanding sprays, mixture motion, and droplet combustion. In lean homogeneous-charge engines, which are designed to operate with a uniform, fuel-lean combustion charge for maximum efficiency, the issues are control of flame speed and quenching. In diesels, which are designed to autoignite and burn in a stratified form, the issues are soot formation and high nitric oxide production. In applications of new fuels, the issues are knock and aromatic emissions. Solutions to these problems will take time, but the new tools are paving the way.

With the accelerated production of diesel engines, public concern for the long-term health effects of particulate emissions has stimulated increased research and necessitated interim control regulations. To design systems which limit soot formation, the origins of particulate emissions and their behavior in combustion environments must be understood. To this end, experiments have been conducted to characterize and measure particulate-formation processes by using laser absorption and scattering as well as optical pyrometry.

The most elusive information sought by engine designers is the detailed mapping of gas-mixture composition and temperature inside the combustion chamber. With increased interest in direct-injection, stratified-charge engines and the widespread introduction of diesels as potentially clean, efficient, and fuel-tolerant engines, industry worldwide is struggling to understand the details of their performance. However, in such fuel-injected engines, the combustion process becomes more complex because of nonuniform fuel distribution and the coupling of chemical and physical processes within the chamber. On the other hand, these engines have great potential for improvement because fuel distribution may be tailored to optimize combustion and minimize pollutants.

In general terms, engine research seeks to find how the air-intake process establishes the controlling fluid motion in the engine and the nature of turbulence set up in the combustion chamber. Then researchers want to know how fuel and air mix inside the combustion chamber, how the mixture burns, and how combustion phenomena resulting from all this lead to generation of undesirable products and to the most efficient burn. Specifically, it is important to understand the formation of nitrogen oxides, carbon monoxide, and unburned hydrocarbons in the combustion chamber—and in diesels, the processes by which particulates are formed and consumed or exhausted.

The evolution of computer modeling has also played a major role in setting goals for the diagnosticians. As large mainframe computers now have the capacity to manipulate complex, multidimensional

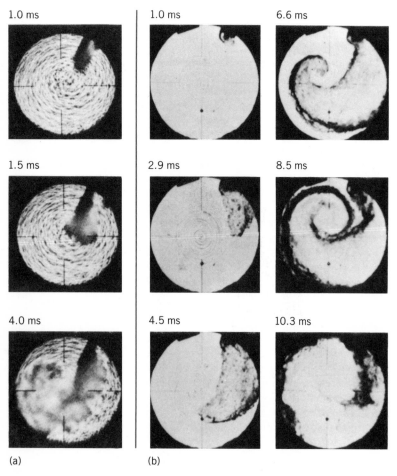

1.0 ms   1.0 ms   6.6 ms

1.5 ms   2.9 ms   8.5 ms

4.0 ms   4.5 ms   10.3 ms

(a)   (b)

Fig. 8.   High-speed photographs taken by the laser shadowgraph technique. Time after ignition is indicated for each frame. (a) Sequence showing gaseous propane fuel injection into a swirling air charge in the engine. (b) Sequence showing premixed methane combustion in the combustion chamber. (*From Improving internal combustion engines, Sandia Technol., pp. 2–11, February 1983*)

computer models of combustion, the pressure exists for the diagnostician to validate (or disprove) these sophisticated models. To do so requires yet more information on temporal and spatial variations of several parameters. New methods of simultaneous, spatially resolved detection have therefore been created, many of which apply to engines.

Optical diagnostic techniques. A variety of these techniques are used to investigate the physical and chemical processes that occur in internal combustion engines: spontaneous and coherent anti-Stokes Raman spectroscopy to measure time- and space-resolved temperature and species concentration; laser Doppler velocimetry to measure fluid flow; pulsed Rayleigh scattering to measure instantaneous turbulent flame structure; Mie scattering and laser absorption to map spatially both temperature and concentration of soot; and various schemes for flow visualization.

Laser shadowgraph. Laser shadowgraph systems have been used to correlate global behavior with detailed point measurements from other diagnostics. In these systems, a laser beam is spatially filtered, expanded, and then collimated before passing through the piston-head window into the combustion chamber. A mirror bonded to the piston surface returns the incident light through a series of collimating lenses to a translucent screen or directly onto the film plane of a camera. High-speed movies and still photographs have been taken of this output (Fig. 8).

Laser Doppler velocimetry. Velocity measurements have been made with a laser Doppler velocimeter. Titanium dioxide particles with nominal 0.2-micrometer diameters are typically used as the light-scattering seed material. A minicomputer records velocity-time pairs, processes the data statistically, and displays velocity histograms for each data set.

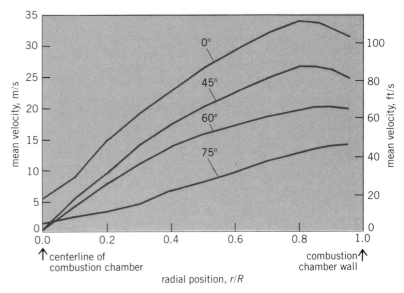

Fig. 9. Turbulence intensities measured inside an internal combustion engine by using laser Doppler velocimetry. Radial position is expressed as ratio of distance from centerline of combustion chamber r to radius of chamber R. Shroud orientation relative to valve is 0, 45, 60, or 75° as indicated. (*After Improving internal combustion engines, Sandia Technol., pp. 2–11, February 1983*)

Exhaustive studies mapping the generation and evolution of mixture motion and turbulence during intake and compression, both with and without fuel injection, have been conducted by using laser Doppler velocimetry in research engines. With changes in the orientation of the shrouded intake valve, swirl rates and turbulence intensities can be varied and their influence on turbulent flame propagation examined (Fig. 9).

Raman techniques. Spontaneous Raman scattering has been used to measure the mixing of fuel and air

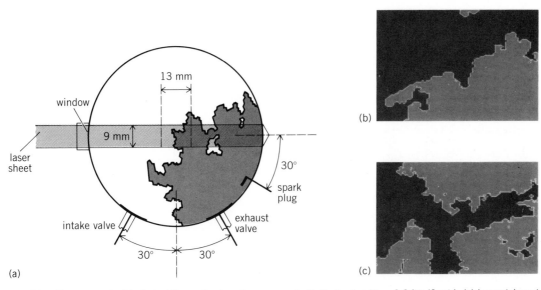

Fig. 10. Measurement of turbulent-flame structure in an engine by imaged Rayleigh scattering. (*a*) Experimental apparatus. 9 mm = 0.4 in.; 13 mm = 0.5 in. (*b*) Image taken at engine speed of 600 revolutions per minute, turbulent velocity fluctuation $V' = 6.6$ ft/s (2 m/s). (*c*) Image taken at engine speed of 1500 revolutions per minute, $V' = 20$ ft/s (6 m/s).

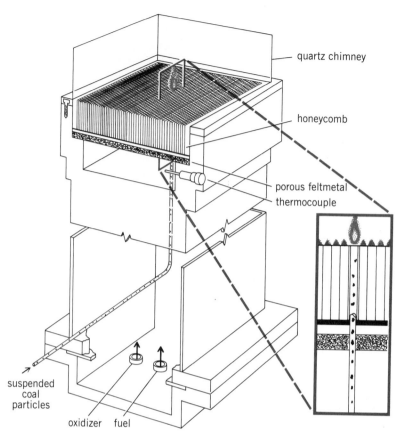

quartz chimney

honeycomb

porous feltmetal

thermocouple

suspended coal particles

oxidizer   fuel

Fig. 11. Laminar flow reactor developed to allow detailed observations of processes occurring during ignition and devolatilization of freely entrained pulverized coal particles in high-temperature oxidizing environments. (*After The science of coal combustion, Sandia Technol., pp. 2–17, August 1985*)

during injection, the fuel-air ratio at the spark plug ignition source, and temperature and density distribution at various points throughout the combustion chamber. Because a smooth, efficient engine is one in which combustion is highly reproducible, techniques have been developed to measure cycle-to-cycle fluctuations.

A recent and very promising development is the application of CARS to the study of diesel combustion. This technique is well suited to making measurements in the highly luminous, soot-laden environments typically found in diesel engines. In addition, it produces a highly concentrated signal beam which allows efficient collection with limited optical access. Spectra of nitrogen, carbon monoxide, and hydrogen have been measured in engines using CARS. Nitrogen spectra are acquired during the compression and power strokes under homogeneous-charge conditions. The gas-phase temperature can then be determined by fitting a calculated spectrum to the data.

Rayleigh scattering. Imaged Rayleigh scattering is also being used to measure instantaneous turbulent-flame thickness in an operating homogeneous-charge engine. The objective of this experimental study is to provide insight into the mechanism of turbulent-

flame propagation and to observe the interaction between turbulence and combustion in an engine (Fig. 10).

Status. These examples represent only a small part of what has now become a worldwide effort to understand engine combustion in finer detail. Time-resolved point measurements of most properties of interest have been demonstrated in research engines. Nevertheless, their application in multidimensional form (imaging) and in multiparameter forms (correlations) is necessary before the computer models can be validated and the model predictions accepted as accurate. Some of the newest techniques show potential to achieve this goal.

COAL COMBUSTION

One feature of the combustion process that has received considerable attention, particularly in response to early environmental concerns over nitrogen oxide emissions, is the mixing rate of the coal and air in the primary combustion zone of a boiler. As in the automotive industry, a major thrust in pulverized coal combustion systems is toward stratified, staged, or off-stoichiometric combustion. It is now accepted that by carefully controlling the distribution and mixing rates of fuel, air, and combustion

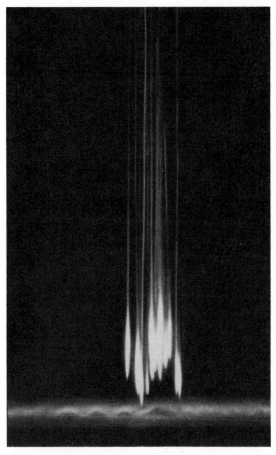

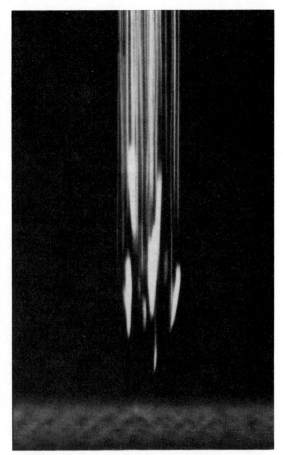

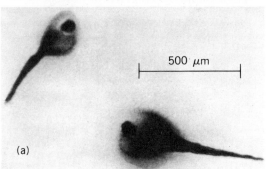

500 µm

(a)

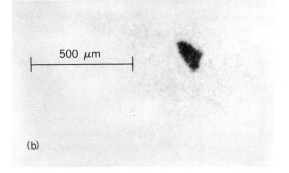

500 µm

(b)

Fig. 12.   Pulverized (a) bituminous coal and (b) lignite undergoing ignition and devolatization in a laminar flow reactor. Upper images are time-integrated views of coal particles burning and moving vertically in reactor. Bright diffuse emissions are attributed to burning of ejected volatile matter. Above the volatile burnout region this emission ceases, and incandescence attributed to heterogeneous char oxidation is observed. Lower images are magnified, time-resolved views of single particles in reactor, showing magnified clouds around particles. (*From The science of coal combustion, Sandia Technol., pp. 2–17, August 1985*)

products, it is possible to reduce emissions, increase fuel flexibility, and increase efficiency (for example, by reducing excess air requirements).

These mixing techniques have been demonstrated to be effective in controlling nitric oxide emissions from coal flames, and some are now being implemented in redesigned burners and combustion chambers. However, because of the strong coupling of the details of coal combustion, the properties of coal itself, and other "outputs" of the combustion process such as the forms and quantities of sulfur and mineral-matter species, interest has been renewed in determining and optimizing the engineering and economic consequences of combustion modification.

Stimulated by these issues, equipment manufacturers are improving designs of combustion equipment by using predictive computer codes to describe and interpret coal-combustion processes in large boilers and furnaces. These more accurate predictive treatments, however, require greater knowledge of local combustion details. Precisely because scientific knowledge of these details is so slight, it is in this area that coal combustion science can have its greatest impact. The key lies in understanding and quantifying the behavior of controlled-mixing combustion processes, including details of the nonlinear processes that govern the time-temperature species history of the reacting fuel particles.

In terms of coal tonnage burned, technology improvements are of unquestioned importance in the United States and other countries. Where direct substitution of coal for oil is the aim, the influence of the liquid used to carry the pulverized coal is of interest. Combustion at high pressures is of concern in any proposal to directly fire coal in large stationary gas turbines. A data base on the combustion of pulverized coal, either neat or in slurried form, is lacking. Fundamental understanding is required about the rates and evolutionary forms of sulfur and nitrogen oxides, mineral matter particle-size distributions, and unburned hydrocarbons during ignition, pyrolysis, devolatilization, and char oxidation of coals.

Laminar-flow reactor studies. To examine the sequential events associated with the burning of pulverized fuels, a variety of laminar-flow reactors (Fig. 11) have been developed with many original features. These reactors permit complete optical and sampling probe access to the reacting fuel particles, allowing processes to be directly observed, photographed, and measured; they also afford control of the material properties of unburned fuel, the rate of addition of this fuel to the reactor, the high-temperature environment to which the fuel is exposed, and the times and rates at which reactions are allowed to occur. The environment is provided by uniform and well-controlled postflame gases of a premixed, laminar flat flame. Coal particles are injected in a dilute single stream on the centerline of the reactor through a 1-mm capillary tube.

Figure 12 shows pulverized coals undergoing ignition and devolatilization in the flow reactor. With this reactor, the early stages of reaction of bituminous coal and lignite were investigated. It was possible to identify soot-formation phenomena during the rapid devolatilization of these coals and to correlate soot-formation propensity with volatile matter content and the porosity of unburned material.

In addition to these phenomenological observations, the flow reactors are instrumented to measure the size, temperature, and velocity of burning coal particles simultaneously. One way to make these measurements is to use an optical system with special apertures to detect scattered and emitted light. This system resolves light emitted by discrete particles of burning coal and char as they pass through the optical sampling region on the centerline of the reactor. To determine particle temperature, this system employs a fast-response optical-electronic system to simultaneously record and process emissions from individual particles at two wavelengths.

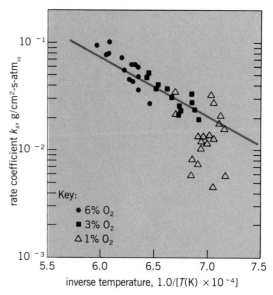

Fig. 13. Chemical reaction rate coefficients derived from heat balances by using particle sizes and temperatures measured with a particle-sizing pyrometry system. The line is a least-squares fit. The slope is indicative of particles whose burning rates are controlled by the combined effects of pore diffusion and the intrinsic chemical reactivity of the particle material. (*After The science of coal combustion, Sandia Technol., pp. 2–17, August 1985*)

Particle size is obtained by using an image-plane coded-aperture technique. Measured temperatures are combined with particle-size information, the properties of the raw coal, and the reactor gas-phase environment to evaluate the absolute burning rates of coals and other solid fuels at very high heating rates (Fig. 13). By using detailed heat, mass, and momentum balances, the effects of heterogeneous reaction kinetics, pore diffusion, and bulk diffusion have been distinguished in determining the dependence of coal burning rate on temperature. The most recent results have demonstrated the important effect of the particle-size distribution and particle temperature on the inferred burning rates.

Although direct measurements of this type are valuable, they are only part of the story. The picture is completed by extractive sampling of hot, partially reacted char and solid ash products in the flow reactor and by in-place measurements of the gas phase with absorption spectroscopy. These efforts have built upon capabilities in electron optics and in Fourier transform infrared spectroscopy. This broad attack on the unknown physics and chemistry of coal combustion will provide the kind of complete data base required for optimizing coal combustion processes.

Laser diagnostics applications. Two recent developments in the laser diagnostics area applied to coal are noteworthy. A long-path-length flow reactor has been designed for direct measurement of temperature and concentrations of gaseous species during coal combustion using CARS. In principle, CARS is especially well suited for such measurements because of its high signal-to-noise characteristics. Conditional sampling is done to eliminate spectral interferences resulting from the occasional breakdown of the intense neodymium:yttrium-aluminum-garnet (Nd:YAG) laser beam on coal particles. However, this breakdown can be used to advantage. Laser spark spectroscopy is being developed to directly measure the elemental composition of fuel droplets, coal particles, and flyash in combustion flows. In this work, three lasers are combined; Mie-scattered argon-ion laser light sizes the particle, a tightly focused helium-neon laser precisely locates the particle and serves as a trigger, and a high-power Nd:YAG pulsed laser vaporizes the particle and excites its elemental constituents to high energy levels. A time-gated optical multichannel analyzer detects the spectrally resolved emission for composition determination.

These examples, as well as the advanced laser diagnostic and computational tools, will be applied to new basic studies of pulverized-coal combustion processes. Such experiments allow control of independent variables, for example, coal type, degree of beneficiation, particle-size distribution, and particle number density. Furthermore, they are designed to permit control and measurement of chemical and physical events that occur during the ignition, pyrolysis, devolatilization, and char-oxidation phases of coal combustion.

[DAN HARTLEY]

Bibliography: Combustion Institute, *18th Symposium (International) on Combustion*, 1981, and *20th Symposium*, 1984; R. M. Fristrom and A. A. Westenberg, *Flame Structure*, 1965; W. C. Gardiner, Jr., The chemistry of flames, *Sci. Amer.*, 246(2):110–124, February 1982.

AUTOMATED SYSTEMS IN MICROBIOLOGY

MICHAEL T. KELLY

Michael T. Kelly is professor of pathology at the University of British Columbia. Holder of a doctor's degree in microbiology from Indiana University, he is certified in medical microbiology by the American Boards of Pathology and of Medical Microbiology. He is a Fellow of the College of American Pathologists and chairperson of the Subcommittee for Automation in Clinical Microbiology of the National Committee for Clinical Laboratory Standards. He is the author of more than 70 publications in the medical and scientific literature.

Microbiological analysis has traditionally been time-consuming and labor-intensive, requiring the isolation of microorganisms from specimens under analysis by culturing on nutrient media. These media, solidified with a gelatinlike material known as agar, are usually contained in covered dishes (petri dishes) or tubes, and the specimens are spread over the media surface. The cultures are then incubated under conditions that favor the growth of the desired microorganisms, and after one or more days colonies appear at sites on the agar surface where the organisms were initially deposited. Individual colonies are then touched with the end of a sterile wire, and the growth is transferred to new media. This process is repeated until the culture consists of only a single type of organism. The microorganism can then be tested to establish its identity and its susceptibility to antimicrobial drugs. The diagnosis of infectious diseases is often delayed for several days while the causative organism is grown and tested.

Within the past 10 years automated systems have been developed that are capable of doing much of the work of identification and antimicrobial susceptibility testing of microorganisms; in some cases automated systems can even replace the culturing of microorganisms. In addition, automated systems often can generate results more rapidly than traditional methods. Such systems have been applied to the detection of organisms in foods and other products, but this article will focus primarily on the applications to the diagnosis of infectious diseases. Automated microbiology systems have been developed for analysis of blood, urine, and respiratory specimens for patients.

ANTIMICROBIAL SUSCEPTIBILITY TESTING

The first commercially successful automated systems for microbiology were designed for testing the susceptibility of bacteria to antibiotics. Antibiotic susceptibility testing is done to identify the drugs that may be most effective for inhibiting the growth of bacteria. Traditionally, as mentioned above, bacteria are cultured from infected areas of the body, and the isolated organism is inoculated over the surface of a large dish of agar. Paper disks impregnated with the antibiotics to be tested are placed on the agar, and the cultures are incubated. As the antibiotics diffuse from the disks into the agar gel, they produce a concentration gradient, and the bacteria growing on the surface of the agar produce a visible film. Disks containing antibiotics to which the organism is susceptible inhibit the growth of the bacteria, and clear zones are observed around them. The degree of susceptibility of the bacterium to each of the antibiotics is then determined from the size of the clear zone around each disk.

This type of testing requires overnight incubation and manual operations for setting up the test and determining the results. Automated susceptibility testing systems generate results in as little as 3 h, and much of the labor is performed by the instrument.

Photometric systems. The first widely available instrument for susceptibility testing of bacteria (Fig. 1) entered the commercial market in the early 1970s. It is essentially a photometric instrument that determines bacterial growth optically. The system consists of an incubator-shaker module and a photometer module. The central part of the system, and its most innovative aspect, is the disposable plastic cuvette that serves as a growth chamber for the bacteria to be tested (Fig. 2). The cuvette consists of an upper chamber with a port for attachment of a tube and a lower chamber with individual compartments. Paper disks impregnated with the antibiotics are dispensed into the individual compartments of the cuvette. Suspensions of the organisms are prepared in tubes of culture broth, and the density of the suspensions is adjusted to a predetermined level so that they appear clear at the time they are dispensed into the cuvettes. The tubes are attached to the cuvettes, which are tilted so that the suspensions first evenly fill the upper compartments of the cuvettes. The cuvettes are then tilted again so that the bacterial suspensions are evenly divided into the individual chambers in the lower compartments. Antibiotics in the paper disks diffuse into the culture broth to achieve predetermined concentrations. The cuvettes are incubated in the incubator-shaker, and they are periodically transferred to the photometer module for growth determinations. The first chamber of each cuvette does not contain antibiotic and serves as a growth control. The suspensions become increasingly turbid as the organisms grow in the control chamber. This turbidity is measured in the instrument by light scattering. Light striking a particle is scattered in several directions, and the instrument measures this scattered light. As the cultures grow, the suspensions contain more and more individual bacteria (particles) and the light scattering is increased. When the instrument detects sufficient growth in the control chamber, it measures the light scattering in each antibiotic-containing chamber of the cuvette. If the light scattering is the same in an antibiotic-containing chamber as in the control chamber, the organism is able to grow in the presence of that antibiotic and is therefore resistant to it. If the light scattering is less in the presence of the antibiotic, the organism is inhibited and is therefore susceptible to the drug.

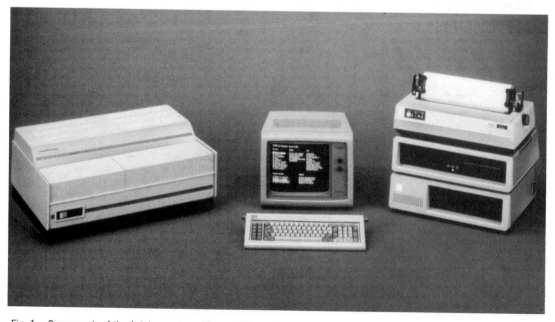

Fig. 1. Components of the Autobac system. (*General Diagnostics*)

After making such determinations for each antibiotic in the cuvette, the information is processed in the instrument's microcomputer and a report is printed listing the susceptibility or resistance to each antibiotic. The testing is completed in 3 to 5 h. The results of antimicrobial susceptibility testing with this system demonstrate approximately 95% agreement with the results obtained by traditional microbiological methods.

Other instruments for testing the susceptibility of bacteria to antibiotics have subsequently become available. Another device similar in concept has several unique features. It uses disposable clear plastic cuvettes and antibiotic-impregnated paper disks, but the design of the cuvette is different. This instrument has a control module, which is essentially a microcomputer for data storage and reporting, and one or more analysis modules. Analysis modules provide a temperature-controlled environment in which the cuvettes are incubated with shaking to provide adequate aeration. Each analysis module accommodates up to eight cuvettes. To use this system, suspensions of the bacteria to be tested are dispensed into cuvettes, and the cuvettes are placed in the analysis module. Growth of the bacteria is monitored by the instrument using light-emitting diodes that shine light through the bacterial suspensions. As the cultures grow, the density increases and less and less light passes through the suspensions. This optical density is measured by photometers in the instrument. Initially, the bacterial suspension is contained in the upper compartment of the cuvette, and the growth of the bacteria is monitored until a predetermined density is achieved. The instrument then automatically transfers the suspension to the individual antibiotic-containing chambers in the lower compartment of the cuvette. The antibiotics diffuse into the culture broth, and growth of the bacteria in the antibiotic-containing chambers is compared to that in the control chamber. Readings are taken automatically by the instrument every 5 min, and the results are stored in the control module. When sufficient growth has occurred, the instrument analyzes the information and reports susceptibility or resistance to each antibiotic. The test is completed in 3 to 6 h depending on the growth rate of the organism. The results of antimicrobial susceptibility testing with this instrument also demonstrate approximately 95% agreement with results obtained by traditional microbiological methods.

Recently, an even more highly automated system has been adapted for antimicrobial susceptibility testing. Originally designed for microbiological testing of lunar samples on the NASA Apollo missions, it was subsequently adapted for culturing urine specimens. This automated system consists of seven instrument components, and it uses a clear plastic card about the size of a credit card for the growth and testing of bacteria. The card has 30 sealed microwells, each connected by a thin channel to an external port on the side of the card. These micro-

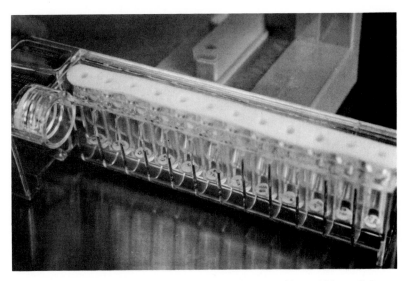

Fig. 2. Disposable plastic cuvette used in the Autobac system. (*General Diagnostics*)

wells contain dried antimicrobial agents; when a bacterial suspension is introduced into the card, the wells fill and the antibiotics dissolve in the culture broth. The instrument components consist of a diluent dispenser that is used to prepare suspensions of the bacteria to be tested, a vacuum-filling chamber that is used to transfer bacterial suspension into the cards, and a sealing module that seals the port on the card through which the suspension is introduced. After sealing, the cards are loaded into a carrier which is placed in the reader-incubator module. After this step, all further operation of the system is automatic, requiring no further input from the operator.

The reader-incubator module provides a controlled growth environment for the inoculated test cards, and optical density readings of each microwell in the cards are taken at frequent intervals by using light-emitting diodes and phototransistor detectors. The status of each test can be checked by the operator at any time with a cathode-ray-tube (CRT) data terminal module. The optical growth information for each microwell is stored and analyzed in the computer control module; at the completion of the analysis, the results are automatically printed on the printer module. As with the other systems, the growth in each antibiotic-containing well is compared with the growth in antibiotic-free control wells, and organisms are reported as susceptible or resistant to the antibiotics tested. Results with this system are available in 8 h. While the results of susceptibility testing with the cards that were available initially did not compare well with standard methods, subsequent generations of the antimicrobial susceptibility testing cards have been improved, and they now demonstrate approximately 95% agreement with traditional methods.

Dilution testing. Other systems for the mechanization of more traditional methods for antimicrobial susceptibility testing have also been developed.

They utilize dilution testing methods in which the antibiotics are serially diluted in broth or agar media in order to achieve a range of antibiotic concentrations. The bacteria are then inoculated to the media and incubated to allow growth. Visible colonies on the agar or turbidity in the broth is provided by the cultures containing less than a critical concentration of each drug. However, no visible growth will occur in the cultures containing an inhibitory concentration of the antibiotic. The lowest concentration of the antibiotic that inhibits growth of the organism is used to determine whether the organism is susceptible or resistant to the drug. The systems for mechanization of this technology provide instruments for inoculation of the antibiotic-containing media and for automatic reading of the end points. In addition, many of these systems have computer modules for interpretation of the results and printers for reporting results.

One of the available systems consists of a replica plating device, a reading device, and a computer module. The replica plating device consists of a seed tray and a multipronged inoculator. The seed tray has multiple small wells to hold suspensions of individual bacteria to be tested. The multipronged inoculator is positioned on an arm that rotates through 180°. The seed tray is placed on one side of the arm, and antibiotic-containing agar plates are placed on the other side of the arm. The arm is lowered so that the prongs of the inoculator enter the bacterial suspensions in the seed tray, and it is rotated through 180° to deposit drops of each bacterial suspension onto the agar plate. The inoculated agar plate is removed, the next plate is put in place, and the process is repeated until all plates in the series have been inoculated. The plates are incubated overnight, in a tray with holes cut out to accommodate the plates, and the tray is placed on a viewing table for recording the results. The presence or absence of growth at each area of inoculation is recorded with a light pen, and the information is stored in the computer. At the completion of reading, the computer analyzes the results and issues reports of the inhibitory concentrations of each antibiotic and whether each organism is susceptible or resistant to each drug tested. Up to 36 organisms can be tested on each plate, making this type of system one of the least expensive to operate. Results can be reported within 18 h, and 95% agreement has been demonstrated between this system and reference testing methods.

Other such systems have been designed to automate broth dilution testing. They differ in the nature of the antibiotic-containing media, but all utilize disposable plastic trays with multiple small wells that contain varied concentrations of the antibiotics. Two systems use trays with frozen antibiotic-containing broth in the wells. Several others use trays with varied concentrations of antibiotics dried in the wells. The antibiotics are reconstituted to the fluid state for testing when the bacterial inoculum is added to each well. Each system also includes multiple inoculation devices to introduce the bacterial suspensions into the wells, and the same organism is introduced into each well of a given tray. Once inoculated, the trays are incubated and then placed in a reader that automatically records the presence or absence of growth in each well. This information is stored and processed by a computer, and reports of inhibitory concentrations and susceptibility or resistance to each antibiotic are provided for each organism tested. The results are available after 18 h for most of these systems, although methods for 4-h reporting have recently been introduced. The accuracy of some of these systems has not been critically evaluated, but those tested have been shown to be highly accurate when compared to conventional testing methods.

BACTERIAL IDENTIFICATION SYSTEMS

All of the antibiotic susceptibility testing instruments so far described also have the ability to identify bacteria and in some cases other organisms as well. Each instrument uses the same growth system for identification as it does for susceptibility testing, and most of the systems utilize standard biochemical tests adapted to the particular instrument for organism identification.

Photometric. One system utilizes a disposable plastic cuvette that has 18 wells for bacterial identification tests. This identification system is unique in that it uses differential growth-inhibitory compounds rather than traditional biochemical tests. Each chamber in the lower compartment of the identification cuvette contains a different inhibitory compound. Some of these are antibiotics, some are dyes, and some are other types of chemicals. Suspensions of the organisms to be identified are prepared, and each is dispensed into a cuvette in the same way as for susceptibility testing. The cuvettes are incubated, and growth readings are taken in the photometer module. The results are stored and analyzed in a computer by a specially designed quadratic discriminant analysis program, and the pattern of differential growth inhibition is used to identify each organism. This system can identify a wide variety of bacteria, including both those that ferment carbohydrates and those that do not. It provides results in 3 h for most organisms, and the system is 95% accurate when compared to reference identification methods.

The susceptibility system that uses a plastic card can be used for bacterial identification also. The cards contain dried biochemical test reagents in the microwells, and these are reconstituted upon addition of bacterial suspensions to be identified. Processing of the identification cards is the same as for the susceptibility testing cards, and once the cards are placed in the reader-incubator all further testing is done by the instrument. The reader-incubator monitors the growth and the biochemical reactions in each microwell of the identification cards, and the information is processed by the computer for identification of each organism being tested. For gram-negative bacteria, results are generated in 4 to 8 h, depending on the organism, and the identifications are highly accurate when compared to those obtained by reference identification methods. Cards are also available for identification of gram-positive bacteria and yeasts, but these have not been thoroughly evaluated and their performance remains to be validated.

Another system uses a disposable plastic cartridge containing dried biochemical test reagents for identification of gram-negative bacilli. The cartridge has a paper cover attached across the top, and the bacterial suspension to be identified is introduced into the cartridge through perforations made in the paper by using a special device. The inoculated cartridge is incubated, and reagents are added to some of the wells to develop the biochemical test reactions. The cartridge is then inserted into a reading compartment in the analysis module, where reactions are read optically; the results are stored and analyzed in the control module. Bacterial identifications are reported after 5 h of incubation. This system is more limited in the groups of organisms that can be identified compared to the other identification systems that have been described, but it does provide approximately 90% accuracy when compared to reference identification methods.

Dilution testing. Systems for automation of dilution susceptibility testing are also used for bacterial identification. Conventional biochemical test media are incorporated in agar plates or in broth in the wells of plastic trays, and the multipronged inoculating devices or instruments used for susceptibility testing are used to inoculate the biochemical media with organisms. Many of these systems offer the advantage that organisms can be identified and tested for antimicrobial susceptibility in the same operation. After incubation, reagents are added to some of the media, and readings of the reactions are taken automatically by the instruments. After computer analysis of the reactions, organism identifications are printed. Most of these systems require overnight incubation, but at least one recently developed system offers identification after only 4 h of incubation. These systems provide 90–95% identification accuracy in the cases where critical performance evaluations have been done.

ANALYSIS OF URINE SPECIMENS

Various methods have been tried in designing instruments for evaluation of urine specimens. Some have been more successful than others.

Photometric. A major advance in microbiological testing is a system for the direct analysis of urine specimens which detects and identifies bacteria without using traditional cultures on agar media. The urine specimen is introduced into a plastic card in the vacuum-filling chamber, the card is placed in the reader-incubator, and growth is monitored auto-

matically. If sufficient growth occurs within 13 h of incubation, a report of bacteria in the urine specimen is issued. Many specimens are reported as positive within only a few hours of incubation depending on the organism involved and the original number of bacteria in the urine. Some of the microwells also contain differential inhibitory substances that allow identification of certain groups of bacteria, and this information is also provided by the instrument. If no growth occurs within 13 h of incubation, the urine specimen is reported as negative for bacteria. This system accurately detects 95% of bacteria present in urine in quantities of at least 100,000 colonies per milliliter.

Other designs are also available for detecting bacteria in urine specimens. To screen urine specimens for bacteria by using the original photometric system, a standard susceptibility testing cuvette is filled with sterile broth so that each chamber in the lower compartment receives an equal volume of broth. A small amount of each urine specimen is then pipetted into an individual chamber, and the cuvette is incubated. Readings are taken at intervals in the photometer module, and if the light-scattering exceeds a particular threshold the specimen is reported as positive for bacteria.

In another photometric system special ampules of broth are used for urine screening. A small volume of each urine specimen is pipetted into an ampule, and the inoculated ampules are inserted into a special holder. This is then placed into the analysis module, where incubation and growth monitoring are carried out automatically. When a preset threshold of optical density is exceeded, the instrument issues a positive report.

With either of these photometric instruments, urine specimens may be reported as positive for

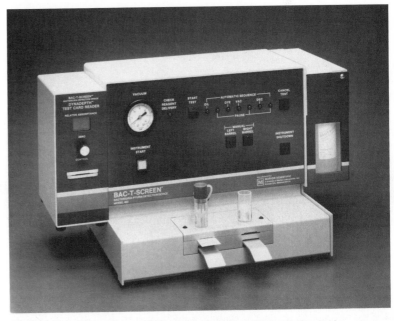

Fig. 3.   Bac-T-Screen urine-screening instrument. (*Marion Laboratories, Inc.*)

bacteria after 3 to 5 h of incubation. If sufficient growth is not detected within 5 h, the specimen is reported as negative for significant numbers of bacteria. Both systems only detect organisms; they do not identify the organisms present. Identification must be done for the positive specimens by culturing some of the positive broth or sampling the original urine specimen for culture on agar growth media. Nevertheless, these systems offer advantages over traditional methods for the detection of bacteria in urine specimens, because the presence or absence of organisms can be reported in a few hours with the instruments whereas traditional cultures require at least overnight incubation. One disadvantage of the automated urinalysis systems is that they do not detect bacteria reliably in quantities of less than 100,000 organisms per milliliter of urine, and such lower concentrations can be significant in certain groups of women with symptomatic urinary tract infections.

Electrical methods. A variety of other instruments have been evaluated for the detection of bacteria in urine specimens with limited success. Electrical impedance has been used, based on the idea that as bacteria grow in a fluid sample the resistance (impedance) to the passage of current across an electrical field is altered. Thus if bacteria are present in sufficient number, a change in impedance will be detectable. However, instruments developed to detect changes in impedance have not proven reliable for detection of bacteria in urine specimens.

Particle counting has also been applied to the detection of bacteria in urine specimens with variable success. The counter instrument draws fluid through a small orifice across an electrical field. As particles, such as bacteria, pass through the orifice, a voltage change occurs, and this change is related to the characteristics of the particle. By using a sophisticated computer analyzer, it is possible to recognize bacteria among other types of particles. Application of this technology to the detection of bacteria in urine specimens at first appeared quite promising, but subsequent experience has not been favorable.

Bioluminescence. Viable bacteria contain adenosinetriphosphate (ATP) which can be detected by light emission in the firefly luciferin-luciferase assay (the same system that makes fireflies glow). In a typical instrument, bacteria are treated so as to release their ATP, the luciferin-luciferase reagents are added, and light is emitted in proportion to the account of bacterial ATP present. A photometer in the instrument detects the light, and if it exceeds a predetermined threshold the specimen is reported positive for bacteria. Instruments of this type have been shown to be 90–95% effective for detection of bacteria in quantities of at least 100,000 organisms per milliliter of urine.

Filtration staining. In this method a device automatically draws a volume of each urine specimen through a filter (Fig. 3). The instrument then stains and washes the filter, and if bacteria are present a pink color develops. This sample instrument provides results in approximately 10 min, and it is 90–95% reliable for the detection of clinically significant numbers of bacteria in urine specimens. As with the photometric instruments, positive urine specimens detected by the filtration staining instrument require culture by traditional methods for the isolation and identification of bacteria. However, the ability to detect bacteria in urine specimens in 10 min is a significant improvement over other methods, and these results can be used to identify those specimens that require culture.

MICROBIOLOGICAL ANALYSIS OF BLOOD SPECIMENS

Detection of bacteria in blood specimens is one of the most important tasks in the clinical microbiology laboratory because of the life-threatening nature of bloodstream infections. Blood cultures have traditionally relied upon inoculation of blood specimens into bottles of broth, which are then incubated to allow growth of any organisms present in the original specimen. Bacteria are usually present in low numbers in bloodstream infections, and often detection requires growth for one or more days in broth. Traditional methods require daily visual inspection of the broth cultures for evidence of bacterial growth, sampling at intervals for subculture to agar media, and sampling for the preparation of stained slides

Fig. 4.   BACTEC NR660 nonradiometric blood-culturing system. (*Johnston Laboratories*)

for detection of bacteria by microscopic examination. These methods require one or more days before organisms can be detected. Fortunately, several new blood culture methods have become available that improve the speed of detection of bacteria in blood specimens.

One instrument performs radiometric detection of bacterial growth in blood cultures. Blood specimens are inoculated into bottles containing a special broth with radiolabeled substrates that can be metabolized by bacteria. If bacteria are present in the blood specimen, they will break down the substrates, releasing radioactive carbon dioxide, which collects in the airspace in the bottle above the broth surface. The instrument is designed with a sampling needle that is automatically introduced into the bottles through a rubber diaphragm. Gas is withdrawn through the needle and analyzed for radioactivity. If the amount of radioactive carbon dioxide in the bottle exceeds a predetermined threshold, the culture is reported as positive and samples can be withdrawn for identification and susceptibility testing.

This method is at least as sensitive as traditional broth culture methods for the detection of bacteria in blood specimens, and it detects positive blood cultures more rapidly. In addition, the instrument performs much of the work that has to be done manually in traditional culture methods, thus reducing the labor requirement for the detection of bacteria in blood specimens. One limitation is the need for radioactive materials. However, a new model (Fig. 4) of the instrument does not require radioactivity. This version is similar to the original instrument except that the carbon dioxide is detected by infrared analysis. Both the original radiometric and the new infrared systems represent major advances in the detection of organisms in blood specimens.

MICROBIOLOGICAL ANALYSIS OF RESPIRATORY SPECIMENS

One of the major contributions of instruments to microbiology has been in the detection of mycobacteria (the organisms that cause tuberculosis) in sputum specimens. Mycobacteria are very slow-growing organisms, and the diagnosis of tuberculosis by traditional methods, requiring culture on special coagulated egg media or agar media, is delayed for 3–6 weeks while the cultures incubate. With a radiometric system, the presence of mycobacteria can be detected in a few days, thus greatly aiding diagnosis.

In the specially adapted radiometric system (Fig. 5), sputum specimens from patients suspected of having tuberculosis are inoculated into bottles of broth culture media specially formulated for the growth of mycobacteria. These media contain radiolabeled substrates that are metabolized by mycobacteria, resulting in release of radioactive carbon dioxide that is detected by the instrument in the same way as for blood cultures.

The same system can be used for testing the sus-

Fig. 5. BACTEC 460TB radiometric mycobacteria-testing system. (*Johnston Laboratories*)

ceptibility of mycobacteria to antituberculous drugs, and the time required for results is again shortened for several weeks to a few days. The culture and susceptibility testing results obtained with the radiometric instrument correlate closely with those obtained by traditional culture methods, and this system is replacing traditional methods in many laboratories.

SUMMARY

Although instruments have been available for microbiological analysis for a relatively short time, they have made a significant impact on the laboratory diagnosis of infectious diseases. Microbiology instruments have proven useful for the detection of organisms in urine and blood specimens in particular, and they are very useful for the detection of mycobacteria in sputum specimens. In addition, instruments have made a valuable contribution to the identification and susceptibility testing of organisms isolated from a variety of clinical specimens. Microbiology instruments offer the potential for more rapid testing which can provide results earlier in the course of a patient's illness and therefore have greater impact on the management of infectious diseases. Many microbiology instruments also offer potential savings in labor costs and may contribute positively to reducing the costs of laboratory diagnosis of infectious diseases.

[MICHAEL T. KELLY]

Bibliography: A. Balows and A. C. Sonnenwirth, *Bacteremia: Laboratory and Clinical Aspects*, 1983; S. M. Finegold and W. J. Martin, *Diagnostic Microbiology*, 6th ed., 1982; J. M. Matsen, Bacterial identification systems, *Clin. Lab. Med.*, 5:1, 1985; R. C. Tilton, *Rapid Methods and Automation in Microbiology*, ASM Publications, 1981.

ADDITIONAL DIMENSIONS OF SPACE-TIME

THOMAS APPELQUIST

Thomas Appelquist is professor and chairperson of the physics department at Yale University. Recipient of a doctor's degree from Cornell University in 1968, he served on the Executive Committee of the Division of Particles and Fields from 1977 to 1980 and on the Experimental Program Advisory Committee of the Stanford Linear Accelerator Center from 1978 to 1981. In 1983, he was a member of the HEPAP Subpanel on New Facilities for the U.S. High Energy Physics Program. He is an associate editor for particle physics on the "Physical Review." Appelquist's research work has centered on gauge field theories of electroweak and strong interactions. Recently he has been working on Kaluza-Klein unified theories of gravity and gauge theories.

During the 1920s, efforts to unify electromagnetism with Einstein's general theory of relativity led to an inspired suggestion by T. Kaluza. His observation, later refined and developed by O. Klein, was that a simple and elegant unification could be obtained by starting with only Einstein's theory in five dimensions. If the extra spatial dimension could somehow become compacitified into a very small circle, it would then not be directly observable. However, the components of the metric associated with the extra dimension would remain as part of the low-energy four-dimensional theory. These could be interpreted as the Maxwell electromagnetic field together with a scalar field of the type considered many years later by C. Brans and R. H. Dicke.

GAUGE THEORIES

From the perspective of the 1980s, it is clear that the intriguing Kaluza-Klein suggestion was an idea whose time had not yet come in the 1920s. The concept of gauge invariance plays a more critical role in this new perspective. Under such an invariance, the laws of physics are unchanged under a group of transformations of the functions which describe the state of the system, the gauge symmetry group. A gauge theory is one whose properties can be derived from the fact that it obeys such a gauge symmetry.

The simplest example of a gauge theory is electromagnetism whose symmetry group is the one-parameter unitary group U(1) of phase rotations of the quantum-mechanical wave function which describes the state of the system. It is found that these phase rotations can be chosen independently at each point of space-time, provided that they are proportional to the charges of the particles, without affecting the physical consequences of the theory; the theory is thus said to have local gauge invariance. Furthermore, the properties of the electromagnetic field described by Maxwell's equations can be derived from this local gauge symmetry.

This local gauge invariance of electrodynamics is just the rotational invariance associated with the periodic extra dimension of the Kaluza-Klein theory. However, it is now believed that electromagnetism is only one part of the Weinberg-Salam electroweak gauge theory, whose symmetry group is the product of U(1) with the unitary group SU(2), and that the strong interactions are described by quantum chromodynamics, an SU(3) gauge theory.

It would seem that present-day efforts to implement the Kaluza-Klein idea should, at the very least, attempt to unify gravitation with all these gauge theories. It is possible to do this by starting with more than one extra dimension, if the extra dimensions curl up into a very small curved, symmetric space (such as a sphere) which then becomes interpreted as the "internal" group space. Thinking along these lines began in the 1960s and started to gain momentum in the mid-1970s after the blossoming of nonabelian gauge field theories. [In such theories the elements of the gauge symmetry group, for example SU(3), do not commute. Physically, this means that the gauge fields interact among themselves, whereas the photon, the quantum of the electromagnetic field, is electrically neutral and does not interact with itself.] The notion of extra dimensions has played an especially important role in supergravity theories, which attempt unification with gravity by incorporating supersymmetry. During the past several years, work on Kaluza-Klein theories has reached a remarkable level of intensity.

These efforts have surely deepened the understanding of Kaluza-Klein theories. Nevertheless, if a realistic unification of gravity with gauge field theories is to emerge from all this, a number of difficult problems will have to be solved.

FIVE-DIMENSIONAL MODEL

Starting in five dimensions, Einstein's general theory of relativity looks essentially the same as it does in four dimensions. There is simply one extra space coordinate. The theory is written in terms of the metric of five space $g_{MN}(x)$. Einstein's equations for this metric have many solutions. Perhaps the simplest is $g_{MN} = \epsilon_{MN}$, the metric of flat, five-dimensional Minkowski space.

Another solution that is of special interest is the product of flat four-dimensional Minkowski space with a closed circle of radius R in the fifth dimension (Fig. 1). [A circle is still flat in the sense that the curvature vanishes.] Within this "vacuum," particles could exist, described by fields which are periodic in the interval $0 \leq x_5 \leq 2\pi R$. It is this Kaluza-Klein solution which might be relevant to the four-dimensional world.

The five-dimensional metric can be written quite generally in the form of Eq. (1), where μ and ν run

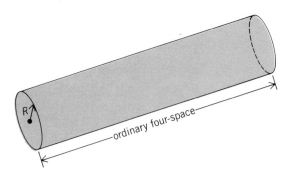

Fig. 1.   Representation of the five-dimensional model. The single long dimension of the cylinder represents the four dimensions of ordinary space-time. The fifth dimension is represented by the closed circle of radius R.

$$g_{MN} = \begin{bmatrix} g_{\mu\nu} + A_\mu A_\nu \phi & A_\nu \phi \\ A_\mu \phi & \phi \end{bmatrix} \qquad (1)$$

from one to four. The fields $g_{\mu\nu}$, and ϕ all depend on x_μ and x_5, with the periodicity requirement in x_5. The "vacuum" referred to above is $g_{\mu\nu} = \epsilon_{\mu\nu}$, $\phi =$ constant, and $A_\mu = 0$. If all these fields are Fourier-expanded on x_5, the x_5 derivatives, in Einstein's equations, will produce masslike terms of order $1/R^2$ for all but the zero term in the Fourier sum (the x_5-independent term).

Since the four-dimensional distances of interest are necessarily much larger than R, the "heavy" modes associated with nonzero terms in the Fourier sum can be expected to effectively decouple. There simply will not be enough energy available to produce them. If this approximation of neglecting the x_5 dependence is made, the five-dimensional theory reduces to Einstein's theory together with Maxwell's electromagnetic theory in four dimensions, where A_μ is the electromagnetic potential. In addition, the field ϕ will be present and can be identifed as the Brans-Dicke field.

In addition to the zero-mode theory, there are charged, massive modes with mass M_n and charge e_n given by Eqs. (2), where n is a positive or nega-

$$M_n = \frac{n}{R\phi^{1/2}}$$
$$e_n = \frac{nG^{1/2}}{2\pi R\phi^{1/2}} \qquad (2)$$

tive integer. Here, a system of units is used in which $\hbar = c = 1$, where $\hbar$ is Planck's constant divided by 2π and c is the speed of light. If the distance $2\pi R\phi^{1/2}$ around the fifth dimension could somehow be fixed relative to the Planck length, $G^{1/2}$, then both the mass and charge of these heavy particles could be computed.

Although the size of the fifth dimension is left undetermined in the classical theory just described, that may no longer be the case in the quantum theory. The vacuum fluctuations of the five-dimensional gravitational field can produce a force which could play a role in determining the value of R relative to the Planck length. This possibility will be discussed after a description of the classical theory in $D > 5$ dimensions.

BEYOND FIVE DIMENSIONS

There was no real reason to extend the Kaluza-Klein idea beyond five dimensions until the emergence of nonabelian gauge field theories. From their invention by C. N. Yang and R. L. Mills in 1954 until their great blossoming in the early 1970s, these theories were studied intermittently by only a few pioneers. In the middle of this era, in 1963, it was suggested that a unification of Yang-Mills theories and gravitation could be achieved in a higher-dimensional Kaluza-Klein framework. A detailed discussion of the Kaluza-Klein unification of gravity and Yang-Mills theories, including the correct form of the $(4 + N)$-dimensional metric, first appeared in 1968. The first complete derivation of the four-dimensional gravitational plus Yang-Mills plus scalar theory from a $(4 + N)$-dimensional Einstein-Hilbert action was finaly given in 1975.

The weakness of this higher-dimensional work was the absence of any good reason as to why any dimensions would compactify, let alone the right number, so as to leave the ordinary four-dimensional "large" world. While the five-dimensional theory at least admitted the compactified fifth dimension along with Minkowski space as a solution to the five-dimensional equations of motion, even this was not true of the higher-dimensional theories.

The essential reason for this is that the higher-dimensional manifolds that give rise to Yang-Mills theories have curvature. If a $(4 + N)$-dimensional Einstein theory is to compactify into the direct product of four-dimensional space-time M_4 and a compact internal space, the internal space will have curvature. But then Einstein's equations also force the other four dimensions to be highly curved. They simply cannot look like the familiar four-dimensional, flat, Minkowski world.

This problem was first addressed by including additional Yang-Mills and scalar matter fields in the higher-dimensional theory. Inclusion of these fields allows classical solutions in which space-time is the direct product of Minkowski space and a compact internal space of constant curvature. This "spontaneous compactification" is achieved, however, by going beyond the pure Kaluza-Klein framework and including extra fields in just such a way as to induce the desired compactification. This program of seeking solutions to the combined Einstein-Yang-Mills equations in $4 + N$ dimensions has been generalized to a larger class of internal spaces.

All this work on classical, higher-dimensional Kaluza-Klein theories provided a springboard for the study of both Kaluza-Klein supergravity and the quantum dynamics of Kaluza-Klein theories.

SUPERGRAVITY IN ELEVEN DIMENSIONS

The basic strong, electroweak and gravitational forces have bosonic quanta. On the other hand, the basic quanta of matter can be both fermionic and bosonic. In quantum theory the exchange of Fermi matter quanta also produces forces which, unlike basic forces, involve a change of the identity of the source (for example, by exchanging an electron between an electron and a photon, the original electron turns into a photon and the original photon into an electron), and these forces as a rule do not follow from a simple gauge principle. This distinction between matter and force bothered Einstein, since in general relativity the field equations involve the curvature, a geometric object on the force side, and the quite arbitrary energy-momentum tensor on the matter-source side. This raises the idea of unifying matter and force as different components of the same agency. Remarkably, this is exactly what happens in certain supergravity theories.

In supergravity theories particles have supersymmetric partners; for example, the gravitino, a fermion, is the partner of the graviton. The maximal, and in some sense the master, supergravity exists in eleven dimensions. Roughly speaking, there are no supergravities beyond eleven dimensions because the gravitino, a spin-vector, in twelve or more dimensions has more than 128 degrees of freedom. When reduced to four dimensions on a torus, this would imply a supergravity containing fields with spin larger than 2; this is generally believed to lead to inconsistencies. In eleven dimensions the gravitino has precisely 128 degrees of freedom, whereas the graviton has only 44. Since the numbers of Bose and Fermi degrees of freedom have to be equal in any supersymmetric theory such as supergravity, this means that Bose fields beyond gravity have to appear in eleven-dimensional supergravity. In fact, supersymmetry dictates that the missing Bose degrees of freedom be supplied in the form of a massless antisymmetric tensor field with three indices A_{MNP} which indeed has $128 - 44$ degrees of freedom, as shown in Eq. (3). Moreover, in eleven di-

$$\binom{11 - 2}{3} = \binom{9}{3} = \frac{9!}{3!6!}$$
$$= 84 = 128 - 44 \quad (3)$$

mensions there exist no matter and no Yang-Mills fields, so that besides gravity only its supersymmetric partner A_{MNP} and gravitino fields serve as matter. The source of gravity is thus fixed by supersymmetry. Force and matter uniquely determine each other; they are but different components of the same supermultiplet.

Several problems, however, stand in the way of producing a realistic theory. First of all, the four-dimensional space has an intolerably large cosmological constant (equivalently, it has far too much curvature) that has to be eliminated somehow. Various ideas have been invoked to produce a flat four-dimensional space-time, though no clear and workable mechanism is yet known. Another serious problem is the lack of chiral fermions (fermions interacting with a left-handedness), which is further discussed below.

KALUZA-KLEIN PHENOMENOLOGY

Eleven-dimensional space-time recommends itself as the habitat of the maximal supergravity theory. Remarkably, there is also a phenomenological argument for eleven as the minimal number of space-time dimensions. If the SU(3) × SU(2) × U(1) gauge symmetry of the electroweak theory combined with quantum chromodynamics in four dimensions is to originate in symmetries of a compact manifold in N "hidden" dimensions [just as the U(1) guage symmetry of electrodynamics originates in the rotational invariance of the single extra dimension of the original Kaluza-Klein theory], then these extra dimensions must be at least seven in number. This follows from the observation that no manifold of dimension three or smaller can have more than six isometries, and thus the eight-parameter group SU(3) can most economically appear as the isometry group of the four-dimensional manifold $CP(2)$. Similarly, an SU(2) × U(1) gauge symmetry in four dimensions is most economically obtained from the isometries of the three-dimensional manifold $S^2 \times S^1$. [Here S^N is the N-dimensional sphere, the locus of points at unit distance from a given point in $(N + 1)$-dimensional euclidean space. Thus, S^1 is the circle and S^2 is the ordinary sphere.] Thus the gauge group of low-energy physics is obtainable from the isometries of the seven-dimensional manifold $CP(2) \times S^2 \times S^1$. However, this manifold is not a solution to the Einstein equations, and as such not relevant as the compactification of eleven-dimensional supergravity. Fortunately, compact manifolds with the isometry group SU(3) × SU(2) × U(1) other than $CP(2) \times S^2 \times S^1$ exists (S^7 is not among them), and some of these are solutions to Einstein's equations.

QUANTUM EFFECTS

It was pointed out by H. B. G. Casimir in 1948 that the zero-point quantum fluctuations of the free electromagnetic field give rise to an attractive force between two parallel, infinite, perfectly conducting

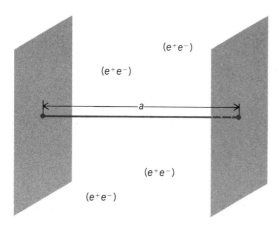

Fig. 2. Casimir effect. The electromagnetic quantum fluctuations (virtual e^+e^- pairs) between two parallel conducting plates separated by distance a give rise to an attractive force proportional to $1/a^3$.

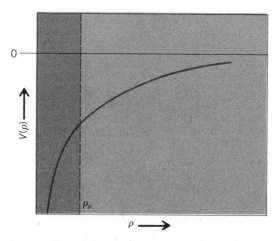

Fig. 3. The quantum potential V in the five-dimensional theory as a function of the radius ρ of the fifth dimension. Below the Planck length ρ_P, the lowest-order computation can no longer be trusted.

plates (Fig. 2). The presence of the plates imposes boundary conditions on the field, which then modify the zero-point energy in a way that depends on the separation between the plates.

The physics of the Casimir effect generalizes to any situation in which boundary conditions are imposed on a quantum field. In all such cases, observable forces are expected to be generated; they may be either attractive or repulsive. For example, if the parallel plates are replaced by a thin, perfectly conducting spherical shell, the Casimir force changes sign.

It is clear that a typical Kaluza-Klein theory, with a space of the form $M_4 \times B$, will naturally exhibit the Casimir effect because of the restriction imposed by the manifold B. For example, in the case that B is a circle of radius ρ, the restriction is just that the momenta are quantized in units of $1/\rho$. In the formulation of perturbative quantum gravity in higher dimensions, it has been found that if the total number of space-time dimensions is odd, then, to lowest order in perturbation theory, no infinities appear in the necessary integrations.

Five-dimensional case. In the five-dimensional case, it was shown, on purely dimensional grounds, that the zero-point (Casimir) energy per unit three-volume must be of the form of Eq. (4), where ρ is

$$V(\rho) = \frac{C}{\rho^4} \tag{4}$$

the radius of the circle and C is a pure number. (Actually, it is proportional to Planck's constant h, which means it can be thought of as induced by one-loop quantum effects.) Because the total space-time dimensionality is odd, C is finite without the need for any subtraction. If a dimensional cutoff is used, then the cutoff dependence can be absorbed in a redefinition of the cosmological constant. The number C turns out to be $-\zeta(5)(2\pi)^4 15/4\pi^2$, where $\zeta(s)$ is the Riemann zeta-function [$\zeta(5) \simeq 1.04$]. Thus there is an attractive force (Fig. 3), just as in the original Casimir effect, which tends to make the extra dimension contract.

This work has been generalized in a variety of ways. The same problem has been studied in the presence of finite temperature. A critical radius ρ_c is found to exist, such that for $\rho > \rho_c$ the extra dimension contracts due to the Casimir effect, whereas for $\rho < \rho_c$ thermal effects make the extra dimension expand. The addition of massive fermions has been shown to produce a repulsive force at short distances that can stabilize the extra dimension; that is, when considered together with the Casimir energy, the fermion contribution produces a potential with a minimum. In the case of N extra dimensions, compactified into a N-dimensional torus, it has been found that one of the dimensions will contract, but that the other $(N - 1)$ dimensions will expand.

Curved internal manifolds. An important step was taken by P. Candelas and S. Weinberg, who considered the case in which the internal manifold has nonvanishing curvature. In particular, they considered the set of manifolds $\mathcal{M}_4 \times S^N$, where $\mathcal{M}_4$ is four-dimensional Minkowski space, and S^N is again the N-dimensional sphere. For the reasons alluded to above, N is taken to be odd. They added a bare cosmological term, as well as a set of matter fields (either scalars or spinors). The potential energy is then taken to consist of three terms: (1) the term arising from the curvature of the N-sphere (not present in the case $N = 1$); (2) the cosmological constant term; and (3) the quantum Casimir energy of the matter fields. They neglect the Casimir energy of the gravitational field itself. Their results are, first, that it is often possible to find a stable solution to the equations of motion. Such a solution requires not only that the potential have a minimum as a function of the radius ρ of the sphere, but also that the potential vanish at this minimum (Fig. 4). This means there is no net cosmological constant, and therefore the scheme in which the four-dimensional space-time is Minkowski space is a consistent one. Furthermore, stability requires the potential to be repulsive at $\rho = 0$, which in turn means that the Casimir force must be opposite in sign to what it was in the case $N = 1$. The fact that this happens is reminiscent of the change of sign in the ordinary electromagnetic Casimir effect in going from parallel plates to a spherical shell.

Perhaps most interesting, the existence of a stable minimum determines the radius of the N-sphere, which in turn, according to standard Kaluza-Klein ideas, fixes the value of the gauge coupling constant [in the case of an N-sphere, the gauge group is 0 $(N + 1)$]. Kaluza-Klein theories are thus possibly unique in offering a dynamical determination of the gauge couplings.

A somewhat disturbing feature of the Candelas-Weinberg work is that the contribution of any one matter field (either scalar or spinor) to the Casimir energy is anomalously small; that is, it contributes to the coefficient of $1/\rho^4$ a number which turns out typically to be of order 10^{-4} or 10^{-5}. As a conse-

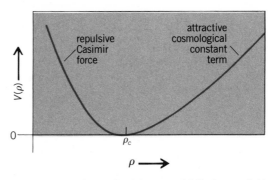

Fig. 4. The Weinberg-Candelas potential V of a manifold with N "internal" dimensions compactified into an N-sphere of radius ρ. The stable point ρ_c can be arranged to have $V(\rho_c) = 0$.

quence, if only a small number of matter fields is included, the minimum of the quantum-corrected potential will be much smaller than the Planck length, implying a coupling constant much larger than unity. To remedy this situation, Candelas and Weinberg are forced to contemplate theories with a large (that is, 10^4 or 10^5) number of matter fields.

Gravitational field contribution. A more natural and physical generalization of the five-dimensional study is to consider the contribution of the gravitational field itself to the Casimir energy with a curved internal manifold. The computation is much more involved than in the case of matter fields, and there are minor discrepancies among the three groups that have published results. Nevertheless, there is general agreement on the following two points: (1) the gravitational contribution per degree of freedom is much larger than the matter-field contribution; (2) however, the gravitational field can also contribute an imaginary part to the potential, indicating an instability in the system, and it seems that there is no minimum of the real part of the effective potential Re V in the region where the imaginary part Im V vanishes. (Minima of Re V have been found in cases where Im $V \neq 0$.) Thus gravity alone at the one-loop level cannot produce a stable, spontaneously compactified solution of the form $M^4 \times S^N$ (N odd). It remains to be seen whether a suitable mix of gravitational and matter fields (such as would be present, for example, in a supergravity theory) can lead to a different result. These questions are actively being studied.

PROBLEM OF CHIRAL FERMIONS

One of the striking properties of the observed quarks and leptons is that they are chiral; that is, the left-handed and right-handed fermions transform under different representations of the $SU(2) \times U(1)$ gauge group. In a Kaluza-Klein theory, these fermions presumably are among the massless modes of some kind of spinor fields in a higher number of dimensions. The question arises as to whether the chiral property descends from the higher-dimensional theory, or whether it is a low-energy manifestation. If the latter, then at some not-too-high energy (that is, some energy insignificant compared to the compactification scale) mirror partners to the observed fermions will be discovered, such that the combined spectrum of original and mirror fermions is no longer chiral.

Arguments against mirror fermions. There are several arguments against the possibility of mirror fermions, among them: (1) the representations of the observed quarks and leptons repeat themselves in identical families; no mirror partners have yet been discovered; and (2) for some miraculous reason, the anomalies that might have infected a chiral theory cancel in the observed spectrum. This removes a potential reason for postulating the existence of the mirror fermions, namely that they would necessarily cancel any anomalies present at the chiral level.

Exclusion of chiral fermions. The prevalent opinion, then, is that within Kaluza-Klein theories the chiral property derives from a chiral spectrum of fermions in higher dimensions. However, it has been found that the spectrum of fermions on any eleven-dimensional manifold $M = M_4 \times B$, where M_4 is four-dimensional space-time and B is a compact seven-dimensional manifold admitting SU(3) $\times$ SU(2) $\times$ U(1) symmetry, cannot be chiral. In fact, it has been shown, using the algebraic properties of the gamma matrices in higher dimensions, that the spectrum of fermions could be chiral only if the dimensionality is 2 mod 8. Finally, using topological arguments, it has been shown that, in any number of dimensions, the Dirac operator cannot admit a chiral spectrum, and neither can the Rarita-Schwinger operator (with the possible exception of internal manifolds that are not coset spaces).

Remedies. The possible evasions of these results within the context of standard Kaluza-Klein ideas are sufficiently unattractive that some rather unconventional remedies have been proposed.

Additional gauge fields. Perhaps the least radical of these is the suggestion that additional gauge fields be included in the higher-dimensional theory. If the compactification involves a topologically nontrivial configuration of these gauge fields, the fermionic spectrum can indeed be chiral. The addition of extra gauge symmetry is precisely the mechanism whereby the now fashionable superstring theories allow for the appearance of chiral fermions.

Noncompact internal manifolds. The proof of the above results requires the internal manifold to be compact. Examples of noncompact internal manifolds of finite volume have been given which admit isometries and also allow chiral fermions. Furthermore, it has been shown that such manifolds can have at least a partially discrete mass spectrum. A further example of the use of noncompact internal manifolds, with relevance to supergravity, has been given.

Quasi-riemannian geometry. The idea of quasi-riemannian geometry has also been proposed as a remedy for the chiral fermion problem. In such a geometry, in D-dimensions the tangent-space group (the flat space that approximates the curved space in the neighborhood of any point) should be chosen to be something other than $0(D - 1, 1)$, the group of rotations in $D - 1$ space and 1 time dimensions. Of course, the tangent-space subgroup corresponding to ordinary space-time must always turn out to be $0(3,1)$. Spinors in a curved manifold actually transform as scalars under general coordinate transformations and spinors under the tangent-space group, so this modification of the tangent-space structure might be expected to wreak havoc with the orderly nonchiral structure of spinors in ordinary riemannian geometry. Analysis of the possible tangent space groups shows that under the appropriate circumstances the fermionic spectrum can indeed be chiral.

PROSPECTS

It now seems that the Kaluza-Klein program has several obstacles to overcome if it is to meet with any real success. The difficulty of obtaining chiral fermions is very troubling, and the role of quantum effects remains mysterious. Attention has largely switched from the traditional local-field framework of Kaluza-Klein theories to the more exotic superstring theories. These, too, postulate the existence of extra dimensions and face many of the same problems in trying to make contact with four-dimensional phenomena. Although presently this goal still seems distant, there seems to be little doubt that theorists will continue to be attracted to higher-dimensional theories. There is a compelling beauty about them, strongly suggesting that the world might really contain extra dimensions.

[THOMAS APPELQUIST]

Bibliography: T. W. Appelquist and A. Chodos, Quantum dynamics of Kaluza-Klein theories, *Phys. Rev.*, D28:772–784, 1983; T. W. Appelquist, A. Chodos, and P. Freund, *Modern Kaluza-Klein Theory*, 1986; N. N. Khuri (ed.), *Proceedings of the Shelter Island Conference*, 1985; A. Salam and J. Strathdee, On Kaluza-Klein theory, *Ann. Phys.*, 141:316–352, 1982; E. Witten, Search for a relativistic Kaluza-Klein theory, *Nucl. Phys.*, B186:412–428, 1981.

A-Z

Abscisic acid

Abscisic acid (ABA), one of the five known plant growth substances, is involved in many aspects of plant growth and development, including stomatal closure, seed development and dormancy, root growth and gravitropism, and senescence. Though it is known that ABA plays a role in many of these physiological processes, the mechanism of regulation of these processes is not known. Various aspects of the physiological role of ABA have been studied, including the biosynthetic pathway, ABA catabolism, regulation of ABA biosynthesis during development of the plant and by the environment, and movement of ABA within the plant.

Biosynthetic pathway. ABA is a chiral molecule, occurring as (+)-(S)-abscisic acid in the natural form (Fig. 1). The chiral center is at carbon 1'.

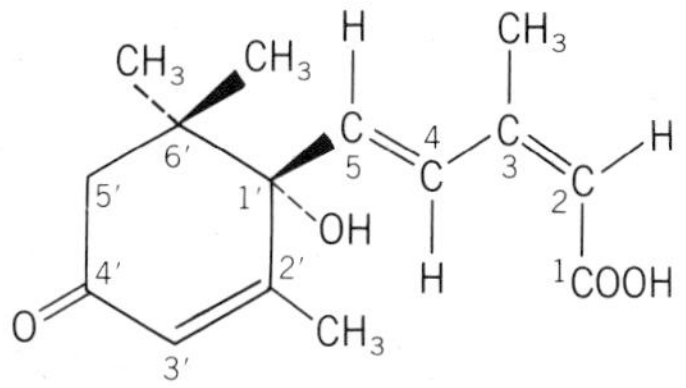

The complete elucidation of the biosynthetic pathway of ABA has yet to be accomplished. Though it is thought that ABA is ultimately derived from mevalonic acid, the immediate precursors to ABA have not been identified. Two possible pathways have been proposed for the biosynthesis of ABA, the direct pathway of biosynthesis and the carotenoid pathway. In the direct pathway ABA is thought to be derived from a C-15 precursor, presumably farnesyl pyrophosphate, although the immediate precursor has not been identified in higher plants. In the carotenoid pathway, the proposed precursors to ABA are C-40 carotenoid-type precursors. Violaxanthin, a carotenoid, can be converted to 2-*cis*-xanthoxin, which in turn may be the direct precursor to abscisic acid.

Recent evidence supports the carotenoid biosynthetic pathway of ABA. Experiments using ^{18}O labeling show that only one labeled oxygen is incorporated into ABA, at carbon 1 of the carboxyl group. If ABA were synthesized directly from mevalonic acid, all of the oxygen atoms in the ABA mol-

Fig. 1. The structure of natural, (+)-(S)-abscisic acid. The numbering scheme of the carbon skeleton is shown.

ecule would be labeled. Therefore, ABA is probably synthesized from a relatively stable pool of precursors, such as the carotenoids.

Site of synthesis. It was once thought that ABA was synthesized in the chloroplast, principally because of the large concentration of ABA found there. In addition, when radiolabeled mevalonic acid was applied to isolated chloroplasts, there was a small amount of the radiolabel incorporated into ABA. Inhibitors of chloroplast protein synthesis do not prevent synthesis of ABA whereas cytoplasmic protein synthesis inhibitors block ABA biosynthesis. Therefore, it is believed that the site of ABA biosynthesis is the cytoplasm.

Catabolism. The ABA molecule is readily metabolized to other compounds, or catabolized, in the plant. Catabolism of ABA can occur through two different pathways, the ABA-ester and the PA-DPA pathways (Fig. 2). Conjugation of ABA to a glucose molecule through the carboxyl group in the side chain results in ABA glucose ester (ABA-ester), which is stable in the plant and localized within the vacuole. The PA-DPA pathway of catabolism results in the rearrangement of ABA to phaseic acid (PA), which in turn may be reduced to dihydrophaseic acid (DPA). These catabolites may also be conjugated to glucose at the side chain. The particular ABA catabolite that accumulates in the plant is spe-

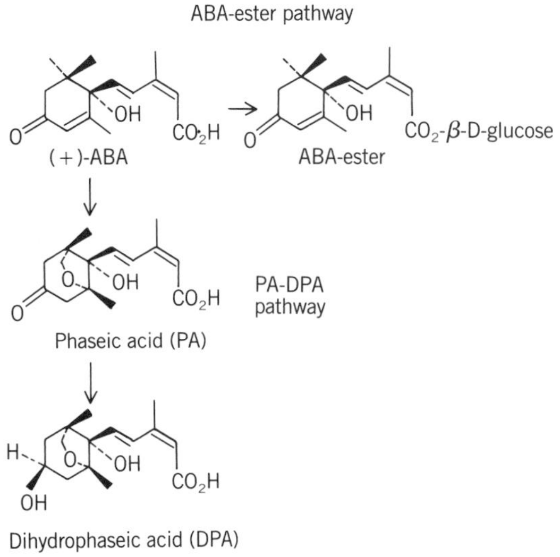

ABA-ester pathway

Fig. 2. Catabolism of abscisic acid, showing the two known pathways.

cies dependent. PA accumulates in tomato, for example, and DPA accumulates in legumes.

The physiological activity of these catabolites is largely unknown. However, in cases where it has been established, any such activity has been less than that of ABA.

Regulation of ABA levels. The biosynthesis of ABA is regulated by several different events during the course of plant growth and development. There are programmed changes in the amount of ABA in different organs of the developing plant. For example, during seed development the amount of ABA within the seed increases during the early stages of development and then decreases during seed desiccation.

The amount of ABA in the plant is also regulated by the environment. During periods of drought, the amount of ABA in the plant increases as much as 40-fold. The loss of water during periods of drought causes the turgor potential of the cells to decrease. This is thought to trigger the increased accumulation of ABA in the plant.

Transport. ABA moves readily within the cell and throughout the plant. It has been shown that the high concentration of ABA in the chloroplast is due to the pH gradients within the cell. ABA, a weak acid, readily moves through the chloroplast membrane when it is protonated. When ABA moves into the higher pH of the chloroplast, the ABA molecule is dissociated. Since the dissociated molecule does not move readily through the chloroplast membrane, the chloroplast is an anionic trap for ABA and could serve as a site of storage. When the pH gradients within the plant change, ABA is released from the chloroplast. Thus changes in pH would result in an immediate supply of ABA.

ABA is readily transported throughout the plant, moving in both xylem and phloem. In order to follow the transport of ABA within the plant, radiolabeled ABA has been injected into the plant and the distri-bution of the radiolabel within the plant has been followed at time intervals. Within several hours of application, the radiolabel can be detected in all parts of the plant; thus ABA can move up or down from the point of application. Frequently, the largest concentration of radiolabeled ABA is found within the youngest leaves and developing buds.

Physiological role. ABA has physiological importance in many fundamental plant processes.

Closure of stomata. ABA promotes closure of the stomata. During periods of drought, stomatal closure helps to prevent loss of water from the plant due to transpiration. Stomatal closure is mediated by movement of ions between the guard cells that make up the stomata and the surrounding free space. Changes in ion concentration result in increases or decreases in turgor, which result in open or closed stomata, respectively. It is hypothesized that the application of ABA can result in stomatal closure by inhibiting the uptake of potassium (K^+) by the guard cells, causing decreased turgor and therefore closed stomata. Three proteins located on the outside of the plasmalemma of the guard cell that specifically bind $(+)$-ABA are beginning to be characterized. These proteins may be involved in the regulation of stomatal aperture by abscisic acid.

Seed development and dormancy. The changes in ABA concentration during the development and maturation of seeds is involved in the prevention of precocious germination. ABA also promotes seed development and the accumulation of seed storage proteins. Work to evaluate the molecular mode of action of ABA during seed development is ongoing.

Root growth and geotropism. ABA has been found to both stimulate and inhibit growth depending on the concentration used, high concentrations (above physiological level) causing inhibition. Changes in ABA synthesis have been correlated with geotropism in the root but there is no conclusive evidence for the regulation of geotropism by ABA. Environmental factors such as water stress affect the ABA level in the root, which results in changes in root growth and uptake of ions.

For background information *see* ABSCISIC ACID; PLANT GROWTH; PLANT HORMONES; PLANT PHYSIOLOGY in the McGraw-Hill Encyclopedia of Science and Technology.

[ELIZABETH A. BRAY]

Bibliography: F. T. Addicott (ed.), *Abscisic Acid*, 1983; R. A. Creelman and J. A. D. Zeevaart, Incorporation of oxygen into abscisic acid and phaseic acid from molecular oxygen, *Plant Physiol.*, 75:166–169, 1984; D. C. Walton, Biochemistry and physiology of abscisic acid, *Annu. Rev. Plant Physiol.*, 31:453–489, 1980.

Agronomy

Information in agronomy has two components. One is knowledge disseminated in printed form. The other, equally important, is heuristic knowledge, gained through years of experience. Heuristic knowledge is the private component, consisting of commonsense principles or working "rules of thumb" that are rarely formalized or published. Thus knowledge is actually the sum of facts plus heuristics, the working body of experience.

Expert systems. A more efficient alternative to traditional computer database management approaches might be provided by expert systems. They have the ability to evaluate the relative importance that each piece of knowledge contributes to the whole. Such systems are not without limits. In order to be successful, an expert system must approach the level of human expertise, within a defined domain.

There are two important components in the development of an expert system. One is the expert who provides the knowledge to the system. The other is the knowledge engineer who, through a series of interviews, extracts knowledge and assists in formalizing the expert's heuristic knowledge. The knowledge engineer then converts the knowledge to a form that can be utilized by the computer.

An expert system is made up of several components that interrelate in a very specific manner (Fig. 1). Knowledge, acquired through an expert by any number of tools, is placed in a knowledge base. The knowledge base is then called on by an inference system which makes decisions based on the input of a nonexpert user. The inference system can also provide the user with advice or explanations.

The knowledge base is not a black box. Information held within the knowledge base is organized in a set of rules. The rules contain a series of conditions that have to be at least partially satisfied for a conclusion to be drawn. Figure 2 shows an example of such a rule, which can be used to identify a weed as Bermuda grass. This rule states that either of two sets of conditions leads to the conclusion. If the six vegetative or the five floral elementary conditions are met or are partially satisfied, then the weed is identified as Bermuda grass. The numbers on the

Bermuda grass		
If 1. Habit is rhizome-&-stolon,	(85%)	
2. Blade-width is fine to medium,	(60%)	
3. Ligule is ciliate,	(50%)	vegetative
4. Sheath is compressed,	(45%)	conditions
5. Collar is narrow,	(25%)	
6. Auricle is absent,	(10%)	
	OR	
1. Flower is spike,	(80%)	
2. Florets = 1,	(75%)	floral
3. Glumes are shorter,	(35%)	conditions
4. Disarticulate is above,	(35%)	
5. Awns are absent,	(15%)	
Then Weed is Bermuda grass.		

Fig. 2. A rule for a knowledge base; to be used for identification of a weed as Bermuda grass.

right are degrees of certainty which indicate the relative importance of each elementary condition toward the conclusion. This degree of certainty represents the weed specialist's (domain expert) degrees of confidence or certainty that the elementary condition supports the conclusion by itself. While the rule is complex, it takes the following general form: If CONDITION, then DECISION.

Potential applications. Current applications of expert systems in agricultural extension are not yet available. The development of expert systems in general is recent, and their proposed use in agriculture is even more recent. Attempts to use this technology began in 1976 with the development of an expert system to diagnose diseased soybean plants, PLANT/ds. PLANT/ds was first developed on a mainframe computer, using the meta-expert system, ADVISE, as the base or shell. One of the goals of this system was to transfer the final expert system to a small, portable microcomputer. This task was completed in 1982 with the successful downloading of PLANT/ds to a personal computer. One of the limitations of the system was that corrections or additions to the knowledge base require that editing take place on a minicomputer where the ADVISE system now resides, then downloading the newly defined expert system to the personal computer.

A second system, PLANT/cd, was also developed to provide assistance in determining the damage from cutworms in corn. The second system was unique in that it used both the surface modeling techniques of expert systems and a deeper simulation model for the development of cutworms. PLANT/cd has not yet been implemented due to limitations in knowledge regarding cutworm damage.

These first two programs provided the emphasis for the current activity on expert systems development, covering two areas associated with the management of turfgrasses. One system being evaluated is WEED, an expert system for identifying unknown weeds found in turfs. A second system, TEA, is also under development to assist in the establishment

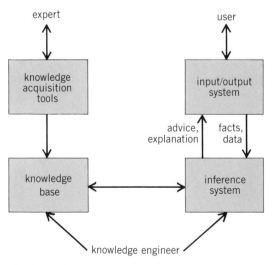

Fig. 1. Basic structure of an expert system. *(After E. A. Feigenbaum and P. McCorduck, The Fifth Generation, Addison-Wesley, 1983)*

and maintenance of turfgrasses under all levels of maintenance. An expert system development environment or shell, AgAssistant, is being constructed on a personal computer to expedite the process of building future systems.

Other expert systems are also under development; however, most programs are in their infancy, and working systems have not yet been developed. While expert systems technology shows great promise for providing quick, efficient answers with a minimum amount of resources, the final evaluation of these techniques is yet to come. It is hoped that within the next 5 years a number of working systems will be available, providing direct answers to current problems. Much research is still needed to develop the appropriate methods of applying this technology to agricultural use.

Knowledge acquisition. The area of knowledge acquisition is a serious bottleneck in the production of new expert systems. The greatest time committed to the development of an expert system is in the identification and coding of appropriate knowledge for use in the system. An approach is needed to speed up the method of interviews between a knowledge engineer and an expert. A data induction program has been developed that uses a matrix of data consisting of various attributes of classes of information and converts them into rules.

An example of the effectiveness of this program was illustrated with the development of PLANT/ds. A list of conditions present in the environment or on the plant was prepared at the time of the diagnosis of diseased soybean plants in a plant clinic. Two hundred ninety diagnosed cases, along with the appropriate descriptions, were utilized in the inductive learning program to develop a set of induced rules. A second set of rules was derived through the interview of plant pathologists. In order to evaluate the performance of each set of rules, those induced versus those derived from experts, each set was used to analyze 340 unknown diseased plants. A summary of the performance of this test indicates that the inductively derived rules operate at a higher level of performance than the expert-derived rules (see the table).

When the first choice of the diagnosis, the one with the highest confidence level, was considered, the inductively derived rules produced the correct answer over 97% of the time. When all suggested

Summary of results from PLANT/ds*, in percent

Type of rules	1st choice	Correct	Not diagnosed	Ratio	Threshold
Expert-derived	71.8	96.9	2.1	2.90	0.65
Inductively derived	97.6	100.0	—	2.64	0.80

*Modified from R. S. Michalski et al., PLANT/ds: An expert consulting system for the diagnosis of soybean diseases, *Proceedings of the 1982 European Conference on Artificial Intelligence*, Orsay, France, July 12–14, 1982.

diagnoses were considered, the inductively derived rules suggested the correct answer for every case. This, by no means, indicates perfection. Both rule bases are operating within a very narrow domain. It does indicate, however, that within specified domains an expert system can offer a high level of performance in comparison to human diagnoses.

While conceptual limitations are still present within expert systems, their utilization on microcomputers is still largely limited by memory or storage capacity. As microcomputer technology expands, the versatility of expert systems will increase.

For background information *see* AGRONOMY; ARTIFICIAL INTELLIGENCE in the McGraw-Hill Encyclopedia of Science and Technology.

[THOMAS W. FERMANIAN]

Bibliography: E. A. Feigenbaum and P. McCorduck, *The Fifth Generation*, 1983; T. W. Fermanian, R. S. Michalski, and B. Katz, An expert system to assist in weed identification for turfgrass management, *Proceedings of 1985 Summer Computer Simulation Conference*, Chicago, Society for Computer Simulation, July 22–26, 1985; R. S. Michalski et al., PLANT/ds: An expert consulting system for the diagnosis of soybean diseases, *Proceedings of the 1982 European Conference on Artificial Intelligence*, Orsay, France, July 12–14 1982.

Air pollution

The study of the effects of air pollution on agricultural crop production has many approaches, including defining the polluted environment, determining the physiological and biochemical responses of plants to pollutants, examining effects on growth and quality of crops, studying interactions of pollutants, and placing pollution responses in perspective with other environmental stresses and crop hazards. It is also important to view the crop production concerns as part of a far wider pollution impact on the plant world, which includes natural ecosystems and forests. New developments in this area include concern about the relationship between gaseous pollutants and acidic precipitation, crop response to trace organic and inorganic contaminants, methods of accurately assessing crop damage, the interactions of pollutants, and methods of protecting plants through genetic manipulation and modification of cultural practices.

Pollutants. Most gaseous and particulate air pollution is a product of industrial and urban activities. These pollutants may be moved by air masses within weather systems and influence crop production over wide areas. The most important gaseous air pollutant affecting agriculture is considered to be ozone, a highly reactive oxidizing agent. Other widely occurring gases include sulfur dioxide, nitrogen oxides (nitric oxide and nitrogen dioxide), and fluoride. Many other gases are toxic to plants, but damage is usually confined to accidental releases in localized areas. Such gases include hydrogen chloride, ethylene, hydrogen sulfide, ammonia, and chlorine. Heavy metals contained in particles may be toxic to plants when the particles fall on plant leaves or enter the soil. Sulfur dioxide and nitrogen oxides dissolve in water droplets to form sulfuric and nitric acid, respectively, resulting in acidic precipitation.

Crop uptake. The response of crop plants to air pollution is complex, and it involves mechanisms of uptake of pollutants and injury to various processes in plant cells and tissues. Crop responses are measured as visible injury, growth retardation, and yield and quality losses (Fig. 1). Responses occur at the cellular, organ, whole-plant, and plant community levels, and the damage done depends upon pollutant dose, which is the combination of pollutant concentration and exposure duration. The effect of a single pollutant may be changed by the presence of other pollutants, external environmental factors, and internal conditions such as physiological status and genetic constitution. Knowledge of the influence of factors such as climatic, soil, and physiological status allows provision of cultural, protectant, and genetic (varietal) manipulations to limit air-pollution damage.

Gaseous pollutants are taken up by plant leaves through the stomatal pores, with the amounts taken up largely determined by the stomatal opening and by air movement that brings polluted air to the leaf surface. The toxic gases diffuse through the stomates to the mesophyll cells, where they dissolve in the water in the cell walls. The dissolved gases pass through membranes and enter the living protoplasm, where injury occurs.

Particles that fall on leaves are trapped on the cuticle, and at least some of the toxic ingredients in the particles migrate into the leaf tissue. While the uptake of gases can be decreased by closing stomata with darkness or chemicals, the uptake of toxicants deposited on the leaf surface is dependent on natural features of the leaf, such as the presence of trichomes or a rough cuticular layer, that foster retention of particles.

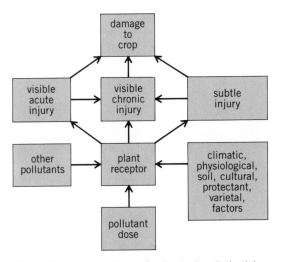

Fig. 1.   Relationship of factors leading to air-pollution injury to plants. (*After W. J. Kender and P. L. Forsline, Remedial measures to reduce air pollution losses in horticulture, HortScience, 18:680–684, 1983*)

Fig. 2.   Visible symptoms of ozone injury on a tobacco leaf. The chlorotic stipple symptom is characteristic of the species, and was called weather fleck because it occurred under bright, warm and humid weather conditions. Controlled exposures to ozone proved that ozone was the causal agent and that the weather conditions were conducive to high ambient ozone levels.

Mechanisms of injury. Air pollutants injure plant tissue by several mechanisms. Cell membranes are injured, enzyme activity is affected, chlorophyll is destroyed, and a host of other primary and secondary effects occur. The first obvious effect of pollution injury is the development of visible injury symptoms on the leaves of affected plants. These injury symptoms generally differ according to the particular pollutant involved. In addition, different plant species may have somewhat different symptoms. Many pollutant injury causes can be diagnosed on the basis of visible symptoms. In general, ozone injury is characterized by chlorotic mottling or flecking on the upper leaf surface (Fig. 2); in many species the chlorosis is associated with browning or bronzing of the tissues. Injury caused by sulfur dioxide and nitrogen oxides often takes the form of interveinal bifacial necrotic lesions, while fluoride injury usually consists of marginal and tip necrosis that takes on a bright orange color. The most obvious visible symptom of injury by other gases, metals, and acidic precipitation is some form of chlorosis associated with the loss of chlorophyll in affected tissues.

Yield. While much information about crop injury symptoms and growth effects in response to gaseous pollutants has been gained through controlled environment research and field observations, little definitive information on yield effects has been available until recently. Now the effects of ozone on yield of economically important crops are much more clearly understood because of the initiation of the National Crop Loss Assessment Network, which is composed of several cooperating programs with field sites in different regions of the United States. Plants are grown in the field by using open-top chambers for the control and manipulation of pollutant concentration around the crop plants. This research has demonstrated that the yield of many crops is reduced by the ozone concentrations occurring outdoors in the

United States. For example, yield losses of up to 20% for cotton, 12% for peanut, 18% for soybean and tomato, and 40% for wheat may be occurring at some locations. These yield reductions have been used in economic analyses to provide estimates of monetary loss to producers and consumers of the crops. A 25% decrease in ozone levels below current levels in the United States would increase the income of producers and result in savings to consumers that combined would be approximately $1.7–1.9 billion.

Much less attention has been given to the effects of particulates on crop growth and yield. Particulates may have a complex chemical composition that makes scientific experimentation difficult. In the particulates that fall on crop plant leaves or on the soil surface, there are many heavy metals known to be injurious to plants. Dose response relationships have been established for visible injury, growth, and yield effects of single and mixed metals in the rooting medium. For example, nickel and copper both decrease the growth of tomato, and there is generally additive joint action when the metals are combined. Nickel is much more injurious than copper. Much less is known of the role of particulate deposited on leaves. The leaf surface will absorb metals from particulates to the detriment of the plant, but definitive information on yield effects is not yet available.

Modifying factors. Crop reponses to air pollutants are readily modified by many external conditions, including environmental and soil factors. Environmental factors that influence response include humidity, light level, daylength, wind speed, and temperature. Soil factors include soil water status and mineral nutrition. Plants grown under water stress usually are less sensitive to air pollutants, while plants with optimum water availability are more readily injured because the stomates tend to be more open. The effects of nutritional status of plants on air-pollutant sensitivity have not been consistent among nutrients and species. Both high- and low-nitrogen status has been shown to increase sensitivity, and phosphorus and sulfur nutrition has been found to affect sensitivity of plants to the commonly occurring gaseous pollutants. The important impacts of environmental and soil factors on plant sensitivity suggest that cultural practices for crops could be modified to minimize crop damage, but there has been little research conducted to explore this concept.

There are major differences among crop species and cultivars in sensitivity to air pollutants, indicating that the genetic makeup of plants also modifies pollutant responses. Laboratory fumigation tests and screening in the field in polluted regions have produced various listings of relative sensitivity of species and cultivars to pollutants. Also, there have been several inheritance studies to determine the genetic systems involved. Heritable differences in sensitivity have been found that could be incorporated in crop breeding programs. Resistance may be recessive or dominant and controlled by single or multiple genes, depending on the species and pollutant. Inheritance studies have also served to identify mechanisms of resistance that can be used as genetic markers, including such variables as stomatal closure patterns and stomate numbers. For example, ozone resistance in onion is determined by closure of stomates, while ozone resistance in bean is related to fewer stomates on the leaves of resistant plants.

Many sources of air pollution emit a combination of two or more major pollutants, so crop plants are likely to be exposed to mixtures of pollutants. The joint action of pollutants may take the form of additive, greater-than-additive (synergistic), or less-than-additive (antagonistic) effects. All three forms have been demonstrated in research on crop growth responses to various mixtures of ozone, sulfur dioxide, nitrogen dioxide, and hydrogen fluoride gases. For example, mixures of ozone and sulfur dioxide may have additive effects on radish, synergistic effects on tobacco, and antagonistic effects on navy bean. Also, the symptoms of injury may be different from those for single gases (Fig. 3). The kind of response is very dependent on the concentration and duration of exposure of each gas. Concurrent and sequential exposures to mixtures also affect the response. There is now a concerted effort to ensure that the experimental conditions used are representative of ambient pollutant concentration, frequency, duration, and time intervals between exposure events. Additional research, using these realistic conditions, is required before a more reliable estimate can be made of the importance of pollutant mixtures to crop growth and yield.

For background information *see* AIR POLLUTION; PLANT GROWTH in the McGraw-Hill Encyclopedia of Science and Technology.

[DOUGLAS P. ORMROD]

Bibliography: American Society for Horticultural Science, Impact of world atmospheric modification

Fig. 3. Visible injury symptoms on navy bean leaves, illustrating the effect of single and mixed gases. The leaf on the left was exposed to ozone and assumed a mottled appearance due to loss of chlorophyll and synthesis of dark-colored pigments. The leaf on the right was exposed to the combination of the same ozone concentration with sulfur dioxide, resulting in a markedly different pattern of chlorosis and no dark pigmentation.

on plant growth and productivity, *HortScience*, 18:665–689, 1983; W. W. Heck et al., Assessing impacts of ozone on agricultural crops: I. Overview, II. Crop yield functions and alternative exposure statistics, *J. Air Pollut. Control Ass.*, 34:729–735, 810–817, 1984; M. Treshow (ed.), *Air Pollution and Plant Life*, 1984; M. H. Unsworth and D. P. Ormrod (eds.), *Effects of Gaseous Air Pollution in Agriculture and Horticulture*, 1982.

Alcoholism

The excessive consumption of alcohol-containing beverages has been associated with the development of sexual dysfunction and, in more severe cases, permanent injury to the organs of the hypothalamic-pituitary-gonadal axis. Such dysfunction and tissue injury can occur independently of the better-known tissue injuries associated with alcohol abuse, such as liver disease, neuropathy, pancreatitis, myopathy, and malnutrition. Sexual dysfunction and tissue damage can be produced experimentally in animals fed ethanol, which is the major constituent, aside from water, of all alcoholic beverages. It is generally believed, therefore, that alcohol per se (ethyl alcohol or ethanol) is the agent responsible for these adverse effects.

Sexual performance. Sexual performance and function are both affected adversely by alcohol abuse. Under experimental circumstances, men with single episodes of moderate alcohol consumption experience a prolonged time to erection, and delayed and less pleasurable ejaculation compared to when no alcohol is consumed. In addition, the frequency and rigidity of erections occurring in response to erotic stimuli are reduced in men who ingest alcohol.

Fewer data are available concerning the sexual responses of women, but available data suggest that women who ingest alcohol experience less vaginal lubrication and have a smaller increase in vaginal blood flow in response to erotic stimuli than do women who do not ingest alcohol.

Although such data are difficult to obtain, and even more difficult to validate, it also appears that chronic alcoholic individuals, both male and female, experience a reduction in libido and in their sexual activity while drinking.

Hormonal effects. Acute short-term alcohol ingestion by normal men, and alcohol administration to male animals, have both been associated with a reduction in testosterone levels and a slight increase in luteinizing hormone levels. These data suggest that alcohol produces a Leydig cell injury. (Leydig cells are located in the testis, and thought to produce testosterone.) Lowered hormone levels can also be demonstrated by using isolated perfused rat testes, and isolated rat Leydig cells maintained in culture.

In contrast to these dramatic effects of acute alcohol ingestion observed in men, similar studies using women have shown no effect of alcohol upon plasma levels of estradiol, progesterone, or the two gonadotropins (follicle-stimulating hormone, FSH, and luteinizing hormone, LH). However, studies using female animals have demonstrated a variety of adverse effects of alcohol ingestion. Alcohol inhibits puberty (delays vaginal opening and the establishment of regular estrous cycles) in female rats; when one ovary has been removed, it inhibits or limits the ovarian hypertrophy that is usually observed; and if alcohol administration is prolonged, it produces atrophy of the uterus and fallopian tubes, loss of ovulatory capacity, and ovarian atrophy. Such effects are associated with reductions in plasma levels of estradiol and progesterone without a compensatory increase in plasma gonadotropin levels.

Similar studies in male animals fed alcohol for prolonged periods and in normal male volunteers fed moderate amounts of alcohol [1 cm³/(kg)(day) or 0.015 fl oz/(lb)/(day)] for 3–4 weeks, resulted in marked reductions in serum testosterone levels, paradoxically normal gonadotropin levels, and, in animals, gross testicular atrophy.

Tissue effects. With prolonged alcohol administration, the testes of experimental animals decrease in volume, the seminiferous tubules shrink, and germinal elements become fewer and of bizarre morphology. In addition, the prostate and seminal vessels atrophy as a consequence of inadequate testosterone stimulation. Similar findings have been reported in chronic alcoholic men, who demonstrate gross testicular atrophy and a histologic pattern that varies from simple spermatogenic arrest to Sertoli cell syndrome (in which no germ cells are present within the seminiferous tubules).

With prolonged alcohol abuse by women or prolonged alcohol administration to female animals, ovulatory failure and loss of corpora lutea lead to a loss of ovarian mass. In contrast to males, germ cell injury is not apparent in females. Rather, the ova present in the ovaries of alcohol-exposed female animals are found predominantly in small primary follicles, and the ovaries demonstrate a marked lack of follicular development. As might be expected, tissue requiring hormonal stimulation for the expression of secondary sex characteristics are atrophic in such animals. Thus the vaginal mucosa of chronic alcohol-exposed female animals is thin; the uterus and fallopian tubes are small and fail to demonstrate endometrial and epithelial development; and the mammary glands become small and atrophic.

Mechanisms. The pathways for the synthesis of sex steroids are similar in males and females. Since androgens are the immediate precursors for estrogens, the effects of alcohol on sex steroid biosynthesis should be quite similar in males and females. Alcohol dehydrogenase, an enzyme which metabolizes alcohol, is found in the gonads of both sexes. The gonadal enzyme, like the alcohol dehydrogenase present in the liver, requires nicotinamide diphosphate (NAD) as a cofactor and is therefore capable of altering cofactor availability and the redox state of the gonads when alcohol is available for metabolism. Such changes would limit steroid synthesis at sites that require NAD, and would enhance the production of reduced steroids over those containing either ketone groups or double bonds within their steroid rings.

Chronic ethanol feeding of rats has been shown also to reduce the testicular content of 3-β-hydroxysteroid dehydrogenase/isomerase which catalyzes the conversion of prognenolone to testosterone, the rate-limiting step in steroid biosynthesis. This reduction of enzyme activity can also be seen when laboratory cultures are subjected to excess cofactor (NAD).

Of equal importance, at least for the adverse effect of alcohol upon the reproductive function of the testes, is the conversion by testis alcohol dehydrogenase of retinol (the circulating form of vitamin A) to retinal (the form of vitamin required by the testes for normal spermatogenesis). The presence of alcohol inhibits retinal generation by the enzyme when the two substrates for the enzyme (retinol and alcohol) are present together.

Chronic alcohol administration to male animals is associated with a reduction in the tissue levels of androgen receptors, and an enhancement of the tissue content of estrogen receptors. Therefore, for any given androgen signal, less androgenic response should be manifested by the tissue, a phenomenon which compounds the effect of the reduction in plasma testosterone levels.

Since alcohol does enhance the tissue content of estrogen receptors in males, it thereby magnifies the effect of circulating estrogens at the tissue level. Such an effect probably accounts, at least in part, for the feminization that is seen occasionally in chronic alcoholic men. It may also account for the relative preservation of secondary sexual characteristics by female chronic alcoholics, who probably have a similar reduction in the biosynthesis of estrogens by their ovaries.

Alcohol, and therefore alcohol abuse, enhances the activity of aromatase in a variety of tissues. Armatase converts androgens to estrogens; its extragonadal activity accounts for about 50% of the estrogens produced daily by normal males. In chronic alcoholics who have reduced steroid (androgen) biosynthesis and advanced alcoholic liver disease with portosystemic shunting, the combination of alcohol's enhancement of extragonadal (but nonhepatic) aromatase activity, increased levels of estrogen receptors, and the shunting of steroids away from the liver enables such men to achieve normal or increased estrogen levels and therefore exacerbate feminization despite androgen deficiency.

For background information *see* ALCOHOLISM; ENDOCRINE MECHANISMS; ENDOCRINE SYSTEM (VERTEBRATE) in the McGraw-Hill Encyclopedia of Science and Technology.

[DAVID H. VAN THIEL]

Bibliography: T. J. Cicero (ed.), *Ethanol Tolerance and Dependence: Endocrinological Aspects*, 1983; D. H. Van Thiel, Ethyl alcohol and gonadal function, *Physiol. Med. Hosp. Pract.*, 19:152–158, 1984.

Alloy

Alloys are usually viewed as solid solutions in which the different atomic species are randomly dispersed on the lattice sites. In fact, this situation is rarely realized. In the case where the interaction between like atoms is stronger than that between unlike ones, clusters of like atoms will predominate. In the case where the interaction between unlike atoms is stronger, local ordering in which atoms tend to be surrounded by unlike neighbors will prevail. If the interaction between unlike pairs is strong enough and the alloy composition is appropriate, the ordering may persist over macroscopic distances (Fig. 1). Such long-range ordered alloys have many unique properties. Because movement of vacancies or of single dislocations destroys local order, their motion is subject to special limitations. Thermally activated processes (such as diffusion) are significantly retarded. In ordered lattices, dislocations must travel in pairs or groups. Alloys having long-range order quite often exhibit yield stresses that increase with increasing temperature rather than decrease as is common in disordered alloys. Many alloys exhibiting long-range order tend to disorder above some critical temperature, although some, such as Ni_3Al, remain ordered up to the melting temperature.

Although long-range ordered alloys are obvious candidates for high-temperature applications, they have attracted relatively little interest because they are often perceived to be inherently brittle. In recent years an effort has been mounted to apply modern alloying concepts to overcome embrittlement in long-range ordered alloys, which are often designated as intermetallic compounds.

Macroalloying and ordered crystal structures. Many alloys having the general composition A_3B exist in ordered crystal structures formed by regular stacking of close-packed layers. The stacking sequence, which is strongly influenced by atomic size and electronic considerations, ranges from cubic or hexagonal packing to more complicated transition structures having unit cells extending over 15 layers. The $(Ni,Co,Fe)_3V$ system illustrates these possibilities and shows how macroalloying can be used to control crystal structure and thereby material properties.

Nickel, cobalt, and iron have roughly the same atomic size, so their influence on crystal structure arises primarily from electronic effects. For nearly 50 years metallurgists have noted correlations between crystal structure and electron concentration, defined as the number of electrons outside filled inert-gas shells per atom (or valence electrons), which is designated as the electron-to-atom ratio. The physical significance of this correlation is obscure, but well-defined relationships exist in many systems. The principle of alloying to control crystal structure can be used to convert a low-symmetry brittle structure into a high-symmetry ductile one.

In Co_3V, stacking of the close-packed layers follows a six-layer sequence in which one-third of the layers has a hexagonal character and two-thirds a

 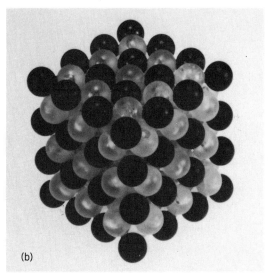

Fig. 1.   Ball models illustrating atomic arrangements in (a) disordered (conventional alloys) and (b) ordered crystal structures (ordered intermetallic alloys).

cubic one. Increasing the electron-to-atom ratio by replacing cobalt with nickel results first in predominantly hexagonal packing and then in fully hexagonal packing. However, reducing the electron-to-atom ratio to a value of 7.89 or less results in an ordered cubic structure. Any combination of nickel, cobalt, and iron that results in an electron density of 7.89 or less produces the cubic ordered structure. This allows considerable scope for alloy development. For instance, by control of electron-to-atom ratio it is possible to replace cobalt, a critical strategic element, with iron and nickel and thus obtain the high-symmetry cubic ordered crystal structure.

The importance of obtaining the cubic ordered structure is illustrated in Fig. 2, in which room-temperature ductility is shown for several hexagonal and cubic ordered alloys. The hexagonal alloys have ductilities too low to permit them to be readily fabricated. All the cubic alloys exhibit excellent ductility and fabricability. The ability to control ordered crystal structure through alloying represents a major breakthrough in alloy design.

Mechanical properties. The cubic ordered alloys have yield stresses that increase with increasing temperature up to the critical ordering temperature T_c, beyond which the strength decreases abruptly (Fig. 3). The decrease in strength is due to the loss of long-range order at temperatures above T_c. The strength levels and the temperature dependence of strength of the commercial solid-solution alloys Hastelloy-X and AISI 316 stainless steel are also shown for comparison. Tensile ductilities of the long-range ordered alloys are excellent, in the range of 30–45%, up to temperatures in the range of 930–1470°F (500–800°C; dependent on alloy composition and T_c), above which they start to drop abruptly because of the onset of grain-boundary cracking. The high-temperature ductility can be substantially improved by the addition of small amounts of titanium and rare-earth elements.

Long-range order has a dramatic effect on the creep properties. In passing through the critical ordering temperature, the creep rate ($\dot{\epsilon}$) changes by two orders of magnitude. The activation energies for creep are little affected by either alloy composition or state of order. The substantial decrease in creep rate on ordering evidently reflects the complex sequence of movements necessary for diffusion in the ordered state. Creep rates for the ordered alloys are roughly three orders of magnitude lower than those

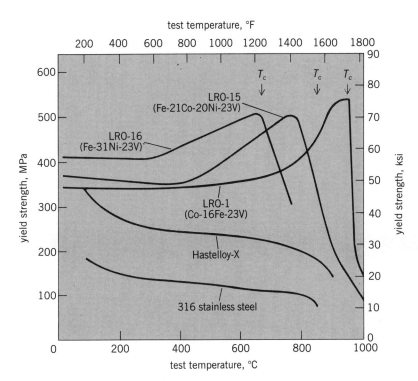

Fig. 2.   Yield strength as a function of temperature for several alloy systems. Yield strengths of long-range ordered alloys (designated LRO) increase with temperatures up to a critical temperature T_c, above which the alloy disorders and the strength decreases. ksi = kips (1000 lb) per square inch.

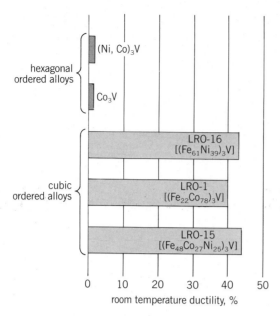

Fig. 3. Room-temperature ductility for several ordered alloys. Alloys that crystallize in the hexagonal crystal structure have low ductility and cannot be fabricated at room temperature. Cubic alloys have excellent ductility and can be readily fabricated.

for commercial solid-solution alloys such as Hastelloy-X and type 316 stainless steel.

An area of composition space has been defined in which ductile cubic long-range ordered alloys based on the system $(Ni,Co,Fe)_3V$ can be produced. Compositions can be manipulated within this space to tailor alloys for specific applications or to create alloys lean in critical strategic materials such as cobalt.

Microalloying and grain-boundary fracture. Low-symmetry crystal structures are not the only source of embrittlement in ordered intermetallic alloys. Single crystals of Ni_3Al, which exists in the cubic form, are highly ductile, but polycrystals are extremely brittle because of grain-boundary weakness. Grain-boundary weakness may arise from two factors; intrinsic, if the boundary is inherently weak relative to the grain interiors, or extrinsic, if impurity segregation to the boundaries weakens them. In overcoming the intrinsic form of embrittlement, dopants are added in part-per-million quantities to react with harmful impurities to bind them in some innocuous form, such as precipitate particles located within the grains. Intrinsic embrittlement is more difficult to deal with, as dopants are required that will segregate to the grain-boundary region and act to strengthen the boundaries, possibly by acting as electron donors. In practice, both embrittlement processes may be operative and both types of dopants may be added.

Very high-purity specimens of Ni_3Al as well as specimens containing dopants introduced to react with harmful impurities such as sulfur exhibit grain-boundary weakness and brittle behavior. It appears that the material is intrinsically brittle. A metamorphosis occurs transforming this useless, brittle sub-

stance to a strong ductile alloy by the addition of boron in a concentration of a few hundred parts per million. Boron-containing alloys can be readily cold-fabricated and have tensile ductilities exceeding 50% at room temperature, while materials without the addition are fully brittle.

The influence of boron is truly astounding, although the mechanism by which it acts is not understood. The solubility limit of boron in Ni_3Al is somewhat in excess of 2000 ppm, so its effect arises from its presence in solution. Boron has been found to segregate strongly to grain boundaries, as evidenced from Auger spectra obtained from freshly fractured grain-boundary surfaces. The signals from boron very nearly disappear after sputtering the surface for 2 min, which removes only a few atom layers. This demonstrates that boron is concentrated in a very narrow region adjacent to the grain boundary. The boron concentration is on the order of several percent, which greatly exceeds the solubility limit, although the boron appears to be in solution in the grain-boundary region. The effect of boron is, however, dependent on alloy stoichiometry; alloys containing less than 25% Al are ductilized by boron additions, while alloys containing greater amounts are not.

Nickel aluminides. Like the long-range ordered alloys, Ni_3Al shows a yield strength that increases with increasing temperature. The decrease in yield strength at very high temperatures is not a result of

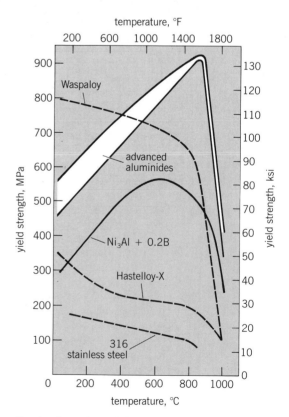

Fig. 4. Comparison of yield strengths of alloys based on Ni_3Al with other high-temperature alloys. The advanced aluminides contain iron and hafnium additions to Ni_3Al. ksi = kips (1000 lb) per square inch.

thermal disordering of the Ni$_3$Al lattice; it remains fully ordered up to the melting temperature, ~2550°F (~1400°C). The decrease in strength is attributed to the appearance of thermally activated slip on cube planes. The strength, therefore, passes through a broad maximum that for boron-doped Ni$_3$Al occurs in the vicinity of 1110°F (600°C).

Unlike many intermetallic compounds, Ni$_3$Al tolerates substantial alloying additions without losing the advantages of the ordered lattice. For example, as much as 10% Fe can be added in solid solution. Ni$_3$Al has been used as a base composition on which to apply macroalloying concepts to enhance properties further, for example, in advanced aluminides. Figure 4 displays yield strength characteristics for an advanced aluminide containing iron and hafnium additions. Note the significant enhancement in strength properties and the displacement of peak strength to higher temperatures, ~1560°F (~850°C) for the case illustrated. The strengthening is entirely a result of solid-solution effects; no precipitates can be detected that could account for even a fraction of the effect.

The presence of aluminum also conveys oxidation resistance to these alloys. In oxidizing atmospheres, compact, adherent, and protective films of oxides were formed on the advanced aluminides. The alloys showed no indication of flaking or scaling of the oxide film during cyclic oxidation for 500 h at temperatures to 2010°F (1100°C) with daily excursions to room temperature. Moreover, the material remained ductile at room temperature after oxidation.

Potential applications. Both the long-range ordered alloys based on (Fe,Co,Ni)$_3$V and Ni$_3$Al can be made highly ductile and fabricable, offering the potential for further macroalloying to produce alloys for specific applications. Because of their high strength and relatively low density, they have potential for turbine disk, vane, nozzle, combustor, and possibly blade applications. They also may be suitable for components for Stirling engines, high-temperature valves, coal conversion systems, steam turbines, and possibly for high-temperature process heat applications such as the gas-cooled reactor.

For background information see ALLOY STRUCTURES; CRYSTAL DEFECTS; GRAIN BOUNDARIES; INTERMETALLIC COMPOUNDS; PLASTIC DEFORMATION OF METAL; SOLID SOLUTION in the McGraw-Hill Encyclopedia of Science and Technology.

[C. T. LIU; J. O. STIEGLER]

Bibliography: C. C. Koch, C. T. Liu, and N. S. Stoloff (eds.), *High-Temperature Ordered Intermetallic Alloys: Materials Research Society Symposia Proceedings*, vol. 39, 1985; C. T. Liu, Physical metallurgy and mechanical properties of ductile ordered alloys (Fe,Co,Ni)$_3$V, *Int. Met. Rev.*, 29:168–194, 1984; C. T. Liu and J. O. Stiegler, Ductile ordered intermetallic alloys, *Science*, 226:636–642, 1984.

Amorphous solid

Semiconductor superlattice structures fabricated from alternating layers of crystalline III–V materials [that is, semiconducting compounds consisting of an element from column III of the periodic table combined with an element from column V, such as gallium arsenide (GaAs)] exhibit many interesting transport and optical properties that are associated with quantum size effects. These superlattices can be fabricated only from materials that have a nearly perfect match in their lattice constants and that can be grown epitaxially on top of one another. Otherwise the density of defects is so great that the phenomena associated with quantum size effects are obscured. It has been recently shown that the range of materials from which superlattices can be fabricated can be extended to hydrogenated amorphous semiconductors (Fig. 1a). Superlattices were made from materials such as hydrogenated amorphous silicon (a-Si:H), germanium (a-Ge:H), silicon nitride (a-SiN$_x$:H), and silicon carbide (a-Si$_{1-x}$C$_x$:H) that are neither lattice-matched nor epitaxial and yet have interfaces that have a low density of defects

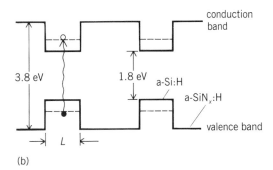

(a)

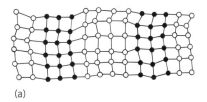

(b)

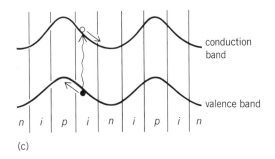

(c)

Fig. 1. Amorphous semiconductor superlattices. (a) Structure of superlattice. (b) Potential well model of a-Si:H/a-SiN$_x$:H. The wavy arrow indicates photoexcitation of an electron-hole pair above the mobility edges. The broken lines indicate the electron and hole mobility edges, shifted relative to the conduction-band and valence-band mobility edges, due to quantum size effects. The width of the a-Si:H layer is indicated by L; also indicated are the mobility gaps (optical energy gaps) of a-Si:H and a-SiN$_x$:H. (c) Energy band diagram of doping superlattice in which a-Si:H is alternately doped n-type and p-type, with undoped i layers in between. The wavy arrow indicates a photoexcited electron-hole pair. The electron and hole become spatially separated because of the built-in fields of the pin junctions.

and are nearly atomically sharp. The stringent requirements for lattice matching in crystalline superlattices are relaxed in the amorphous case because of the nonperiodic structure and the ability of hydrogen to passivate coordination defects.

Film growth and structure. The amorphous superlattice materials are made by a plasma-assisted chemical vapor deposition process in which the composition of the reactive gas in the plasma reactor is changed periodically. For instance, silane gas (SiH_4) is used for the a-Si:H layers and a 5:1 mixture of ammonia (NH_3) and silane for the silicon nitride layers. To achieve abrupt interfaces between the layers, the gas exchange time in the reactor must be short compared to the time it takes to grow a monolayer—a condition that is readily satisfied. The deposition technique has the capability of scaling up to a large volume and economic production process.

To probe the uniformity of the layers and the structure of the interfaces, a number of different techniques have been used. Figure 2 shows an example of an electron micrograph of a hydrogenated amorphous Ge:H/Si:H superlattice in which the individual layers are only 1.3 nanometers wide or the equivalent of four monolayers. The layers are smooth to better then 0.5 nm. The fact that this degree of perfection is achievable in the uppermost layers of a 440-layer-thick film indicates that the amorphous superlattice can be grown without cumulative roughening.

Because the amorphous superlattices are periodic in one dimension, they diffract x-rays and neutrons. From the intensities of the diffracted Bragg peaks, it is possible to derive information on how smooth the layers are; in the case of a-Si:H/a-SiN$_x$:H superlattices, a root-mean-square roughness of 0.5 nm was determined. The ability to make high-quality

superlattices over large planar or curved surfaces makes these materials attractive for application as x-ray or neutron mirrors.

From measurements of bulk properties of the superlattice such as infrared absorption, Raman scattering, and atomic composition, it is possible to deduce information on the structure of the interfaces on an atomic scale. This is because in superlattices which are made with a large density of interfaces the material in the interface region occupies significant volume fraction of the material. Infrared spectroscopy of the Si-H stretch vibrations and nitrogen-15 nuclear resonant reaction measurements of the hydrogen concentration in a-Si:H/SiN$_x$:H superlattices show that the first 2 nm of a-Si:H deposited on a-SiN$_x$:H has an appreciably higher hydrogen concentration than the bulk of the layer. The extra hydrogen is believed to relieve the strain caused by the large mismatch between the two materials. In the amorphous Ge:H/Si:H superlattices the mismatch is much smaller, and no extra hydrogen is apparent in the interface region. Raman spectroscopy has shown that the interface consists of a monolayer-wide disordered silicon-germanium alloy.

The information obtained from the structure of the superlattices provides new insights into the growth of amorphous films. The fact that the amorphous superlattices can be grown with nearly atomically sharp interfaces suggests that there must be some smoothing mechanism that prevents roughening of the growing surface and enables the film to grow one monolayer at a time, analogously to the expitaxial growth of crystalline materials.

Electronic properties. Quantum confinement of charge carriers (electrons and holes) by potential barriers in crystalline semiconductors raises the allowed electron and hole energy levels and gives rise to a density of states that increases in discrete steps. This structure in the density of states is reflected in an increase in the optical band gap and fine structure in the optical absorption spectrum. This picture of quantum confinement can be carried over to amorphous superlattices for charge carriers that are above the mobility edges. The mobility edge in amorphous semiconductors defines the demarcation between extended states (above the mobility edge) and localized states (below the mobility edge). Figure 3 shows the optical absorption spectrum of a-Si:H/a-SiN$_x$:H superlattices. The fine structure in the optical absorption that is observed in crystalline superlattices is absent presumably because of disorder broadening. The evidence for quantum size effects come from the large increase in the optical band gap which is manifested by the blue shift in the absorption edge with decreasing a-Si:H layer thickness (Fig. 3). The increase in band gap can be explained quantitatively on the basis of the one-dimensional periodic potential in Fig. 1b, using an effective mass of 0.2 m for the electrons and 1.0 m for the holes (where m is the free electron mass). The shift in the mobility edges with decreasing a-Si:H layer thickness also explains the observed

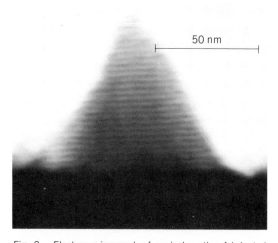

50 nm

Fig. 2. Electron micrograph of conical section fabricated from the top surface of a 440-layer-thick a-Si:H/a-Ge:H superlattice having a 2.7-nm periodicity. Dark bands in the image are the a-Ge:H layers. (*After H. W. Deckman, J. H. Dunsmuir, and B. Abeles, Transmission electron microscopy of hydrogenated amorphous semiconductor superlattices, Appl. Phys. Lett., 46:171–173, 1985*)

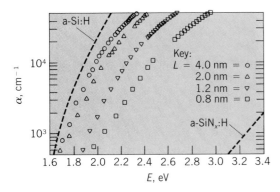

Fig. 3. Optical absorption coefficient α versus photon energy E for a series of a-Si:H/a-SiN$_x$:H superlattices for different a-Si:H layer thicknesses L and a constant a-SiN$_x$:H thickness of 3.5 nm. Also given is α for thick (unlayered) a-Si:H and a-SiN$_x$:H films. (After B. Abeles and T. Tiedje, Amorphous semiconductor superlattices, Phys. Rev. Lett., 51:2003–2006, 1983)

large decrease in the electrical conductivity in the plane of the layers. Quantum size effects in layers as thick as 4 nm indicate that the coherence of the wave functions is 4 nm or larger. The ability to tune the optical band gap of the superlattice has potential application, for instance in solar cells, where it is desirable to grade the band gap of the semiconductor to match the solar spectrum.

Electroabsorption. Another interesting property of amorphous superlattices is the large electroabsorption effect they exhibit. Electroabsorption is the increase in optical absorption when an electric field is applied normal to the film. In ordinary amorphous materials, the effect is vanishingly small. In superlattices, large electroabsorption results from an asymmetry in the two interfaces (a-Si:H deposited on a-SiN$_x$:H and a-SiN$_x$:H deposited on a-Si:H) which give rise to large built-in electrical fields (up to 4×10^5 V/cm have been observed in a-Si:H/a-SiN$_x$:H superlattices), with the sign of the field alternating between the layers. These fields are caused by electrons, transferred from the nitride to the silicon; the interface charge involved (6×10^{12} cm^{-2}) is concentrated within 2 nm of the hydrogen-rich interface region formed when silicon is deposited onto silicon nitride. This charge has been attributed to strain-relieving defects induced by lattice mismatch at the interfaces.

Conductivity and photoconductivity. The charge spillover from the a-SiN$_x$:H to the a-Si:H layers causes a large increase in the electrical conductivity. When the a-Si:H layers are thick enough (greater than 4 nm), so that quantum size effects are not significant, the conductivity becomes five or six orders of magnitude higher than it is in intrinsic undoped bulk a-Si:H films. Unlike conventional impurity-doped a-Si:H, transfer-doped a-Si:H is highly photoconductive. The new doping mechanism is analogous to modulation doping in crystalline III–V compound semiconductors in which a substitutionally doped wide-band-gap semiconductor, joined in a heterostructure with a narrow-band-gap semicon-

ductor, will dope the narrow-gap material by a charge spillover. More recently another kind of amorphous modulation-doping superlattice was synthesized in which alternating layers of a-Si:H were doped p-type with boron and n-type with phosphorus. This superlattice is the analog of the crystalline *nipi* or *npnp* structure. The energy band diagram of the *nipi* superlattice is shown in Fig. 1c. The materials show enhanced photoconductivity resulting from the spatial charge separation of the photogenerated carriers at the *pin* junctions.

Photoluminescence. Photoluminescence measurements in amorphous semiconductor superlattices show that interface defects can have a profound effect on the luminescence efficiency. This is clearly demonstrated by comparing the photoluminescence of silicon layers (optical band gap 1.8 eV) in a-Si:H/SiN$_x$:H superlattices with the photoluminescence of germanium layers (optical band gap 1.1 eV) in a-Ge:H/Si:H superlattices. In the a-Si:H/a-SiN$_x$:H superlattice the interfaces are the major source of nonradiative recombination centers (approximately 10^{11} centers per square centimeter per interface) as evidenced by the drop in photoluminescence efficiency when the silicon layer thickness decreases and becomes comparable to the capture radius R_c (approximately 7 nm) of the photoexcited electron-hole pairs. The capture radius R_c can be regarded as defining a sphere surrounding each photogenerated electron-hole pair such that if the sphere contains a nonradiative recombination center the pair recombines nonradiatively; otherwise it recombines radiatively. On the other hand, in the a-Ge:H/a-Si:H superlattices nonradiative recombination is dominated by centers which are in the bulk of the a-Ge:H layers while the contribution of the interfaces is negligible. Consequently when the a-Ge:H layer thickness is reduced below $2R_c$, the number of accessible recombination centers in the germanium layers decreases and the photoluminescence efficiency increases by more than two orders of magnitude over that of bulk unlayered a-Ge:H films. This enhancement of the photoluminescence in germanium layers together with the high quality of the germanium/silicon interfaces opens up new opportunities for understanding luminescence phenomena in a-Ge:H in particular and energy transport processes in amorphous semiconductors in general.

The ability to synthesize layered amorphous semiconductors with interesting, novel electronic properties is expected to have a major impact on the science and technology of amorphous semiconductors.

For background information *see* AMORPHOUS SOLID; SEMICONDUCTOR HETEROSTRUCTURES in the McGraw-Hill Encyclopedia of Science and Technology. [BENJAMIN ABELES]

Bibliography: G. H. Döhler, Solid state superlattices, *Sci. Amer.*, 249(5):144–151, November 1983; J. Pankove (ed.), *Semiconductors and Semimetals*, vol. 21C: *Hydrogenated Amorphous Silicon: Electronic and Transport Properties*, 1984.

Anxiety states

There is no consensus for a definition of anxiety, and there are few empirical data to distinguish it from fear and other emotional states. Nonetheless, there is agreement that the anticipation of pain, bodily injury, or death usually produces subjective discomfort and physiological changes that are recognizably different from other emotions. It seems reasonable to suppose that the neurobiological substrates associated with normal fear and anxiety in humans also underlie pathological fear, anxiety, or panic, although these disorders may have additional pathophysiologies. Intensive studies in humans and other animals have examined the role of central noradrenergic systems in the production or reduction of anxiety or alarm using drugs, electrical stimulation, or lesions of the brain. It has been postulated that the major brain noradrenergic system innervated by the nucleus locus coeruleus mediates an alarm function that may include anxiety or fear.

Locus coeruleus–noradrenergic system. The nucleus locus coeruleus is a compact cell group located in the caudal pontine portion of the central gray matter. In monkeys the locus coeruleus contains approximately 70% of the brain's noradrenergic neurons. It contains five well-defined tracts that innervate all the brain cortices, specific thalamic nuclei, hypothalamic nuclei, olfactory bulbs, cerebellar cortex, and spinal cord. The afferent and efferent connections of the locus coeruleus–noradrenergic system are well suited to mediate the rather global effects associated with emotion. Notably, locus coeruleus neurons project to areas of the limbic system that are classically thought to be involved in emotion. The locus coeruleus receives innervation from every major sensory modality, from pain-sensitive neurons, and from the limbic system. Locus coeruleus projections form complete feedback loops to and from many brain areas, including those diverse regions that mediate physiological responses to pain and fear. Biochemically the locus coeruleus–noradrenergic system has presynaptic and probably somatodendritic receptors that allow modulation of activity by a wide variety of neurochemicals.

The hypothesis that overactivity of the locus coeruleus–noradrenergic system may be a neural mechanism for mediating fear and related emotions has been studied in nonhuman primates. In the stump-tailed monkey *(Macaca arctoides)*, low-intensity electrical stimulation of the locus coeruleus, which activates noradrenergic neurons, results in behavioral responses essentially identical to those observed in natural fear states. Such evidence is consistent with earlier work in rodents indicating that stress responses and anxiety were associated with increased noradrenergic functioning. Activation of the locus coeruleus, either electrically or by drugs, produces profound alterations in behavior and affects nearly every major brain area and autonomic region that is altered by fear. The behavioral effects of stimulation are similar to those found following the administration of drugs that activate neurons in the locus coeruleus. Lower levels of locus coeruleus activation are correlated with attentiveness and vigilance to physiologically relevant stimuli; progressively reduced levels are correlated with inattentiveness, impulsiveness, and fearlessness. Lesions of the locus coeruleus and pharmacological manipulations that decrease locus coeruleus activity also decrease fearlike behaviors, partially antagonize the effects of electrical stimulation, and decrease the frequency of some behavioral responses in social situations.

Several clinical studies have attempted to determine whether central noradrenergic activity is increased during anxiety-fear states in humans. Alterations of peripheral sympathetic function (heart rate, blood pressure, respiration, and so on) are found during fear or anxiety, but are probably nonspecific physiological correlates of arousal and stress. Increases found in the concentrations of noradrenaline, adrenaline, or their metabolites in the blood plasma of normal individuals may reflect postural changes or stress; but since they are largely measurements of the sympathetic nervous system, they do not provide evidence for the existence of central noradrenergic or adrenergic hyperactivity. However, a metabolite of noradrenaline, 3-methoxy-4-hydroxyphenylethyleneglycol (MHPG), is partially derived from central noradrenergic metabolism and is the major metabolite of noradrenaline in brain. When noradrenergic activity is increased, the amount of this metabolite that leaves the brain is greater than that measured in the arterial supply to the brain. In direct studies in monkeys, there is very high correlation between the amounts measured in various brain regions and the amounts in cerebrospinal fluid and plasma, suggesting that MHPG concentrations in plasma or cerebrospinal fluid might be useful estimators of central noradrenergic activity. Anxiety ratings have been correlated with increased MHPG concentrations in cerebrospinal fluid of depressed individuals and with increased plasma concentrations of MHPG in phobic individuals exposed to a phobic stimulus. These changes may reflect both increased brain noradrenergic activity and increased peripheral noradrenergic function.

Pharmacological evidence. To determine the biological significance of these correlations, it would be important to know whether experimentally increased noradrenergic function produces anxiety in humans. Although an older study reported that electrical stimulation in the region of the locus coeruleus produced severe fear in humans, the location of the electrodes could not be determined precisely, and this procedure is not feasible under most experimental conditions. However, certain alpha-2-adrenergic antagonists (such as yohimbine and piperoxane) cross the blood-brain barrier, and at low doses increase brain noradrenergic neuronal activity without directly producing peripheral or postsynaptic effects. These compounds thus provide a means of increasing noradrenergic system activity in a noninvasive and somewhat physiological manner. Doses of yohimbine, which preferentially activate brain noradrenergic systems, have been found to induce anxiety states and even panic in psychiatric patients. Yohimbine also significantly increased the levels of MHPG in plasma; this increase was significantly correlated with the anxiety induced by the drug. Conversely, low doses of alpha-2-agonists (such as clonidine) acutely decrease net noradrenergic neuronal functioning by stimulating the effects of noradrenaline and adrenaline both at alpha-2-mediated somatodendritic autoreceptors (this decreases neuronal firing) and at presynaptic sites (this decreases noradrenaline release). The prediction that clonidine should reduce anxiety has received clinical support from studies of people with panic attacks. Clonidine is also correlated with decreased plasma MHPG and decreased responses to phobic-panic stimuli.

Noradrenergic hyperactivity has also been suggested to be an important component of drug withdrawal syndromes that include increased anxiety or fear. Most important of these is the opiate abstinence syndrome which, in primates, resembles stimulation of the locus coerulus and reproduces many of the signs and symptoms of anxiety and fear. Several investigations have found that abrupt discontinuation from methadone and antagonist-precipitated withdrawal (using naltrexone) in individuals addicted to methadone result in increased levels of plasma MHPG. The efficacy of clonidine in the treatment of opiate withdrawal also supports the role of noradrenergic hyperactivity in the symptoms of opiate withdrawal. Some preliminary data also suggest the effectiveness of clonidine in reducing the symptoms of benzodiazepine, alcohol, and nicotine withdrawal.

Several other compounds also increase anxiety or precipitate panic in individuals with anxiety disorders. Lactate infusions precipitate anxiety attacks in anxious neurotics, but not usually in others. Carbon dioxide has been found to produce anxiety and panic attacks in anxious and normal individuals; this may result from dose-dependent increases in locus coeruleus firing, or from a behavioral sensitivity to this or other effects of carbon dioxide. Caffeine and a benzodiazepine "inverse agonist" (which activates benzodiazepine receptor–mediated systems) also increase anxiety or panic. Some evidence suggests that these effects also may result partially from activating noradrenergic neurons.

Investigations of common biochemical properties of pharmacologic agents effective in relieving anxiety and related affective disorders may provide additional clues to the mechanisms underlying anxiety. Many anxiolytic drugs have been associated with effects on a variety of brain neurochemical systems, which may be responsible for sedative, hypnotic, muscle-relaxant, anticonvulsant, and other effects of the drugs, in addition to their anxiolytic properties. Substantial evidence suggests that the anxiolytic properties of the opiates, barbiturates, and ethanol may be explained by the inhibition of noradrenergic neurotransmission. The tricyclic and monoamine oxidase–inhibiting antidepressants have efficacy for reducing panic attacks, and both classes of drug are known to reduce net noradrenergic function after chronic administration.

Opioids are well known to produce anxiolytic and mood-altering effects in humans. Although low doses of opiates can initially be dysphoric, the most consistent effect of heroin is an anxiolytic one, in addition to its attenuation of pain. Opiate agonists inhibit locus coeruleus activity in animals, and block painful stimulation-induced increases in locus coeruleus activity. Barbiturates also have been extensively used for their anxiety-reducing properties. Consistent with a unified noradrenergic hypothesis of anxiety, one of their numerous biochemical effects is to alter noradrenergic metabolism, which reflects a change in locus coeruleus activity. Their anxiolytic effects may be due to their ability to augment the inhibitory effects of GABA and also to antagonize excitatory afferents to the locus coeruleus. Pharmacological, biochemical, and electrophysiological studies have shown that benzodiazepines also produce a decrease in locus coeruleus activity. The specific mechanism by which this occurs remains elusive. Specific benzodiazepine receptors in the mammalian brain have recently been described, and are presumed to be the most direct site of action of benzodiazepines. However, these receptors are not located on the locus coeruleus, and the benzodiazepines do not interact directly with the alpha-2-adrenergic receptor, the GABA receptor, or the opiate receptor all located in that vicinity. Considerable evidence does exist to indicate that benzodiazepines may activate GABA receptors on locus coeruleus neurons, and in turn inhibit locus coeruleus neuronal activity.

Some relevant literature derived from studies of rodents contradicts a hypothesis of noradrenergic hyperactivity as crucial to anxiety or fear. Human and nonhuman primate data indicating the involvement of noradrenergic systems and the interaction of other neurotransmitter and peptide systems on the functioning of the locus coeruleus have utilized biochemical correlates of anxiety or fear, experimental pharmacological induction or reduction of anxiety in normals and psychiatric patients, and biochemical actions of antianxiety agents. Possible alterations of the locus coeruleus and other brain areas and neurotransmitter systems that impinge upon its functioning might lead to pathological anxiety and to other fear-panic-related affective disorders.

For background information *see* Abnormal behavior; Phobic reaction; Psychopharmacologic drugs in the McGraw-Hill Encyclopedia of Science and Technology.

[Jane R. Taylor; D. Eugene Redmond, Jr.]

Bibliography: D. S. Charney et al., Yohimbine-induced anxiety and increased noradernergic functions in humans: Effects of diazepam and clonidine, *Life Sci.*, 33:19–29, 1983; D. S. Charney and D. E. Redmond, Jr., Neurobiological mechanisms in human anxiety: Evidence supporting central noradrenergic hyperactivity, *Neuropharmacology*, 22(12B):

1531–1536, 1983; A. H. Tuma and J. D. Maser (eds.), *Anxiety and the Anxiety Disorders*, 1985.

Apical meristem

The role played by the shoot apical meristem is central in the life of a plant. Located at the tip of each growing shoot, the few hundred cells that make up the meristem are enclosed and protected by young, growing leaves (Fig. 1). Although hidden from view, this single meristem functions to produce all the new cells of the shoot system. Numerous descriptive studies have followed the structural changes that this meristem undergoes during the production of leaves. But once structural changes can be detected, the physiological and biochemical events that control these changes have already occurred.

Phosphofructokinase (PFK) activity in *Dianthus* tissue on dry-weight, cell, and cell cubic micrometer bases

Tissue	PFK activity* (substrate consumed per hour)		
	millimoles per kg dry wt	femtomoles per cell	attomoles per cell (μm^3)
Apex	1056	75	115
Leaf pair 1	1013	81	124
Leaf pair 2	812	65	77
Leaf pair 6			
Epidermis	140	666	10
Mesophyll	160	1454	11
Leaf pair 9	80	—	—

*Original determinations of activity were based on dry weight of samples; data were converted to cell and cell cubic micrometer bases according to previous determinations of cell volume and number of cells per unit dry weight of sample types.

The physiological events associated with these processes are not as well understood, since, because of the meristem's small size, it has not been possible to collect enough apices (without damaging them) to routinely employ standard biochemical and physiological methods of investigation.

Most current knowledge on meristem function is qualitative and has come from the use of stains and dyes on sectioned material. However, the development of quantitative biochemical techniques that are applicable to small tissue samples permits entirely new questions to be asked about the apical meristem and its products. With these methods the activity of enzymes or the level of metabolites can now be determined on a mole per dry weight basis in small samples, single cells, parts of cells, or small groups of cells (see table). Since enzymes are an expression of gene activity, the results from this technique can be used as an effective bridge between knowledge of structure and molecular biology, and permit important biochemical questions to be addressed.

Basic technique. The basic technique consists of measuring the metabolite or enzyme of interest by utilizing an enzymatic reaction that employs a pyridine nucleotide (nicotinamide adenine dinucleotide, NAD, or nicotinamide adenine dinucleotide phosphate, NADP). The advantage of using pyridine nucleotides lies in their fluorescence properties and their selective stability in either acid or base. The sensitivity of fluorometry easily permits measurement of less than a nanomole of pyridine nucleotide. Although many biological reactions employ the reduced forms NAD(H) or NADP(H) directly, others do not but nonetheless usually generate a product that ultimately can be tied to such a reaction. The amount of pyridine nucleotide present must be amplified through two coupled, enzymatic reactions. Amplification of 400 million times is possible and permits measurement of as little as 10^{-18} moles of a substance within an individual sample. It is this high sensitivity that enables quantitative measurements to be made in samples within the apical meristem or leaf primordia.

For analysis of small samples from regions of the

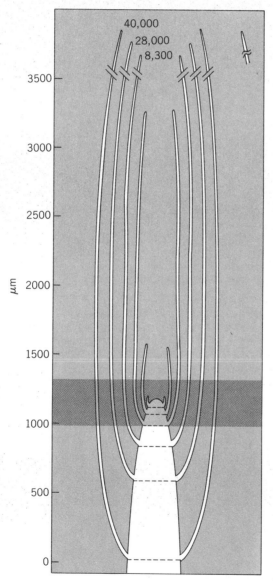

Fig. 1.   Diagram of the shoot tip of *Dianthus chinensis* which illustrates the relationship between the apical meristem and the length and point of insertion of the youngest six pairs of leaves. The shaded region is where the samples are taken for biochemical determinations. The values given represent the lengths of the leaves in micrometers. (*Copyright 1985 by Judith Croxdale*)

Fig. 2. Apical meristem and youngest pair of leaves viewed by scanning electron microscope. *(From J. G. Croxdale and W. H. Outlaw, Jr., Glucose-6-phosphate-dehydrogenase activity in the shoot apical meristem, leaf primordia, and leaf tissues of Dianthus chinensis L., Planta, 157:289–297, 1983)*

apical meristem or the youngest leaf primordia, the tissue must be specially prepared. First, shoot tips are rapidly frozen. In species where only the meristem is of interest and young leaves can be removed easily and completely, entire terminal buds are dried and the young leaves are removed by dissection. The resulting meristem is in full view, and samples from the desired areas of the meristem are taken. In other species, however, pairs of leaves interlock along their length and grow vertically for several centimeters. If one attempts to remove whole leaves from the dried shoot, they break off above their attachment point to the stem and obscure access to the younger, shorter leaves and the meristem itself. Therefore, frozen shoots of these species are encased in a nearly solid gelatin mixture, refrozen, and serially sectioned to fully expose the meristem and provide samples of all leaf primordia. The frozen sections are then freeze-dried. It is from dried sections or whole pieces of tissue that samples are cut by means of hand-held knives. It is possible to routinely cut precise samples, including incipient leaf primordia and the youngest leaf (Fig. 2) as well as different cell types within the leaf such as the epidermis, mesophyll, xylem, or phloem.

Vegetative growth. Since apical meristems produce all the new cells of the shoot and are enclosed by young leaves, they are unable to manufacture their own food by photosynthesis, but instead must import carbohydrate. Carbohydrates can be oxidized by two main pathways, glycolysis or the pentose phosphate shunt. The products of the latter pathway would be useful to actively dividing cells in providing a source of reducing power for biosynthesis, the precursor for nucleic acid synthesis, and an amino acid essential in the formation of lignin, which provides the rigidity for aerial structures. Thus it might be anticipated that a greater proportion of carbohy-

drate would be oxidized by this route. By comparing the activities of the two key enzymes that control the two pathways, one can predict the pathway with the greater capacity. By using *Dianthus chinensis* shoots of known developmental age, it has been shown that within the apical meristem the capacity for glycolysis exceeds that of the pentose phosphate shunt. This also holds true for the first two pairs of primordia. With the sixth leaf pair, however, the capacity for carbohydrate oxidation by the pentose phosphate shunt is greater than by glycolysis. These results were unexpected and may indicate that the apical meristem and leaf primordia are supplied with the compounds of the pentose phosphate pathway from nearby but older cells, similar to the situation hypothesized to occur in a similar region of roots.

The pattern of the quantitative values reveals a great deal about structure. For example, measurements of four enzymes involved in carbohydrate oxidation show that incipient primordia are not biochemically different from other cells within the apical meristem. Enzymatic activity is homogeneous throughout the meristem and, prior to emergence, leaf primordia are not biochemically specialized. Thus, whatever signals the cells in a particular region to become a leaf is not expressed by differential activity of these four enzymes. Although one would not anticipate that enzymes and metabolites of carbon metabolism were the signals per se, one might expect to find these compounds within the cells of an incipient primordium due to its growth rate. However, once a primordium is cut off from the meristem, enzymatic activity may diverge in the primordium. Thus, primordia are biochemically specialized early, but not when within the meristem proper.

Hexokinase activity can also be used to elucidate structure. It has been demonstrated that the changes in hexokinase activity that occur throughout the morphogenesis of leaves is paralleled along the length of a single developing leaf. Thus, as leaf cells age and differentiate, they undergo changes in enzymatic activity that occur due to their position and state of differentiation within an individual leaf.

Changes in enzymatic activity are correlated with the structural differentiation of leaves. Marked changes in enzymatic activity normally take place between leaf pair 2 and leaf pair 6. Measurements taken from the fourth leaf pair show a gradient of activity depending upon the area within the leaf that is measured. These areas within the leaf correspond to regions with known structural differences.

Even mature cells within an individual leaf show biochemical heterogeneity. In fact, when mixed values are obtained for a metabolite or enzyme in the different cell types (epidermis, guard cells, mesophyll, phloem, xylem) of a leaf, it can be used as a signpost that differentiation and specialization have occurred.

Studies using *Dianthus* shoots are under way to determine when the plastids of developing leaves become photosynthetically competent in a light-grown plant and where within the leaf these chloroplasts are found. This will be determined by measuring relevant enzymes of the light and dark reactions and metabolites of carbon fixation. These results will be compared to the fine structure of plastids by establishing ultrastructural bases, for example, volume of cytosol, organelle number, suborganellar volume, and organellar membrane area, on which to base activity and metabolite level. These results will indicate the pattern of gene expression during the development of the photosynthetic apparatus.

Transition to flowering. This biochemical technique has also been applied to studies on the carbohydrate status of the meristem during the floral transition. Research on the floral evocation of *Sinapis alba* apical meristems indicates that a 50% increase in sucrose concentration occurs within the meristem 10 h after an inductive photoperiod. Similar to *Dianthus*, no biochemical specialization could be found within different regions of the meristem. For this work, samples were taken from the margins of the meristem where mitotic activity is known to be higher than in the larger, central cells of the meristem, the other area sampled. These results indicate that an early physiological event in floral transition is the accumulation of sucrose. Further work is in progress on the measurement of invertase and sucrose synthase as it relates to flowering and the nutritional status of the meristem.

For background information *see* APICAL MERISTEM; DEVELOPMENTAL BIOLOGY; PLANT MORPHOGENESIS in the McGraw-Hill Encyclopedia of Science and Technology. [JUDITH G. CROXDALE]

Bibliography: M. Bodson and W. H. Outlaw, Jr., Elevation in the sucrose content of the shoot apical meristem of *Sinapis alba* at floral evocation, *Plant Physiol.*, 77S:111, 1985; J. Croxdale, Quantitative measurements of phosphofructokinase in the shoot apical meristem, leaf primordia, and leaf tissues of *Dianthus chinensis* L., *Plant Physiol.*, 73:66–70, 1983; O. H. Lowry and J. V. Passonneau, *A Flexible System of Enzymatic Analysis*, 1972.

Arboriculture

Arboriculture, a branch of horticulture, concerns the selection, planting, and care of trees, shrubs, vines, and ground-cover plants in natural and designed landscapes. A tree usually has one vertical trunk and is at least 15 ft (5 m) tall. A shrub has several vertical or semiupright branches arising at or near the ground and is usually shorter than a tree. A vine or wall shrub has slender, flexible stems that need support. Ground-cover plants include prostrate, spreading shrubs and vines. These terms are somewhat arbitrary but they help characterize plant growth and form. Knowing the potential form and size of plants and their stages of growth is essential in order to plan landscapes or to care for plants.

Arborists are concerned primarily with trees since they become large, they are long-lived, and they dominate landscapes visually and functionally. Although research has often verified the wisdom of early practitioners, new findings and experience now suggest changes in some commonly accepted practices. These include selecting quality plants, planting techniques, training trees, response to pruning, fertilization, and chemical control of growth.

Landscape functions. Plants serve several landscape functions. They can provide privacy, define space, and progressively reveal vistas. Plants can be used to reduce glare, direct traffic, reduce soil erosion, filter air, and attenuate noise. They influence the microclimate by evaporative cooling; interception of the Sun's rays, reflection, and reradiation; and modification of rain, fog, and snow deposition. Plants can be placed to decrease, increase, or direct wind. Even so, certain plants irritate some people with their pollen, leaf pubescence, toxic sap, and smelly flowers and fruit. Additionally, because of their large size and extensive root systems, trees can be dangerous and costly: their branches can fall, injuring people and damaging property, and their roots can clog sewers and break curbs and sidewalks.

Plant selection. Selection of an appropriate plant species and even the cultivar (variety) is a critical factor. A plant's growth habit and its mature size are usually the first criteria in selection. Whether the plant is meant to shade, screen, enclose, accentuate, direct, or protect will determine what plant is selected. Unfortunately, mature plants that are too large for their space are more the rule than the exception. Sturdy root systems tolerant of poor soil conditions protect plants against storms and pests. Leaves, flowers, and fruit not only are visually important but also can have considerable impact on maintenance. Time and duration of leaf and fruit fall and the ease of removal are often key considerations when selecting plants, particularly trees.

The lowest temperatures on record determine in large measure which plants will survive and which will thrive in a specific site. In the autumn before plants have sufficiently hardened, they can be seriously injured or killed by low temperatures they could easily tolerate in midwinter. Spring and fall frosts can kill flowers and new shoots. Plants, particularly trees, should be selected to withstand the lowest temperatures expected.

Although most plants grow best in a deep, loam soil, some are able to withstand unfavorable soils, such as those that have poor drainage, strata of different textures, extremes of acidity or alkalinity, or chemical toxicities. Species that use less water or have extensive root systems are favored in arid regions.

Species native to the area are the ones most often recommended. In many situations, however, particularly around buildings, the microclimate and the soil are different from what they were before construction, and native plants may do quite poorly. Many plants, once planted, grow well even though they may not be able to reproduce naturally. In addition, tolerance to air pollutants is becoming increasingly important in many urban areas. The urban environment can be quite harsh on most landscape plants.

Plant quality can be as important as the kind of plant. Roots free from kinks and circling are essential if plants are to perform well and withstand the elements. It is desirable that trees be able to stand without support and have small branches along the trunk to protect and nourish the trunk during establishment years.

Planting and initial care. Unless the soil has been compacted, the planting hole need only be deep enough to take the roots; in fact, usually it is best that the plant be slightly higher than it was in the nursery. The soil should be firmed around the roots and watered to settle the soil and remove air pockets. Plants transplanted from containers will initially need to be irrigated more frequently than when they were in the container.

Unless the plant is overgrown, the only pruning needed will be some thinning to begin tree structure development, to shape shrubs, or to select vine canes to trellis. The more young trees are free to move in the wind, the stronger they will become.

Establishment and care. Little pruning is necessary for central-leader trees, conifers and some hardwoods for them to grow into strong, well-shaped landscape specimens. Species that become round-headed, however, may need considerable pruning in their early years in order to ensure the desired height of branching and to develop a strong branch structure.

Even though pruning invigorates the growth of individual branches, the less a branch or tree is pruned the larger it will become. Therefore, only large-growing branches that are too low or will compete or interfere with more desirable branches should be removed. Permanent branches, particularly of large-growing trees, should be smaller in diameter than the trunk where they arise, and vertically spaced at least 1 ft (30 cm) apart along the trunk (2–4 ft or 60–120 cm is even better).

A tree will be more open and retain its natural form if branches are removed completely (thinned) in contrast to being headed, or stubbed back (see illus.). Heading concentrates subsequent growth just below the pruning cut, and results in dense foliage with weakly attached branches. In removing a branch, the final cut should be just to the outside of the branch bark ridge in the crotch and the collar below. Such a pruning cut minimizes the size of the wound and reduces the possibility that the trunk will become infected if the branch base decays.

Fertilization of young plants is necessary for rapid growth; mature plants, however, may need little or no added nutrients to remain healthy. Nitrogen is almost universally deficient in soils and usually is the only element to which trees and other large-growing plants will respond. Nitrogen fertilizers are water-soluble and can be applied to the soil or lawn surface and watered in. In alkaline soils, the availability of iron and manganese may be so low for certain plants that they exhibit the typical pale leaves

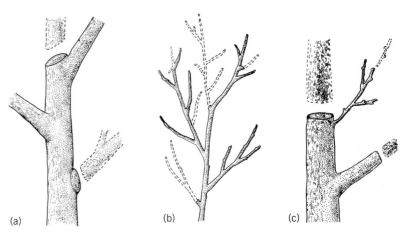

Methods of plant pruning. (a) Thinning (removing a branch at its origin) of a mature tree. (b) Thinning of a young tree. (c) Heading (pruning to a small lateral branch or to a stub) of a mature tree. (*After R. W. Harris, Arboriculture: Care of Trees, Shrubs, and Vines in the Landscape, Prentice-Hall, 1983*)

with narrow (due to iron deficiency) or wide (manganese deficiency) darker green bands along the veins. Increasing soil acidity or applying chelated nutrients should ameliorate these problems.

Irrigation can ensure establishment of young plants, the attractive appearance of foliage, and even survival. Most plants perform best when adequate soil moisture is available. Many mature plants can, however, endure long periods without rain or irrigation if a large volume of moist soil is present and plants have extensive root systems. Water use by plants is greatest during the few weeks before and after the summer solstice. If the water supply is limited, it is important for the soil to be fully moist at the beginning of the growing season; subsequently, normal rainfall or irrigations will usually be adequate. A few heavy irrigations are more efficient than frequent light ones. In irrigated landscapes, however, more plants do poorly due to too much water than not enough.

Mulch, material placed on the soil surface, has many advantages in addition to controlling weeds: it can protect the soil from compaction and erosion, conserve moisture, moderate soil temperature, provide an all-weather surface for traffic, and allow plants to root in the most fertile and well-aerated soil close to the surface. A wide range of organic and inorganic or synthetic materials can be used as mulch. Mulch should be kept 5 cm (2 in.) away from the trunks of plants in order to prevent the trunk from being kept too moist or from being damaged by rodents.

Chemical control of growth, particularly of trees and shrubs, helps in keeping plants within bounds. A number of chemicals are used by arborists to reduce the growth of trees primarily under utility lines. Chemical effectiveness varies with plant species, stage of growth, method of application, and, to a certain extent, weather.

For background information *see* AGRICULTURAL SOIL AND CROP PRACTICES; LANDSCAPE ARCHITEC-

TURE; ORNAMENTAL PLANTS in the McGraw-Hill Encyclopedia of Science and Technology.

[RICHARD W. HARRIS]

Bibliography: B. Ferguson, *All About Trees*, 1982; R. W. Harris, *Arboriculture: Trees, Shrubs, and Vines in the Landscape*, 1983; A. L. Shigo, Compartmentalization of decay in trees, *Sci. Amer.*, 252:96–103, 1985; B. F. Wilson, *The Growing Tree*, 1984.

Archeology

Past human subsistence is the subject of one of the most important questions in archeological research. The quest for food directly involves many aspects of prehistoric human society, including group size and social organization, residence patterns, technology, and transportation. Information on past diet is essential to characterize the trophic position of prehistoric human populations, their utilization of the environment, the determinants of site location, the nature of subsistence activities, and even the social status of some individuals. Most methods for the reconstruction of past diets have not provided reliable results, but new studies of the chemical composition of prehistoric human bone may provide more substantive information.

Current evidence for diet comes from a number of lines of research: analysis of human and animal bone, of preserved plant remains and fecal matter, and of dental wear and disease, and through predictive modeling. There are serious deficiencies with each of these methods, however. The vast majority of archeological sites do not contain preserved organic materials or any trace of food remains. Even in situations of excellent preservation, food remains may not provide an accurate representation of what was actually consumed. Studies of stature and other skeletal indicators can provide information on the quality of the diet but not its components. Dental investigations provide limited information on certain aspects of diet (such as gritty versus soft foods) but do not reveal the actual foods consumed. Predictive models provide intriguing suggestions of potential subsistence patterns, but these projections are dependent upon available archeological evidence.

Fortunately, the chemical analysis of human bone offers a means for obtaining more reliable information on important aspects of paleonutrition, such as marine versus terrestrial components in the diet, the importance of plants versus animals, and the presence of certain species of plants.

Bone tissue is essentially a network of mineralized fibers, composed of a matrix of organic collagen filled with inorganic calcium phosphate (hydroxyapatite) crystals. Bone contains three major components: a mineral fraction (bone ash), an organic matrix (collagen), and water. In addition to these major components, a number of minor and trace elements are incorporated during the manufacture of bone tissue.

Both the elemental and isotopic composition of bone is considered in chemical studies. Isotopic studies concentrate on the organic part of bone (collagen) while elemental analyses focus on the mineral portion (apatite). By dry weight, organic materials constitute about 30% and minerals about 70% of bone. Collagen, a protein, composes 90% of the organic portion of dry, fat-free bone.

Isotopic analysis. Isotopic analysis has concentrated on carbon and nitrogen. While the amount of these elements in bone is under strict metabolic control, the ratio of stable isotopes ($^{13}C/^{12}C$ and $^{15}N/^{14}N$) in bone collagen reflects the ratio found in the diet. A mass spectrometer is used to measure these isotopic ratios in collagen. Carbon isotope ratios are reported in terms of $\delta^{13}C$, which is the difference between the ratio of $^{13}C/^{12}C$ in the sample and the same ratio in a reference standard of marine carbonate. $\delta^{13}C$ values generally become more positive along the continuum from plants, to herbivores, to carnivores in both marine and terrestrial regimes.

The interpretation of carbon isotopic ratios in bone collagen is relatively straightforward. Individual variability in isotope ratios is due largely to diet, since the reservoirs of carbon isotopes in the sea and in the atmosphere are constant. Diagenesis—postdepositional chemical change in bone—does not appear to alter isotopic ratios.

Carbon isotope ratios have been used to distinguish the consumption of marine versus terrestrial organisms in human diet. The $^{13}C/^{12}C$ ratio in seawater bicarbonate is higher than in atmospheric carbon dioxide. This difference is also seen in the plants that inhabit the two regimes, as well as in the bone collagen of animals that feed on these plant species. In addition, many tropical terrestrial grasses (called C_4 species) utilize a photosynthetic pathway that efficiently metabolizes carbon dioxide by initial conversion to a four-carbon compound which incorporates more available ^{13}C. C_3 plants, more common in temperature areas, produce a three-carbon compound. The carbon isotope ratios in the bone collagen of animals feeding on three- or four-carbon plants reflect the differences in the two categories. This principle has been used to study the introduction of corn into prehistoric North America (see illus.).

Staple isotope ratios of nitrogen ($^{15}N/^{14}N$) in animal bone also reflect the ratio from diet. Three major classes of organisms can be distinguished by using nitrogen isotopes: nitrogen-fixing plants and the animals that feed on those plants; terrestrial food chains not involved in nitrogen fixation; and marine foods not based on nitrogen fixation. Fresh-water systems may constitute another class but are not yet well documented. Given this information, it should be possible to examine human bone to distinguish groups whose diets are based largely on leguminous plants, marine foods, or terrestrial (nonleguminous) foods. Studies of nitrogen isotope ratios in archeological material are, however, rare. In one successful application, examination of both carbon and nitrogen isotopes in human bone from Tehuacan, Mexico, has indicated that consumption of tropical grasses began much earlier than is evidenced in the archeological remains.

Elemental analyses. Concern with the by-products from nuclear testing during the 1950s initiated intensive investigations of the relationship between bone chemistry and diet. Strontium-90, a harmful radioactive isotope produced by fission weapons testing, appeared in substantial quantities in milk and other foods and in the human skeleton. Studies of the movement of strontium through the food chain indicated that the element was differentially distributed (see table). For example, the amount of strontium in vegetation varies by plant part and species. This information was used initially to distinguish browsing and grazing diets among fossil herbivores.

In vertebrates, approximately 99% of all body strontium is deposited in bone tissue. Clear differences in bone strontium levels can be seen along the food chain. Herbivores incorporate in bone tissue only about 20–25% of the strontium from the plants they ingest. Carnivores consume less strontium in their diets and discriminate against it in favor of calcium in the manufacture of bone tissue. Thus, the bones of herbivores and carnivores can be distinguished by their strontium content. Omnivores, such as humans, fall between these two extremes, depending upon the amount of plant food in the diet.

Marine fish and shellfish exhibit quite high levels of strontium due to the higher mineral content of ocean waters. Low strontium levels are found in fresh-water fish and shellfish, compared to marine species. Plants concentrate strontium in roots and lower leaves, but some species of nuts also contain high levels of strontium.

Human strontium levels should reflect a composite strontium value from various foods, with marine

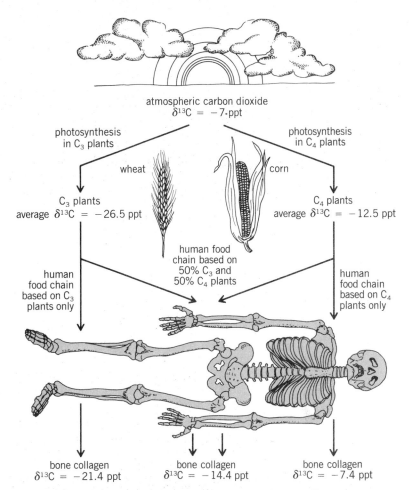

Carbon isotope ratios in C_3 and C_4 plants and in human bone. The $\delta^{13}C$ ratio is the difference in parts per thousand of the ratio between ^{13}C and ^{12}C isotopes in the sample and the ratio in a reference standard of marine carbonate. (*After P. S. Zurer, Archaeological chemistry, Chem. Eng. News, vol. 61, no. 8, 1983*)

Table 1. Strontium contents of various foodstuffs*

Food type	Sr^{2+}, µg/g	Food type	Sr^{2+}, µg/g	Food type	Sr^{2+}, µg/g
Seafood		Root vegetables		Fresh fruits	
Clam	25.9	Yam	1.1	Grape	1.8
Shrimp	6.2	Beet	2.4	Banana	0.9
Halibut	4.0	Carrot	2.6	Blueberry	0.6
Fish flour	280.0	Turnip	2.0	Strawberry	2.3
Anchovy	16.5	Parsnip	4.2	Nuts	
Meat		Onion	0.6	Brazil	107.0
Round steak	1.9	Garlic	3.5	Pecan	14.0
Lamb chop	1.8	Legume vegetables		Fats and oils	
Pork liver	0.8	Lima bean	4.4	Butter	0.8
Beef kidney	1.2	Lentils, dry	5.1	Corn oil	<0.1
Chicken breast	0.5	Green beans	1.1	Cottonseed oil	0.1
Chicken leg	1.2	Red beans	0.6	Safflower oil	<0.1
Dairy		Leafy vegetables		Condiments and	
Cheese	5.0	Lettuce	3.4	spices (dry)	14.0
Milk	0.5	Celery	1.7	Nutmeg	119.0
Grains		Brussels sprouts	0.06	Cinnamon	79.0
Wheat	3.5	Broccoli	1.0	Allspice	30.0
Millet	1.3	Cauliflower	0.5	Cocoa	23.0
Barley	1.0	Fleshy vegetables		Tea leaves	
All Bran	8.6	Green pepper	0.3		
Fresh corn	0.5	Squash	0.9		
		Tomato	0.5		
		Mushroom	0.5		

*After H. L. Rosenthal, in S. C. Skoryina (ed.), Content of stable strontium in man and animal biota, *Handbook of Stable Strontium*, pp. 503–514, Plenum Press, 1981.

animals and plants contributing higher levels of strontium and terrestrial animals providing only low amounts. Human bone is completely remodeled over a period of 5–10 years, so that chemical information on diet represents a composite summary of diet.

Strontium analysis is also relatively straightforward. Diet and the environment are the major sources of strontium in human bone. Local environmental levels of natural strontium determine the total amount of strontium available to the food chain. The composition of diet, such as the proportions of plants versus meat, determines the amount of strontium entering the bloodstream and bone tissue.

For prehistoric human populations, strontium analysis has been used to examine the relative contribution of plants and animals to the diet, marine versus terrestrial foods, the transformation from hunting-gathering to agriculture, the relationship between status and diet, and the age of weaning. In a seminal study, Margaret Schoeninger examined a large burial population from the site of Chalcatzingo, Mexico, correlating differences in diet with positions of status or prestige in society. Higher-status burials, indicated by the presence of jade in the graves, had lower bone strontium levels than individuals provided only with pottery or lacking grave furnishings. Greater meat consumption by the higher-status individuals could explain the observed differences.

Despite the positive results of this study, strontium analysis is still in an experimental stage; a number of questions remain to be resolved regarding other possible sources of variation, including individual differences and diagenesis. Individual variability within a population is relatively high, with a coefficient of variation ranging between 20% and 35%. Sources for this variability include age, sex, reproductive status, and individual metabolism. The effects of postdepositional chemical changes in bone (diagenesis) are not well understood, and studies to date are often contradictory. Strontium levels in bone appear to be unaffected in most depositional contexts, but are dramatically changed in others. The potential effects of diagenesis and environmental differences make the comparison of populations from different localities questionable.

For background information see ARCHEOLOGY; FOSSIL MAN; PALEONTOLOGY in the McGraw-Hill Encylopedia of Science and Technology.

[T. DOUGLAS PRICE]

Bibliography: L. L. Klepinger, Nutritional assessment from bone, *Annu. Rev. Anthropol.*, 13:75–96, 1984; T. D. Price, M. J. Schoeninger, and G. J. Armelagos, Bone chemistry and past behavior: An overview, *J. Human Evol.*, 14:419–447, 1985; M. J. Schoeninger, *Dietary Reconstruction at Chalcatzingo, a Formative Period Site in Morelos, Mexico*, Univ. Mich. Mus. Anthropol. Tech. Rep. 9, 1979; A. Sillen and M. Kavanagh, Strontium and paleodietary research: A review, *Yearbook of Physical Anthropology*, 25:67–90, 1982.

Atom clusters

An atom cluster (or molecular cluster) is an assembly of atoms (or molecules) which are weakly bound together; they display properties intermediate between those of isolated gas-phase atoms and bulk condensed media. There is no precise definition for the number of entities included in the range termed "cluster." Such systems may contain as few as two and perhaps as many as several thousand individual components, their unique distinction being that an appreciable number of them is present on the surface of the cluster at any time. Although their existence has been known for many years, atom and molecular clusters have become the subject of intense investigation in recent years due to the promise that results will serve to bridge the gap between atomic and molecular physics on one hand, and that of condensed material and surface physics on the other.

Classification and importance. Characteristically, matter is thought to be in one of three states, solid, liquid, or gas, with possible altered properties if the system is highly ionized (a plasma is sometimes referred to as the fourth state of matter). In view of the fact that their properties lie between those of the gas and bulk condensed phase, clusters have been referred to as the fifth, or aggregated, state of matter. Clusters are ubiquitous in the lower and upper atmosphere and in the effluent from virtually every combustion process. Potential application of research on clusters includes the development of new methods in materials processing, photography, and microelectronics, and understanding the physical basis for catalysis and nature of small grains in interstellar media.

Production and study. The study of clusters in the gas phase depends on suitable methods for effecting the desired degree of aggregation. For producing neutral clusters of simple atoms and molecules from substances with high vapor pressures, adiabatic expansion from high pressure into vacuum is a common technique. In the case of highly refractory metals such as copper, iron, nickel, tungsten, and molybdenum, vaporization is not easily carried out and different techniques are required. An especially useful method involves pulsed laser vaporization (Fig. 1). A neodymium–yttrium-aluminum-garnet (YAG) laser is employed as the source of vaporization; the laser beam is focused onto a target rod made of the desired material which is rotated within the throat of a pulsed, supersonic helium expansion nozzle. As the vapor cools and the high-density helium gas expands, clustering of metal atoms is accomplished.

Although several methods have been utilized to study cluster ions, the most common one employs thermal ion sources, where clustering is effected through the electrostatic interaction of the ion with other atoms or molecules at controlled reaction temperatures. Such methods have provided a wealth of thermochemical information, including direct mea-

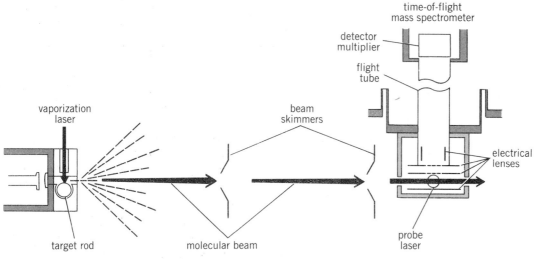

Fig. 1. Pulsed laser vaporization equipment and supersonic molecular beam two-photon ionization mass spectrometer for analyzing vaporization products. Cluster ionization is accomplished by the probe laser. Electrical lenses introduce cluster ions into the flight tube, where they are temporally separated based on velocity differences, hence enabling mass analysis.

surements of cluster bond energies (and inferences of cluster structure) from data acquired through the use of combinations of these sources with mass spectrometers.

Metal atom clusters. The large fraction of atoms existing on a cluster surface make clusters excellent prototypes in understanding the properties of surfaces on the molecular level, including elucidating questions about size effects on metallic conductivity. Additionally, new insight into the nature of metal bonding is derived from such studies.

Alkali metal clusters and coclusters with other metal atoms such as magnesium and zinc have been studied. Ionization potentials of smaller-sized clusters, determined by photoionization techniques, display oscillatory trends in relative values, with lower ionization potentials generally observed for the odd-numbered clusters.

Interesting comparisons can be made with the equation below, based on classical electrostatics,

$$W(R) = W_\infty + \tfrac{3}{8} (e^2/R)$$

which describes the influence of particle size on the ion image potential contributions to the work function of a system with spherical symmetry. Here, W represents the work function, R the radius of the equivalent sphere supporting elementary charge e, and W_∞ the bulk work function of the polycrystalline metal. There is a remarkable correspondence of the experimental data with theory at all but the smallest cluster sizes (Fig. 2a).

Interesting oscillations in their ionization potentials have also been found for several large metal clusters including ones composed of iron, nickel, and aluminum. Nevertheless, many species do still correlate reasonably well about an average line derived from the classical theory if the fine details of the oscillations are ignored (Fig. 2b). The results do

not necessarily imply that atoms are already metallic at small degrees of aggregation, but only that, excluding curvature, to a first approximation the intrinsic part of the ionization potential (the energy necessary to remove an electron to infinity) is roughly independent of the size of the clusters.

Detailed spectroscopic studies of a number of metal dimers have shown that such molecules as Cr_2, Mo_2, and Be_2 have unusually short bond lengths with high vibrational frequencies and large force constants. New insights into metal bonding are derived. For example, $3d$ (and $4d$) electrons have been found to have an important role in the metal-metal bond in Cr_2, Mo_2, and V_2. Comparison of V_2 and Cr_2 has revealed the importance of the $3d$ contraction with increasing nuclear charge.

Among the most interesting findings for metal clusters is evidence for a size effect in reactions occurring on transition-metal atom clusters. Dissociative chemisorption of molecular hydrogen and deuterium on cobalt and niobium clusters has a dramatic dependence on the number of atoms contained in the cluster. For instance, in cobalt, the trimer, pentamer, and all clusters with ten or more atoms react almost completely, while the atom, dimer, and clusters with six to nine atoms exhibit little or no reactivity. Particularly intriguing are correspondences between variations in electron binding energy and chemisorption reactivity of iron clusters with respect to molecular hydrogen. The oxidation of a sodium tetramer cluster lowers its work function, an effect similar to the reduction of the bulk work function of the metal by surface contamination.

Magic numbers. An important focus of research has been investigation of factors influencing anomalies in the abundances of cluster distributions at certain cluster sizes, commonly referred to as magic numbers. Abrupt discontinuities in the abundances

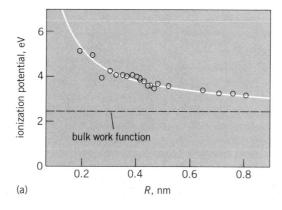

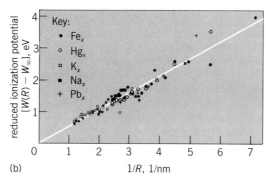

Fig. 2. Comparison of ionization potentials of metal atom clusters with the classical electrostatic equation discussed in the text. (a) Plot of experimentally determined photoionization potentials for sodium clusters (Na$_x$) against R, representing the radius of a sphere with the same volume as an n-atomic metal cluster. The solid curve is based on the classical electrostatic equation. The bulk work function for polycrystalline sodium (2.75 eV) is shown for comparison. (b) "Reduced" ionization potentials, representing the difference between measured ionization potential $W(R)$ and extrapolated bulk work function W_∞, based on a best fit of the data to the classical electrostatic equation (solid line), plotted versus $1/R$ for five metal systems. (After M. M. Kappes et al., On the manifestation of electronic structure effects in metal clusters, J. Chem. Phys., 1985)

of neighboring cluster sizes have been studied for rare-gas atom clusters as well as metals (for example, lead and bismuth), salt systems such as sodium chloride (NaCl), and systems of hydrogen-bonded molecules such as water, alcohols, and ammonia. The role of fragmentation and cluster ion stabilities rather than neutral structures in promoting the observed magic numbers has been established in all but perhaps the metal cluster systems, where the evidence is presently ambiguous. For instance, as many as six molecules are lost from an ammonia cluster following its threshold photoionization.

Organic molecules in rare-gas clusters.

Pioneering investigations of the spectroscopy of organic molecules coclustered with rare-gas atoms have been carried out. A combination of photoexcitation and laser-induced fluorescence has enabled a study of the shifts in the electronic states of the organic probe particle as a function of degree of aggregation. Most of the organic molecules which have been

studied involve ring compounds, and the data show a strong general preference for the bonding of the rare-gas atoms at central positions above a given ring. Conformers have been found where the bonding sites may be doubly occupied, with others vacant. In most cases studied, the rare-gas atoms lead to a redshift in the electronic state of the organic molecule. Shifts linear in polarizability of the clustering atom for attachment of both the first and second atoms to ringed compounds such as substituted benzenes have been found.

Typically, the spectral line widths of the spectral features are very narrow for the first few clusters. Van der Waals modes, located to the blue side of the main spectral resonances, are due to the vibrations arising from the rare-gas atom "beating" against the organic ring. Especially interesting are data for large clusters such as the example shown in Fig. 3 for argon clusters attached to phenylacetylene. At high degrees of aggregation, the main spec-

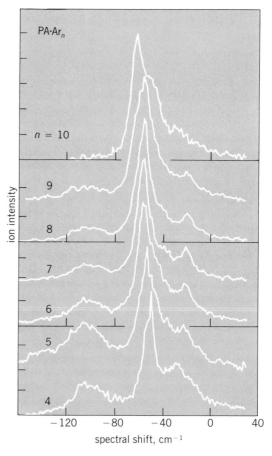

Fig. 3. Resonance-enhanced two-photon ionization current versus one-photon energy for clusters composed of phenylacetylene (PA) bound to argon. The ion currents are recorded at the mass/charge ratios corresponding to PA·Ar$_n$ ($4 \leq n \leq$ 10). The energy scale is relative to the S_1 electronic origin of phenylacetylene. The ion current is relative and different for each spectrum. (After P. D. Dao, S. Morgan, and A. W. Castleman, Jr., Resonance enhanced multiphoton ionization of van der Waals molecules: Studies of spectroscopic shifts of phenyl acetylene clustered with molecules and atoms, Chem. Phys. Lett., 111:38–46, 1984)

tral feature redshifts asymptotically to a limiting value; the spectra also display the onset of additional modes to both the red and blue of the main resonance; the latter bear some resemblance to phonon modes expected when a cluster system begins to develop properties associated with a solid.

Despite the similarities of the spectral features to those of a condensed phase molecule, the ionization potentials of such systems differ dramatically from those of the condensed phase. For example, in the case of ionization potentials for argon clustered to paraxylene, the total shifts for the hexamer are only one-tenth the value measured for similar species isolated in rare-gas matrices (thin films of condensed matter). Evidently, for systems composed of dielectrics in contrast to metallic systems, a very large number of coclustered atoms is needed for the ionization potential to approach that of the bulk phase.

For background information *see* ATOMIC STRUCTURE AND SPECTRA; CHEMICAL STRUCTURES; INTERMOLECULAR FORCES; IONIZATION POTENTIAL; MOLECULAR STRUCTURE AND SPECTRA; WORK FUNCTION (ELECTRONICS) in the McGraw-Hill Encyclopedia of Science and Technology.

[A. WELFORD CASTLEMAN, JR.]

Bibliography: J. Jortner, Level structure and dynamics of clusters, *Ber. Bunsenges. Phys. Chem.*, 88:188–201, 1984; T. C. Märk and A. W. Castleman, Jr., Experimental studies on cluster ions, *Adv. Atom. Mol. Phys.*, 20:65–172, 1984; M. D. Morse and R. E. Smalley, Supersonic metal clusters, *Ber. Bunsenges. Phys. Chem.*, 88:228–233, 1984; E. Schmacher et al., On metal-atom clusters, IV. Photoionization thresholds and multiphoton ionization spectra of alkali-metal molecules, *Helv. Chim. Acta*, 61:453–487, 1978.

Atomic beams

The thermal motion of atoms is often a significant obstacle in experiments that determine atomic properties. Problems arise both from the nonzero velocity and from the variation of velocity from one atom to the next. For atoms in a gas, this velocity spread is comparable to the average speed. In spectroscopy, the energy difference between atomic energy levels is measured by using the frequency of radiation. The precision of the energy determination is limited by the uncertainty principle: the motion of the atoms allows only a limited time for observation, and this limitation implies an uncertainty or broadening of the energy measurement. Furthermore, even though Doppler-free techniques have nearly eliminated the influence of Doppler effects, residual relativistic effects can be eliminated only by reducing the kinetic energy of the atoms. These effects shift the apparent frequency differently for each atom depending on its velocity, causing further spread in the measurement.

In experiments involving atomic collisions, or deflection of atoms by external forces, the observations would be simplified by using atoms of a single, well-defined velocity. The spread of velocities in thermal motion makes this difficult to achieve.

For the above reasons, atomic physicists have developed methods to reduce the temperature of gas-phase atoms, and thus to improve the accuracy of measurements. Recently several groups have succeeded in laser-cooling neutral atoms to very low temperatures. These very slow atoms have now been trapped as well.

Laser cooling. The basic idea behind laser cooling is quite simple. Each massless photon of laser light carries a momentum $p = h\nu/c$ in the direction of the laser beam (where h is Planck's constant, ν is the laser frequency, and c is the speed of light). When an atom absorbs light, it receives a slight "kick" from the momentum of the photon, and makes a transition to a higher state. The atom's momentum Mv changes by the photon momentum $h\nu/c$, so that the velocity v of an atom with mass M changes by $\Delta v = h\nu/Mc$. For sodium atoms illuminated by light of the yellow color of low-pressure sodium lamps, the velocity change is a mere 0.03 m/s (0.1 ft/s) out of a typical thermal velocity of 1000 m/s (3000 ft/s). Obviously, each atom must absorb many photons in order to undergo a significant reduction in velocity.

After absorbing light, an atom must subsequently emit it before it can absorb again. The emitted light also carries momentum and gives the atom a kick as it leaves, but the direction of these kicks is random and symmetric. Thus, after many absorption-emission cycles (scatterings) the kicks from the emitted light average nearly to zero, while the kicks from the absorbed light, being all in the direction of the laser beam, can add up to a large change in velocity.

Doppler-shift compensation. Since atoms in a beam all travel in essentially the same direction, reduction of their velocity is achieved simply by directing properly tuned light opposite to the beam. However, an atom can absorb light and be slowed only if its transition energy matches the light energy $h\nu$. For sodium this means that the frequency of the laser must be correct to within about 10 MHz out of an optical frequency of about 5×10^{14} Hz. Unfortunately the atoms see the frequency of the laser light Doppler-shifted by much more than 10 MHz. In fact, the spread of Doppler shifts for a thermal sodium beam is about 2000 MHz. This has two important consequences: First, only a small fraction of atoms in the beam have the right velocity to be in resonance to absorb light from the laser. Second, even those atoms that start with the right velocity will soon slow down to the wrong velocity: 200 scatterings change the velocity by 6 m/s (20 ft/s) and the Doppler shift by 10 MHz.

Without some compensation of the Doppler shift, only a small fraction of the atoms can be decelerated by a small fraction of their average thermal velocity. Two methods have been used to provide such compensation. The first involves changing the frequency of the laser and was suggested by V. S. Letokhov in 1976. The second involves changing the energy of the atomic states and was suggested by W. D. Phillips in 1980. This discussion will concentrate on the second method.

Atoms in a magnetic field experience energy-level changes (Zeeman shifts) that can compensate for the Doppler shifts. The apparatus to accomplish this is shown in the illustration. A sodium atomic beam is directed along the axis of a solenoid, or wire helix, with more windings at one end than at the other. This winding pattern causes the coil's magnetic field to vary along its length. The laser, which is coaxial with both the atomic beam and the solenoid, is resonant with atoms of a specific velocity v_0 at the solenoid entrance where the field is strongest, taking into account both the Doppler shift and the Zeeman shift. As the atoms scatter the light they slow down and move toward lower magnetic field. The smaller Doppler shift is thus compensated by the smaller Zeeman shift, and the atoms stay in resonance. For a properly wound solenoid, the process continues until the atoms come to rest near its low-field end. Also, another atom with an initial velocity smaller than v_0 eventually reaches a point in the solenoid where the magnetic field has the right strength to Zeeman-shift it into resonance with the laser. It then starts to scatter light and slow down along with the initially faster atoms. Thus all of the atoms in the beam with velocities smaller than v_0 can be slowed to a stop. The final spread of velocities of about 10 m/s (30 ft/s) is equivalent to a temperature of 0.1 K (0.2°F above absolute zero). This is laser cooling of an atomic beam.

Minimum distance and time. For sodium atoms with $v_0 = 1000$ m/s (3000 ft/s), the minimum distance required to bring the atoms to rest is about 0.5 m (1.5 ft), and the minimum time is 1 ms. These limits apply because of a limit on how fast the atoms can decelerate: the atoms can absorb light only as fast as they emit it, and the average time required for such spontaneous emission is 16 ns for sodium. This leads to a maximum deceleration of 10^6 m/s^2 (3×10^6 ft/s^2).

Trapping of neutral atoms. One of the major motivations for the work on laser cooling of atomic beams has been the possibility of trapping atoms in electromagnetic fields. Atomic and molecular ions are routinely held in such "bottles," but because the electromagnetic forces on neutral atoms are much smaller than those on charged ions, it was impossible to trap neutral atoms until the development of laser cooling of atomic beams. Now atoms are available which move so slowly that even quite weak forces can confine them.

The first successful trap for neutral atoms relies on the energy of interaction between an atomic magnetic dipole moment and an externally applied magnetic field. Atoms whose magnetic moments are aligned antiparallel to the magnetic field experience an increase in potential energy as the field increases. Thus, they are repelled from regions of high field and can be confined in a region of low field.

A trap that produces such a region consists of two identical coaxial coils of wire carrying current in opposite directions (see illus.). Midway between the coils, on axis, the magnetic field is zero. The magnitude of the field increases linearly for any displacement away from the trap center. The trap has a maximum field of 0.025 tesla and can contain sodium atoms as fast as 3.5 m/s (11.5 ft/s). This corresponds to a temperature of 17 mK (3×10^{-2} °F above absolute zero).

Laser-cooled atoms are put into the trap as follows: Atoms are decelerated and brought to rest near the low-field end of the solenoid as described above. Atoms which are slightly closer to the high-field end of the solenoid are still moving slowly toward the trap. When the cooling laser beam is shut off (by a chopping wheel in the illustration), these atoms leave the solenoid and move into the trap. Then the cooling laser beam is turned on again for just long enough to stop them, the trap current is turned on, and all those atoms which are slow enough and close enough to the center of the trap are confined. A mechanical shutter (see illus.) prevents additional atoms from entering the trapping region after the trap is on.

Atoms have been confined in the magnetic trap for longer than 1 s, nearly 10^5 times longer than uncooled atoms and 100 times longer than cold but untrapped atoms would stay in a similar region. The 1-s trapping time is limited by the rate at which atoms are knocked out of the trap by collisions with fast atoms of background gas. With better vacuum, considerably longer trapping times should be possible.

Laser cooling and trapping of atoms is a new and rapidly changing field. Among the developments that can be expected are cooling in traps to microkelvin temperatures, traps using lasers to both confine and cool atoms, and the use of laser-cooled or trapped atoms in ultrahigh-resolution spectroscopy, collision experiments, measurement of long lifetimes, observation of quantum collective effects, and tests of fundamental symmetry principles.

For background information *see* LASER SPECTROSCOPY; MOLECULAR BEAMS in the McGraw-Hill En-

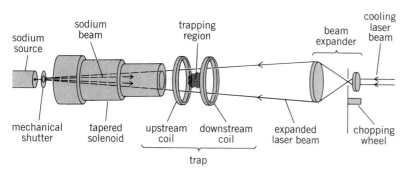

Apparatus for cooling and trapping sodium atoms.

cyclopedia of Science and Technology.

[HAROLD METCALF; WILLIAM D. PHILLIPS]
Bibliography: W. D. Phillips (ed.), Special issue on laser-cooled and trapped atoms, *Prog. Quant. Electr.*, 8:115–259, 1984; S. Stenholm and P. Meystre (eds.), Special issue on mechanical effects of light, *J. Opt. Soc. Amer.*, pt. B, 2:1751, November 1985.

Atomic structure and spectra

Electrons in the innermost shells of atoms exhibit very different behavior from the outer electrons. The latter are rather loosely bound and give rise to optical transitions and chemical bonding; their study through traditional spectroscopy dates back to the turn of the century and has played a pivotal role in the development of quantum mechanics. Inner-shell electrons, by contrast, have only recently been subject to detailed study: these electrons are bound so tightly (with an energy that can exceed 100 keV in atoms heavier than radon) that it takes modern accelerator techniques to excite them in a controlled manner.

Properties of inner shells. Because of their high energy, inner-shell electrons exhibit strong relativistic effects. In particular, the Breit interaction, which comprises magnetic and retardation effects, can dominate over the static Coulomb force. Quantum-electrodynamic phenomena can become very pronounced in atomic inner shells. For example, self-energy and vacuum polarization, which together produce the Lamb shift, can contribute as much as 0.5 keV to the K-shell binding energy of heavy atoms.

An important difference exists between the ways in which inner- and outer-shell excitations will decay. When an outer-shell vacancy is filled by an electron that falls into the empty state, a photon is probably emitted. The frequency of the emitted light is equal to the released energy divided by Planck's constant (Bohr's frequency rule). Such radiative transitions are quite unlikely in inner shells. When an electron falls into a hole in, say, the M shell of a heavy atom, the probability of an (x-ray) photon being emitted is less than 1 part per thousand. Instead, the dominant decay mode of most inner-shell vacancies is through radiationless or Auger transitions: an electron falls into the hole, and the released energy is carried (by a virtual photon) to another bound atomic electron, which is ejected.

The high intensity of Auger transitions is due to two factors: large matrix elements of the r^{-1} operator that mediates these transitions, and a very large number of channels (a typical L vacancy in a heavy atom can be filled through more than 2000 different radiationless transitions). Because radiationless transitions are so intense, inner-shell vacancies have very short lifetimes and large widths (Fig. 1). A K-shell vacancy in uranium lives only for a few attoseconds, and even in copper the K-vacancy lifetime is less than a femtosecond. These lifetimes constitute an important consideration for the design

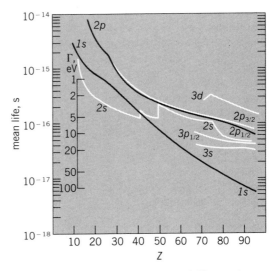

Fig. 1. Mean lives of atomic K, L, and M vacancies, as functions of atomic number Z. The inset scale shows corresponding level widths Γ. (*After B. Crasemann, M. H. Chen, and H. Mark, Atomic inner-shell transitions, J. Opt. Soc. Amer. B, 1:224–231, 1984*)

of possible x-ray laser schemes; for example, to maintain a population inversion with a K hole in copper would require a pumping power of the order of 10^{15} times the 8-keV binding energy, or 1 W per atom, far in excess of any power levels available under ordinary circumstances.

Clearly, the properties of atomic inner shells are unusual and a challenge to explore, both from an applied point of view and to learn more about fundamental aspects of relativity and quantum electrodynamics. The most important tool for probing inner shells has been synchrotron radiation.

Synchrotron radiation. The electromagnetic radiation emitted by relativistic electrons under transverse acceleration in storage rings has properties that make it an ideal probe of atomic structure. Intensity and tunability are the most important of these properties.

Electrons traversing the bending magnets of storage rings radiate considerable power. Thus, 100 mA of 3-GeV electrons confined to a circular path of 33-ft (10-m) radius emit more than 70 kW of electromagnetic radiation. The electrons radiate in the forward direction in a tight cone of aperture approximately mc^2/E radians, where m and E are the electron's mass and energy, and c is the speed of light, or 0.2 milliradians for electrons of energy $E = 3$ GeV. It is thus possible to bring substantial flux to experimental apparatus well removed from the ring. Formidable as this flux is, it can be enhanced yet further by several orders of magnitude through use of insertion devices, wigglers, and undulators.

The synchrotron-radiation spectrum extends continually from the infrared to the hard x-ray regime; there is no other known source which has this property. For virtually all applications, experimenters select a monochromatic slice of the synchrotron-radiation spectrum, using diffraction gratings at the longer wavelengths and Bragg diffraction for x-rays. These methods give resolution better than 1 eV at 5 keV and allow computer-controlled tuning over a wide range of the spectrum.

Spectrometries. Absorption spectrometry is the simplest and oldest method for determining atomic energy levels. The first atomic absorption spectrum obtained with synchrotron radiation was that of helium in the 20-nanometer region, measured in the early 1960s; this pioneering work led to the discovery of doubly excited states of helium that autoionize: as one electron returns to the ground state, the other electron is ejected. This finding was of crucial importance for the understanding of electron correlation effects because it represents the simplest possible example of the highly correlated motion of two excited electrons. More recently, x-ray absorption measurements have revealed the structure of inner-shell edges in rare gases, leading to precise determinations of hole-state energies (Fig. 2). Yet, the scope of absorption spectrometry is limited because it generally reflects only the undifferentiated response of the atomic electrons as a whole.

Photoelectron spectrometry provides the necessary discrimination among final states. With tunable synchrotron radiation, this technique has proven an extremely valuable tool for the study of electronic structure and dynamics. By measuring photoelectron intensity as a function of angle and polarization of the incident light, and even determining the spin of the photoelectrons, it is possible to achieve a complete determination of the photoionization process and gain information that is very sensitive to many-body effects such as coupling between channels of excitation. Most recently, it has even been possible to use synchrotron radiation to photoionize atoms that have been placed in an excited state through simultaneous laser irradiation; this technique provides access to a whole manifold of previously unknown states.

Dynamics of inner-shell transitions. One of the most interesting insights recently gained through synchrotron-radiation spectrometry of atomic inner shells concerns the dynamics of the excitation of deephole states and their decay. If an inner atomic electron is removed through absorption of a very energetic x-ray photon, far above threshold, then this excitation process and the subsequent filling of the vacancy through photon or Auger-electron emission can be considered a two-step process. The decay is separated from the excitation by the intervening relaxation of the surrounding electrons, and the photon and electron emissions can be treated as independent processes by means of first-order time-dependent perturbation theory. This traditional two-step model of atomic excitation and deexcitation has now been found to break down near threshold. Synchrotron radiation provides the means for measuring the threshold-excitation and interference effects that signal the breakdown of the two-step model.

Resonant Raman effect. This effect epitomizes the failure of the two-step approximation: here a single second-order matrix element describes both excitation and deexcitation. Resonant Raman scattering can be thought of as the inelastic analog of resonance fluorescence. Both the radiative (x-ray) and radiationless (Auger) versions of the process have been seen with synchrotron light. In the resonant Raman effect, a photon of exactly the right energy (within the natural lifetime width) promotes an atomic electron to an excited state, which decays in one and the same step. The emission line exhibits linear dispersion and subnatural linewidth at resonance.

Postcollision interaction. The two opposite regimes, resonant Raman scattering at threshold and wholly distinct excitation and deexcitation in the high-energy limit, are not separated by a drastic discontinuity. The phenomenon that links them and leads gradually from one into the other is postcollision interaction.

The postcollision interaction mechanism bridges excitation and deexcitation by feeding energy from one to the other, smoothly linking resonant Raman scattering with the opposite two-step extreme. The phenomenon arises near the threshold of photoionization when the slowly receding photoelectron is still within, or near, the residual singly charged ion as the decay occurs. If this decay consists of the emission of an Auger electron, the ion suddenly becomes doubly charged, and the receding photoelectron occupies a deeper potential well. Because the photoelectron's energy cannot increase instantaneously, the energy difference is transferred to the Auger electron. By slowly tuning highly monochromatic x-rays through an atomic inner-shell threshold and observing the energy of a suitable Auger peak, the resonant Raman effect and its continuation in the form of postcollision interaction can be traced (Fig. 3).

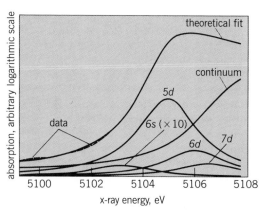

Fig. 2. L_2 absorption edge of xenon, measured under high resolution with synchrotron radiation at the Stanford Synchrotron Radiation Laboratory. Also shown are the theoretical fit to these data and the contributions to this fit from partial cross sections for the excitation of $2p_{1/2}$ electrons to the continuum and to the $6s$ and nd states. The limit of the Rydberg series (the energies of the absorption peaks for the $5d$, $6d$, . . ., nd, . . ., states) leads to an accurate value for the L_2 energy level. (*After M. Breinig et al., Atomic inner-shell level energies determined by absorption spectrometry with synchrotron radiation, Phys. Rev. A, 22:520–528, 1980*)

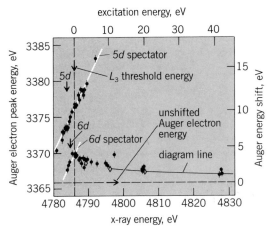

Fig. 3. Peak energies of xenon L_3-M_4M_5 Auger electrons (emitted from the M_5 level when an electron falls from the M_4 level to a vacancy in the L_3 level), as functions of excitation energy. The L_3 photoionization threshold energy and the energies required to excite L_3 electrons to the $5d$ and $6d$ levels are indicated by arrows. The resonant Raman regime with linear dispersion occurs near the 4786-eV L_3 threshold energy, and linear fits to the Auger-electron energies are shown in this region. The line labeled $5d$ spectator arises from atoms in which the photoelectron is trapped in the (previously empty) $5d$ orbit; in the $6d$ spectator case, the photoelectron is caught in a $6d$ orbit, rather than escaping into the continuum. Above threshold, the Auger energy shift due to postcollision interaction gradually subsides. The diagram line consists of Auger electrons emitted when the photoelectron escapes. Barred points indicate measurements performed in the Stanford Synchrotron Radiation Laboratory; open diamonds are theoretical predictions based on resonance scattering theory. (*After G. B. Armen et al., Threshold excitation of short-lived atomic inner-shell hole states with synchrotron radiation, Phys. Rev. Lett., 54:1142–1145, 1985*)

Unified description. A very challenging theoretical task has been the development of a unified description of atomic excitation and deexcitation that consistently encompasses the resonant Raman, postcollision-interaction, and two-step regimes. This aim now appears to have been attained by T. Åberg and his collaborators, in terms of resonance scattering theory.

For background information *see* ATOMIC STRUCTURE AND SPECTRA; AUGER EFFECT; QUANTUM ELECTRODYNAMICS; RAMAN EFFECT; SYNCHROTRON RADIATION in the McGraw-Hill Encyclopedia of Science and Technology.

[BERND CRASEMANN]

Bibliography: B. Crasemann (ed.), *Atomic Inner-Shell Physics*, 1985; B. Crasemann (ed.), *X-Ray and Atomic Inner-Shell Physics—1982*, 1982; D. J. Fabian et al. (eds.), *Inner-Shell and X-Ray Physics of Atoms and Solids*, 1981; H. Winick and S. Doniach (eds.), *Synchrotron Radiation Research*, 1980.

Barnacle

Barnacles are a group of crustaceans that have developed biological glues which enable them to attach firmly to hard substrates. In recent years the adhesion mechanisms of barnacles have been intensively studied; this research was prompted by practical efforts to control the biological fouling of engineered structures such as ships, docks, and drilling rigs in marine environments.

Cement apparatus. The body of the adult barnacle is enclosed by a series of tough calcareous plates and is firmly attached by the anterior end to the substrate. Charles Darwin first described the cement apparatus of adult barnacles in 1854, but is was not until the early 1970s that the nature of the glue was identified. Figure 1*a* shows the cement apparatus of the common intertidal barnacle *Balanus balanoides*. The base of this barnacle is thin and membranous; running in interlinked concentric rings within the base are the cement ducts. These ducts transport the glue (usually referred to as cement) from discrete gland cells to openings under the base. When first produced, the cement is a clear, runny liquid which passes easily through the ducts and spreads under the base, coming into intimate contact with both the base and the hard substrate. The runny cement is mostly protein and water, but it also contains polyphenol oxidase and polyphenols which eventually bring about cross-linking of the proteins and thus tanning of the cement (that is, curing of the glue). The cement turns white and rubbery when fully cured and bears more resemblance to an extremely viscous fluid than to a solid cement. The base increases in size as the calcareous shell of the barnacle grows, and the cement duct system keeps pace with this expansion by laying down a fresh ring of organic cement for each growth increment (Fig. 2).

Adult barnacle adhesion. When adult barnacles are forcefully pulled straight upward from a substrate, the cement itself invariably fails, leaving some cement on both the animal's base and the substrate. The force required to do this from a fairly smooth slate surface is about 900,000 pascals (130 lb/in.^{2}) of the area of failure. Thus for a barnacle with a cement area under the base of 1 cm^2 (0.15 in.^{2}), a force equivalent to a weight of roughly 9 kg (20 lb) would be needed to completely remove the barnacle. The biochemical composition of the cements of the barnacle species studied to date appear to be very similar; the forces required to cleave them are likewise similar. Some acorn barnacles are very large; *B. nubilis*, for example, can have a basal area in excess of 20 cm^2 (3 in.^{2}) and would require a force equivalent to some 180 kg (roughly 400 lb) to remove it from the substrate.

Factors affecting adhesion. Siliceous rocks possess (as do glass, many metals, and some synthetic plastics) a fairly high free surface energy. Such free surface energy implies high levels of either ionic forces (those which bind chemicals) or the weaker dispersion forces (more akin to gravity or magnetism) which promote various types of adhesive bond. The adhesion of barnacles has been tested on some synthetic plastic surfaces with low free surface energies, such as polymethylmethacrylate (Plexiglass) and polytetrafluorethylene (Teflon). Barnacles find

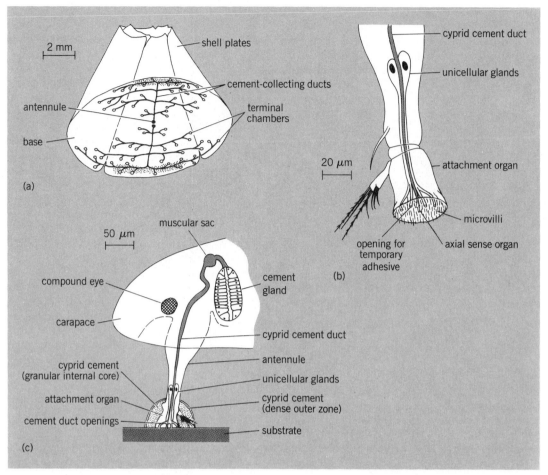

Fig. 1.   Acorn barnacles. (a) Diagram of an adult barnacle showing the proliferation of the cement ducts over the base; shaded area shows the position of the bulk of the most recent cement gland cells (*after G. Walker, The early development of the cement apparatus in the barnacle, Balanus balanoides (L.) (Crustacea; Cirripedia), J. Exp. Mar. Biol. Ecol., 12:305–314, 1973*). (b) Distal end of the cypris antennule of *Balanus balanoides* (*after J. A. Nott and B. A. Foster, On the structure of the antennular attachment organ of the cypris larvae of Balanus balanoides (L.), Phil. Trans. Roy. Soc. (B), 256:115–134, 1969*). (c) Anterior of a settled *B. balanoides* cyprid (*after G. Walker, The adhesion of barnacles, J. Adhesion, 12:57–58, 1981*)

these surfaces difficult to adhere to; forces recorded when they are forcefully removed are as low as one-tenth of the values obtained on surfaces with higher free energy. Furthermore, there is considerably less failure within the cement itself; rather, it is the adhesive cement-substratum bond that is broken. Of the nonstick surfaces used for these experiments, those with the lowest contribution of ionic forces to the free surface energy produced the least adhesion. These results may prove useful in designing antifouling coatings for submerged structures.

The roughness of a surface also has some effect on the adhesion of barnacles, but the exact nature of that effect is uncertain. On surfaces such as slate, moderate roughness induces stronger adhesion, while some of the strongest adhesion measurements have been obtained from very smooth, high free energy surfaces, such as well-cleaned glass or resin-reinforced fabric laminates (Tufnol).

The results of adhesion testing described above offer some clues about how barnacle glues work. On natural surfaces the cement itself fails during testing, due to the rupturing of chemical cross-links between protein molecules within the cement. Such failure of what is in general a thin film of a very viscous fluid, the cement, is reminiscent of the failure of bonds created by the viscous effects of classical Stefan adhesion. Viscous adhesion alone is not sufficient explanation since the keying-in effect utilized by many commercial crack fillers is also exhibited by barnacle cements on certain roughened surfaces. The uncured cement readily flows into most of the nooks and crannies under the barnacle base, where it eventually cures to gain a reasonable mechanical hold on the substrate. Finally, ionic forces also play a part in barnacle adhesion. Charged regions of the proteinaceous cement show strong attraction for similarly charged areas of the substrate. Such bonding may well play a large part in the adhesion between the cement and the cuticle of the barnacle base since both have similar protein composition.

Barnacle larva adhesion. Sessile organisms, especially those that live in the intertidal zone, face many problems of survival. Barnacles overcome the problem of "staying put" by sticking themselves very effectively to hard substrates with a protein-aceous glue. They overcome the problem of coloniz-ing and recolonizing substrates by producing larvae which are carried by currents in the sea. These larvae often feed and develop in the water column, but eventually they attach themselves to hard substrates for further development.

The settlement-stage larva is known as a cyprid. It is torpedo-shaped, with six posterior propulsive limbs and two relatively large anterior appendages, the antennules, which terminate in the attachment organs. The bell-shaped attachment organ carries many very small cuticular projections, or villi, over its disklike surface (Fig. 1b). These villi are covered in yet another sticky proteinaceous secretion which is produced by unicellular glands in the antennule. When a cyprid makes contact with a hard substrate, the attachment organs are applied to the surface, and adhere with sufficient force to prevent detach-ment by moving sea water.

Cypris larvae do not turn into adult barnacles the instant they attach. They move about the surface, apparently assessing various chemical and physical parameters which indicate the suitability of the site for adult survival. The initial attachment of the cy-pris larva must therefore be of a temporary nature. The adhesive secretion used, although from glands very similar in nature to those producing the adult cement, does not cure like the adult cement. Pre-sumably, it is like the adult cement but without the tanning agents polyphenol and polyphenol oxidase. The mechanism is very similar to that for adult ce-ment, but with a cohesive strength some quarter to half that of the adult cement.

Gregarious settlement. Sessile organisms tend to be gregarious (that is, they settle and live close to-gether). For most barnacles this is a biological ne-cessity since, though they may be hermaphrodites, few are able to self-fertilize; they must settle, there-fore, within penis reach of each other.

The problem of how barnacle cyprids detect the presence of other cyprids and adult barnacles when they setle has been studied since the 1950s. The traditional belief was that cyprids were stimulated to settle after they had touched an adult barnacle or a recently metamorphosed cyprid. The recognition was considered as a chemical sense activated on contact with barnacle cuticle or with the similar at-tachment proteins. More recently, however, it has been suggested that cyprids recognize these proteins by how hard their antennules stick to them. Barna-cle cyprids explore hard substrates, using their an-tennules to "walk around." Each time the antennu-lar disk is applied to the surface and detached to walk on a pace, the adhesive secretion is cleaved, leaving a small round "footprint" of secretion on the surface. If many cyprids walk over the same sur-face, these footprints result in an extensive, if dis-

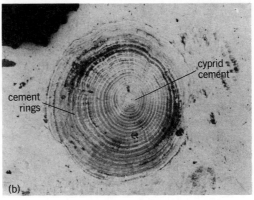

Fig. 2. Cement deposits on (a) the base of a *Balanus cren-atus* and (b) the substrate from which it was pulled. Both have been stained with bromphenol blue. The dark concen-tric rings of cement are clearly visible, as is the small area of cyprid cement in the center of the rings of adult cement left on the substrate.

continuous, layer of protein over the surface. It has been shown that cyprids can be encouraged to settle by their own proteinaceous adhesive as well as by the cuticular proteins of adults. The presence of adult barnacles is, therefore, not a prerequisite for the gregarious settlement of barnacle cyprids.

Permanent cyprid attachment. When the cyprid has eventually selected a settlement site, the tem-porary attachment afforded by the antennulary disks is augmented by the mass release of a more perma-nent cement, which is identical in chemical com-position to that of the adult cement. This cement is released from two large cement glands in the body of the cyprid (Fig. 1c), and passes down a single cement duct in each antennule and out through a network of smaller ducts in the attachment disk. This cement takes some 1.5–2 h to cure and attains a final strength only slightly higher than that of adult cement (some 950,000 Pa or 138 lb/in.). Adhesion testing shows that the mechanisms operating for cy-pris permanent cements are identical to those for the adult.

For background information *see* BALANOMORPHA; BARNACLE; CIRRIPEDIA in the McGraw-Hill Ency-clopedia of Science and Technology.

[A. B. YULE]

Bibliography: P. T. Grant and A. M. Mackie

(eds.), *Chemoreception in Marine Organisms*, 1974; A. J. Southward (ed.), *The Biology of Barnacles*, 1986; A. B. Yule and D. J. Crisp, Adhesion of the cypris larvae of the barnacle, *Balanus balanoides*, to clean and arthropodin treated surfaces, *J. Mar. Biol. Ass. U.K.*, 63:261–271, 1983.

Basement rock

An adequate understanding of the nature of the crystalline continental crust in North America must include that large portion which lies mostly buried beneath younger sedimentary rocks in the midcontinent region of the United States. This includes the region between the Appalachian Mountains on the east, the Great Lakes on the north, and the Rocky Mountains on the west. Crystalline rocks—that is, those of igneous or metamorphic origin—in this region are almost entirely of Precambrian age and are exposed in only three relatively small areas: (1) the St. Francois Mountains of southeastern Missouri, where about 350 mi^2 (900 km^2) are underlain by granite and rhyolite; (2) Mayes County in northeastern Oklahoma, where Precambrian granite is exposed in a small region along Spavinaw Creek; and (3) the Eastern Arbuckle Mountains of southern Oklahoma, where granite, granitic gneiss, and related rocks are exposed in the Belton Anticline. All other knowledge of the crystalline basement in this region has been obtained from study of cuttings and cores returned from deep drilling in the search for oil, gas, and minerals.

Continental crystalline terranes. Serious study of basement rocks in the midcontinent began in the early 1960s, when workers began collecting samples from earlier drillings to determine types and ages of the rocks recovered. However, only the rubidium-strontium (Rb-Sr) and potassium-argon (K-Ar) methods of radiometric age determination were applicable to small samples at that time.

At the University of Kansas accumulation of samples from the basement rocks of Kansas and Missouri was begun in 1968 and was complementary to a major study of exposed rocks in the St. Francois Mountains. In 1973 the uranium-lead (U-Pb) method of age determination was applied to the mineral zircon ($ZrSiO_4$) separated from basement rock samples and containing trace amounts of uranium and thorium. The results led to the discovery of two major terranes, or areas, of the continental crust that are composed almost exclusively of the volcanic rock rhyolite and the chemically similar shallow intrusive rock granite. These terranes are known from Ohio across Indiana, Illinois, southern Missouri, southern Kansas, most of Oklahoma, and into the Panhandle region of Texas. Thus a large region of the central part of the continent, almost 1200 mi (2000 km) long and 120 to 180 mi (200 to 300 km) wide, is underlain by these distinctive rock types. Their formation was clearly an important event in the evolution of the continent (see illus.).

Geologic record. To put these granite-rhyolite terranes into proper context, it is necessary to review some of the geologic history of the midcontinent region. The oldest rocks known are very ancient, or Archean, granitic rocks exposed in the Canadian Shield and extending southward into Wisconsin, Michigan, and Minnesota; rocks of similar age and type are also known in Wyoming and Montana (see illus.). These rocks range in age from about 3.6 Ga (1 Ga = 10^9 years) in southwestern Minnesota to 2.6–3.0 Ga elsewhere.

About 1.85 Ga ago, an event occurred along the southern margin of the Canadian Shield that involved the eruption of volcanic rocks, the deposition of sediment in basins associated with volcanoes, and the emplacement of magma bodies (plutons) in the upper part of the crust. The rocks formed are quite similar to those now observed in the arcuate volcanic island chains and mountains, such as the Aleutian Islands and the Andes Mountains, that now ring the Pacific Ocean and are known to mark the convergence of great plates of the lithosphere. Thus it would appear that the southern edge of the Canadian Shield was the locus of such a convergent boundary 1.85 Ga ago and the volcanic processes then formed new continental crust, just as new crust is evidently forming in volcanic island arcs and volcanic mountain chains now. Entirely similar processes producing similar rock assemblages occurred along the northwestern margin of the Canadian Shield, in what is now the Canadian Northwest Territories, and in a great belt from northern Alberta and Saskatchewan across Canada to Greenland at virtually the same time. These events may have signaled the assembly of the continent from more ancient Archean blocks.

Somewhat later, between 1.63 and 1.78 Ga ago, a series of similar volcanic-plutonic terranes were formed to the south of the Archean block in Wyoming and Montana. These rocks, now exposed in Colorado, New Mexico, and Arizona, were formed as yet more convergent plate boundaries formed along the southern edge of the growing continent. Drill-hole data indicate that rocks of this age and type extend across the midcontinent area, where they are known in the subsurface of eastern Colorado, northern Kansas, Nebraska, and northwestern Iowa. The rock record indicates that the continent grew around its edges, mostly by volcanic processes related to convergent plate boundaries; either this process occurred mostly on the southern edge of the continent or that is where the record of growth is preserved.

The presence of the extensive granite-rhyolite terranes, lying to the south and east of the Archean blocks and the volcanic and plutonic rocks that accreted between 1.63 and 1.85 Ga ago, evidently records quite a different kind of continent-building event. These rocks are distinctly younger, having formed between 1.48 and 1.34 Ga ago (see illus.), and they are chemically quite different, consisting of material much higher in SiO_2 and the alkali metals Na and K than the accreted volcanic and plutonic rocks. Age determinations by the U-Pb method

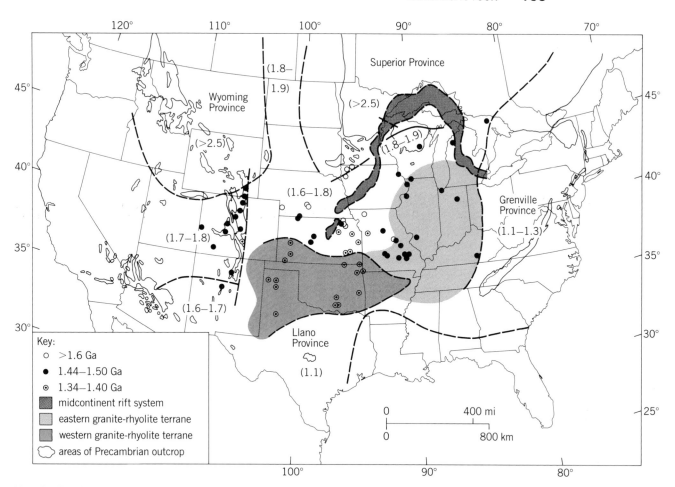

Map of United States representing distribution of crystalline rocks and their ages. Circles are the locations of dated samples, most of which are drill core or drill cuttings. Age range of most of the crystalline crust in specific regions is indicated by values in parentheses. (*After M. E. Bickford and W. R. Van Schmus, Discovery of two Proterozoic granite-rhyolite terranes in the buried midcontinent basement: The case for shallow drill holes, in B. Raleigh, ed., Proc. 1st Int. Symp. on Continental Drilling, Springer-Verlag, 1985*)

on separated zircons have shown that the granite and rhyolite that underlie the eastern midcontinent (Indiana, Illinois, Kentucky, southeastern Missouri) was formed 1.45 to 1.48 Ga ago, whereas entirely similar rocks that underlie the western midcontinent (southwestern Missouri, southern Kansas, Oklahoma, Texas Panhandle) were formed 1.34 to 1.40 Ga ago. Apparently, isolated plutons, formed during both of these igneous events, occur to the north of the granite-rhyolite terranes, where they intrude older rocks of the 1.63- to 1.85-Ga events. Thus there are at least two terranes of these distinctive rocks, and the events that formed them occurred first in the eastern region and then about 100 Ma (1 Ma = 10^6 years) later in the western region.

Terrane characteristics. The key to understanding the origin of the granite-rhyolite terranes lies in their distinctive chemical composition, their lack of deformation, and in what may be called their mantle separation age— the time the rock material was separated from the Earth's mantle below the crust. Mantle separation ages may be determined by study of the radioactive parent-daughter pair ^{147}Sm-^{143}Nd.

The granite-rhyolite terranes are notable for the uniformity of rock types. Other than minor occurrences of basaltic rocks (for example, as late dikes in the exposed St. Francois Mountains), granite and rhyolite are essentially ubiquitous. Intermediate volcanic rocks like andesite, which are characteristic of volcanic island arcs, are essentially absent, as are sedimentary rocks, metamorphic rocks, or any kind of deep-seated plutonic rocks. Rhyolitic volcanic rocks, characterized by high SiO_2, are typical of regions in which rifting of the continental crust has been associated with crustal melting to yield surface volcanism and shallow plutonism; a more recent example would be the widespread rhyolitic flows formed during the last 17 Ma in the Great Basin of Nevada, Arizona, and Utah in response to crustal extension. The lack of deformation in the granite-rhyolite terranes is also suggestive of formation in a rift environment; the rocks are probably extensively faulted, as can be seen in the exposed St. Francois Mountains, but the penetrative deformation and metamorphism that accompanies folding in convergent settings is lacking.

Rhyolite veneer model. Recent Sm-Nd studies have shed much light on the origin of the granite-rhyolite terranes. This work revealed that the Sm-Nd mantle separation ages of rocks of the granite-rhyolite terranes are about 1.8 Ga regardless of the crystallization ages of the rocks as indicated by U-Pb ages of zircons. The implication is that a major part of the continental crust of North America, including that south of the northern tier of states (Wyoming to Michigan) and extending as far south as central Texas, was formed about 1.8 Ga ago by processes involving separation of material from the mantle. Rocks of the granite-rhyolite terranes, therefore, must have formed from partial melting of previously formed crust and probably lie upon it as a relatively thin veneer. These relationships are consistent with the other features of the granite-rhyolite terranes and support formation and emplacement in a rift environment.

An interesting occurrence to the west, in the Wet Mountains of southern Colorado, also supports the model presented above. In the Wet Mountains, the San Isabel batholith, a large granitic pluton whose U-Pb age is 1.36 Ga, was emplaced at midcrustal levels into preexisting crust that is at least 1.7 Ga old. This pluton is exposed because of uplift in the southern Rocky Mountains that occurred within the last 100 Ma. Presumably the San Isabel batholith represents a frozen magma body that was originally emplaced at intermediate crustal levels, perhaps at a depth of about 6 to 9 mi (10 to 15 km); higher-level plutons and perhaps surface volcanic rocks may have originally been above the San Isabel batholith but have now been removed by erosion. These relationships suggest that the volcanic terranes that now underlie the plains may have extended at least as far west as central Colorado and that midcrustal plutons like the San Isabel probably occur below the granite-rhyolite terranes of the midcontinent region. *See* MASSIF.

For background information *see* BASEMENT ROCK; CONTINENTS, EVOLUTION OF; GRANITE; LITHOSPHERE; PLUTON; RHYOLITE in the McGraw-Hill Encyclopedia of Science and Technology.

[M. E. BICKFORD]

Bibliography: W. R. Muehlberger et al., Geochronology of the midcontinent region, United States, pt. 3: Southern area, *J. Geophys. Res.*, 72:5409–5426, 1966; B. K. Nelson and D. J. DePaolo, Rapid production of continental crust 1.7 to 1.9 b.y. ago: Nd isotopic evidence from the basement of the North American mid-continent, *Geol. Soc. Amer. Bull.*, 96:746–754, 1985; J. J. Thomas, R. D. Shuster, and M. E. Bickford, A terrane of 1350–1400 m.y. old silicic volcanic and plutonic rocks in the buried Proterozoic of the midcontinent and in the Wet Mountains, Colorado, *Geol. Soc. Amer. Bull.*, 95:1150–1157, 1984; W. R. Van Schmus and M. E. Bickford, Proterozoic chronology and evolution of the midcontinent region, North America, in A. Kroner (ed.), *Precambrian Plate Tectonics*, pp. 261–296, 1981.

Biopotentials and ionic currents

Biopotentials are voltage differences that can be measured between separated points in living cells and tissues. There are ionic charge transfers, or currents, associated with these biopotentials. Membrane channel currents and their related potentials are of great interest to those concerned with the operations of active neural systems.

The propagation of the conducted action potential results from fast, accurately controlled, voltage-dependent permeability changes to sodium and potassium ions that take place across the axon membrane. The permeation of sodium ions occurs in pores in the cell membrane that have been called sodium channels. Similarly, potassium ion permeation takes place through pores called potassium channels. These channels have been identified as molecular entities with distinct operational characteristics.

It has now become possible to compare the precise chemical composition and configuration of channels at the atomic level with the electrical characteristics of the currents flowing through these molecules.

Now it is widely accepted that a two-state ion-selective membrane channel, capable of being gated between closed and open states, is the molecular mechanism responsible for excitation in neuromuscular systems. These channels are specific protein molecules capable of undergoing conformational changes in their structure that account for the observed gating properties.

Channels. There are two general types of channels, and these are classified according to the way in which they respond to stimuli. Electrically excitable channels have opening and closing rates that are dependent on the transmembrane electric field. Chemically excitable channels (usually found in synaptic membranes) are controlled by the specific binding of certain activating molecules (agonists) to receptor sites associated with the channel molecule.

The long conducting parts of nerve cells, such as axons, transmit information by means of trains of constant-amplitude nerve impulses or action potentials. The information content is encoded into the frequency of impulse transmission. The generation of any one of these impulses is brought about by a sequence of movements of specific ions through an ensemble of protein molecules bridging the nerve cell membrane.

In order to describe the sequence of events that occur during an action potential, and to explain the macroscopic properties of electrically excitable membranes (threshold, propagation velocity, refractoriness, anesthetic blockade, and so on), it is necessary to understand that these properties arise from the ensemble characteristics of many channels activated in some form of recruited synchrony.

The most accurate measurement of these ensemble characteristics has involved voltage-clamping a reasonably large area of excitable membrane and recording the ionic current flow through thousands of parallel channels.

These membrane proteins or channels can exist in either of two states. In the open state they admit ion flow, and in the closed state ion flow is prohibited. Transitions between resting, open and closed states, or vice versa, are referred to as channel gating. Such gating is the result of channel molecule conformational changes.

The ionic currents flowing through a large area of nerve membrane containing an ensemble of many channels were described quantitatively in 1952. From this description, it was possible to reconstruct the action potential and predict many of its characteristics. One such prediction concerned the gating mechanism of the channels that were only inferred at that time. It was proposed that the channels possessed charged molecular entities, and that these entities moved in response to the electrical field across the membrane.

Gating currents. By 1973 such a gating current in the sodium channels of the squid giant axon had been measured. Recently, measurements were made of currents generated in response to sinusoidal clamped voltages applied across sodium channels in the squid axon membrane. By subjecting these sinusoidal gating currents to a Fourier analysis, the harmonic content of the currents was determined. In addition, the same methods and analysis were applied to computer-generated model schemes for sodium channel gating. None of the existing models for gating fits the gating current harmonic data, and a new gating current model was proposed.

Sequencing. Not only did this new model fit the gating current data, but it made very interesting predictions as to the molecular conformations taking place in the sodium channel structure.

One of the major advances in the application of the techniques of molecular biology and genetic engineering to the study of channels took place in the early 1980s in Japan. This research, following the successful sequencing of several pituitary hormones and significant active peptides such as beta lipotropin, determined the gene sequence for the different subunits of the acetylcholine receptor channel. From the nucleic acid sequence the precise amino acid sequence of this protein was determined.

Following this, the sodium channel protein, the transducer protein in visual cells which is linked to activated rhodopsin, and the sodium-potassium adenosine triphosphatase protein (the sodium pump molecule) were sequenced.

Several neurotoxins have been shown to bind specifically to sodium channels in chick cardiac muscle cells, skeletal muscle fibers of the rat, rat brain cells, and the electric organ of the electric eel, *Electrophorus electricus*. By making use of these specific bindings, highly purified sodium channel protein has been obtained. The molecular weight of the polypeptide sodium channel protein ranges from 200,000 to 300,000 daltons, depending on the species.

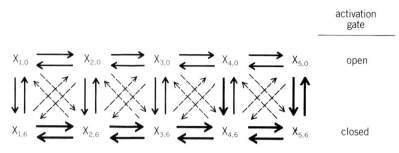

Fig. 1. Kinetic diagram for model conformational transitions. (*After J. G. Fohlmeister and W. J. Adelman, Jr., Gating current harmonics, II. Model simulations of axonal gating currents, Biophys. J., 48:391–400, September 1985*)

Recombinant DNA methods were used to clone DNA (cDNA) sequences complementary to messenger RNA coding for the sodium channel polypeptide of the eel electroplaque. The complete amino acid sequence of this polypeptide was obtained by making use of a nucleotide sequence analysis of this cloned DNA. The eel electroplaque sodium channel protein was shown to contain 1820 amino acid residues with a molecular weight of 208,321 daltons.

Assuming that the amino acid sequences in sodium channels in eel are similar to those in squid, subgroups in the channel structure were identified as being responsible for channel gating. On this basis, a tentative three-dimensional model for the sodium channel was constructed which would be consistent with both the amino acid sequence in the eel electroplaque sodium channel protein structure and the gating current harmonic data. Figure 1 shows the gating current kinetic model, and Fig. 2 shows the hypothetical arrangement of transmembrane polypeptide segments which can be constructed from the primary channel structure.

The kinetic scheme shown in Fig. 1 describes two molecular processes moving with independent degrees of freedom. Each of the X's corresponds to a molecular conformation, where $X_{i,j}$ is the conformation corresponding to molecular substates X_i ($i = 1$, . . ., 5) of the primary process and X_j ($j = 0,6$) of the secondary process. Horizontal transitions are strongly voltage-dependent; vertical transitions are weakly voltage-dependent. The broken arrows indicate coincidental but rare transitions. Major transitions between substates are indicated by heavy arrows. As $X_{5,0}$ is the most likely open channel state giving rise to a conducting channel, there are five likely closed states $X_{1,6}$, . . ., $X_{5,6}$ which precede open activation. Furthermore, all upwardly (or downwardly) pointing arrows are equal once the substate populations of the primary process have settled into their steady-state values, preventing net circulation in any kinetic loop and avoiding perpetual motion. With the addition of kinetics describing the sodium inactivation process, the sets of equations derived for the behavior of the sodium channel are extremely robust, and have been used to describe most observable biopotential phenomena associated with the generation of the nerve impulse.

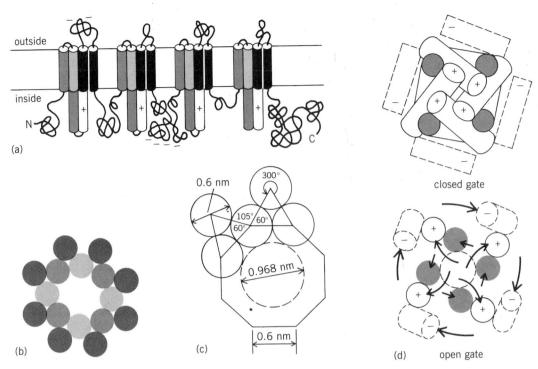

Fig. 2. Transmembrane polypeptide segments. (*a*) Hypothetical arrangement of the transmembrane polypeptide segments. (*b*) Sodium channel as viewed face-on to the membrane. Circular elements are coded as in *a*. (*c*) Dimensions of the hypothetical channel assuming α-helical segments whose axes are spaced by 0.6 nanometer. The pore diameter may be reduced to between 0.5 and 0.6 nm due to the extension of amino acid residues from the inner layer of segments. (*d*) A hypothetical activation gate extending from the model channel (given in *b*) toward the cytoplasm (out of the page). Uprighting motion of the gate elements is indicated by the arrows in the open gate diagram. Motion of ($-$) elements constitutes the primary kinetic process; motion of ($+$) elements constitutes the secondary kinetic process. (*After J. G. Fohlmeister and W. J. Adelman, Jr., Gating current harmonics, II. Model simulations of axonal gating currents, Biophys. J., 48:391–400, September 1985*)

For background information *see* BIOELECTRIC MODEL in the McGraw-Hill Encyclopedia of Science and Technology.

[WILLIAM J. ADELMAN, JR.]

Bibliography: C. de Duve, *A Guided Tour of the Living Cell*, vol. 2, 1984; J. F. Fohlmeister and W. J. Adelman, Jr., Gating current harmonics: I. Sodium channel activation gating in dynamic steady states, and II. Model simulations of axonal gating currents, *Biophys. J.*, 48:375–390 and 391–400, 1985; M. Noda et al., Primary structure of *Electrophorus electricus* sodium channel deduced from cDNA sequence, *Nature (Lond.)*, 312:121–127, 1984.

Bordetella pertussis

Whooping cough is a contagious disease of the respiratory tract characterized by coughing, with the whooping sound of an inspired breath following a coughing spasm. The bacterium *Bordetella pertussis* causes whooping cough by growing upon the ciliated cells of the respiratory tract and producing damaging substances there. These damaging toxins and enzymes have only recently been understood. Effective pertussis vaccines consisting of whole bacterial cells have been available since the 1940s, and have reduced the incidence of whooping cough. However, pertussis whole-cell vaccines contain many irrelevant substances, tend to cause local reactions, and rarely may even induce serious reactions. Japan is using its recently developed purified vaccine consisting of only a few proteins from *B. pertussis*.

Pathophysiology. Whooping cough in humans is characterized by the adherence of *B. pertussis* cells to the cilia of the cells of the nasopharynx, trachea, and other structures of the respiratory tract. Colonized cells die within a few days and are extruded from the epithelial layer. In some instances the lungs are infected, probably after mucosal surface fails to move mucus and particulates upward.

The exact nature of the substances that induce the cough is not known, but *B. pertussis* makes several noxious products. These toxins may affect mice and humans differently.

Pertussis toxin. Pertussis toxin is an ADP-ribosylating toxin (that is, it attaches an adenosinediphosphate-ribose group) which covalently modifies a regulatory protein of the human cell membrane, so that the cell no longer responds to normal hormone signals that cause a decrease in the intracellular level of cyclic adenosinemonophosphate (cAMP). Such a change in cell regulation affects the movement and effectiveness of human white blood cells, and causes pancreatic beta cells to secrete insulin, which can lead to low blood sugar levels. Effects on white blood cells include elevation of small lymphocyte numbers in the blood, sometimes to very high levels.

Pertussis toxin has several additional actions, one of which is in attaching the bacteria to the cilia of susceptible human cells. The toxin interacts with another *B. pertussis* surface protein called the filamentous hemagglutinin, which mediates adhesion to ciliated respiratory cells. In rodents, pertussis toxin induces sensitivity to histamine shock, depresses blood glucose, and elevates blood leukocytes, but is not active in adhesion to ciliated cells.

Adenylate cyclase. Adenylate cyclase, an enzyme which is excreted by *B. pertussis*, is ingested by human white blood cells. Once inside the cells, the enzyme leads to formation of more cAMP and to altered cell properties. White cells may be paralyzed, which keeps them from eliminating invading bacteria. The adenylate cyclase is present in some vaccines.

Lipopolysaccharide. A third noxious product is a structural component of the bacterial cells called lipopolysaccharide or endotoxin. Lipopolysaccharide is found as part of the cell envelope of gram-negative bacteria; it consists of lipid molecules inserted into the outer membrane, a carbohydrate core, and carbohydrate chains that extend toward the outer environment. This toxin is stable at high temperatures, unlike most protein toxins.

Vaccines. The standard DTP vaccine consists of diphtheria toxoid, tetanus toxoid, and pertussis vaccine (thus the acronym). (A toxoid is a toxin that is chemically treated so that toxicity is destroyed while ability to induce protective antibodies is retained.) Three doses of DTP vaccine for infants, then one dose at age 2 years, and one dose at age 5 comprise the standard vaccine regimen in the United States, which has markedly reduced the incidence of whooping cough but has not eliminated the disease.

Pertussis vaccine, which usually consists of whole cells, is a crude mixture of *B. pertussis* components. At least two toxic components are originally present: lipopolysaccharide, which is not destroyed by heating, and a substance which is destroyed by heating at 133°F (56°C) or by aging in the presence of Merthiolate. The lipopolysaccharide in the pertussis component of DTP tends to cause local redness, pain, and swelling at the injection site, as well as fever. More serious reactions, which are rare, may include a shock syndrome and neurologic disease. It is impossible to determine with certainty which vaccine component is responsible for any adverse effect; most apparent adverse effects are coincidental. A further complication in establishing causation is that infants may have convulsions or other neurologic symptoms from causes unrelated to vaccination. However, many scientists believe that the shock syndrome results from pertussis toxin in the vaccine, and its elevation of blood insulin.

Japanese scientists have developed a semipurified vaccine which is now in use in Japan. The major components are filamentous hemagglutinin and pertussis toxoid. Infants are not immunized, and vaccine is given only after age 2. This vaccine contains almost no lipopolysaccharide, and is being tested for its protective capability in several countries.

Bordetella pertussis surface antigens, called agglutinogens, are likely to be important in immunity, and should be included in new vaccine tests. The adenylate cyclase is another candidate for inclusion in vaccines.

As one last complication, *B. pertussis* components, including pertussis toxin and lipopolysaccharide, are immune adjuvants, that is, they cause the body to mount a stronger immune response to other substances. DTP vaccine may need to be reformulated and retested if the nature of the pertussis component is altered.

For background information *see* MEDICAL BACTERIOLOGY; TOXIN; VACCINATION; WHOOPING COUGH in the McGraw-Hill Encyclopedia of Science and Technology.

[CHARLOTTE PARKER]

Bibliography: D. W. Confer and J. W. Eaton, Phagocyte impotence caused by an invasive bacterial adenylate cyclase, *Science*, 217:948–950, 1982; M. Pittman, Pertussis toxin: The cause of harmful effects and prolonged immunity of whooping cough—A hypothesis, *Rev. Infect. Dis.*, 1:401–412, 1979; E. Tuomanan and A. Weiss, Characterization of two adhesins of *Bordetella pertussis* for human ciliated respiratory-epithelial cells, *J. Infect. Dis.*, 152:118–125, 1985; A. C. Wardlaw and R. Parton, *Bordetella pertussis* toxins, *Pharmaceut. Ther.*, 19:1–53, 1983.

Boring and drilling (mineral)

A superdeep drill hole is one which penetrates to depths in excess of 20,000 ft (6 km). In the United States, superdeep drilling is conducted primarily by the petroleum industry in exploration for natural gas in sedimentary basins of the Gulf Coast, West Texas, Oklahoma, California, and the Rocky Mountain overthrust belt. The deepest hole in the United States is Bertha Rogers No. 1, drilled to a depth of 31,441 ft (9.58 km) in sedimentary rocks of the Anadarko Basin, western Oklahoma; it ended when the bit encountered molten sulfur. The Kola Peninsula Superdeep Hole, located in the northwestern Soviet Union, is the deepest in the world at greater than 39,000 ft (12 km), most of it drilled into Precambrian granite gneiss.

Superdeep drilling technology. To accomplish superdeep drilling, new technology was required to construct heavy-duty drill rigs and fluid pumping systems, design new drill bits with longer life and strong but light drill strings, and develop new materials and tools to withstand high temperatures and pressures and corrosive-reactive formation fluids. Additionally, new techniques have been developed for high-speed automatic drill pipe–handling systems and downhole drilling motors. Superdeep drilling is conducted or proposed by several countries for the purpose of scientific investigations of the Earth's continental crust. This drilling, with coring and special geophysical logging at great depth, requires new technology in coring bits and core recovery systems, and in materials for seals, electronic components, and conducting cables.

To drill under superdeep conditions, a prerequisite is to maintain the hole as straight as possible to reduce wear on the well bore, casing, and drill string. Another important factor is drilling time (equated to cost). Since pulling the entire drill string out of, and replacing it back into, the hole (a technique known as round tripping) at great depths to change a worn bit is time-consuming, bit design must be directed toward increased life. Finally, the drill string and bottom-hole assembly must be regularly inspected and maintained or replaced to avoid loss of materials downhole that require time and effort to recover. Such work, known as fishing, is sometimes unsuccessful, causing loss of the hole.

Materials that can withstand the harsh environment of superdeep drilling must be specially developed, with attention given to tools and instruments made of composite materials. For a normal geothermal gradient, temperatures at 33,000 ft (10 km) can range from 480 to 570°F (250 to 300°C). Pressures at that depth, which may exceed 16,000 psi (1100 kg/cm^2), are such that the rock making up the borehole wall spalls into the hole; cores recovered from these depths break into poker chip–shaped disks due to the pressure release. Formation fluids that contain acid gases (H_2S or CO_2) may cause continuous corrosion on the drilling tools and react with drilling fluids (muds) and cement, causing them to lose their original properties. Many materials presently being used by the drilling industry are not designed to withstand the superdeep environment.

Drilling industry. Petroleum industry deep-drilling practice calls for rotary drilling predominantly through sedimentary rocks. In mining, although drilling depths are generally limited to about 15,000 ft (4.6 km) by the depths to which underground mining may be conducted economically, the rocks penetrated are often harder metamorphic and igneous types. Scientific research drilling is conducted most often in igneous and metamorphic rocks, and it requires extensive time for downhole sampling and measurements.

In petroleum exploration holes, little core sampling is required because of the industry's capability for interpreting downhole environments from analysis of geophysical logs and drill cuttings. Heavy-duty rotary rigs are used to drill large-diameter holes and handle the heavy casing loads required to protect the hole from rock instability caused by high pressures. Tungsten carbide insert roller-cone bits turned at low speeds are most effective for cutting rock. To control high formation downhole rock pressures and rock instability while drilling, heavy drilling muds are used. Special additives to oil-based muds are used to reduce the effects of acid gases. Coring at various depth intervals requires extra

round trips of the drill string.

The mining industry drills small-diameter holes and uses only limited casing, allowing for use of lighter rigs. Core is recovered almost continuously by using high-speed diamond core bits, with water as a drilling fluid. Wire-line core barrels are used. These fit inside the drill string and, after capturing a core, are pulled to the surface inside the string without round-tripping the string, thus saving time. Although the mining industry does not do superdeep drilling, this technique is related to the technology used in scientific research drilling.

Scientific drilling. Superdeep scientific drilling presents a challenge to technology. The Soviet Union has drilled and cored one hole over 39,000 ft (12 km) deep in the Kola Peninsula and has one in progress in the Caucasus Mountains at about 26,000 ft (8 km). West Germany is planning a cored hole to depths of 46,000 to 49,000 ft (14 to 15 km), and the United States is developing a program of scientific drilling that will include holes to depths in excess of 33,000 ft (10 km). These drilling projects are proposed by scientists to solve critical questions about the Earth's crust that can be answered only with a combination of surface and downhole geological data. Unique to the projects are the requirements for collecting continuous core samples to the total depth, and for developing deep observatories in some holes to monitor long-term changes in the Earth's stress field, seismicity, and other physical and chemical properties. The Soviet Union's Kola Superdeep Hole was started in 1970, and it has taken over 15 years to reach 39,000 ft (12 km), making it a very costly project. With new technological development, core drilling to 49,000 ft (15 km) should be accomplished in about 7 to 10 years.

Rotary drilling. In rotary drilling, the drill string is turned from the surface, causing abrasion of the string, casing, and well bore. Downhole motors located at the end of the drill string are driven by high-pressure drilling fluids pumped down the hole to rotate the bit. Thus, the drill string itself barely rotates. The Soviet superdeep project developed a downhole turbine to drive its bit, but the entire drill string must be tripped out of the hole every 33 ft (10 m) of advance to replace the wornout bit. Using automatic drill string–handling techniques, the Russians are able to accomplish this round trip in about 18 to 20 h from 39,000 ft (12 km). The challenge is to develop a long-lived bit on a downhole motor, and to recover core with a wire-line core barrel between round trips for bit changes and motor maintenance.

Drill strings. Most drill strings for deep drilling use high-strength steel, requiring a heavy-duty rig at the surface. Aluminum alloy drill strings used by the Russians, being much lighter, allow for the use of lighter rigs. However, aluminum is rapidly corroded by H_2S, which may be present in superdeep holes. Composite high-strength aluminum and steel alloy drill strings need to be developed. To use lighter rigs with heavy and high-strength drill strings, hydraulic jacks may be used on the rig to handle the weight of the drill string and casing loads. The drill rods and couples must be inspected at frequent intervals during the drilling project to detect possible material failure before a long and costly fishing job is required.

Drill hole logging. Little understanding is presently available concerning the proper interpretation of geophysical logs collected from igneous and metamorphic rather than sedimentary lithologies. Careful correlations between logs and the properties and features of rocks collected by coring are essential. Most current logging tools have a temperature limitation of between 482 and 527°F (250 and 275°C), and most conducting cable for geophysical logging is limited by its strength and its resistance to high temperatures and corrosive fluids. Measurement-while-drilling techniques enable certain geophysical measurements to be made downhole during actual drilling. The number and type of tools that can be used for measurement-while-drilling are limited, as are the means of transmitting the needed data to the surface. Development of this technilogy is vital to superdeep drilling.

Hybrid systems. Hybrid drilling systems using a combination of petroleum and mining industry practice are proposed for superdeep drilling. One system being developed for ocean drilling collects samples in soft sediments as well as hard rocks. An inner core barrel with a thin-walled diamond coring bit is contained coaxially within the barrel of a larger roller cone bit. The diamond bit and barrel is extended through the roller bit and rotated independently with a downhole motor to cut core while the roller bit is stationary. The outer roller cone bit and barrel provide an arbor for the inner diamond core bit, which cuts core. The hydraulic drive is then switched to the roller cone bit, and the hole is advanced to the end of the inner core barrel stroke. The inner core barrel, core, diamond bit, and downhole motor are then retrieved with a wire line as a package.

Kola Superdeep Hole. The Kola Superdeep Hole is an example of present capabilities and limitations in superdeep drilling and coring. The purposes of this scientific drilling project are to understand the local structure, to test models derived from surface geological and geophysical surveys, to discover the nature of geological and geophysical discontinuities at depth, and to determine ore formation mechanisms and their source depths. One reason for the success of this drilling project is that the bottom-hole temperature at about 39,000 ft (12 km) is only 403°F (206°C). The Soviet advanced open borehole drilling method used by the Soviets for this drilling project has the following strategy: drill and case a large-diameter hole into solid rock; install sacrificial casing to protect the permanent casing from abrasion of the drill string; and drill to total depth with a downhole turbo-drive motor using very lightweight mud, reaming the hole and setting additional casing only when necessary.

For the Kola Superdeep Hole, casing has been set to 6600 ft (2 km), the rest of the hole being open. An 8½-in. outer-diameter (21.6-cm) four-cone tungsten carbide roller bit is driven by a downhole turbine, collecting a 2½-in. diameter (6.4-cm) core. At 39,000 ft (12 km) between 20 and 45% of the core is recovered. About 26 to 33 ft (8 to 10 m) of penetration is made before the bit is worn, so the core pieces are collected in an inner magazine above the bit and recovered during a round trip to change bits. The upper section of the drill string consists of about 6600 ft (2 km) of steel rod; the remainder is made of an aluminum alloy with steel couplings. The drill string and couplings are inspected frequently at an on-site ultrasonic test facility; to date, no failures of the drill string have occurred. The rig derrick has a hook-load capacity of 440 tons (400 metric tons) and stands 223 ft (68 m) high, capable of core drilling to at least 49,000 ft (15 km). *See the feature article* CONTINENTAL DRILLING.

For background information *see* BORING AND DRILLING (MINERAL); WELL LOGGING (MINERAL) in the McGraw-Hill Encyclopedia of Science and Technology.

[ROBERT S. ANDREWS]

Bibliography: P. Britton, Ultradeep drilling, in L. Leroy, D. Leroy, and J. Raese (eds.), *Subsurface Geology: Petroleum, Mining, Construction*, 4th ed., Colorado School of Mines, 1977; Ye. A. Kozlovsky, The world's deepest well, *Sci. Amer.*, 251(6):98–104, 1984.

Bridge

This article discusses two areas of recent improvements in bridge construction: incremental launching and the use of glued-laminated deck panels in timber bridges.

Incremental launching method.

Incremental launching is a novel method of constructing medium- and short-span bridges without using temporary supports between piers. This technique, developed by F. Leonhardt of West Germany in 1962, is especially suited to sites where environmental constraints make temporary supports difficult and expensive. In the United States it is coming into widespread use. While usually applied to straight bridges or bridges with constant horizontal or vertical curves, incremental launching has also been used to construct spiral-curved bridges.

Incremental launching is done by constructing the superstructure (the bridge girder) at one abutment and sliding it out toward the opposite abutment as each new segment is added. Alternatively, it can be pushed from both abutments toward an intermediate span, or from the intermediate span to the abutments. The bridge superstructure is constructed of segments, precast or cast in place, which are launched as they are completed.

The length of the incrementally launched bridge is limited by the capacity of the available equipment, with a length of 2000 ft (600 m) considered comfortable. A bridge of this type requires slightly deeper girders than one constructed by standard

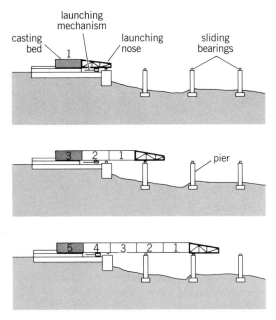

Fig. 1. Incremental launching operation, showing several stages. The individual segments are numbered in the order of fabrication and launching.

methods; 15 is the usual span-to-depth ratio. Temporary piers may be required by bridges that are longer.

Four items of auxiliary equipment are required for the launching operation (Fig. 1): (1) a launching mechanism (jacks) to push or pull the bridge superstructure; (2) special sliding bearings to allow the girder to slide across the top of the piers with minimum friction; (3) the lightweight launching nose to reduce the cantilever load at the front end of the superstructure; and (4) the casting bed.

Launching mechanism. A classic launching mechanism (Fig. 2a) consists of a horizontal hydraulic jack to push to a vertical jack with a Teflon bottom that slides on a stainless steel pad. The launching operation is started by using the vertical jack to lift the bridge girder from the formwork supports at the abutment. The horizontal jack next pushes the girder forward. The vertical jack is retracted, lowering the bridge back to the supports, and the horizontal jack is repositioned, and the cycle is repeated. This equipment is efficient and has been applied in many incrementally launched bridges. However, the maximum length of girder that such a mechanism can push is limited in that the friction restraint between the vertical jack and the bridge girder must be larger than the combined friction resistance of all other supports, or else the vertical jack will skid. For small bridges, the launching mechanism may consist simply of center-hole jacks which pull steel rods attached to the far abutment. For longer spans such a pulling device attached to the piers or the opposite abutment may also be used to assist the main launching mechanism.

Sliding bearings. Because the soffit of the bridge is usually not smooth enough to slide, special bear-

ings are necessary to reduce friction resistance during launching (Fig. 2b). In most cases the bearing consists of a concrete block covered with a stainless steel sheet. Neoprene pads coated with Teflon are inserted between the concrete blocks and the soffit of the girder. This method makes it possible to reduce the friction resistance to about 2% of the vertical reaction.

Lateral restraint is achieved by using rollers or simply attaching steel angles to the bearing to assure that the bridge girder will remain in proper alignment as it travels. Finally the temporary bearings are replaced by the final bearings after the launching operation is completed.

Launching nose. During launching, the front end of the bridge girder is cantilevered over one full span just before it reaches the next bearing. This usually creates an unacceptably large bending moment in the girder. To reduce this bending moment a light steel launching nose (Fig. 1) can be attached to the tip of the cantilever. The launching nose is also slanted upward to compensate for the deflection of the cantilever so that when its front end reaches the sliding bearing on top of the next pier it will engage the bearings easily.

Casting bed. For a cast-in-place concrete bridge a local casting bed is required to form and pour the concrete. A typical casting bed is two segments long. This length permits the webs and top slab (flange) of one segment to be poured at the same time as the bottom slab of the following segment,

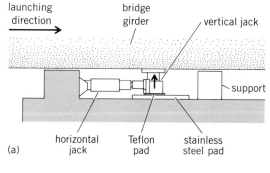

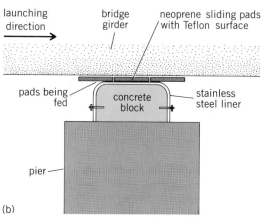

Fig. 2. Auxiliary equipment for the incremental launching operation. (a) Launching mechanism. (b) Sliding bearing.

and the formwork for the deck slab and the web can be supported on the hardened bottom slab.

As the bridge is launched from pier to pier, small uneven areas at the soffit of the girder can create large bending moments in the girder. The tolerance in the design is usually stated as 0.2 to 0.4 in. (5 to 10 mm). Therefore, the casting bed must be supported solidly by either spread footings or on piles so that it is very sturdy and unyielding.

Floating bearings. In cases where a tight tolerance is very difficult to achieve or if the spans are so short that even a 0.2-in. (5-mm) tolerance would create unacceptable bending moments during launching, it is possible to support the sliding bearings on adjustable hydraulic jacks. By adjusting the pressure in the jack and therefore regulating the reaction at the bearings to specified forces, it is possible to control the bending moment in the girder during launching.

[MAN-CHUNG TANG]

Glued-laminated timber bridges. The national highway bridge infrastructure of the United States is in serious disrepair. There are some 500,800 bridges on the public road system that are 20 ft (6 m) and longer, not including bridges on private roads, such as driveways, and roads into logging areas. Of those half million bridges, about 200,000 are estimated to be in disrepair and are deficient either functionally, in that they are not wide enough, or structurally, in that they do not possess the load capacity they are required to carry. A good percentage would be apt candidates for glued-laminated (glulam) repair.

Old-style bridges. A typical U.S. Forest Service bridge of the 1960s had a nail-laminated deck consisting of vertical 2-in. (5-cm) dimension lumber laminations nailed to each other and to the bridge stringers, or main supporting members. Decks were originally used with closely spaced stringers where the deflection of the decking system was not a major factor. The result was a tight, good-looking deck system.

When the stringers came to be glued-laminated in the mid-1960s, stringers could be spaced farther apart with consequent greater deflection of the decking that spans between the stringers. Prior to glulam decking, vehicles passing over the bridge caused working out of the nails and loosening of the deck. This, combined with the shrinking or swelling of the individual laminations due to the pickup of moisture from the elements and drying, caused excessive loosening. Not only does this cause the integrity of the deck to suffer, but it permits moisture to move between the loosened laminations onto the stringers, which increases the possibility of decay in the stringer system.

Glued-laminated decks. Quite recently, glued-laminated decks were developed to solve both of the problems with nail-lamination, that is, weakening of the deck over time, and decay of the stringers due to moisture. Deck weakening would be prevented since there are no nails to loosen, while stringer de-

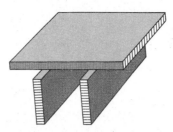

Fig. 3.   Conventional glued-laminated deck and stringers.

cay would be prevented by providing a tight roof for stringers and thus keeping moisture from reaching them.

Glued-laminated timber is an engineered, stress-rated product of a timber laminating plant. It is composed of specially selected and prepared wood laminations securely bonded together with adhesives. The grain of the wood in each lamination is approximately parallel to the length of the member. Individual laminations may be joined end to end to make longer lengths, or placed edge to edge to make wider pieces. Laminations may be bent to a curved form during manufacture.

A glulam deck panel is, in effect, a laminated beam laid on its side perpendicular to the stringers (Fig. 3). A design procedure for "glulam" deck panel bridges was published by the American Institute of Timber Construction (AITC) in 1975. AITC also obtained approval for use of the glulam deck system design procedure from the American Association of State Highway and Transportation Officials (AASHTO) in its latest specification for highway bridge design in the United States.

Glulam procedure. In the conventional glulam deck system, panels are laid perpendicular to the stringers. This can be done with lightweight equipment, which saves on construction costs and permits construction in remote areas. When first developed, the panels were connected by dowels so that their loads were shared. The dowel holes were predrilled at laminating plants, after which the panels were pressure-treated with wood preservatives to increase the durability of the wood. In the field, panels were connected to the stringers by spikes, and wood preservative was poured into the holes drilled at the job site to protect untreated wood.

Great care is required in fabricating the dowel holes and assembling deck units in the field. If not enough care is taken, or dimensional changes occurred after the panels were treated with preservatives and after the fabrication of the holes, there can be field assembly problems caused by misalignment of dowel holes, or by holes of the wrong diameter.

An alternate method of attaching the glulam deck panels to the stringers has now been developed which uses angular clips bolted in place in the field. Load tests on this clip system showed results comparable to the dowel-connected deck panels. AITC obtained approval from AASHTO for the design procedure for the clip system deck panels. This connection system is now more widely used than the dowel system.

When deck panels are treated with preservatives such as creosote, or pentachlorophenol in heavy oil, the migration of moisture into the wood is minimized. Furthermore, the higher working stresses permitted for wood used under dry conditions may be counted on to minimize the required size of the panels. Other important advantages are that glulam deck panels can be installed even in inclement weather, and they can be used as replacement decks on steel stringer bridges.

Finished glulam panel deck bridges have guard rails attached and, typically, have an asphalt wearing surface applied; glulam decking lends itself to this design.

All of the deck panel systems described above were placed perpendicular to the stringer systems. Another method is to eliminate the stringers and to use thicker and longer glulam deck panels placed longitudinally over the supporting structure frame. This is known as a longitudinal deck panel bridge. Advantages of this system are that less clearance is required over the water or whatever is beneath the bridge, stringers are eliminated, and erection is simplified. An acceptable uniform design procedure for the longitudinal deck bridge is under development.

For background information *see* BRIDGE; LUMBER MANUFACTURE; PRESTRESSED CONCRETE in the McGraw-Hill Encyclopedia of Science and Technology.

[RUSSELL P. WIBBENS]

Bibliography: *Erection Procedure for Glued-Laminated Timber Bridge Decks with Dowel Connectors*, USDA For. Serv. Res. Pap. 263, 1976; *Procedure for Design of Glued-Laminated Orthotropic Bridge Decks*, USDA For. Serv. Res. Pap. FPL 210, 1973; *Simplified Design Procedure for Glued-Laminated Bridge Decks*, USDA For. Serv. Res. Pap. FPL 233, 1974; M. C. Tang, Recent development of construction techniques in concrete bridges, *Transport. Res. Board Rec.*, no. 665, 1978; *Weyerhaeuser Glulam Wood Bridge Systems*, Weyerhaeuser Co., 1980; H. Wittfoht, *Building Bridges*, Beton-Verlag, 1984.

Cancer (medicine)

Techniques of gene cloning have advanced so rapidly that it has become possible to produce in pure form large quantities of some of the substances that the human body produces in its own defense. Of special interest in cancer therapy are lymphotoxin and tumor necrosis factor, products of the immune system. These substances are distinguished from most other products of the immune system by their profound toxic effects on various cell types, including fibrosarcomas, mammary carcinomas, lymphomas, and other tumor cells in tissue culture. They also inhibit the growth in mice of a methylcholanthrene-induced (Meth A) sarcoma by causing a hemorrhagic necrosis.

Lymphotoxin. This substance, first recognized in 1968, is produced by the lymphocytes of every mammalian species that has been examined, including human, hamster, rat, and mouse. Toxic against cells in tissue culture, lymphotoxin is one of several factors known as lymphokines that are released from T cells (thymus-derived lymphocytes) after stimulation with a specific antigen or mitogen such as concanavalin A or phytohemagglutinin. [Other lymphokines include immune interferon, interleukin 2 (T-cell growth factor), and migration inhibitory factor.]

Lymphotoxin is produced both by helper and cytotoxic T cells. It also appears in supernatants of some lymphoid tumors adapted to growth in tissue culture. The amino acid composition of the lymphotoxin produced by the human B lymphoblastoid cell line RPMI 1788 has been determined and the protein sequenced. Measured by molecular sieve chromatography, the protein has a relative molecular mass (M_r) of 64,000 daltons, and by sodium dodecyl sulfate (SDS) polyacrylamide gel electrophoresis the monomeric lymphotoxin has a molecular mass of 25,000 daltons. The protein has 171 amino acids and is glycosylated. This glycosylation is essential neither for the in vitro toxic effect against the mouse tissue culture cell line L929 nor for its in vivo effect against the Meth A sarcoma in mice. The results of recent research suggest that lymphotoxin produced by human peripheral blood T cells is identical to that produced by cell line RPMI 1788. The existence of additional molecular forms of lymphotoxin has been suggested; however, these have not yet been sequenced, nor have the genes been cloned, and the relationship to the product described above is unclear.

The gene for lymphotoxin produced by the human lymphoid line has been cloned and recent studies have demonstrated that mitogen-stimulated mouse T cells which produce lymphotoxin synthesize large quantities of a ribonucleic acid (RNA) homologous to the cloned gene for human lymphotoxin. This suggests that the murine form of lymphotoxin is very similar to that of the human.

Lymphotoxin probably plays a role in defense against foreign invaders. It may also participate in fighting malaria and several viral diseases. Its production by T cells of both the helper and killer lineages suggests a diversity of roles, including a role in killing by cytolytic T cells. Possibly lymphotoxin helps to regulate the immune system, as it is moderately toxic to both B and T cells, and it may also be involved in the pathology of T-cell-mediated autoimmune diseases. Lymphotoxin is produced by antigen-specific T cells in response to tumor antigens and thus participates in tumor immunity. It is an excellent candidate for a mediator of defense against tumors, as most reports suggest that a lower dose of lymphotoxin is needed to kill tumor cells than normal cells. Another possibility is that lymphotoxin or a closely related molecule, leukoregulin, actually prevents the development of tumors.

Though the original analyses of lymphotoxin were

based on its ability to kill in tissue culture the mouse cell line L929, it is also effective in causing necrosis of other tumors as well. The mechanism of the tumor cytotoxicity caused by lymphotoxin is under investigation. One known effect of lymphotoxin treatment is an increased synthesis of target cell lysosomal enzymes. Another involves damage of target cell deoxyribonucleic acid (DNA).

Tumor necrosis factor (TNF). This substance was first reported in 1975 in the serum of mice and rabbits which had been injected first with *Mycobacterium bovis* strain bacillus Calmette-Guérin (BCG) and then with endotoxin. When the serum substance was injected into mice bearing the transplanted Meth A sarcoma, a hemorrhagic necrosis of the tumor occurred. Hence the factor was called tumor necrosis factor. It also inhibits mouse cell line L929 and thus can be studied in a less cumbersome and less expensive assay in tissue culture. Whereas lymphotoxin has always been studied as the product of lymphocytes in tissue culture, tumor necrosis factor was originally isolated from serum or plasma and appears to be the product of activated macrophages. The human promyelocyte leukemia cell line HL60 was used to purify the protein and to clone the gene. The mouse macrophage cell line J7774 can also be induced to make tumor necrosis factor. The structure of tumor necrosis factor as isolated from the human HL60 cell line is a protein with a relative molecular weight of approximately 17,000 daltons; it has 157 amino acids, an isoelectric point of 5.3, and contains two cysteines which form an internal disulfide bridge. It is not glycosylated. Tumor necrosis factors of several molecular weights were originally described in serum; those from rabbits are 39,000 daltons and 67,000 daltons; those from mouse serum are 40,000 daltons and 150,000 daltons. It is possible that in the natural state aggregates or tumor necrosis factor occur or that several different species are produced in the body.

It has been difficult to assign a function for tumor necrosis factor in the normal animal. It is possible that T cells, after stimulation with antigen, induce macrophages to produce tumor necrosis factor which could then participate in necrotic lesions associated with delayed hypersensitivity and certain cases of viral immunity. It can kill malarial parasites, suggesting that tumor necrosis factor may play a role in defense against parasitic infections. Its cytotoxic effect against tumors suggests a natural role in tumor immunity. Recent evidence indicates that it is similar or identical to cachetin, a substance which inhibits lipid metabolism and has been associated with wasting in parasitic diseases and cancer.

Comparisons. The most striking similarity between lymphotoxin and tumor necrosis factor is a functional one. Both kill L929 cells in tissue culture and cause hemorrhagic necrosis of the Meth A sarcoma in the animal. Nevertheless, there are profound differences in the method of induction, the cell of origin, and the biochemistry of the molecules. When the structure of the products of cloned genes for the two proteins are compared, they are clearly different, though certain areas of homology are apparent. The products of the cloned genes obtained from the lymphoblastoid cell line RPMI 1788 lymphotoxin and the promyelocytic cell line HL60 tumor necrosis factor vary in important ways. Though their apparent molecular weights are similar, striking differences in structure are apparent. These include the presence of a disulfide bridge in tumor necrosis factor and its absence in lymphotoxin. Furthermore, lymphotoxin contains three methionines and tumor necrosis factor contains none. These differences in amino acid sequence and three-dimensional structure are reflected as antigenic differences. Antibodies made to human tumor necrosis factor do not cross-react with human lymphotoxin. Another difference is that lymphotoxin is a heterogeneous glycoprotein, whereas tumor necrosis factor is not subject to significant posttranslational processing. Some areas of homology in the two molecules may represent regions which are crucial to function. As much as 28% of the molecules are homologous and, in particular regions, the homology is much higher. Some workers believe that tumor necrosis factor and lymphotoxin should be considered as members of a multigene family, and should be called alpha and beta tumor necrosis factor. However, further discussion and study are warranted before the nomenclature is changed.

Therapeutic applications. The ideal anticancer drug should discriminate between normal and transformed cells. It must be effective if administered systemically, be nonimmunogenic, and have minimal side effects. Until now, one of the problems in immunotherapy has been that the antigenic pattern of tumors can change significantly as they evolve. A perfect anticancer treatment would be effective regardless of the tumor's antigenic display.

There is considerable hope that lymphotoxin and tumor necrosis factor can be used in tumor therapy. First, studies of both cytokines indicate that they usually kill transformed cells more effectively than normal cells. Furthermore, effects on Meth A tumor growth have already been noted when the substances are injected directly into the tumors or even intravenously. Another positive aspect is that a synergistic effect in the killing of some tumors has been noted at least in tissue culture combinations of immune interferon and lymphotoxin. This suggests that small doses of several individual cytokines may be very effective.

The wide range of sensitivity of different tumors to the factors presents a problem in using lymphotoxin and tumor necrosis factor as anticancer agents; so far no pattern (for example, by tumor type) has been apparent. Until the basis of sensitivity to the factors is understood, it will be difficult to predict whether a tumor in a particular patient will be affected or not. Long-term studies of systemic administration of cytokines must be carried out to identify potential side effects. Because lymphotoxin does inhibit cells of the lymphoid lineage in tissue culture, immunosuppressive effects must be seriously considered. The identity of tumor necrosis factor and cachectin suggests that only a limited systemic dose range may be useable for treatment of cancer patients.

In spite of possible problems, there seems to be a tremendous potential for lymphotoxin and tumor necrosis factor in the fight against cancer. Their effectiveness at very low concentrations, the fact that their activity does not depend upon a continuous expression of a particular antigen by the tumor cell, and the fact that they are natural human products are all advantages. Furthermore, human-derived products are less likely to be immunogenic in humans and thus more readily tolerated. That lymphotoxin acts, at least in part, by fragmenting target cell DNA suggests that a combined modality of toxin administration and radiation therapy may be particularly useful. Another approach is the targeting of these toxins directly to tumors by using monoclonal antibodies.

For background information *see* CANCER (MEDICINE); NEOPLASIA; ONCOLOGY in the McGraw-Hill Encyclopedia of Science and Technology.

[NANCY H. RUDDLE]

Bibliography: B. Aggarwal et al., Human tumor necrosis factor: Production, purification, and characterization, *J. Biol. Chem.*, 260:2345–2354, 1985; E. A. Carswell et al., An endotoxin-induced serum factor that causes necrosis of tumors, *Proc. Nat. Acad. Sci.*, 72:3666–3670, 1975; G. A. Granger and T. W. Williams, Lymphocyte cytotoxicity in vitro: Activation and release of a cytotoxic factor, *Nature*, 218:1253–1254, 1968. N. H. Ruddle, M. B. Powell, and B. S. Conta, Lymphotoxin, a biologically relevant model lymphokine, *Lymphokine Res.*, 2:23–31, 1983.

Cell (biology)

About 4 billion years ago, the first organisms evolved on Earth. They were probably similar to modern-day prokaryotes (bacteria). From 3.7 billion years ago until about 2 billion years ago, Earth was in an "Age of Prokaryotes" covered by bacterial communities which now may be found in the fossil record. Two billion years ago, specific events occurred that resulted in the origin of a new type of cell, the eukaryotic cell, which is present in all protoctists (such as amebas and ciliates), fungi, plants, and animals.

Prokaryotes and eukaryotes. The cells of prokaryotes and eukaryotes differ in their organization. In prokaryotes, the genome, which consists of a loop of deoxyribonucleic acid (DNA), is in the cytoplasm. In eukaryotes, the genome (chromosomes consisting of tightly coiled DNA with histone proteins) is enclosed within a membrane, forming the nucleus. Major metabolic pathways of prokaryotes, such as respiration and photosynthesis, may occur throughout the cell. In contrast, these major pathways in eukaryotes occur in distinct membrane-bound packages—mitochondria for respiration, and plastids,

such as chloroplasts, for photosynthesis. Many prokaryotes are motile by the use of simple whiplike flagella, composed of flagellin protein and attached to wheellike structures (basal bodies) that rotate, and flick the flagella. Many eukaryotes, in contrast, are motile with complex motility organelles (cilia or undulipodia), which are composed of at least 200 different proteins including tubulin, the protein that forms microtubules. Motility is accomplished by the sliding of the microtubules.

Symbiosis. All organisms are associated with other organisms of different species; some of the more intimate associations are designated as symbioses. In symbiotic associations the organisms are usually in close physical proximity and may share products or physical protection. The classic examples of this phenomenon are lichen symbioses, in which algae are protected from desiccation within the body of a fungus and the fungus receives photosynthetic product from the algae. *See* LICHENS.

There is considerable evidence that eukaryotic cells are the products of specific symbiotic associations between prokaryotes, which probably began to occur 2 billion years ago. According to this idea, the mitochondria, the plastids, the motility organelles (undulipodia), and the nucleocytoplasm of eukaryotes were once separate organisms. Their association is, however, obligate in modern eukaryotes. Essentially, multicellular eukaryotes such as humans are walking communities of symbiotic bacteria. Although the components of the symbiosis can no longer survive on their own, there are many clues to their former independent identities (Fig. 1).

Advantages of symbioses. The atmosphere of early Earth probably contained only trace amounts of oxygen; the first organisms were anaerobic. Slowly oxygen began to accumulate as the waste product of photosynthetic cyanobacteria, and by about 2 billion years ago enough oxygen had accumulated in the atmosphere that anaerobic organisms were being poisoned. Oxygen is a toxic, corrosive gas, and only aerobic organisms with special oxygen-detoxifying enzyme systems can tolerate it. One of the most innovative detoxification systems is respiration, which actually uses oxygen in energy-generating metabolism. It would have been a great advantage for an anaerobic prokaryote host to acquire a symbiont with a respiratory pathway (that is, premitochondria). In addition, photosynthetic organisms as symbionts (preplastids) would provide the host with photosynthetic products, and motile episymbionts would provide motility. The advantages to the endosymbionts (internal) and episymbionts (attached) may be removal of waste products by the host, provision of some nutrient by the host, and a uniform, protected environment.

Nucleocytoplasm. The main body of modern eukaryotic cells, containing the cytoplasm and the nucleus (which encloses the chromosomal genome), represents the original prokaryotic host organism which acquired the mitochondria and plastids as endosymbionts and the motility organelles (undulipodia) as episymbionts. The prokaryotic host or pre-nucleocytoplasm probably had no cell wall, which allowed easy uptake of the mitochondria and plastids, or perhaps active penetration by the mitochondria and plastids. The host prokaryote probably had proteins similar to histone proteins (modern eukaryotic DNA-binding proteins) associated with its DNA. A modern group of bacteria, perhaps descendants of the original host bacteria, are the mycoplasms, which have no cell wall and have histonelike proteins.

Mitochondria. Respiration occurs on membranes within mitochondria. One of the products of respiration, energy-rich adenosinetriphosphate (ATP), is distributed to the rest of the cell for use as an energy source for metabolism. Certain modern bacteria, such as *Paracoccus*, perform a similar set of reactions on their internal membranes, and thus are considered to be possible descendants of the bacteria that first entered symbioses as mitochondria. Modern mitochondria still retain some of their bacterial characteristics: they are enclosed in a double membrane, they have their own DNA and ribonucleic acid (RNA), and they are sensitive to some antibiotics that typically affect prokaryotes. Mitochondria still have a semblance of autonomy in that they can undergo divisions as the rest of the cell divides, and in that they have their own independent genome. The genomes of mitochondria, however, have become greatly reduced over the last 2 billion years, and mitochondria are no longer able to grow as independent organisms.

Plastids. Photosynthesis occurs on membranes within plastids. Different types of plastids have different sets of light-capturing pigments: for example, chloroplasts of plants and green algae have chlorophylls *a* and *b*; phaeoplasts of brown algae have chlorophylls *a* and *c1*; and rhodoplasts of red algae have chlorophyl and phycobiliproteins. Each type of plastid may have had a different symbiotic origin. Chloroplasts seem to be related to a green bacteria, *Prochloron*, which has chlorophylls *a* and *b*, and rhodoplasts seem to be related to cyanobacteria, which have chlorophyll *a* and phycobiliproteins. The origin of phaeoplasts remains unknown. Plastids have their own DNA, and have RNA that is very similar to bacterial RNA. Plastids are sensitive to several antibiotics that affect bacteria and, as do mitochondria, they undergo divisions and have a genome separate from that of the nucleus, but greatly reduced.

Motility organelles. Spirochetes (long, thin, highly motile bacteria) are possible candidates for descendants of the common ancestors of eukaryotic motility organelles (undulipodia). Modern spirochetes have a tendency to form motility symbioses with other organisms—for example, a protoctist found in the hindguts of termites, *Mixotricha*, is completely covered with millions of moving spirochetes (Fig. 2). Of the major eukaryotic organelles, however, the origin of motility organelles remains most obscure. The membrane-bound motility organelles do have an intrinsic motility and are very com-

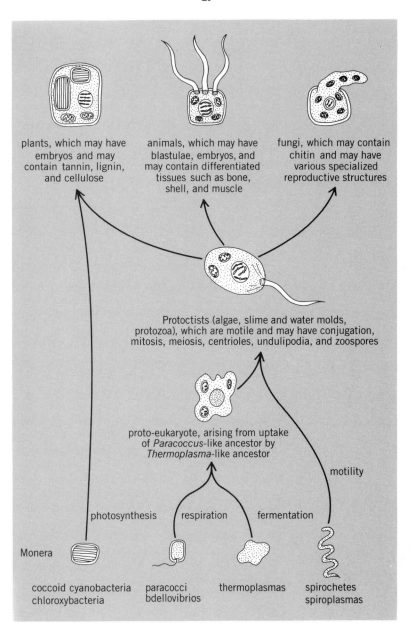

Fig. 1.   Representation of model for the origin of eukaryotic cells by symbiosis. (*After L. Margulis, Symbiosis in Cell Evolution, W. H. Freeman, 1981*)

plex (composed of tubulin and at least 200 other proteins). There is no DNA in motility organelles, though there is a small piece of RNA at the base of each one, possibly a remnant of an ancient prokaryotic genome system. The most important evidence for the symbiotic origin of motility organelles is the presence of tubulinlike proteins and small tubules (similar to the primary structural feature of motility organelles) in some spirochetes.

Genome reduction. When organisms become associated, their needs become different from when they were independent. Some of their gene products may become redundant, or unnecessary because of a different environment (for example, inside a host a protective cell wall may no longer be useful). Since it requires energy to maintain all of the genes

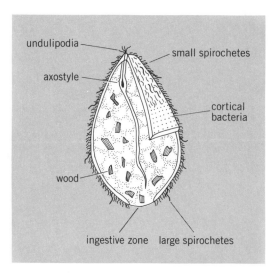

Fig. 2. *Mixotricha paradoxa*, a protoctist found in the gut of termites, with symbiotic spirochetes. (*After L. Margulis, Symbiosis in Cell Evolution, W. H. Freeman, 1981*)

in a genome, organisms that carry redundant and unnecessary genes are at a disadvantage. Thus there is a selection pressure for the loss of extra genes.

It has been demonstrated that some genes have traveled from the mitochondria to the nucleus and from the mitochondria to the plastid. The mechanism of transport is not well understood, but there may be several possibilities: viruses and viruslike entities can transport pieces of DNA, and some genes seem to have intrinsic abilities to move from one DNA strand to another. The selective advantage of moving genes, particularly genes that are useful to a symbiosis, is that a centralized location for the genes (that is, the nucleus) may be more efficient in coordinating gene products and replicating genes in cell division.

For background information *see* CELL (BIOLOGY); CELL PLASTIDS; CILIA AND FLAGELLA; LIFE, ORIGIN OF; MITOCHONDRIA in the McGraw-Hill Encyclopedia of Science and Technology.

[BETSEY DEXTER DYER]
Bibliography: L. Margulis, *Symbiosis in Cell Evolution*, 1981; S. Wolfe, *Biology of the Cell*, 1981.

Cell differentiation

Cell differentiation is broadly defined as a set of regulatory processes that converts the egg into a mature organism composed of many different kinds of tissues and organs. In mammals and many other vertebrates, the initial steps of this process (often called cell determination) are not evident until the egg has cleaved into a large number of cells. In some invertebrates, however, determination begins as early as the two-cell stage, although the biochemical and structural features associated with the later stages of differentiation are still invisible. Precocious cell determination in invertebrates appears to be caused by the regionalization and unequal segregation of cytoplasmic regulatory factors (cyto-

plasmic determinants) between the embryonic cells. Recent studies have shown that some of these factors may be maternal messenger ribonucleic acid (mRNA) molecules that are synthesized during oogenesis, localized in specific regions of the egg cytoplasm, and partitioned between the embryonic cells during cleavage. The localization and segregation of maternal mRNAs appears to be mediated by their association with regionalized cytoskeletal domains underlying specific cytoplasmic areas (ooplasms) of the egg.

Maternal RNA as cytoplasmic determinant. Maternal mRNAs have been suspected as cytoplasmic determinants, but only recently has convincing evidence been obtained for this role. The first experimental evidence was obtained in relation to the condition known as double abdomen, in which the cephalic segments of an insect (*Smittia*, a midge) embryo are converted to a partial mirror image of the abdominal segments by ultraviolet (UV) irradiation of the anterior pole of the egg. Repair of ultraviolet effects by white light, an enzymatic process known to occur only with nucleic acid substrates, and mimicry of ultraviolet effects by localized application of ribonuclease (RNase) are consistent with the idea that the ultraviolet-damaged materials are RNA molecules.

The second kind of experimental evidence implicating RNA molecules in cell determination is derived from rescue studies of maternal-effect mutations in genes that govern embryonic dorsal-ventral polarity, one of the first expressions of cell determination in fruit fly (*Drosophila melanogaster*) embryos. Because important gene products are accumulated in the egg during oogenesis, maternal-effect mutations are expressed as recessive alleles even after fertilization by sperm carrying wild-type dominant alleles. Several maternal-effect mutations, including *snake*, *easter*, *Toll*, and *spatzle*, show altered dorsal-ventral polarity and are lethal, but can be partially or completely rescued by microinjection of RNA containing polyadenylate [poly (A)] and derived from wild-type embryos. Since poly (A) sequences are restricted to mRNAs, this is strong evidence that maternal mRNAs are involved in the determination of dorsal-ventral polarity.

It has been proposed that maternal mRNAs serve as cytoplasmic determinants by directing the synthesis of proteins which in turn serve either as regulatory molecules that govern the nature of gene activity in different cell types, or as some of the structural or enzymatic proteins characteristic of differentiated cells. Several means can be envisioned in which the differential distribution of mRNA molecules to different types of embryonic cells can be achieved. First, all mRNA species could be segregated to each blastomere, but certain species might be degraded in some of the cells. Second, certain mRNA species could be actively segregated to individual cells during cleavage. Third, certain mRNAs could already be differentially localized in the egg and partitioned to different cells by cleavage. These possibilities have been difficult to test by biochemical means because they require methods for the mass separation of ooplasms. It has been possible, however, to use another approach: determination of the position of individual mRNAs in sections of eggs and embryos by in-place hybridization. For in-place hybridization, eggs or embryos are fixed and sectioned, after which the sections are allowed to react with radioactive nucleic acid probes which exhibit sequence complementarity to maternal mRNA. The location of hybrids formed between the radioactive probes and the mRNA molecules in the sections is determined by autoradiography, a technique that identifies the cytological position of radioactive emissions by their translation into visible silver grains in a photographic emulsion. In-place hybridization studies have provided the first indication that maternal mRNAs are unevenly distributed in the egg cytoplasm.

Spatial distribution of maternal mRNAs. Several model systems have been used to explore the possibility of differential spatial distribution of maternal mRNAs. Ascidian (tunicate) eggs and embryos are particularly good subjects to map the location of mRNAs, since the eggs contain three different colored ooplasms which exhibit specific developmental fates and can serve as markers for identifying the localization of mRNAs. After fertilization the ooplasms are subject to dramatic spatial rearrangements, and eventually become fixed in positions where they will enter different embryonic cells during the subsequent cleavages. The ectoplasm (a cytoplasm originally derived from the oocyte nucleus or germinal vesicle) enters the epidermal cells, the endoplasm (a yolk-filled ooplasm) enters the brain, notochord, and gut cells, and the yellow crescent cytoplasm enters the muscle cells. In-place hybridization with poly (U), a probe that detects all mRNAs by binding to their poly (A) sequences, and with cloned DNA probes recognizing histone and actin mRNAs, has shown that mRNAs exhibit different spatial distributions in the egg and early embryo. Most of the mRNA molecules are localized in the ectoplasm (Fig. 1). Histone mRNA, however, is evenly distributed within the ooplasms, and actin mRNA is concentrated in the ectoplasm and yellow crescent cytoplasm. The ectoplasmic and yellow crescent actin mRNAs migrate with these ooplasms during the cytoplasmic movements that follow fertilization, become fixed in position prior to the first cleavage, and are subsequently partitioned to specific embryonic cells. The epidermal and muscle cells obtain most of the maternal actin mRNA.

The localized actin mRNAs of ascidian eggs also show distinct origins during oogenesis. The ectoplasmic actin mRNA is originally present in the germinal vesicle (oocyte nucleus), while the yellow crescent actin mRNA is initially located in the oocyte cortex. This spatial arrangement suggests that yellow crescent actin mRNA may be transcribed and transported to the cytoplasm earlier during oogenesis than the ectoplasmic actin mRNA.

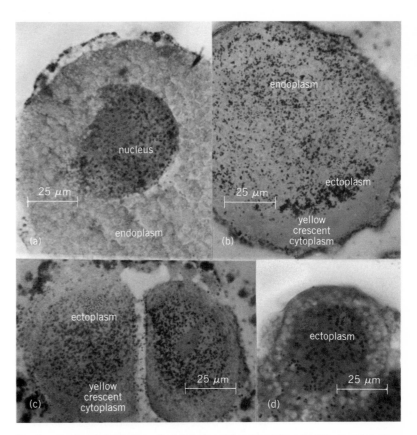

Fig. 1.   In-place hybridization of sectioned ascidian eggs with poly (U), in which silver grains show the localization of total mRNA. (a) Mature oocyte, in which most messenger ribonucleic acid is in the nucleus. (b) Fertilized egg, showing comparative labeling of the three major cytoplasmic regions. (c) Two-cell embryo. (d) Four-cell embryo, showing an ectoplasmic blastomere. (*From W. R. Jeffery and D. G. Capco, Differential accumulation and localization of poly-A containing RNA during early development of the ascidian Styela, Dev. Biol., 67:152–166, 1978*)

Localized mRNA molecules have also been described in annelid (*Chaetopterus* and *Nereis*) eggs, in which ooplasms exhibiting unique developmental fates are also present. As in ascidian eggs, the localized mRNAs migrate with their respective ooplasms after fertilization, become fixed in certain locations, and are partitioned between different embryonic cells.

The pattern of mRNA distribution observed in ascidian and annelid eggs supports the idea that mRNAs are prelocalized in specific regions of the egg and partitioned into different embryonic cells during cleavage. This situation may not be true for all eggs, however. Evidence for uneven mRNA distributions is lacking in mammalian and sea urchin eggs, each of which exhibits a relatively homogenous cytoplasm and is not subject to precocious cell determination.

Messenger RNA cytoskeletal associations. The rearrangement of mRNA during ooplasmic segregation suggests these molecules may be tenaciously associated with regionalized cell structures. The structural matrix could be membrane systems, cytoskeletal elements, or components associated with either of these entities. The involvement of mem-

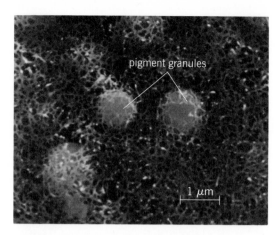

Fig. 2. A portion of the cytoskeleton which is present near the plasma membrane in the yellow crescent region of ascidian eggs. The pigment granules are enmeshed in a filamentous network. (*From W. R. Jeffery and S. Meier, A yellow crescent cytoskeletal domain in ascidian eggs and its role in early development, Dev., Biol., 96:125–143, 1983*)

branes or cytoskeletal elements in mRNA localization has been tested by the extraction of eggs with non-ionic detergents, which extract membrane lipids and other soluble components but leave the egg cytoskeletal system intact. Ascidian eggs have been shown by such detergent extraction to contain a complex cytoskeletal framework consisting of a superficial network of actin filaments (Fig. 2) and a more internal lattice of filaments associated with particular cytoplasmic organelles. This cytoskeletal domain is very well developed in the yellow crescent cytoplasm; it segregates with this region during early development and eventually enters the embryonic muscle cells. A similar cytoskeletal framework is present in the cortex of other eggs.

Most of the mRNA is retained in the cytoskeletal fraction when ascidian or annelid eggs are detergent-extracted. Moreover, in-place hybridization of sections of eggs treated with detergent shows that mRNAs are localized in the same regions of the cytoskeletal framework as they are in intact eggs (Fig. 3). This brings up the possibility that the egg cyto-

skeleton is involved in mRNA localization: the localization, rearrangement, and partitioning of maternal mRNAs may be dependent on their recognition and strong interaction with egg cytoskeletal components.

Conclusions and prospectus. Maternal mRNA molecules have been found to be unevenly distributed in the egg cytoplasm and segregated to the embryonic cells during early development. This distribution may be mediated by the association of mRNA with cytoskeletal domains which are themselves localized in the egg. The discoveries of the localization and differential partitioning of mRNA to specific embryonic cells and the ability of mRNA to rescue maternal-effect mutations which govern embryonic polarity suggest that maternal mRNA plays an important role in cell determination. Further progress in this area will require analysis of the spatial distribution of many different species of mRNA, particularly those which code for proteins involved in developmental regulation, and the development of microinjection assays which can be used to test the determinant activity of mRNAs. Ultimately it will be necessary to see if specific maternal mRNAs can affect developmental features after introduction into embryonic cells where they do not normally exist.

For background information *see* BLASTULATION; EMBRYOLOGY; GENE ACTION in the McGraw-Hill Encyclopedia of Science and Technology.

[WILLIAM R. JEFFERY]

Bibliography: K. V. Anderson and C. Nuselein-Volhard, Information for the dorsal-ventral pattern of the *Drosophila* embryo is stored as maternal mRNA, *Nature*, 311:223–227, 1984; W. R. Jeffery, Spatial distribution of messenger RNA in the cytoskeletal framework of ascidian eggs, *Dev. Biol.*, 103:482–492, 1984; W. R. Jeffery, C. R. Tomlinson, and R. D. Brodeur, Localization of messenger RNA during early ascidian development, *Dev. Biol.*, 99:408–420, 1983; W. R. Jeffery and L. Wilson, Localization of messenger RNA in the cortex of *Chaetopterus* eggs and early embryos, *J. Embryol. Exp. Morphol.*, 75:225–239, 1983.

Cellular immunology

Recent research has provided deeper understanding of the nature and activity of receptors for antigen on the surface of T lymphocytes and of cell-surface complement receptors.

Antigen-specific receptors. The immune response to foreign invaders such as bacteria, viruses, or tumor cells is mediated by antibodies or by specialized cells of the immune system. In the so-called humoral immune response, B lymphocytes (B cells) differentiate into antigens, plasma cells which secrete antibodies that are specific for foreign substances. In contrast, T lymphocytes (T cells) mediate the antigen-specific cellular immune response. Subsets of T cells play different roles in the immune defense. Effector T cells function as cytotoxic T cells in the rejection of foreign tissue grafts or tumors and in the elimination of virus-infected cells,

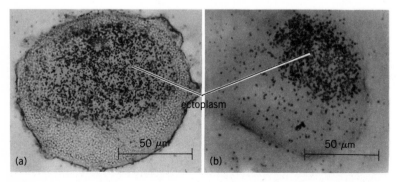

Fig. 3. In-place hybridization with poly (U) of (a) sectioned ascidian egg, and (b) cytoskeletal framework showing the localization of total mRNA. Note the position of grains in the ectoplasm of the egg and its cytoskeleton. (*From W. R. Jeffery, Spatial distribution of messenger RNA in the cytoskeletal framework of ascidian eggs, Dev. Biol., 103:482–492, 1984*)

and are also involved in the delayed hypersensitivity response. Regulatory T lymphocytes can amplify (as helper T cells) or suppress (as suppressor T cells) the responses of other T or B lymphocytes in an antigen-specific fashion. In addition to T and B cells, other lymphoid cells (for example, natural killer cells) and myeloid cells (for example, macrophages and granulocytes) play a role in the immune response. Yet only T and B cells appear to carry antigen-specific cell-surface receptors.

B-cell receptors are membrane-bound immunoglobulins that bind to soluble antigen. In contrast, antigen receptors on the surface of T cells appear not to bind to soluble antigen. Rather, the recognition of antigen by T-cell surface receptors depends on the expression of the antigen on the surface of antigen-presenting cells. Eventual antigen and receptor interaction triggers T-cell proliferation and thus results in an increase of the number of antigen-specific T-cells. In the case of the cytotoxic T cells, receptor-antigen interaction also induces the lysis of target cells.

Because of the absence of a typical free-ligand/receptor interaction, the T-cell receptor for antigen remained elusive for almost 20 years after the discovery of T lymphocytes. Since 1982, however, methodology to isolate and characterize T-cell receptors has been developed, and genes coding for T-cell receptors have been isolated and identified. Sufficient insight into function and repertoire of T-cell receptors will lead to an understanding of their role in a variety of disease states such as leukemias and lymphomas, autoimmune diseases, malignant disorders, and allergies.

Functionally unique cloned T lymphocytes. The discovery of T-cell growth factor (interleukin 2 or IL 2) allowed for cloning and long-term test-tube propagation of functionally unique T lymphocytes. Thus, helper, suppressor, and cytotoxic T cells have been cultured while maintaining their specificity. In addition, antigen-specific T cells have been fused with a tumor cell line (BW5147). The fusion products, which can grow in the absence of IL 2, are the so-called T-T hybridomas. The studies with T-cell clones and T-T hybridomas had two significant outcomes. They showed that these cells maintain exquisite specificity for a given antigenic determinant after long-term test-tube culturing, and they proved that T-cell receptors recognize antigen only in cell-cell contact. In the case of helper or suppressor T cells, these intercellular interactions occur with antigen-presenting cells, B cells, or macrophages that present processed antigen at their surface to T cells. Cytotoxic T cells detect antigen on the surface of the cells which they destroy specifically, the target cells.

Receptor-antigen recognition in all T cells is restricted by interaction between T-cell receptors and specific products of the so-called Major Histocompatibility Complex (MHC) locus. The phenomenon of MHC restriction, in which T-cell receptors recognize a combined specificity made of a foreign an-

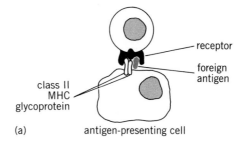

(a)   antigen-presenting cell

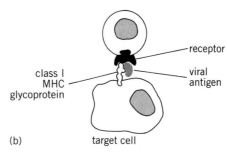

(b)   target cell

Fig. 1. Major histocompatibility (MHC) restriction. Large gray components are nuclei. *(a)* Helper T cell. Antigen receptors on helper T cells recognize a combined specificity made of a foreign antigen and a class II MHC glycoprotein on the surface of an antigen-presenting cell. *(b)* Cytotoxic T cell. These cells recognize foreign (usually viral) antigens on the surface of a target cell in association with class I MHC glycoproteins.

tigen and MHC products, was confirmed by experiments with cloned T lymphocytes. For instance, cytotoxic T-cell receptor-antigen interaction is restricted by MHC class I antigens, and helper and occasionally suppressor T-cell receptor and antigen interaction by MHC class II antigens. The phenomenon of MHC restriction is not well understood on the molecular level; whether antigen and MHC product are recognized by more than one receptor is still uncertain (Fig. 1).

Studies with T-cell clones and T-T hybridomas also revealed that cell-surface molecules other than T-cell receptor, antigen, and MHC antigens are necessary for intercellular interaction. The principal role of the so-called accessory molecules (for instance T4, T8, LFA1, or T11) is to establish adhesion between the T cell and target or antigen-presenting cell in order to facilitate subsequent recognition of the antigen and MHC product by the T-cell receptor. Some of these structures (for example, T11) may also be involved in the T-cell proliferative responses. However, the major breakthrough in the last 4 years has been the molecular analysis of the T-cell receptors for antigen and MHC products.

T-cell receptor for antigen. T-cell receptors on helper and cytotoxic cells could be described only after both functionally active antigen-specific T-cell clones or T-T hybridomas and monoclonal antibodies specific for the given T-cell clone had been generated. In studies using human or mouse T-cell clones or T-T hybridomas, such clone specific (clonotypic) antibodies are shown to affect the function of that clone. For instance, the cytotoxicity of an antigen-specific effector T-cell clone could be blocked by a specific reagent. Similarly, it could be shown that helper T-cell functions were stimulated or blocked by specific clonotypic antibodies.

Clonotypic antibodies react with two proteins on the surface of T cells: the glycosylated proteins, alpha (α) [50,000–55,000 daltons molecular weight] and beta (β) [40,000–45,000 daltons molecular weight], which are connected by disulfide bridges (S-S links), forming a protein heterodimer. Comparative protein analysis such as peptide mapping and two-dimensional gel electrophoresis have shown that these glycoproteins are unique to an individual clone. The structural heterogeneity of the clonotypic heterodimer resides in the protein backbone (32K) of the glycoprotein and not in the oligosaccharide side chains. Accordingly, this variability was later found to be reflected in the genes coding for the clonotypic heterodimer.

Together, the high level of variability of the clonotypic structure and the specific effect of the anti-clonotypic antibodies on the function of T-cell clones have suggested that the clonotypic heterodimer was the T-cell receptor. This notion was further supported by the apparent homology of these structures to immunoglobulins.

Unlike B-cell antigen receptors, which comprise two immunoglobulin heavy and light chains, the T-cell receptor heterodimer is associated with at least three invariable membrane proteins (T3). Together, they form the so-called T-cell receptor/T3 complex (Fig. 2). These T3 proteins, gamma (γ), delta (δ), and epsilon (ε), are not involved in antigen binding, but may serve a role in signal transduction through the plasma membrane. In general, antigen binding triggers T-cell proliferation, and this may be transmitted through a change in the interaction between the α/β heterodimer and the T3 chains. How this transduction occurs on a molecular level is uncertain. This question is currently being studied by several laboratories.

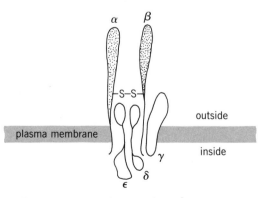

Fig. 2. T-cell receptor/T3 complex in the plasma membrane of human T lymphocytes. The T-cell receptor glycoproteins (α and β) form a disulfide-linked heterodimer, which varies between T-cell clones. The three proteins T3-γ, -δ, and -ε are associated with the α and β chains via protein-protein interactions. Stippled areas represent variable elements.

Coding for T-cell receptor/T3 complex. Recently, human and murine T-cell-specific cDNA clones have been isolated that contain the α and β chain of the T-cell antigen receptor. The first cDNA coding for a T-cell receptor protein (β chain) was cloned by Mark Davis and Steven Hedrick and their collaborators at the National Institutes of Health. They reasoned that given the predicted variability of the T-cell receptor genes, the genomic DNA may be rearranged in much the same fashion as immunoglobulin genes. Indeed, a T-cell-specific cDNA that rearranged its chromosomal DNA was isolated. Upon nucleotide sequence analysis, it was found that this cDNA coded for a protein structure which was homologous to immunoglobulin. Meanwhile similarities had been found between the predicted amino acid sequence based on cDNAs isolated by the Davis-Hedrick approach and limited protein sequences of the clonotypic α and β chains. Thus, conclusive evidence was gathered to prove that cDNAs coding for the clonotypic heterodimer had been isolated. Analysis of a series of cDNAs quickly assured that the α and β chains, like immunoglobulins, contain variable and constant regions. Together with earlier-mentioned data, the cDNA analysis provided strong evidence for the idea that these clonotypic structures are the T-cell receptors.

The genes coding for the α and β chains are similar to immunoglobulin genes in their structure, in their unique recombination mechanism, and in the regulation of their expression. Much still needs to be learned about the molecular mechanisms involved in these processes. Another major task for immunologists will be to estimate the extent of the T-cell receptor repertoire. Thus far only limited information about the variable-region genes has been obtained, but it appears that the T-cell repertoire is much smaller than that of the B cell. Based on the current estimate of the number of variable-region genes for the α and β chains, the T-cell receptor may consist of approximately 10^5 different variable regions. The B-cell receptor repertoire could be up to two to three orders of magnitude larger. This implies that the level of antigenic discrimination by T cells may be quite different from that in B cells.

The cDNA clones coding for one of the T3 polypeptide chains (T3-δ) of mouse and human have recently been isolated. Based on sequence and nucleotide comparisons, this structure is quite different from immunoglobulins or the T-cell receptor. It has a relatively long intracellular domain which may play a role in signal transduction. Studies to clone the genes coding for T3-γ and T3-ε are well under way. Clones of these genes will enable study of gene expression during thymic differentiation (maturation and progressive diversification of T cells within the thymus gland) and will aid in the attempts to determine the precise function of the corresponding proteins.

Summary. The T-cell receptor/T3 complex is a protein ensemble inserted in the plasma membrane of T lymphocytes. Whereas the α and β chains vary between different T-cell clones, that is, the chains are related but distinct, the γ, and δ, and ε chains are invariant between T-cell clones. Recently, genes coding for the α, β, and δ chains have been isolated. Functional studies in several laboratories suggest a division of labor between the α- and β-chain heterodimer (the T-cell receptor) and the three other polypeptide chains (T3-γ, -δ, and -ε): the T-cell-receptor α and β chains together recognize antigen and the MHC product on the antigen-presenting cell, after which an activation signal is transduced to the cytoplasmic side of the plasma membrane via the T3 complex.   [COX TERHORST]

Complement receptors. Complement receptors are cell-surface structures which possess the ability to specifically bind certain proteins of the complement system. The complement system plays an important role in the organism in recognizing and destroying infectious agents, including many viruses, bacteria, fungi, and parasites. Most of these actions are mediated through the various complement receptors.

The complement system consists of 20 different plasma proteins which circulate throughout the body in an inactive, native form. The system is activated when it encounters many pathogens, especially when these foreign substances are coated with antibody. After activation, the various complement factors bind to each other and to the substance responsible for their activation in a precisely regulated manner. This cascade-type activation sequence is characterized by the formation of indigenous complement enzymes able to cleave certain other complement molecules, thereby generating fragments.

Activation. Complement may be activated through either of two activation pathways, termed the classical and alternative pathways, each of which involves the reaction steps of several complement proteins. Classical pathway activation involves activation, binding, and cleavage of complement proteins termed C1, C2, C3, and C4. Alternative pathway activation involves activation of proteins termed Factor B, D, H, I, and properdin, as well as C3, which is common to both activation pathways. The remainder of the complement sequence includes the reaction steps of proteins called C5, C6, C7, C8, and C9, and the actions of regulatory proteins.

When complement is initiated, for example, by a virus-antibody complex via the classical pathway, the viral particles become coated with fragments of C1, C2, C3, C4, and C5 termed C1q, C2a, C3b, C4b, and C5b. These bound complement fragments posses other sites through which they bind to complement receptors located on leukocytes and other cells. In this process, the virus becomes adherent to the surface of a complement receptor-bearing cell via a complement molecule bridge. Occupancy of the complement receptors triggers various responses by the receptor-bearing cell depending on the cell type involved. In addition, other complement fragments diffuse away from the area of activation and bind to cells bearing specific receptors for these molecules. The binding of these fragments to the surface receptors of the cells also triggers various responses by the cell.

Interactions. One or more distinct complement receptors are present on the surface of most of the types of cells found in human blood. Certain tissue cells also express complement receptors. These various receptors interact only with the activated forms of specific complement components and have little, if any, ability to bind the native, circulating forms of these factors. Thus, complement activation is a prerequisite for complement-receptor interactions, and accordingly, for the manifestation of the various biological activities of the complement system which are largely mediated through such contacts.

Receptors for the C1q portion of C1 are found on human B lymphocytes and on polymorphonuclear leukocytes, monocytes, and macrophages. The receptor on B lymphocytes is a glycoprotein-proteoglycan complex. The C1q receptor interacts with sites on the C1q molecule which are exposed only following complement activation. The C1q receptor mediates the attachment of classical pathway activators to the surface of the various C1q receptor-bearing cell types via a C1q bridge. In this manner, the C1q receptor could facilitate antibody formation or destruction of the complement activator by the C1q receptor-bearing cell, though the exact functions of the receptor are not yet known.

There are six distinct receptors for C3 activation fragments located on many different types of cells. Each of these preferentially reacts with a single C3 fragment. The C3a receptor, which binds the small C3a activation peptide, is located on mast cells, polymorphonuclear leukocytes, and macrophages. Although the requirements for binding have been analyzed, the receptor has not been purified. Binding of C3a to the receptor triggers release of histamine and other vasoactive amines from mast cells and basophils, secretion of lysosomal enzymes by polymorphonuclear leukocytes, and stimulation of arachadonic acid metabolism in macrophages. These mediators stimulate other cellular responses such as smooth muscle contraction and increase vascular permeability with resulting exudation of fluid into tissues.

The C3 receptor termed CR1 preferentially binds the initial C3 activation product, C3b. CR1 is found on human red blood cells, B lymphocytes, some T lymphocytes, and on polymorphonuclear leukocytes, monocytes, and macrophages. It is also present on glomerular podocytes, a cell type in human kidneys and on dendritic reticular cells in lymphoid tissue germinal centers. CR1 is a single glycosylated polypeptide chain which exists in allotypic forms with molecular weights ranging from 160,000 to 240,000 daltons. The consequences of engagement of CR1 by activator-bound C3b differ depending on the cell type involved. Erythrocyte CR1 regulates complement activation and aids in the clearance of immune complexes. On certain lymphoid cells and phagocytic cells, CR1 synergizes with antibody to augment destruction of C3b bearing complement activators. The functions of kidney and dendritic reticular cell CR1 receptors is unknown.

The C3 receptor known as CR2 binds the terminal C3 cleavage products C3d,g and C3d. This receptor is confined to B lymphocytes. It is a single-chain glycoprotein with a molecular weight of 145,000 daltons. In addition to its role in binding C3d,g and C3, CR2 has been shown to be the structure used by Epstein-Barr virus, a human herpesvirus, to attach to and infect human B lymphocytes. The physiologic functions of CR2 have not been fully elucidated, but current evidence suggests that it is involved in triggering B lymphocytes to proliferate and differentiate.

The CR3 complement receptor binds the intermediate C3 cleavage product, C3bi. CR3, which is confined to certain cytolytic lymphocytes and phagocytic cells, is composed of two noncovalently linked glycosylated polypeptide chains with molecular weights of 180,000 and 90,000 daltons. CR3 is a member of a family of cell-surface glycoproteins, which share the same smaller polypeptide chain; the other members of the family do not interact with complement activation fragments. Interactions of complement activator-associated C3bi with CR3 on phagocytic cells enhances antibody-dependent responses leading to destruction of the activator. Individuals genetically lacking CR3 are predisposed to recurrent life-threatening bacterial infections.

CR4 is a newly described C3 receptor which binds C3bi and C3d,g and under certain circumstances, C3d. It is present on polymorphonuclear leukocytes and monocytes. Although not yet well characterized biochemically or functionally, CR4 is distinct from CR1 and CR3.

The sixth C3 receptor is the C3e receptor which selectively reacts with a small degradation product of C3 termed C3e. It is confined to polymorphonuclear leukocytes. Engagement of this receptor in the organism releases leukocytes from the bone marrow, thereby producing a transient increase in their level in the circulation.

Both primary cleavage fragments of C4, namely C4a and C4b, bind to the surface of a number of types of cells. Studies indicate that C4a binds to the C3a and C4b receptors, respectively, located on various cells. The affinity of the C4 fragments for these receptors is lower than the affinity for the C3 fragments.

One of the most important complement receptors is the C5a receptor which binds only C5a, the primary C5 activation fragment. This receptor is found on the surface of mast cells, polymorphonuclear leukocytes, monocytes, and macrophages. Although many aspects of the binding reaction have been documented, and it has been clearly shown to differ from the C3a receptor located on the same cells, the structure of the C5a receptor has not been determined. Engagement of the C5a receptor leads to the biological actions noted for the C3a receptor stimulation. In addition, however, C5a binding to the C5a receptor induces polymorphonuclear leukocytes and monocytes to migrate rapidly toward the area where complement activation is occurring. This process,

termed chemotaxis, leads to an influx of such cells and is an important constituent of the inflammatory response. In addition, C5a binding to C5a receptors and polymorphonuclear leukocytes leads to aggregation or clumping of these cells. In tissues this process probably augments the destruction of microorganisms and other complement activators.

Pathogen elimination. The biologic response elicited upon engagement of the complement receptors by their respective ligands either produce or enhance inflammatory processes or augment host defense mechanisms operative against human pathogens which activate complement. These various biologic activities together form an integrated system which aids in the elimination of such pathogens. Thus, complement activation produces C3a, C4a, and C5a which engage the relevant receptors and induce the release from the receptor-bearing cells of histamine and other vasoactive amines as well as products of arachadonic acid metabolism (leukotrines and prostaglandins). These mediators contract smooth muscles, increase vascular permeability, and produce tissue swelling. Similar activity in the context of the inflammatory response tends to localize infectious agents or injurious ducts, which have caused the complement activation, and to prevent their spread throughout the body. The chemotactic properties of C5a bring in leukocytes which release enzymes, oxidants, and other substances that aid in the destruction of the pathogen. In addition, these phagocytic cells as well as lymphocytes interact via specific receptors with C1q, C3b, C3bi, C3d,g, C3d, and C4b attached to the surface of the complement activator. This process triggers further release of various substances which destroy the pathogens and also, in the case of phagocytic cells, it facilitates ingestion and intracellular destruction. This scenario describes the major biological functions of complement in organisms as presently understood. In addition to these actions, which are all concerned with the inflammatory response and with the destruction of pathogens, recent studies suggest that complement receptors fulfill other functions concerned with cellular regulation and with the induction of and modulation of several types of immune responses. These additional properties of the receptors are under active investigation. *See* Immunogenetics.

For background information *see* Antigen; Cellular immunology; Compliment; Immunoelectrophoresis; Immunology in the McGraw-Hill Encyclopedia of Science and Technology.

[NEIL R. COOPER]

Bibliography: M. M. Davis, Molecular genetics of the T-cell receptor β chain, *Annu. Rev. Immunol.*, 3:537–560, 1985; D. T. Fearon and W. W. Wong, Complement ligand to receptor interactions that mediate biological responses, *Annu. Rev. Immunol.*, 1:243, 1983; T. E. Hugli, Structure and function of anaphylatoxins, *Springer Semin. Immunopathol.*, 7:193–219, 1984; G. D. Ross and M. E. Medof, Membrane complement receptors specific for bound fragments of C3, *Adv. Immunol.*, 37:217–267, 1985; R. D. Schreiber, The chemistry and biology of complement receptors, *Springer Semin. Immunopathol.*, 7:221–249, 1984; C. Terhorst and P. van den Elsen, in J. M. Cruse and R. E. Lewis, Jr. (eds.), The T-cell receptor/T3 complex, *The Year in Immunology 1984–1985*, 1985.

Chaetognatha

Drifting aquatic organisms belong to an ecological category called plankton. The majority of planktonic animals, the zooplankton, are herbivorous crustaceans. However, zooplankton also include carnivores such as medusae, jellyfish, and wormlike animals called chaetognaths (arrowworms). In the open ocean, chaetognaths are the most abundant planktonic carnivore. Recent studies of chaetognath behavior have led to new understanding of the biology of this small group of ubiquitous animals.

The phylum Chaetognatha is composed of nearly 70 species, of which 60 are planktonic. The remaining species attach to underwater substrates as adults. Planktonic chaetognaths are found in all oceans, from near the surface to depths of thousands of meters, depending on the species. Adult chaetognaths range in length from 6 to 100 mm (0.09 to 1.55 in.), with a diameter of 1 to 4 mm (0.02 to 0.06 in.; Fig. 1). Fins are situated both along the sides of their cylindrical bodies and around the tail. Internally, the body is divided by transverse septa into three regions: head, trunk, and tail. The head is equipped with grasping spines (chaetae) for seizing prey (other zooplankton), which are swallowed whole. The trunk comprises 50–80% of the body length, and encloses a coelom (body cavity) and a straight gut tube. Chaetognaths are hermaphrodites, with ovaries located in the central trunk, and testes in the tail. No respiratory or excretory organs have been identified in these animals.

The known sense organs of chaetognaths are mechanoreceptors and eyes. Hair-fans, some of which are sensitive to waterborne vibrations, project from the body surface and fins. Chaetognaths are thus able to perceive the vibrations of swimming prey (usually copepods), and this permits feeding in darkness. Chaetognath eyes are small, simple structures, consisting of a multilobed pigmented cell surrounded by several hundred light-sensitive cells. The eye has no specialized focusing apparatus, and since the receptive fields of many of the receptor cells overlap, the eye cannot form a high-resolution image. Such simple eyes are probably used only for detecting the intensity distribution of underwater light and shadow.

Vertical migration. Among zooplankton, there are several behaviors that can be mediated by the type of rudimentary light receptor possessed by chaetognaths. These important behaviors include startle responses and diel (daily) vertical migrations. Startle responses are rapid evasive movements elicited either by shadows or by mechanical disturbances (such as those caused by potential predators). In

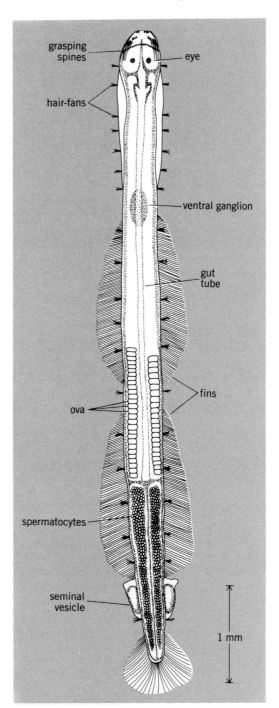

Fig. 1. Dorsal view of the planktonic chaetognath *Sagitta hispida.*

contrast, vertical migrations involve sustained swimming, and accompanying mass movements of zooplankton over vertical distances from a few meters to several hundred meters. The speed and extent of the migration vary with species and water depth. The most common migration pattern consists of movement of zooplankton closest to the surface at night, and descent to their greatest depth during daytime. As the ascent phase of the migration usually begins near sunset, it is generally believed that initiation of the migration depends on a light cue. Vertical migration probably occurs in all phyla with planktonic representatives.

Chaetognaths are delicate animals, easily injured during capture and laboratory manipulations. However, a few inshore species can withstand handling, and most studies on chaetognath behavior involve to these resilient species. Off the coast of the southeastern United States, the chaetognath of choice for experimental work is *Sagitta hispida.*

In nature, diel vertical migration is usually documented by repeatedly sampling at predetermined depths for animals of interest. A typical study might involve obtaining plankton samples simultaneously from four depths at 3-h intervals for a period of 2 days. Analysis of the species abundance in each of the 51 samples would provide information on the vertical distribution of the animals at particular times of day. By using this sampling technique, it was found that, near Beaufort, North Carolina, adult *S. hispida* congregated near the bottom (at 6–7 m or 20–23 ft depth) until shortly after sunset. As light levels decreased, the animals rose en masse toward the surface. By sunrise, this species was again virtually absent from the upper water column. Subsequent laboratory work has dealt with the light-receptive capabilities and migration-related behavior of this species.

Light responses. Two important aspects of any visually mediated behavior are its dependence on light wavelength and intensity. For chaetognaths, a convenient behavior for quantifying light sensitivity is phototaxis—the tendency to swim toward or away from a light source. In experiments in which chae-

tognaths are given a choice of swimming toward or away from a light directed at them horizontally, the proportion of animals swimming toward the light can be quantified. This has been done for a range of light intensities and wavelengths, yielding a series of response-intensity functions. From this experiment, the spectral sensitivity of the animals has been determined, assuming that the strength of phototaxis depends only on how many photons are absorbed by the chaetognath eye. For wavelengths that are absorbed at high efficiency by the photoreceptors, fewer photons are required in the stimulus beam to elicit a given level of phototaxis.

Sagitta hispida performed only a positive phototaxis (that is, swam horizontally toward the light), and was most sensitive to light of wavelengths 440–520 nanometers (Fig. 2). These wavelengths cover the blue-green region of the visible spectrum. This spectral sensitivity is typical of zooplankton living in coastal areas, and maximizes photosensitivity to underwater light in the blue and blue-green seas in which *S. hispida* is normally found.

In order to study light responses that might be involved in the ascent phase of diel vertical migration, it was necessary to examine the vertical movements of chaetognaths, since responses to gravity could also be involved in the migration. To do this, *S. hispida* was maintained under overhead lights having the approximate intensity and wavelength found in their natural habitat, and were then exposed to reductions in light intensity. Animals were placed in a tall chamber, and isolated near the bottom beneath a removable transparent partition. The intensity of the overhead light was then reduced or the light was switched off. After a brief waiting period, the chamber partition was removed, and the proportion of animals swimming upward was measured. The instantaneous experimental changes in light intensity were intended to mimic intensity reductions occurring at sunset, though the natural reductions would be more gradual.

The results of this experiment indicated that whenever *S. hispida* was left in darkness, the animals swam upward vigorously. This upswimming behavior occurred whenever the light intensity was reduced below a particular level [5×10^{16} photons/(m^2) (s)], regardless of the intensity of the original adapting light. The intensity at which the animals displayed strong upswimming was very close to intensities measured near the bottom only at sunset, just prior to the animals' natural ascent (Fig. 3). It would appear that this migration intensity, which lies above the threshold for vision in this species, represents a cue to the animals to begin sustained upward movement. Intensities higher than the migration intensity depress the upswimming activity.

A subsequent experiment dealt with the interaction between light direction and intensity in controlling the behavior of migrating chaetognaths. This experiment was performed as above, except that the stimulus light was presented from below the animals. In this case, when light intensity was lowered

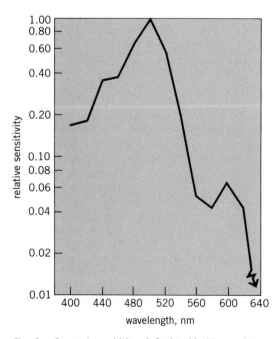

Fig. 2. Spectral sensitivity of *Sagitta hispida,* as determined by measuring the strength of positive phototaxis to light of different wavelengths. *(After A. J. Sweatt and R. B. Forward, Jr., Spectral sensitivity of the chaetognath Sagitta hispida Conant, Biol. Bull., 168:32–38, 1985)*

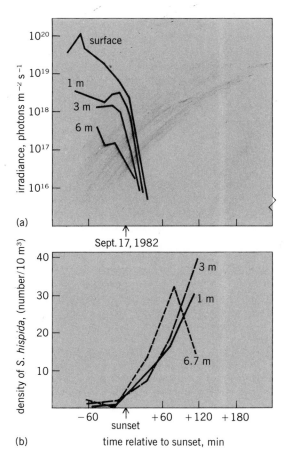

Sept. 17, 1982

(a)

(b)

Fig. 3.   The ascent phase of the vertical migration of *Sagitta hispida*. *(a)* Light-intensity profiles at the sampling site near Beaufort, North Carolina. *(b)* The evening increase in abundance of adult chaetognaths in the water column. Vertical distribution was determined by repeatedly collecting chaetognaths simultaneously from three depths, at 40-min intervals during sunset. 1 m = 3.3 ft. *(After A. J. Sweatt and R. B. Forward, Jr., Diel vertical migration and photoresponses of the chaetognath Sagitta hispida Conant, Biol. Bull., 168:18–31, 1985)*

longer periods of passive sinking.

In the absence of a cue for sustaining upward swimming, any negatively buoyant zooplankter in the open ocean runs the risk of sinking out of the productive upper layers. For *S. hispida*, the ascent cue can be provided at any time of day, due to the depth-dependent decrease in light intensity in the water column. A sinking chaetognath should stop its descent when it reaches a depth where the light intensity is low enough to stimulate negative geotaxis. This could explain why *S. hispida* is not often found in the open ocean below 200 m (650 ft) depth, where the light intensity approximates its migration (ascent) intensity. There are also many chaetognaths and other zooplankton which rarely come near the ocean surface at any time of day. The contributions of light responses, buoyancy factors, and feeding requirements to the distribution of these deep-living species remain to be studied.

For background information *see* CHAETOGNATHA; MARINE ECOSYSTEM; TAXIS in the McGraw-Hill Encyclopedia of Science and Technology.

[ANDREW J. SWEATT; RICHARD B. FORWARD, JR.]

Bibliography: D. L. Feigenbaum and M. R. Reeve, Prey detection in the Chaetognatha: Response to a vibrating probe and determination of attack distance in large aquaria, *Limnol. Oceanogr.*, 22:1052–1058, 1977; R. B. Forward, Jr., Light and diurnal vertical migration: Photobehavior and photophysiology of plankton, *Photochem. Photobiol. Rev.*, 1:157–209, 1976; T. Goto and M. Yoshida, The role of the eye and CNS components in phototaxis of the arrow worm *Sagitta crass* Tokioka, *Biol. Bull.*, 164:82–92, 1983.

Chelation

A chelate is a ligand which binds to a metal through at least two electron donor groups, normally forming a five- or six-membered ring system. The number of donors in a particular chelating ligand is described by its denticity. Metals require multiple donor groups (typically 4 to 10) to satisfy their coordination requirements, so in dilute solution a chelating ligand which has the requisite number of donor groups will be less dissociated than its nonchelate analogs. The enhanced stability of the chelated metal complex is a particularly important property for metal-binding pharmaceuticals, since the introduction of free metal ions into biological systems often results in toxic side effects. Chelating ligands now occupy an important role in medicine and are used as antidotes to metal poisoning and as diagnostic probes for the analysis of bodily tissues and fluids.

Therapeutic agents. Therapeutic chealting agents are used as antidotes for metal poisoning which can result from accidental ingestion or from breakdown of the body's assimilation system. Nonessential elements such as mercury, lead, or plutonium which are ingested tend to accumulate in specific body tissues where they interfere with normal biochemical pathways. In such cases, it is desirable to be able

to the level for the migratory ascent, the animals again swam upward, even though they were able to perceive the light below them. This indicates that the ascent at low light intensity is oriented primarily with respect to gravity, and not with respect to light direction. This type of ascent is termed a negative geotaxis, in that it moves animals away from the center of the Earth. Thus, while *S. hispida* is capable of positive phototaxis, its migratory ascent results from a light-dependent negative geotaxis.

Purpose of migration. There are several hypotheses concerning the adaptive advantage of diel vertical migration among the zooplankton. One major idea is that zooplankton escape visual predators by avoiding well-lit surface waters during the day, and approach the surface only at night, when they feed on planktonic algae and other organisms. While such an advantage may accrue to migrating chaetognaths, examination of the swimming mode of *S. hispida* may lead to an alternative explanation. This species is negatively buoyant, and swims by repeated upward darting motions, interspersed with

to treat the patient with a chelating agent which binds the offending metal specifically and strongly, forming a soluble complex which can rapidly be excreted.

In practice it has been extraordinarily difficult to design chelating drugs with the combination of selectivity and lack of toxicity. Relatively few chelating drugs have advanced to clinical trials since the development of 2,3-dimercaptopropanol (BAL, or British anti-lewsite) during World War II for the treatment of arsenic poisoning produced by the war gas lewsite, Cl—CH=CH—AsCl$_2$. One class of compounds which has received a great deal of attention, the aminocarboxylates, is exemplified by ethylenediaminetetraacetic acid (EDTA; shown as its cobalt(III) complex in structure I). Although EDTA

(I)

has been tested as a therapeutic metal-decorporating agent for nearly every type of metal poisoning, it is currently recommended (in conjunction with BAL) only as the antidote for the treatment of acute lead poisoning. EDTA has limitations as a chelating agent because it is not selective, and binds many divalent ions, including calcium and magnesium. In order to design specificity into chelating drugs, it is necessary to understand the coordination chemistry of the metal in question and apply this knowledge to the synthesis of new ligands. This approach is being taken for the preparation of new ligands for iron chelation therapy.

Iron. Although iron is an essential element, it is toxic when present in excess, partly due to the fact that the human body has no mechanism for the elimination of the metal. There are two major aspects of iron poisoning. The accidental ingestion of iron supplements by children in the United States remains a serious form of poisoning. Iron overload, or hemochromatosis, results from the treatment of the genetic disease beta-thalassemia (Cooley's anemia). This disease affects nearly 5 million people worldwide and renders the body unable to synthesize the beta chain of hemoglobin, the iron-containing oxygen carrier in the blood. Treatment requires repeated transfusions of whole blood through the lifetime of the patient, leading to an accumulation of toxic levels of iron in the form of insoluble hyroxides.

The National Institutes of Health has sponsored a significant research effort toward the development of iron-specific chelators for use in the treatment of Cooley's anemia. For clues to the design of such a chelator, some researchers have chosen to study siderophores—low-molecular-weight iron chelators secreted by microorganisms specifically to sequester iron(III) at physiological pH, conditions in which Fe(III) normally exists as an insoluble hydroxide (solubility product equilibrium constant $K_{sp} \cong 10^{-39}$). These compounds typically contain three hydroxamic acid or catechol groups, each of which chelates the iron in a bidentate fashion so that an octahedral complex is formed. Examples of siderophores include desferrioxamine B (DFO; II), a tri-

(II)

hydroxamic ligand produced by fungi, and enterobactin (III), a cyclic triester of 2,3-dihydroxyben-

(III)

zoylserine, secreted by enteric bacteria such as *Escherichia coli*. Enterobactin has the distinction of forming the world's strongest iron(III) complex, with a formation constant (K_f) of 10^{52}.

Despite its large formation constant and selectivity for iron(III), enterobactin is not a useful drug since it promotes bacterial growth and infection in the patient. Desferrioxamine B, which also forms a strong iron complex ($K_f = 10^{31}$) and is highly selective for iron(III) over divalent metal ions, is not metabolized by bacteria in the body. It is currently used to treat iron-overloaded patients. Judging by these criteria, it would be expected to perform adequately, but the drug has several drawbacks, including the fact that high doses and prolonged treatments are required to achieve negative iron balance. Also, oral administration is not effective and DFO must be administered by subcutaneous infusion.

One possible reason for the lack of effectiveness of DFO is its inability to remove iron quickly from the iron-transport protein transferrin, the source of the largest labile iron pool in blood plasma. Presumably, iron removed from transferrin would be replenished via biochemical pathways from the relatively inaccessible iron in ferritin, the iron-storage protein. Thus, overall iron levels should be more efficiently reduced by using a chelating drug which is capable of removing iron from transferrin. Since it was known that catechols are more effective than hydroxamic acids in removing iron from transferrin, a derivative of DFO was prepared in which the amine terminus was connected to a catechol moiety (IV). This compound, DFOCAM-C (V), was shown to remove iron from transferrin approximately 100 times as rapidly as DFO. Studies indicated that DFOCAM-C was indeed more effective at removing iron from animals than DFO itself, lending credence to the hypothesis. Recently, another derivative of DFO, DFOHOPO (VI), has been synthesized and it incorporates a hydroxypyridinone group. The hydroxypyridinone ligands are monoprotic and are more effective chelating agents than catechols under mildly acidic conditions. DFOHOPO is approximately four times as effective per mole ligand as DFO itself in removing iron in animals, making it one of the best compounds yet tested.

Plutonium. Plutonium-239 contamination poses a special environmental hazard because of toxicity and long half-life. Iron(III) and plutonium(IV) have similar affinities for the protein transferrin, suggesting that iron(III)-specific chelating ligands might be modified to accommodate the eight-coordinate Pu(IV) ion. Ligands containing four catechol groups have been synthesized. The ligand 3,4,3-LICAM-C (VII) was tested in animals, and caused the elimination of 88% of the amount of injected plutonium, compared with 10% loss in control animals and 70% removed by the currently used plutonium-decorporation drug, diethylenetriaminepentacetic acid (DTPA; VIII).

Radiopharmaceuticals. Radiopharmaceutical compounds, containing a radioactive nuclide, are useful for diagnosis or treatment of disease. After administration to the subject, the distribution of the nuclide is determined and quantified by detecting the emitted radiation. These results yield a density profile indicating into which tissues the isotope has been localized. The usefulness of this procedure depends upon the pharmacological behavior of the complex which is administered, but the radiopharmaceuticals typically are used to search for tumors or abnormalities in heart function.

One widely used nuclide, technetium-99, is formed by the radioactive decay of molybdenum-99. Various chelates of ^{99}Tc are used to obtain different tissue specificities. For example, *N*-dimethylphenylcarbamoylmethyliminodiacetic acid (HIDA; IX) forms a 2:1 complex with ^{99}Tc and has been shown to be effective in diagnosing a variety of liver dysfunctions in humans. The phosphonate complexes of ^{99}Tc [methylenediphosphonate (MDP; X) and hydroxymethylenediphosphonate (HMDP; XI)] are bone seekers, since the coordinated phosphon-

ate retains much of its affinity for calcium, and provides useful skeletal and myocardial infarct imaging agents.

NMR imaging agents. One of the most exciting medical developments in recent years is the clinical application of nuclear magnetic resonance (NMR) technology. Hydrogen is an NMR active nucleus and, as part of the water molecule, a predominant component of the human body. Different images depending on the type of tissues in which the water is localized can be observed, and these images of the human body can reveal abnormal tissues.

The inherent NMR contrast between different tissues is dependent on the water concentration, temperature, viscosity, and local magnetic environment. Contrast-enhancing pharmaceuticals have the potential of being selective for a specific type of tissue. One way to achieve differential enhancement is to use a paramagnetic pharmaceutical which would change the local magnetic field around the water molecules, and thus the image.

Paramagnetic $3d$-transition metals (for example, Fe^{3+}, Mn^{2+}) and lanthanide ions (for example, Gd^{3+}) may be useful agents, provided they can be administered in a nontoxic manner. Gadolinium(III), with seven unpaired electrons in its $4f$ orbitals and a magnetic moment of 7.9 Bohr magnetons, is an ideal candidate for an imaging agent. Unfortunately, when administered as the chloride, Gd^{3+} produces toxic effects in animals ($LD_{50} = 0.26$ mmol/kg) and is associated with a long-term biological risk. However, the introduction of Gd^{3+} as its stable octadentate DTPA chelate ($K_f = 10^{23}$) significantly lowers its toxicity ($LD_{50} = 20$ mmol/kg) and allows for the enhancement of spontaneous central nervous system tumors. Thus, in spite of the fact that the DTPA complexes gadolinium in solution, the water molecules in the second coordination sphere are sufficiently magnetically perturbed to provide an enhanced signal.

For background information *see* CHELATION; CHEMICAL BONDING; COMPLEX COMPOUNDS; COORDINATION CHEMISTRY; ETHYLENEDIAMINETETRAACETIC ACID; NUCLEAR MAGNETIC RESONANCE (NMR) in the McGraw-Hill Encyclopedia of Science and Technology.

[THOMAS J. MC MURRY; KENNETH N. RAYMOND]

Bibliography: A. E. Martell (ed.), *Inorganic Chemistry in Biology and Medicine*, 1980; A. E. Martell, W. Anderson, and D. Badman (eds.), *Development of Iron Chelators for Clinical Use*, 1981; P. M. May, The present status of chelating agents in medicine, *Prog. Med. Chem.*, 20:225, 1983; *Phys. Chem. Phys. Med. NMR*, 16:97–104, 105–113, 145–155, 1984, and 17:113–122, 1985; K. N. Raymond, Specific sequestering agents for iron and actinides, *Environmental Inorganic Chemistry*, 1985; S. J. Rodgers and K. N. Raymond, Ferric iron sequestering agents 11, *J. Med. Chem.*, 26:439–442, 1983.

Chemical dynamics

Electron-transfer reactions are a particular type of oxidation-reduction (redox) reactions. In a redox reaction the oxidation numbers of two or more elements change as reactants are transformed into products. Redox reactions are well defined in a stoichiometric sense, but the term oxidation-reduction has no mechanistic connotations. Electron transfer is one type of redox process, but in addition to the usual stoichiometric significance, the term electron transfer has a mechanistic implication, namely that somewhere along the reaction pathway connecting reactants to products there is an elementary step in which an electron localized in an orbital of the reducing agent is transferred to an orbital localized in the oxidizing agent.

Electron-transfer reactions: elementary steps.

Electron-transfer reactions of metal complexes, whether inner-sphere or outer-sphere, proceed via a

sequence of elementary steps: formation of precursor complex, intramolecular electron transfer within the precursor complex, and dissociation of the successor complex. The oxidation states of the metal centers in the precursor complex are those characteristic of the reactants, whereas the oxidation states in the successor complexes are those characteristic of the products. The reacting complexes maintain the integrity of their coordination shells in the precursor and successor complexes of outer-sphere reactions. For inner-sphere reactions, the precursor and successor complexes are binuclear complexes in which a common ligand is bound to the two metal centers undergoing electron transfer. Reactions (1) and (2)

$$Co^{III}(NH_3)_5OH_2{}^{3+} \mid Fe^{II}(CN)_6{}^{4-} \rightleftharpoons$$

$$Co^{II}(NH_3)_5OH_2{}^{2+} \mid Fe^{III}(CN)_6{}^{3-} \quad (1)$$

$$(NH_3)_5Ru^{III}N \bigcirc NFe^{II}(CN)_5 \rightleftharpoons$$

$$(NH_3)_5Ru^{II}N \bigcirc NFe^{III}(CN)_5 \quad (2)$$

are examples of intramolecular electron-transfer reactions which transform precursor complexes into successor complexes. Reaction (1), where the vertical lines indicate intimate ion pairs, represents an outer-sphere process, whereas reaction (2) illustrates an inner-sphere process.

Intramolecular electron transfer. According to current concepts in electron-transfer theory, the rate of electron transfer is governed by nuclear and electronic factors. The precursor and successor complexes have different sizes and charge distributions, and their most stable configurations will occur in different regions of nuclear configuration space. The intramolecular electron-transfer process that transforms the precursor complex into its electronic iso-

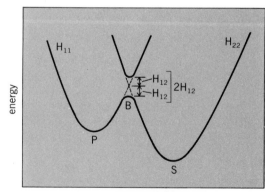

Fig. 1. Schematic representation of zero-order and first-order surfaces for electron transfer. P = precursor complex; B = transition state for intramolecular electron transfer; S = successor complex; H_{11} = precursor complex surface; H_{22} = successor complex surface; H_{12} = splitting at intersection region. (*After A. Haim, Intramolecular electron-transfer reactions of ion pairs: Thermal, optical, and photochemical pathways, Comments Inorg. Chem., 4:113–149, 1985*)

mer, the successor complex, corresponds to the crossing of the system, as depicted in Fig. 1, from one surface to the other. The nuclear factor corresponds to changes in the positions of the nuclei, both metal-to-ligand bond distances and orientation-of-solvent dipoles, necessary to bring the precursor complex from its equilibrium configuration to that appropriate to the intersection region. In the latter configuration, the energies of the system before (precursor) and after (successor) electron transfer are equal. The electronic factor corresponds to the probability of electron transfer once the intersection region has been reached, and depends on the magnitude of H_{12}, the electronic coupling between the metal centers. If H_{12} is not too small (>0.5 kcal), every time that the required nuclear configuration is reached the electron is transferred; that is, the system follows the path given by the lower surface and the process is referred to as adiabatic. If H_{12} is too small, the system reaches the intersection region with enough momentum to jump to the upper surface, and only occasionally is electron transfer consummated. This is referred to as nonadiabatic electron transfer. The magnitude of H_{12} depends on the distance between the metal centers, the symmetry of donor, acceptor, and carrier orbitals, and perhaps the exoergonicity of the reaction.

Electron exchange and cross-reactions. A very important class of electron-transfer reactions are electron-exchange or self-exchange reactions, shown in reactions (3) and (4), where an asterisk is used to

$$*Fe(OH_2)_6^{3+} + Fe(OH_2)_6^{2+} \rightleftharpoons$$
$$*Fe(OH_2)_6^{2+} + Fe(OH_2)_6^{3+} \quad (3)$$

$$*Ru(NH_3)_6^{3+} + Ru(NH_3)_6^{2+} \rightleftharpoons$$
$$*Ru(NH_3)_6^{2+} + Ru(NH_3)_6^{3+} \quad (4)$$

denote one of each pair of a species that have identical composition but different charge and are exchanging electrons. Since no net chemical changes occur in an electron-exchange reaction, special techniques must be devised to measure the rates of such reactions. A common technique involves isotope labeling of one of the two oxidation states of a redox couple and measurement of the rate at which the label appears in the other oxidation state. Reactions between one partner of a redox couple and one partner of another redox couple are referred to as cross-reactions. The cross-reaction between the $Ru(NH_3)_6^{3+/2+}$ and $Fe(OH_2)_6^{3+/2+}$ couples is depicted in reaction (5).

$$Fe(OH_2)_6^{3+} + Ru(NH_3)_6^{2+} \rightleftharpoons$$
$$Fe(OH_2)_6^{2+} + Ru(NH_3)_6^{3+} \quad (5)$$

Two of the properties of interest when a cross-reaction takes place relate to thermodynamics and kinetics, that is, how far the reaction proceeds and the rate of approach to equilibrium. The first property can be determined by simply subtracting the standard reduction potentials of the two relevant redox couples. As far as kinetic predictions are concerned, one of the most useful consequences of cur-

rent theories of electron-transfer reactions is the Marcus cross-relationship, given in Eq. (6), which

$$k_{12} = (k_{11}k_{22}K_{12}f_{12})^{1/2} \quad (6)$$

allows the calculation of rate constants for cross-reactions from properties of the individual couples. In Eq. (6), k_{11} and k_{22} are the rate constants for the electron-exchange reactions of the two couples, k_{12} is the rate constant for the cross-reaction between the two couples, K_{12} is the equilibrium constant for the reaction, f_{12} is a correction factor which is nearly 1 for mildly exoergonic reactions, all reactions are taken to be adiabatic, and electrostatic corrections have been neglected. In the above example, k_{11}, k_{22}, and k_{12} are the rate constants for reactions (3), (4), and (5), respectively, and K_{12} is the equilibrium constant for reaction (5).

Biological electron transfer. For several years, chemists have been probing the reactivity of redox-active metalloproteins by studying their electron-transfer reactions with small inorganic molecules or ions. These bioinorganic cross-reactions represent an intermediate situation between the purely biological electron-transfer reactions and the simple inorganic systems. In particular, the reactions of cytochrome c with several inorganic couples of known electron-exchange rate and reduction potential have been studied kinetically. By utilization of the known

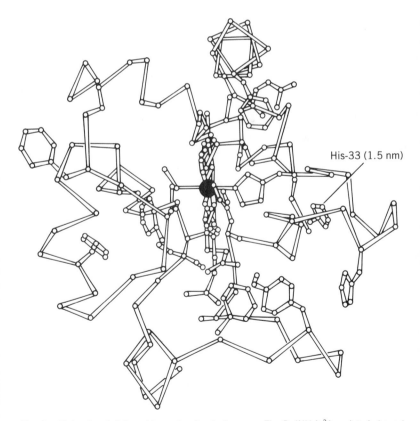

His-33 (1.5 nm)

Fig. 2. Molecular skeleton of horse heart cytochrome c. The $Ru(NH_3)_5^{3+}$ moiety is bound at His-33 which is 15 angstroms (1.5 nanometers) from the iron center. (*After J. R. Winkler et al., Electron-transfer kinetics of pentaammineruthenium(III) (histidine-33)-ferricytochrome C: Measurement of the rate of intramolecular electron transfer between redox centers separated by 15Å in a protein, J. Amer. Chem. Soc., 104:5798–5800, 1982*)

k_{22} values for the inorganic couples, the value of K_{12} calculated from the reduction potentials of the biological and inorganic couples, and the measured values of k_{12}, it is possible to calculate values of k_{11}, the rate constant for the electron-exchange reaction of the ferricytochrome-ferrocytochrome couple. The calculations yield an average value of $1.3 \times 10^3 \ M^{-1} \ s^{-1}$, in excellent agreement with the value of approximately $1 \times 10^3 \ M^{-1} \ s^{-1}$ measured directly by nuclear magnetic resonance techniques. The agreement is encouraging and suggests that the treatment of nuclear and electronic factors which successfully correlates electron-transfer reactions of inorganic couples may also be applicable to more complicated biological electron-transfer reactions.

Another area under active investigation relates to the question of the distance dependence of rates of biological electron transfer. By appropriate modifications of metalloproteins, intramolecular electron transfer in modified cytochrome c (Fig. 2), myoglobin, azurin, and hybrid hemoglobin has been shown to proceed with significant rates across 1.0 to 2.5 nanometers. Such long-range electron transfer raises the question of the detailed mechanism of the transfer. Proposed mechanisms involve direct tunneling from one metal to the other and orbital mediation by the backbone that connects the metal centers.

For background information *see* BIOLOGICAL OXIDATION; COMPLEX COMPOUNDS; COORDINATION CHEMISTRY; OXIDATION-REDUCTION in the McGraw-Hill Encyclopedia of Science and Technology.

[ALBERT HAIM]

Bibliography: A. Haim, Intramolecular electron transfer reactions of ion pairs: Thermal, optical, and photochemical pathways, *Comments Inorg. Chem.*, 4:113–149, 1985; A. Haim, Mechanisms of electron transfer reactions, *Prog. Inorg. Chem.*, 30:273–357, 1983; R. A. Marcus and N. Sutin, Electron transfer in chemistry and biology, *Biochem. Biophys. Acta*, 811:265–322, 1985; H. Taube, Electron transfer between metal complexes: Retrospective, *Science*, 226:1028–1036, 1984.

Chemical ecology

Marine organisms possess many types of defensive mechanisms, of which use of diverse toxins is very common. Soles of the genus *Pardachirus*, for example, use chemical means to repel predatory attacks, in addition to camouflaging themselves on the sandy bottom. They release a toxic secretion to repel would-be predators, such as sharks. The Red Sea Moses sole *(P. marmoratus)* has been shown to be sharkproof; when it is disturbed, its peculiar toxin glands discharge a soapy secretion that contains pardaxin, a peptidic toxin long believed to be the only repellent factor. Toxic steroid glycosides have recently been discovered in the secretion of an Indo-Pacific sole *(P. pavoninus)*, suggesting that the repellent has a more complex chemical composition. Similar glycosidic toxins were subsequently found in the Moses sole secretion.

Toxic steroid monoglycosides. When hydrophilic components such as peptides are removed from the lyophilized secretions of *Pardachirus* spp. by acetone precipitation, steroid monoglycosides are found to be the only major organic components, that is, the secretion is relatively free of fats. The glycosides isolated from the *P. pavoninus* secretion were named pavoninins. The major pavoninins are pavoninin-1 (structure I*a*), pavoninin-3 (I*b*), and pavoninin-5

(I*a*)

pavoninin-3 (R = ·ııılll OH)
pavoninin-5 (R = ◀ OH)

(I*b*)

(I*b*). Each glycoside has *N*-acetyl-β-D-glucosamine (β-GlcNac) attached to a glycosidic linkage at the 7α or 15α position of the steroid aglycon of the cholestane skeleton. The glycoside mixture exhibited moderate shark repellent activity against lemon sharks *(Negaprion brevirostris)* in captivity. Aside from repellent activity, pavoninins are ichthyotoxic (toxic to fishes) and hemolytic. Despite differences in the position of glycosidation as well as variation at the A/B ring moiety of the steroid aglycon, each pavoninin possesses a similar level of ichthyotoxicity, that is, it is lethal to killifish *(Oryzias latipes)* at concentrations of 5–10 ppm.

Analogous monoglycosides isolated from the secretion of *P. marmoratus*, named mosesins, have comparable ichthyotoxicity but possess somewhat stronger shark repellent activity than pavoninins in the same bioassay systems. Although not identical to pavoninins, they share many structural characteristics with them in that each mosesin also has a cholestane skeleton that is acetoxylated at the terminus of the side chain, and is monoglycosylated at the 7α position, as is pavoninin-1. The sugar in mosesins, however, is β-D-galactose (β-Gal) or its 6-acetate derivative (β-Gal-6-Ac), and their steroid aglycons are oxygenated at one more position than those of pavoninins. No genins (aglycons) are found in common between the two series, but there are like pairs such as pavoninin-1 (I*a*) and mosesin-1 (II*a*), or pavoninin-5 (I*b*) and mosesin-2 (II*b*). The oxygenated terminus of the side chain suggests that soles utilize

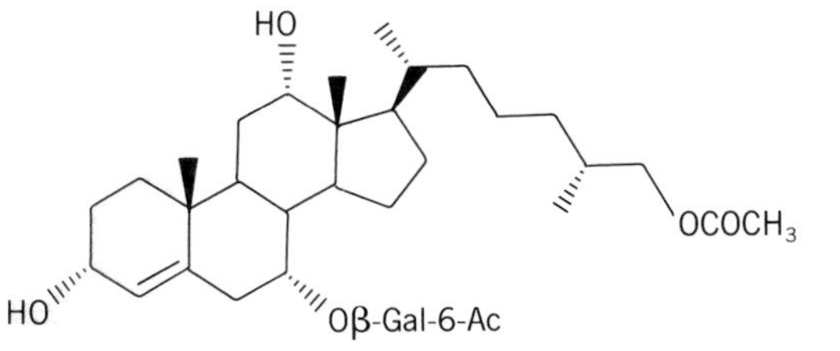

(IIa)

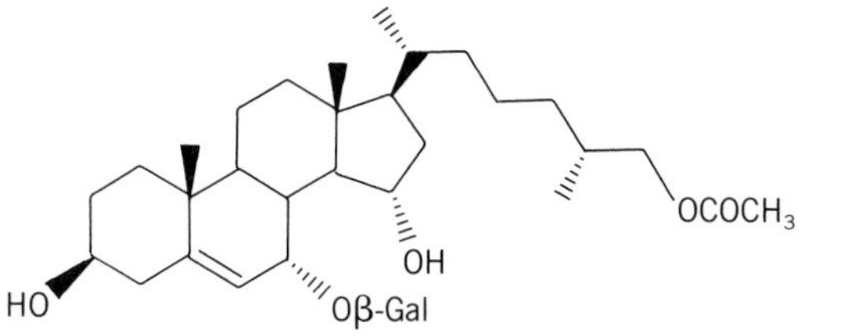

(IIb)

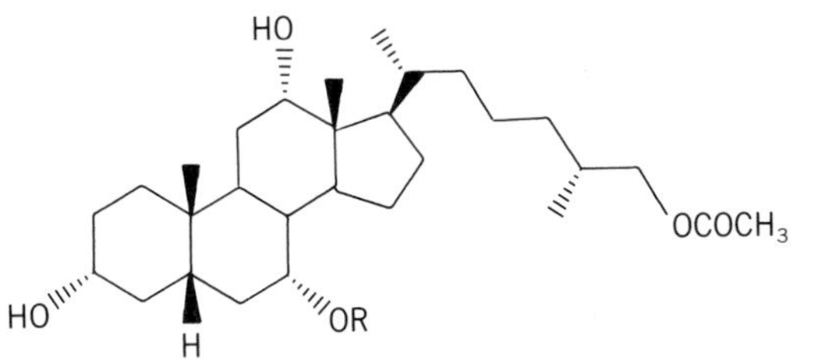

mosesin-3 (R = β-Gal-6-Ac)
mosesin-4 (R = β-Gal)

(IIc)

catabolic products of cholesterol for their defense. Indeed, the genin of mosesin-3 or -4 (IIc) is a likely precursor to cholic acid, a major bile acid in vertebrates.

Due to the amphiphilic nature of the chemical structures, these molecules have strong surfactant (surface-active agents such as the wetting and cleaning components of soaps) properties, similar to other defensive substances of fishes, ichthyocrinotoxins, with the possible exception of tetrodotoxin in puffers. Amphiphilicity of mosesins-1 and -3 arises from a particularly elaborate architecture in which all the hydrophilic hydroxyl groups are oriented toward the α face of the molecules. While steroid or triterpenoid glycosides are well-known defensive agents of echinoderm starfish and sea cucumbers, respectively, pavoninins and mosesins are unprecedented as isoprenoidal ichthyocrinotoxins of chordates.

Toxic peptides (pardaxins). While there have been many studies of the pharmacology of *P. marmoratus* pardaxin, its chemical characterization had been delayed by difficulties in purification caused by its being invisible at the conventional ultraviolet wavelength of 280 nanometers and by its inconsistent response to analytical techniques such as chromatography and electrophoresis. However, recent application of reverse-phase liquid chromatography to peptide chemistry has allowed purification of the peptidic toxins from the acetone precipitate of the *P. pavoninus* secretion. The ichthyotoxic fraction obtained by gel filtration of the precipitate exhibited weak shark repellent activity against the white-tip reef shark *(Triaenodon obesus)*. This fraction was further separated into two toxic fractions by anion-exchange chromatography; the active component in each was purified through a reverse-phase column and the amino acid sequence was determined. Both are linear peptides of 33 amino acids differing from each other only at positions 14 and 31, where leucine and glutamic acid in the major peptide are replaced by isoleucine and glycine respectively in the minor peptide. Preliminary reinvestigations of the *P. marmoratus* secretion indicated that it has a peptide that is very similar, though not identical, to the major toxic peptide of *P. pavoninus*. These peptides are all eluted from a gel filtration column with a retention volume corresponding to a molecular weight of around 13,000 daltons, matching the reported value of *P. marmoratus* pardaxin and indicating their tetramer formation in aqueous media. The primary structure of *P. pavoninus* pardaxin is

10
Gly-Phe-Phe-Ala-Leu-Ile-Pro-Lys-Ile-Ile-
20
Ser-Ser-Pro-Leu-Phe-Lys-Thr-Leu-Leu-Ser-
30
Ala-Val-Gly-Ser-Ala-Leu-Ser-Ser-Ser-Gly-
Glu-Gln-Glu

The strong surfactant property of the pardaxin match their amphiphilic sequences, where the seven C-terminal residues are hydrophilic while the rest, particularly the positions 2–6, are largely lipophilic. The toxins found in *Pardachirus* spp. resemble melittin, a well-documented hemolytic peptide in bee venoms, in the molecular size and linear amphiphilicity. Bioactivities and proposed modes of action of pardaxins also coincide well with those reported for melittin. Thus, pardaxins and melitten probably share a similar role in the two completely different ecosystems.

Although both the steroid glycosides and the peptides have shown perceivable shark repellent activity, their efficiency adds up to less than that known for the whole secretion. A degree of synergism has been noted between the two factors in the ichthyotoxicity assay, but syntheses of these compounds in large quantity must be achieved in order to clarify the role of each component in repelling sharks and their possible synergistic interaction in the natural secretion.

For background information *see* CHEMICAL ECOLOGY in the McGraw-Hill Encyclopedia of Science and Technology.

[KAZUO TACHIBANA]

Bibliography: L. Bolis, J. Zadunaisky, and R. Gilles (eds.), *Toxins, Drugs, and Pollutants in Marine Animals*, pp. 2–42, 1984; E. Clark, Shark repellent effect of the Red Sea Moses sole, in B. J. Zahuranec (ed.), *Shark Repellents from the Sea, New Perspectives*, pp. 135–150, 1983; N. Primor, J. Parness, and E. Zlotkin, Pardaxin: The toxic factor from the skin secretion of the flatfish *Pardachirus marmoratus* (Soleidae), in P. Rosenberg (ed.), *Toxins, Animal, Plant, and Microbial*, pp. 539–547, 1978; K. Tachibana, K. Nakanishi, and S. H. Gruber, Shark-repelling ichthyotoxins of the Red Sea Moses sole, *Pardachirus marmoratus*, in preparation; K. Tachibana, M. Sakaitani, and K. Nakanishi, Pavoninins, shark-repelling and ichthyotoxic steroid *N*-acetylglucosaminides from the defense secretion of the sole *Pardachirus pavoninus* (Soleidae), *Tetrahedron*, 41:1027–1037, 1985; S. A. Thompson et al., Melittin-like peptides from the shark-repelling defense secretion of the sole, *Pardachirus pavoninus*, submitted to *Science*.

Chemotherapy

Although early diagnosis and surgical removal remain the most effective means of treating cancer, chemotherapy has also proved to be valuable for the management and, in some instances, eradication of this disease. Among the newer chemotherapeutic agents, *cis*-diamminedichloroplatinum(II), or cisplatin (structure I), has been the most effective of an-

$$
\begin{array}{ccc}
H_3N & & Cl \\
& \diagdown \ \diagup & \\
& Pt & \\
& \diagup \ \diagdown & \\
H_3N & & Cl
\end{array}
$$

(I)

titumor drugs. This platinum-based compound is especially active against testicular cancer, and has been successfully employed in the treatment of various other tumors including ovarian, head and neck, bladder, lung, and cervical. The trans isomer (structure II) has no chemotherapeutic activity, but a

$$
\begin{array}{ccc}
H_3N & & Cl \\
& \diagdown \ \diagup & \\
& Pt & \\
& \diagup \ \diagdown & \\
Cl & & NH_3
\end{array}
$$

(II)

number of analogs of cisplatin have been undergoing clinical trials. The compound diammine(1,1-cyclobutanedicarboxylato)platinum(II), or carboplatin (structure III), has much the same chemotherapeutic

$$
\begin{array}{c}
O \quad\quad H_2 \\
\parallel \quad\quad C \\
H_3N \quad O-C \quad C \\
\diagdown \quad\quad\quad \diagdown \\
Pt \quad\quad C \quad CH_2 \\
\diagup \quad\quad\quad \diagup \\
H_3N \quad O-C \quad C \\
\parallel \quad\quad H_2 \\
O
\end{array}
$$

(III)

activity as cisplatin but with somewhat less toxic side effects. Carboplatin has just been released for use in the United Kingdom. Studies of the molecular mechanism of action of cisplatin have revealed that it binds to deoxyribonucleic acid (DNA) in the cancer cell in a manner different from that of the inactive trans isomer. Understanding this difference may ultimately lead to the rational design of even more effective anticancer drugs.

History of platinum drugs. The remarkable biological activity of cisplatin was accidentally discovered in 1965, when it was reported that electrolysis products from a platinum electrode inhibited cell division in the common laboratory bacterium *Escherichia coli*. Subsequent studies with a mouse tumor model system revealed potent antitumor activity. By 1973, clinical trials clearly demonstrated cisplatin to be very effective against testicular and ovarian cancers that had been unresponsive to the best prior therapies. For a while it appeared as though the gastrointestinal and, especially, the renal toxicity of the heavy-metal drug would prohibit its widespread use, but it was discovered that simply giving patients large amounts of water prior to injection greatly reduced kidney toxicity without diminishing anticancer activity. In 1979 the U.S. Food and Drug Administration approved cisplatin for the treatment of human cancer, and by 1983 it had become the leading anticancer drug sold.

Pharmacology and toxicology. Solutions of cisplatin dissolved in physiological saline are given as an intravenous injection. Chloride ion must be present since the compound slowly reacts in water ultimately to form toxic or inactive platinum compounds. Although experimental protocols vary widely, one commonly employed procedure is to administer the compound by injection every 3 or 4 weeks. Cisplatin rapidly clears the plasma after injection, and most of the platinum is excreted into the urine within a few days. Cisplatin induces nausea and vomiting in nearly all patients, but since the early 1980s various antiemetic agents have had an impact on this most severe, dose-limiting side effect. (With sufficient pretreatment hydration, nephrotoxicity is no longer the major dose-limiting toxicity.) Some patients experience a ringing in the ears, peripheral neuropathy, myelosuppression, and occasionally allergic reactions.

Clinical picture. Cisplatin is usually given in combination with other anticancer drugs. Its most spectacular success is in the area of testicular cancer where, when incorporated with vinblastine and bleomycin, cisplatin treatment cures nearly all patients having stage A (testis alone) or stage B (metastasis or retroperitoneal lymph nodes) testicular carcinoma. Both combination and single-agent protocols have also shown responses in patients with, especially, ovarian cancer as well as bladder, head and neck, non-small-cell lung, and cervical cancer. In addition to combination chemotherapy, cisplatin has also been shown to be effective when combined with radiation therapy, with some evidence for a synergistic effect between the two in killing tumor cells. Several platinum analogs are in various stages of development and screening in the hope of producing agents with reduced toxic side effects.

Mechanism of action. When *cis*-diamminedichloroplatinum(II) dissolves in water, the chloride ions are slowly released over many hours in a stepwise fashion to form chemically more reactive, positively charged aqua complexes. The relatively high chloride ion concentration in blood serum suppresses these hydrolysis reactions, however, maintaining the drug in its relatively unreactive parent form. Cisplatin passively diffuses across the cell membrane; there is no evidence for its selective uptake into tumor cells or specific tissues. Once inside the cell, the drug encounters a markedly reduced chloride ion concentration, and hydrolysis commences. Structure-activity studies of a variety of platinum compounds have shown that antitumor activity is best when the complex is neutral, presumably to facilitate its transport into cells, and when the hydrolyzable leaving groups are adjacent to each other in the square-planar coordination environment of the platinum complex.

The key to understanding this profound cis stereochemical requirement came following numerous experiments pointing to DNA as the primary target in the cancer cell responsible for the activity of cisplatin. Binding of platinum to DNA inhibits DNA replication, which is necessary for the growth and survival of a tumor. It is also possible that tumor cells are deficient, relative to normal cells of the same tissue, in their ability to repair cisplatin lesions from their DNA. Thus there could be a window of concentration where the platinum is toxic to the tumor cells while being tolerated relatively well by normal cells. The relative toxicities of *cis*- versus *trans*-DDP may similarly be the consequence of differential repair. Recent experiments using a model system in which intact African green monkey kidney cells were infected with Simian Virus 40 have shown that, whereas equal amounts of the two isomers bound to the viral DNA in the host cells are equally effective at inhibiting DNA replication, much more *trans*- than *cis*-DDP must be added to produce equitoxic doses. The experiments showed that the adducts made by the trans isomer to DNA in the cell were repaired much more efficiently than the adducts formed by the cis isomer.

Test-tube studies of the adducts formed by cisplatin with DNA revealed a preference for sequences containing adjacent guanosine nucleosides and, to a lesser extent, adjacent adenosine-guanosine pairs. Crystal structure determination of the major adduct, *cis*-[Pt(NH₃)₂{d(pGpG)}], has demonstrated binding of the N(7) nitrogen atoms of the two guanine rings at the adjacent positions in the platinum coordination sphere originally occupied by the chloride ions in cisplatin (see illus.). Nuclear magnetic resonance studies in solution and molecular mechanics calculations indicated that this adduct could be incorporated into longer, double-stranded DNA with only relatively local disruption of base pairing and helix structure. The trans isomer, however, is stereochemically incapable of linking adjacent nucleotides in this manner and is known to form intrastrand

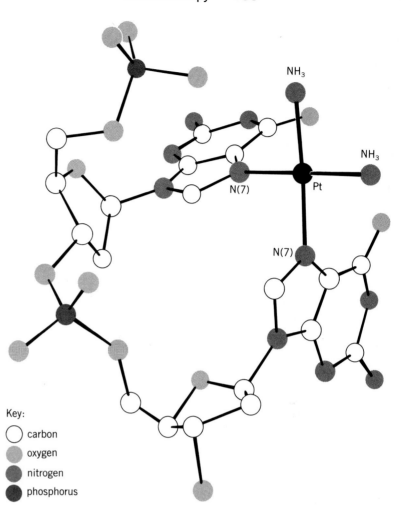

Key:
○ carbon
● oxygen
● nitrogen
● phosphorus

Molecular structure of the major adduct of cisplatin with DNA, *cis*-[Pt(NH₃)₂{d(pGpG)}], based on crystal structure determination. (*After S. E. Sherman et al., Science, 230:412, 1985*)

cross-links with one or more intervening nucleotides. Such a structure would be more disruptive, possibly accounting for the greater recognition and repair of *trans*- versus *cis*-diamminedichloroplatinum(II) adducts with DNA in the cell.

Future prospects. Intensive ongoing research at both the fundamental and clinical levels should produce new platinum and other heavy-metal-based antitumor drugs in the near future. Antibodies raised against platinum-DNA adducts have already been useful clinically to monitor drug clearance and patient response. Methods for selective drug delivery to tumor cells and for chemically minimizing the toxic side effects of cisplatin are being actively pursued.

For background information *see* CANCER (MEDICINE); CHEMOTHERAPY; PHARMACEUTICAL CHEMISTRY in the McGraw-Hill Encyclopedia of Science and Technology.

[STEPHEN J. LIPPARD]

Bibliography: M. P. Hacker, E. B. Douple, and I. H. Krakoff (eds.), *Platinum Coordination Complexes in Cancer Chemotherapy*, 1984; S. J. Lippard, *Science*, 218:1075, 1982; P. J. Loehrer and L. H. Einhorn, *Ann. Intern. Med.*, 100:704, 1984; A. L. Pinto and S. J. Lippard, *Biochim. Biophys. Acta*, 780:167, 1984.

Chromatography

Affinity chromatography is a purification method based on biological recognition; it has become one of the major means for the purification of biologically active molecules. Hundreds of proteins have been purified to homogeneity by affinity chromatography, and the method has also been instrumental in solving biological problems such as the mechanism of action of many enzymes and hormones, the nature of cell-cell interactions, and isolations and purifications in genetic engineering. Affinity chromatography has been applied in industry, particularly in biotechnology and medicine.

Principle. The basis of this technique is simple. Biomolecules usually have either natural or artificial molecules that they recognize. If one of these partners can be covalently bound to a polymeric carrier, the immobilized compound can then be used to isolate its complement by simply passing an extract containing the latter through the immobilized matrix. Molecules not possessing an appreciable affinity for the immobilized ligand will pass unretarded through the column, whereas molecules capable of binding the ligand will be adsorbed. The desired biomolecule can then be eluted by a high concentration of the soluble or native form of the immobilized partner or by changing external conditions so that the complex between the biomolecules will no longer be stable and the desired molecule will be eluted in a purified form. External conditions that can be changed are pH, ionic strength, solvents, and temperature.

Biomolecules purified. The various classes of compounds that have been purified by this method are discussed in the following paragraphs. These include antibodies and antigens, enzymes and inhibitors, regulatory enzymes, dehydrogenases, transaminases, hormone-binding proteins, vitamin-binding proteins, receptors, glycoproteins, lectins, genes, cells, viruses and phages, bacteria, and miscellaneous others.

Antibodies and antigens (immunoaffinity chromatography). One of the major and earliest uses of affinity chromatography was in immunology for the purification of antibodies on immobilized antigens. The reverse approach, namely the use of immobilized antibodies for the isolation of their corresponding antigens, has become popular in recent years. The antigens can be peptides, proteins, enzymes, receptors, nucleic acids, and other binding biomolecules.

The use of synthetic peptides for preparing antibodies has also found widespread application. These antibodies have been used subsequently to purify the parent protein from which the peptide was modeled. This approach has been applied particularly in molecular biology and may also serve as a basis for preparation of synthetic vaccines.

Monoclonal antibodies are being adopted rapidly for purification purposes in all fields of biology and biotechnology. If the trend continues, a shift from affinity chromatography to immunoaffinity chromatography for purification of biologically active compounds may ensue.

Enzymes. Enzymes represent the largest category of proteins thus far purified by affinity chromatography. By virtue of their discrete specificity, a large variety of ligands are available. The purification of enzymes can be divided into two categories: enzymes of narrow specificity (such as inhibitors for proteolytic enzymes or for nucleases) and class-specific enzymes, involving use of general ligands (such as coenzymes). The coenzymes widely in use are nicotinamide adenine dinucleotide (NAD), nicotinamide adenine dinucleotide phosphate (NADP), and adenosinetriphosphate (ATP), and other derivatives of adenine for dehydrogenases, and pyridoxal phosphate for transaminases. Another category of general ligands (pseudo-affinity chromatography) are the immobilized triazine dyes, for example Cibacron Blue. These colorful columns are also gaining popularity because of their facile preparation, stability, low cost, and commercial availability.

Lectins and glycoproteins. Lectins are sugar-binding, cell-agglutinating proteins of nonimmune origin which have been used extensively as macromolecular carbohydrate-specific reagents. Lectins can be purified by affinity chromatography on columns containing sugars and, alternatively, immobilized lectins are being used to isolate glycoproteins, glycopeptides, and glycolipids. The use of lectins for affinity chromatography of glycoconjugates is readily accomplished due to the low affinity of the lectins to sugars. Elutions can therefore be performed easily. Among the best-known lectins are concanavalin A (specific for D-mannose and D-glucose), wheat germ agglutinin (specific for N-acetyl-D-glucosamine), and peanut agglutinin (specific for D-galactose). Most of the proteins which are associated with membranes and which are exposed to the cell exterior are glycoproteins. Therefore lectin columns have been used in the early stages of purification of membrane-associated glycoproteins.

Receptors. Affinity chromatography has been crucial for the study and purification of the category of proteins comprising the receptors. A receptor is a molecule that recognizes a specific chemical entity (an effector), binds to it, and initiates a series of biochemical events resulting in a characteristic physiological response. Receptors exist for hormones, neurotransmitters, vitamins, and drugs. A variety of receptors have been purified on columns containing the corresponding effector. Another approach to isolate receptors is to use antibodies, as described above.

Biotechnology—recombinant DNA. The rapid development of the field of genetic engineering has depended on the concept of bioselective adsorption.

Without the development of affinity chromatography, progress in genetic engineering would have been very slow, if not impossible. Affinity chromatography has been used in all stages of recombinant DNA technology. Specific messenger RNA can be obtained on columns of immobilized chromosomal DNA in a single-step procedure. The enzymes used (DNA polymerase and reverse transcriptase) in DNA processing steps have all been affinity-purified, and the final products are usually identified and purified on monoclonal antibody columns.

Miscellaneous. Affinity chromatography can also be used for the isolation of cells, viruses, phages, bacteria, and so on. The isolation of cells which bear a given surface receptor can be accomplished by an approach similar to that used for the isolation of the free receptors; that is, cells, containing known receptors, can be isolated on columns containing the appropriate ligand. Lectins are also major tools for isolation or separation of cell populations, particularly for cells of the immune system (for example, lymphocytes) for medical application.

Related areas of affinity purification. The broad scope of actual and potential applications in affinity chromatography has caused the development of subspecialty adaptations, many of which are now recognized by their own nomenclature. These include covalent affinity chromatography, hydrophobic chromatography, metal chelate chromatography, affinity electrophoresis, affinity partitioning, affinity density perturbation, transition-state affinity chromatography, subzero affinity chromatography, high-performance affinity chromatography, filter affinity transfer, lectin affinity chromatography, immunoaffinity chromatography, affinity therapy, and affinity cytochemistry.

Some of these subcategories have become generally accepted and are widely used: hydrophobic chromatography for early-step purification, covalent chromatography for protein and peptides containing thiol groups, affinity electrophoresis for lectins and glycoproteins, and affinity partitioning for large-scale purifications.

Future perspectives. Today, affinity chromatography is the most potent of all separation methods and is widely used. The procedures are now being scaled up and automated for the purification of proteins prepared by genetic engineering. Eventually, genetic engineering may render affinity chromatography obsolete since it may be possible to introduce the means for purification of the products at the level of the gene.

For background information *see* CHROMATOGRAPHY in the McGraw-Hill Encyclopedia of Science and Technology.

[MEIR WILCHEK]

Bibliography: I. M. Chaiken, M. Wilchek, and I. Parikh (eds.), *Affinity Chromatography and Biological Recognition*, 1983; W. H. Scouten, *Affinity Chromatography*, 1981; W. H. Scouten (ed.), *Solid Phase Biochemistry*, 1983; M. Wilchek, T. Miron, and J. Kohn, Affinity chromatography, *Meth. Enzymol.*, 104:3, 1984.

Circuit (electronics)

The conventional methods of assembling electronic components onto printed wiring boards, in which wire leads are inserted through holes in the boards, have essentially reached their limits with respect to cost, weight, volume, and reliability. For increased board density, the current trend is toward the use of surface-mount technology (SMT), in which usually smaller components are mounted on the surfaces of the boards. This technology makes it possible to produce more reliable assemblies at reduced weight, volume, and cost. The weight of the printed wiring assemblies using surface-mount technology is reduced because surface-mounted components can weigh up to 10 times less than their conventional counterparts and can occupy about one-half to one-third as much space on the printed wiring board surface.

Surface-mount technology. Surface-mount technology is a revolutionary packaging technology in which through-hole mounted components are replaced by surface-mounted components. The assembly is soldered by reflow (vapor phase, infrared) or wave soldering processes, depending upon the mix

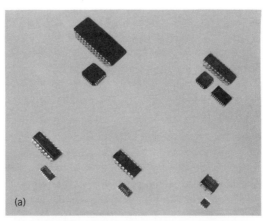
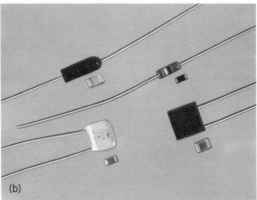

Fig. 1. Surface-mounted components and their conventional equivalents. *(a)* Active components. *(b)* Passive components.

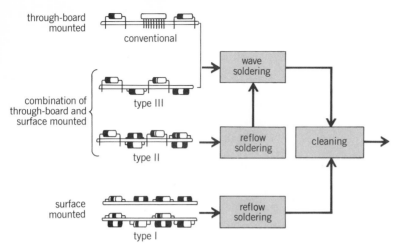

Fig. 2. Configuration of components in conventional and surface-mounted (types I, II, and III) assemblies, and the soldering processes required for each type.

of surface-mounted and through-hole-mounted components. Examples of surface-mounted active components include small outline integrated circuits (SOIC) and plastic leaded chip carriers (PLCC). The smaller size of surface-mounted components, compared with their conventional equivalents (Fig. 1), and the option of mounting them on either or both sides of the printed wiring board can reduce board area from 20 to 80%. The surface-mounted components, actives and passives, when attached to printed wiring boards, form three major types of surface-mount technology assemblies, commonly referred to as type I, type II, and type III (Fig. 2). Type I assemblies consist entirely of surface-mounted components, on either one or both sides of the board. Type II and type III assemblies have both surface-mounted and leaded components; in type II assemblies the surface-mounted components are on both sides of the board, while in type III assemblies they are only on the bottom of the board.

Component identification. Surface-mounted components, especially resistors and capacitors, are very small, making it very difficult to provide part markings on them. This requires a system of part handling and control to prevent part mixing. If parts do get mixed, they must be either positively identified or discarded. In most cases, discarding such parts is the cheaper alternative. For this reason, it is of critical importance to use surface-mounted chip resistors and capacitors only on tapes and reels that not only provide positive part markings (on the reel cover) but also prevent loss of part solderability of the component terminations during handling.

Component placement. The requirements for accuracy in component placement make it almost mandatory to use autoplacement machines to place surface-mounted components on the board. Selection of the appropriate autoplacement machine is dictated by the types of components to be placed and their volume. There are basically four types of

autoplacement machines available: in-line placement equipment, simultaneous placement equipment, sequential placement equipment, and sequential/simultaneous placement equipment.

There are many autoplacement equipment machines available in each of the four categories. Guidelines must be established for the selection of a machine specifying, for example, what kind of parts are to be handled; whether they come in bulk, magazine, or on a tape, and whether the machine can accommodate future changes in tape sizes. Selection and evaluation of tapes from various vendors for compatibility with the selected machine is very important. Special features, such as adhesive application, component testing, board handling, and reserve capability for further expansion in a machine, may be of special interest for many applications. Reliability, accuracy of placement, and easy maintenance are important to all users.

Soldering. Like the selection of autoplacement machines, the soldering process selection depends upon the type of components to be soldered and whether or not they will be used in combination with leaded parts. There is no one best method of soldering, and the sequence and process of soldering depend upon the type of surface-mounted assembly (Fig. 2). For example, if all components are of surface-mounted types, the reflow method (vapor phase, infrared) may be desirable. However, for type III surface-mount technology, wave soldering is most desirable. No matter which soldering method is chosen, some process issues will require resolution.

Wave soldering. In wave soldering, adhesive is used to glue the passive surface-mounted components onto the bottom of the board, and the whole assembly is wave-soldered in one operation. The surface-mounted components go through the wave.

Outgassing and solder skips are two main concerns in wave soldering. The outgassing or gas evolution occurs on the trailing terminations of chip resistors and capacitors. It is believed to be caused by insufficient drying of flux, and can be corrected by raising the printed wiring assembly preheat temperature. The other concern, solder skips, is caused by the shadow effect of the part body on the trailing terminations. Orienting the part in such a way that

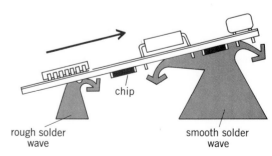

Fig. 3. Dual-wave soldering system for surface-mounted assemblies.

both terminations are soldered simultaneously solves shadow effect problems.

The most common method of solving both outgassing and shadow effect problems is to use the dual-wave system, where the first wave is a rough wave and the second wave is smooth (Fig. 3).

Adhesive application. In wave soldering, selection and application of adhesive plays the most critical role. With too much adhesive, no fillets or poor solder fillets may result if adhesive gets on pads. Too little adhesive will fail to accomplish its objective of holding parts through the solder wave. A good adhesive should form a single part, be colored, have a long shelf life, be easy to apply, and have an adequate bond strength with short cure time. In addition, after curing and soldering, the adhesive should remain moisture-resistant, nonconductive, noncorrosive, and be reworkable.

Vapor-phase soldering. Vapor-phase soldering is used for type II and type I assemblies (Fig. 2). Also known as condensation soldering, it uses the latent heat of vaporization of an inert liquid. Solder paste is screened on the board, the part is placed by an autoplacement machine, and the whole assembly is soldered in the vapor of the inert liquid at 419°F (204°C). The vapor is used only as a heating medium. A secondary vapor and cooling coils (Fig. 4) are used to minimize the vapor loss of the primary fluid, which is fairly expensive. The main concerns in vapor-phase soldering are part movement and formation of solder balls which can short conductors or component leads. Such problems can be avoided by proper selection of solder paste, pad design, and process control. Both solder joint defects and reliability depend to a great degree on component footprint design (pad size and orientation).

Infrared reflow soldering. In infrared reflow soldering, radiant or convective energy is used to heat the assembly. There are basically two types of infrared reflow processes: focused (radiant) and nonfocused (convective). The latter has proven more desirable for surface-mount technology. The focused infrared process radiates heat directly on the parts and may unevenly heat assemblies. The heat input on the part may also be color-dependent. In the nonfocused or diffused infrared process, the heating medium can be air or an inert gas or simply the convection energy. A gradual heating of the assembly is necessary to drive off volatiles from the solder paste. After an appropriate time in preheat, the assembly is raised to the reflow temperature for soldering and then cooled.

Cleaning. Cleaning of surface mount technology assemblies, in general, is harder than that of conventional assemblies because of smaller gaps between surface-mounted components and the printed wiring board surface. The smaller gap may entrap flux which may cause potential reliability problems if the printed wiring assembly is not properly cleaned. Hence, the cleaning process to be used is dependent upon the flux used for wave soldering.

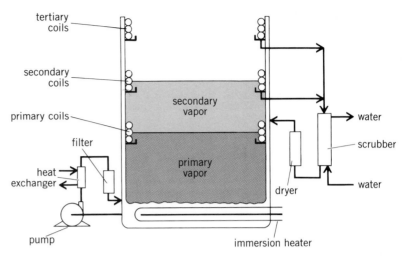

Fig. 4. Vapor-phase soldering system for surface-mounted assemblies.

Most companies use rosin-based fluxes generally known as rosin-activated (RA) or rosin-mildly-activated (RMA). For rosin-activated fluxes both solvent and aqueous cleaning are required, whereas for rosin-mildly-activated flux, solvent cleaning alone is sufficient.

Repair and rework. Due to the absence of plated through-holes, surface-mounted components are easier to remove and the possibility of thermal damage is nonexistent. There are various tools available for removing components. One of the most common is a forklike tip attachment made for the conventional soldering iron to fit the part being removed (Fig. 5). Heat is provided by the soldering iron, and the part is removed by twisting the tool. Hot air equipment is also used to remove components. One of the main concerns when this equipment is being used is preventing damage to adjacent components.

Testability. Most companies use "bed-of-nails" in-circuit testing for conventional assemblies. Use of surface-mounted components does not impact testability if the rule for testability of assemblies is not violated. The rule that must be followed is to provide through-hole passages or test pads to allow electrical access to each test node during in-circuit testing. If possible, this electrical access should be provided both at top and bottom of the wiring board,

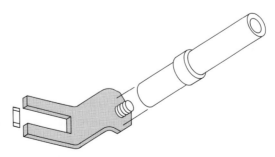

Fig. 5. Soldering iron tip attachment for removing surface-mounted components.

but the bottom access is a necessity.

For background information *see* CIRCUIT (ELEC-TRONICS); INTEGRATED CIRCUITS; PRINTED CIRCUIT in the McGraw-Hill Encyclopedia of Science and Technology.

[RAY PRASAD]

Bibliography: C. Capillo, How to design reliability into surface mount assemblies, *Electr. Pack. Prod.*, 25(7):74–80, July 1985; H. W. Marstein (ed.), Automation thrives as companies strive for SMT capability, *Electr. Pack. Prod.*, 25(5):68–74, May 1985; T. Ormond (ed.), Surface mount technology: A special report, *EDN*, 9:104–124, April 18, 1985; R. P. Prasad, Critical issues in implementation of SMD technology, *IPC-Tech. Rev.*, 24(10):15–18, December 1983.

Climatic change

A decade ago the research submarine *Alvin* first found and sampled ocean-floor hydrothermal springs near the Galapagos spreading center. Since that time oceanographers have come to realize that this process of hydrothermal circulation in ocean-floor rocks exerts a strong, and in some cases even controlling, effect on the chemistry of the oceans. Recent efforts to discern the timing and consequences of sea-floor hydrothermal activity have further suggested that this process may be linked to important changes in the Earth's climate.

Investigations of the relationship between climate and sea-floor hydrothermal systems not only should aid in understanding the history of the Earth's climate but also may provide clues to the consequences of future climate changes. For example, at present much concern exists that the Earth will experience a greenhouse effect during the next several decades, resulting from the build-up of CO_2 in the atmosphere caused by deforestation and fossil-fuel burning. Various mathematical models have predicted that this effect will result in an increase in the average surface temperature of the Earth by about 2.7–8.1°F (1.5–4.5°C). Such an increase could be significant, inasmuch as the present average Earth surface temperature is only about 59°F (15°C). One means of assessing the possible consequences of the predicted warming trend would be to study analogous periods in the Earth's history, such as times when increased atmospheric CO_2 concentrations caused climatic changes. Recent investigation of sea-floor hydrothermal activity indicates that such a period may have occurred during the Eocene, about 50 million years (m.y.) ago.

Chemical processes in the ocean. The direct chemical link between sea-floor processes and CO_2 fluxes is provided by the widespread hydrothermal activity along oceanic ridges. Fissures and fractures formed in fresh ridge-crest basalt as it cools and is rifted apart provide conduits for the circulation of cold bottom waters into the underlying crust. The circulating sea water penetrates to depths of a few kilometers, reacts chemically with hot basalt at temperatures in excess of 570°F (300°C) and completes

its convective cycle by emerging as hot springs along the ocean floor. The chemical exchanges that occur during this process include the removal of magnesium and sulfate from sea water and a concomitant enrichment of calcium, potassium, silica, iron, manganese, and other trace elements within the hydrothermal solution.

The chemical exchange of calcium for magnesium in this process is particularly relevant to the CO_2 cycle. Both laboratory and field investigations have shown that each Mg^{2+} ion which is removed from sea water is replaced by a chemically equivalent amount of Ca^{2+} ion. The only significant process that balances Ca^{2+} inputs to the oceans is the precipitation of biogenic $CaCO_3$, which is accompanied by the formation of CO_2, as shown in the reaction below.

$$Ca^{2+} + 2HCO_3^- \rightarrow CaCO_3 + CO_2 + H_2O$$

Once these calcium inputs were discovered, the next important step was to determine whether the amount of Ca^{2+} entering the oceans from the hydrothermal vents is great enough to influence the chemical balances of the whole ocean. Estimates of this amount have been calculated on the basis of three different geochemical data sets, and all give essentially the same result: the input of hydrothermal Ca^{2+} to the ocean is about 5×10^{12} moles per year. Rivers constitute the only other major source of Ca^{2+} to the oceans, accounting for an input of about 12.5×10^{12} moles per year. The chemical reaction shown above indicates that for each mole of Ca^{2+} input, one mole of CO_2 is produced. Consequently, it can be demonstrated that present-day sea-floor hydrothermal activity accounts for about 29% of the total ocean contribution to atmospheric CO_2. Finally, recent budget calculations estimate that the oceans contribute somewhere between 47 and 75% of the total global atmospheric CO_2 input; thus the calculations indicate that sea-floor hydrothermal activity plays a significant role in the present-day CO_2 budget, accounting for about 14 to 22% of the total atmospheric CO_2 input.

Ridge-axis hydrothermal history. If the intensity of hydrothermal activity during certain times in the geological past was significantly greater than today, then it is reasonable to argue that these periods represent examples of counterparts to CO_2-induced climatic changes. Leg 92 of the Deep Sea Drilling Project, which crossed the East Pacific Rise (EPR) at 19°S during early 1983, was undertaken to provide the first direct documentation of the geologic history of ridge-axis hydrothermal activity. That expedition showed that hydrothermal activity in the southeast Pacific reaches relative maxima of five to ten times present values, based on mass-accumulation rates of Fe-rich sediment, in the earliest Miocene and the late Miocene. These episodes of hydrothermal activity documented by the Leg 92 scientific party therefore do not correspond to known spreading-rate fluctuations, but do coincide with the two periods of ridge jumping and reorganization of di-

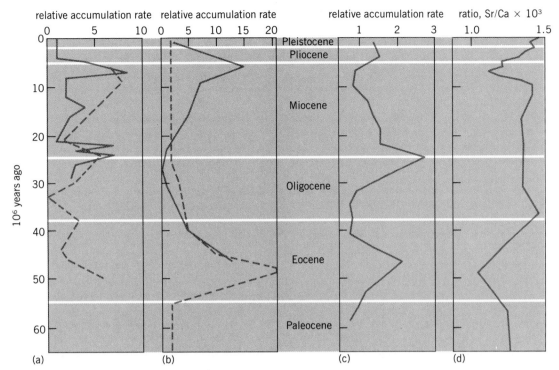

Relative mass accumulation rates of (a) iron, (b) opal, (c) calcium carbonate, and (d) the Sr/Ca ratio of foraminiferans. Accumulation rates increase and Sr/Ca ratio decreases about 50 m. y. ago during early to middle Eocene time. Data compiled from several sources; two sets of data are plotted in a and b.

vergent and transform plate boundaries along the East Pacific Rise that occurred approximately 18–22 and 6–8 m.y. ago. These results and the discovery of large-scale hydrothermal activity along the tectonically complex but slowly spreading Juan de Fuca Ridge lead to the suggestion that ridge-axis tectonism, in the form of axis jumps and reorganizations, rather than spreading rate, controls the amount of hydrothermal activity that occurs through geologic time.

The early to middle Eocene was the last time of greatly enhanced accumulation of hydrothermally derived material on a global scale. These indications of enhanced hydrothermal activity are important because the early Eocene is the time of the last worldwide episode of ridge jumping and general reorganization of sea-floor spreading centers, and has long been recognized as experiencing the warmest and most humid Cenozoic climate. These two events, increased tectonism and warm climate, may be linked by cause and effect by the process described above. Recalculation of the carbonate mass balance model developed by R. A. Berner and colleagues with a fourfold increase of hydrothermal calcium input to the oceans results in a doubling of CO_2 in the oceans and thus a doubling of CO_2 in the atmosphere, assuming the two concentrations are related linearly. Computer-based climatic simulations predict a global warming of 2.7–8.1°F (1.5–4.5°C) in response to a modern doubling of atmospheric CO_2.

This hypothesis describes the occurrence of the following events approximately 50 m.y. ago: (1) a global warming above preexisting temperatures of a magnitude similar to or greater than that predicted for the modern case; (2) a possible indication of increased humidity; (3) a decreased pole-to-equator temperature gradient; (4) increased accumulation of hydrothermal Fe in deep-sea sediments; (5) increased accumulation of Si in deep-sea sediments; and (6) increased Ca concentrations in sea water and increased accumulation of $CaCO_3$ in deep-sea sediments. Strong evidence exists for all six of these apparently related phenomena.

The Eocene global warming about 9°F (5°C) above Paleocene values has long been recognized from both terrestrial and oceanic records. Humid Eocene climates are suggested by the nature of the prevailing flora. The early to middle Eocene is also the time of the warmest high-latitude temperatures. The input of both hydrothermal and opaline materials to the sea floor increased severalfold 50 m.y. ago (see illus.). Most of the Eocene opal has undergone diagenetic conversion to the widespread chert deposits of that age, so accurate opaline flux values for that time are difficult to construct. The Sr/Ca ratios from samples of well-preserved fossil planktonic foraminifera show a significant decrease of 15 to 25% during the Eocene, about 50 m.y. ago (see illus.), a finding which has been attributed to an increased hydrothermal supply of Ca^{2+} during these periods. Summaries of carbonate accumulation indicate a maxima during the Eocene when accumulation rates more than doubled (see illus.). Deep waters of the ocean are more corrosive toward solid $CaCO_3$ because they contain greater amounts of dissolved CO_2. Variations in the amount of CO_2 in deep waters will affect the position of both the lysocline (the depth horizon at which there is observed a marked increase in the rate of carbonate dissolution), and the carbonate compensation depth, or CCD (the depth horizon at which the rate of carbonate input to the sediments is exactly balanced by the rate of carbonate dissolution, and below which there is no net accumulation of solid carbonate). A significant increase in hydrothermal activity should cause increased CO_2 levels in deep waters and thus a shallower carbonate compensation depth (CCD) and lysocline, which in turn would shift the concomitant increase in carbonate production toward shallower depositional environments. This is the pattern observed for the Eocene.

The data indicate that a severalfold increase in sea-floor hydrothermal activity, occasioned by early Eocene tectonic activity, caused a CO_2-induced global greenhouse effect. The several other predicted results of increased hydrothermal activity also occurred: changes in global climate and increased deposition of ferrigenous, opaline, and calcareous sediments. In summary, on the basis of the results assembled thus far, there appears to be a clear link between sea-floor hydrothermal activity, tectonism, and climate.

For background information *see* CLIMATIC CHANGE; GREENHOUSE EFFECT, TERRESTRIAL; SEA WATER in the McGraw Hill Encyclopedia of Science and Technology.

[ROBERT M. OWEN; DAVID K. REA]

Bibliography: R. A. Berner, A. Lasaga, and R. M. Garrels, The carbonate-silicate geochemical cycle and its effect on atmospheric CO_2 over the past 100 million years, *Amer. J. Sci.*, 283:641–683, 1983; Leg 92 staff, Advection in the East Pacific, *Nature*, 304:16, 1983; National Academy of Sciences, *Report of the Carbon Dioxide Assessment Committee*, 1983; R. M. Owen and D. K. Rea, Sea-floor hydrothermal activity links climate to tectonics: The Eocene carbon dioxide greenhouse, *Science*, 227:166–169, 1985.

Communications satellite

Recent advances in communications satellite service include (1) the operation of the *Arabsat* satellites (serving countries in North Africa and the Near East) and the *MORELOS* satellite (serving Mexico) and (2) the development of private international communication satellites.

Arabsat satellite. The *Arabsat* communications satellite (Fig. 1) was designed and manufactured by Aerospatiale of France in partnership with Ford Aerospace of the United States for the Arab Satellite Communications Organization (Arabsat), which comprises 22 countries and owns and operates the satellite system. The spacecraft is designed to provide service to the Arabsat member states, located in North Africa and the Near East, for telephone, data, telex, and television transmission (Fig. 2).

Operational service of the two-satellite system started early in 1985. The satellites are located in geosynchronous orbit at 19° and 26°E longitude. Design life of each satellite is 7 years. The initial on-station weight of the *Arabsat* satellite was 1305 lb (592 kg). The first *Arabsat* satellite was launched by the Ariane launch vehicle and the second by the U.S. Space Transportation System (STS).

The spacecraft configuration (Fig. 1) is of the momentum-bias three-axis stabilized type with a main body box and solar power panels extending from the north and south sides. The solar panels rotate once per day relative to the spacecraft body to face the Sun. Antennas are located on three sides of the main body, oriented to face the Earth.

The satellite receives and transmits in the C-band and transmits also in the S-band frequency ranges. The transponders receive right-hand and left-hand circularly polarized signals at 6 GHz from the receiving antenna. The wideband signals are amplified and down-converted to 4 GHz. The down-converted signals are then transmitted by 8.5-W traveling-wave-tube amplifiers. A signal from one channel is translated to the S-band and routed through two parallel 50-W traveling-wave tubes to the S-band (2-GHz) transmitting antenna.

The C-band antennas are located on the east and west sides of the spacecraft. Each antenna can receive and transmit both circular polarizations. The S-band antenna is a slotted waveguide array located on the Earth-facing panel of the spacecraft.

The tracking, telemetry, and command subsystem provides command reception, telemetry transmission, and ranging functions. The telemetry unit provides satellite subsystem operational status information. The satellites in orbit are controlled from two ground stations. The primary station is located in Dihrab, near Riyadh, Saudi Arabia, and the secondary station is located in Tunis, Tunisia.

The attitude determination and control system uses a momentum-bias three-axis stabilization design. Two independently powered skewed momentum wheels provide gyroscopic rigidity and torque control. Redundant infrared Earth sensors provide the reference data.

The primary structure consists of a graphite-epoxy/light-alloy honeycomb central cylinder which transmits the loads and houses the propellant tanks. Equipment for both the communication and service modules is placed on panels made from light-alloy honeycomb with either light-alloy or graphite-epoxy skins.

The electrical power subsystem is a dual-bus, direct-energy transfer system relying on the solar generator in sunlight and nickel-cadmium batteries during eclipse. Two rigid, Sun-oriented solar array wings, consisting of 20,000 solar cells bonded to a flexible kapton skin reinforced with graphite-epoxy frames, supply 1.3 kW.

The thermal control system uses passive techniques. The highly dissipative equipment is located

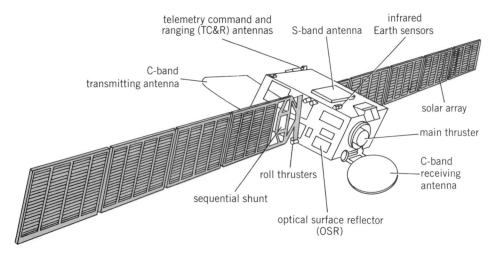

Fig. 1. Configuration of the *Arabsat* communications satellite.

on the inner side of the north-south panels, and surface coatings (paints, optical surface reflectors, superinsulation) control the heat exchanges. Heater elements are used for components having relatively small allowable temperature ranges such as propellant and batteries.

An integrated bipropellant [nitrogen tetroxide (N_2O_4) and monomethyl hydrazine (MMH)] propulsion subsystem provides both the apogee motor function and reaction control. Apogee maneuvers are executed with a 110-lbf (490-N) main thruster. A redundant system composed of 5-lbf (12-N) thrusters is used for orbit and attitude control.

Morelos satellite. The *MORELOS* communications satellite (Fig. 3) was produced by the Hughes Aircraft Company for the Secretaria de Comunicaciones y Transportes of Mexico. The spacecraft is designed to provide domestic service in Mexico (Fig. 4) for telephone, data, and television transmission. Operational service of the first of two satellites started in mid-1985. The satellites will be located in geosynchronous orbit at 113.5° and 116.5°W longitude. Design life of each satellite is 9 years. The initial on-station weight of the *MORELOS* satellite was 1440 lb (653 kg), and it was designed to be launched by the U.S. Space Transportation System.

The two main elements of the Hughes gyrostat configuration spacecraft (Fig. 3) are the spinning rotor and the despun, Earth-oriented platform containing the communications repeater and the antenna. A bearing and power transfer assembly, consisting of a brushless dc motor, precision ball bearings, and signal and power slip ring assemblies supplies the rotating interface.

The satellite receives and transmits in both the C-band and the K_u-band frequency ranges. The C-band repeater is a single-conversion channelized design that receives at 6 GHz and transmits at 4 GHz. The receiver is an all-solid-state microwave integrated-circuit design. Final channel amplification is provided by traveling-wave-tube amplifiers. Each of the six wideband traveling-wave-tube amplifiers has

a saturated output power of 10.5 W; each of the 12 narrowband traveling-wave-tube amplifiers has a saturated output power of 7.0 W.

The K_u-band repeater is a single-conversion channelized design that receives at 14 GHz and transmits at 12 GHz. The receiver is an all-solid-state microwave integrated-circuit design. Final channel amplification is provided by four 19.4-W traveling-wave-tube amplifiers.

The structure above the bearing and power trans-

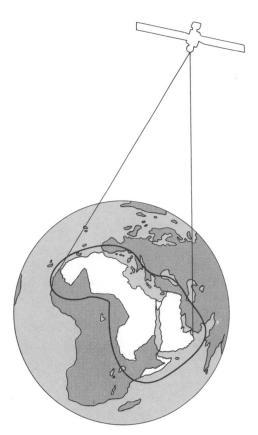

Fig. 2. Typical antenna coverage of the *Arabsat* satellite.

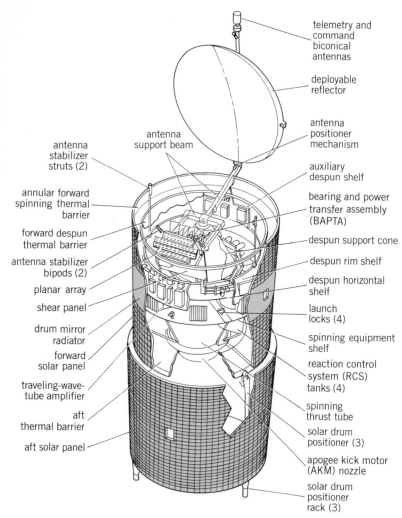

antenna
stabilizer
struts (2)

annular forward
spinning thermal
barrier

forward despun
thermal barrier

antenna stabilizer
bipods (2)

planar array

shear panel

drum mirror
radiator

forward
solar panel

traveling-wave-
tube amplifier

aft
thermal barrier

aft solar panel

antenna
support beam

telemetry and
command
biconical
antennas

deployable
reflector

antenna
positioner
mechanism

auxiliary
despun shelf

bearing and power
transfer assembly
(BAPTA)

despun support cone

despun rim shelf

despun horizontal
shelf

launch
locks (4)

spinning equipment
shelf

reaction control
system (RCS)
tanks (4)

spinning
thrust tube

solar drum
positioner (3)

apogee kick motor
(AKM) nozzle

solar drum
positioner
rack (3)

Fig. 3. Configuration of the *MORELOS* communications satellite.

fer assembly supports parabolic and planar array communications antennas and the telemetry and command antennas. The parabolic communications antenna is composed of two orthogonally polarized offset reflectors, one behind the other, sharing a high-gain 71-in.-diameter (180-cm) aperture. The front, horizontally polarized reflector is transparent to vertically polarized signals. Each reflector is fed by an independent dual-mode multihorn feed array.

Telemetry and command omnidirectional antennas are deployed after the spacecraft is ejected from the space shuttle. Command of the spacecraft is performed at C-band, and telemetry information from the spacecraft is available as pulse-code modulated and real-time frequency-modulated (FM) data. The telemetry transmitters feed a toroidal-beam omnidirectional antenna through two of the communications traveling-wave tubes or, for regional coverage, through the communications antenna.

A two-axis radio-frequency beacon tracker maintains elevation and azimuth beam pointing errors within ±0.05°. Error signals generated by the command track receiver are sent to the antenna posi-

tioner which processes the north-south error to control the antenna boresight in elevation. If beacon presence is lost, despin control reverts to the Earth-sensor reference mode.

The sunlight-illuminated spacecraft receives primary electrical power from forward and aft cylindrical solar arrays. Two nickel-cadmium batteries located on the spinning section receive power from charge arrays and supply secondary electrical power to the spacecraft for eclipse operation.

The spacecraft's primary heat rejection path is through a cylindrical drum radiator. Units with high thermal dissipation, including the traveling-wave-tube amplifiers and spacecraft batteries, have direct radiative coupling with this radiator.

The reaction control subsystem, located on the spinning section, consists of one radial and two axial thrusters to control spacecraft attitude and orbit position. All thrusters are canted to provide spin speed control. Four titanium tanks have a maximum hydrazine propellant load capability of 465 lb (211 kg). A solid-propellant apogee kick motor provides the impulse required to inject the spacecraft into a near-geosynchronous equatorial orbit from the intermediate transfer orbit established by the perigee kick motor which is fired after satellite ejection from the Space Transportation System. [WILLIAM J. KECK]

Private international satellites. During 1983 the United States embarked upon the development of private international communications satellites, an international satellite system separate from Intelsat.

Five such satellite systems have been proposed.

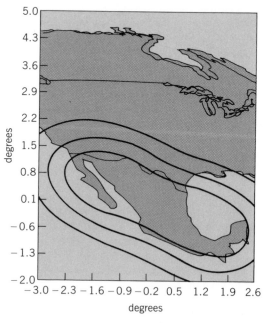

Fig. 4. Typical antenna coverage of the *MORELOS* satellite. The axes are dimensioned in degrees north and south and east and west of the antenna aiming point near Mexico City; the origin of these angles is the satellite. Double-line boundaries result from composite of views from the two different satellite positions. The three contours represent constant levels of effective radiated power from the satellite antennas.

They include, in order of application: ORION, International Satellite, Inc. (ISI), Cygnus, Pan Amsat, and Finsat. These systems intend to offer communication services to a unique community of users, with system designs different from those of Intelsat or other regional systems. In addition, a United States company, Pacific Satellite, Inc., has joined with Papua New Guinea to propose the provision of private communication satellite services in the Pacific.

Policy. The prospect of these private international communication satellite systems precipitated a major policy review by the United States government, led by the Senior Interagency Committee on International Communications Policy. This committee issued a report recommending the approval of these private systems under certain conditions. The criteria were:

1. Each system would be restricted to providing services through the sale or long-term lease of transponders or space segment capacity only for communications that are not interconnected with public-switched message networks (except for emergency restoration service).

2. One or more foreign authorities must authorize use of each system and enter into consultation procedures with the United States to ensure technical compatibility and to avoid significant economic harm to Intelsat.

The committee recommended that these systems should be authorized in accordance with these criteria, provided that the President, in accordance with provisions of the Communications Satellite Act of 1962, determined that international communications satellite systems separate from Intelsat were "required in the national interest." The President made this finding on November 28, 1984.

Subsequently, the Federal Communications Commission exercised its regulatory responsibility by seeking comments from the public. The United States systems received initial approval in July 1985.

System description. A typical private system being implemented under this arrangement is that of International Satellite, Inc. This private system provides service from anywhere in the continental United States at frequencies in the K_u-band (14 GHz uplink and 12 GHz downlink) to almost anywhere in western Europe (Fig. 5). The system would place two communications satellites (Fig. 6) with specialized capability in orbit so as to provide high capacity between these two geographic sectors flanking the Atlantic. Because of the power in the satellites and the narrow focus of the beams, this satellite system can provide service to customers at their places of business, with small antennas (10 ft or 3 m).

Service. These private systems, by virtue of policy and regulatory decisions, are not permitted to offer international telephone service, which is the backbone of the services provided by Intelsat. The purpose of the policy is to protect Intelsat from economic harm. However, part of the rationale behind

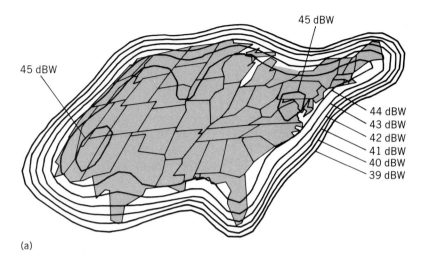

(a)

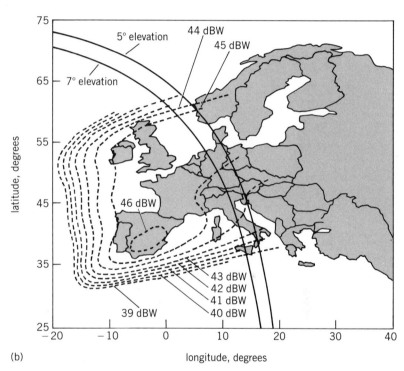

(b)

Fig. 5. Proposed satellite system coverage by International Satellite, Inc. (a) Continental United States coverage. (b) West European coverage.

the authorization of these private systems is that they will be able to develop and offer new services. In the case of International Satellite, Inc., for example, these include:

1. Europe-originated video to small stations in the continental United States, and conversely video from centers in the United States to western Europe.

2. Fully interconnected, intracorporate data communication networks operating from antennas located in customer premises throughout the continental United States and western Europe.

3. High-speed facsimile distribution to and among business offices in the United States and western Europe in either bilateral or multilateral (broadcast) mode.

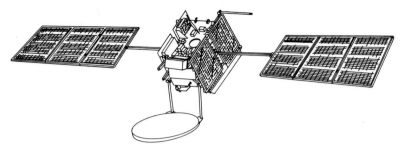

Fig. 6.   Proposed configuration of International Satellite, Inc. (ISI), communications satellite.

New capabilities. Private communication satellite systems offer capabilities for communication over the Atlantic Ocean that may not be available otherwise. This is accomplished by taking advantage of the unique properties of available satellite technology not now being used for international communication satellite service. Table 1 compares some of these differences in the capabilities of International Systems Inc. and Intelsat.

Table 1. Qualitative comparison of private and Intelsat systems at K_u-band

Characteristic	Intelsat	International Satellite, Inc.
Coverage		
United States	East coast only	Full continental United States
W. Europe	80%	All
Earth station size (antenna diameter)	16–32 ft (5–10 m)	10 ft (3 m)
Spectrum	750 MHz	2000 MHz
Satellite power	45 W	480 W

Efficient use of geostationary orbit. Private communication satellite systems must coexist in the same orbit and spectrum as the United States domestic communication satellites, as well as those of Intelsat. The orbit is the arc of geostationary orbit between 30° and 60°W longitude. The spectrum is those radio-frequency allocations assigned for use to communication satellites at 4/6 GHz and 11–12/14 GHz.

There are over 50 Intelsat member countries in the Atlantic basin, and the orbit positions used by Intelsat must be such as to provide service to them all. Intelsat uses orbit positions at 47°, 50°, and 53°W longitude, among other orbit locations between 30° and 60°W. In addition, it has requested additional orbit positions in this arc. The private systems have requested orbit locations in similar parts of this orbit, at positions indicated in Table 2.

The situation is further complicated by the highly competitive use of geostationary orbit by United States domestic communication satellites. Approximately 50 orbit positions have been assigned to these satellites between 62° and 146°W longitude. This includes a number of organizations which have been newly granted two orbit positions each.

The problems are posed by the electromagnetic proximity of the private international satellite communication system with both Intelsat and United States domestic satellites. Generally, satellites operating on the same frequencies at 4/6 GHz require at least 3° of orbital spacing, and those at 11–12/14 GHz 2° of orbital spacing. It can be seen from Table 2 that some of the requested private-system orbit positions are in direct conflict with either Intelsat positions or those of United States systems. Some of the conflicts include ORION with both Intelsat and Videosat, and Cygnus with Intelsat.

The mechanism for resolving these conflicts is established in the International Radio Regulations of

Table 2. Location of private and Intelsat satellites in geostationary orbit*

System	Longitude 60°W ——————————— 45°W ——————————— 30°W										
ISI	58° / X	56° / X									
ORION							47° / K			37.5° / K	
Cygnus								45° / K	43° / K		
Pan Amsat			57° / C/K					45° / C/K			
Intelsat				56° / K/C	53° / K/C	50° / K/C					34.5° / K/C
Videosat									43.5 / K	37.5° / K	

*Letters indicate frequency bands.

the International Telecommunication Union. The basis for revising these regulations was established at a treaty conference of this organization dealing specifically with the geostationary orbit in 1985.

For background information *see* COMMUNICATIONS SATELLITE in the McGraw-Hill Encyclopedia of Science and Technology.

[DONALD M. JANSKY]

Bibliography: Aerospatiale Space and Ballistic Systems Civilian Programmes, 1983; American Institute of Aeronautics and Astronautics, *9th Communications Satellite System Conference*, San Diego, AIAACP 821, March 7–11, 1982; Hughes Aircraft Co., *MORELOS Satellite System Summary*, Ref. F2017, 1983; International Astronautical Federation, *Congress*, Budapest, Hungary, October 10–15, 1983.

Computer networking

The widespread use of local area networks (LANs) for every possible configuration of computers and terminals has brought about a shift in the focus of research and development efforts from hardware design issues, such as proper topologies and media access methods, to open networks, network management, and integration of data with voice and video on one transmission medium. Whereas proprietary LANs are proliferating and are becoming more functionally sophisticated for their particular application, a current trend is to design the networks to accept any vendor's equipment and make them truly open. At the same time, management issues such as fault resiliency and tolerance are being addressed as personal computer LANs become more sophisticated. Design work is also proceeding on creating the chips and the accompanying software to integrate data with voice and video on the integrated services digital network (ISDN).

Network compatibility. Standardization of the appropriate signals for the individual topologies and media access methods is being completed by various committees of the Institute of Electrical and Electronics Engineers (IEEE) and the International Standards Organization (ISO), and thus any vendor wanting to develop a LAN can now do so by designing the network around one of the standards. This has led to an oversupply of proprietary baseband and broadband LAN designs that for the most part are not compatible. The compatibility issues are being addressed from different perspectives: wideband backbone networks that tie LANs together, operating systems that make different LANs seem transparent to one another, and higher-level protocol solutions using the well-established protocols of System Network Architecture (SNA), XNS (Xerox Network Systems), and Transmission Control Protocol and Internet Protocol (TCP/IP).

Operating system. The widespread business use of personal computers has brought about the development of personal-computer LANs. A family of these LANs has evolved to support the dominant microcomputer operating system, MS-DOS, for which over 40,000 application packages have been written. It has been possible to connect computers based on this operating system to only a limited extent and, for their part, personal-computer LANs have followed their own rules. This lack of standardization has had the greatest effect on software developers who were inhibited from writing network applications.

These problems are addressed by the MS-DOS 3.1 operating system for personal computers, which has many built-in fundamental multiuser functions built in as high-level network primitives. At the highest level of the open systems interconnection (OSI) model—the applications level—every LAN that supports MS-DOS 3.1 functions presents a standardized interface to applications software. This means that any multiuser software package written for MS-DOS 3.1 runs on any LAN that supports MS-DOS 3.1. Thus software developers can write one version of a network application for all MS-DOS networks and need not face the high cost and speculation of writing a separate version for each LAN.

Wideband network. An example of advanced LAN technology is ProNET-80 (see illus.), an 80-megabit/s network with a completely decentralized token passing star-shaped ring configuration. The network allows connectivity between different makes of personal computers, minicomputers, and mainframe computers. Up to 255 Unibus or Multibus workstations can be connected, using any combination of shielded twisted pairs of coaxial fiber-optic cables, or infrared links. Software-compatible to the 10-megabit/s ProNET-10, this superhighway for communications can be used in a variety of applications, such as a backbone network for several LANs, a high-speed transmitter of complex high-resolution graphic data, or a superfast host-to-host network in a distributed processing environment. ProNET-80 uses the same passive wire centers for connections to all stations as did ProNET-10. The wire centers add fault tolerance to the system, and automatically bypass a failed node. Also the network contains an error detection system to ensure data integrity by correcting parity errors and determining the exact location of the error. This is done by inserting a format error message in the host interface. By polling the network, an operator can determine exactly where the error occurred.

The network's token-passing ring method assures real-time response and guaranteed access to the media. Another similar token-passing ring network has been introduced recently, but is scheduled to operate at a 4-megabit/s rate, as specified by the IEEE 802.5 standard. However, a five-chip set developed for this network can be manipulated to operate at 10 times that rate, thus leaving room for expanding the network in the future.

Optical fiber transmission. While ProNET-80 can operate on a twisted-pair system, its basic advantage is that the LAN can be connected via fiber optics. Token rings, in general, are well suited to optical transmissions as they depend on point to point as opposed to broadcast connections such as in a bus configuration. Fiber-optic cables can now be ex-

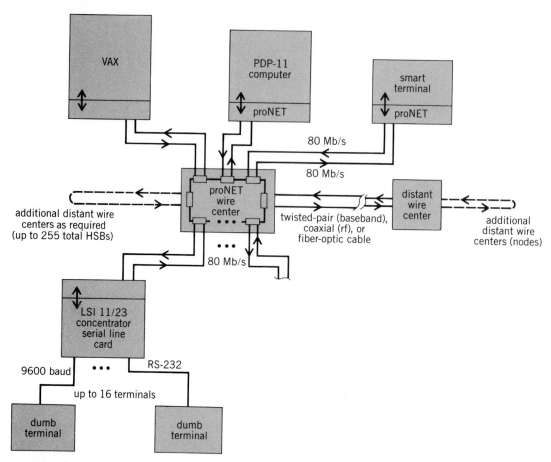

Configuration of ProNET-80, an advanced local area network. HSB = host-specific interface board. Concentrator serial line card allows interface between many terminals to the proNET wire center by concentrating data serially. (*After H. C. Salwen, In praise of ring architecture for local area networks, Comput. Des., 22(3):183–192, March 1983*)

pected to operate over distances of 1 to 2 mi (2 to 3 km). The maximum distance that a high-data-rate LAN can reach without resorting to long-haul technology or cumbersome lower-speed bottlenecks is thus greatly enlarged by the use of fiber-optic technology. In an optical-fiber-ring LAN, for example, a 250-node network could extend through a circle with a diameter over 120 mi (200 km). With fiber-optic repeaters or bridges that connect ring LANs together, the range of optical-fiber LANs is limited only by the maximum acceptable total delay around the ring. Compared to the delay limit in a 10-megabit/s CSMA/CD (carrier sense multiple access/collision detection) LAN of roughly 50 microseconds, token rings have no intrinsic delay limit. Thus the number of repeaters in any given network is much lower, thereby increasing reliability by reducing parts counts and shortening the length of the critical path.

The table summarizes the types of transmission media used in LANs and highlights the advantages of optical fiber technology. The values presented are averages, and such parameters as range, loss, and cost are, of course, dependent on bandwidth and signal quality.

Protocols. Adherence to popular communication protocols is becoming critical for LAN vendors. As LANs are connected to one another and to backbone networks to form global area networks, protocols such as TCP/IP, XNS, and the entire collection of SNA protocols have to be dealt with. Several vendors provide TCT/IP protocol software packages for various LAN configurations. TCP/IP has been in daily use since 1982, and became the Department of Defense packet network protocols in 1983. The TCP/IP protocols, which were developed by the Defense Advanced Research Projects Agency (DARPA) as a vendor-independent network protocol, are actually a superset of the original protocols used by DARPA's ARPANET, which is a nationwide network for Department of Defense–sponsored researchers.

Token-ring configuration. Token-ring LANs are unique among the three standards that have been developed for these networks (the other two being CSMA/CD on baseband cable and token-bus on broadband cable) in that the nodes are physically connected serially by a transmission medium, such as twisted pairs or optical fiber. Several advantages exist in choosing a token-ring configuration for a LAN. These include ease of fault isolation, performance stability under load, the use of predomi-

Comparison of transmission media used in local area networks

Medium	Range	Bandwidth	Loss at 20 MHz	Immunity to electromagnetic interference	Security	Cost	Comments
Twisted pair (shielded)	<4000 ft (1200 m)	<100 kHz	High	Very good	Fair	Low	Inexpensive, limited range
Twinax	<0.6 mi (1 km)	<200 MHz	80–160 dB/mi (50–100 dB/km)	Very good	Fair	High	"Controlled impedance," can be 802-compatible*
Multiconductor flat cable	<300 ft (100 m)	<10 MHz	High	Poor	Fair	High	Shielding available, can be byte parallel
Coaxial cable	<1.2 mi (2 km)	<400 MHz	10.5 dB/mi (6.5 dB/km) 100 dB/mi (60 dB/km)	Fair	Fair	Very high Low	Can be 802-compatible*
Fiber optics	1–2 mi (2–3 km)	>500 MHz-km	1.6–13 dB/mi (1–8 dB/km)	Excellent	Very good	High	Costs continue down
Telephone line (PABX, PBX)	—	<60 kHz	High	Poor	Poor	Low	Low cost, installed base
Infrared radiation	1000 ft (300 m)	<20 MHz	High	Good	Fair	Currently high, potentially low	Free space line of sight, reduced range in fog (moisture-dependent)
Microwave radiation	<30 mi (50 km) [horizon]	>100 MHz	Inversely proportional to distance	Poor	Fair	High	Free space line of sight, subject to scattering, link loss depends on antenna size

*Refers to standards developed by IEEE Committee 802.

nantly digital rather than analog engineering, and the promise of fiber-optic technology. The effect of the token-ring LAN on SNA protocols is not disruptive, and very few changes would be required to attach an SNA node to the ring. Because SNA is a layered architecture, supporting a new physical and data link control layer has little effect on the rest of the architecture. Thus, path control, for example, could remain unchanged, thereby allowing SNA nodes attached to the token-ring LAN to participate in an SNA network immediately.

ISDN and Digital Multiplexed Interface. Because token-ring configurations lend themselves to fiber-optic implementations which can support a high bandwidth, and this in turn means that the medium can accommodate more than data, including voice and video transmission, token-rings could become primary contenders for parts of the integrated services digital network (ISDN). Conceived by the International Consultative Committee on Telegraphy and Telephony (CCITT), ISDN is a public end-to-end digital communication network with signaling, switching, and transport capabilities. It will support a wide range of services, such as voice, data, video, facsimile, and music over standard interfaces. The initial chips are being developed that allow potential

users to interface to the network. Only a small amount of integrated circuits with well-defined functions will be required to satisfy a broad spectrum of telecommunication design applications.

To accommodate the data portion in the ISDN, several computer manufacturers have proposed a Digital Multiplexed Interface to provide a cost-effective, high-speed interconnection between terminals and host computers in a private-branch-exchange environment. Because it is based on the use of standard T-carrier telephone transmission facilities and 64-kilobit/s transmission and switching, applications of Digital Multiplexed Interface can provide data connectivity over geographically dispersed private networks. For local applications, Digital Multiplexed Interface can operate at distances up to 1300 ft (400 m) between the private branch exchange and host computer without signal regeneration. For distances greater than 1300 ft (400 m), Digital Multiplexed Interface makes use of the standard T-carrier repeater line or a fiber-optic option.

Manufacturers implementing the Digital Multiplexed Interface specification will replace the individual single-channel data lines and data modules that are currently used for private-branch-exchange networking with the high-bandwidth, multichannel interface. A Digital Multiplexed Interface has been implemented between a Hewlitt-Packard HP 3000 computer and AT&T's 75/85 private branch exchange and systems of other private-branch-exchange manufacturers. Immediate implementation of the Digital Multiplexed Interface and the ability to upgrade to ISDN compatibility via firmware were important objectives that were balanced against the need to make the product inexpensive enough to provide substantial cost advantages over existing private-branch-exchange solutions. While achieving these objectives was possible because good approaches to design were taken by using existing integrated circuits, very large-scale integrated circuit (VLSI) implementations of the current design, now under development, should lead to further improvement.

The proliferation of proprietary LANs has kept the connection cost to these networks relatively high. While standards being developed for the various configurations by IEEE Committee 802 will reduce these costs substantially, even greater cost reductions will result from VLSI implementation of the ISDN. Connection costs per node for ISDN are expected to drop below those for local area network standards by 1987.

For background information *see* DATA COMMUNICATIONS; MICROCOMPUTER; OPTICAL COMMUNICATIONS in the McGraw-Hill Encyclopedia of Science and Technology.   [NICOLAS MOKHOFF]

Bibliography: N. Mokhoff, LANs team up to widen the network connection, *Comput. Des.*, 24(2):96–115, February 1985; W. Stallings, *Local Network Technology Tutorial*, IEEE Catalog EHD 208–1983, 1983; R. J. Sundstrom et al., SNA directions—A 1985 perspective, in A. Wojcik (ed.), *1985 National Computer Conference Proceedings*, 54:589–644, July 1985.

Computer software

Computer hardware description languages (CHDL) are notations and computer languages that facilitate the documentation, design, and manufacturing of digital systems. Extending beyond early applications in logic synthesis and simulation, these languages are now used to drive very large-scale integrated (VLSI) circuits, and combine program verification techniques with expert system design methodologies.

Hierarchies of levels of detail. Digital system descriptions use a hierarchy of levels of detail. A system is depicted as a network of components of a lower level, each of which is in turn a "system" composed of yet simpler components at an even lower level (see table). At some point, components are taken as primitive and their properties are not described in terms of lower-level elements.

Evolution of CHDLs. Computer hardware description languages have been used since the 1950s, and their evolution has been tied to the maturity of the computer design process. Simpler notations were used to describe early systems at the gate and register transfer levels, and the primitive components

Hierarchy of levels of design

Component	Connection	Behavior
Gate	Layout	Timing diagrams
Flip-flop	Wire list	Boolean equation Truth tables
Register	Block diagram	State diagram
Arithmetic-logic unit (ALU)		Register transfer
Processor	?	Instruction set
Network	Processor-memory-switch (PMS) diagram	?

were gates, registers, and so forth (transistors and capacitors being in some sense at a lower level but of more concern to electrical engineers and physicists than to most computer designers).

Starting in the late 1960s, instruction set processor description languages were introduced to aid the design activities at a higher level. This was in response to the needs of computer architects who were more concerned with the behavior of the instruction set of the computer than with the implementation or realization of the data paths and control units. To a large extent, this evolution was triggered by the introduction of computer families, sharing the same instruction set but implemented with widely diverse technologies and cost-performance trade-offs. The notations borrow heavily from regular Algol-like programming languages, although the languages are tuned to the description of purely behavioral aspects of the system, ignoring implementation details. An instruction set processor description of an early minicomputer, the DEC PDP-8, is shown in the illustration.

Recently, processors, memories, and switches have appeared as primitive elements used to build multiprocessors and array processors of different kinds, leading to geographically distributed computer networks. There are few computer hardware description languages in use at this level because formal design methods are still evolving.

Concurrently with the appearance of languages appropriate for the different levels of design, the domain of applications for the languages within each design level has also grown. This evolution is brought about by advances in computer programming technology. While the notations for early systems tended to be designed mostly for documentation and simulation, more formal applications of computer hardware description languages have become increasingly common.

As the complexity of hardware designs increases, hardware and software design activities tend to blur. Current problems being addressed by designers of computer hardware description languages include correctness and verification, separate compilation facilities, access to program libraries, version control, and configuration management. Many good ideas have been developed to enhance the productivity of programmers and the correctness of the program development process, and a similar array of tools may soon be developed to aid the hardware design process. The commonality of activities has led some researchers to propose the use of regular programming languages, such as Ada, as computer hardware description languages. The migration of computer science results into hardware design is further illustrated by current approaches to very large-scale integration (VLSI) synthesis.

VLSI synthesis. The use of very large-scale integration has made complete processors and memories available in a single silicon chip. Two aspects of this technological revolution are, first, that primitive components continue to increase in complexity and, second, that the rate of introduction of new components continues to increase.

While the increased power of the primitive components simplifies the design task, the rapid introduction of new, more complex components requires an acceleration in the design process if a new technology is to realize its potentiality—some estimates indicate an exponential growth in worker-hours per month required to design and lay out complex integrated circuits. This can be achieved only through automatic means.

Early design automation systems were constrained by a fixed, built-in set of components and by a direct or canonical implementation philosophy. That is, the designer's specification would be translated into hardware in a manner very similar to a macroexpansion in an assembly programming language. A better approach is to eliminate the first constraint by taking as inputs both the designer's specification and the description of the components or building blocks from some design data base, thus speeding up the incorporation of new technologies into the design process. This works well when designing with standard packages, subject to common interconnection rules.

For a variety of reasons, a direct translation of the initial specification might not lead to an acceptable implementation. For example, the specification might have been written in a style that made it easy to read and perhaps extend by other designers. Thus, a more intelligent handling of the designer's specification is required. This is particularly true in the design of integrated circuits, where geometric constraints must be taken into account.

The design activity can be characterized as a constrained optimization problem along multiple dimensions, such as cost, speed, power dissipation, shape, size (or area), and component count, and the number of possible solutions (that is, implementations of a given behavioral specification) can be very large. Unfortunately, there are no algorithms that could lead a design automation program through the entire space of designs, evaluating them and retaining only those that meet some criteria of optimality. Emerging design automation systems rely instead on heuristic techniques to explore the design space and often borrow sophisticated methods developed in the field of artificial intelligence. At present, however,

```
PDP8 := begin
** Mp.state **
M\Memory[0:4095]<0:11>,

** Pc.state **
PC\Program.Counter<0:11>,
cpage\current.page<0:4>,
lac<0:12>,
        L\Link<>                  := lac<0>,
        AC\Accumulator<0:11>      := lac<1:12>,
** Instruction.Format **
i\instruction<0:11>,
        op\operation.code<0:2>    := i<0:2>,
        ib\indirect.bit<>         := i<3>,
        pb\page.0.bit<>           := i<4>,
        pa\page.address<0:6>      := i<5:11>,
        ................          ! declaration of I/O
                                     and micro-operation bits.

** Address.Calculation **
eadd\effective.address<0:11> :=
        begin
        DECODE pb =>
                begin
                0 := eadd = '00000 @ pa,
                1 := eadd = cpage @ pa
                end next
        IF ib =>
                begin
                IF eadd<0:8> eql #001 => M[eadd] =
                  M[eadd] + 1 next
                eadd = M[eadd]
                end
        end,

** Interpretation.Process **
interpret :=
        begin
        REPEAT    begin
                  i = M[PC]; cpage = PC<0:4> next
                  PC = PC + 1 next
                  execute()
                  end
        end,

execute :=
        begin
        DECODE op =>
                begin
        #0\AND    :=      AC = AC and M[eadd()],
        #1\TAD    :=      lac = lac + ('0 @ M[eadd()]),
        #2\ISZ    :=      begin
                          M[eadd] = M[eadd()] + 1 next
                          IF M[eadd] eql 0 => PC = PC + 1
                          end,
        #3\DCA    :=      begin
                          M[eadd()] = AC next
                          AC = 0
                          end,
        #4\JMS    :=      begin
                          M[eadd()] = PC next
                          PC = eadd + 1
                          end,
        #5\JMP    :=      PC = eadd(),
        #6\iot    :=      input.output(),
        #7\opr    :=      operate()
                  end
        end,
...................................
end
```

Instruction set processor description of DEC PDP-8.

chip design remains an expensive, time-consuming labor-intensive activity. It will be a few years before the performance of artificial-intelligence-based tools becomes adequate for mass-produced designs in a truly automated silicon foundry.

For background information *see* ARTIFICIAL INTELLIGENCE; DIGITAL COMPUTER; DIGITAL COMPUTER PROGRAMMING; INTEGRATED CIRCUITS; OPTIMIZATION in the McGraw-Hill Encyclopedia of Science and Technology.

[MARIO R. BARBACCI]

Bibliography: Computer hardware description languages, *Computer*, special issue, vol. 18, no. 2, February 1985; G. Moore, VLSI: Some fundamental challenges, *IEEE Spectrum*, 16(4):30–37, April 1979; New VLSI tools, *Computer*, special issue, vol. 16, no. 12, December 1983; *Proceedings of the IFIP 6th International Symposium on Hardware Description Languages and Their Applications*, Pittsburgh, May 23–25, 1983; *Proceedings of the IFIP 7th International Symposium on Computer Hardware Description Languages and Their Applications*, Tokyo, August 29–31, 1985.

Computerized tomography

The introduction of computerized tomographic (CT) scanning during the early 1970s had a revolutionary impact on the use of x-rays in diagnostic imaging. The cross-sectional image produced by computerized tomography solved the problem of superposition in ordinary projection imaging and thus greatly improved the visualization of soft tissues in the body. During the first several years of development, dramatic technological improvements in both image quality and scanning speed were introduced. Today computerized tomographic scanners can resolve structures of about 1 mm and have exposure speeds in the range of 1–5 s. Recently, computerized tomography technology has stabilized at this level which appears to be the limit of technology based on the use of rotating x-ray tubes.

Although computerized tomography has achieved an important role in diagnostic radiology, its applications are restricted by the performance level of conventional machines. For example, the resolution and speed of computerized tomography are more than a factor of 10 worse than the corresponding capability of conventional x-ray film imaging. Thus computerized tomography cannot be used in applications which require fine-detail resolution such as in imaging blood vessels or where there is rapid motion such as in heart imaging.

Recently a new approach to computerized tomography has been introduced, based on the use of scanning electron beams. This breakthrough has already led to a commercial computerized tomographic scanner for cardiac imaging with a scan speed of 50 ms. Resolution competitive with conventional scanners has been achieved, and there are indications that advances in this area are feasible in the future as well. This development has depended on innovations in accelerator physics, detector technology, tungsten metallurgy, and fast computerized data acquisition and control systems.

Scanning electron beam system. Computerized tomography requires the rotation of a point source of x-rays about the body. In conventional systems the x-rays are produced in an x-ray tube that is mechanically rotated within a large gantry mechanism. Within the tube the x-rays are produced by directing a beam of electrons from a small cathode onto a nearby tungsten anode. A voltage of 130 kV and a current of 100 mA are typical. Speed limitations occur due to the centripetal forces acting on a heavy rotating x-ray tube and the limited capacity of such tubes to dissipate the electrical energy generated.

In scanning electron beam systems, a large evacuated scan tube containing a ring-shaped tungsten target is used. Motion of the x-ray beam is produced by deflecting a beam of electrons along a circular track on the tungsten ring by using magnetic deflection similar to that used in the cathode-ray tube. Since no mechanical motion is required, high scanning speeds can easily be achieved. Since the scan tube must be large enough to partially surround the body, the tungsten rings are relatively massive and have a much higher heat capacity than conventional x-ray tubes. Thus, beam currents on the order of 1000 mA are possible, and scans can be taken in rapid sequence without the need for a cooling interval.

Electron beam transport and focusing. Electron beams of this power represent a major challenge and have required the development of new theoretical accelerator physics. In such beams the self-force due to electrostatic repulsion of the electrons is an important effect that tends to defocus the beam. The space charge distribution within the beam is also dependent on ionization of residual gas in the vacuum system. The design of the magnetic focusing system that transports and focuses the electron beam over the required distance must explicitly take these factors into account, and this added complexity has been the major obstacle to the development of electron beam systems of this type. The successful development of the Imatron Cine-CT scanner illustrates the existence of at least one solution to this problem.

The Imatron Cine-CT scanner is based on an electron beam scan tube that was intended for use in cardiac computerized tomography imaging (Fig. 1). An electron gun is used to produce a beam at 130 kV and 800 mA. After exiting the gun, the beam expands due to space charge effects and is refocused by magnetic lenses. The beam is then bent through an angle of 33 to 37° by a pair of orthogonal dipole electromagnets. The plane of bend is rotated by applying currents to these magnets that are approximately sinusoidal and 90° out of phase. Thus the deflected beam is caused to sweep along the circular tungsten target rings at the focus. Since the scanner is intended to image multiple slices, four target rings are provided and can be selected in sequence by varying the bend angle.

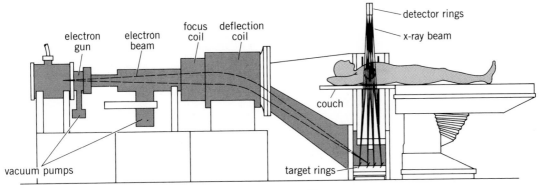

Fig. 1. Side view of a scanning electron beam computerized tomographic scanner.

X-ray collimation and detection. An x-ray-opaque housing (collimator) defines the x-ray beam to a fan shape that rotates with the motion of the electron beam spot on the tungsten rings. Under computer control, the focal spot is precisely steered through 210° in 50 ms. The collimated x-ray beam penetrates the body and is recorded by an adjacent pair of ring-shaped detector arrays above the body. Each detector ring consists of 432 crystal-photodiode detectors in a 210° arc. A 30° sector of the fan intercepts the body, and the corresponding detectors record the distribution of transmitted radiation at each of 1250 positions along the target ring.

Data processing and scanning. A data acquisition system digitizes and stores the detector data at a rate of 16 million bytes per second in a random-access memory that stores the data for 80 images. The data are then processed in an array processor that synthesizes the corresponding images by using a mathematical technique called reconstruction from projections. The digital image data are then displayed on a video screen by using a variety of image-processing techniques that aid diagnosis and analysis.

Since this scanner operates completely under computer control, a wide variety of scanning sequences and operational modes are available. Approximately 40 megabytes of user software support various scanning applications and data analysis techniques. Two basic scanning speeds are provided, a 50-ms speed that gives a resolution of 0.3 line per millimeter and a 100-ms speed that gives 0.75 line per millimeter when combined with a double-density detector ring. Typically up to 20 or more of these scans can be taken in rapid sequence with an 8-ms interscan delay. For dynamic studies, this gives a maximum speed of 17 images per second at each of two simultaneous slices. For higher-exposure, low-noise images, scans can be averaged. By using 10-frame averages, an object with a contrast of approximately 0.3% and a diameter of 4 mm can be detected.

Applications. With the Imatron Cine-CT scanner, volume images consisting of multiple slices can be obtained in order to examine anatomy in three dimensions. The 3-in. (8-cm) length of the heart can

be scanned in 224 ms by using four 50-ms scans on sequential target rings. Figure 2 illustrates such an eight-slice study of the heart obtained during the diastolic phase of a single beat. The fast scan speed freezes the cardiac motion, enabling exceptional clarity of the entire internal anatomy. Clinical applications of these images include sizing of infarcted and ischemic tissue, visualization of aneurysms and associated thrombus, estimation of myocardial mass and wall thickness, and assessment of coronary anatomy.

Study of heart motion. In the scanner's continuous mode at 17 frames per second, the motion of the heart can be studied. In this mode 10-frame movies at eight levels can be acquired by using four successive beats. Applications include evaluation of ventricular volumes, stroke volume, ejection fractions, regional wall thickening, regional wall motion, and cardiac output. Figure 3 illustrates an analysis of these parameters using a semiautomated program. The endo- and epicardial borders of the left ventricle have been outlined, with the half-height contour method, for the diastolic and systolic frame at each of the seven levels, and the quantitative results displayed. The computer calculates wall thickening and radial motion as a function of angle about the center of mass, and global parameters such as left ventricular area and myocardial area at each level are then determined and summated to yield volumes, ejection fractions, and myocardial mass. Preliminary research indicates that fast computerized tomography is the most accurate available technology for such studies.

Flow mode. The Cine-CT scanner was designed to measure blood flow in vessels, cardiac chambers, and tissue by using its "flow mode" of operation. In this mode a series of multislice scans is obtained, each triggered at a fixed time interval after the electrocardiographic R-wave, following rapid intravenous injection of a small amount of contrast medium. By using time delays corresponding to diastole, a series of scans is obtained on successive beats at the same cardiac phase. When these images are displayed as a movie, the heart appears stationary, and density changes in the ventricular cavities and myocardium are seen during passage of the con-

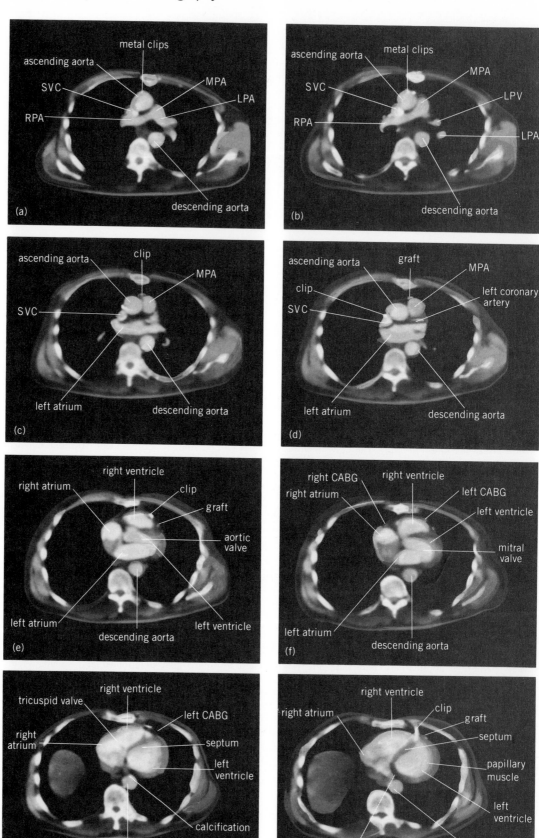

Fig. 2. Eight-level volume study obtained during one heart-beat of a patient with coronary artery bypass grafts. The grafts are seen in cross section at several levels. The native coronary arteries and other cardiac structures are labeled.

Parts *a–h* show successive levels. SVC = superior vena cava; RPA = right pulmonary artery; MPA = main pulmonary artery; LPA = left pulmonary artery; LPV = left pulmonary vein; CABG = coronary artery bypass graft.

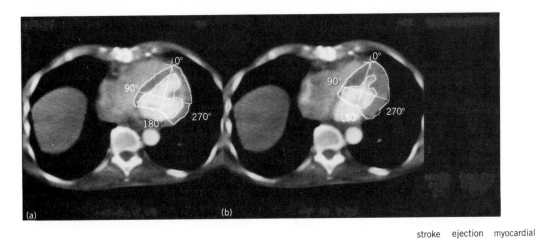

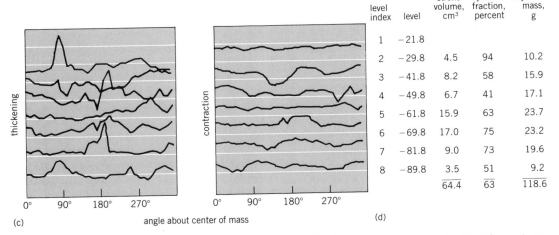

level index	level	stroke volume, cm³	ejection fraction, percent	myocardial mass, g
1	−21.8			
2	−29.8	4.5	94	10.2
3	−41.8	8.2	58	15.9
4	−49.8	6.7	41	17.1
5	−61.8	15.9	63	23.7
6	−69.8	17.0	75	23.2
7	−81.8	9.0	73	19.6
8	−89.8	3.5	51	9.2
		64.4	63	118.6

(c)   angle about center of mass   (d)

Fig. 3.   Computer-assisted analysis of left ventricular function. (a) Cross section through heart at midlevel of left ventricle, taken at 50-ms scan speed and acquired at diastolic phase of the cardiac cycle. Operator-assisted tracing of the inner and outer borders of the left ventricle is shown. (b) Cross section at the same level of the systolic phase. Seven additional levels were recorded simultaneously and analyzed. (c) Computer-calculated wall thickening and radial wall motion (contraction) as a function of angle about the center of mass for each of seven levels. (d) Table of global parameters for these levels, and their summation.

trast bolus through structures of interest. Among applications for this technique are measurement of cardiac output, patency and quality of coronary arteries and bypass grafts, and tissue flow in the myocardium and other organs.

The capability of fast computerized tomography to improve detection and quantitation of a wide variety of cardiac diseases represents a significant advance. Fast computerized tomography is also being applied to a variety of noncardiac diagnostic problems where dynamic studies or blood flow measurement are needed. As electron beam scanning technology is perfected, it appears likely that fast, multislice scanners will gradually displace conventional computerized tomographic scanners and will widen the fraction of x-ray examinations that can be performed by computerized tomography.

For background information *see* COMPUTERIZED TOMOGRAPHY; PARTICLE ACCELERATOR in the McGraw-Hill Encyclopedia of Science and Technology.

[DOUGLAS P. BOYD]

Bibliography: D. W. Farmer et al., In vivo assessment of left ventricular wall and chamber dynamics during transient myocardial ischemia using cine-computed tomography, *Amer. J. Cardiol.*, 55:560–569, 1985; G. T. Herman, *Image Reconstruction from Projections: The Fundamentals of Computerized Tomography*, 1980; C. Higgins (ed.), *CTT of the Heart: Experimental Evaluation and Clinical Application*, 1983; G. N. Hounsfield, Computed medical imaging, *Science*, 210:22–28, 1980; M. J. Lipton, C. B. Higgins, and D. P. Boyd, Computed tomography of the heart: Evaluation of anatomy, *J. Amer. Coll. Cardiol.*, 55:555–695, 1985.

Cosmology

Vast spaces exist between the galaxies of the observable universe. It is possible that most of the matter in the universe is thinly distributed in intergalactic space and has escaped detection due to its very low density. Galaxies are believed to have formed a few billion years after the big bang, but it is very likely that not all the primordial matter went into forming galaxies, and that a very large amount of this matter has been scattered through the observable universe. In addition, the motions of galaxies in clusters, deduced from Doppler velocity measurements of the spectral lines they produce, indicate that, according to accepted laws of physics, there must be far more mass in the clusters of galaxies than the total sum of the masses of the individual galaxies observed. Hence, astronomers are trying to find and explain this "missing mass."

Evidence for missing mass. In 1933 F. Zwicky applied the viral theorem (a derivation of the mass of a stable system from its observed dynamical properties) to the Coma cluster of galaxies and provided the first evidence for the missing mass. In order to describe the amount of missing mass, astronomers use the mass-to-luminosity ratio M/L of objects such as stars. This ratio indicates how much more mass is measured dynamically than is found from visible starlight alone. The larger is M/L, the more nonluminous matter exists. By definition $M = 1$ and $L = 1$ for the Sun. As larger structures are examined, M/L is found to have larger values. Study of the motions of stars in the solar vicinity yields a value of M/L of approximately 3. One part of the mass contributing to this ratio is in visible stars, one part is in detectable interstellar material, and one part remains unknown as missing mass, even in the vicinity of the Sun. Toward the outer edges of the Galaxy, the discrepancy becomes larger and M/L increases to values as large as 20. Such values are derived by observing that the rotation curves of the Galaxy, and other galaxies, remain flat out to very large distances.

The relative motions of galaxies in small groups and in large galaxy clusters also demonstrate the existence of nonluminous matter; application of the virial theorem to such systems yields values of M/L ranging from 50 to 300.

Relation to critical density. According to the big bang cosmology, the expansion of the universe will eventually be reversed if the gravitational attraction among its parts is stronger than the force of the initial explosion. If the value of M/L for the whole universe is about 1000 to 2000, the universe will eventually stop expanding and will collapse to its initial state (a "closed" universe). If M/L is less than this critical value, then the universe will expand forever (an "open" universe).

Cosmologists define the quantity Ω as the ratio of actual mean density of the universe to the critical density, where the critical density is the one just sufficient to close the universe. Both the observed mass and the assumed mass derived from virial considerations suggest that $\Omega = 0.1$ to 0.2, and that only about 10% of the mass contributing to this value (corresponding to $\Omega = 0.01$ to 0.02) is baryonic.

Recent ideas suggesting modifications of the big bang model, termed inflationary and quantum cosmology, provide reasons for favoring the value $\Omega = 1$. This value would require 10 times more unseen mass than is inferred from virial-type considerations. At present, there is no dynamical evidence that $\Omega = 1$; that is, no systems are known with M/L around 1000 to 2000, which would raise the value of Ω to 1.

Searches for missing mass. Early searches for the missing mass attempted to find large amounts of intergalactic dust or gas, but neither component seems to be present in sufficiently large amounts to solve the missing mass problem. Observations throughout the electromagnetic spectrum have revealed small amounts of unsuspected matter, observable at optical or radio wavelengths, such as faint streams, jets, and tails extending into intergalactic space from peculiar or colliding galaxies; neutral hydrogen interstellar gas in large galaxy halos, and galaxy extensions; and even a suspected neutral hydrogen intergalactic cloud found in the Leo group of galaxies. X-ray telescopes in space have detected extended x-ray emission from the intracluster space in galaxy clusters. This high-energy emission is due to a very low-density gas at a temperature of tens of millions of degrees. It is assumed that this hot intracluster gas was ejected from the interstellar media of the galaxy members of the clusters by stellar explosions or by the gravitational interaction (collisions and tidal effects) between galaxies. All these observations of additional matter in the universe add up to only 10 to 12% of the mass derived from application of the virial theorem to galaxy clusters. The rest of the mass, if it exists, remains unknown.

There has been considerable speculation on the possible kinds of unseen mass in the universe. Some suggestions call for an extremely large number of late type-M dwarf stars (very faint and approximately 0.1 solar mass each) populating the halos of galaxies, or even smaller planet size objects; very massive black holes (approximately 10^6 solar masses each) populating the universe have also been suggested. Other suggestions postulate nonbaryonic unseen mass. If the neutrino, an elementary particle, has mass, then neutrinos may constitute the entire missing mass of the universe, since they were presumably produced in great quantities in the big bang. Other highly speculative suggestions invoke hypothetical particles called axions and photinos.

For background information *see* BIG BANG THEORY; BLACK HOLE; COSMOLOGY; GALAXY; GALAXY (EXTERNAL); INTERSTELLAR MATTER; NEUTRINO; STAR in McGraw-Hill Encyclopedia of Science and Technology.   [YERVANT TERZIAN]

Bibliography: J. N. Bahcall, Self-consistent determinations of the total amount of matter near the Sun, *Astrophys. J.*, 276:169–181, 1984; W. Forman and C. Jones, X-ray imaging observations of clusters of galaxies, *Annu. Rev. Astron. Astrophys.*, 20:547–85, 1982; M. P. Haynes, R. Giovanelli, and M. S. Roberts, A detailed examination of the neutral hydrogen distribution in the Leo triplet, *Astrophys. J.*, 229:83–90, 1979; M. J. Rees, Is the universe flat?, *J. Astrophys. Astron.*, 5:331–348, 1984; V. C. Rubin, Dark matter in spiral galaxies, *Sci. Amer.*, 248(6):96–108, 1983; S. E. Schneider et al., Discovery of a large intergalactic HI cloud in the M96 Group, *Astrophys. J.*, 273:L1–L5, 1983; S. E. Schneider and Y. Terzian, Between the galaxies, *Amer. Sci.*, 72:574–581, 1984.

Developmental biology

Throughout life the rates at which various proteins are synthesized are controlled to meet the changing needs of cells. The mechanisms by which translation (protein synthesis) is regulated are as varied and complex as the protein synthesis mechanism itself. In all cells, protein synthesis requires the transcription of nuclear deoxyribonucleic acid (DNA) into messenger ribonucleic acids (mRNAs), which are then transported from the nucleus to the cytoplasm where they direct the synthesis of specific proteins. In the cytoplasm, ribosomes, which are made from RNA and proteins, read and translate the mRNAs into proteins. Additional protein factors are essential: initiation factors stimulate the binding of mRNAs to ribosomes, and elongation factors stimulate the movement of ribosomes along the messages. There are many possible points at which cells might regulate the production of proteins. Modification or regulation of any step in the process of synthesis in the nucleus or in the cytoplasm could, in principle, be employed to regulate protein synthesis.

During oogenesis in many animals, the developing eggs, or oocytes, grow enormously in size and acquire all the materials needed to sustain the organism through the embryonic stages. These materials include ribosomes and mRNAs needed to produce new proteins after the egg is fertilized. At fertilization the male gamete (sperm) donates its half of the genetic material and activates the dormant cytoplasm of the egg to produce both metabolic energy and the proteins required to stimulate cell replication and cleavage.

During cleavage, some of the proteins and mRNAs become segregated in groups of cells and appear to function as determinants that direct the specialization of these cells during embryogenesis. One example of specialized cytoplasm which contains a localized determinant is the germ plasm found in the oocytes of some insects and amphibians. This germ plasm contains granules containing RNA and protein. Cells which came to contain this germ plasm during the cleavage stage eventually form gametes. Thus, as in the case of the nucleus, materials such as ribosomes and germ plasm are passed along to the next generation in the cytoplasm, giving a continuity from one generation to the next.

Protein synthesis rates. The rates of protein synthesis in the cytoplasm vary with each phase of an organism's life. During oogenesis, ribosomal RNA and ribosomal proteins are synthesized actively as the oocyte produces and stores the translational machinery needed after fertilization. Messenger RNAs, which will direct the synthesis of the specific proteins needed during the cleavage stage, are made and stored, as are many other proteins, including tubulin, needed for cell division, and necessary chromosomal proteins such as histones. The storage of these RNAs and proteins facilitates the rapid cell division during the cleavage stage of the early embryo.

At the end of oogenesis, most eggs enter a dormant stage until the moment of fertilization. In many animal species, factors in the cytoplasm of the mature oocytes depress energy metabolism during this stage, and prevent the stored mRNAs and ribosomes from synthesizing proteins.

After fertilization, ionic changes in the cytoplasm trigger the awakening of these stored components. Energy metabolism and protein synthesis become activated and, in minutes, are active at from two to fifty times the rates seen in the unfertilized egg. Cleavage is very rapid until, depending on the species, several hundred to several thousand cells are produced. At the end of the cleavage stage, the rate of cell division slows again, and specific genes are activated to transcribe new mRNAs that will direct the differentiation of the cells as tissues and organs are formed.

Protein regulation in oocyte. How the rate of protein synthesis is regulated in the developing oocyte, in the mature egg, and in the fertilized egg during the cleavage stages comprises important questions for developmental biologists. In the oocyte, protein synthesis is limited by the amount of mRNAs that can be transcribed on its DNA. Many animal species circumvent this limitation by selectively replicating some genes in order to produce during oogenesis large amounts of the RNAs which are coded for by these replicated genes. Genes for ribosomal proteins and ribosomal RNA are in this group. Some mRNAs are regulated in amphibians in the early phases of oogenesis by a protein that masks mRNA and thereby prevents its translation. This masking factor is gradually reduced in amount and disappears at the end of oogenesis, enabling mRNA to be translated. The reason for this oogenic repression of mRNA translation is not understood. Perhaps it preserves messages until they have reached the proper location in the cytoplasm or until sufficient ribosomes are available. Another type of mRNA masking that may occur in oocytes and eggs is modification of the primary and secondary structure of the mRNA. About half of the mRNAs of echinoderm and amphibian eggs contain repetitive RNA sequences that bond to each other to form loops. Messenger RNAs in such states are not translatable. It is not known whether looped mRNAs serve as a storage form with regulatory function or are simply on a degradative pathway.

Increase in synthesis rate. At the time of fertilization of many eggs, including those of mice, frogs (*Xenopus*), clams (*Spisula*), and starfish, the rates of protein synthesis increase 2- to 4-fold. In sea urchins, the rates of protein synthesis increase 20- to 50-fold at fertilization. Many researchers seek to determine the regulatory mechanism for these increases in synthesis. No new gene products are required. If the nucleus of an egg is removed and the egg is artificially activated by chemicals or by pricking, the rates of protein synthesis increase as rapidly as with the nucleus intact. Thus, the change occurs entirely within the cytoplasm by ionic or molecular cues.

An important ionic cue is a transient increase in calcium ion concentration, either by the release of Ca^{2+} from intracellular stores or by the uptake of Ca^{2+} from sea water (in marine species). In most animals this increase in calcium ions occurs during oocyte maturation. This change triggers the begin-

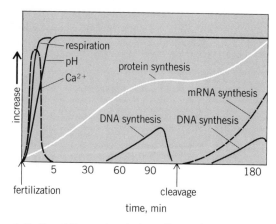

Activation of the zygote cytoplasm. The earliest events after fertilization of sea urchin eggs include a temporal increase in calcium ion which triggers an increase in respiration, and an increase in the pH of the egg cytoplasm. Protein synthesis increases dramatically when the pH becomes alkaline (about pH 7.4).

ning of a number of metabolic processes, including an increase in the rate of protein synthesis.

Translational control. Translational control has been studied extensively in the amphibian *Xenopus*, in the surf clam *Spisula*, and in various species of sea urchins. In *Xenopus*, protein synthesis increases several-fold immediately after hormones signal the maturation of the oocyte in preparation for fertilization. Although it is not known what component of the protein synthetic machinery limits the rate of synthesis before this signal, it has been shown that mRNA is not the rate-limiting factor. When mRNA is injected into oocytes, the rate of protein synthesis does not increase.

In *Spisula*, the rates of protein synthesis increase about 4-fold, and different proteins are synthesized when the oocytes undergo the final maturation events. These changes are correlated with changes in the structure of the mRNAs being translated. Before maturation the mRNAs being translated contain polyadenylated [poly(A)] tails. After maturation these mRNAs loose their poly(A) tails and are no longer translated, while other mRNAs become polyadenylated and are translated. How the poly(A) tails signal the translation of the mRNAs is not known.

Regulation by pH level. In sea urchin eggs, a large, transient increase in calcium ions is found immediately after fertilization; simultaneously, alkalinization of the cytoplasm occurs. It is the increase of the cytoplasmic pH from about 7.0 to 7.4 that initiates the joining of ribosomes and stored mRNAs, and thus the increase in the rate of protein synthesis (see illus.). How the pH change activates the protein synthesis machinery is currently under investigation. Any molecule that is part of the protein synthesis machinery could be blocked in the unfertilized eggs, or several different molecules could be partially inhibited and together contribute a multiple effect to decrease the rates of protein synthesis in the unfertilized eggs. So far, two types of

molecules have been implicated as regulatory molecules. Sea urchin ribosomes appear to contain an inhibitory protein that prevents the translation of mRNAs by ribosomes in test tubes. In addition, certain types of initiation factors may be in short supply in the active regions of the cytoplasm, and thus limit translation. Messenger RNAs, although stored, are not rate-limiting in the unfertilized eggs, or after fertilization when the rates of protein synthesis are rapidly increasing. Analogous to the experiment in amphibians, mRNAs injected into unfertilized and fertilized sea urchin eggs are translated, but do not increase the rate of protein synthesis. Thus, some component other than mRNA limits protein synthesis.

In mammalian eggs the rate of protein synthesis increases less rapidly after fertilization than in the eggs described above. There is less need for these embryos to develop rapidly or to store massive quantities of materials, as the embryos are protected from the environment and provided with nutrients externally. Less is known, however, about the regulation of protein synthesis in mammalian eggs and embryos because of the small numbers of eggs produced compared to those produced by other types of animals. The studies described suggest that various species have adapted the protein synthesis machinery in different ways to meet their particular pattern of embryo development.

For background information *see* DEVELOPMENTAL BIOLOGY; EMBRYOLOGY; GENE ACTION in the McGraw-Hill Encyclopedia of Science and Technology.

[MERRILL B. HILLE]

Bibliography: M. B. Hille et al., Translational control in echinoid eggs and early embryos, in R. Sawyer and R. Showman (eds.), *Belle W. Baruch Library in Marine Science*, vol. 12, 1985; R. A. Laskey et al., Protein synthesis in oocytes of *Xenopus laevis* is not regulated by the supply of messenger RNA, *Cell*, 11:345–351, 1977; J. D. Richter and L. D. Smith, Developmentally regulated RNA binding proteins during oogenesis in *Xenopus laevis*, *J. Biol. Chem.*, 258:4864–4869, 1984; E. T. Rosenthal, T. Hunt, and J. V. Rudderman, Selective translation of mRNA controls the pattern of protein synthesis during early development of the surf clam, *Spisula solidissima*, *Cell*, 20:487–494, 1980.

Diagenesis

The diagenesis of clastic sediments involves a combination of physical and chemical processes operating on a complex detrital mixture of aluminosilicate, carbonate, and silicate minerals as well as other inorganic and organic components. Temperatures encountered during diagenesis range from the Earth's surface temperature up to the onset of metamorphism (about 250°C or 480°F), and depths range from the Earth's surface to less than 30,000 ft (about 10,000 m). Present-day porosity and permeability of a clastic rock is a consequence of the diagenetic modification of the original (primary) porosity. As depth of burial increases, the general result of these processes is to decrease porosity and permeability via compaction and the formation of intergranular cements (commonly quartz and calcite). However, other processes operative during diagenesis can act to reverse this trend and actually increase porosity and permeability in the subsurface (termed porosity enhancement). The dissolution of intergranular cements as well as framework grains is the major process resulting in porosity enhancement. The development of enhanced porosity by mineral dissolution requires a mechanism for the mass transfer of the dissolved materials. Without such a mechanism, pores would become filled with reaction products such as clay, zeolite, or carbonate minerals, and quartz overgrowths. Predicting the occurrence of these porosity-enhancing processes in time and space is of particular importance to the petroleum industry, as the pores in clastic rocks commonly form the reservoir for liquid hydrocarbons.

Studies of the maturation reactions of organic matter indicate that water-soluble organic compounds are released at the molecular level from a substance known as kerogen prior to the generation of liquid hydrocarbons. The presence of these water-soluble organic compounds increases the solubility of aluminosilicate and carbonate minerals significantly by forming organic complexes with the cations released from the minerals. Organic compounds are also effective pH buffers, and affect the stability of carbonate phases in the subsurface. An understanding of the interaction of organic and inorganic constituents during diagenesis is essential to the prediction of the occurrence of enhanced porosity in the subsurface.

Kerogen maturation. Because heat is constantly flowing out of the Earth, the temperature of a sediment increases as the sediment is buried to greater depths. The actual change in temperature with depth depends on heat flow at the particular location and the thermal conductivity of the various rock layers. The organic matter undergoes chemical changes as the temperature (depth of burial) increases. The original organic compounds combine to form a high-molecular-weight material called kerogen. No unique kerogen molecular structure exists because kerogens are not all formed from the same original materials and because the kerogens themselves change as they mature. Kerogen undergoes numerous changes as temperature continues to increase. Water-soluble organic compounds, liquid hydrocarbons, and natural gas are all products of kerogen maturation. Kerogen composition is conveniently described by the relationship between two ratios, the atomic hydrogen:carbon ratio (H/C) and the atomic oxygen:carbon ratio (O/C). These ratios are usually displayed on a graph known as a van Krevelen diagram (Fig. 1a). The position of a given kerogen sample on the diagram will depend on both the original type of organic matter present in the sediment and the extent to which the organic matter has matured (a function of both time and temperature). Maturation pathways of the three general types of organic matter are shown in Fig. 1a. The type I

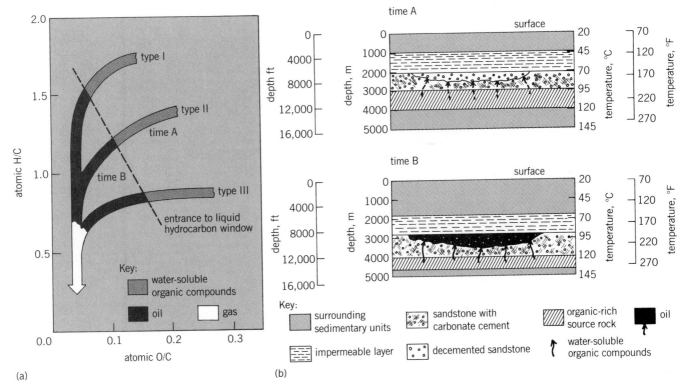

(a)   (b)

Figure 1.   Kerogen maturation. (a) Van Krevelen diagram showing maturation pathways. Water-soluble organic compounds are produced prior to liquid hydrocarbons regardless of kerogen type. (b) Schematic burial relationships corresponding to the maturation levels indicated by the van Krevelen diagram.

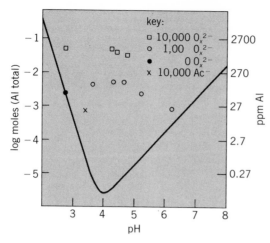

Fig. 2.   Results of a 2-week dissolution experiment run at 100°C (212°F). Aluminum values measured in postrun solutions are plotted versus pH. The mineral separate used was andesine. The solubility curve for gibbsite (aluminum hydroxide) is plotted for reference; these aluminum concentrations would be expected in the absence of any organic aluminum complexation. O_x^{2-} is the concentration of oxalic acid in solution in mg/liter (ppm). Ac^- is the concentration of acetic acid in solution in mg/liter (ppm). These measured aluminum values are not equilibrium values; in nearly all experimental runs an aluminosilicate gel phase formed during cooling. (*After R. C. Surdam, S. W. Boese, and L. J. Crossey, Chemistry of Secondary Porosity, Amer. Ass. Petrol. Geol. Mem. 37, pp. 127–151, 1984*)

classification represents kerogen originating from an algal source, type II indicates a normal marine source, and type III indicates a terrestrial (or woody) source material. Also shown schematically in Fig. 1a are the points where the different kerogens enter the liquid hydrocarbon window. These points represent the onset of liquid hydrocarbon generation. The positions of these points must be viewed as very approximate, as the actual temperature corresponding to a given stage of maturation will vary considerably even within a given kerogen type. Water-soluble organic compounds are produced prior to the liquid hydrocarbons regardless of kerogen type.

Water-soluble organic compounds. A number of studies have been done on water-soluble organic compounds which consider their characterization and distributions in the diagenetic environment.

Characterization. Water-soluble organic compounds have been measured in oil field waters in concentrations up to 10,000 mg/liter (ppm). Among the diverse types of organic compounds observed are short-chained carboxylic acids (both monofunctional and difunctional forms) and phenols. Most abundant is acetic acid, a monofunctional carboxylic acid. Commonly, acetic acid dominates the alkalinity of oil field waters produced in the temperature range 80–120°C (180–250°F). This means that the acetate system will buffer the oil field waters rather than the carbonate system. Propionic acid and formic acid (both monofunctional carboxylic acids) are also common, but at somewhat lower concentration levels. Oxalic acid (a difunctional carboxylic acid) as well as other difunctional forms with carbon numbers ranging up to 10 are present in concentrations generally less than 100 mg/liter (ppm). Phenols (a benzene ring with attached hydroxyl groups) are also observed at the sub-hundred-ppm level.

Distribution. The highest concentrations of the water-soluble organic compounds are observed in the temperature range of 80–120°C (180–250°F). At temperatures less than 80°C (180°F), it is hypothesized that the low concentrations are due to bacterial consumption of the simple organic compounds in the subsurface. At temperatures greater than 120°C (250°F), thermal decarboxylation begins to occur and CO_2 and water or methane are the products. At temperature exceeding 180°C (360°F), carboxylic acids are virtually absent.

Enhanced mineral solubilities. Two different groups of minerals were studied, the aluminosilicate minerals and the carbonate minerals.

Aluminosilicate minerals. A series of dissolution experiments were performed using the types of organic compounds observed in oil field waters, a variety of minerals, and a range of pH values. Typical results are shown in Fig. 2. Aluminosilicates include feldspars, clays, and zeolites as well as lithic fragments. As aluminum is the least soluble major component in the aluminosilicate minerals, an increase in aluminum solubility will enhance the solubility of the mineral. Oxalic acid (a difunctional carboxylic acid) increases the solubility of aluminum significantly. A solubility increase of four orders of magnitude is observed at a pH value of 4. Experiments with various phenols indicate that they are as effective as oxalic acid in mobilizing aluminum. The aluminum is mobilized by the formation of a water-soluble organic complex. The mobilization of aluminum is critical for the development of enhanced porosity: in the absence of a complexing agent, a reaction product (most likely an aluminum-

Fig. 3.   Scanning electron photomicrograph showing framework grain dissolution of a dissolved feldspar grain with an authigenic albite rim from a sandstone at a present depth of 400 m (1312 ft) from the Bighorn Basin, Montana. Maximum burial is estimated at 2000 m (6560 ft). Note the absence of reaction products in the vicinity of the dissolved grain. (*E. Sven Hagen, University of Wyoming*)

bearing clay mineral such as kaolinite) would form in the immediate vicinity of the grain, and no effective porosity increase would result. Thus the water-soluble organic compounds produced from the kerogen molecule during diagenesis enable waters expulsed from the source rock to dissolve aluminosilicate framework grains as the waters migrate through adjacent sandstones.

Carbonate minerals. The organic acid anions also have an affinity for the calcium, magnesium, and iron found in carbonate minerals, and should destabilize these minerals as well as the aluminosilicate minerals. The system is complicated by the interaction of organic buffering, the carbonate minerals, and dissolved CO_2 (gas). In the absence of another buffer, increasing the partial pressure of CO_2 (g) in a fluid will lower the pH (via formation of carbonic acid), and carbonate minerals will dissolve. If the fluids are buffered by acetate, an increase in the partial pressure of CO_2 (g) will actually stabilize the carbonate minerals. The increase in the partial pressure of CO_2 (g) is also closely related to the maturation of the kerogen and the onset of the thermal decarboxylation of the organic solvents. The complex interaction of water-soluble organic buffers (such as acetate) and CO_2 (g) produced from the organic matter by thermal decarboxylation will determine the stability of carbonate minerals in the subsurface.

Prediction of porosity enhancement. The processes of decementation and framework grain dissolution account for the overwhelming majority of enhanced porosity in clastic rocks. Data from oil field waters and dissolution experiments indicate the water-soluble organic compounds can affect the stability of both aluminosilicate framework grains and carbonate minerals. An understanding of how and when these organic compounds are produced from kerogen during diagenesis allows the type and quantity of the available organic compounds to be estimated. This information can be used to estimate the amount of porosity enhancement to be expected from the dissolution of aluminosilicate framework grains (Fig. 3) or carbonate cements. Porosity enhancement can be viewed as part of an overall process involving the redistribution of porosity in the subsurface on a basin-wide scale. Although porosity enhancement occurs in regions proximal to source rocks, reaction products may precipitate as chemical conditions or temperatures may change along the fluid migration pathway. It is only through a process-oriented approach that porosity enhancement can be understood, and consequently predicted. The presence of porosity in clastic rocks is a prerequisite for a hydrocarbon reservoir, and thus the economic implications of this predictive capability are significant.

For background information *see* DIAGENESIS; KEROGEN; OIL FIELD WATERS; PETROLEUM GEOLOGY; SANDSTONE in the McGraw-Hill Encyclopedia of Science and Technology.

[LAURA J. CROSSEY]

Bibliography: B. Durand (ed.), *Kerogen: Insoluble Organic Matter from Sedimentary Rocks*, 1980; D. L. Gautier, Y. K. Kharaka, and R. C. Surdam, *Relationship of Organic Matter and Mineral Diagenesis*, Soc. Econ. Paleontol. Mineralog. Short Course 17, 1985; D. A. McDonald and R. C. Surdam (eds.), *Clastic Diagenesis*, Amer. Ass. Petrol. Geol. Mem. 37, 1984; P. A. Scholle and P. R. Schlugar (eds.), *Aspects of Diagenesis*, Soc. Econ. Paleontol. Mineralog. Spec. Publ. 26, 1979; R. C. Surdam, S. W. Boese, and L. J. Crossey, *Chemistry of Secondary Porosity*, Amer. Ass. Petrol. Geol. Mem. 37, pp. 127–151, 1984.

Ecological interactions

Although brachiopods are only a minor component of modern benthic faunas, they were the dominant shelled suspension feeders during the Paleozoic. Their ecological interactions are of interest both in illuminating the paleoecology of this historically important group and for the insight they provide into the potential evolutionary diversity possible in bivalved suspension feeders (a body plan shared by brachiopods, bivalved mollusks, and some crustaceans). The phylum Brachiopoda is divided into two classes: the Articulata, in which interlocking teeth and sockets permanently join the two valves of the shell, and the Inarticulata, which lack hinge structures. Numerous other features distinguish the two classes, including dramatic differences in embryological development, shell composition, and the presence or absence of a complete gut. Some workers argue that the two classes each deserve status as a phylum. Regardless of the merits of this proposal, it is important to distinguish ecology of the two classes.

The vast majority of both inarticulate and articulate brachiopods permanently attach to the substrate at larval metamorphosis, either by cementation of one valve of the shell, or through a fleshy stalk, the pedicle. As permanently attached epifauna, they compete for space on the substrate and this is an important component of their ecology. Studies in both New Zealand and the west coast of North America have shown that articulate brachiopods are poor spatial competitors, particularly when competing with sponges and mussels. In contrast, the inarticulate brachiopod *Discinisca strigata* from the Pacific coast of Panama has proven to be dominant in spatial competition, routinely defending its own space and coopting space occupied by neighboring polychaete worms (spirorbids and serpulids), bryozoans, and sponges.

Discinisca strigata characteristics. *Discinisca strigata* (Fig. 1) is a small (approximately 0.6 in. or 1.5 cm long) animal with a limpet-shaped dorsal valve and a flat ventral valve, the latter pierced by a hole through which the pedicle extends. The inorganic component of both valves is calcium phosphate. The shell is fringed by setae, chitinous outgrowths of the mantle, which are differentiated into long anterior setae ornamented with threadlike outgrowths (Fig. 1) and shorter lateral and posterior setae ornamented with short thornlike processes. The

Fig. 1. A group of *Discinisca strigata* on their natural rock substrate. Note the long, prominent anterior setae on each animal; these setae function as an incurrent siphon for the brachiopods' feeding currents.

threadlike processes on the anterior setae are entangled, linking the anterior setae to form a functional siphon; water is pumped into the mantle cavity anteriorly and exits laterally (the inverse of the flow pattern seen in articulates). When *D. strigata* closes its valves, it exhibits a stereotyped behavior pattern consisting of rotating the valves both individually (the valves, since they lack hinge structures, can be disarticulated) and together around the pedicle, sweeping the setae through an arc around the shell. The competitive dominance of *D. strigata* involves interactions of features of the animal's morphology, such as the setal structure and the shell composition, and aspects of its behavior such as the rotation of the shell on closure and the direction of water flow through the shell gapes.

Competitive mechanisms. When *D. strigata* is observed on its natural rock substrate, each individual is surrounded by a distinct zone or halo extending around the shell for a distance equal to the length of the lateral or posterior setae, and from which sponges and sometimes bryozoans have been excluded. When adjacent sponges are thick and fleshy, the sponge may arch over (but never touch) the shell of the brachiopod. Neighboring bryozoan colonies often show a pronounced thickening of the colony at the tips of the lateral and anterior setae caused by the superposition of bryozoan zooids; this frontal budding pattern in bryozoans is known to be induced when growth of the colony is blocked by some external agent. The cleared zones surrounding the brachiopods are produced by the sweeping action of the lateral setae when the animals rotate on their pedicles; the thornlike protuberances on the lateral setae abrade surrounding sponges and the weakly calcified buds produced on the edges of bryozoan colonies, and induce frontal budding in more mature zooids. This mechanism is effective only near the substrate, but sponges are prevented from arching over the setae and overgrowing the brachio-

pod's shell because the shell is surrounded by a pool of particle-depleted water that exits from the brachiopod's lateral gapes and within which the sponge's growth is inhibited.

If the brachiopods are removed from the substrate, serpulid and spirorbid tubes and bryozoan colonies previously hidden under each shell can be seen to be highly worn and eroded (Fig. 2); the bryozoan zooids are often worn down to their base, and the worm tubes are worn to the point where the entire length of the internal cavity of the tube is exposed. Spirorbid tubes that lie at the edge of the brachiopod's shell, and bryozoan colonies that extend under the shell, are often bisected, with those portions beneath the brachiopod worn nearly to the substrate while portions extending beyond the shell are untouched. Underlying calcareous epifauna are abraded by the brachiopod's ventral valve, an effect possible only because the calcium phosphate of the brachiopod's shell is much harder than the calcium carbonate that makes up the skelton of other epifaunal organisms. This physical abrasion of calcareous competitors as a mechanism of spatial competition is unique to *D. strigata*.

Juvenile *D. strigata* are sometimes attached to zooids in the middle of living bryozoan colonies, implying that these brachiopods can metamorphose directly on a living bryozoan. The presence of a functional siphon (that is, the anterior setae) is vital to this ability, allowing the brachiopods to feed from water above the level of the surrounding bryozoan zooids' lophophores. The pedicles of some adult *D. strigata* are attached to and completely underlain by dead, worn bryozoan colonies; by abrading the zooids' skeletons, *D. strigata* which metamorphose on bryozoan colonies can coopt the space previously occupied by some or all of the bryozoan colony.

Other epifaunal organisms. Even though *D. strigata* uniformly wins competitive interactions with other epifauna, this brachiopod is not the commonest organism on substrates where it is found. This

Fig. 2. Two spirorbids, one of which (right) lay under the valves of a *Discinisca strigata*. Due to the slope of the substrate, the brachiopod lay at an angle, producing the beveled abrasion of the spirorbid.

discrepancy between dominance in competitive interactions and its lack of dominance in the fauna is most likely due to this brachiopod's low recruitment rate and determinate growth.

The mechanisms that allow *D. strigata* to win competitive interactions for space on the substrate are, for the most part, not available to articulate brachiopods. Articulate brachiopods uniformly exhibit an anterior incurrent–lateral excurrent flow pattern of their feeding current through the shell gapes, and so cannot protect their shell and attachment site with a pool of particle-depleted water. In addition, the setae of articulate brachiopods are simple straight shafts lacking the thornlike projections of the lateral setae of *D. strigata*, and so are less effective in abrading surrounding epifauna. Some articulates can prune surrounding sponges by rotating their shell on the pedicle, but they cannot abrade calcareous epifauna since the shell of articulates is also composed of calcium carbonate—their shell would be abraded to the same extent as the epifauna.

The competitive abilities of *D. strigata* far exceed those of articulate brachiopods and are much greater than would be expected for a noncolonial animal with determinate growth; its ability to physically abrade underlying calcareous animals is unique. These abilities have undoubtedly been a prime contributor to the genus's success since it arose in the Jurassic, and present an object lesson to ecologists on the importance of investigating the mechanisms which underlie competitive interactions.

For background information *see* ARTICULATA (BRACHIOPODA); BRACHIOPODA; ECOLOGICAL INTERACTIONS; INARTICULATA; MARINE ECOSYSTEM in the McGraw-Hill Encyclopedia of Science and Technology.

[MICHAEL LABARBERA]

Bibliography: P. J. Doherty, A demographic study of a subtidal population of the New Zealand articulate brachiopod *Terebratella inconspicua*, *Mar. Biol.*, 52:331–342, 1979; M. LaBarbera, Mechanisms of spatial competition of *Discinisca strigata* (Inarticulata: Brachiopoda) in the intertidal of Panama, *Biol. Bull.*, 168:91–105, 1985; M. J. S. Tevesz and P. L. McCall (eds.), *Biotic Interactions in Recent and Fossil Benthic Communities*, 1983; C. W. Thayer, Brachiopods versus mussels: Competition, predation, and palatability, *Science*, 228:1527–1528, 1985.

Electrical power engineering

For many applications, ranging from thermonuclear fusion research and high-energy particle accelerators to lasers and electromagnetic launchers, power must be delivered in short, intense bursts. To deliver this pulsed power, energy must be collected at lower power, stored, and then released almost instantaneously.

Pulsed-power systems can now deliver gigajoules of energy, megamperes of current, or terawatts of power. Pulse widths range from microseconds at the

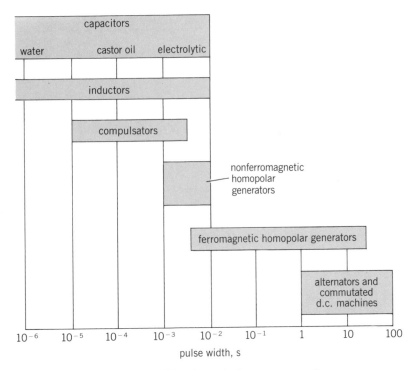

Fig. 1. Pulse widths compatible with various pulsed-power components.

highest power levels to several seconds at the highest energy levels. In all pulsed-power mechanisms, energy is first stored, then compressed to the appropriate pulse duration, and finally switched, or released. Electric energy can be stored directly in electric fields (capacitors) or magnetic fields (inductors). Energy stored in inertial systems based on flywheels can be converted to electric pulses (see table). Figure 1 summarizes the pulse widths compatible with the various pulsed power components.

Capacitors. Capacitors are one of the most highly developed energy-storage technologies. Energy-storage capacitors typically use castor oil and kraft paper for the dielectric; some modern units use polymer films. Higher-voltage systems operating at megavolt levels use pure water as a dielectric, taking advantage of its high dielectric constant. But water cannot sustain a high electric stress for a long time, since the polarization of the water molecules allows the charge to leak between the conductive plates; water capacitors must be charged more rapidly than other types. Work on low-temperature water-glycol mixtures has significantly increased the intrinsic time constant of such capacitors.

Capacitors are available at voltages up to 100 kV, and their modular construction allows them to be connected either in series, to achieve megavolt levels, or parallel, to reduce source impedance and deliver currents of up to 10 MA. Their primary limitations are relatively high internal impedance and low energy density (about 0.2 ohm and 0.2 kilojoule/kg).

Inductors. Energy may also be stored in a magnetic field, usually a solenoidal or coaxial configu-

Typical characteristics of pulsed-power systems

Storage technique and device	Storage capacity	Energy density, kJ/kg	Power density, kW/kg	Pulse width	Module voltage	Source impedance, ohms	Short-circuit current, kA	Storage time, s
Electric field								
Capacitor	15 kJ	0.2	8000	0.1–0.5 ms	10,000	0.212	50	1000
Magnetic field								
Inductor								
Room temperature	5 MJ	1.2	324	1–5 ms	3,000	0.002	1500	0.45
Cryogenic	3 MJ	3.1	1000	1–100 ms	5,000	0.005	1000	1.2
Superconducting	500 kJ	2.2	50	1 ms	10,000	3142	3	10^{12}
Inertial								
Flywheel								
DC generator	0.8 MJ	0.32	0.3	1 s	1,800	0.0142	1	100
Homopolar generator	6 MJ	8.5	70	0.1–0.5 s	100	10^{-5}	2000	415
Alternator	185 MJ	1.3	0.7	1 s	6,900	1.12	6	3000
Compulsator	200 kJ	3.8	250	0.1–2 ms	6,000	0.084	71	254

ration. Whereas capacitors can amplify current, being charged at low current and discharging at high current but operating at the same voltage in both cases, inductors can amplify voltage while holding current essentially constant. Thus, capacitors are best suited to uses in which adequate voltage is the prime concern—gas breakdown, for example—and inductors are best for applications in which current is the chief consideration, such as force generation. Inductors can store higher energy densities than capacitors but typically have shorter time constants. Except for superconducting inductors, they typically store energy for less than 1 s, whereas capacitors can store energy for 1000 s or more.

In addition, inductors require a dc opening switch to extract the stored energy, whereas a capacitor requires only a closing switch. Largely because of the lack of adequate opening switches, use of inductive energy-storage devices is not widespread.

Networks. Pulsed-power systems generally consist of networks of capacitors or inductors connected by appropriate switches and designed to compress pulses of electric energy sequentially into successively shorter time intervals. These pulse-compression networks amplify power by accepting input power at a low level and storing the energy sequentially in several stages, with each stage discharging the stored energy at a higher power level (and thus in a shorter time) than that at which it was received. At each capacitor stage, a switch is closed and the stored energy is rapidly discharged into the next load. Similarly, an inductor is discharged when a switch is opened to transfer the pulse of current into the next load. In the capacitive circuit, the switch may be as simple as a spark gap that breaks down when the capacitor reaches a preset peak voltage. In the inductive circuit, a simple switch may be a fuse designed to fail after the inductor reaches peak current. Accurate timing requirements, repetitive operation, or efficiency considerations often require more elaborate switches.

Inertial devices. As requirements for larger pulsed-power energy stores have grown, the high energy density and low cost of inertial energy-storage systems have made them increasingly attractive. In-

ertial devices, which convert mechanical energy into electrical energy, can couple directly to primary power sources. A generator can be directly joined to the shaft of a gas turbine or diesel engine, for example, to provide primary energy conversion as well as energy storage. As larger systems are contemplated and as laboratory applications of pulsed power move into the field, this advantage of direct connection to primary power sources becomes more important.

Homopolar generator. In recent years the homopolar generator, a simple low-impedance dc machine, has been developed specifically for pulsed-power applications (Fig. 2). As the only dc rotating machine without a commutator, it avoids most of the problems associated with commutated dc machines.

In the simplest homopolar generator (Fig. 3), a monolithic conducting disk is rotated in an axial magnetic field, producing voltage between its shaft and outer radius. Sliding-contact technology currently limits the surface speeds of the homopolar rotor to a range of 720–820 ft/s (220–250 m/s).

Fig. 2. A 6.2-MJ compact homopolar generator at the University of Texas at Austin, connected to a 3.1-MJ cryogenically cooled inductor, which further compresses the generator pulse for an output current of 1 MA.

The homopolar machine can be modeled electrically as a capacitor. This means that the machine can be used as a capacitor in various pulse-compression circuits, the differences being that it can be "charged" mechanically rather than electrically and that it is a capacitor of low voltage (tens to hundreds of volts) and extremely high capacitance (thousands of farads).

Since the entire monolithic rotor acts as the armature conductor, internal impedance is extremely low, less than 10 microhms. However, the rotor makes only one pass through the magnetic field and therefore produces relatively low voltage. In pulsed applications, the homopolar generator, because of its low impedance, is capable of operating at high-output current (1–2 MA) into very low-impedance loads. Many topological variants have been explored, including disks, drums, and spools.

Compulsator. The conventional ac machine—the synchronous machine or alternator—is by far the most highly developed of the rotating machines because of its use in central-plant generating stations. It combines the high-voltage capability of the wound armature with greatly simplified current collection. A variation of the synchronous machine, the compensated pulsed alternator or compulsator, invented in 1978, is essentially an alternator with reduced internal impedance, which is more suitable for sub-millisecond pulse generation.

The compulsator differs from the conventional synchronous machine in having an additional stationary winding in series with the rotating armature winding. The function of this additional winding is to compensate the internal inductance of the machine at one point in the cycle, usually at peak voltage, through the principle of flux compression. The inductance thus goes through a minimum when the fixed windings are opposed, and a sudden pulse of current results (Fig. 4).

Both the rotating and stationary windings are of the air-gap type, and they are bonded through the ground-plane insulation to the smooth surface of the laminated steel rotor and stator; they do not fit into slots, as in conventional machines. This construction reduces the leakage inductance of the windings by minimizing the exposure of ferromagnetic mate-

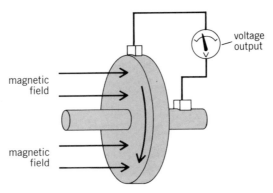

Fig. 4. Variation of voltage, inductance, and current during a compulsator cycle.

rial to the fringing field of the conductor.

Only a few compulsators have been built. It seems clear, however, that they work best in the range 6–15 kV and that typical inductance variations of 25:1 to 200:1 can be achieved. Pulse widths from 100 microseconds to 10 milliseconds appear possible.

The compulsator is easier to drive with most prime movers since it slows only slightly during a pulse, typically 5–20%, instead of stopping as the homopolar generator does. In addition, its self-commutation, because of the cyclical inductance variation, makes it more suitable for burst or continuous operation in the repetitive pulse mode (Fig 4).

Applications. Controlled thermonuclear fusion is a major application of pulsed-power technology. High-power pulsed-power supplies play key roles in high-voltage, current, and magnetic-field research, as well as in powering particle accelerators for high-energy physics research. A prime example is the advanced test accelerator at the Lawrence Livermore National Laboratory, which generates 250-kV, 40-kA bursts with 20-nanosecond rise times at a repetition rate of 1 kHz. High-power pulsed supplies are also essential for x-ray generators used as diagnostic tools in a variety of applications. Another major driving force in pulsed-power technology is the production of electromagnetic pulses to simulate the effects of nuclear weapons. *See* MAGNETIC FIELD; PARTICLE ACCELERATOR.

The use of pulsed-power systems for launching projectiles is emerging. Potential applications of this technology range from launching gram-sized projectiles to tens of kilometers per second, for equation-of-state research and impact-fusion experiments, to materials processing and even ballistic launching of a ton of payload into space from the Earth's surface.

Homopolar generators have been adapted to weld cross sections of several square inches (tens of square centimeters), making it possible to join large metal sections in the field in a fraction of a second without fluxes, filler metal, or contaminants. Post-weld heat-treating of the weld zone can be accomplished in the same pulse. Other materials-processing applications of pulsed-power technology include the heating of metal billets for rolling and forging,

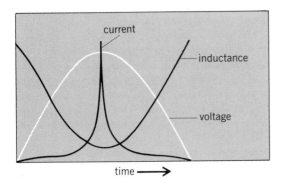

Fig. 3. Components of simple homopolar generator.

electrical sintering of metallic powders, in-place generation of amorphous metal coatings, and the spraying of metal powders.

For background information *see* ALTERNATING-CURRENT GENERATOR; CAPACITOR; ENERGY STORAGE; IMPULSE GENERATOR; INDUCTOR in the McGraw-Hill Encyclopedia of Science and Technology.

[WILLIAM F. WELDON]

Bibliography: T. A. Aanstoos and J. M. Weldon, Homopolar pulse welding of API X-60 high strength line pipe, *Welding J.*, 63(7):23–28, July 1984; J. H. Gully, Assembly and testing of a compact, lightweight homopolar generator power supply, in T. H. Martin and M. F. Rose (eds.), *4th IEEE International Pulsed Power Conference, Albuquerque, June 6–8, 1983*, pp. 411–414, 1983; W. F. Weldon, Pulsed power packs a punch, *IEEE Spectrum*, 22(3):59–66, March 1985; W. F. Weldon et al., Compulsator: A high power compensated pulsed alternator, in V. Nardi, H. Sahlin, and W. H. Bostick (eds.), *2d International Conference on Energy Storage, Compression, and Switching, Venice, December 5–8, 1978*, pp. 925–938, 1983.

Electrical utility industry

The year 1985 witnessed a continued hiatus in orders for new generating equipment, despite substantial growth in both demand and energy sales. Uncertainty about the form of legislation designed to reduce acid rain complicated utilities' decisions about expanding capacity, and the coming into service of long-delayed generating units again boosted reserve margins to extraordinary levels.

The consensus growth rate in demand of 2–2.9% per year is expected to create capacity shortages in some regions, and utilities began to plan on substantial installations of combustion turbines in the early 1990s to address this situation. A new phenomenon, generating capacity built by nonutility consortia, appeared. A private developer has proposed construction of two 150-MW steam turbine generators in Vermont, the total output to be sold to local utilities. In New Mexico, a consortium of utilities, builders, and an indigenous Indian tribe are planning a new station to be fired by coal mined on the Indian reservation. These activities are a response to the forecast need for additional capacity and the utilities' unwillingness to commit the resources to build that capacity without assurances of an adequate return on the investment.

Ownership. The industry is dominated by the investor-owned corporations, which serve 76.5% of the 98.9 million electric customers in the United States. Municipal, state, and district publicly owned entities serve only 13.4% of the customers, while cooperatives serve 10.1%. Federal utilities are basically wholesalers and do not serve retail customers directly.

Investor-owned utilities also own and operate 76.8% of the nation's installed generating capacity. Publicly owned utilities own 10.2% of all installed capacity, and cooperatively owned utilities 3.4%.

Federally owned capacity is roughly comparable to that of other publicly owned utilities, about 9.6%.

The seeming discrepancy between the percentage of customers served by cooperatives and the much smaller percentage of capacity owned by them reflects the fact that most such organizations are primarily distribution companies that buy their power from others at wholesale rates and distribute it to their individual customers.

Cooperatives are the fastest-growing sector of the industry for at least two reasons: (1) much of the residential, commercial, and industrial expansion is occurring in the exurban areas that they tend to serve; and (2) they are able to raise financing on more advantageous terms than can the private utilities. The result of these factors is that cooperatives—and, to a lesser extent, publicly owned utilities—increasingly either have bought heavily into capacity constructed by the larger investor-owned utilities or have built such facilities themselves with the intent of selling the power to capacity-short private utilities. Rarely is a plant now constructed by a private utility that does not have cooperatives or publicly owned utilities as substantive partners in ownership.

Capacity additions. Utilities had a total generating installed capacity at the end of 1984 of 658,017 MW, having added 10,187 MW during that year. Utilities added 15,671 MW during 1985, raising total industry installed capacity to 673,688 MW (see table). Capacity represents the actual aggregate design rating of all generating units connected to utility systems. In actuality, these units may not be capable of achieving their design rating for various reasons. These include such factors as lower-than-normal reservoir levels at hydroelectric installations, boiler tube leaks, or full or partial outage due to maintenance requirements or forced shutdowns. For this reason, the North American Electric Reliability Council prefers to list capability—the ability to perform at peak demand periods—rather than capacity as the true measure of a system to supply load.

The composition of the new capacity additions during 1985 was 218 MW of conventional hydroelectric, 1404 MW of pumped storage hydroelectric, 9235 MW of fossil-fueled steam, 4646 MW of nuclear steam, and 168 MW of combustion turbines. The composition of total plant by type of generation at the end of 1985 is given in the table (see also illus.).

Fossil-fueled capacity. Since 1980, all new fossil-fired units entering service have been fueled by coal as dictated by the Fuels Use Act of 1974. In 1985, one unit entering service in Texas was designed to burn natural gas, and a number of others are planned for future years. This reflects a certainty that the Fuels Use Act's prohibition of gas as a fuel will fall victim to the projected surplus of gas in future years. Utilities plan to install an addition 56,200 MW of fossil-fired capacity between 1984 and 1993.

Geothermal capacity, strictly speaking, is not fos-

United States electric power industry statistics for 1985*

Parameter	Amount	Increase compared with 1984, %
Generating capacity, MW		
Conventional hydro	66,893 (10%)	0.33
Pumped storage hydro	13,696 (2%)	11.4
Fossil-fueled steam	464,742 (69%)	2.0
Nuclear steam	71,720 (10.6%)	6.9
Combustion turbine	51,795 (7.7%)	0.3
Internal combustion (diesel engine)	4,841 (0.7%)	−3.1
TOTAL	673,688	2.3
Noncoincident demand	451,200	0.2
Energy production, GWh	2,472,500	4.7
Energe sales, GWh		
Residential	777,400	3.5
Commercial	578,200	5.8
Industrial	838,700	7.0
Miscellaneous	86,500	7.7
TOTAL	2,279,900	5.7
Revenues, total, $\times 10^6$ dollars	118,700	8.5
Capital expenditures, total, $\times 10^6$ dollars	38,579	9.6
Customers, $\times 10^3$		
Residential	87,431	1.8
TOTAL	98,901	1.8
Residential usage, kWh/customer	8,965	1.6
Residential bill, units/kWh (average)	7.42	0.9

After 36th annual electric utility industry forecast, *Elec. World*, 199(9):51–58, September 1985; 1985 statistical report, *Elec. World*, 199(4):53–76, April 1985; and extrapolations from monthly data of the Edison Electric Institute–Association of Electric Companies.

sil-fired, but is included in that category for convenience. Utilities commissioned into service an additional four units in 1985, totaling 170 MW.

Utilities spent $7.8 billion on construction of fossil-fired units in 1985. Investor-owned utilities spent $5.9 billion, publicly owned utilities spent $1.1 billion, cooperatives spent $0.5 billion, and federal agencies spent $0.2 billion.

Nuclear power. The hiatus in new orders for nuclear plants that began in 1978 remained unbroken in 1985, and there are no plans for additional orders for the foreseeable future. However, units on which construction began as many as 14 years ago continue to enter service. Four new units came on line in 1985, bringing the total to 80 units spread among 55 individual stations. Three of the units added were pressurized-water reactors (PWR); the remaining unit was of the boiling-water reactor (BWR) design. The total number of each type now in service is 51 and 29 units, respectively.

Current utility plans call for another 55,100 MW of nuclear capacity to enter service by 1995, which would raise nuclear power's percentage of total capacity to about 20%.

The damaged Three-Mile Island #2 reactor (near Harrisberg, Pennsylvania) which experienced a partial core meltdown in 1979 remains out of service. Clean-up has proceeded steadily, however, and removal of the damaged core is now expected during 1986. The twin #1 unit, which was shut down after the accident through legal and regulatory action, was cleared for a return to service in 1985. It was successfully brought to full power and re-entered commercial service late in the year.

Combustion turbines. Combustion turbines have historically been used to supply about 10–15% of the capacity required to meet the highest yearly peak demand. When they are used in this way for about 200 h/year or less, their low capital costs of about $325/kW, versus $950/kW for a coal-fired unit, more than offsets their higher fuel consumption.

Utilities plan an additional 2600 MW of combus-

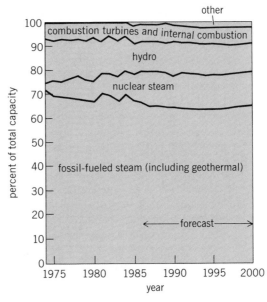

Probable net mix of generating capacity. (*After 36th annual electric utility industry forecast, Elec. World, 199(9):51–58, September 1985*)

tion turbines between now and 1993. Capital expenditures for this type of capacity amounted to $43.7 million in 1985.

Combustion turbines are also used in tandem with steam turbines in highly efficient combined cycle operation. Utilities have 5800 MW of combined-cycle capacity currently operating on their systems.

Hydroelectric installations. Future plans call for an additional 4000 MW of conventional hydroelectric capacity to be constructed over the next 10 years, although appropriate locations for dams are becoming increasingly difficult to find and license for development. Utilities spent $483 million for such units in 1985.

The future of hydroelectric energy may rest with small-scale projects carrying ratings of 1–50 MW. Such units can be prepackaged and installed in locations where modest heads of 15–20 ft (4.5–7.5 m) are available. The special incentives for the development of these small units provided by the Public Utilities Regulatory Policy Act continue to spur the installation of small units by private developers who sell the output to utilities.

Pumped storage. Pumped storage represents one of the few possible methods for storing large amounts of electric energy. Utilities spent $178 million on pumped storage capacity during 1985, of which $176 million was by investor-owned companies.

Demand. After a substantial growth of 6.3% in peak demand in 1984, growth sagged to only 0.2% in the summer of 1985. It is normal for a year of high growth to be followed by one of subnormal growth, but the flat response in the summer of 1985 was primarily attributable to widespread, abnormally cool weather throughout the heavily loaded Midwest. Rates for electricity continue to rise, and the effect of negative elasticity is also a factor. Long-term growth in summer peak demand now has reached a consensus level of 2.0–2.9% per year, compounded anually.

The long-term trend shows a slight decrease over the forecast period. This decline arises naturally from a mix of demographic factors that characterize the United States. Population is growing at a decreasing rate and is aging beyond the years of peak consumption. This effect is compounded by price-induced conservation. Because of the high technological content of the utilities' plant, the preponderance of skilled labor employed, the high cost of capital engendered by unresponsive regulation, and escalating fuel and pollution-control costs, electricity prices should rise slightly more than inflation over the next decade.

The pattern of load growth for utilities must be analyzed in terms of regions, rather than nationally. In 1985, for example, utilities in the Northwest experienced demand growth in the 10–13% area, while the entire midwestern region suffered a loss compared to 1984. New England, which experienced severe heat, had a second year of 4% growth,

and seems to be increasingly insulated against the effects of economic recession.

The overall national pattern of slowly declining load growth will continue to have for some years a major effect on reserve margin—that is, the excess of installed capacity over actual demand—on a national basis. A rule of thumb is that national reserve margin should be about 25%, and it is now about 34%. Although utilities are, in response, canceling or delaying construction of additional capacity, construction of units started years ago and now nearing completion will support this high reserve for the next few years. Despite the high national margin, individual utilities, especially those in areas of high growth, will fall below margins required for reliable operation by the early 1990s.

The North American Electric Reliability Council prefers to use a different measure of adequacy of capacity, a factor called capacity margin. This is calculated by dividing the difference between actual peak demand and available capacity at time of peak by the available capacity. This capacity factor reflects a more realistic picture of spare capacity by accounting for capacity that is out of service at time of peak for various reasons.

Usage. Sales of electricity rose at a much more substantial rate than did peak demand. Overall sales to all classes of customers gained 5.7% over 1983, boosted by the continued strong economy and the recovery of housing starts. Commercial sales held up well, rising 5.8%; this was the second year in a row that sales grew at about this rate, after a series of slow years. Industrial sales also did well, surging 7.4% over 1983, the greatest single-year increase since 1976. Residential sales shared in the general increase, though at the lower rate of 3.5%. (Sales in each category are given in the table.)

Despite the continual increases in the cost of electricity, electric heating for residences, driven by the growing popularity of the heat pump, continues to gain. Over half of all new residences built in the United States for more than a decade have been electrically heated. Heating energy sales for the year 1985 rose to 172×10^9 kWh.

Residential use per customer reversed its decline and rose slightly from 8827 kWh/customer in 1984 to 8965 kWh in 1985. This produced revenues of 7.4 cents/kWh and an average annual residential bill of $665. In contrast, each cooperative utility customer used 10,995 kWh for which they paid only 6.93 cents/kWh. Total revenues from the residential segment of the industry for the investor-owned utilities was $118.7 billion, and $10.7 billion for the cooperative.

Fuels. Utilities have shifted to coal from oil and gas as fuels ever since the energy crisis of the early 1970s and, in fact, have been forced to do so by the provisions of the Fuels Use Act of 1974. The consumption of coal in 1985 rose 6.3% to 666.4×10^6 tons (602.8×10^6 metric tons). Oil use again declined, by 16.7%, to 204.5×10^6 bbl (32.5×10^3

m³). With gas currently in good supply, gas usage increased 6.9%, to $3,110 \times 10^{12}$ ft³ (88.0×10^{12} m³).

The same pattern repeated in energy generated. Coal generated 1341.7×10^{12} kWh, accounting for 64% of the total. Oil was used for 119.8×10^{12} kWh, or 5.7%. Gas was the fuel for 297.4×10^{12} kWh, or 14.2% of the total. Nuclear generation continued to gain as new stations came in line for full base-load operation, and generation from this source rose to 15.8% of total energy produced, or 327.6×10^{12} kWh. The remainder was contributed primarily by hydroelectric installations, with minor contributions from other sources such as wind, waste, and solar.

Distribution. Distribution capital expenditures for 1985 amounted to $6.1 billion, with an additional $2.5 billion expended in maintenance. During the year, 11,700 mi (18,847 km) of three-phase equivalent overhead lines and 6830 three-phase equivalent miles (11,002 km) of underground came into service at voltages ranging from 4.16 to 35 kV. The majority of this construction was at 15 kV, which accounted for 7228 three-phase equivalent miles (11,643 km) of overhead and 4717 three-phase equivalent miles (7598 km) of underground circuitry. The percents of total construction held by the other voltages were 14.0%, 21.8%, and 1.9% for 35 kV, 25 kV, and 4 kV, respectively. For underground construction, the equivalent percentages were 3.9%, 26.9%, and 14%, respectively.

During 1985, utilities energized 15,375 MVA of distribution substation capacity and expended $28.2 million in capital dollars for distribution substation construction.

Transmission. Utilities spent $3.1 billion in capital accounts for transmission lines in 1985. They spent $999 million for overhead lines operating at 345 kV and above, and $698 million on lines operating at voltages of 220 kV and below. For underground construction, which can cost on average eight times more than equivalent overhead construction, capital expenditures were $1.7 million at voltages of 220 kV and above, and $28.2 million for circuits of 161 kV and below. Utilities installed 2929 mi (4718 km) of overhead lines at 345 kV and above, but 5402 mi (8702 km) at 220 kV and below. Looking at the power-transmitting capability of these lines rather than just length gives a different perspective. New lines at 345 kV and above had an estimated capability of 2908 GW-mi (4685 GW-km), compared to only 735 GW-mi (1183 GW-km) for those operating at lower voltages. The proportions of lines at various voltages are different for underground cable because of the technology involved. In 1985, only 5 mi (8 km) of underground circuitry operating at or above 230 kV was built. There was 57 mi (92 km) at or below 161 kV.

Utilities brought into service 33.6 GVA of transmission substation capacity in 1985 and expended $891 million for substation construction. Maintain-

ing existing transmission plant cost $702 million.

Capital expenditures. Total capital expenditures in 1985 were $43.4 billion. Of this total, $29.5 billion was paid for generating facilities, $3.7 billion for transmission, $7.5 billion for distribution, and $2.6 billion for miscellaneous facilities and equipment.

Total assets held by investor-owned utilities at the end of 1985 were $377 billion. Electric cooperatives had assets of $10.7 billion.

For background information see ELECTRIC POWER GENERATION: ELECTRIC POWER SYSTEMS; ENERGY SOURCES; TRANSMISSION LINES in the Mc Graw-Hill Encyclopedia of Science and Technology.

WILLIAM C. HAYES

Bibliography: Edison Electric Institute, *Statistical Yearbook of the Electric Utility Industry*, 1984; *1985 Annual Data Summary Report*, North American Electric Reliability Council, 1985; 1985 annual statistical report, *Elec. World*, 199(4):53–76, April 1985; 36th annual electric utility industry forecast, *Elec. World*, 199(9):51–58, September 1985.

Electroencephalography

The utilization of topographic mapping of the brain's electrical signals is predicated on two long-established clinical methods of recording brain activity, electroencephalography (EEG) and evoked potentials (EP). In these techniques, multiple electrodes are affixed to an individual's scalp in a grid pattern. The frequency and intensity of electrical activity over time at each electrode location is recorded on a continuous graph. Electroencephalograms register neurological activity in the resting state. In recording evoked potentials, electrical activation in various regions of the brain is elicited by physical and mental stimulation. This technique has extended the diagnostic power of electroencephalography.

The clinical utility of both electroencephalography and evoked potentials has been limited by the complexity of interpreting their results. The neurophysiologist is faced with the task of mentally converting the multiple erratic line recordings into a meaningful series of maps depicting various aspects of brain function. Electroencephalograms contain an immense amount of information which must be analyzed for four general dimensionalities: (1) spatial (physical extent and movement); (2) spectral (amplitude of electrical charge within the frequency band); (3) temporal (duration of each given electrical event); and (4) statistical (comparison of a given individual's brain electrical activity with that of normal subjects).

The diagnostic limitations of electroencephalography and evoked-potential analysis are not due to too little information in the recordings, but rather to their wealth of information and their complexity. This complexity can create problems in reliability of interpretation between different readings, and even between different readers of the same electroencephalographic data.

Topographic mapping development. Beginning in the 1950s, researchers sought ways to improve the reliability and sensitivity of interpreting electroencephalogram tracings. But it was not until the 1970s that the development of micro- and minicomputers allowed topographic mapping of brain electrical activity. These computers are capable of performing all four analytic modes described above and translating the resulting data into instant or serial maps of brain activity displayed on computer terminals using color graphics. The computer can use Fourier analysis and the fast Fourier transform (FFT) algorithm to decompose electroencephalographic signals into spectral (wavelength) components. This analytic technique may be likened to extracting and quantifying the sound of a particular instrument or section of instruments from the total sound of an orchestra. Such complex analysis can be crucial to identification of neurological pathology. In the analysis of evoked-potential data, the computer is also able to localize the point at which electrical activity begins and the trajectory of this activity as it spreads and interacts with other electrical responses in the brain.

Topographic mapping technique. To create topographic images of electroencephalography, numeric values are assigned to frequency bands and these are translated into a gray scale or a rainbow scale. Data are collected from a number of electrodes, usually 20, and an interpolation algorithm is used to develop intermediate values between electrodes (Fig. 1).

Topographic images of evoked potentials can be made at millisecond intervals over a period of time and then viewed, like a movie or cartoon, as the computer displays these images on screen at the terminal. The spectral, temporal, and spatial analyses required in electroencephalography and evoked-potential interpretation are performed by the computer. When reading the traditional electroencephalogram, the neurophysiologist may take advantage of these imaging techniques to support an interpretation. Figure 2 provides a simple block diagram of the elements used in generating topographic images of brain electrical activity.

When viewing a topographic map, the clinician is still confronted with the critical question of whether the image (or electroencephalographic recordings associated with the image) represents the neurological activity of a normal, healthy individual or that of one with a pathological condition. A statistical technique called significance probability mapping (SPM) has been developed to aid in this discrimination. Significance probability mapping compares an individual's topographic image to an image based on neurological data of a comparable reference group, and generates a new image that depicts the deviation of the individual from the normal group. Such comparisons provide the final complicated step in clinical evaluation of electroencephalographic evoked-potential data, and have proven extremely valuable in diagnosing neurologic abnormalities. Additionally, significance-probability mapping techniques are valuable in research applications comparing two or more groups of subjects under a number of resting or activated (evoked) conditions.

Research applications. Much of the technical groundwork for topographic mapping was set down

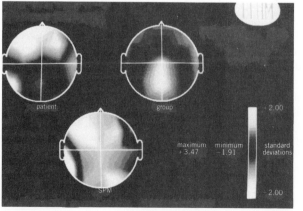
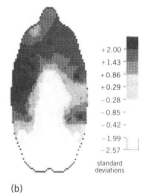
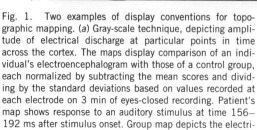

(a)

(b)

Fig. 1. Two examples of display conventions for topographic mapping. (*a*) Gray-scale technique, depicting amplitude of electrical discharge at particular points in time across the cortex. The maps display comparison of an individual's electroencephalogram with those of a control group, each normalized by subtracting the mean scores and dividing by the standard deviations based on values recorded at each electrode on 3 min of eyes-closed recording. Patient's map shows response to an auditory stimulus at time 156–192 ms after stimulus onset. Group map depicts the electrical response of a comparable control group to the same stimulus at the same point in time. Significance probability map depicts regional differences between patient and control group based upon standard deviation from mean amplitude (*courtesy of F. H. Duffy*). (*b*) Modification of the standard 10–20-electrode placement system, developed by M. Buchsbaum, using 32 electrodes and a four-point linear interpolation. Map shown is a top-down view, with a gray scale depicting brain activity within a specified frequency range.

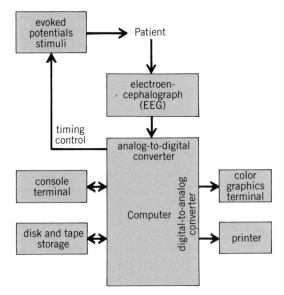

Fig. 2. Primary elements of equipment required for topographic mapping of brain electrical activity.

in Europe, with increased activity in the United States and Japan over the past several years. The emphasis is shifting to establishment of applications of this technique to clinical diagnosis and laboratory research. A substantial number of research applications have also resulted, many of which are described below.

Dyslexia. In the late 1970s studies were done to investigate the application of brain electrical activity mapping (BEAM) to a better understanding of aberrant brain function in specific reading retardation (dyslexia). By using the significance probabilty mapping technique, clearly defined regional neurological differences between dyslexic boys and a carefully selected comparison group without dyslexia were found. Such a physiological "signature" of dyslexia may provide the basis for earlier diagnosis and remediation of this language disorder.

Schizophrenia. Various studies have confirmed previous clinical findings on the pathological condition of regions of the brain associated with schizophrenia. In addition to the traditional frontal lobe abnormalities, deficiencies have been observed, under certain circumstances, in the parietal and temporal regions of the brain's left hemisphere. Further research may lead to enhanced knowledge and treatment of this psychological debility.

Epilepsy. A number of research laboratories have applied topographic imaging techniques to the study of epileptic seizures. Topographic imaging may be a more sensitive detector in the diagnosis of covert epilepsy, aiding in discrimination of this condition from similar syndromes and potentially pointing the way to the underlying cause of this condition. The topographic distribution of the epileptic spike itself has been mapped, and new information as to its origins is appearing.

Stroke. Japanese researchers have combined topographic mapping analysis and the study of regional blood flow in the brain in order to study the effects of stroke. Such studies permit more accurate delineation of the functional extent of such brain lesions.

Alzheimer's disease and aging. The brain electrical activity of patients with Alzheimer's disease, both presenile and senile, has been studied, in comparison to control subjects of similar age. The analysis defined specific areas of the brain which were dysfunctional in senile patients (right posterior temporal lobes) when compared to control groups. A related study addressed changes over time of brain wave characteristics associated with the normal aging process. These studies were designed to provide the basis for further research into the pathologies of brain function in Alzheimer's disease. They provide firm evidence that Alzheimer's disease should not be thought of as premature aging—in medically healthy subjects the effect of normative aging or brain function is surprisingly small.

Brain development in early childhood. Neurological imaging techniques have also been applied to the study of cerebral development or newborn babies and premature babies. The mapping techniques were used to locate areas of the newborn brain associated with higher functioning. This research was applied to analyzing dysfunction in premature babies, who have been shown to have a higher risk for subsequent learning disabilities. Ongoing research should lead to diagnostic methods for very early identification of "at-risk" babies for learning problems, and thus provide the basis for specialized strategies to deal with developmental deficiencies.

Tumor detection. A number of applications of brain electrical activity mapping address early detection of brain lesions or tumors. Such techniques as the computerizd tomography (CT) scan and nuclear magnetic resonance (NMR) imaging are also available for such diagnosis, but topographic imaging has been demonstrated as a highly sensitive detection device for tumor recurrence and so provides the basis for prompt treatment of tumors.

Clinical applications. The analyses described above are examples of research applications of topographic mapping. In the clinic, the most extensive use to date has been in the resolution of neurological "mystery" cases, in which a patient has elicited any of a number of abnormal behaviors but has been tested as normal in clinical, neurological, and psychiatric evaluations, and has also tested as normal on radiographic examinations such as the computerized tomography scan. The clinician must establish whether such a patient is organically diseased or emotionally ill. It is common for organic brain disease to produce disturbances in behavior that often mimic more classical "functional" or emotional illness. In children, such cases often appear to be behavioral disorders; in adults the illness may be psychiatric or, for instance, putative epilepsy. The sensitivity of topographic mapping in interpreting electroencephalography and evoked-potential results has proven extremely useful in the detection of organic pathologies and their related neurological characterizations.

For background information *see* BIOPOTENTIALS AND IONIC CURRENTS; BRAIN; ELECTROENCEPHALOGRAPHY; NEUROBIOLOGY in the McGraw-Hill Encyclopedia of Science and Technology.

[FRANK H. DUFFY]

Bibliography: F. H. Duffy (ed.), *Topographic Mapping of Brain Electrical Activity*, 1986; F. H. Duffy et al., Significance probability mapping: An aid in the topographic analysis of brain electrical activity, *Electroenceph. Clin. Neurophysiol.*, 51: 455–462, 1981; F. H. Duffy and N. Geschwind (eds.), *Dyslexia: A Neuroscientific Approach to Clinical Evaluation*, 1985; A. Remond and F. Offner, A new method for EEG display, *Electroenceph. Clin. Neurophysiol.*, 7:453, 1952.

Electromagnetic field

For several decades there has been concern that human exposure to the electric and magnetic fields associated with high-voltage power lines might lead to adverse health consequences. Research on this question has demonstrated that under some circumstances power-line frequency fields, even rather weak fields, can interact with and produce changes in certain types of living tissue, especially nerve tissue. Given the major role that natural electromagnetic processes play at the cellular and molecular level, this is not surprising. However, biological effects are not the same thing as adverse health consequences. Despite substantial research, no persuasive evidence that power-frequency (50 or 60 Hz) fields can lead to adverse health impacts has yet emerged. However, given the state of scientific understanding, it is not possible to draw definitive conclusions.

In the 1960s a number of workers in Eastern Bloc countries published reports suggesting that people working in substations of very high-voltage power systems were experiencing a variety of nonspecific complaints such as headaches and fatigue. It was suggested that the causative agent might be the electromagnetic fields associated with these lines, but no systematic attempt was reported to exclude other factors such as noise, spark discharges, or chemical leaks from transformers or circuit breakers. While attempts to replicate these findings in the West were essentially unsuccessful, they did have the effect of stimulating interest in the subject. Interest was further stimulated by a proposal to bury a large extremely low-frequency (ELF; 30–300 Hz) antenna in several midwestern states in order to communicate to submarines at sea. These concerns, as well as more recent concerns about power line fields, have resulted in a substantial program of high-quality research.

Research on adverse health consequences. While much of the early research on 50- or 60-Hz fields suffered from severe problems of quality control and experimental design, a large volume of high-quality research results are now available. This research is of four broad varieties: (1) laboratory screening studies to search for potential problems that warrant careful examination; (2) laboratory studies of specific effects or mechanisms of interest, carried out with whole animals; (3) laboratory studies of processes and mechanisms performed at the level of individual organs, of cells, or at the subcellular level; and (4) epidemiological studies of human populations.

Screening studies. A number of screening studies that exposed rats, mice, and other laboratory animals to strong electric fields (for example, 1–100 kV/m) have examined a wide variety of biological variables. With few exceptions, no effects from brief or prolonged exposures have been observed. Only a few screening studies have been conducted for magnetic fields; they have also yielded largely negative results. Studies of possible effects on natural and farm plants and animals have clearly demonstrated that, with the exception of a problem with bees, there are no ecologically or agriculturally significant consequences from power-line field exposure. In very strong electric fields bees can receive shocks as they walk around in the hive, and this disrupts their behavior. The problem is easily solved by shielding the hive with grounded chicken wire.

Studies of specific effects. A number of specific effects have been identified and studied in detail. At electric field strengths above several kilovolts per meter, people and animals can sense the presence

Approximate range of (a) electric and (b) magnetic fields encountered in various common environments. The horizontal bands indicate the range of values. There is a higher likelihood that values lie in the more darkly shaded regions. 1 milligauss = 10^{-7} tesla.

of electric fields. A variety of experiments have explored the behavioral responses of animals in such fields. Chronic stresslike responses appear not to occur. Long-term exposure to strong electric fields do appear to produce changes in circadian fluctuations in the level of certain endocrine hormones, especially those associated with the pineal gland. There is no agreement on the potential health implications, if any, of these changes. Studies of reproduction in mammals have yielded largely negative results; one study of miniature swine produced ambiguous results that could be interpreted as indicating an increased incidence of developmental abnormalities, but these results could be due to a number of problems with the study. There have been far fewer studies conducted with magnetic fields. Pulsed low-frequency magnetic fields have been found to accelerate the regrowth of broken bones and are used clinically. Studies of the development of chicken eggs in pulsed magnetic fields have shown significant developmental abnormalities, but only for certain pulse waveforms not typically associated with power-line fields.

Organ, cell, and subcellular studies. A variety of studies at the level of individual living organs and cells and at the subcellular level have yielded strong evidence that certain kinds of low-frequency electric and magnetic field exposure can affect at least a few biological processes. Some studies suggest that the effect involves processes that occur at the outer surfaces of cells, either in the lipid bilayer or in the complex receptor proteins which float in that layer. Little theoretical understanding is available. Some of the effects show response only at certain frequencies and field strengths, which suggests some form of resonant process. Other experiments appear to show transient effects when fields are turned on. Several studies have suggested that the Earth's constant magnetic field may interact with some low-frequency fields to produce an effect.

Epidemiological studies. Epidemiological studies of occupationally exposed people such as power-line workers have shown no significant health impacts. Several studies have reported an association between the incidence of certain cancers and proximity to power distribution circuits (the wires on poles in the street) or employment in so-called electrically related occupations. The workers who carried out some of these studies have suggested that exposure to power-line-frequency magnetic fields may promote the growth of cancer. However, these studies suffer from a variety of problems which raise doubts about the validity of this suggestion. These problems include ambiguity about whether the groups that showed effects actually received more field exposure, and a failure to control statistically the effects of a variety of factors such as smoking which are known to affect cancer rates. A number of additional studies of this kind are under way.

Sources and measures of field exposure. People are exposed to 50- or 60-Hz fields from many sources. The illustration summarizes the approxi-

mate range of field strength associated with several common sources. However, if field exposure is a source of concern, it is not clear that stronger fields should necessarily be of greater concern than weaker fields. Thus, a scientifically based program of field exposure reduction is not now possible. One strategy to attempt to reduce possible hazards might be to try to make transmission-line exposures similar to those that people receive from other sources. Without explaining their rationale, a few states have imposed limits on the strengths of electric fields from high-voltage lines. No limits have been imposed on other sources such as electric blankets, and no limits have been imposed on magnetic fields.

Comparison with other risks. Thousands of environmental agents are being explored as possible sources of risk. These include a wide variety of common "natural" foods and seasonings, and chemicals that have been used for years. Many common activities like driving and sun bathing carry known risks that could be better controlled. In dealing with possible risks from 50 or 60 Hz fields, it is important to keep some perspective on such other risks as well as on the considerable benefits of electric power.

For background information *see* BIOPHYSICS; ELECTROMAGNETIC FIELD; TRANSMISSION LINES in the McGraw-Hill Encyclopedia of Science and Technology.   [M. GRANGER MORGAN]

Bibliography: W. R. Adey, Tissue interactions with non-ionizing electromagnetic fields, *Physiol. Rev.*, 61:435–514, 1981; *Bioelectromagnetics* (journal); H. B. Graves et al., *Biological Effects of 60-Hz Power Transmission Lines*, report to the Florida Department of Environmental Regulation of the Florida Electric and Magnetic Fields Science Advisory Commission, Tallahassee, 1985; M. G. Morgan et al., Power-line fields and human health, *IEEE Spectrum*, 22(20):62–68, February 1985; W. T. Norris et al., People in alternating electric and magnetic fields near electric power equipment, *Electr. Power*, 31:137–141, 1985; A. R. Sheppard, *Biological Effects of 60-Hz Power Transmission Lines*, report to the Montana Department of Natural Resources and Conservation, Helena, 1983.

Electron microscope

A major goal in the semiconductor industry is the reduction in size of integrated circuit components to a submicrometer level. Direct examination of the features and defects in such devices would then be beyond the capabilities of optical microscopy, and would require scanning electron microscopy, a technique commonly used throughout the industry. Concomitant with this scaling-down of the lateral dimensions, device improvements are equally dependent on decreasing the thickness of films which lie stacked on one another on the semiconductor substrate crystal (Fig. 1). Such films, which are crucial to the operation of the device, may be even thinner than the resolution limit of the most advanced scanning electron microscopes (approximately 5 nano-

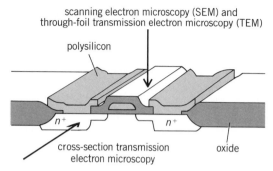

scanning electron microscopy (SEM) and
through-foil transmission electron microscopy (TEM)

polysilicon

n^+ n^+

cross-section transmission
electron microscopy

oxide

Fig. 1. Cross-section view of various layers which can make up part of a simple integrated circuit. The viewing directions for cross-section transmission electron microscopy (TEM) and for conventional microscopy are indicated.

meters). Thus a more powerful instrument, the transmission electron microscope, must be employed to investigate their nature and their influence on the circuit's electrical characteristics. There has been a proliferation in the application of this approach in basic semiconductor research, which will undoubtedly impact technological developments in the future.

While transmission electron microscopy has been used extensively in other areas of materials science, it is only now beginning to play a major role in semiconductor investigations. Examination of a device in cross section (Fig. 1) is essential to uncover the relationship of the thin films and their interfaces with their associated electrical performance. A major, but somewhat mundane, problem which has inhibited progress concerns specimen preparation. Samples can be fabricated from virtually any starting material, but they must be made sufficiently thin that an electron beam can be transmitted through them without significant absorption. For the highest resolution and clarity (currently a little less than 0.2 nm), thicknesses of 10 nm or less are required. Procedures for bringing this about have now been established.

HREM technique. The most useful way to reveal the structure is by high-resolution electron microscopy, also known as the lattice imaging technique. The specimen is carefully oriented in the microscope, using externally driven tilt controls, so that the atoms in the crystal are aligned exactly parallel to the imaging electron beam. By positioning suitable apertures, the transmitted electron beam is allowed to interfere with diffracted beams in producing the final image. When the microscope is suitably focused at high magnification, a regular array of image spots is seen which can be directly related to the lattice periodicity of the crystalline sample. Under special imaging circumstances the actual atomic positions in the structure are displayed. Of course, an amorphous, or noncrystalline, material produces an irregular picture.

Studies of silicon-based materials. Figure 2a shows a typical example, the interface between a silicon substrate crystal and the oxide which has

been grown on it by a simple chemical reaction. The regular pattern associated with the silicon (Si) crystal is readily apparent, so long as high enough magnification is employed (Fig. 2a is at approximately 10^7 times). The silicon dioxide (SiO_2) layer (essentially glass) is amorphous and so its image has no regularity. The transition from one material to the other is noticeably abrupt. Oxidation of silicon to produce an overlying oxide layer is a procedure used extensively during semiconductor manufacturing. The oxide can be employed in a variety of tasks from being an insulating layer, to preventing dopant implantation and diffusion, to even playing an active role in the operation of a circuit. Thus many pictures such as Fig. 2a have been taken. It is the reproducibility and integrity of the silicon–silicon dioxide interface, so dramatically exposed by lattice imaging, which is responsible to a great degree for the wide application of silicon integrated circuits.

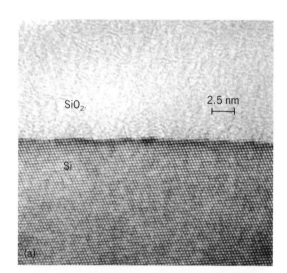

SiO_2 2.5 nm

Si

(a)

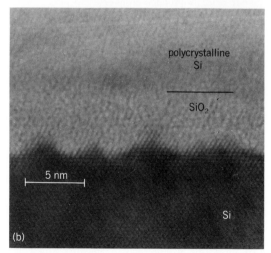

polycrystalline Si

SiO_2

5 nm

Si

(b)

Fig. 2. High-resolution electron microscope images of the silicon–silicon dioxide (Si-SiO_2) interface. (a) Conventional, thick oxide, with topographically smooth interface. (b) Thin oxide, with rough interface, separating silicon substrate from polycrystalline silicon contact layer. (*Courtesy of J. C. Bravman and A. H. Carim*)

An interesting case study illustrates the power of high-resolution electron microscopy for semiconductor diagnostic work. Extremely thin oxides (less than 10 nm), separating the silicon substrate from a polycrystalline silicon contact layer, are being explored for use as tunnel dielectrics and as thin gate oxides for metal-oxide-semiconductor (MOS) devices. In order to produce a well-controlled oxide thickness, oxidation is carried out at low temperature. Unfortunately this leads to a layer with inferior insulating properties, with electrical breakdown occurring prematurely. Several possible causes could explain this phenomenon. High-resolution electron microscopy, however, pinpointed a topographically rough silicon–silicon dioxide interface (Fig. 2b), with silicon protrusions only 1–2 nm in height, as the source of this effect. Such information could not have been obtained by alternative means.

These examples represent only a small fraction of the work performed on silicon-based materials. Interfaces have been examined at this level for silicon grown on insulators (for example, silicon on sapphire), insulators grown on silicon, metallic and other electrically conducting contacts to silicon, silicide formation on silicon, epitaxial silicon growths, silicon intercrystalline grain boundaries, and even lattice imperfections and precipitations within silicon crystals themselves. Not only are topographic effects sought (as in Fig. 2b), but the nature and extent of chemical reactions which have occurred at interfaces, the presence or absence of intermediate layers between adjoining thin films, the manner by which one crystal phase grows on top of its substrate, and even the atomic arrangements at the interface itself. The latter requires careful imaging under well-known experimental conditions, an extensive series of image-matching calculations, and the fortunate circumstance that the interface is atomically smooth and exactly parallel to the imaging electron beam. When these conditions are achieved, the analysis can be quite revealing. Much valuable knowledge is thus derived to help researchers toward their goal of producing better materials combinations and thereby more advanced microcircuits.

Studies of compound semiconductors. The high-resolution electron microscopy technique has proved equally profitable in the examination of compound semiconductor systems, especially those based on gallium arsenide (GaAs). The lattice structure is similar to that of silicon, and whereas the specimens may be somewhat more delicate, the larger lattice spacings and heavier average atomic number often result in pictures of even greater clarity and contrast. Figure 3 shows an example in which a layer of gallium arsenide has been grown by molecular beam epitaxy (MBE) onto a parent gallium arsenide single crystal. This type of growth can now be sufficiently controlled that perfect continuity of the gallium arsenide is retained (epitaxy) as well as a film free from imperfections. However, in the present example, the procedure has resulted in a high-defect-density gallium arsenide overlayer. High-resolution

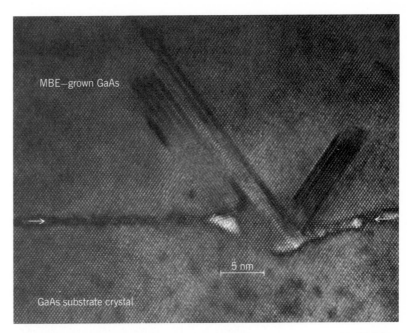

Fig. 3.   Lattice image of molecular beam epitaxy (MBE)–grown gallium arsenide (GaAs) on a gallium arsenide substrate crystal, showing that defects in the deposited growth can originate at small surface agglomerates. Arrows indicate the growth interface. (*Courtesy of T. Yamashita*)

electron microscopy of the interface (Fig. 3) shows that the imperfections originate at small noncrystalline agglomerates (some as small as 1 nm) which were present on the substrate crystal surface. While their identity is not established, it is clear that clean surfaces are essential for the proper epitaxial growth. The surface preparation was suitably altered following the present observations, accompanied by improvement in the structure and properties of the material. This work serves to illustrate that not only can such small entities alter the electrical characteristics of a device material, but they can have long-range influence on the quality of thin-film growth (in this case by the introduction of many faults). The transmission electron microscopy approach provides a direct means for identifying the causes of a problem, from which a remedy is often obvious.

Studies of superlattices. As well as studying the integrity of interfaces, high-resolution electron microscopy has provided much insight in the characterization of multilayer superlattices, which are employed in quantum-well and other devices (for example, modulation-doped field-effect transistors). In these materials, thin layers of two compounds, or semiconductor alloys, are deposited in an alternating sequence. The layers are generally only a few atomic planes thick, and so again high-resolution electron microscopy is essential for studying their structure. It is important that the lattice mismatch (due to the simple fact that there are small differences in the atomic diameters) is not accommodated by lattice faults, particularly dislocations. Conditions to bring this about, as well as limitations on

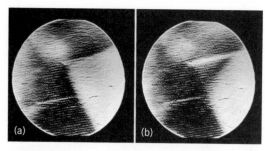

Fig. 4. Individual frames from a television recording of silicon regrowth at 1330°F (730°C). About 3 s elapsed between frame *a* and frame *b*, but the sudden growth burst, involving the apparently simultaneous crystallization of hundreds of atoms, occurred between two successive videotape frames 1/30 s apart. (*Courtesy of M. A. Parker*)

material combinations and layer thicknesses, are determined from structural information provided by high-resolution electron microscopy.

Time-resolution studies. High-resolution electron microscopy is experiencing a significant enhancement in capability. Several commercial instruments are becoming available with greater resolving power and a wider range of accessories. For instance, it has now been shown that it is possible to heat a specimen to quite high temperatures in the transmission electron microscope and still retain sufficient control to record changes at lattice resolution. Figure 4 shows pictures from a television video recording, taken only 3 s apart, of the growth of crystalline silicon into an amorphous silicon overlayer. This process is employed after doping the semiconductor by ion implantation, which destroys the crystallinity of the silicon, to reproduce perfectly crystalline silicon once more. Similarly, reaction chambers can be added to transmission electron microscopes so that thin-film depositions can be observed as they are carried out. It seems likely therefore that transmission electron microscopy will take on the form of a minilaboratory in which materials are observed, fabricated, and perhaps even tested, all at the same time.

In summary, as microelectronic devices are made increasingly smaller, higher-power microscopy such as transmission electron microscopy is required to locate and identify subtle, minute features which are responsible for modifying electrical performance. The contribution of the high-resolution electron microscopy technique should become more direct over the course of time, assisting the development of advanced technologies.

For background information *see* ELECTRON MICROSCOPE; INTEGRATED CIRCUITS; SEMICONDUCTOR HETEROSTRUCTURES in the McGraw-Hill Encyclopedia of Science and Technology.

[ROBERT SINCLAIR]

Bibliography: J. C. Bravman and R. Sinclair, The preparation of cross-section specimens for transmission electron microscopy, *J. Electr. Microsc. Tech.*, 1:53–63, 1984; A. H. Carim and A. Bhattacharyya, Si/SiO$_2$ interface roughness: Structural observations and electrical consequences, *Appl. Phys. Lett.*, 46:872–874, 1985; J. M. Gibson and L. R. Dawson (eds.), *Layered Structures, Epitaxy and Interfaces*, vol. 37 of *Materials Research Society Symposium Proceedings*, 1985; J. C. H. Spence, *Experimental High Resolution Electron Microscopy*, 1980.

Electronics

Ceramic- and glass-metal packages were originally used for the interconnections and environmental protection of silicon and germanium devices, such as diodes and transistors. They are now used for intraconnections, substrates, power distribution, cooling, and inputs and outputs as well. As the number of circuits per chip increases, each of these functions takes on greater importance.

Current technology. There are three basic methods for joining silicon chips to their substrates: back bonding with gold connector wires, flip-chip bonding with solder pads, and decal bonding.

Back bonding, the original process which is still in use, bonds the silicon chips to the substrates with either a gold-silicon eutectic braze or a silver-loaded epoxy cement. The connections from the chip pads to those on the substrate are made with 0.001- to 0.003-in.-diameter (0.025- to 0.075-mm) gold wires which are ultrasonically or thermally bonded to both sets of pads.

In flip-chip bonding, up to 121 0.005-in.-diameter (0.125-mm) 95% lead–5% tin solder pads are deposited on each silicon chip. The pads on these chips are then placed on the vias in the ceramic substrate, held in place with a solder flux by surface tension, and then passed through a belt furnace at about 750°F (400°C). (Vias are holes in the ceramic layers, filled with metallic paste, which electrically connect the conductive pattern from layer to layer.) This method is used in manufacturing the thermal conduction module (see illus.).

Decal bonding uses a thin gold-plated copper foil which has been etched to mate with the pads on the chips and substrate. These patterns are the decals, and they are ultrasonically or thermally bonded to the chips and substrate. The chips are either back-bonded to a substrate or encapsulated in an epoxy resin.

Environmental protection. Silicon integrated circuit chips are usually coated with thick silica layers which provide adequate protection from the environment. In order to protect the aluminum conductors from humidity, the chips must be hermetically sealed. There are several approaches to achieve this. A thick plastic coating can be used which will delay the penetration of moisture, or glass-metal or ceramic-metal seals can be used. In the latter case a flange is brazed to the metallization on the ceramic and then a cover is welded or brazed to this flange. Another method, used with the newest version of the thermal conduction module (discussed below), is to directly seal the cover to the bare ceramic with the use of a C ring.

Intraconnections. The intraconnection wiring determines the transmission speed of the signals within the substrate. Table 1 lists several of the ceramics being used and those which are being considered for future use. It is important to reduce the permittivity of the ceramic because it determines the speed of the signal. As the dielectric constant of the ceramic is reduced, the characteristic impedance of the line Z has to be maintained between 50 and 100 ohms, as determined by Eq. (1), where L is the in-

$$Z = \sqrt{L/C} \tag{1}$$

ductance of the line, and C is the capacitance of the line to ground. Therefore, the thickness of the ceramic layers must be reduced to provide these values. As the layers become thinner, there is less space for the metallic components. Thus, the entire structure is subject to constraints which depend on the conductivity of the metal.

Table 2 lists the properties of the metals in use now and those being considered for the future. Molybdenum and tungsten are the best compromises for adhesion, electrical conductivity, and thermal expansion for alumina substrates. As silicon chips increase in size, however, the thermal expansion of the ceramic should be reduced in order to minimize the stresses in the joints between the silicon and the ceramic. The Coffin-Manson equation (2) gives the

$$N_f = c/\epsilon^2 \tag{2}$$

number of cycles to failure by fatigue, N_f, where c is a constant for the system and ϵ is the strain. N_f can be over 20,000 cycles with current ceramic materials and silicon. As the silicon chips get larger, however, fatigue failures will become a more serious problem due to the larger stresses.

Power distribution. Higher-conductivity metals will be needed in multilayer ceramic structures for power distribution. Moybdenum and tungsten may be replaced with copper or gold when low-firing ceramic materials are used. Other alternatives could be the use of decoupling capacitors or bus-bars on the substrates.

Cooling. There are two methods for removing the heat from chips on a module. Chips which are back-bonded to the substrate can have their heat removed through the substrate from the back of the chip. In the flip-chip technology, chips are cooled by aluminum pistons or ceramic or metal covers in contact with the back of the chips. Each of these configurations provides a direct path for heat to leave the module through a water or air exchange arrangement.

Inputs and outputs. The leads on the substrates are used to provide a means of communication to the rest of the system. The lead reduction ratio parameter is a measure of the efficiency of the package. It is calculated from the total number of pads on all the chips divided by the total number of leads on the substrate. Early device packages had a lead reduction ratio of 1. As levels of integration grew, this value increased to 9 for the thermal conduction module.

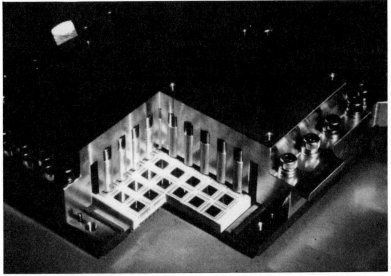

(a)

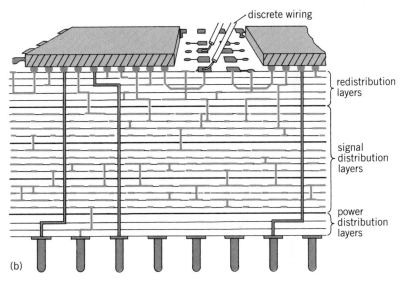

(b)

Thermal conduction module. (a) Cutaway view. (b) Schematic cross section of multilayer ceramic substrate with chips and pins attached. The discrete wires are used to make engineering changes.

Advanced packages. The thermal conduction module (see illus.) represents the most advanced ceramic package in the electronics industry. A new version of this module will be used in the next IBM series of large-scale computers. As in earlier versions, this substrate supports 133 chips, has 1800 pins, and is 3.5 in. (90 mm) square, but each chip now dissipates up to 6 W. The other major change is the elimination of the flange on the substrate, which has been replaced with a C-ring direct seal to the bare ceramic. This substrate also uses a 92% alumina ceramic with molybdenum.

Another type of substrate can have up to 41 ceramic layers, which are made from 55% alumina and 45% lead borosilicate glass. Each substrate can accommodate up to 76 chips, which are decal-bonded to the substrate by an ultrasonic thermal

Table 1. Properties of ceramic materials

	Strength, kpsi (MPa)	Thermal conductivity, W/(m·°C)	Thermal expansion 10^{-6}/°C (10^{-6}/°F)	Relative permittivity (K)
Alumina (Al_2O_3) compositions				
92%	46 (317)	16.7	7.7 (4.3)	9.5
96%	40 (276)	25.1	7.8 (4.3)	10.2
55% + 45% glass	43 (296)	5.0	4.2 (2.3)	7.5
Future materials				
Aluminum nitride (AlN)	40 (276)	210	4.5 (2.5)	8.8
Silicon nitride (Si_3N_4)	85 (586)	33.5	3.1 (1.7)	6.0
Silicon carbide (SiC)	20 (138)	340	3.1 (1.7)	40
Mullite	24 (165)	10	3.8 (2.1)	5
Cordierite	15 (103)	10	1.6 (0.9)	6
Glass-ceramic mixtures	28 (193)	3.5	6.5–7.5 (3.6–4.2)	6.5–7

compression bonder. The metal is gold or 85% silver–15% palladium. These substrates are sintered at 1650°F (900°C) in order to prevent melting of the gold.

Another manufacturer uses a 96%-alumina ceramic with tungsten, which sinters about 2800°F (1550°C). Molybdenum-manganese pastes are used on the surface of the modules, which are subsequently coated with up to five layers of polyimide and copper. The lines are 0.01 in. (0.25 mm) wide, with 0.02-in. (0.50-mm) spacings, and vias in the ceramic substrates have 0.008-in. (0.20-mm) diameters. The thin-film metal structures have 0.002-in. (0.05-mm) line widths with 0.005-in. (0.13-mm) spacings and vias with 0.0032-in. (0.08-mm) diameters in the polyimide.

Future materials. Ceramic materials being considered for future use include aluminum nitride (AlN), silicon nitride (Si_3N_4), silicon carbide (SiC), mullite-cordierite mixtures, and glass-ceramic mixtures.

Pure aluminum nitride has an extremely high thermal conductivity, a thermal expansion close to that of silicon, and a strength of 40,000 psi (276 megapascals). The problem with aluminum nitride is that it has to be sintered at very high temperatures. For example, with a 3% yttria addition as a sintering aid, this material has to be sintered at 3300°F (1800°C). Furthermore, this addition reduces its thermal conductivity.

Silicon nitride has a low thermal conductivity; however, it has a strength of 85,000 psi (586 MPa), a thermal expansion closer to that of silicon, and a dielectric constant of about 6. However, it also must

be sintered at high temperatures and pressures to produce a dense material. Thin films can be produced by chemical vapor techniques.

Silicon carbide has the highest thermal conductivity of all the ceramic materials being considered, as well as a thermal expansion very close to that of silicon. Its dielectric constant, however, is about 40, and it has to be hot-pressed or hot-isostatically-pressed to obtain a high density.

Mixtures of mullite ($3Al_2O_3 \cdot 2SiO_2$) and cordierite ($2MgO \cdot 2Al_2O_3 \cdot 5SiO_2$) are also being studied for future substrate materials. They can be sintered in oxidizing or reducing atmospheres in the temperature ranges of 1800 to 2650°F (1000 to 1450°C). Their thermal expansions can be adjusted to match that of silicon, and dielectric constants are in an acceptable range of 5 to 6.

Various glass-ceramic mixtures are being developed in order to obtain low-firing materials which can be used with copper or noble metal pastes, such as gold or silver-palladium. A mixture of 35% alumina, 40% borosilicate glass, and 25% forsterite ($2MgO \cdot SiO_2$) ceramic has been developed and is being used with silver-palladium conductors for multilayer ceramic substrates. A low-temperature ceramic in the ZnO-Al_2O_3-MgO-SiO_2 system, cofired with gold, is also being developed for multilayer substrates. Finally, a 50% alumina–50% borosilicate glass multilayer substrate using copper metallurgy is being studied.

New processes. Doctor blading, the process now used to manufacture ceramic layers, can be extended to produce plasticized ceramic sheets as thin as 0.001 in. (0.025 mm). At this point other tech-

Table 2. Properties of metals

Element	Melting point, °C (°F)	Resistivity, 10^{-8} ohm-m	Thermal expansion, 10^{-6}/°C (10^{-6}/°F)
Tungsten (W)	3415 (6179)	5.5	4.5 (2.5)
Molybdenum (Mo)	2625 (4757)	5.2	3.5 (1.9)
Palladium (Pd)	1552 (2826)	10.8	11.0 (6.1)
Copper (Cu)	1083 (1981)	1.7	17.0 (9.4)
Gold (Au)	1063 (1945)	2.2	14.2 (7.9)
Silver (Ag)	960 (1760)	1.6	19.7 (10.9)
Aluminum (Al)	660 (1220)	3.7	23.9 (13.3)

nologies, such as sputtering, evaporation, chemical vapor deposition, and sol-gel chemical synthesis, could be implemented in order to manufacture the next generation of ceramic or polymer substrates. Aluminum conductors could then be used with these processes because of their low operating temperatures. Also, these technologies are very suitable for making thin films, which will permit the use of a much broader range of materials.

For background information *see* CERAMIC TECHNOLOGY; CIRCUIT (ELECTRONICS); INTEGRATED CIRCUITS; PACKAGING OF EQUIPMENT; PRINTED CIRCUIT in the McGraw-Hill Encyclopedia of Science and Technology.

[BERNARD SCHWARTZ]

Bibliography: B. Schwartz, Microelectronics packaging, in *Electronic Ceramics*, Amer. Ceram. Soc. Spec. Publ. 3, pp. 12–15, May 3, 1969; B. Schwartz, Review of multilayer ceramics for microelectronics packaging, *J. Phys. Chem. Solids*, 45:1051–1068, 1984.

Elementary particle

A major share of the scientific endeavor throughout history has been devoted to the search to uncover the fundamental laws and building blocks of nature. This collective quest has culminated in the identification of various fundamental forces, or interactions, which together govern the behavior of everything known. Two of them, gravity and electromagnetism, are quite familiar in terms of daily experience and application. The other nuclear and subnuclear forces have a measurable strength only over distances of 10^{-15} m or less.

Extensive experimental studies have also revealed a large number of particles that seem to be elementary, meaning that they are not made out of anything else. These include 45 particles, called quarks and leptons, arranged in three very similar "families" of 15 particles each. (The electron is an example of a lepton, whereas the proton is made out of three quarks.) There are additional elementary particles, called gauge particles, whose main role is to transmit the forces described above.

The first section of this article discusses superstring theory which gives promise of understanding the properties of these particles and forces from first principles. The second section discusses lattice gauge theory, which introduces a space-time lattice to study the interactions between quarks and their gauge fields.

Superstrings. The overriding concern of elementary particle physics is to understand the various particles and forces in terms of a simple and elegant mathematical principle. However, tables that list properties of particles (masses, lifetimes, decay modes, and so forth) display few discernible patterns. The hope for finding an explanation for particle properties in terms of a simple principle lies in the mathematical fact that symmetrical equations can have asymmetrical solutions. This spontaneous symmetry breaking means that the equations that

are sought may bear little resemblance to the physics they are supposed to describe. The solutions of the equations can be very complicated and messy, subtly concealing the underlying symmetry.

The most severe obstacle that has arisen in attempts to construct a unified theory of nature is an apparent clash between the requirements of gravity and quantum mechanics. There is a very beautiful and successful theory of gravity, Einstein's general theory of relativity, but it falls apart when one tries to interpret it quantum-mechanically. Quantum calculations based on Einstein's gravity theory give infinite answers that have no sensible interpretation. In the quantum context, Einstein's theory can be viewed as the theory of a particular kind of elementary particle, called a graviton. In the mathematical theory, the graviton and other elementary particles are described as points having no spatial extension whatsoever. This could be the source of the problem.

String theories. In 1974 it was proposed that elementary particles are not points, but rather are one-dimensional curves called strings, which have zero thickness and whose length is typically the Planck length, 10^{-35} m, 20 orders of magnitude smaller than the atomic nucleus. Any quantum theory in which gravitons and other elementary particles are described as points seems to have unacceptable infinities, whereas the string theories not only need gravity in order to be consistent but are also apparently free from these infinities.

The known string theories are only consistent if, in addition to the usual time dimension, there are nine space dimensions. Invoking more than the usual three space dimensions may seem bizarre, but is nonetheless a perfectly sensible possibility in the context of a quantum string theory. The basic idea is that six dimensions are unobserved because they are too small. They are curled up into some sort of six-dimensional ball. This compactification is an example of spontaneous symmetry breaking: All nine space dimensions occur symmetrically in the fundamental equations, but the physically relevant solution has six of them very small and three very large. In principle, the existence of extra dimensions could be demonstrated by experiments in which particle collisions at high energies produce new kinds of particles with very specific properties. Unfortunately, the size of the extra dimensions that seems most likely is so small that the required energy is very far beyond what will be available in the foreseeable future.

The simplest and most promising theories involve only closed strings with no free ends. In these theories the basic force is described by a single fundamental interaction in which either two strings join to give a single string or one string divides into two. By contrast, the general theory of relativity contains an infinite number of fundamental interactions according to which any number of gravitons can interact at a point. The string theory replaces this complicated structure by the one basic interaction described above. This simplification is one of the novel features that make possible the elimination of infinities from the quantum theory.

Supersymmetry. Another key to the cancellation of infinities is a special kind of symmetry called supersymmetry. Supersymmetry first arose in 1971 in the context of string theory. It was independently introduced for point particles (in four-dimensional space-time) at about the same time. Supersymmetry can be defined as the symmetry of a theory under a mathematical transformation that interchanges bosons and fermions. Bosons and fermions are two fundamental categories into which all particles are divided. Bosons have an integral spin angular momentum, whereas fermions have half-integral spin. Quarks and leptons are examples of fermions, whereas gauge particles are bosons. Supersymmetry implies the existence of new, so far unobserved, particles to serve as partners for each observed particle, that is, a fermionic partner for each known boson and a bosonic partner for each known fermion. The experimental observation of such particles, which could take place in the near future, would be one of the most important discoveries in the history of science.

One of the fundamental attractions here is that all of matter, as it is presently known, is composed of fermions and all the natural forces are transmitted by bosons. A theory that brings fermions and bosons together on an equal footing thus, at least in principle, represents a step in the direction of a complete unification of both matter and forces.

In supersymmetric string theories, called superstring theories, there are cancellations that apparently are sufficient to ensure that well-defined finite results are obtained. (A complete proof is still required.) Analogous cancellations also take place in supersymmetric point-particle theories containing gravity (supergravity theories). However, in those theories the cancellations almost certainly are insufficient to obtain meaningful finite results.

Uniqueness. Even though previous theories successfully describe a wide range of phenomena, they are not entirely satisfactory in a number of regards. For one thing they are incomplete since gravity is omitted. Also, they are not as unique as would be desirable. Their construction requires choosing a particular symmetry group (there are an infinite number of possibilities), choosing how many particles of each spin to include, and assigning them to particular symmetry patterns. It is also necessary to adjust a large number of parameters in the equations. All of these choices are made so that the theory fits the experimental facts as closely as possible. Ideally, a fundamental theory should uniquely determine the symmetry, the types of elementary particles and all their properties, without having to "dial knobs" in the equations. Superstring theories show a great deal of promise in these regards.

In August 1984 it was discovered that supersymmetric theories in 10-dimensional space-time possess quantum-mechanical inconsistencies called anomalies unless the symmetry group is restricted to one of just two possible choices. The allowed symmetry groups turned out to be (in standard mathematical terminology) $SO(32)$ and $E_8 \times E_8$. These are very large symmetry structures, each having 496 generators. There is no problem embedding all known elementary-particle symmetries in either one of them. A superstring theory was developed containing both open and closed strings, with the $SO(32)$ symmetry, and evidence was found that it is free from anomalies and infinities. A few months later, two new superstring theories containing only closed strings, one with $E_8 \times E_8$ symmetry and a second example with $SO(32)$ symmetry, were formulated. These developments have excited much interest. For the first time, it has been demonstrated that requiring a consistent unification with gravity can resolve many of the ambiguities that are present in previous theories. Not only is the symmetry group determined (up to two distinct alternatives), but so is everything else including the complete spectrum of elementary particles (interpreted as oscillation modes of the string), all their interactions, and even the dimensionality of space and time.

Consequences of the theory. A great deal of work is still required to relate superstring theories to phenomena observed in the laboratory. However, the phenomenology of superstrings is progressing rapidly. The prospects, especially for the $E_8 \times E_8$ theory, appear extremely bright. A crucial step required for making contact with experiment is to deduce a detailed geometric description of the six compact dimensions. One possibility that is currently receiving much attention is that they form a Calabi-Yau space, a particular type of six-dimensional space with a number of favorable mathematical properties. These studies are pushing at the frontiers of mathematics, and are resulting in an unprecedented level of communication between physicists and mathematicians.

The $E_8 \times E_8$ superstring theory incorporates many features that were previously identified as important but could not be implemented in a single theory. In this theory they appear automatically. One of these features is an understanding from fundamental principles of why the basic laws are not symmetrical under mirror reflection (parity violation). Another encouraging feature is that the experimentally observed symmetries can readily emerge from one of the two E_8's when spontaneous symmetry breaking occurs as a consequence of six dimensions being compactified.

[JOHN H. SCHWARZ]

Lattice gauge theory. Computer simulations of field theory formulated on a discrete space-time lattice have become an important tool in the study of the interactions between quarks and gauge fields. Recent calculations confirm the phenomenon of confinement, wherein an isolated quark would have infinite energy.

Quark confinement. Quarks are the conjectured fundamental constituents of hadrons, that class of elementary particles which sense the nuclear interaction. The class includes the proton, the neutron, and strongly interacting mesons. A theory where the quarks are bound together by gluonic fields is the prime candidate for an underlying dynamics of the nuclear force.

Although there are several rather compelling reasons to believe in quarks as a new layer of substructure in matter, an isolated quark has never been observed. This is in striking contrast to the empirical observation that in hadronic collisions anything that can be created will be.

A resolution of this enigma is referred to as confinement; as a quark is pulled from a hadron, it experiences a restoring force which remains nonvanishing even for asymptotically large separations. Thus it would take an infinite amount of energy to fully isolate the quark.

The confinement mechanism is believed to involve nonlinear interactions among the gluon fields. In this theory, called a gauge theory because it generalizes the gauge invariance of electromagnetism, the gluon field lines form into tubes of flux connecting the quarks. These flux tubes have a finite energy per unit length; consequently the energy of a quark grows linearly with separation as it is pulled from the other quarks which provide a sink for this flux. Investigations of high angular momentum states yield the strength of this long-range force; in popular units it is about 14 tons (1.3×10^4 newtons).

Space-time lattice. The best evidence that confinement is more than a conjecture comes by way of the formulation of the gauge theory on a space-time lattice. In this approach the continuum of space-time is replaced by a discrete set of points or sites. Quarks move through the structure via sequential hops from site to nearby site (see illus.). In the course of these hops they interact with the gauge gluons which are fields located on the "bonds" of the lattice.

At first the lattice formulation seems quite perverse because there is no evidence that the physical vacuum has any discrete structure, crystalline or other. Theorists are using a mathematical trick to remove some of the divergences which are rampant in quantum field theory. The lattice constitutes an ultraviolet cutoff in that all wavelengths less than twice the lattice spacing cannot propagate. The field theory becomes a well-defined mathematical system, amenable to various calculational schemes, including numerical simulation. Nevertheless, the lattice is just a temporary artifact, and physics requires a final continuum limit, wherein the lattice spacing is taken to zero.

The infinities of field theory are an old problem which has been treated quite successfully by a variety of means for quantum electrodynamics. With the strong nuclear interactions, however, the need for the lattice arises from some rather unique features of confinement. This phenomenon cannot be treated as a perturbation of free quarks and gluons. The unconfined theory with vanishing gauge couplings has no resemblance to the observed physical

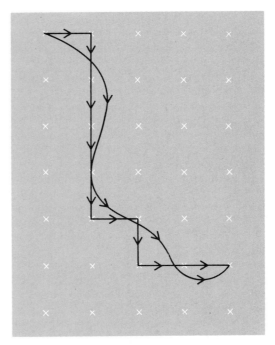

Path of a quark through space and time in lattice gauge theory, approximated by a sequence of discrete steps. Results are to be extrapolated to the continuum limit of vanishing lattice spacing. (*After M. Creutz, High-energy physics, Phys. Today, 36(5):35–42, May 1983*)

world. This contrasts sharply with electrodynamics, where the diagrammatic approach of R. Feynman has played a central role. Most conventional schemes for dealing with the ultraviolet divergences of a field theory are based on this perturbative expansion. To go beyond and study nonperturbative phenomena requires a nonperturbative cutoff. Herein lies the main virtue of the lattice approach, which makes the theory well defined before any expansions or approximations are begun.

The use of a space-time lattice makes manifest the deep connections between field theory and statistical mechanics. In this analogy, the field theoretical coupling constant corresponds directly to temperature. Thus the particle physicist has available all the machinery of statistical mechanics, including high-temperature expansions, mean field theory, and numerical simulation via Monte Carlo methods. It is the latter approach which has generated considerable interest, as it opens the possibility of calculating hadronic properties from first principles.

Monte Carlo simulation. The lattice formulation reduces the solution of a field theory to a problem of evaluating multiple integrals, the dimension of these integrals being proportional to the number of lattice sites under consideration. This suggests, at least for a finite system, attempting a numerical evaluation. The large dimensionality of the integrals, however, makes any direct attack impractical; a 16^4 lattice, typical for many current simulations, has 262,144 bonds, the fields on each of which are to be integrated over. The appearance of large numbers immediately suggests a statistical treatment. The goal of a Monte Carlo simulation is to provide a small number of field configurations which are typical of thermal equilibrium in the statistical analog.

In this context, a Monte Carlo program begins with some initial configuration of the fields, stored as an array in a computer memory. The processor then loops through these variables while making pseudorandom changes in such a manner that configurations are ultimately encountered with a Boltzmann weighting. Effectively, the computer serves as a heat bath at a temperature corresponding to the field theoretical coupling constant.

The Monte Carlo technique provides an opportunity to perform numerical experiments on virtual systems in which the physicist can arbitrarily choose the underlying dynamics. Thus various dynamical features can be isolated and their role studied in such bulk phenomena as phase transitions. Furthermore, as entire field configurations are stored, in principle any desired function of the fields is available for study. The method converges well in both high- and low-temperature regimes and interpolates well in between.

Use of supercomputers. To apply these methods to lattice gauge theory and strong interaction physics, much use is made of supercomputers. The fastest computing machines available are necessary to overcome certain inherent problems. In a Monte Carlo calculation, statistical fluctuations are always present, giving uncertainties which only decrease with the square root of the computer time. An improvement of one order of magnitude in accuracy requires an increase in computer power of two orders of magnitude. In addition to high-speed computations, large computer memories are also necessary. As space-time is four-dimensional, the linear sizes of lattices which will fit on modern supercomputers are still quite limited, with 10 to 20 sites on a side being typical.

Results. Despite these limitations, the simulations of lattice gauge theory have given rather compelling evidence for quark confinement and have produced some remarkable quantitative results on the characteristics of quark interactions. The first calculation to come from this approach was of the numerical strength of the constant force between asymptotically separated quarks. The result is a ratio of the relative strength of this long-range force to the strength of the perturbative interaction of quarks at short distances. The next important result of such simulations was the observation and determination of the physical temperature of a phase transition between conventional matter composed of normal hadrons, and a high-temperature plasma consisting of a thermal gas of quarks and gluons. It is hoped that this phase will be directly observed in high-energy collisions of heavy nuclei. Finally, there have been predictions of new particles which are bound states of the gluons alone. Searches for these quarkless particles, sometimes called glueballs, form an active area of experimental work.

Most of the simulations consider the full dynamics of the gluon fields while treating the quarks as external sources. Calculations that go on to include the kinetic effects of valence quarks, with accuracies of order 10%, have reproduced much of the hadronic spectrum. This is rather remarkable since it involves an approximation of unknown validity. In particular, processes involving excitation of quark-antiquark pairs from the vacuum are ignored.

Advanced calculations. To go beyond this so-called valence approximation is an area of intense ongoing research. Severe technical difficulties arise in numerically treating fields which exhibit the Pauli exclusion principle. Although there are algorithms known which correctly take these effects into account, they typically require several orders of magnitude more computer time than the gluon simulations. Such calculations are a severe strain on existing computer capabilities. Indeed, some theorists have recently begun designing special-purpose processors to obtain the needed computational power at a reasonable cost.

To summarize, lattice gauge theory coupled with Monte Carlo simulation has provided the particle physicist with a powerful tool to study the interactions of quarks and gluons. It has enabled first-principles calculations of hadronic properties. Indeed, at present it seems to be the most promising technique applicable to intermediate and long-range effects. The approach has also played a key role in increasing interdisciplinary activity between specialists in particle physics, statistical mechanics, and computer science.

For background information *see* ELEMENTARY PARTICLE; FUNDAMENTAL INTERACTIONS; GLUONS; GROUP THEORY; MONTE CARLO METHOD; QUANTUM CHROMODYNAMICS; QUANTUM FIELD THEORY; QUARKS; STATISTICAL MECHANICS; SYMMETRY LAWS (PHYSICS) in the McGraw-Hill Encyclopedia of Science and Technology.

[MICHAEL CREUTZ]

Bibliography: P. Candelas et al., Vacuum configuratons for superstrings, *Nucl. Phys.*, B258:46–74, 1985; M. Creutz, *Quarks, Gluons, and Lattices*, 1983; M. B. Green and J. H. Schwarz, Anomaly cancellations in supersymmetric D = 10 gauge theory and superstring theory, *Phys. Lett.*, 149B:117–122, 1984; D. J. Gross et al., Heterotic string theory (I): The free heterotic string, *Nucl. Phys.*, B256:253–284, 1985; C. Rebbi, The lattice theory of quark confinement, *Sci. Amer.*, 248(2):54–65, February 1983; J. H. Schwarz, *Superstrings: The First 15 Years of Superstrings Theory*, 2 vols., 1985; K. Wilson, Confinement of quarks, *Phys. Rev.*, D10:2445–2459, 1974.

Endocrine mechanisms

Over the past few decades, significant progress has been made in understanding the function of the pineal gland. There is a firm experimental basis for stating that the pineal plays a central role in regulating the reproductive activity of seasonally breeding mammals. Beyond that, little is known for certain, but there is a host of intriguing possibilities.

The pineal gland is unique in the vertebrate endocrine system. Endocrine glands function by regulating the internal operation and composition of the body via chemical secretions called hormones. In general, hormones act by either stimulating or inhibiting the functions of their target tissues, and each hormone has specific effects on its targets which are similar in all vertebrates. The pineal, however, provides the body with a hormonal signal that is neither stimulatory nor inhibitory, and that may vary in its effects depending upon species, physiological status, and time of year. Thus, the pineal gland does not function like other endocrine glands and cannot be studied by using traditional endocrine methods. This realization has contributed significantly to recent advances in understanding pineal physiology.

The pineal gland is a neuroendocrine transducer, that is, it receives environmental information from the nervous system and relays it to the body by means of a daily endocrine signal. This internal signal reflects variations in the day-night cycle (photoperiod) and thus provides individuals with a means of coordinating activities, for example, reproduction, with changes in the environment. This signal may also allow the body to synchronize its internal operations. The bulk of pineal research has been done on mammals; therefore, this discussion will deal exclusively with mammalian pineal physiology.

Biological clock. The pineal contains many biologically active compounds. Only one of these, the indoleamine melatonin, has been clearly demonstrated to have a role in pineal physiology. Melatonin levels in the pineal and in body fluids increase at night and decrease during the day in all species studied, not only in diurnally active animals, but in nocturnally active ones as well. This daily rhythm in melatonin levels serves as an internal time signal to the body. However, the biological clock that drives the melatonin rhythm resides in the brain, not in the pineal gland.

The pathways that carry nerve impulses from the brain to the pineal are illustrated in Fig. 1. Impulses generated by light striking the retina are transmitted along nerve tracts to the suprachiasmatic nuclei which function as a central biological clock. Here, photoperiodic information from the environment is evaluated and the operation of the clock adjusted accordingly. Impulses then pass through the paraventricular nuclei and deeper brain structures to the spinal cord. The spinal cord routes these impulses to the sympathetic nervous system, where they pass to the superior cervical ganglia. Sympathetic fibers carry the impulses to their final destination, the pineal gland.

In the pineal, cells synthesizing melatonin receive relatively few impulses from the brain during the day, so they produce relatively little melatonin. At night, however, the number of nerve impulses received increases substantially. This stimulates the

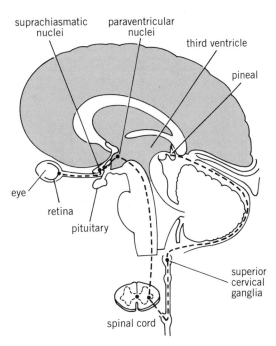

Fig. 1. Midsagittal section of the human brain showing the neural pathway (broken line) from the eye to the pineal gland. (*After L. Tamarkin, C. J. Baird, and O. F. X. Almeida, Melatonin: A coordinating signal for mammalian reproduction?, Science, 227:714–720, 1985*)

melatonin-producing cells to increase their levels of key enzymes used in melatonin synthesis, and this, in turn, causes a large increase in melatonin production. The 24-h melatonin rhythm thus generated functions much like the hands of a clock, that is, it indicates to the body whether it is day or night. Figure 2 illustrates typical daily melatonin profiles for monkeys, rats, and humans.

Variations in the photoperiod affect the daily melatonin rhythm because light has two effects on pineal function. Light of sufficient intensity suppresses melatonin production in all mammals, including humans; constant light abolishes the melatonin rhythm. In marked contrast, a melatonin rhythm persists indefinitely in animals kept in the dark; however, animals living in constant darkness for extended periods, without daily exposure to

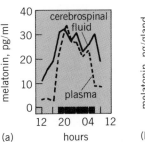 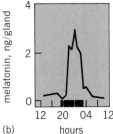 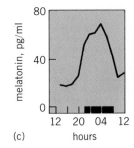

Fig. 2. Daily melatonin profiles in (a) rhesus monkey plasma and cerebrospinal fluid, (b) rat pineal, and (c) human plasma. The bars on the horizontal axes represent the daily period of darkness. (*After L. Tamarkin, C. J. Baird, and O. F. X. Almeida, Melatonin: A coordinating signal for mammalian reproduction?, Science, 227:714–720, 1985*)

light, gradually lose synchrony with the natural photoperiod. The duration and timing of the photoperiod that an animal sees, then, affects the duration and timing of the nocturnal increase in melatonin levels.

Regulation of reproduction. Many animals from temperate and polar latitudes have seasonal life cycles. They experience annual changes in body weight, coat color, physical activity (such as hibernation), and reproduction. Seasonally breeding animals have adapted their physiology to sense annual changes in the environment. They use this information to determine where they are in their annual cycle and when it is time to mate. The timing of breeding is especially important since offspring must be born only during that time of the year that favors survival.

The environmental cue most widely used for a seasonally breeding animal to determine the time of the year is the annual variation in the daily photoperiod. In temperate regions, summer days typically have 14 h of light and 10 h of dark; the reverse is true in the winter. This is reflected by the daily melatonin rhythm. Figure 3 shows how melatonin profiles for two seasonal breeders, the Djungarian or Siberian hamster and the sheep, change with time of year. In both, the duration of the nocturnal increase in melatonin is short when the night is short, and long when the night is long. Other mammals

reflect seasonal changes in their melatonin rhythms in other ways. By mechanisms which are still not understood, these changes in the pattern of melatonin secretion can alter reproductive activity.

There is one important difference between hamsters and the sheep, however. Hamsters are reproductively active during long days (short nights) while sheep are inactive, and vice versa. Each animal responds to the environmental information provided to it via the melatonin signal in the way that it is genetically programmed to respond. Thus melatonin is neutral in its effect, and it neither stimulates nor inhibits. The physiological response of an organism to melatonin depends on species, physiological status, time of year, and other factors.

Other functions of melatonin. Unlike seasonal breeders, many species of mammals, including humans and the laboratory rat, breed all year round. These nonseasonal breeders show no obvious physiological changes at any time during the year, even though they exhibit a daily melatonin rhythm. Further, some strains of laboratory mice carry a mutation that abolishes their melatonin rhythm, and they survive quite well without it. One of the major unsolved problems in understanding pineal physiology is whether nonseasonal breeders need a melatonin rhythm or even a pineal gland at all. Ideas on this subject abound, but some are better supported by experimental evidence than others.

There is evidence that a daily melatonin rhythm is important in regulating the routine activity of the reproductive system. The maintainance of normal ovarian function and the timing of ovulation within the female reproductive cycle are two processes which may be regulated by the daily melatonin rhythm. Melatonin has been shown to regulate the responsiveness of the mammary glands (breasts) and uterus to the female sex hormones. The pineal plays a role in timing the onset of puberty in some mammals; however, no causal relationship between pineal function and the onset of puberty in humans has been demonstrated. Other observations suggest that the pineal may play a role in timing the onset of parturition (birth).

The pineal is hypothesized to influence the function of several endocrine glands in addition to those of the reproductive system, but it is not known whether this effect is direct or indirect via the influence of melatonin on the brain. These glands include the pituitary gland, the thyroid gland, and the adrenal glands. There is also evidence that the thymus may be subject to regulation by the pineal. If this is true, the pineal could play a role in modulating the body's specific, cellular immune defenses. There is even speculation that the pineal may play a role in the body's resistance to cancer.

The possibility that melatonin may regulate behavior has stimulated a great deal of popular interest in pineal physiology. Many people, including some scientists doing research on the pineal gland, believe that "jet lag" is due to the body's melatonin rhythm being out of synchrony with the local day-

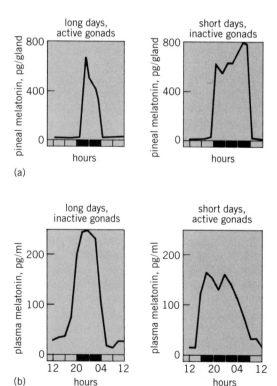

Fig. 3.   Daily profiles of (a) pineal melatonin of Djungarian or Siberian hamsters in long days and short days and (b) plasma melatonin of sheep in long days and short days. Bars on horizontal axes represent the daily period of darkness. (*After L. Tamarkin, C. J. Baird, and O. F. X. Almeida, Melatonin: A coordinating signal for mammalian reproduction?, Science, 227:714–720, 1985*)

night cycle. While a loss of synchrony does occur when a person travels across several time zones, it is not known whether this is the cause of the jet lag. The same people also feel that a person recovers from jet lag when the brain's biological clock resets itself to the new local time. As attractive as this hypothesis sounds, it also has no experimental basis in fact.

Another aspect of behavior which may be controlled by the pineal is seasonal affective disorder. Behavioral scientists have noted that in the fall of the year, when days grow shorter (and nights get longer), some individuals get severely depressed. This depression typically persists until spring, when the days get longer (and the nights shorter). Many seasonally depressed people seem to benefit from a winter vacation to warmer climates where, perhaps not coincidentally, the days are longer and the nights shorter. Research is currently under way to establish whether seasonal alterations in melatonin rhythms occur in humans and if such changes are responsible for seasonal affective disorder. As yet, there are no firm experimental data on the subject.

For background *see* ENDOCRINE MECHANISMS; PINEAL BODY in the McGraw-Hill Encyclopedia of Science and Technology.

[CURTIS J. BAIRD]

Bibliography: B. Fellman, A clockwork gland, *Science 85*, 6(4):76–81, 1985; R. J. Reiter (ed.), *The Pineal Gland*, vols. 1–3, 1981–1982; R. Relkin (ed.), *The Pineal Gland*, 1983; L. Tamarkin, C. J. Baird, and O. F. X. Almeida, Melatonin: A coordinating signal for mammalian reproduction?, *Science*, 227:714–720, 1985.

Energy storage

Superconducting magnetic energy storage is unique among the technologies proposed for diurnal energy storage for the electric utilities in that there is no conversion of the electrical energy, which is stored directly as a circulating current in a large superconducting coil or magnet, into another energy form such as mechanical, thermal, or chemical. Thus, one advantage of superconducting magnetic energy storage is the inherent high storage efficiency that is possible because energy conversion processes are avoided. The actual round-trip efficiency of a large unit is expected to be 90% or greater. The fast response (approximately 10 ms) of the system to power demand means that a diurnal storage unit can also function as a swing generator or provide system stabilization. The major components of a superconducting magnetic energy storage system are a large superconducting coil cooled by liquid helium, an ac-to-dc converter, and a refrigerator that maintains the temperature of the helium coolant (see illus.).

Superconducting magnetic energy storage for diurnal load leveling still requires major technology development to gain practical application. Design studies and analyses continue to improve the engineering feasibility of this type of energy storage. The best estimated costs, combined with the high efficiency, make superconducting magnetic energy storage a strong contender in the economics of future utility energy storage.

Storage requirements and technologies. Variations in electric power usage have required utilities and manufacturers to design power-generating units capable of cycling and to develop methods of storing energy to meet varying power demands. Most of these demands are periodic, but the cycle time may vary from a few seconds to a year. The annual variation is usually accommodated by scheduling power equipment outage and major maintenance for low-demand seasons. The daily and weekly variations, however, are the most important because of the sheer magnitude of the power variations that occur during periods as short as an hour. The change in power demand is often from 60% of the peak load at 7 A.M. to 90% at 9 A.M. On a system with a

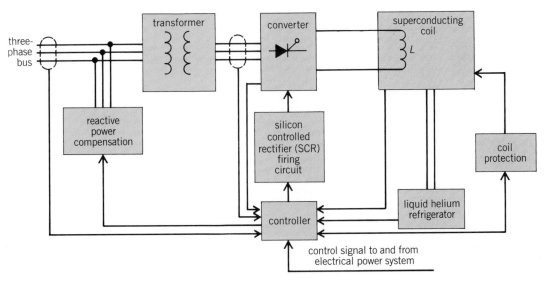

Components of a superconducting magnetic energy storage system.

2000-MW peak load, this variation is about 600 MW. Large coal-fired and nuclear power plants, which are the most efficient generating units, are normally designed to operate at full or nearly full capacity with little or no power variation. Their expected life is considerably decreased when forced to cycle by large fractions of total output capacity. Gas turbines, old and intermediate-sized power plants, and energy storage units, which are designed for cycling or are less affected by changing power levels, are cycled to meet the daily and weekly power variations.

Superconducting magnetic energy storage systems interface with the grid through transformers and the silicon controlled rectifier Graetz bridge converter (see illus.). The charging-discharging rate and the power flow between the three-phase power line or grid and the coil are determined by the amplitude and polarity of the bridge voltage. A phase-controlled converter requires reactive power from the ac bus.

Stabilization versus load leveling. The same concepts and broadly similar circuits as first proposed for diurnal load leveling can be used for small superconducting magnetic energy storage units utilized solely to stabilize power transmission grids to prevent undesirable power oscillations. The distinction between the two kinds of systems is one of scale. Those for stabilization are small, with 8.4 kWh (30 MJ) of energy stored, a 10-MW power rating, and an average coil diameter of 10 ft (3 m), compared to 1–10 GWh (3.6–36 × 10 TJ), 300–1000 MW, and 1000–3000 ft (300–900 m) diameter for a large diurnal load-leveling system. The frequency of power input-output for operation is 0.35 Hz for the small stabilizer in contrast to once a day for a large load-leveling system.

Stabilization unit. A superconducting magnetic energy storage stabilization system was built by the Los Alamos National Laboratory to damp potentially destructive power oscillations on the Western U.S. Power System, particularly on the Pacific AC Intertie that is used to transmit power from the Northwest to southern California. The 30-MJ superconducting inductor that stored energy for this purpose was contained in a nonconducting dewar and supported by a helium refrigerator and a gas-handling system mounted on trailers. Energy flowed in and out of the inductor at frequencies from 0.1 to 1.0 Hz with power amplitudes up to 11 MW. The principal oscillation to be damped has a characteristic frequency of 0.35 Hz. The maximum current of the superconducting coil was 5 kA with terminal voltages up to 2.2 kV. The coil interfaced with the Bonneville Power Administration 13.8-kV bus at the Tacoma Substation through a converter and transformers. The system could be operated with the converter either in parallel bridge mode or for constant reactive power control with the bridges in an opposed mode. An analysis of the Western U.S. Power System response and dynamic characteristics was successfully performed with the system. As the first

large-scale utility application of superconductivity in the United States, the stabiliztion unit proved to be compatible with utility service.

A variety of power-generating and energy storage technologies can satisfy the cyclic power demand. Some, such as gas turbines, hydroelectric, and pumped hydroelectric, have been used widely. Several new technologies, including compressed air, underground pumped hydro, batteries, and superconducting coils, show promise for possible future applications. With adequate development, and assuming reasonable costs can be achieved for these technologies, utilities will be able to select the type of plant to optimize power generation capability in terms of cost and performance.

Principle of superconducting storage. Superconductivity was first observed by H. Kamerlingh Onnes in 1911. Though there were schemes to apply this discovery, none had any chance of success until the discovery of the "hard" or type II superconductors. Superconductors can carry very large currents with no measurable resistive component, and this trait gives superconductors a unique capability for wide application in the electrical and electronics industries.

Though several superconductors are commercially available, alloys of niobium and titanium appear to be the most effective for superconducting magnetic energy storage. This material when operated at 1.8 K ($-456.4°F$) can carry currents up to 5900 A/mm^2 at a field of 5 teslas, which is 300 times greater than the typical operating current density in copper. Energy for superconducting magnetic energy storage is stored in superconducting coils in keeping with the equation below, where E is the energy in joules, L

$$E = 0.5LI^2$$

the coil inductance in henrys, and I the current in amperes in the superconducting winding.

Superconducting system characteristics. Because superconducting magnetic energy storage is inherently very efficient, it has the potential of application in systems with large energy storage requirements and meets many of the utilities' requirements for diurnal storage. An unusual feature of superconducting magnetic energy storage is the cost scaling with size, which is different from that for other storage devices. For a given design, the cost of a superconducting magnetic energy storage unit is roughly proportional to its surface area and the required quantity of superconductor. The cost per unit of stored energy (megajoule or kilowatthour) decreases as storage capacity increases.

The charge and discharge of a superconducting magnetic energy storage unit through a multiphase converter or Graetz bridge allows the system to respond within tens of milliseconds to power demands that could include a change from maximum charge rate to maximum discharge power. This rapid response allows a diurnal storage unit to provide spinning reserve and to improve system stability. Both the converter and the energy storage in the coil are

Design parameters for a diurnal load-leveling superconducting magnetic energy storage system

Parameter	Value
Energy stored	5250 MWh (18.9 TJ)
Power	1000 MW
Maximum current	200 kA
Maximum voltage	10 kV
Coil diameter	3300 ft (1000 m)
Coil height	62 ft (19 m)
Width of trench	23 ft (7 m)
Height of trench	82 ft (25 m)
Number of coil turns	556
Number of coil winding layers	4
Maximum magnetic field	5 T
Liquid helium temperature	1.8 K ($-456.4°F$)
Liquid helium volume	8×10^5 gallons (3×10^6 liters)
Refrigeration power	4.1 MW
Efficiency	90–93%
Cost	$960,800,000

highly efficient because there is no conversion of energy from one form to another. The major loss during storage is the energy required to operate the refrigerator that maintains the superconducting coil at 1.8 K ($-456.4°F$).

Design of load leveling systems. Large diurnal load-leveling superconducting magnetic energy storage systems have received considerable attention since 1971 in a series of evolutionary design concepts. Common to these designs has been the use of superfluid [1.8 K ($-456.4°F$)] liquid helium for maintaining the coil at low temperature, and structural support of the large magnetic loads experienced by the superconducting windings by subsurface load-bearing rock. Superconductors carry more current at lower temperatures—hence, the choice of the 1.8 K ($-456.4°F$) helium bath instead of 4.2 K ($-452°F$), the normal boiling point of helium, to reduce the amount and cost of the superconductor. Ordinary magnets and inductive coils are physically constrained mostly by steel structure; however, because of the huge size of a diurnal load-leveling superconducting magnetic energy storage system, the only economically feasible support is against a rock wall in a trench, just below the Earth's surface. Dimensions and other parameters for the most advanced conceptual design of a large superconducting magnetic energy storage system are presented in the table.

For background information *see* ELECTRIC POWER SYSTEMS; ENERGY STORAGE; SUPERCONDUCTING DEVICES in the McGraw-Hill Encyclopedia of Science and Technology.

[JOHN D. ROGERS]

Bibliography: W. V. Hassenzahl, Superconducting magnetic energy storage, *Proc. IEEE*, 71:1089–1098, 1983; R. J. Loyd, T. Nakamura, and J. R. Purcell, Design improvements and cost reductions for a 5000 MWh superconducting magnetic energy storage plant, *Los Alamos Nat. Lab. Rep.*, LA-10320-MS, 1985; J. D. Rogers et al., 30-MJ superconducting magnetic energy storage system for electric utility transmission stabilization, *Proc. IEEE*, 71:1099–1107, 1983.

Ethylene

Ethylene is a simple alkene gas ($H_2C{=}CH_2$) and plant growth regulator. In recent years its biosynthesis from methionine, including a recycling pathway for sulfur and carbons, has been elucidated. Studies using antagonists of ethylene action indicate that ethylene regulates its own production by autocatalysis and autoinhibition. Conjugation of its precursor 1-aminocyclopropane-1-carboxylic acid (ACC) may also control ethylene synthesis.

Regulation of ethylene biosynthesis. In a pathway that recycles methionine, ethylene is produced from the amino acid methionine via S-adenosyl methionine (SAM), a common biological methyl (CH_3) donor. From SAM the plant derives both ACC, the immediate precursor of ethylene, and methylthioadenosine (MTA). The adenine portion of MTA is hydrolyzed off, leaving methylthioribose (MTR). Studies with radioactive labels have shown that the CH_3—S of the old methionine molecule and the adjacent two carbons of the ribose of MTR [which come from adenosinetriphosphate (ATP) via SAM] are then recycled into a new molecule of methionine. Such conservation enables the plant to maintain high ethylene production even when its supplies of methionine and sulfur are small. The two carbons from ribose go on to become the carbons at the base of the ACC triangle (its unusual cyclopropyl ring) and subsequently those in ethylene. The course of ethylene biosynthesis in all known systems proceeds according to the reaction pathway: methionine → SAM → ACC → ethylene. Synthesis can be regulated by changes in ACC production, in the conversion of ACC to ethylene, or both (see illus.). Differences in ethylene emanation do not necessarily indicate differences in ethylene synthesis; they could represent changes in hormonal binding or metabolism.

ACC synthase. ACC synthase, the enzyme which converts SAM to ACC, is probably cytoplasmic. It is very pH-sensitive, but has been partially purified from wounded tomato fruits and is believed to weigh about 57,000 daltons. Studies with inhibitors and effectors indicate that ACC synthase is a pyridoxal phosphase enzyme. When ethylene production is induced, new synthesis of the enzyme is often initiated, making formation of ACC the rate-limiting step. Since polyamines are also made from SAM, their synthesis occurs at the expense of ACC synthesis, and vice versa.

Ethylene-forming enzyme. The existence of an ethylene-forming enzyme catalyzing conversion of ACC to ethylene was formerly doubted, but of the four stereoisomers of the ACC analog 1-amino-2-ethylcyclopropane-1-carboxylic acid, only one is oxidized in plants to 1-butene, the ethylene analog. Stereospecificity seems to indicate enzymatic con-

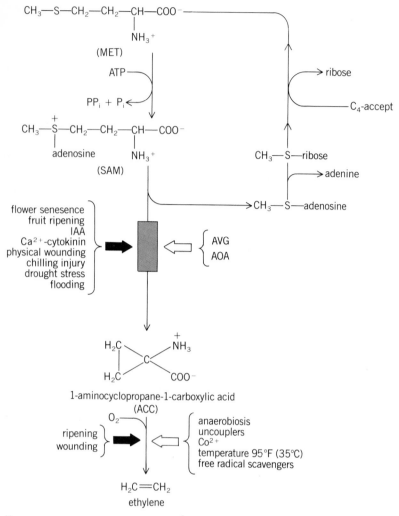

Key:

⇦ induction of synthesis of the enzyme

⬛➡ inhibition of the reaction

Regulation of ethylene biosynthesis. Shaded area indicates a reaction that is normally suppressed and is the rate-limiting step in the pathway. MET = methionine; PP$_i$ = inorganic pyrophosphate; P$_i$ = inorganic phosphate; AVG = aminoethoxyvinylglcine; AOA = aminoxyacetic acid; IAA = indoleacetic acid (auxin); SAM = S-adenosyl methionine. (*After S. F. Yang, Biosynthesis of ethylene and its regulation, in J. Friend and M. J. C. Rhodes, eds., Recent Advances in the Biochemistry of Fruits and Vegetables, pp. 89–106, Academic Press, 1981*)

in the vacuole, perhaps on the inner tonoplast. Isolated vacuoles are not able to form ACC, but can convert ACC to ethylene. It could also be on the inner mitochondrial membrane; pea mitochondria make ethylene from ACC at a rate 10 times that of whole tissue. When ionophores increase the entry of ACC into the organelles, ACC-dependent ethylene production rises 60- to 70-fold. Submitochondrial particles have even higher ethylene-forming enzyme activity.

Conjugation of ACC. Not all ACC becomes ethylene. The N atom of the hormone precursor is also conjugated to a malonyl group from malonyl CoA by a peptide linkage to become *N*-malonylamino-1-cyclopropane-1-carboxylic acid. MACC is not a source of ethylene but represents a dead-end metabolite since the addition of this malonyl group to ACC is essentially irreversible. Conjugation of ACC may be a by-product of diverting otherwise toxic D-amino acids, which also undergo malonylation. (ACC is recognized as either D or L, since it has no asymmetric carbon.) D-Alanine and D-methionine inhibit conversion of ACC to MACC, stimulating ethylene evolution. Since ethylene production can be promoted by reducing malonylation, biosynthesis may be partially regulated by the conjugation of ACC.

Translocation of ACC. ACC is sometimes synthesized in a tissue distant from the target of ethylene action. It is produced in flooded roots and transported to the aerated shoot, where it is converted to ethylene, causing formation of aerenchyma (a hollow stem), initiation of crown roots, and epinasty (drooping leaves). Carnation petals produce ethylene, wilt, and die immediately after pollination. ACC transmitted on the pollen may be the substrate for early ethylene production. Wound-induced ACC then forms in the pollen tube. Some ACC is oxidized by the ethylene-forming enzyme there, but removal of that ethylene does not prevent senescence or petal ethylene production. ACC is also transported to the petals, where it is converted into more ethylene, beginning autocatalytic synthesis. The rise in ethylene evolution precedes an increase in the concentration of ACC, indicating translocation of ACC.

Control of synthesis. Ethylene synthesis is naturally regulated by ethylene concentration. Both promotive and inhibitory effects can be observed by using silver ions (Ag$^+$), norbornadiene (NBD), or carbon dioxide—competitive inhibitors of ethylene binding and action but not of ethylene synthesis. Differences in the amount of ACC or ethylene produced with and without Ag$^+$ or NBD indicate the extent of autocatalysis or autoinhibition. In closed systems, these strong effects may overwhelm effects of other treatments.

Autocatalysis (where a small amount of exogenous ethylene causes a tremendous increase in endogenous ethylene formation) is common in fruits, flowers, and leaves. In unripe tomato and cantaloupe, exogenous ethylene markedly enhances ACC-dependent ethylene production; ethylene promotion of the ethylene-forming enzyme activity precedes an in-

version conclusively, since chemical means do not prefer any configuration.

The ethylene-forming enzyme has not been isolated, but is thought to be membrane-bound because agents such as detergents, osmotic shock, cold shock, and high temperatures destroy its activity. The conversion requires oxygen; anaerobic systems show enhanced ACC production, but will form no ethylene unless provided with some O$_2$. The ethylene-forming enzyme appears to require energy, since uncouplers inhibit its activity. Although some peroxidase inhibitors also limit ethylene production, the enzyme is probably not a peroxidase, since peroxidases do not convert ACC to ethylene in a cell-free system. The ethylene-forming enzyme may be

crease in ACC synthase.

Exogenous ethylene inhibits auxin-induced ethylene biosynthesis from methionine in tobacco leaf disks by reducing the concentration of free auxins and of free ACC. The latter is mediated through suppression of ACC formation and increased conjugation of ACC to MACC. In unripe tomatoes, ethylene again markedly stimulates malonylation of ACC and D-amino acids. White light reversibly inhibits ACC conversion to ethylene in senescing or wilting leaves. In the light, photosynthesis lowers the internal CO_2 concentration. In the dark, CO_2 could bind to the ethylene binding site, preventing autoinhibition while causing ethylene emanation to increase.

Effect of other chemicals. Ethylene production is affected by all other plant hormones. In most systems, application of auxin induces or increases ethylene evolution, particularly by initiating the de novo synthesis of ACC synthase and enhancing its activity. Auxin sustains ethylene-forming enzyme activity in cultured pear cells, perhaps by maintaining membrane integrity. Sugars alone enhance both steps of ethylene production in tobacco leaf disks, but sucrose also stimulates ethylene emanation by increasing auxin uptake and hydrolysis of auxin conjugates.

Calcium ions (Ca^{2+}) promote the formation of ACC in mung bean hypocotyls, senescing cucumber cotyledons, and ripening apples. In the latter tissue, Ca^{2+} also increases ethylene-forming enzyme activity, and may stimulate uptake of exogenous ACC.

Conditions inducing evolution. Fruit ripening is the best-known developmental event triggering ethylene formation. Climacteric fruits give off a burst of ethylene just before a transient rise in respiration that precedes ripening. Seed germination, pollination, senescence of flowers and leaves, and periods of active cell division all feature heightened ethylene production.

Generally, environmental factors inducing ethylene synthesis are stressful. These include both drought and flooding; chilling; air pollutants such as SO_2; pathological infections; cutting, crushing, rubbing, shaking, or bending plant parts; and transfer of cultured cells. The amount of ethylene evolved varies with genotype; a drought-resistant wheat gives off more ethylene when wilted than does a drought-sensitive one.

Application of light prior to or during the induction of ethylene evolution reduces the amount of ethylene observed. Wilted wheat leaves give off levels of ethylene similar to those detected in darkness for at least 4.5 h after being transferred to light. Biosynthesis of a precursor or release of the hormone may be enhanced by darkness.

Activity. Plants' responses to ethylene vary with species, tissue, and stage of development. Inhibitory effects are well known, but the phytohormone also promotes growth. Although ethylene inhibits elongation in pea stems, it stimulates internodal elongation of deep-water rice.

Ethylene has many effects on cell walls. In pea stems, it limits longitudinal growth but promotes radial growth (thickening). Cytoplasmic microtubules reverse directions, causing cellulose to be deposited in the wall in a transverse rather than longitudinal orientation. At the same time, longitudinal walls become less sensitive to acid-induced loosening, but transverse walls increase in sensitivity. Ethylene can decrease the extensibility of the wall, inhibiting lateral expansion of leaves as well as elongation. When cell enlargement is prevented, cells take up less water. The resulting lack of dilution raises the cell sap osmolality. Ethylene enhances synthesis of the hydroxyproline-rich glycoprotein of the cell wall, and modifies amounts of other wall constituents and activities of enzymes.

Ethylene often turns on catabolic enzymes, especially cellulases. This accelerates storage-protein hydrolysis, chlorophyll deterioration, fruit softening, and complete cellular breakdown in abscission layers and aerenchyma. However, in anabolic systems it induces synthesis of RNA, proteins, and pigments.

Ethylene causes seeds and buds to be released from dormancy or apical dominance, initiation of crown roots, formation of hypocotyl hooks in emerging legumes, induction of flowering in pineapple, and floral sex changes in cucumber. The hormone is involved in epinasty, leaf movements of thigmonastic plants, and changes in geotropic responses. Commercially, ethylene is used to promote fruit ripening and abscission at harvest. Its antagonists prevent the untimely ripening of produce.

Binding and metabolism. Ethylene uptake by tissues is a linear function of concentration from 1 to 1000 nanoliters per liter. The hormone has reported binding sites on the inner and outer sides of cell walls and on endoplasmic reticulum and protein body membranes. Recently an ethylene-binding site was partially purified from bean membranes. It is a hydrophobic, asymmetrical protein with a shielded negative charge. The fact that ethylene-induced effects are inhibited by silver ions implicates a possible sulfhydryl binding site. Ethylene is oxidized by plants to CO_2, ethylene oxide, or ethylene glycol. Its metabolism is inhibited by cold temperatures or CO, but not by CO_2.

For background information *see* ETHYLENE; PLANT GROWTH; PLANT HORMONES; PLANT METABOLISM in the McGraw-Hill Encyclopedia of Science and Technology.

[ELISE ROSE]

Bibliography: E. M. Beyer, P. W. Morgan, and S. F. Yang, Ethylene, in M. B. Wilkins (ed.), *Advanced Plant Physiology*, pp. 111–124, 1984; S. Philosoph-Hadas, S. Meir, and N. Aharoni, Autoinhibition of ethylene production in tobacco leaf disks: Enhancement of 1-aminocyclopropane-1-carboxylic acid conjugation, *Physiol. Plants* (Copenhagen), 63:431–437, 1985; C. J. R. Thomas, A. R. Smith, and M. A. Hall, Partial purification of an ethylene-binding site from *Phaseolus vulgaris* L. cotyledons, *Planta*, 164:272–277, 1985; S. F. Yang and N. E. Hoffman, Ethylene biosynthesis and its regulation in higher plants, *Annu. Rev. Plant Physiol.*, 35:155–189, 1984.

Euglenida

Historically, the unicellular euglenoid flagellates have presented an evolutionary enigma. The group contains green organisms with chloroplasts (phototrophs), with a nutrition based on photosynthesis, as well as colorless organisms which depend on capture of prey (phagotrophs) or absorption of nutrients directly from the environments (osmotrophs). As a result, euglenoids are classified both as algae and as protozoa—the phototrophs in the algal class Euglenophyceae and the phagotrophs in the protozoan order Euglenida. However, very distinctive ultrastructural characters in both green and colorless groups establish that they are closely related and must therefore be grouped together.

Recently, new ultrastructural evidence has established the presence of a vestigial phagotrophic apparatus in two phototrophic genera, *Colacium* and *Euglena*. Its position in these organisms as well as its association with the cytoskeleton supports the proposal that the phagotrophic euglenoids are related to primitive protozoa, the kinetoplastid flagellates. This suggests that the phototrophic euglenoids are derived from a phagotrophic ancestor which acquired chloroplasts, by ingesting and establishing a permanent relationship either with photosynthetic green algal cells or with green algal chloroplasts. The osmotrophic forms are believed to be derived subsequently from either of the phagotrophs or from the phototrophs and will not be considered here.

General structure. Euglenoids generally are elongate cells, although some genera may be spherical or flattened. Two flagella are attached at the anterior end, where they are inserted in a flagellar pocket. The cell membrane is associated with a thick, proteinaceous layer forming a pellicle. The pellicle is organized in discrete strips or ridges which are underlain by, and presumably supported by, characteristic bands of microtubules. The nucleus is located near the center of the cell and, characteristically, contains an endosome and small chromosomes which remain condensed throughout interphase.

Phagotrophic euglenoids. These euglenoids lack chloroplasts. Their pellicular strips tend to be broad. The flagellar pocket opens at the anterior end of the cell close to the opening of the phagotrophic apparatus. The two openings lie together in a depression or groove, the vestibulum. Both flagella may extrude from the flagellar pocket as in *Peranema* (Fig. 1*a*). One flagellum (the leading flagellum) generally extends straight forward from the cell. It provides the swimming motion and may also be tactile in function. The other flagellum (the recurrent flagellum) turns posteriorly and lies along the side of the cell. This flagellum lies between the cell and the substrate when the cell is gliding along a surface. In phagotrophs such as *Calycimonas*, only one flagellum (the leading flagellum) is emergent from the flagellar pocket (Fig. 1*b*). The other flagellum is very short and lies within the flagellar pocket. The contractile vacuole, if present, empties into the flagellar pocket.

Phototrophic euglenoids. These euglenoids contain chloroplasts. The pellicle is made up of narrow strips. The flagellar pocket of the phototrophs (the reservoir) does not open at the vestibulum, as in the phagotrophs (compare Fig. 1*c* with *b*). It lies deeper in the cell and is connected to the outside by a tubular "canal." The canal is lined by microtubules which can be traced to the microtubules underlying the pellicle. Therefore, the canal is an extension which is formed by the invagination of the pellicle surface. The reservoir of the phototrophs is homologous with the flagellar pocket of the phagotrophs. It contains the bases of the two flagella as well as the photosensory apparatus—the eyespot granules (stigma) which lie in the cytoplasm just outside the reservoir membrane, as well as the paraflagellar

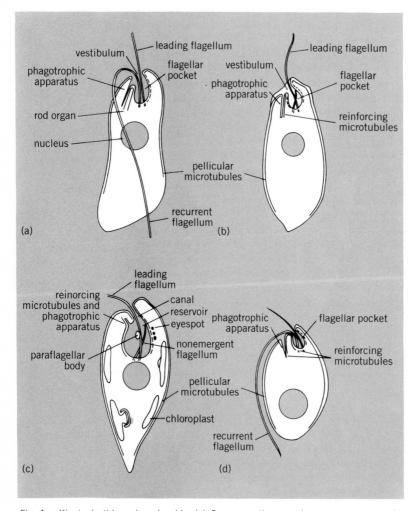

Fig. 1. Kinetoplastids and euglenoids. (*a*) *Peranema*; the nemadesm may serve as reinforcing microtubules. (*b*) *Calycimonas*. (*c*) *Colacium*. (*d*) *Bodo*. (*After R. L. Willey and R. G. Wibel, A cytostome/cytopharynx in green euglenoid flagellates (Euglanales) and its phylogenetic implications, BioSystems, 18:369–376, 1985)*

body which is attached to the base of the leading flagellum on a level with the eyespot. The two flagella of the phototrophic euglenoids may both emerge from the reservoir (through the canal), as in *Eutreptia*, or as in the commonly known genus *Euglena*, one flagellum is emergent and locomotory whereas the other is restricted to the reservoir. The contractile vacuole, if present, empties into the reservoir.

Phagotrophic apparatus: euglenoid. In the phagotrophic euglenoids, the phagotropic apparatus is a membrane-lined pocket which opens at the vestibulum close to the opening of the flagellar pocket (Fig. 1a and c). Associated with the membrane of the phagotrophic apparatus are four "vanes" of microfilamentous material which radiate into the cytoplasm, and a set of microtubules which may be in the form of a single band, as in *Calycimonas* (Fig. 2a), or one or more complex rods (rod organs) as in *Peranema* and *Entosiphon* (Figs. 1a and 2d). During feeding, in *Peranema* the rods extend forward to expand and brace the opening of the phagotrophic apparatus. The enlargement of the opening serves to aid in engulfing prey but, at the same time, occludes the opening of the flagellar pocket.

Until recently, there was no structure known in the phototrophic euglenoids which might represent a primitive or vestigial homolog of the phagotrophic apparatus. Occasionally, the reservoir has been mistakenly identified with phagotrophic activity and the canal described as a gullet. However, the reservoir is, instead, homologous with the flagellar pocket and is not known to take in, much less digest, particulate matter or prey. However, in the green phototrophic genera, *Colacium* and *Euglena*, there is a small pocket which opens into the anterior end of the reservoir at the reservoir-canal transition region (Fig. 1c). Since this transition region is equivalent to the phagotrophic vestibulum, the position of the pocket opening is directly equivalent with the phagotrophic apparatus. The organization of the pocket is not unlike that of the phagotrophic *Calycimonas*—membrane-bound, with microfilamentous strands and a band of reinforcing microtubules (compare Fig. 2b and a). The same pattern is seen in the phagotrophic apparatus of *Peranema* if the microtubules of the rod organs are considered homologous with the reinforcing microtubules (Fig. 2d).

Phagotrophic apparatus: kinetoplastid. The protozoa with which the euglenoids have been frequently associated on ultrastructural and biochemical grounds are the kinetoplastid flagellates, Kinetoplastida. The kinetoplastids include the parasitic trypanosomatids (one of which causes sleeping sickness in humans), and the free-living bodonids (tiny, biflagellated cells common in fresh-water ponds). Comparison of the pocket of *Colacium* with the phagotrphic apparatus of *Bodo* and *Cryptobia* reveals extraordinary similarities of structure (compare Fig. 2c and b). The phagotrophic apparatus of the kinetoplastid is membrane-bound, braced by a microtu-

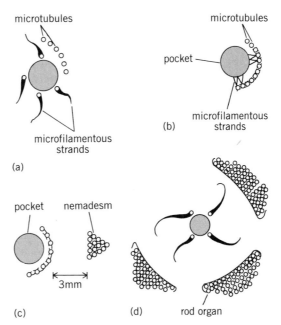

Fig. 2. Cross sections (diagrammatic) through the phagotrophic apparatus of (a) *Calycimonas*, (b) *Colacium*, (c) *Bodo*, and (d) *Entosiphon* to illustrate the similarities of structure. (*After R. L. Willey and R. G. Wibel, A cytostome/cytopharynx in green euglenoid flagellates (Euglanales) and its phylogenetic implications, BioSystems, 18:369–376, 1985*)

bular band which appears, in *Cryptobia*, to be connected with the membrane by microfilamentous strands and, in *Bodo*, has a small rod organ (also known as nemadesm) similar to those of *Entosiphon* and *Peranema*. In the bodonids, the phagotrophic apparatus opens at the anterior end of the cell a short distance from the flagellar pocket rather than in a common opening, such as the vestibulum of the phagotrophic euglenoids (Fig. 1c).

Further evidence of homology of the phagotrophic apparatus of the bodonid and euglenoid flagellates comes from the organization of the cytoskeleton. In the bodonids, one of the microtubular bands lining the flagellar pocket curves back to act as a support for the phagotrophic apparatus—the reinforcing microtubules (Fig. 1d). The homology of this band in *Bodo* and *Colacium* is sufficiently close to permit use of the same name for it in both genera (compare Fig. 1d and c). The pathways of the microtubular bands in the phagotrophic euglenoids are not yet known and are under active investigation. However, the complex microtubular rod organs in *Peranema* and *Entosiphon* are close to the position in which the reinforcing microtubules are expected to be found; temporarily, the nemadesms can substitute for the reinforcing microtubules until more information is available. However, it is in the simple reinforcing microtubules of *Colacium* that the close relationships of the bodonid and euglenoid phagotrophic apparatus can be demonstrated.

Phylogenetic implications. What does the existence of a small, microtubular-braced pocket at the reservoir-canal transition region in phototrophic euglenoids suggest concerning phylogenetic relationships in the group as a whole? First, the ultrastructural homologies of the pocket with the phagotrophic apparatus of the phagotrophic euglenoids and the kinetoplastid zooflagellates support the recent linkage of the euglenoids with the kinetoplastids in a common phyletic group known as the Euglenozoa. Second, the position of the pocket at the reservoir-canal transition region suggests an evolutionary direction which has, until now, been lacking. The opening of the phagotrophic apparatus of both the kinetoplastids and the phagotrophic euglenoids independent of the flagellar pocket suggests that this character is primitive. The opening within the reservoir of the phototrophic euglenoids is probably derived. It is highly unlikely that the inner opening gave rise to the outer opening by evagination since the inner opening is restricted to only two known genera and is found in no other group. Indeed, an evolutionary sequence can be constructed in which the phototrophic euglenoid pocket structure can be derived from the phagotrophic apparatus of the kinetoplastids by (1) rotation of the phagotrophic apparatus opening toward the opening of the flagellar pocket until they both lie in a common vestibulum (as in the phagotrophic euglenoids; Fig. 1a and b), and then (2) invagination of the vestibulum resulting in a deeply situated flagellar pocket connected to the external environment by a canal lined with microtubules derived from the pellicle (compare Fig. 1a–d). As a result, the opening of the vestigial phagotrophic apparatus would open at the reservoir-canal transition region as observed in *Colacium* (Fig. 1c).

The selective pressures leading to reservoir and canal formation appear to be associated with phototrophic nutrition. It can only be speculated that the development of the photosensory apparatus as well as increased water (perhaps, also, protein and carbohydrate) transport associated with the contractile vacuole region, resulting from increased osmoregulatory activity, may have contributed to the observed size increase and invagination of the reservoir.

If the phototrophic euglenoids are derived from the phagotrophic euglenoids, then the chloroplasts of the phototrophs must be secondarily acquired, and must not be a primitive character developed during early prokaryote-eukaryote evolution. Several biochemical and ultrastructural characters of the euglenoid chloroplasts are similar to those of the green algae. It had already been suggested that the chloroplasts of *Euglena* may have originated by the ingestion by a phagotrophic cell of either (1) green algal chloroplasts or (2) green algal cells, with a resulting integration of the chloroplasts or cells (reduced over evolutionary time) into the host (phagotroph) physiology. The implication that phototrophic euglenoids are derived from the phagotrophic euglenoids, established by the presence and position of the vestigial phagotrophic apparatus, supports strongly this endosymbiotic hypothesis for the origin of euglenoid chloroplasts. Therefore, the euglenoid

flagellates can be considered to have evolved from the kinetoplastid protozoa with the subsequent derivation of the green, phototrophic euglenoids by some endosymbiotic event.

For background information *see* ALGAE; EUGLENIDA; PLANT EVOLUTION; PROTOZOA in the McGraw-Hill Encyclopedia of Science and Technology.

[RUTH L. WILLEY]

Bibliography: J. O. Corliss, The kingdom Protista and its 45 phyla, *BioSystems*, 17:87–126, 1984; J. F. Frederick (ed.), Origins and evolution of eukaryotic intracellular organelles, *Ann. N. Y. Acad. Sci.*, 361:193–208, 1981; P. A. Kivic and P. L. Walne, An evaluation of a possible phylogenetic relationship between the Euglenophyta and Kinetoplastida, *Orig. Life*, 13:269–288, 1984; G. F. Leedale, Phylogenetic criteria in euglenoid flagellates, *BioSystems*, 10:183–187, 1978; W. H. R. Lumsden and D. A. Evans (ed.), *Biology of the Kinetoplastida*, 1976; R. L. Willey and R. G. Wibel, A cytostome/cytopharynx in green euglenoid flagellates (Euglenales) and its phylogenetic implications, *BioSystems*, 18:369–376, 1985; H. W. Woolhouse (ed.), *Advances in Botanical Research*, 1983.

Facies (geology)

The blueschist facies is defined to include both glaucophane-bearing rocks (commonly metavolcanics) and associated isofacial nonglaucophanic metamorphic rocks. Transitional assemblages from a typical blueschist assemblage of glaucophane + lawsonite + chlorite + sodic pyroxene + sphene to a greenschist assemblage of actinolite + epidote + chlorite + albite + quartz + sphene are common in major high-pressure metamorphic terranes. Recent experimental and field studies indicate that both blueschist and greenschist assemblages may be stable over a considerable pressure-temperature (P-T) range. Hence the blueschist-greenschist facies boundary is not a unique line on a P-T diagram; instead, depending on the bulk rock composition, the transition may occur over a range of P-T conditions.

Tectonic settings. Blueschist and related metamorphic rocks occur in elongate, narrow zones where they are generally associated with rocks of oceanic affinities such as ophiolites and abyssal sea sediments. Blueschists have a restricted world distribution and invariably are located within tectonic belts that occur principally around the margins of the Pacific Ocean, in the Caribbean and Indonesian regions, and the Alpine-Himalayan regions. For example, in western America, blueschist and related rocks are exposed discontinuously in the coastal belts of Baja California, California, southwestern Oregon, Washington, and Alaska. The most commonly encountered protoliths for blueschist facies rocks are mafic volcanics, graywacke, shale, and cherts characteristic of deep-basin accumulations.

The pressure-temperature conditions of blueschist facies metamorphism are distinct and separate from other types of metamorphism; an extremely low geothermal gradient on the order of 8–15°C/km (23–44°F/mi) is required. Minimum depths of formation are about 10–15 km (3–4 kilobars), or 6–9 mi (300–400 megapascals), and temperatures about 250°C (480°F). Maximum inferred depths are approximately 30–45 km (9–13 kilobars), or 18–81 mi (900–1300 megapascals), for blueschist belts in Japan, California, and the Alps, and highest inferred temperatures are on the order of 450–550°C (840–1020°F). These low-temperature, high-pressure metamorphic suites are confined to Paleozoic and younger orogenic terranes.

The occurrence of high-pressure and low-temperature minerals, lithology of mixed oceanic and deep-basin protoliths, and restricted distribution suggests that blueschist facies metamorphism may develop within a geosuture (a boundary zone between contrasting tectonic units) between pairs of convergent lithosphere plates by either great depths of burial in a subduction zone or tectonic thickening related to the emplacement of ophiolite by obduction. Slow uplift of these high-pressure complexes from great depth provides a retrograde P-T path that ensures partial or complete conversion of blueschist assemblages to greenschists, as in the Alps. On the other hand, rapid uplift of blueschist facies rocks along a convergent plate junction may preclude the back reaction to more "normal" greenschist assemblages, as in western California. Postmetamorphic thrusting and recumbent folding commonly obscure evidence of thermal- or pressure-gradient trends, and the tectonic details during blueschist facies metamorphism remain obscure. The common change of blueschists to higher-temperature and lower-pressure metamorphic facies within some geosutures shows that the preservation of blueschists within Precambrian cratons is quite unlikely, as later metamorphism or plutonism tends to destroy blueschist minerals.

Occurrences. High-pressure metamorphic terranes with associated blueschist and greenschist facies assemblages can be interpreted as (1) greenschist protoliths for blueschist facies metamorphism, such as in Anglesey, United Kingdom; (2) blueschist retrograded to later greenschist during emplacement, such as in the Alps and Japan; or (3) contemporaneous formation of blueschist and greenschist at depths, such as those in the western United States and many other blueschist belts elsewhere in the world. Textural relations and age dating are evidence for the distinction of these three different origins.

Blueschist, greenschist, and transitional assemblages occur as intimate interlayers in foliated schists or in the rims and cores of unfoliated pillow basalts in some blueschist terranes. Subtle changes in color and in the occurrence of blue sodic amphibole and green actinolite as small-scale intercalations are attributed to the difference in total iron, Fe_2O_3 content, and Na/Ca ratio (all high in blueschists). The chemical features are in part established by magmatic processes but largely by differences in the sea water/rock ratio during postigneous alterations. Heterogenous alteration of pillow lavas and other fragmental deposits, followed by intense flattening-shearing during metamorphism, provides a mechanism for generating blueschists and greenschist interlayered on the centimeter scale.

Experiments. Pressure-temperature conditions for metamorphic reactions can be experimentally determined. Arguments favoring deep burial for formation of blueschists on the order of 20 to 30 km (12 to 18 mi) or more are derived from a comparison of the observed mineral parageneses with experimentally determined phase equilibria. The stability of glaucophane with end-member composition $Na_2Mg_3Al_2Si_8O_{22}(OH)_2$ has been extensively investigated; nearly all experiments consistently indicate that glaucophane requires high pressure and low temperature for its formation. However, application of the stability of glaucophane in its own bulk composition to natural blueschists is significantly limited by the compositional complexity of sodic amphibole, by the variation in stability of glaucophane with different compositions, and by the variation in coexisting phases in natural blueschists.

Pressure-temperature positions of two reactions have been recently determined in order to evaluate the nature of the blueschist-greenschist transition boundary. The results as shown in Fig. 1 locate the boundary for the first reaction at about 350°C, 7.8 kilobars (660°F, 780 MPa) and 450°C, 8.2 kilobars (840°F, 820 MPa), and that for the second reaction

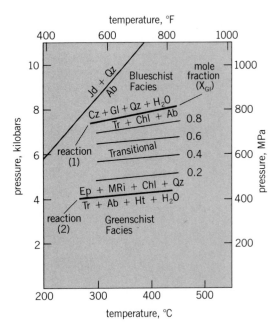

Fig. 1. A pressure-temperature (P-T) diagram showing experimentally determined stability fields of blueschist, transitional, and greenschist facies assemblages. Reaction (1) is for Fe-free basaltic composition and reaction (2) is for Fe^{3+}-saturated system. Within the wide P-T transitional field, reaction (1) displaces continuously toward lower pressure as Fe^{3+} increases in the system as expressed by the mole fraction of glaucophane component in sodic amphibole (X_{Gl}). Ab = albite; Chl = chlorite; Cz = clinozoisite; Ep = epidote; Gl = glaucophane; Ht = hematite; Jd = jadeite; MRi = Mg-riebeckite; Qz = quartz; Tr = tremolite.

at about 300°C, 4 kilobars (570°F, 400 MPa); both reactions have similar gentle P-T slopes. The data derived from this work emphasize that the stability of glaucophane is highly dependent on starting materials and bulk composition (hence mineral assemblage) and that the boundary for the blueschist-greenschist transition is governed by a continuous reaction. In the P-T region between these two reactions, there is a blueschist-greenschist transitional assemblage of sodic amphibole + actinolite + epidote (+ albite + chlorite + quartz). Compositions and proportions of these phases for a given basaltic bulk composition change systematically as a function of P and T. Reaction (1) shown in Fig. 1 defines the maximum stability for the greenschist assemblage in the Fe-free basaltic system. With gradual introduction of Fe_2O_3 into the system, this reaction is displaced continuously toward lower pressure, and both epidote and sodic amphibole increase their Fe^{3+} contents (Fig. 2). In the Fe^{3+}-saturated system, reaction (2) occurs at pressures of about 4 kilobars (400 MPa) and 300°C (570°F). So long as both hematite and magnetite occur, this re-

action remains at fixed pressure isothermally and has fixed mineral compositions for a given bulk composition.

Significance. The pressure-Fe^{3+} content relations (Fig. 2) indicate that the composition (Al_2O_3 content) of sodic amphibole coexisting with epidote + actinolite + chlorite + albite + quartz decreases systematically with decreasing pressure and can be used as a pressure indicator (for example, a geobarometer) for metamorphism. Pressure estimates for selected blueschist terranes in California, Japan, New Zealand, and New Caledonia using this method are in good agreement with those derived from sodic pyroxene geobarometry.

Mineral assemblages and approximate compositions of sodic amphiboles at the pressures P_1, P_2, and P_3 shown in Fig. 2, and at temperature of 300°C (570°F) are illustrated in Fig. 3. Tie lines are schematically drawn for the coexisting phases. Complete solid solution is assumed for glaucophane–Mg-riebeckite and for clinozoisite-epidote. These three diagrams (Fig. 3) illustrate the paragenetic and compositional variations of blueschist-greenschist transitional assemblages as a function of bulk composition. For example, common basaltic rocks have compositions between those of epidote and actinolite, whereas iron-stones may be very oxidized and contain abundant hematite. At pressure P_3 where reaction (1) occurs, basaltic rocks contain the typical blueschist assemblage epidote + glaucophane (+ Ab + Chl + Qz + sphene), whereas iron-stones may contain epidote + Mg-riebeckite + hematite. At intermediate pressures, common metabasites contain the transitional assemblage epidote + actinolite + sodic amphibole, whereas low-Fe_2O_3-bearing basalts have only a greenschist assemblage. Hence blueschist, transitional, and greenschist assemblages with decreasing bulk rock Fe^{3+}/Al ratio could be interlayered and could be formed at identical pressure-temperature conditions. If pressure continuously decreases, both epidote and sodic amphibole may systematically vary their Fe^{3+}/Al ratios. At pressures lower than P_1, basaltic rocks with various Fe^{3+}/Al ratios contain only greenschist facies assemblages.

Thus, the blueschist-greenschist facies boundary is a typical example of sliding equilibria for metamorphic facies transitions that are defined by continuous reactions. Depending on the bulk composition, blueschist, greenschist, and transitional assemblages are stable over a considerable P-T range and may form during one-stage recrystallization. Coexistence of these assemblages in the same outcrop or in a single specimen has been documented in many blueschist terranes. Such one-stage metamorphism should be differentiated from the two-stage recrystallization with greenschist after blueschist facies metamorphism or vice versa in certain orogenic belts.

For background information see BLUESCHIST METAMORPHISM; FACIES (GEOLOGY); METAMORPHIC

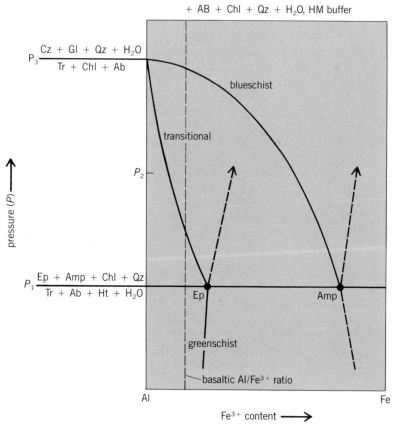

Fig. 2. Pressure-Fe^{3+} content plot at temperature of 300°C (570°F) showing qualitative change in compositions of sodic amphiboles and epidote for buffered assemblage in the transitional zone: (1) the solid lines are for epidote–tremolite–sodic amphibole and (2) the broken lines for epidote–hematite–sodic amphibole. P_1 and P_3 are respectively the equilibrium pressure for the reactions (2) and (1). Ab = albite; Amp = amphibole; Chl = chlorite; Cz = clinozoisite; Ep = epidote; Gl = glaucophane; Ht = hematite; Qz = quartz; Tr = tremolite. (After S. Maruyama, M. Cho, and J. G. Liou, Experimental investigations of blueschist-greenschist transition equilibria, Geol. Soc. Amer. Mem. 164, 1986)

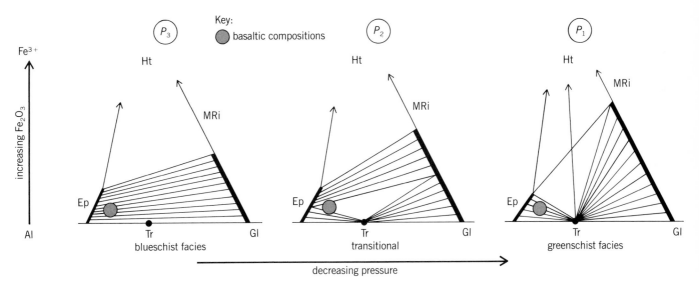

Fig. 3. Three schematic isobaric diagrams at temperature (T) = 300°C (570°F) showing variation of mineral assemblage and composition of sodic amphibole as a function of pressure. Ep = epidote; Gl = glaucophane; Ht = hematite; MRi = Mg-riebeckite; Tr = tremolite. *(After S. Maruyama, M. Cho, and J. G. Liou, Experimental investigations of blueschist-greenschist transition equilibria, Geol. Soc. Amer. Mem. 164, 1986)*

ROCKS; METAMORPHISM in the McGraw-Hill Encyclopedia of Science and Technology. [J. G. LIOU]

Bibliography: J. H. Carman and M. C. Gilbert, *Amer. J. Sci.*, 283A:414–437, 1983; M. A. Dungan, J. A. Vance, and D. P. Blanchard, *Contrib. Mineral. Petrol.*, 82:131–146, 1983; W. G. Ernst, *Benchmark Papers in Geology*, vols. 17 and 19, 1975; S. Maruyama, M. Cho, and J. G. Liou, *Geol. Soc. Amer. Mem. 164*, 1986.

Fault and fault structures

The Meers Fault recently and unexpectedly has been recognized as a potential source for a moderate-to-large earthquake, even though it is an old Paleozoic fault and is in a seismically inactive part of Oklahoma. The earthquake potential was first reported by M. C. Gilbert in 1983. His observations were surprising, because the fault is in the central United States, a region of minimal earthquake activity. This discovery adds another exception to the general rule that earthquakes of the central and eastern United States possess magnitudes less than 5½ to 6.

History. The Meers or Thomas Fault was first described formally in 1951 when it was recognized as being part of the Frontal Fault System, a west-northwest-trending boundary zone between the mainly Cambrian igneous rocks of the Wichita Mountains and the partly petroleum-bearing early-to-late Paleozoic sediments of the Anadarko Basin. In 1985 these boundary faults were shown to possess over 3 to 5 mi (5 to 8 km) of vertical displacement and to show large- to small-scale structures indicating large left-lateral offsets. Total length is not well defined, but tectonic maps suggest the zone is more than 300 mi (500 km) long between Amarillo, Texas, to or

beyond the Ouachita Mountains (Fig. 1).

In 1983 evidence was presented for a fresh-appearing fault scarp of 16 mi (26 km) length and 16 ft (5 m) height along a section of the Meers Fault. The fault has little or no historical earthquake activity, but the geological character indicates a potential for future surface rupturing from a magnitude 6–7 earthquake. This interpretation suggests that the Meers Fault, and perhaps other faults in the central and eastern United States, could represent an unrecognized earthquake potential.

Although there are no nuclear reactors that could be affected by the fault, its unusual location in the central-eastern United States region of generally minimal earthquake activity led the U.S. Nuclear Regulatory Commission (NRC) to initiate investigations to evaluate its seismogenic character. Subsequently, other groups studied fault style, rates and ages of displacement, geomorphic expression, seismicity, and structural setting.

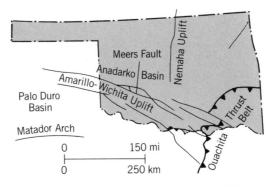

Fig. 1. Map showing location of Meers Fault, Oklahoma, between the Amarillo-Wichita Uplift and the Anadarko Basin.

Fig. 2.   Meers Fault scarp showing left-lateral offset (features on the far side of the fault are offset to the left) of ridgelines and streams.

Youthful expression. The fault scarp is spectacularly straight, steep, and conspicuous with early morning illumination (Fig. 2), even though it shows a maximum height of only 16 ft (5 m). Detailed studies of the scarp have revealed scarplets only 1 ft (0.3 m) high with over 25° slope in alluvium and soil. It has been estimated that the most recent scarp is less than one or two thousand years old, an estimate confirmed by carbon-14 dating, which bracketed the time of last movement between 500 and 2000 years ago. The NRC defined a capable fault as one that has moved once in the last 35,000 years, or more than once in the last 500,000 years. The Meers Fault fits this definition both from the recent age of last movement and from the beveling of the young scarp which suggests repeated displacements.

The scarp is sharp and narrow, both in the western part where it breaks brittle conglomerates of the

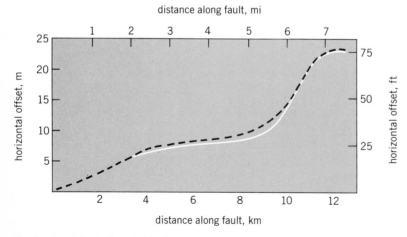

Fig. 3.   Left-lateral offset of ridgelines and valley floors crossing the northern part of the Meers Fault. The broken-line portion of the curve represents places with no quantitative measurements. (*After A. R. Ramelli and D. B. Slemmons, Surface offsets and scarp morphology, Meers fault, Seismol. Soc. Amer. Abstr., 80th Annual Meeting, 1985*)

Post Oak Formation, and in the partly faulted monocline of the eastern section where it crosses the plastically deformed shales of the Hennessey Formation.

Recurrent activity. At least three different slants or bevels in alluvial sediments are seen in the scarp slope, which indicates a recurrent character for faulting events. This is shown by a lower, younger, and steeper bevel of 25° near the base of the scarp, a 15° slope on the middle part, and a 10° slope along the top. The 30–72 ft (10–22 m) lateral offset of a ridgeline also implies recurrent displacements, because this is greater than the 30 ft (10 m) observed from the world's largest earthquakes of above magnitude 8. The recurrent activity has been well demonstrated by recent geomorphic studies.

Scarp length. Field studies confirmed the 16-mi (26-km) length, but low-sun-angle aerial photographs of the fault zone show an extension of another 7.5 mi (12.5 km) to the southeast. Surface faulting for a zone of this length corresponds to an earthquake magnitude of 7 to 7½.

Strike-slip component. Detailed geomorphologic studies show the left-lateral offset (features on the far side of the fault are shifted toward the left) to be four times the vertical component on the western part of the scarp. Measurements indicate left-lateral offsets of up to 80 ft (25 m) along a segment with a scarp up to 16-ft (5-m) height. Figure 3 shows offsets that cause many consistent jogs in alignment of ridges and streams crossing the fault. The scarps are also eccentric, with exaggerated heights on the left side of the ridges and little or no height on the right side viewed from the south.

The westernmost part of the fault in the Post Oak Formation clearly shows left-lateral offsets (Fig. 2), but the partly folded and partly faulted rocks of the eastern part of the fault, in shales of the Hennessy Formation, appear to have striations that indicate mainly vertical displacements.

Earthquake activity. The Meers Fault is in an extremely quiet area with no known associated historical earthquakes. It has been demonstrated that although the Meers Fault is quiet seismically, the fault zone (Fig. 1) shows a weak alignment of epicenters between Amarillo and the Mississippi Valley in Arkansas.

Prehistorical earthquake magnitudes. The total length of the Meers Fault was compared with worldwide historical earthquakes that possess surface rupturing. A left-lateral fault with a 25-mi (40-km) length would be expected to cause an earthquake of magnitude 6½ to 7½. Similar estimates using maximum displacement are not satisfactory for present analysis, because the true number of episodes was not determined. The exploratory trenches that have been studied indicate the total vertical component of rupture is 10 ft (3 m) in a monoclinal flexure, but the amount of strike-slip component is not known (it may be as much as four times greater). Also, scarps or deformed sediments with monoclinal folding may not show clearly the number of faulting events. An additional problem of using displacements is that the magnitude may be sensitive to the drop in stress. Many seismologists believe that intraplate earthquakes possess large stress drips and, for a given earthquake magnitude, the rupture lengths are shorter and the maximum displacements larger than interplate earthquakes. Although the frequency of large earthquakes may differ from that of the New Madrid region (magnitudes about 8 in 1811–1812) or the Charleston, South Carolina, region (magnitude about 7 in 1887), the Meers Fault appears to be capable of causing large-magnitude earthquakes and strong ground motions.

Significance. Potential earthquake impact on dams, lifelines, structures, and facilities in this area has not been evaluated for earthquakes of magnitude 6½ to 7½, because such earthquakes previously would not have been reasonably expected. Tectonic rupturing here or on extensions of the zone could cause unknown effects on petroleum and natural gas migration and accumulation. An even more important possibility is that the Meers Fault is part of a larger system of active faults or the only known example of other unrecognized faults in the eastern or central United States. The fault and topographic patterns in the region east of the Rio Grande Rift in New Mexico suggest connections to this system. The regional stress measurements and compatible earthquake focal mechanisms may imply that other parts of this zone are also active. This clearly expressed fault, almost 360 mi (600 km) east of the Rocky Mountains, is important because it is within the central and eastern United States region where earthquake risk and hazard is generally determined by historical earthquake activity, not by active fault investigations used near plate tectonic boundaries. The growing number of exceptions to this approach—New Madrid, Charleston, St. Lawrence River, Canada, and now the Meers Fault—indicate possible complications in this method.

For background information *see* EARTHQUAKE; FAULT AND FAULT STRUCTURES; SEISMIC RISK in the McGraw-Hill Encyclopedia of Science and Technology.

[DAVID B. SLEMMONS]

Bibliography: M. C. Gilbert, Possible Quaternary movement on the Meers fault, southwestern Oklahoma, *EOS,* 64:313, May 3, 1983; A. R. Ramelli and D. B. Slemmons, Surface offsets and scarp morphology, Meers fault, *Seismol. Soc. Amer. Abstr., 80th Annu. Meet.,* 1985; D. B. Slemmons, A. R. Ramelli, and S. Brocoum, Earthquake potential for recent multiple surface ruptures along the Meers fault, southwestern Oklahoma, *Seismol. Soc. Amer. Abstr., 80th Annu. Meet.,* 1985; S. Weisburd, A fault of youth, *Sci. News,* 127:23, 363–365.

Fertilization

Reproduction in angiosperms, that is, the flowering plants, presents a stark contrast to reproduction in the animal kingdom. First, genetic selection occurs through copious production of multicelled pollen; intense competition occurs between the pollen tubes that deliver the gametes. Second, the gametes—two per pollen grain—appear rather unspecialized and are nonmotile. Finally, the successful pollen tube delivers two sperm cells accurately and directly to the egg, where each of the sperm cells fuses with different cells in a process unique to angiosperms known as double fertilization. The product of sperm cell fusion with the egg results in an embryo; the fusion of the other sperm cell with the polar nuclei of the central cell results in the formation of a genetically distinct nutritive tissue known as endosperm. This article discusses recent research on the development of flowering plant gametes, the mechanism of gametic fusion, and the possibility that sperm cells are predestined to fuse with specific female cell types during double fertilization.

Male gamete. The male gamete is produced by synchronous meiotic divisions in specialized cells known as microsporocytes. Each division results in four genetically distinct microspores. Each microspore then divides mitotically to produce a vegetative cell, which later forms the pollen tube, and a generative cell, which divides mitotically once more to form two sperm cells. Division of the generative cell occurs after pollination in some plants and before pollination in others, leading to a distinction between bicellular and tricellular pollen, respectively.

Upon pollination (deposit of the pollen grain on the stigma) a tube forms through a hole already present in the resistant outer wall of the pollen grain. The pollen tube then elongates through specialized stylar transmitting tissue, and approaches the female gamete, which in different flowers may be as near as several micrometers or as far as several inches away from the tip of the stigma. Pollen tube growth may take only 45 min, at rates as rapid as 35 mm/h (0.14 in./h), in a relative of the dandelion, or growth may take several months, as in witchhazel. Nonetheless, the male gametes, the sperm, appear to travel in the tube as passive participants in the process of gametic delivery, without independent means of locomotion. The pollen tube has stringent recognition requirements in many flowering plants and, if compatible, is directed successfully to the ovary and female gamete by chemotropic substances. Although it is believed that calcium may act directly as a chemotropic agent in some species, other pollen tubes appear to require different factors or even proteins. Ion distributions in the pollen tube and surrounding tissue may also be essential to its physiology. Pollen tube growth stops when the tip becomes precisely located next to the egg cell, where it deposits the sperm cells. The sole purpose of the pollen is thus to deliver the sperm cells to the egg cell.

Recent research on the mechanism of fertilization has led to the concept of a male germ unit consisting of two sperm cells and the vegetative (tube) nucleus. According to this concept, the sperm cells remain in close association throughout their journey within the pollen tube and are apparently linked throughout their passage. The grasses are an apparent exception to this pattern, since their sperm cells have a looser form of association. Although some sperm cells appear identical, other, dimorphic, sperm cells may differ greatly in organelle content and thereby may differ in their potential to transmit specific organelles during fertilization.

Female gamete. The female gamete, the egg, is initiated within the ovule by a single cell, known as the megasporocyte. This cell undergoes meiosis to produce four genetically distinct megaspores, of which one, two, or four may provide the genetic material required in the formation of the female plant, the megagametophyte, loosely known as the embryo sac. In more than 70% of flowering plant species, the embryo sac develops from a single megaspore with the other meiotic products simply aborting. Such a sequence is known as monosporic development and results in the megaspore nucleus dividing in three cycles of mitosis to form eight nuclei, followed by cytokinesis at the end of the last division. The egg in this type of embryo sac is just one of seven cells in the embryo sac derived from a single meiotic product. The other embryo sac cells include two synergids, three antipodals, and one typically binucleate central cell. The nuclei of the central cells originate at the poles of the megagametophyte and are therefore known as polar nuclei, although they may often fuse prior to fertilization to form a single secondary nucleus.

Four of the embryo sac cells function directly in double fertilization: the egg, two synergids, and the central cell with its polar nuclei form which may be termed the female germ unit. Angiosperm embryo sacs of different species frequently vary with respect to the number of megaspore nuclei participating in the formation of the megagametophyte, the exact number of polar nuclei in the central cell, and the presence and number of antipodals.

Sperm deposition. The end of the pollen tube's growth occurs when the pollen tube enters one of two specialized cells, the synergids, within the embryo sac. The passively conveyed sperm cells are delivered as the tip of the pollen tube bursts and osmotic pressure within the tube pushes the contents of the tube though the small hole. By virtue of the synergid's organization at fertilization, the sperm cells gain entrance to an area of the egg and central cell that lacks a cell wall and is receptive to cellular fusion. The naked, wall-less sperm cells exit the synergid and become appressed to these wall-less segments of the egg and central cell, allowing direct membrane apposition (Fig. 1).

The sperm cells outwardly appear to have an equal chance of fertilizing the egg, but research on the genus *Plumbago* and earlier work on corn suggest that fertilization may be preferential, with one sperm cell having a distinct advantage in fertilizing the egg. The remaining sperm cell fuses with the central cell and expresses its genetic potential in the production of the endosperm, which will provide nutrition for the growth of the embryo.

Gametic and nuclear fusion. Actual fertilization is the result of two distinct fusions: gametic fusion of the cytoplasm of the sperm and egg, and nuclear fusion of the gametic nuclei into one. Cellular fusion is accomplished during a rapid sequence of events lasting perhaps less than a minute. Fusion bridges form at points between membranes of male and female gametes until they become so numerous that isolated membrane fragments are formed (Fig. 2). These membrane fragments then rapidly dissipate by forming vesicles. The sperm nucleus and its cytoplasm are thus directly incorporated into the female cell. However, only the cytoplasm contiguous with the nucleus at the time of fusion is incorporated; the rest remains unfused and eventually degenerates.

Nuclear fusion takes much longer than gametic fusion. After the male nucleus is aligned with its female counterpart, fusion is initiated by the joining of the outer leaflet of the nuclear envelopes, followed quickly by the joining of the inner leaflet. The fusion causes bridges to form in the nucleoplasm and may be accompanied by additional fusions elsewhere in the nuclei until the two nuclei clearly become one. The processes of physical fusion end at this point, and the genetic and physiological processes of development begin. The endosperm typically develops sooner than the embryo.

Sperm cells account for half of the embryo's nuclear heredity, but for a smaller amount of its cytoplasmic heredity. Two classes of cytoplasmic organelles containing deoxyribonucleic acid (DNA) have been observed in the sperm cells: mitochondria in 100% of those species studied, and plastids in 30% of the species studied. Both are important for cell

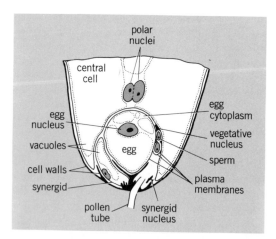

Fig. 1. Fertilization in a typical angiosperm embryo sac viewed in longitudinal section. The sperm cells, released from the tip of the pollen tube into one synergid, reach an area between the egg and central cell where they may contact both cells simultaneously. During double fertilization, one sperm will fuse with the egg to produce the embryo, and the other will fuse with the central cell to form the endosperm. (*After C. Dumas et al., Emerging physiological concepts in fertilization, What's New Plant Physiol., 15:17–20, 1984*)

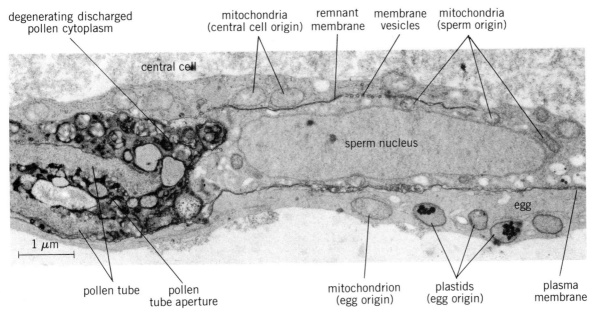

degenerating discharged pollen cytoplasm

central cell

mitochondria (central cell origin)

remnant membrane

membrane vesicles

mitochondria (sperm origin)

sperm nucleus

egg

1 μm

pollen tube

pollen tube aperture

mitochondrion (egg origin)

plastids (egg origin)

plasma membrane

Fig. 2. Transmission electronmicrograph of the pollen tube and sperm of *Plumbago zeylanica* within 1 min of gametic fusion in the central cell, the earliest after fusion that has been observed in any angiosperm. Evidence of sperm membranes remaining from gametic fusion and sperm mitochondria can still be observed in the central cell after fusion. The membranes will dissipate within minutes of fusion. Sperm plastids remaining from the fusion of the other sperm cell with the egg are evident below. (*From S. D. Russell, Fertilization in Plumbago zeylanica: Gametic fusion and the fate of the male cytoplasm, Amer. J. Bot., 70:416–434, 1983*)

function: the mitochondria provide energy through respiration, and the plastids provide energy through photosynthesis in the functioning chloroplast. Organellar DNA transmitted into the zygote may compete or recombine with the organellar DNA of the egg, or alternatively, it may be eliminated after transmission. The importance of the paternal organellar DNA complement is presently a central point of controversy in mechanisms of fertilization.

Control of polyspermy. Since the fusion of more than one sperm cell with the egg, an event known as polyspermy, can be detrimental to the offspring, and since polyspermy seems as infrequent in plants as in animals, it is probable that some mechanism permits only one sperm cell to fuse with the egg and inhibits any further fusions. Although animals possess an external fertilization membrane to the outside of the zygotic cell membrane, there is no obvious counterpart in higher plants. Evidently, the polyspermic block is rather the result of physiological changes in the existing plasma membrane that makes it unreceptive to additional fusions.

For the two sperm cells deposited by a single pollen tube, preferential fusion patterns may represent a significant block to polyspermy. Also, the occurrence of more than one pollen tube in the embryo sac is such a rare event that one might suggest that polyspermy in angiosperms may be predominantly controlled at the pollen tube level. The final block to polyspermy is the eventual formation of a cell wall around the zygote, which inhibits any further membrane surface contacts and potential fusion with surrounding cells.   [SCOTT D. RUSSELL]

Bibliography: C. Dumas et al., Emerging physiological concepts in fertilization, *What's New Plant Physiol.*, 15:17–20, 1984; S. D. Russell, Fertilization in *Plumbago zeylanica*: Gametic fusion and fate of the male cytoplasm, *Amer. J. Bot.*, 70:416–434, 1983; S. D. Russell, Preferential fertilization in *Plumbago*: Ultrastructural evidence for gamete-level recognition in an angiosperm, *Proc. Nat. Acad. Sci.*, 82:6129–6132, 1985; S. D. Russell, Ultrastructure of the sperm of *Plumbago zeylanica*, II. Quantitative cytology and three-dimensional organization, *Planta*, 162:385–391, 1984.

Fire detection

Electronic fire detection systems have been used for many years. The widespread use of microprocessors permits more accurate fire detection and allows significant improvement in discriminating against unwanted alarms. New systems are able to make an intelligent appraisal of a situation and to decide whether a false or true fire exists. They are also able to maintain the system in a calibrated state for long periods and indicate when fire detecting elements are likely to reach a false alarm condition.

Conventional systems. The most commonly used fire detection systems divide the area to be protected into zones, generally of about 2000 yd^2 (2000 m^2) in area. A number of detectors (usually up to 20) are connected to a single wire in a zone (Fig. 1), and during a fire one of these devices switches on, effectively shorting the line. The voltage applied to the line then falls to a much lower level, which is sensed at the fire control panel. This is annuciated

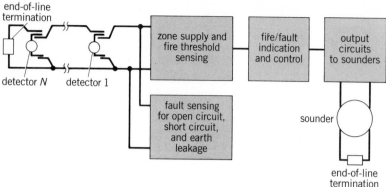

Fig. 1. Conventional fire detection system. (*After M. D. Stephenson, Automatic fire-detection systems, Electr. Power, 31:239–243, 1985*)

with indicators which are normally red and show the location of the fire only to the nearest zone.

To ensure that the line is unbroken, an end-of-line termination is connected after the last detector. This is generally a resistor or Zener diode that passes a continuous current. If the line becomes open-circuited by either breakage or removal of a detector, the end-of-line current falls to zero and alerts the fire control panel.

The fire control panel makes it possible to silence sounders or reset the system after a fire has been announced. The strict manner in which this is done is controlled by different regulations in various countries.

Detectors and sensors. There are three main types of fire detectors in use: heat (using a rate of rise of temperature, a fixed temperature, or a combination of both), smoke (using ionization, light scatter, or light obscuration), or flame (which may be ultraviolet-sensitive, infrared-sensitive, or a combination of both). The detectors sense one of these physical parameters and switch into a conducting state at a predetermined level of heat, smoke, or flame.

On the other hand, sensors give signals of relative levels of heat, smoke, or flame. In this case the determination and decision as to whether a fire is present may be made either close to the sensor or at a central point such as the fire panel.

Modern systems. Modern fire detection systems combine several features to assist in (1) locating the fire to the nearest detector, (2) detecting a fire earlier and discriminating against false fires such as cigarette smoke, and (3) keeping the fire detection system in a fully calibrated condition as contamination phenomena begin to degrade its operational state. These fire systems (commonly described as intelligent) utilize microprocessors to perform these functions. Addressability of each sensor or detector (outstation) point can be achieved by a switch for each outstation, or by using the numerical position of the outstation on the line as the address. Very often each outstation uses a single-chip microprocessor so that functions other than addressability can be realized. These include commands from the

fire panel to the outstation to initiate fire checking and calibration sequences at the sensor or detector, or to switch on an indicator at the outstation in the event of a fire.

Early detection and discrimination. As buildings grow larger, the time taken to evacuate in the event of a fire or false alarm becomes greater. It is therefore essential that the building (or floor) is evacuated at the earliest opportunity, and that an alarm indicates an actual fire rather than a misinterpretation by the fire sensor or detector.

Fires grow at various rates (Fig. 2) which can generally be classified as very fast (such as gasoline fires), medium (wood fires), and slow-smoldering (cotton, or cigarettes in a waste bin). The *N*-heptane fire is very fast-growing and, although it is very easily detected at an early stage, evacuation before a dangerous smoke level is reached becomes difficult. The medium-growth-rate fire illustrated by flaming wood allows earlier evacuation decisions to be based on the rate of growth of smoke. Smoke densities are often measured by the obscuration of a light beam. A smoke obscuration of 15%/m (5%/ft) is considered to be a point where people cannot survive. Time to evacuate a building (or floor) sensibly may be as much as 10 min. Therefore, if the rate of growth of smoke is 1.5%/(m)(min) [0.5%/(ft)(min)], a fire can be announced at as low as 2.5%/m (0.75%/ft), allowing extra time to evacuate. Normal optical smoke detectors are set to produce a fire signal at about 5%/m (1.5%/ft) on a typical test fire. A smoldering fire may stay undetected for some time and then may quickly burst into flames and produce a raging fire. A great deal of research is being carried out to separate signals generated by these fires from nonfire signals to enable detection in the early stages of fire.

The basic fire growth signals appear substantially monotonic. Closer examination of these signals, par-

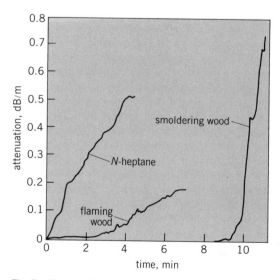

Fig. 2. Fire growth patterns monitored by an optical-scatter sensor. (*After M. D. Stephenson, Automatic fire-detection systems, Electr. Power, 31:239–243, 1985*)

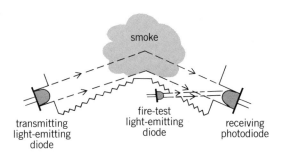

smoke

transmitting
light-emitting
diode

fire-test
light-emitting
diode

receiving
photodiode

Fig. 3. Light-scatter smoke detector or sensor. (*After M. D. Stephenson, Automatic fire-detection systems, Electr. Power, 31:239–243, 1985*)

ticularly at lower smoke levels, shows violent fluctuations. Signals from cigarette smoke can show large peaks as clouds of particles pass through the sensing chamber. These can be filtered out to a large entent to give a fairly smooth output signal.

Signal processing. Signals from fire sensors can be processed either at the outstation individually or at the fire panel collectively. Current trends favor the latter approach. Signals are stored in digital format and are processed with digital filtering techniques. Central processing makes it possible to examine the states of more than one sensor in order to make a decision based on collective information. It also allows for the easy control of calibration or checking processes.

Calibration. Many of the fire detecting elements now in use are smoke sensors. In such devices, smoke and other gases must be allowed to freely flow into the sensing chamber. Therefore, from the time a fire detecting system is installed, there is continuous deterioration in efficiency as the sensing chamber becomes contaminated. In conventional systems, detectors are replaced on a regular but random basis to overcome the problem of them becoming either oversensitive (prone to false alarms) or insensitive (not signaling a fire or signaling too late). Modern systems use various methods to ensure that a sensor is within its required parameters and, where possible, to implement corrections to return it to its original parameter values. These methods ensure that the system stays in a fully calibrated condition for longer periods and that sensors are replaced when they become contaminated, rather than according to a random replacement approach.

An example of a calibration method for an optical scatter sensor is shown in Fig. 3. Optical scatter sensors work by a light-emitting source (usually a light-emitting diode) projecting a beam toward a point x. A light receiver aimed at point x normally sees no light. When smoke is introduced at point x, light is scattered by the smoke particles and registered by the light receiver. This type of sensor operates particularly well with white and blue smoke.

The optical system is not perfect, and a leakage signal is present that will vary with the amount of contamination or light output from the light transmitter. A second light transmitter is switched on to produce a signal at the receiver equivalent to a known level of smoke in the sensing chamber. This provides a signal that is dependent on contamination levels. The contamination changes are generally measured in days, weeks, and months, whereas smoke level changes caused by fires are measured in minutes and hours. It is fairly simple, therefore, to separate out drift changes, due to contamination, and fire signals.

Subtracting the background signal from the known fire level signal gives the span of the device. The percentage of span at which a fire is decided can then be calculated almost irrespective of the contamination level. It is possible to introduce fixed constants outside of which a sensor is considered to be too contaminated to allow reliable fire detection. These contaminated sensors can be recorded before they reach a critical state, allowing time for cleaning before a false alarm occurs.

Products of combustion. Earlier detection of fire can be obtained by sensing the atmosphere for the products of combustion. At this time, sensors able to detect these gases are easily contaminated and are prohibitively expensive for general use. Work is being carried out in this area by using infrared techniques. These will be introduced in the petrochemical industries, where risks of fires, explosions, and poisonous gases are most prevalent.

For background information *see* ALARM SYSTEMS; FIRE TECHNOLOGY in the McGraw-Hill Encyclopedia of Science and Technology.

[MICHAEL D. STEPHENSON]

Bibliography: S. Daws, Intelligent approach to fire protection, *Elect. Rev.*, 215(7):27–28, 1984; R. G. Gower, *Developments in the Fire Protection of Offshore Platforms*, vol. 1, 1978; M. D. Stephenson, Automatic fire-detection systems, *Electr. Power*, 31(3):239–243, 1985.

Fluids

The term electrically active fluids is used to include both electrorheological fluids, which exhibit their properties under electric fields, and ferrofluids, which respond only to magnetic fields.

Electrorheological fluids have been known for 30 years but have only recently attracted intense interest in several countries, such as Britain, the Soviet Union, and Japan. Research has provided better fluids, while development work on electrorheological fluid devices, such as activators, dampers, and clutches, indicates that sizable engineering loads can be managed.

In the early 1960s ferrofluids were developed as a means for controlling fuel flow under conditions of weightlessness in space. Since then, many areas have been suggested for their use. Among the better-known applications are dynamic feedthrough seals, bearings and lubricants, damping for loudspeaker coils, separation of scrap metals, and dust exclusion seals for computer disk drives.

Electrorheological fluids. An electrorheological fluid is basically a colloidal suspension of finely di-

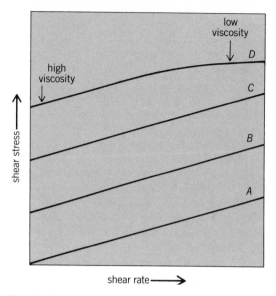

Fig. 1. Schematic diagram of shear stress against shear rate for increasing electric fields. A: zero field and newtonian behavior. B and C: idealized Bingham fluid with yield point increasing with field. D: Bingham fluid showing thixotropy.

vided particles (dispersed phase) in a carrier liquid (continuous phase) that is usually an insulating oil. When an electric field is applied, the rheological, or flow, properties of an electrorheological fluid are changed dramatically through a marked increase in resistance (viscosity) to flow of the fluid. Above a critical field, in the region of 3–4 kV/mm (75–100 kV/in.), it behaves as a solid, and ceases to flow until a certain shear stress (the force that makes one section of fluid slide past an adjacent one) is reached. The process is reversible so that the fluid can be switched almost instantly to a solid and back again.

The variety of applications of such fluids was immediately obvious, and attempts were made to use them in clutches, dampers, and vibrators. However, the initial efforts were futile because of the abrasive nature of the slurries, and interest in the subject lapsed for nearly 20 years.

Composition. Some of the slurry systems originally used consisted of particles of starch or activated silica gel suspended in kerosene. Similar compositions having a consistency of paste are still being examined, but the emphasis has changed to a search for colloids which exhibit, and preserve, their liquidity. Modern electrorheological fluids usually contain polymeric particles, such as lithium polymethylmethacrylate, in a chlorinated hydrocarbon oil, but their precise compositions are proprietary knowledge or held in patents. The particles range in size from 5 to 20 micrometers, while their concentration varies between 20 and 30% by volume. The requirements for suitable fluids are demanding; density matching is necessary to prevent settling out of the suspended particles, while high boiling points and low freezing points are necessary to sustain liquidity and give a

reasonable working range of temperature.

Flow behavior in an electric field. In the absence of a field, the flow characteristics of modern electrorheological fluids can be considered as newtonian (curve A in Fig. 1). However, when a field is applied across the flow direction, startling non-newtonian behavior occurs: the fluid appears to become "solid" and a finite shear stress (yield point) is needed to break down its structure and to initiate flow (curves B and C). Moreover, the higher the field, the higher the yield point. This behavior is characteristic of a so-called Bingham plastic, where above the yield stress the shear rate increases linearly with stress. The familiar thixotropic (nondrip) paints are everyday examples of Bingham fluids (curve D): brushing (shear stress) reduces viscosity and the paint flows easily, whereas the paint will not drip when left undisturbed.

The potential importance of electrorheological fluids for a wide range of applications stems from two features of this liquid-solid transition. First, the structural change to a "solid" occurs in about 10^{-3} s, and when the field is switched off, the substance relaxes to the liquid state just as quickly. Second, although high fields are required, typically 3 kV per millimeter (75 kV per inch) of fluid gap, the conduction currents are only in the microampere range. Consequently, the power requirements of any switching operation are still low enough to be handled by solid-state power supplies. Therefore, electrorheological fluids make feasible a cheaper, faster, and simpler alternative to the relatively slow conventional hydraulic or pneumatic control systems.

Potential application. Among the potential applications of electrorheological fluids are devices such as clutches, fluid drives and couplings, pressure and flow valves, dampers, shock absorbers, and actuators. When a voltage is applied to an electrorheological fluid between two plates in a sealed chamber (Fig. 2), the fluid acts as a coupling medium, and torque can be transmitted at a magnitude which is a direct function of the field across the plates. If the field is high enough so that the fluid appears to become "solid," the two plates will then rotate together with no slip, just as does a normal dry-plate clutch. The torque transmitted can be increased by using groups of interleaved plates. The automotive indus-

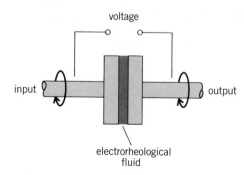

Fig. 2. Elementary parallel-plate clutch arrangement.

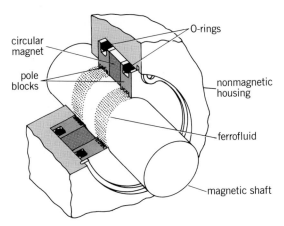

circular magnet
pole blocks
O-rings
nonmagnetic housing
ferrofluid
magnetic shaft

Fig. 3. Four-stage rotating shaft seal. O-rings provide static seal. Ferrofluid focused between magnetic poles and shaft forms nonrubbing circular seal. (*Ferrofluidics Limited*)

try is conducting research in this area.

Ferrofluids. Ferrofluids, like electrorheological fluids, are colloidal suspensions, but the similarity between the two types of fluid ends here. The particles in a ferrofluid are magnetic and ultramicroscopic in size, and their concentration is only 1 or 2% by volume. Typically their average diameter is 10 nanometers, or about 10–20 times the size of a small molecule. To prevent their settling, the particles are coated with a long-chain surfactant such as oleic acid. Iron, cobalt, and nickel, which are all ferromagnetic, and the ferrite magnetite (lodestone) are commonly employed, but some magnetic alloys have been used. Various carrier liquids are suitable such as water, diesters, hydrocarbons, fluorocarbons, and more recently liquid metals such as mercury.

Behavior in a magnetic field. Bulk specimens of iron, cobalt, and nickel consist of many individual magnetic domains, each magnetized to saturation but randomly oriented, so that their overall magnetization is zero in the absence of an external magnetic field. The average particle size in a ferrofluid is about 100 times smaller than a domain so that each particle can be considered as a small single domain of a ferromagnetic material. When an external magnetic field of low-to-moderate strength is applied, all the particles do not fall into alignment, and the field must be increased considerably before full saturation is achieved. The magnetic strength of a ferrofluid has not exceeded 15% of that of the bulk material. Unlike an electrorheological fluid, a ferrofluid does not go "solid" or congeal when it is subjected to a field; it becomes magnetized but it still remains a low-viscosity liquid.

Applications. The most developed application of ferrofluids is that of sealing regions of differing pressure that are connected by rotating shaft seals (Fig. 3). Ferrofluid is retained in each ring "tooth" on the soft-iron pole blocks by the focusing effect of the magnetic field from the circular permanent magnet.

The combination of tooth and ferrofluid constitutes a stage that can support a pressure differential of about 5 lb/in.^{2} (35 kilopascals). Figure 3 shows a typical four-stage arrangement on each pole block. Multistage seals can withstand much higher pressures, each ring of fluid automatically taking its equal share of the pressure difference. The performance of ferrofluid shaft seals is impressive. They can sustain pressures of 600 lb/in.^{2} (4 megapascals) at a continuous rating of 10,000 revolutions per minute, or 100,000 revolutions per minute for shorter periods. This compares most favorably with conventional O-ring seals, which are limited to rotational speeds of about 300 rpm.

For background information *see* COLLOID; NONNEWTONIAN FLUID; RHEOLOGY in the McGraw-Hill Encyclopedia of Science and Technology.

[T. J. GALLAGHER]

Bibliography: D. A. Brooks, Electrorheological effects adds muscle, *Control and Instrumentation*, 14:57–58, October 1982; R. E. Rosensweig, Magnetic fluids, *Sci. Amer.*, 247(4):136–145, October 1982; W. M. Winslow, Induced fibration of suspensions, *J. Appl. Phys.*, 20:1137–1140, 1949; M. Zahn and K. E. Shenton, Magnetic fluids bibliography, *IEEE Trans. Mag.*, MAG-16:387–409, 1980.

Food engineering

Freeze drying has a long history of use in the preservation of delicate heat-sensitive materials such as biological and food products. The use of low-temperature dehydration results in a lightweight, chemically stable product with an extended shelf life at room temperature. Long-term preservation of consumable products without the use of additives makes this a valuable processing technique. However, the advantages of the storage stability are offset by the processing charges. For bulky food products, the dehydration times are lengthy (≈ 11 h), and the cost is a large fraction of the value of the product.

For bulky products such as meat and vegetables, the sublimation rate is controlled by the conduction of heat to the ice through the dried outer layer. To avoid thermal degradation of the food, it is necessary to place a limit on the temperature gradient in the dried layer, which restricts the rate of drying. The use of microwave power, on the other hand, offers a way to circumvent the problem of energy transport through the dried layer. Microwave power, with its relatively long wavelength, can pass through the dried layer with little attenuation and be absorbed within the frozen region. Energy may be supplied to the sublimation interface by conduction through the frozen core while the dried layer remains cool. With this approach, the drying rate can be accelerated beyond the conventional limits without an associated loss in product quality. However, despite this apparent substantial advantage over conventional techniques, which has been known for a long time, microwave power has not been widely applied commercially to freeze drying. The general

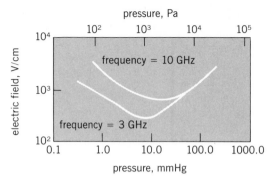

Fig. 1. Breakdown of air in a radio-frequency electric field. (*After A. D. McDonald, Breakdown in Gases, John Wiley and Sons, 1966*)

consensus has been that the savings in freeze dryer size, labor, and floor space, due to reduction in drying time, were offset by the higher energy costs. Moreover, it was felt that technical problems involved in the utilization of microwave energy would hinder its adoption commercially. However, in the light of recent economic and technical developments and indications that microwave processing may provide freeze-dried food products of a higher quality, the interest in applying microwave to freeze drying should rise.

Technology. Proper design for microwave freeze dehydration requires solving problems of corona discharge, nonuniform heating, impedance mismatch, and applicator efficiency.

Corona discharge. The ionization of the gas in the vacuum chamber, variously known as breakdown, discharge, or corona, has been a major barrier to the development of microwave freeze drying. At a given frequency, the electric field necessary to initiate breakdown in a specific gas is a function of pressure. Typical breakdown curves are shown in Fig. 1, where the region above a given curve indicates

that the electric field is strong enough to induce breakdown. For sublimation to occur in the freeze drying chamber, pressure-temperature conditions must be below the triple point of water. On the breakdown curve, this pressure (610 pascals or 4.6 mmHg), lies to the left of the minimum breakdown electric field. Maintaining the pressure below the breakdown range in a freeze dryer is difficult, since water vapor is generated during the freeze drying process. In a system with a limited condensation rate, the added water vapor produces an increase in the chamber pressure, which in turn increases the probability of breakdown within the cavity. Smooth operation of the freeze dryer requires that a balance must be found between the power fed to the product and the corresponding chamber pressure. In addition, the design of the vacuum barrier and of the transducer coupling the feed to the applicator appears crucial, but an adequate design can be achieved.

Uniform heating. Nonuniform heating with microwaves has long been considered a major potential difficulty in the freeze drying application. It is indeed a critical parameter. Nonuniform microwave energy distribution in the applicator may result in unwanted early corona discharge, meltback, or overheating in the high-energy zones. Use of a multimodes cavity should overcome this problem.

If the microwave source is to provide a uniform energy distribution in the load in a multimodes cavity, the theory indicates that the generator should have a nonzero emission band width. This can be realized through frequency modulation of the microwave generator. Sweep of the microwave frequency over a certain band width should be a powerful means of achieving or improving uniform heating. Use of mechanical mode stirrers, as sometimes proposed, does not seem to be efficient in improving the uniformity of the field.

Impedance mismatch. The difficulty of matching the microwave applicator to the transmission line due to the variable dielectric load (drying food product) is also considered a major problem. Good impedance match between the applicator and the line is needed in order to achieve low standing waves in the line and low reflections back toward the generator. Standing waves in the line result in a loss in power-handling capacity of the line, while high reflected power toward the generator represents a loss in efficiency of the applicator and is detrimental to the life of the tube, if the generator is not protected by a circulator. However, this drawback can be overcome by proper design.

Applicator efficiency. The microwave applicator can be made almost 100% efficient. Experiments have shown that the total microwave power input can be transferred to the food load, within the accuracy of the experiment, provided that the empty applicator is characterized by low losses. Low losses can be accomplished by decreasing wall losses (good conductor and high volume to surface ratio) and reducing leakage.

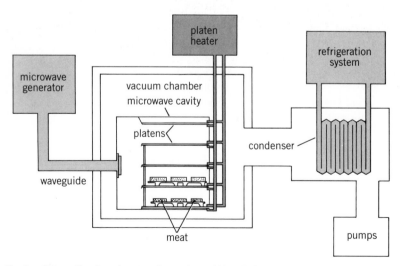

Fig. 2.   Schematic of a microwave freeze dryer. (*After H. B. Arsem and Y. H. Ma, Aerosol formation during the microwave dehydration of beef, Biotech. Prog., 1(2):104–110, June 1985*)

Combined microwave and radiant freeze drying.
Recent work has demonstrated that advantages such as lower microwave power levels, improved product quality, and favorable overall economics could be gained by a combined microwave and radiant freeze dryer (Fig. 2). It is believed that existing large-scale conventional freeze drying systems can easily be converted to combined microwave and radiant drying systems. This would provide considerable reduction in initial capital costs to make the overall economics more favorable.

For background information *see* FOOD ENGINEERING; MICROWAVE in the McGraw-Hill Encyclopedia of Science and Technology.

[Y. H. MA]

Bibliography: H. B. Arsem and Y. H. Ma, Aerosol formation during the microwave dehydration of beef, *Biotech. Prog.*, 1(2):104–110, 1985; Y. H. Ma and P. R. Peltre, Freeze dehydration by microwave energy: Part II. Experimental study, *AIChE J.*, 21:344–350, 1975; A. D. McDonald, *Microwave Breakdown in Gases*, 1966; P. R. Peltre, H. B. Arsem, and Y. H. Ma, Applications of microwave heating to freeze dehydration: Perspectives, *AIChE Symp. Ser.*, 73(163):131–133, 1975.

Food manufacturing

Most food products can be classified as perishable or semiperishable: quality deteriorates during storage, and the rate of deterioration is usually temperature-dependent. Conventional inventory management techniques allow for recognized quality changes due to time but not those due to temperature. A suitable time-temperature indicator would permit the inventory management criterion to be changed from age to remaining shelf life based on cumulative thermal exposure. This change would reduce waste and prevent thermally abused product from reaching the consumer.

Time-temperature indicator requirements. A useful time-temperature indicator must satisfy several conditions. First, it must undergo an observable temperature-dependent change, that is, change must be irreversible under field conditions and easily and quickly detectable. Finally, it must be possible to correlate the change with changes in product freshness.

Second, a commercially viable monitoring system must be applicable to many products, a condition that imposes other requirements. Ideally, the monitoring system should be able to cover a wide range of time-temperature exposure. Food product shelf lives range from days at 32°F (0°C) to years at room temperature. Diverse products are likely to have different temperature sensitivities and deterioration mechanisms. For an indicator reading to be used as a direct measure of product quality, indicator change must imitate product change in several respects. Not only must the indicator and product reaction rate equations have the same form, but the reaction rate constants must also have the same temperature dependence.

Fig. 1. Components of time-temperature indicator label: (*a*) product information, (*b*) indicator information, and (*c*) indicator material. (*Allied Corporation, LifeLines Technology*)

Shelf-life monitoring system. A system has been developed that can monitor a variety of products by using indicator change as an indirect measure of product quality change. Computer software is used to translate indicator change into product quality change. A wide time-temperature range is covered by using a family of color-changing indicator materials.

Figure 1 shows a typical label printed with bar codes and an indicator material. The bar codes contain product and indicator identification and other information. The indicator material can be any one of a family of acetylenic monomers that undergo solid-state polymerization, producing a color change. The polymerization reaction follows first-order kinetics, so reflectance decreases exponentially as time increases. Reaction temperature dependence follows the Arrhenius relationship; a plot of log (reaction rate) against inverse absolute temperature is linear, with a slope proportional to the activation energy. Monomer composition and concentration can be varied to cover a wide time-temperature range.

Bar codes and polymer reflectance are read by a portable hand-held microcomputer with an attached optical wand. Software within the microcomputer uses this information to calculate product freshness or remaining shelf life. This software can use product and indicator data to estimate remaining shelf life of products following different rate equations with different activation energies. The bar code information is needed so that the software can locate the particular indicator and product files it needs to translate indicator reflectance into product quality. Only data files have to be changed to monitor a different product or use a different indicator. The calculation procedure, and therefore the software structure, is unchanged. Product quality can be displayed directly on the hand-held computer, and data can be downloaded periodically to a remote host computer, where they can be analyzed and stored.

Test results. Product quality typically decreases monotonically over time, with the reaction rate doubling for each 10°C (18°F) increase in temperature. Results of tests of two products showing this typical behavior are given in Fig. 2*a–b*. The system has been tested on a variety of food products, under constant and variable temperature conditions. Figure 2*a* shows the results of ultrahigh-temperature milk tests. Ultrahigh-temperature processing in-

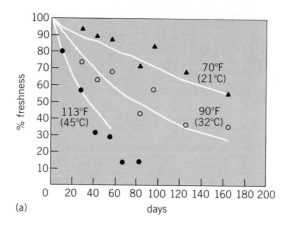

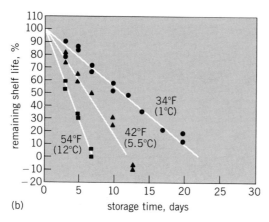

Fig 2. Freshness and shelf life tests. (a) Ultrahigh-temperature (UHT) milk; comparison of measured freshness (data points) with estimates from time-temperature indicators (curves) (*after R. Zall et al., Evaluation of automated time-temperature monitoring system in measuring freshness of UHT milk, 9th Annual Eastern Research Highlights Conference, Washington, D.C., November 7–8, 1984*). (b) Comparison of estimated and measured remaining shelf life of irradiated cod fillets. Points are averaged estimates from 10 indicators; lines are average measured shelf lives (*after B. L. Tinker et al., Evaluation of automated time-temperature monitoring system in measuring the freshness of chilled fish, IIR Conference on Stored Lives of Chilled and Frozen Fish Products, Aberdeen, Scotland, October 1985*).

volves heating to a higher temperature for a shorter time than conventional pasteurization. The product is relatively stable and need not be refrigerated. Tests therefore lasted several months at temperatures up to 113°F (45°C). Product freshness was evaluated by a taste panel whose averaged ratings were then put on a 0–100% basis. It appears to follow first-order kinetics with an activation energy of about 15 kcal (63 kilojoules) per gram-mole. The lines representing calculated freshness are a good approximation to measured values. In this case an indicator material with an activation energy of 28 kcal (118 kJ) per gram-mole, almost twice that of the product, was used successfully. This difference does not introduce any significant estimation error; accuracy is limited by uncertainty in the organoleptic measurements.

Results of shelf life tests on a more perishable food product, irradiated cod fillets, are shown in

Fig. 2b. This product has a relatively short shelf life, and must be refrigerated. Tests therefore lasted over 1–3 weeks at 34–54°F (1–12°C). Shelf life decreases linearly over time, and can be represented by zero-order kinetics with an activation energy of about 17 kcal (71 kJ) per gram-mole. In this graph the curves show average remaining shelf life at three temperatures, while the data points are estimates made by using reflectance readings of indicators with an activation energy of 23 kcal (96 kJ) per gram-mole. Estimates agree closely with the averaged data. The software compensated for the difference between product and indicator activation energies, and also for the difference between the zero-order kinetics of the product and the first-order kinetics of the indicator material.

Field implementation. Indicators can be automatically attached to product shipping containers on a packing line. This ensures that both the indicator and the product start at known freshness levels. Strategic scanning of indicator labels attached to these products can be used to elucidate the status of inventory anywhere in the distribution chain. Troublesome distribution paths can be easily monitored, and thermally stressed products can be identified. Appropriate actions can be taken, and inventories can be rotated to extend the useful life of a product. This will reduce product spoilage and optimize product quality during distribution. Telecommunications allow easy transfer of freshness data from remote locations to a host computer so that an entire inventory control system can be integrated and monitored from a single location.

For background information *see* CHEMICAL THERMODYNAMICS; FOOD MANUFACTURING in the McGraw-Hill Encyclopedia of Science and Technology.

[PAUL G. FRIEDMANN; THADDEUS PRUSIK]

Bibliography: S. C. Fields and T. Prusik, Shelf life estimation of beverage and food products using bar coded time-temperature indicator labels, *4th International Flavor Conference*, Rhodes, Greece, July 23–27, 1985; T. P. Labuza, *Shelf Life Dating of Foods*, Food and Nutrition Press Inc., Westport, Connecticut, 1982; J. H. Wells and R. P. Singh, Performance evaluation of time-temperature indicators for frozen food transport, *J. Food Sci.*, 50(2): 369–371, 378, March-April 1985.

Forest management and organization

Forests may change slowly, but many techniques behind their management have seen rapid innovations since the 1970s. This article reports on two innovations: methods to measure scenic beauty, and the advent of portable data collectors. Accurate scenic beauty measurement is essential if environmental standards are to be upheld, while electronic portable data collectors allow rapid and accurate assessment of measurable forest attributes.

SCENIC BEAUTY MEASUREMENT

Forest scenic beauty is an intangible phenomenon and has historically not been managed equally with quantifiable forest products such as timber and water. The recent application of psychophysical techniques to the scenic beauty measurement problem, however, has permitted evaluation and prediction of this important forest product. Mathematical models can be formulated which predict quantities of change in scenic beauty as a function of changes in levels or amounts of actual physical landscape objects such as tree density, topographic relief, and insect damage to tree crowns.

Need for scenic beauty assessment. Forestry's traditional neglect of scenic beauty has been strongly challenged as a result of tree harvesting controversies on some National Forest lands, and by the emergence of environmentalist political strength. On the basis of such legislation as the National Environmental Policy Act and the National Forest Management Act, the forestry industry is now faced with the problem of assessing values for scenic beauty as well as for timber, forage, water, wildlife, and recreation. All except scenic beauty have been easily quantifiable.

Assessment systems, based upon the artistic judgments of scenic quality experts such as landscape architects, have emerged as one answer to the problem of scenic beauty evaluation. Such assessment procedures, however, raise new problems. First, deference to experts is often not acceptable to the public, particularly when the public is the beneficiary or user of the product in question. Second, such assessment systems are often inefficient approaches to evaluating the esthetic consequences of forest management decisions across a changing set of forest management situations. Finally, these systems are based in the art-design lexicon of the expert. Color, line, texture, and contrast mean little to the forest manager who must deal with species composition, tree density, insect damage, and harvest prescription. It is difficult, if not impossible, to give equal weight to timber and scenic beauty when they are evaluated in different languages or predicted by different forest characteristics and units of measure.

Scenic beauty measurement. Forestry research, in combination with psychophysical methods, has provided a reliable and valid method for measuring and predicting scenic beauty. Psychophysics deals with the problems of defining and calibrating the relationships between physical phenomena, such as loudness and brightness, with human perceptual responses to them. The problem is one of correlating observable physical properties of stimuli with unobservable psychological responses.

The essential idea behind psychophysical scaling is that people can differentiate among stimuli along some given dimension (such as scenic beauty), but that for any given stimulus (such as a forest scene) their evaluations will vary with time and will also depend upon many other factors such as their experience with the stimuli or their mood. Thus, repeated evaluations of any single stimulus along a perceptual dimension such as scenic beauty will be distributed as a normal curve of evaluation scores around a mean value. The more similar two stimuli are perceived to be, the greater the overlap in these

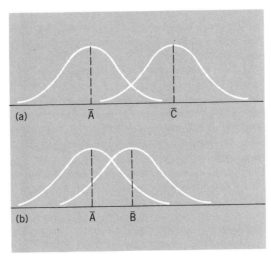

Fig. 1. The overlap of scenic beauty distributions can be used to represent the perceived similarity between landscapes with (a) less overlap and (b) more overlap. As a consequence of the overlap, the scenic beauty sampling distribution means $\overline{A}$ and $\overline{B}$ in b are more similar than the mean values $\overline{A}$ and $\overline{C}$ in a. (After R. B. Hull et al., Measurement of scenic beauty: *The law of comparative judgment and scenic beauty estimation procedures, Forest Sci.*, 30:1084–1096, 1984)

distributions (Fig. 1). While the underlying psychological dimension (such as scenic beauty) of each stimulus cannot be directly observed, the mathematical or statistical positioning of these response distributions can be. In essence, it is the amount of confusion between stimuli which is measured (distributional overlap). These amounts of overlap of the distributions can be mathematically translated into an interval scale of a psychological dimension such as scenic beauty—a scale which measures the magnitude of scenic beauty results. This interval scale, or metric, is the key to developing predictive scenic beauty models which can relate scenic beauty of forests to the composition of physical features in those forests.

The most used and tested of the psychophysical methods for computing scenic beauty values are the Scenic Beauty Estimation Method and the Law of Comparative Judgment. In the Scenic Beauty Estimation method, panels of human observers make judgments about the amount of scenic beauty of each of a series of photographs which represent the forest scenes of interest. In the Law of Comparative Judgment approach, the observers view all possible pairs of forest scenes as represented by photographs and make judgments about which of each pair has the most scenic beauty. Observer evaluations, from either the Scenic Beauty Estimation or the Law of Comparative Judgment method, are subjected to a series of mathematical transformations, defined by psychophysical methods such as signal detection theory, and a scenic beauty metric for the photographs results. Both methods have been extensively tested for reliability and validity, and both have performed well. A separate validity issue concerns the use of two-dimensional photographs as surrogates for a three-dimensional world. The results of approximately a dozen studies indicate, however, that photographs are valid substitutes for actually assessing scenic beauty on-site.

Statistical models. Once scenic beauty values are computed for an array of forest scenes which vary along some measurable dimension of interest (for example, tree density), measurements of this dimension and others in the visual field are made. The intent is to use these measurements, called landscape object predictors, as predictors of scenic beauty. The measurements of the landscape object predictors can be made either from the photographs, which the subjects evaluated, or from on-the-ground measurements. For example, if photographic measurements are used, in the case of tree density the proportion of the photograph in trees would be used as the predictor measurement. If, on the other hand, ground measurements were used, the numbers of trees per unit area (or the cross-sectional area of the tree stems per ground unit area) would be used as the potential predictor measure.

The modeling process is a linear regression in which scenic beauty is predicted by the landscape object measurements for all the photographs which were evaluated by the panel of human observers. Using statistical methods to evaluate the degree of data fit and predictive power results in models similar to the equation below, where SBE_i = scenic

$$SBE_i = a + b_iX_i + \cdots + b_nX_n + e$$

beauty estimate in psychological units, a = Y-axis intercept, b_i = slope (unit change in SBE, per unit change in X_i), X_i = predictor (landscape object measure), and e = error resulting from lack of perfect prediction. Such equations permit scenic beauty to be predicted based upon changes or differences in predictor variable measures. By substituting different values for the predictors in such an equation, scenic beauty estimates for different forest landscapes can be predicted. If on-the-ground measurements of the landscape objects were used as predictors, then the same type of measures that are used to assess timber and other tangible resources are used in the scenic beauty models. Thus, forest management can assess trade-offs in product output from forests as a result of physical modifications of those environments.

Predictors of scenic beauty. Landscape variables that have been documented to be valuable predictors of scenic beauty include: forest harvesting types (clearcutting vs. selective cutting), tree density, tree size, topography, viewing distance, ground vegetation, insect damage to tree crowns, and dead and downed tree stems. The relationships may be positive (increasing tree size increases scenic beauty) or negative (increasing insect damage decreases scenic beauty), and they provide information not only on the direction of the change in scenic beauty, but also on the magnitude of such change. Forest management can decide, therefore, if the changes in scenic beauty are proportionately large or small as a result of various environmental modifications.

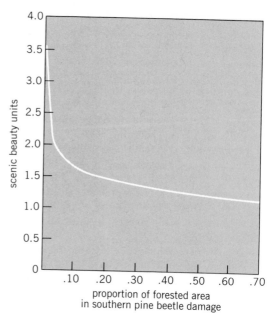

Fig. 2. Predicted loss of scenic quality for forests with different levels of insect damage. The curve resulted from a predictive model specification of: scenic beauty units = 1.03 − 0.28 ln X (ln X = the natural logarithm of the proportion of forest area in insect damage). (*After G. J. Buhyoff and W. A. Leuschner, Estimating psychological disutility from damaged forest stands, Forest Sci., 24:424–432, 1978*)

Current efforts. Much of the early work on these models assumed that the relationships between scenic beauty and the predictor variables were linear. However, the phenomena of decreasing marginal utility from economics, and sensory thresholds from psychology, imply that better model forms may often be curvilinear. In the case of decreasing marginal utility, as more and more of a certain landscape object is added to the forest view, the effect on scenic beauty of adding another unit of that variable becomes smaller and smaller until the change in scenic beauty is mathematically negligible. This type of relationship has been found in the southeast United States for the scenic beauty predictor of insect damage to tree crowns. A logarithmic relationship between southern pine beetle damage and scenic beauty is the most predictive (Fig. 2). In this case, scenic beauty decreases are minimal after just 10% of the forest canopy shows signs of insect damage. Such models are less likely to be grossly biased at data extremes. Similarly, the concept of a sensory threshold suggests that some predictors may have extremely small visual impacts at low levels, but have very large effects on scenic beauty once a threshold amount is exceeded. These curvilinear relationships not only are theoretically sound but are intuitively pleasing as well; they have proved to be very good descriptions of the behavior of scenic beauty in forest settings.

[GREGORY J. BUHYOFF]

PORTABLE DATA COLLECTORS

Foresters collect and analyze large amounts of data to support management decisions, including information on land, timber, wildlife, and other resource characteristics. Electronic portable data collection terminals are now being used in these field inventories to enhance productivity and accuracy by speeding up the process of data collection, validation, and transfer to computers for further processing.

Forest inventory methods. Forest inventories are accomplished by randomly sampling key parameters that measure resource quality and quantity. The parameters that are sampled vary depending on the information needs of the resource manager. For example, managers concerned with producing timber will want information on timber type, size, quality, and rate of growth. A timber inventory, therefore, will sample individual tree parameters such as species, diameter, height, past diameter and height growth, percent defect (soundness), form, tree or log grades, and type of product that can be produced. Sampling intensity also varies; the desired level of statistical reliability will determine the number of sample measurement plots on a given area (stand), size of a sample plot, and intensity of sampling within each plot.

Regardless of inventory type or sample intensity, data collected on parameters measured in the field have traditionally been recorded on paper, using a recording sheet, or tally card, in which data are recorded with a combination of alphabetic or numeric codes and actual numeric measurements. Before processing, the data are verified for accuracy and completeness, usually by someone other than the person who recorded the measurements. Inaccuracies or missing data may require the forester to return to the field to remeasure some of the sample plots. After the data have been completely verified, they are summarized, either manually or by computer, using a combination of mathematical and statistical equations. With the advent of microcomputers, most foresters can automatically, rather than manually, process their data. Computer-aided processing does, however, require an additional step: data must be typed onto punch cards or directly entered into the computer via the keyboard of an interactive terminal. After entry, the data must again be verified for accuracy and completeness.

Portable data collection terminals. Electronic portable data collectors (PDCs) have been used extensively in the retail industry to conduct product inventories. Since the mid-1970s, these devices have begun to appear in forest research experiments and in timber inventory applications.

A portable data collector is a self-contained unit that includes a power supply, keyboard, semiconductor memory, display screen, communication ports, and controlling (operating system) software (Fig. 3). Physical size varies; some collectors are comparable to hand-held electronic calculators, 5.75 × 3.5 × 2.0 in. (14.6 × 8.9 × 5.1 cm), while others are the size of a clipboard, 10.5 × 10.5 × 1.4 in. (26.7 × 26.7 × 3.5 cm). There are usually two types of power supply. The main supply consists of four to six AA rechargeable bat-

Fig. 3. Electronic portable data collector. This unit measures 7.8 × 3.6 × 2.2 in. (19.7 × 9.1 × 5.7 cm).

teries that provide for continuous operation over a period of 15 to 50 h. A backup lithium or NiCad battery prevents data loss in the event the main power supply is accidentally depleted.

Data and application software is stored in dynamic random-access memory (RAM), while the software that controls operation is stored in nonvolatile read-only memory (ROM). RAM storage capacity may vary from 8 kilobytes (Kb) to 256 Kb or more. The keyboard of a portable data collector, which partly determines its size, may offer 20 to 60 keys that can be used for full alphanumeric and special-symbol data entry. Smaller keyboards combine several functions on each key. Display screens use either liquid-crystal display (LCD) or light-emitting diode (LED) technology to display one to four lines of 16 to 40 alphanumeric characters.

Most portable data collectors provide an RS-232-C serial communications port for transmitting data to a host computer. Some also provide an internal modem that can be used with an acoustic coupler to transmit data over telephone lines. Other communication options are available to link portable data collectors to laser scanners, bar code wands, and analog devices such as calipers and thermometers.

Some portable data collectors can be programmed by using standard high-level languages such as BASIC. Programs can be entered by using the data collector's own keyboard, or transferred (downloaded) from a host computer. Others may be programmed by using proprietary operating system commands.

Because of harsh environmental conditions encountered in forestry data collection, portable data collectors suitable for forestry work are sealed to prevent dust and moisture damage. Some may also be built to withstand great physical shock and temperature extremes.

Benefits in forest inventory work. A portable data collector can be programmed to record and store data in the same manner used to record and store data on a paper tally card. In addition, it can be programmed to automatically verify the accuracy of each item entered. Because a constant sequence is followed for each record entered, measurements cannot be inadvertently missed. Thus, a data set brought back from the field is complete and accurate: it does not have to be manually inspected.

Data stored in a portable data collector are electronically transmitted to a host computer by direct connection or by indirect connection via telephone. Transfer times are a fraction of the time required when using manual keyboard entry systems. To verify accurate data transmission, each data set is transmitted twice and automatically compared by the computer to detect discrepancies.

The increased productivity possible through a reduction in time spent preparing timber inventory summaries can be significant. For example, a 1984 feasibility analysis of portable data collectors for timber sale inventories on the Ozark–St. Francis National Forest in Arkansas reported a projected savings of 12.6 h per timber sale. Over one year it was estimated that forest managers could save approximately $6500 net of expenditures by using portable data collectors.

Other applications. There are many other applications for portable data collectors in forestry. In addition to inventories of timber, portable data collectors can be used to inventory other resources such as wildlife habitat, nontimber vegetation, soil, and water. Portable data collectors are also being used in wood product manufacturing applications such as log scaling, log inventory, and lumber inventory. Because many portable data collectors can be programmed for mathematical operations, they can also be used to process and summarize data onsite. This increases their usefulness in many applications, including surveying, road engineering, and harvesting productivity analysis.

For background information *see* FOREST MANAGEMENT AND ORGANIZATION; FOREST MEASUREMENT; PSYCHOPHYSICAL METHODS; STATISTICS in the McGraw-Hill Encyclopedia of Science and Technology.

[TIMOTHY M. COONEY]

Bibliography: G. J. Buhyoff and W. A. Leuschner, Estimating psychological disutility from damaged forest stands, *For. Sci.*, 24:424–432, 1978; T. M. Cooney, Portable data collectors and how they're becoming useful, *J. Forest.*, 83:18–23, 1985; T. C. Daniel and J. Vining, Methodological issues in the assessment of landscape quality, in I. Altman and J. F. Wohwill (eds.), *Human Behavior and Environment*, pp. 40–84, 1984; R. B. Hull, G. J. Buhyoff, and T. C. Daniel, Measurement of scenic beauty: The law of comparative judgment and scenic beauty estimation procedures, *For. Sci.*, 30:1084–1096, 1984; USDA Forest Service, *Automating the Timber Sale Tally System*, prepared by James R. Flanders and R. Bruce Taylor, Evaluation Rep. 2, Ozark–St. Francis National Forests, October 1984; E. H. Zube, J. L. Sell, and J. G. Taylor, Landscape perception: Research, application and theory, *Landscape Plan.*, 9:1–34, 1982.

Fractals

Most natural growth proceeds irreversibly; once it occurs, it cannot be undone. Furthermore, many irreversible growth processes lead to disorderly structures. Some examples are soot, smoke, colloids, gels, complex fluid-flow patterns, lightning, and electrolytic deposits. Traditionally, scientists have avoided analyzing such complex structures, but recent discoveries have given remarkable insights into these objects. First, simple theoretical models of the growth patterns have shown that many apparently disorderly growths display a nontrivial scaling symmetry; that is, they are fractals. Second, experimenters have demonstrated that the specific objects mentioned above, among others, are in fact fractal. Thus, a simplifying, unifying feature has been found in a complex, seemingly chaotic situation.

Scaling symmetry and the mass dimension. The pattern shown in Fig. 1 (called a Vicsek snowflake, after its inventor) is an example of a fractal. It looks as if it might have grown by adding units from outside, a process discussed below, but it also has a special structure: each part is identical to the whole seen at a different magnification. This structure has self-similarity or scaling symmetry.

B. Mandelbrot has precisely characterized such symmetry, pointing out that the abstract definitions of dimensionality used in pure mathematics are relevant to natural processes. Mandelbrot's fractal dimension (in this case, what is known as the mass dimension) characterizes the scaling symmetry of Fig. 1 as follows. If each unit of the pattern is considered to have a mass, then the dependence of the total mass M on the size of the object R can be expressed by Eq. (1), where f is a constant. For an

$$M = fR^D \tag{1}$$

ordinary shape, D is easy to write down.

For a circular pattern, D is 2; for a chain of units, D is 1. But for Fig. 1, tripling the radius only multiplies the number of units by 5. This means that $D = \log 5/\log 3 = 1.46$. When D is not a simple integer like a spatial dimension, the object in question is a fractal.

This definition of the dimension D, together with the notion of scaling symmetry, leads to many patterns, some of great beauty. It is also physically important because it characterizes the object in terms of the relationship of many scales. Thus, the details of any particular scale, such as the microscopic nature of the units in Fig. 1, are not important. Any physical property which depends only on scaling should be universal, that is, common to any system of the same scaling symmetry. If two objects have the same fractal dimension D, they will share many properties independent of their detailed makeup. Such universality is invaluable in the study of critical phenomena. The new work extends this idea to growth processes, with similarly useful results. Characterization of fractal symmetry in terms of mass dimension is particularly appropriate for growing structures: in practice, mass and size can be tracked as the object grows.

Growth models. The simplest growth models that lead to fractals are all variations on one theme: a large structure that grows by adding small subunits which stick and aggregate whenever they touch. Two models of this type will be discussed: particle aggregation and cluster aggregation.

Particle aggregation models. Particle aggregation models may be used in the study of the growth of a metal deposit in an electrochemical cell. Under the proper conditions, ions wander in the solution and attach to the electrodes or to the previously deposited metal when they touch it. A computer simulation model lets a particle walk randomly until it arrives at a certain fixed distance from the current aggregate, where it sticks. Then another particle is released, and so on.

Computer simulations of this process, known as diffusion-limited aggregation (DLA), have been carried out (Fig. 2). The surprising result is the open, tenuous nature of the growth, which is a random fractal. The growth is wispy for two reasons. First, in irreversible growth a particle stays where it sticks; it never (or in real systems, rarely) seeks a more stable location. Also, sticking tends to occur at tips because particles wander in from the outside. Fractal matter always has large holes. The fractal dimension D, measured on a series of pictures like Fig. 2, is 1.7. For a three-dimensional deposit it is 2.4.

The wispy structure of electrolytic deposits has long been known (and considered a nuisance). Measurements of the growth of a copper deposit have found D to be in agreement with computer models. Two-dimensional patterns of zinc leaves have been produced which look almost exactly like Fig. 2.

Other objects with diffusion-controlled growth (that is, random walking from outside) are very common in nature. As discussed above, they can all be expected to have very similar scaling properties. An example of great technological interest is a fluid-flow system which exhibits so-called viscous fingering. Such an object occurs when a relatively nonviscous fluid (like water) is pumped into a viscous fluid (like oil) in a porous medium. Exactly this method is used to enhance petroleum recovery. The water tends to break into a complex branching pattern; oil flows more easily away from protruding tips, or fingers (Fig. 3). This branching pattern can also be produced by letting air intrude into glycerine held between glass plates (Fig. 3).

The equations governing viscous fingering are identical to those of diffusion-limited aggregation. Thus the pattern of the less viscous fluid should resemble Fig. 2, as verified by several experiments.

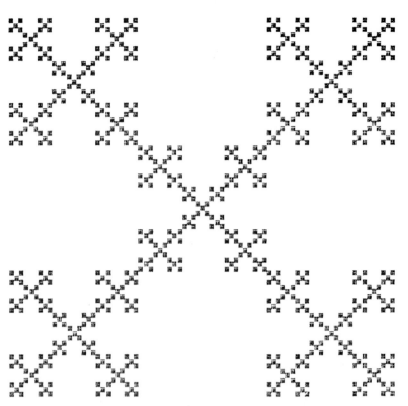

Fig. 1. Vicsek snowflake, an exact fractal.

Another even more dramatic example is dielectric breakdown, for example, lightning. The form of the electric potential near a growing breakdown also resembles diffusion-limited aggregation. In fact, the tendency for lightning to form near a sharp tip is exactly the principle of the lightning rod. Measurements on two-dimensional patterns of sparks known as Lichtenberg figures showed, once more, a fractal dimension D of about 1.7 and a geometry resembling Fig. 2.

Fig. 2. Diffusion-limited aggregate in two dimensions.

Fig. 3.   Complex branching pattern of viscous fingering, produced by letting air intrude into glycerine held between glass plates. (*Courtesy of E. Ben-Jacob et al.*)

Cluster aggregation models. Cluster aggregation models are one step more complicated than particle aggregation. They consider a collection of particles all moving randomly at once. Whenever two particles meet, they stick. But in contrast to the case of particle aggregation, the two-particle cluster continues to move, forming three-particle clusters, and so forth. This is a realistic description of what happens in a coagulation process. For example, soot forms and grows to macroscopic size because tiny carbon particles stick together in a flame or exhaust. Computer simulations have been performed showing that the process forms fractals.

These cluster aggregates are different from diffusion-limited aggregation. They are even more tenuous and wispy because it is even harder for a diffusing cluster to penetrate another cluster than it is for a single particle to enter an aggregate. As a result, the two clusters stick at the surface. The open nature of cluster aggregates is reflected in their fractal dimensions: 1.4 for two dimensions of space and 1.8 for three, compared to 1.7 and 2.4, respectively, for diffusion-limited aggregation.

A series of experiments on aqueous gold colloids have shown that a suspension of tiny gold spheres in water stick together in a way that is described in great detail by the cluster aggregation model. The fractal symmetry is measured directly on electron micrographs. Other experiments have shown that certain types of smoke are cluster aggregates.

The class of growth models so far studied is probably a small subset of those which will eventually be important for describing nature. It is remarkable, however, what a wide range of phenomena can already be treated, and how ideas taken from fractal geometry simplify and unify the treatment.

For background information *see* CRYSTAL GROWTH; DIFFUSION IN GASES AND LIQUIDS; LIGHTNING in the McGraw-Hill Encyclopedia of Science and Technology.

[LEONARD SANDER]

Bibliography: F. Family and D. P. Landau, *Kinetics of Aggregation and Gelation*, 1984; B. Mandelbrot, *The Fractal Geometry of Nature*, 1982; T.

Witten and L. M. Sander, Diffusion-limited aggregation, *Phys. Rev.*, B27:5686–5697, 1983.

Galaxy

Radio pictures of a large region centered on the nucleus of the Galaxy reveal a striking system of parallel, filamentary structures exceeding 100 light-years (6×10^{14} mi or 10^{15} km) in length. Their geometry implies that they are shaped by a relatively strong magnetic field oriented perpendicular to the disk of the Galaxy (a poloidal field), unlike the magnetic field known to exist elsewhere in the Galaxy, which lies in the galactic disk and wraps around the center (an azimuthal field). A dynamo of galactic scale might play a role in generating the poloidal field.

The Sun is located in the outskirts of the Milky Way Galaxy. It is therefore difficult to peer into the central region of this swarm of stars, not only because the nucleus is located some 30,000 light-years (1.8×10^{17} mi or 2.8×10^{17} km) away, but also because tiny interstellar dust grains distributed along the line of sight between the Sun and the nucleus effectively shroud the activity there from the view of optical astronomers. Galactic nuclei can sometimes generate enormous quantities of energy by enigmatic processes. In the most extreme cases—Seyfert galaxies, radio galaxies, and quasars—the energy output from the nucleus may rival or even exceed the luminous energy emitted by the entire galaxy. In contrast, the nucleus of the Milky Way Galaxy is relatively calm, at least at present. But because it is much nearer than its more energetic counterparts, it is much easier to study in detail and may yield the key to understanding them.

The problem of obscuration is circumvented by observing at infrared or radio wavelengths where the dust grains are rather transparent. In recent years, infrared and radio astronomers, observing predominantly the gas and dust near the galactic center, have synthesized a picture in which the variety of known phenomena result from a complex interplay between giant molecular and atomic clouds, hot plasma, magnetic and gravitational fields, and a bright, central source of luminous and mechanical energy that may surround a massive black hole.

Sagittarius A. The large-scale radio picture of the galactic center is dominated by Sagittarius A, the brightest radio source in the constellation of Sagittarius (Fig. 1). This source, which coincides with the nucleus, has an overall extent of about 50 light-years (3×10^{14} mi or 5×10^{14} km), and encompasses a radio halo surrounding a 20-light-year (1.2×10^{14} mi or 2×10^{14} km) center of activity which displays a curiously detailed structure on every scale. Observations made with very high angular resolution, employing the technique of very long-baseline interferometry (VLBI), reveal that a tiny but very bright radio source lies at a position close to, and possibly coincident with, the precise dynamical center of the Galaxy. Its size is less than about 20 times the separation between the Earth and the

Fig. 1. Radiograph of Sagittarius A at a wavelength of 6 cm. (*From R. D. Ekers et al., The radio structure of Sgr A, Astron. Astrophys., 122:143–150, 1983*)

Sun, which, by astronomical standards, is minuscule for something that radiates so intensely. The nature of this radio source remains a mystery, although infrared astronomers have independently concluded from their observations that a compact concentration of matter—probably a black hole having a mass about a million times the mass of the Sun—must be present along the same line of sight (given the uncertainties in position). This and other evidence suggest that the tiny nuclear radio source is a manifestation of energy released as intersellar matter spirals inexorably into a central black hole under the combined effects of gravitational and viscous forces.

Most of this activity occurs on a very small scale (several light-hours, or less). Other details of Sagittarius A and its environment have been observed on a much larger scale (from about one light-month up to a few hundred light-years), primarily with the Very Large Array (VLA), a network of 27 radio telescopes located near Socorro, New Mexico, and operated by the National Radio Astronomy Observatory. The Very Large Array employs the principle of interferometry to form radio images in which the angular resolution is comparable to that of the best optical telescopes. Very Large Array images of Sagittarius A show a complex, multiarmed structure about 10 light-years (6×10^{15} mi or 10^{14} km) across, and centered on the compact source observed by very long-baseline interferometry. Here the radio emission originates in a plasma which has presumably been ionized by radiative energy emanating from the mysterious nuclear object. Some of the arms of this structure may be tongues of gas which have been pulled out of their orbits by tidal forces and are now streaming toward the central mass concentration, while othes may be plumes or jets of gas which are being spewed away from the

nucleus along the rotation axis of the Milky Way. Surrounding this is a 20-light-year (1.2×10^{14} mi or 2×10^{14} km) ring of radio emission, which appears to represent the projected surface of a bubble blown by some past explosion, possibly a supernova which occurred fortuitously close to the galactic nucleus.

The Arc. Radio images of the galactic center region made with a field of view exceeding about 100 light-years (6×10^{14} mi or 10^{15} km) reveal what is called the Arc (Fig. 2). Known for over two decades, the Arc has long appeared to be a hook-shaped appendage to Sagittarius A. Recent Very Large Array observations, however, show that it consists of two bundles of remarkable radio filaments having no known counterparts elsewhere in the sky. One of the bundles contains about a dozen parallel filaments oriented perpendicular to the disk of the Galaxy, and maintains a surprising uniformity over their 100-light-year (6×10^{14} mi or 10^{15} km) length. This geometry, coupled with the observation that portions of these filaments are linearly polarized, strongly indicates that the emission process is synchrotron radiation, whereby relativistically moving electrons (those moving at almost the speed of light) are accelerated by a magnetic field, and in the process are forced to emit electromagnetic radiation which, in general, is linearly polarized. The division of the Arc into filaments is further evidence supporting their magnetic nature: relativistic particles would naturally stream along the magnetic field lines, so the filaments can be understood as tubes of magnetic flux in which the particles are constrained to move. Indeed, the streaming electrons

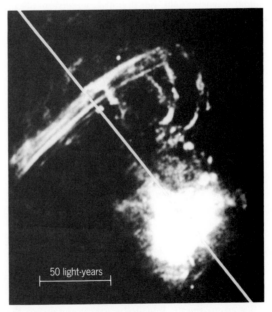

Fig. 2. The Arc at 20 cm wavelength. Sagittarius A and its surrounding radio halo constitute the large feature at the lower right. The superimposed line indicates the approximate location of the galactic plane, that is, the midline of the Milky Way.

might even generate the magnetic field in which they move if they carry a substantial current. This notion is strengthened by the presence of faint radio loops wrapped in an apparently helical fashion around the filaments.

The second bundle of filaments has a substantially different character in that they are less uniform, are more arched, and do not emit by the synchrotron mechanism. Observations of radio emission lines from recombining hydrogen atoms and of far-infrared emission from dust surrounding the filaments reveal that the filaments of the second bundle are composed of a thermal (nonrelativistic) plasma. At their extremity, they join the ends of the filaments in the first bundle at a right angle, so that the two bundles appear to be connected. The significance of the connection is an open question, although it is possible that the second bundle of filaments represents a continuation of the electric current that may be present in the first bundle. If so, ohmic (resistive) heating might play a role in generating the ionized plasma filaments within the large cloud of cool molecular gas that is known to coincide with these filaments.

Poloidal field. Any profound understanding of the physical processes occurring in the Arc must surely involve a complex magnetohydrodynamical treatment. Some important global implications might be more readily extracted, however. The magnetic field inferred from the parallel, nonthermal filaments (the first bundle) is oriented perpendicular to the galactic disk. It is thus a poloidal field, like that of the Earth's dipole field, in which the polar axis is roughly aligned with the Galaxy's rotation axis. It is possible to extrapolate from this field to imagine that the entire galactic center region is laced with a poloidal magnetic field, but that it is perceived only in the Arc where it is "illuminated." On a galactic scale, a poloidal field might be generated by the rotation of the Galaxy itself, in a manner akin to that of the dynamos hypothesized to occur in the interiors of the Earth and Sun. By combining galactic rotation with some form of noncircular motion (large-scale turbulence, for example) the galactic dynamo would tap the Galaxy's well-established azimuthal magnetic field to generate the poloidal one.

Asymmetry and energy source. If this picture is generally applicable, then one must solve the problem of why the poloidal field is illuminated by radio emission at a position which is offset from the galactic center by 100 light-years (6×10^{14} mi or 10^{15} km), that is, why the Galaxy is not symmetric about its rotation axis. The answer is probably linked to the fact that the distribution of interstellar matter is asymmetric: There is much more interstellar molecular gas on the Arc's side of the galactic center than on the opposite side, and since the magnetic field lines are tied to the interstellar gas, it should not be surprising that it is more concentrated, and thus more likely to manifest itself, on the side where the gas is found. Also, there is the question of what supplies the energy by which the Arc is illuminated.

The very interesting possibility that electric fields are responsible is under consideration. Clearly, many profound and challenging questions have been raised by the new view of the galactic center Arc.

For background information *see* BLACK HOLE; GALAXY; QUASARS; RADIO ASTRONOMY; SYNCHROTRON RADIATION in the McGraw-Hill Encyclopedia of Science and Technology.

[MARK R. MORRIS]

Bibliography: R. D. Ekers et al., The radio structure of Sgr A, *Astron. Astrophys.*, 122:143–150, 1983; K. Y. Lo et al., On the size of the galactic center compact radio source: Diameter < 20 a.u., *Nature*, 315:124–126, 1985; E. Serabyn and J. H. Lacy, Observations of the galactic center: Evidence for a massive black hole, *Astrophys. J.*, in press; F. Yusef-Zadeh, M. Morris, and D. Chance, Large, highly organized radio structures near the galactic centre, *Nature*, 310:557–561, 1984.

Gene

Transposable P elements provide a means for introducing cloned genes into the germ line chromosomes of the fruit fly, *Drosophila melanogaster*. With the recent development of this transformation technique, molecular geneticists are learning about the requirements for the mobilization of P elements. Investigators can follow individual transformation events by including visible genetic markers within the P-element vectors. This technique is extremely useful because researchers can insert genes or regulatory deoxyribonucleic acid (DNA) fragments into marked P-element vectors and ask questions about the sequences that control the developmental expression or regulation of genes. Transformation studies have begun to reveal the importance of large chromosomal domains, which strongly affect the workings of individual genes. In addition, P-element mutagenesis has become an important method for generating mutation in intact cells and for cloning genes.

P family of transposable elements. The P element is a DNA sequence capable of moving around the *Drosophila* genome and causing other elements to jump, but only when male fruit flies that carry P elements in their chromosomes (P strains) mate with females that lack them (M strains). The mobilization of the elements occurs in the germ line of the developing embryo; therefore, it is not until the second-generation offspring of the mated pair are born that the effects of this transposition are seen. The second generation shows many abnormalities, including chromosomal rearrangements, increased sterility, and a high mutation rate. These phenomena are collectively referred to as hybrid dysgenesis. When females containing P elements are crossed with M-strain males, or when crosses are made within strains, hybrid dysgenesis does not occur. Germ line transformation simulates hybrid dysgensis by injecting cloned P-element DNA molecules into the developing germ line of M-strain embryos.

Molecular biologists have learned about the struc-

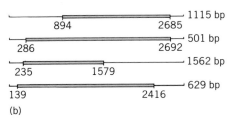

(a)

(b)

Fig. 1. Comparison of (a) complete P-element DNA with (b) four defective P-element sequences cloned from the *white* locus. The shaded areas mark the deleted portion from each defective element. Numbers at right refer to the number of nucleotide pairs that are present in each element. Open reading frame (ORF) regions designate sequences that have the capability of encoding polypeptides that may serve to regulate the transposition of the elements. (*After K. O'Hare and G. M. Rubin, Structures of P transposable elements and their sites of insertion and excision in the Drosophila melanogaster genome, Cell, 34:25–35, 1983*)

ture of P elements by cloning them out of P strains and determining their nucleotide sequence, a necessary first step for developing a transformation system. There are two types of P elements, complete and defective ones. The larger, complete elements are about 2.9 kilobases (1 kb = 1000 bases) in length and can move themselves as well as other elements around the genome. The defective elements vary in size, but are smaller than the 2.9-kb element, and can only mobilize in the presence of a complete P element. The defective elements lack variable amounts of DNA from the middle portion of the complete element (Fig. 1), which suggests that the middle portion of the complete element contains genes whose products control transposition. Since both classes maintain the same DNA sequences at their ends, it appears that the ends are important for the integration of P elements into the chromosomes during P-element transposition. Each end is an inverted repeat of the other, a common feature of transposable elements.

Genetic engineering of P elements. With the knowledge of the structural features just outlined, G. Rubin and A. Spradling engineered P-element molecules to make them suitable vehicles for introducing cloned genes into the chromosomes of the fruit fly. They used a gene called *rosy* (ry). Normally flies have brick-red eyes (ry^+), but ry^- mutants have brown eyes due to a genetic defect in pigment production. The molecular biologists inserted the ry^+ gene into the middle of a defective P element by using recombinant DNA technology, thus creating a transposon. They assumed that if they could get these transposons into the chromosomes of the developing germ cells and create an environment that mimicked hybrid dysgenesis, the DNA would integrate into the chromosomes. The proper dysgenic conditions were simulated by co-injecting

complete P elements with the defective P elements into the germ line of M-strain embryos. The complete elements contributed the genes required for transposing the defective P-element transposons. Rubin and Spradling reasoned that since integration events should occur in the germ line, the transformed genes should be stably inherited like any other gene.

The experimenters selected the progeny of the injected embryos with red eyes (putative transformants) and mated them to study the inherited DNA. Molecular hybridization to the DNA of their descendants demonstrated that most of the injected molecules had integrated without gross rearrangements into random chromosomal sites, usually averaging one or two integrations per transformant. These experiments showed that P elements could be used as vehicles for introducing other genes into the chromosomes, and that the ry^+ gene could be used as a visible marker for transformation events.

Gene regulation studies. Many investigators now routinely employ P-element transformation to ask specific questions about the regulation of gene expression. A wide variety of individual genes, including genes that control the fly's behavior, genes under strict developmental control, and ubiquitous enzyme genes, have been studied by this technique.

The general design of these experiments has been to insert the gene of interest into the marked P-element vector, microinject these molecules as described above, and choose ry^+ transformants (those with red eyes) from the progeny. Stable lines containing one copy of the inserted gene are established, and the sites of integration are determined by hybridization of the transposon to the polytene chromosomes, a technique that labels the locus on the chromosome where the transposon has integrated, using radioisotopically labeled complementary DNA sequences. The integrated copies are analyzed at the molecular level to ensure that no rearrangements have occurred during transposition, and favorable lines are selected for studies of expression of the transformed genes in their diverse new chromosomal positions.

The expression of transformed genes can be assayed in a number of ways. If there are mutants available that lack the gene product of the newly introduced gene (such as the ry^- mutant) one can simply look at the final fly for "rescue" of the mutant phenotype (red eyes). If no mutants exist, one can alter the gene before it is injected by inserting heterologous DNA (DNA from a different gene) into the transcribed portion of the gene. Such an insertion results in a larger RNA which can be distinguished in size from the normal RNA made by the endogenous gene (Fig. 2). The transformed RNA can also be specifically identified by its complementarity to the heterologous inserted DNA. At the RNA level, time-specific and tissue-specific patterns of transcription from transformed genes can be analyzed and compared with those from the endogenous genes, as can the quantities of RNAs produced. Alternatively, the final gene product can be assayed if

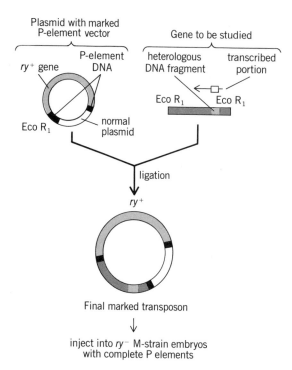

Fig. 2. Structure and construction of a marked transposon. A plasmid which contains the *ry*⁺ gene inserted between P-element ends is constructed, and a section of heterologous DNA is inserted into a DNA fragment containing the gene to be studied. The increase in size of the transcribed portion of DNA of the gene to be studied (as a result of the heterologous DNA fragment) is shown. When both the plasmid and the gene are digested with the restriction endonuclease Eco R_1, the two fragments can join, via their cohesive Eco R_1 termini, which produces the final marked transposon.

there is a known enzymatic activity or phenotype associated with it that can be distinguished from that of the host fly.

All of the above methods have shown that genes carry sequences required for their proper developmental regulation very close to the transcribed sequences. Usually these regulatory regions are just upstream (within 500 nucleotides) from the transcribed region; sequences dictating different tissue specificities for the same gene have been separated by deletion analysis of this region. Several studies suggest, however, that to obtain the normal quantitative levels of RNA produced, sequences farther upstream are needed. Since the same fragment tested in five different places often shows different amounts of transcription from each site, biologists have concluded that chromosomal sites of integration influence quantitative transcription.

Another approach to understanding gene regulation is to look for fragments containing controlling elements, rather than genes, to see if they can assert their control over new chromosomal domains. An example of this is the chorion (eggshell) amplification control element. The chorion genes are organized in two clusters in the chromosomes. The DNA in these clusters undergoes amplification prior to the transcription of the genes, resulting in maximal amounts of RNA and protein synthesis in a very short time period. Transformation studies testing DNA fragments from the amplified regions have identified centrally located fragments (amplification control elements) which autonomously initiate amplification when placed in novel chromosomal sites with the proper developmental timing and tissue specificity.

Another general control mechanism, dosage compensation, regulates the levels of transcription of many X-chromosome–linked genes so that males, with only one X chromosome, produce the same amounts of RNA as females, with two X chromosomes. The control sequences for the process of dosage compensation of the X-chromosome–linked *white* gene have been shown to be carried on a small DNA fragment located adjacent to the *white* gene.

Other uses. In addition to being useful tools for studying genetic regulation, P elements can be used to generate new mutations and to clone genes. As mentioned earlier, hybrid dysgenesis induces mutations by the jumping in (or out) of P elements to genes. If the P element integrates into a desired gene, the gene can be fished out of the genome of the mutated animal by using hybridization to the P-element DNA. Similarly, if a P element is close to a gene of interest, hybrid dysgenesis can induce imperfect excisions of the element from this gene, resulting in deletions which include the flanking material containing the gene of interest.

For background information *see* GENE; GENE ACTION; GENETIC ENGINEERING in the McGraw-Hill Encyclopedia of Science and Technology.

[LAURA KALFAYAN]

Bibliography: D. V. DeCicco and A. C. Spradling, Localization of a cis-acting element responsible for the developmentally regulated amplification of *Drosophila* chorion genes, *Cell*, 38:45–54, 1984; T. Hazelrigg, R. Levis, and G. M. Rubin, White Locus DNA transformation in *Drosophila*: Dosage compensation, *zeste* interaction and position effects, *Cell*, 36:469–481, 1984; K. O'Hare and G. M. Rubin, Structures of P transposable elements and their sites of insertion and excision in the *Drosophila melanogaster* genome, *Cell*, 34:25–35, 1983; G. M. Rubin and A. C. Spradling, Genetic transformation of *Drosophila* with transposable element vectors, *Science*, 218:348–353, 1982.

Genetic engineering

In the past decade, understanding of how living organisms synthesize proteins has progressed to the point where molecular biologists can use this synthesis machinery in the manufacture of proteins of interest—particularly for pharmaceutical use. Fundamentally, this genetic engineering involves first the isolation from human deoxyribonucleic acid (DNA) of a gene encoding a particular protein. This isolation step is accomplished by cloning the DNA and amplifying it in the bacteria *Escherichia coli*. Second, the human regulatory signals preceding and following the gene are removed and replaced with heterologous regulatory signals from bacteria, yeast, or other microbial or eukaryotic sources. The assembled human gene and heterologous regulatory signals are then transferred from *E. coli* to the heterologous cell culture system to synthesize the human protein. The two steps in this process are known respectively as the cloning and the expression of a gene.

Hemophilia. Hemophilia is a sex-linked genetic disorder that affects about 1 in 10,000 males worldwide. About 90% of the hemophilia cases are due to a defect in the protein Factor VIII, a component of the blood coagulation cascade. Current treatments of hemophilia utilize concentrates of Factor VIII derived from human plasma. Since Factor VIII is a trace protein in plasma, these concentrates are made from the pooled blood of thousands of donors. Thus, hemophiliacs risk exposure to various blood-borne viral diseases such as hepatitis and acquired immune deficiency syndrome (AIDS). Genetic engineering provides the opportunity to manufacture Factor VIII in much larger quantities and in an environment where such viral risks should be eliminated.

Cloning Factor VIII. Cloning of the Factor VIII gene begins with the purification and characterization of the protein. This work had progressed slowly for a number of years because Factor VIII is a trace protein and because it is easily degraded by various enzymes during the isolation procedure. Recent large-scale purifications beginning with 100 liters (25 gallons) of human plasma have succeeded in purifying up to 500 micrograms of the protein. Characterization of this purified material showed that Factor VIII is a very large protein containing some 2000–3000 amino acids—one of the largest proteins known.

The large-scale purification has yielded sufficient quantities of Factor VIII so that the amino acid sequence in several short regions of the molecule could be determined. These amino acid sequences were used to deduce the corresponding DNA sequence which codes for these short regions. Oligonucleotides of 14–50 bases with the DNA sequence encoding these sequenced regions of Factor VIII were then chemically synthesized and radioactively labeled with phosphorus-32 for use as probes for the Factor VIII gene.

A heterogeneous mixture of fragments derived from human chromosomal DNA was then joined to DNA molecules of the bacteriophage λ so that a library of cloned human DNA could be replicated by *E. coli*. This library, containing about 1 million independent clones, was screened by hybridization with the ^{32}P-labeled oligonucleotide probes for Factor VIII. Hybridization is a technique in which the self-complementary strands of the DNA are separated and allowed to find the corresponding strand once again. When a radioactive probe sequence is present, it can substitute for the native complementary strand and provide a radioactive tag for the

clone of interest. In this manner the first clones of a short portion of the Factor VIII gene were isolated and their DNA sequence determined.

The chromosomal genes of eukaryotes are usually split up into short exon regions (typically of 50–350 base pairs) which are separated by intron sequences that can sometimes be quite long (up to 30,000 base pairs). After a ribonucleic acid (RNA) copy of the chromosomal gene is made in the nucleus, the introns are removed by a process known as RNA splicing to give a cytoplasmic messenger RNA (mRNA) which codes for the protein and which contains only exon sequences. To express Factor VIII in a heterologous system, it was this spliced mRNA sequence that was desired. With the Factor VIII exon probe that was cloned from the chromosomal DNA, a number of human cell lines were screened by the hybridization technique to find a source of Factor VIII mRNA. One of about 40 cell lines showed low but detectable levels of Factor VIII mRNA. A DNA copy of this RNA was made by a series of steps starting with the viral enzyme, reverse transcriptase, which synthesizes a DNA strand starting with a mRNA template. Several libraries of such DNA copies (cDNA libraries) cloned into bacteriophage λ were screened with the Factor VIII probe to isolate a series of overlapping clones that represent the entire Factor VIII mRNA. The DNA sequence of these clones was determined, and from this sequence the entire amino acid sequence of Factor VIII was derived.

With the cDNA clones for Factor VIII it was possible to clone and characterize the entire chromosomal copy of the Factor VIII gene (Fig. 1). The gene contains 26 exons and encompasses 185,000 base pairs of the X chromosome—the largest gene known to date. This gene codes for a mRNA of about 9000 base pairs, including about 170 base pairs of 5′ untranslated RNA, a protein coding region of 7053 base pairs beginning with an initiation

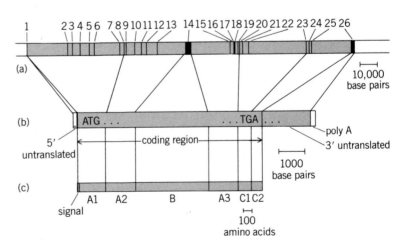

Fig. 1. Diagram of (a) Factor VIII gene, (b) messenger RNA, and (c) protein. The gene has 26 exons and encompasses 185,000 base pairs of the X chromosome. The messenger RNA has the exons spliced together and is about 9000 base pairs in length. The protein coding region of the messenger RNA is 2351 amino acids long, including a 19-amino-acid signal peptide that is cleaved off when Factor VIII is secreted from the cell.

methionine codon (ATG) and ending with a stop codon (TGA), and a 3′ untranslated region of 1805 base pairs. The protein begins with a 19-amino-acid signal sequence specifying that the protein is to be secreted from the cell. This sequence is cleaved off in the course of this secretion process, leaving a full-length protein of 2332 amino acids. A number of internal homologies in the Factor VIII protein can be found, suggesting that various gene duplications have occurred in the course of the evolution of the protein. There is a triplicated region of about 375 amino acids (the A repeat), a large unique region of 930 amino acids (the B region), and a 150-amino-acid duplication (the C repeat).

The cloned DNA probes for the Factor VIII gene can also be used to localize and determine the precise molecular defects that are the cause of hemophilia in some cases. In four patients whose DNA have been analyzed in detail, the defects differ, showing that various genetic mutations can lead to the same disease phenotype. Two of these four patients have a short portion of the Factor VIII gene deleted from the chromosome. The other two have apparently only a single nucleotide change which converts the normal codon specifying an amino acid into a stop codon. All four of these defective Factor VIII genes would be expected to make truncated forms of the protein which would be inactive.

Expression of Factor VIII. Since Factor VIII is a large and therefore complicated protein, the cloned Factor VIII DNA was engineered for expression in mammalian tissue culture cells. The Factor VIII cDNA clones were assembled into an expression plasmid that could be replicated in *E. coli* to produce large amounts of the plasmid DNA. This plasmid also contains eukaryotic control signals, so that Factor VIII can be produced when the plasmid is introduced into mammalian tissue culture cells. This expression plasmid (Fig. 2) contains a bacterial

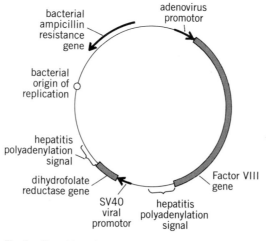

Fig. 2. Plasmid used to express Factor VIII in tissue culture cells. This plasmid contains the DNA sequences necessary to grow large quantities of the DNA in *Escherichia coli*, as well as the Factor VIII and dihydrofolate reductase genes that function in mammalian tissue culture cells.

origin of replication and a selectable marker gene (ampicillin) so that it can be replicated in *E. coli*. It also contains a eukaryotic promoter or RNA transcription start signal from adenovirus, followed by the assembled Factor VIII gene from the cDNA clones (about 7000 base pairs). The Factor VIII mRNA is terminated by a polyadenylation signal from hepatitis virus. This signal is necessary for proper mRNA processing in eukaryotic cells. Following this is a second transcription unit which provides a selectable marker gene in eukaryotic cells. This consists of another viral promoter, the coding sequence for the enzyme dihydrofolate reductase and a second polyadenylation signal. When this expression plasmid is introduced into mammalian tissue culture cells, occasionally (with a frequency of 1 per 1000 to 1 per 1,000,000) the plasmid integrates into a chromosome in the cell lines. Cells that stably integrate the plasmid can then be selected by virtue of the dihydrofolate reductase gene with the use of a selective medium. When medium from these selected cells is assayed for Factor VIII activity, it can be demonstrated by a number of criteria that these cells now produce active, human Factor VIII. A variety of techniques can be used to increase the yield of Factor VIII synthesis by these cell lines to the point where sufficient quantities can be purified for animal and human testing of the protein.

For background information *see* DEOXYRIBONUCLEIC ACID (DNA); GENETIC ENGINEERING; RIBONUCLEIC ACID (RNA) in the McGraw-Hill Encyclopedia of Science and Technology.

[WILLIAM I. WOOD]

Bibliography: J. Gitschier et al., Characterization of the human Factor VIII gene, *Nature*, 312:326–330, 1984; J. Gitschier et al., Detection and sequence of mutations in the Factor VIII gene of hemophiliacs, *Nature*, 315:427–430, 1985; G. A. Vehar et al., Structure of human Factor VIII. *Nature*, 312:337–342, 1984; W. I. Wood et al., Expression of active human Factor VIII from recombinant DNA clones, *Nature*, 312:330–337, 1984.

Genetics

The term C value refers to the number of nucleotides in the haploid genome of a particular organism. For several decades, the relationship between C value, information content, and morphological complexity has been clouded by apparent inconsistencies leading to the "C-value paradox." The C-value paradox consists of several distinct but related observations: (1) morphologically similar organisms sometimes have widely different C values; (2) the amount of deoxyribonucleic acid (DNA) in the genome seems to correlate poorly with organismal complexity; and (3) there appears to be vastly more DNA in many eukaryotic genomes than necessary to code for all the proteins believed to be made.

Measured genome site. Figure 1 illustrates the first two points. Variations of two orders of magnitude are evident among relatively closely related organisms, such as the amphibians. Moreover, some relatively primitive organisms, such as gastropods, have a higher C value than the mammals. Some amphibians have 25 times the genomic information content of humans! For illustration of the third point, consider humans. Humans are thought to make approximately 100,000 proteins. Since the size of an average protein is about 500 amino acids, and since each amino acid is specified by three nucleotides of DNA, it takes 1500 nucleotides to code for the average protein. Thus, if 100,000 proteins were made, about $1500 \times 100,000 = 1.5 \times 10^8$ nucleotides of DNA are used to code for protein. But, as shown in Fig. 1, the size of the human genome is at least 2×10^9 nucleotides. Thus, there appears to be at least 10 times more DNA in the human genome than needed; this observation has been extended to many other organisms as well.

Minimum genome size. The lack of correlation of genome size with either morphological or informational considerations presents a confusing picture. The situation has been somewhat clarified, however, by more accurate measurements and more thoughtful treatment of the data. First, it should be noted that genome size estimates of the kind summarized in Fig. 1 were generated by different workers using different techniques, and are subject to considerable error. When it is realized that the measurements are accurate only within a factor of two or three, the variability among related groups shown in Fig. 1 is reduced. Second, it is now known that some of the really extreme cases of variation of genome size are unusual and probably arise from large-scale amplifications of extensive regions of the genome, or even repeated rounds of amplification of the entire genome. Observations such as the wide variation in C value among the amphibians can probably be explained by such evolutionarily unusual occurrences. Finally, the relationship between organismal (that is, morphological) complexity and genome size is simplified if the total genome size, which may contain extraneous as well as essential DNA, is not considered, but rather the minimum known genome size among organisms in particular taxonomic groups. This yields a clearer picture of the minimum amount of DNA needed to specify particular morphologies. As shown in Fig. 2, when the data are presented in this form, a relationship between genome size and organismal complexity is evident.

Eukaryotic genome organization. Using minimum genome size enables some rationalization for the variation in C value observable in the animal world. However, the third aspect of the C-value paradox—why eukaryotic genomes are so large relative to the information needed to code for protein—remains to be addressed. New information generated by the use of recombinant DNA techniques sheds light on genomic organization which helps to understand the situation.

Many genes do not occur only once in the genome. Rather, they are repeated and organized into multigene families. Often, each of the multiple cop-

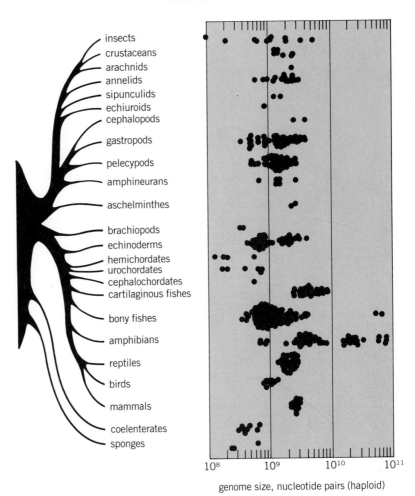

Fig. 1.   Distribution of measured genome sizes among animals. (*After R. J. Britten and E. H. Davidson, Repetitive and non-repetitive DNA sequences and a speculation on the origins of evolutionary novelty, Quart. Rev. Biol., 46:111–138, 1971*)

ies of the gene is subtly different and performs a slightly different function, or is expressed differently during development. For example, a cluster of genes codes for the beta subunit of human hemoglobin (Fig. 3). Some of the members of this particular multigene family are apparently nonfunctional (they have been called pseudo-genes). A large number of genes, previously thought to occur only once, turn out to be members of such multigene families. Well-studied examples include genes coding for globin, actin, tubulin, histones, and ovalbumin. So many genes are known to be thus organized that multigene families appear to be the rule rather than the exception. Since many proteins are encoded by multiple copies of genes, more DNA of the genome must be committed to coding for each protein than was previously thought.

Further, fine-scale mapping of eukaryotic genomes reveals that there are spaces between genes that are often as large or larger than the genes themselves. For example, the beta globin multigene family is scattered over almost 6×10^7 nucleotides (Fig. 3). Taken together, the repetitiveness of many

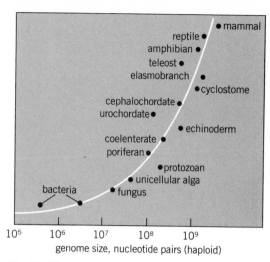

Fig. 2. Minimum known genome sizes among representatives of various taxonomic groups. (*After R. J. Britten and E. H. Davidson, Gene regulation for higher cells: A theory, Science, 165:349–357, 1969*)

genes and the spaces between them account for a significant fraction of the "extra" DNA of eukaryotic genomes.

Still more genomic DNA contributes directly to gene structure, but not to the coding of proteins. Examples are leader and trailer sequences in messenger ribonucleic acid (RNA) molecules. These are continuous with the coding region of the message and participate in initiation and termination of translation, ribosome binding, message stability, and transport to the cytoplasm, but do not specify amino acids to be incorporated into protein.

Surprisingly, a large amount of nonprotein-coding DNA is found within structural genes themselves. These regions of noncoding DNA, called introns, interrupt the coding sequence of most eukaryotic genes. Introns are transcribed along with the coding DNA, but are spliced out of the nuclear transcript prior to transport to the ribosomes for translation into protein. The physiological function of introns (if any) is unknown. They range in size from about 50 to several thousand nucleotides, and some genes contain a large number of them. The collagen gene,

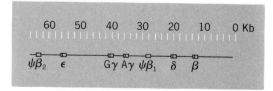

Fig. 3. Arrangement of the human betalike globin genes. The scale indicates length of the chromosomal segment in thousands of nucleotides. The genes are shown as boxes wherein the shaded regions indicate coding DNA and the open segments denote introns. Nonfunctional genes (pseudogenes) are represented as open boxes. (*After L. I. Grossman and G. P. Moore, Recombinant DNA and basic research: What we have learned, L.S.A. Mag., University of Michigan Press, 6:15–20, 1983*)

for example, has more than 50 introns making up more than 90% of its length.

Noncoding DNA. In spite of the considerations enumerated above, there still exists a large amount of DNA in eukaryotes whose function is unknown; it does not code for protein. It is possible that this DNA has no function. It may even be parasitic, in a sense, in that it confers no benefit to the organism and persists because of its own propensity for replication or amplification. This idea has been termed the theory of selfish DNA. However, several lines of evidence suggest that most nonprotein-coding DNA does have function.

Some nonprotein-coding DNA functions as a regulator or modulator of protein-coding DNA, which is often directly adjacent or nearby. Examples are sequences which promote polymerase binding, or enhance transcription efficiency, or play a role in processing of nuclear RNA. All genes probably need promoters, and there may be other regulatory signals generally required for correct gene function. Some genes have special regulatory signals, often in adjacent DNA, which modulate the gene in response to a particular external signal. Genes that are activated by heat shock provide an example of a group of genes controlled by a common regulatory signal.

Some nuclear RNA transcripts may have regulatory rather than coding roles. Theories regarding mechanisms of gene regulation have been proposed suggesting that RNA-RNA or RNA-DNA hybridization may activate genes or affect their processing. Such regulatory mechanisms, if they exist, could explain the function of some noncoding DNA.

DNA sequences that function as regulators often carry out their function by physical, sequence-specific interaction with proteins. Noncoding DNA may also be involved in sequence-specific protein interactions, not directly for gene regulation, but rather to maintain chromosomal structure. DNA exists in cells complexed with proteins which function in transcription, replication, condensation, and chromosome pairing. The necessity for proper interaction with these chromosomal proteins may be the function of a large amount of the noncoding DNA. In addition to sequences required for interaction with protein, noncoding DNA may affect structure by influencing the three-dimensional conformation of DNA itself. This can happen through intrastrand base pairing to form stem-loop structures, or by changes in conformation of the DNA helix such as the Z conformation. These sequence-specific, three-dimensional changes may be important in chromosome structure, or regulation of coding DNA, and may be controlled by nonprotein-coding DNA sequences.

Summary. Some aspects of the C-value paradox have been resolved by more accurate measurement, by further information which allows reinterpretation of the data to discern patterns where none was previously evident, and by sorting out of exceptions from the general rule. Much of the "extra" DNA of eukaryotic genomes appears to be functional, although the full range of what these functions may be is as yet unknown.

For background information *see* Deoxyribonucleic acid (DNA); Gene; Gene action in the McGraw-Hill Encyclopedia of Science and Technology.

[Gordon P. Moore]

Bibliography: R. J. Britten and E. H. Davidson, Repetitive and non-repetitive DNA sequences and a speculation on the origins of evolutionary novelty, *Quart Rev. Biol.*, 46:111–138, 1971; G. P. Moore, The C-value paradox, *Biol. Sci.*, 34:425–429, 1984; R. A. Raff and T. C. Kaufman, *Embryos, Genes and Evolution*, 1983; M. Rosbash et al., Analysis of the C-value paradox by molecular hybridization, *Proc. Nat. Acad. Sci. USA*, 71:3746–3750, 1974.

Glass and glass products

For thousands of years silicate glass has been made by melting powders of silica (SiO_2) together with fluxes such as oxides of sodium (Na_2O), potassium (K_2O), calcium (CaO), magnesium (MgO), lead (PbO), or boron (B_2O_3). Melting and forming temperatures for such traditional glass-making operations range from 1800°F (1000°C) to over 3600°F (2000°C) depending on glass composition (Fig. 1). Newly developed sol-gel processing methods make it possible to produce optically transparent amorphous silica and silicate materials at temperatures below 400°F (200°C). Low-temperature gel-glasses are only one-half the density of traditional vitreous materials of the same composition. Thus, large, lightweight optics is potentially one of the first applications. Also, because of the low density of gel-glasses, their strength is only one-half that of traditional silicate glasses. The strength and density can be increased, however, by heating the material to temperatures ranging from 840°F (450°C) to 2400°F (1300°C), depending upon gel-glass composition (Fig. 1). At the densification temperature, voids or pores (with diameters of 1–10 nanometers) in the gel-glass network collapse, the material shrinks, and the density, strength, hardness, and index of refraction increases. Consequently, it is possible to "tailor-make" various combinations of physical properties of gel-glasses by varying their thermal history. When all pores are eliminated, the properties of gel-glasses become equivalent to those of traditional melt glasses.

Sol-gel processing. Sol-gel processing involves six steps. Soluble metal organic precursors such as tetraethyloxysilane (TEOS) or tetramethyloxysilane (TMOS) are mixed with water (step 1). A hydrolysis reaction (1) occurs which forms Si—O—H bonds. Si—O—H bonds react to form Si—O—Si bonds, and release water in a condensation reaction. These Si—O—Si structural units bond with neighbors in a polycondensation reaction (2) which creates colloidal, submicrometer particles of silica and the beginning of a silica network, termed a sol (Fig. 2a).

The sol, which behaves like a fluid, can be cast

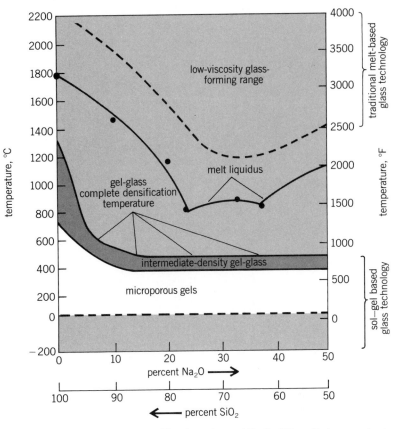

Fig. 1. Thermal processing-composition dependence of Na_2O–SiO_2 melt glasses and sol-gel glasses.

into a mold (step 2) in the shape of the desired object.

Interconnections form within the sol and the viscosity increases. At a specific time, termed the gel point, the network becomes completely interconnected and rigid; that is, gelation occurs (step 3). The sol is now a brittle solid termed a gel (Fig. 2b).

To make the gel useful, it must be aged (step 4) and dried (step 5). During aging the number of Si—O—Si bonds composing the solid gel network increases while the remaining liquid [methanol (CH_3OH) and water (H_2O)] migrates from the pores. Thus, the gel shrinks and its density and strength increase during aging.

Drying involves removal of the liquid from the pores of the gel. The pore size is usually very small, in the range of 1 to 10 nm. Consequently, capillary stresses developed during drying can be very high. If the drying stresses are greater than the strength of the gel network, cracks develop and the gel breaks into a powder. Such powders can be used commercially, for example, as abrasive grains.

In order to make monolithic objects, such as optical components, drying stresses must be controlled by: (1) very slow evaporation, (2) hypercritical evaporation which converts the pore liquid to a gas before it is removed and thereby eliminates the solid-liquid interface which causes the capillary stresses, (3) decreasing the pore-liquid surface energy by

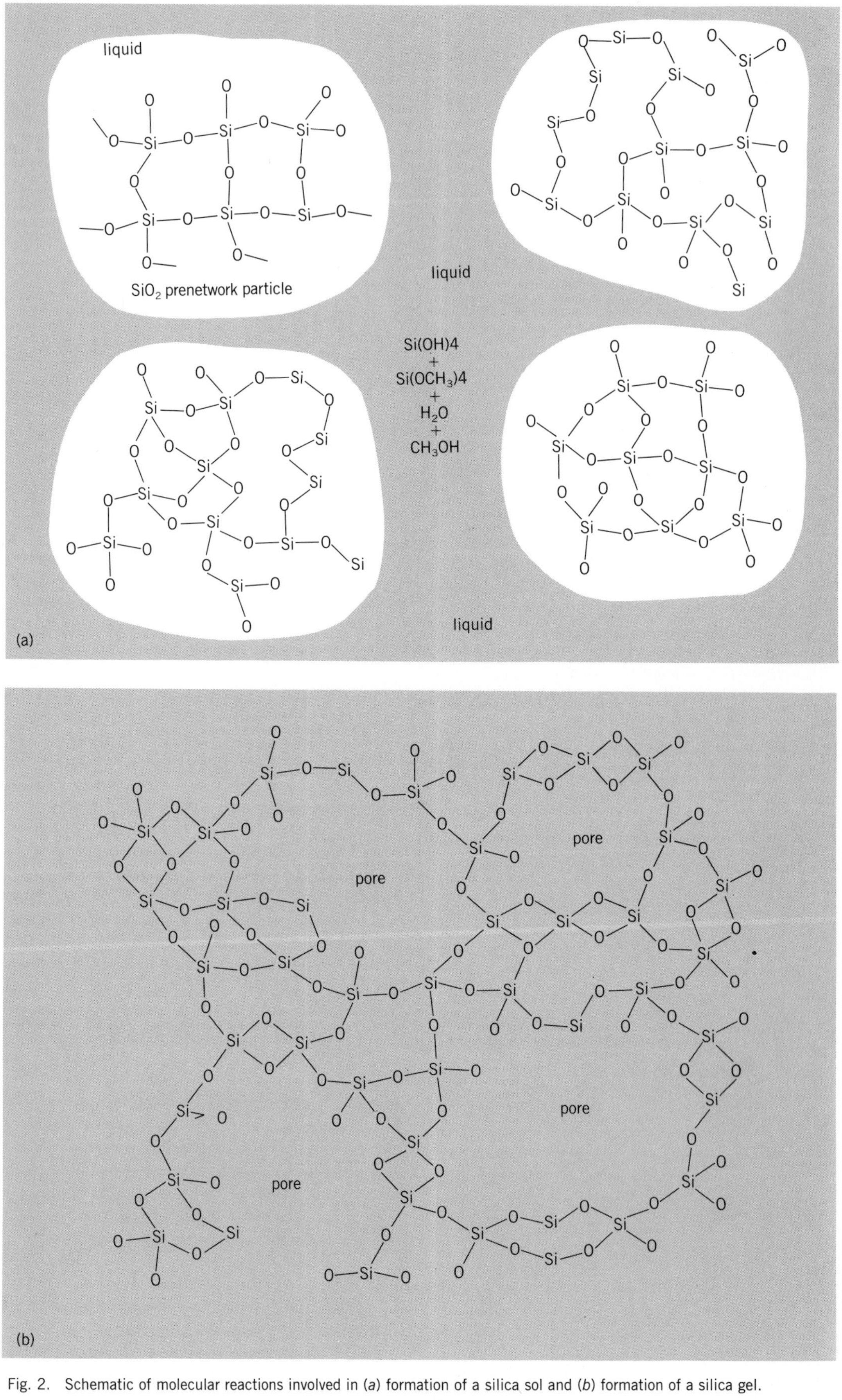

Fig. 2. Schematic of molecular reactions involved in (*a*) formation of a silica sol and (*b*) formation of a silica gel.

$$CH_3-O-\underset{\underset{CH_3}{\overset{\overset{CH_3}{O}}{|}}}{\overset{\overset{O}{|}}{Si}}-O-CH_3 + 4(H_2O) \rightarrow$$

$$HO-\underset{\underset{H}{\overset{\overset{H}{O}}{|}}}{\overset{\overset{O}{|}}{Si}}-OH + 4(CH_3OH) \quad (1)$$

$$HO-\underset{\underset{H}{\overset{\overset{H}{O}}{|}}}{\overset{\overset{O}{|}}{Si}}-OH + HO-\underset{\underset{H}{\overset{\overset{H}{O}}{|}}}{\overset{\overset{O}{|}}{Si}}-OH + HO-\underset{\underset{H}{\overset{\overset{H}{O}}{|}}}{\overset{\overset{O}{|}}{Si}}-OH \rightarrow$$

$$HO-\underset{\underset{H}{\overset{\overset{H}{O}}{|}}}{\overset{\overset{O}{|}}{Si}}-O-Si-O-\underset{\underset{H}{\overset{\overset{H}{O}}{|}}}{\overset{\overset{O}{|}}{Si}}-OH + 2H_2^+O \quad (2)$$

adding surfactants, (4) elimination of very small pores, or (5) obtaining monodisperse pore sizes which eliminates differential capillary stresses. Adding specific organic chemicals to the sol, termed drying-control chemical additives, can aid in controlling pore size distributions very effectively and can yield large dried silica gels with only several days of total processing time. Similar results can be obtained in even shorter times by using high-pressure autoclaves and a hypercritical drying process.

Properties and applications. The density of dried gels ranges from as low as 5% of the density of a melt-derived glass of the same composition to as much as 60% of theoretical density. The very low-density gels are termed aerogels, whereas the higher-density gels are called xerogels. Low densities and associated large surface areas (greater than $500 \text{ m}^2/\text{g}$) make dried gels attractive for catalyst substrates, thermal barriers, lightweight fillers for plastics, microballoons to contain deuterium oxide for nuclear fusion targets, lightweight large-scale optics, and optical waveguides.

Densified gels are useful as abrasive grains and can be prepared as fibers. Sols can also be applied as coatings on substrates such as window glass to improve their reflection properties.

For background information *see* GEL; GLASS AND GLASS PRODUCTS in the McGraw-Hill Encyclopedia of Science and Technology.

[LARRY L. HENCH]

Bibliography: C. J. Brinker, D. E. Clark, and D. R. Ulrich (eds.), *Better Ceramics Through Chemistry*, vol. 32, Materials Research Society, 1985; L. L. Hench and D. R. Ulrich (eds.), *Science of Ceramic Chemical Processing*, 1986; L. L. Hench and D. R. Ulrich (eds.), *Ultrastructure Processing of Ceramics, Glasses and Composites*, 1984; R. K. Iler, *The Chemistry of Silica*, 1979.

Gold metallurgy

Heap leaching allows the commercial exploitation of marginal gold deposits. In a heap leaching operation (see illus. *a*) ore crushed to 2-in. (5-cm) diameter or finer is placed on an impermeable leaching pad of hyperlon plastic, asphalt, or packed clay. Sodium cyanide (NaCN) leaching solution at pH 11 is sprayed onto the ore to dissolve the gold. It is essential for the pH to be high in order to maintain protective alkalinity both for gold recovery and for prevention of deadly cyanide gas (HCN). After percolating through the ore heap, the solution is pumped to a processing plant for gold removal (illus. *a*).

In the processing plant, gold is removed by passing the solution upward through columns of granular activated carbon. Gold and silver adsorb onto the carbon, and the barren cyanide solution is fortified with makeup solution and returned to the heap. Over a hundred small heaps (<100 tons or 90 metric tons), medium heaps, and large heaps (>100,000 tons or 90,000 metric tons) have been constructed in the last decade. This technology has also been applied to old tailings piles and mined overburden dumps (illus. *b*).

One of the major innovations responsible for economical heap leaching has been the use of activated carbon for gold recovery from solution. Another has been constructing more permeable heaps. Future innovations in leach solution chemistry and in-place or underground leaching hold even more promise (illus. *c*). In-place or underground leaching could save the cost of mining and environmental surface disturbance. Using acid thiourea or thiosulfate solutions instead of cyanide could reduce environmental risks of underground leaching, and the costly environmental protection bond money required for cyanide.

Gold recovery from solution. Heap leaching saves money over established vat leaching practice because the ore is neither ground so fine nor moved so far. In addition, there is no need for expensive filtering of the ore from the gold-bearing solution after leaching. Even though heap leaching gives lower gold recovery (sometimes lower than 60%), marginal gold ores (<0.05 oz/ton or 1.57 g/metric ton of gold) can be treated. However, without the advent of activated carbon adsorption, heap leaching would still be uneconomical, because the gold-bearing solutions from a heap contain so little gold (1–3 ppm).

Activated carbon adsorption is the process of recovering gold from gold-bearing leach solutions. Activated carbon is made by burning coconut shell in air-starved conditions. Used as ⅛-in. (0.3-cm) granules, it will adsorb gold from dilute cyanide solutions until it becomes saturated at several hundred oz/ton.

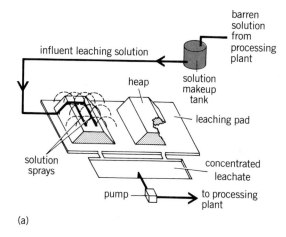

(a)

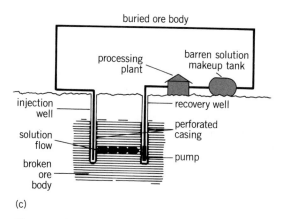

(b)

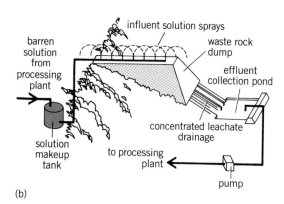

(c)

Diagrams of three gold recovery leaching operations. (a) Heap leaching. (b) Dump leaching. (c) In-place leaching of a buried ore body.

The real innovation is the discovery of how to rinse or strip the gold off the activated carbon so that the carbon could be reused for gold adsorption. In addition, the rinse solutions are highly concentrated with gold. This reduces the cost of the next step: electroplating the gold onto steel wool electrodes for metal recovery. In comparison to the dilute leach solutions from the heap, the highly concentrated rinse solutions from the carbon require smaller vessel volumes and have higher electrowinning current efficiencies.

Before the development of gold desorption from carbon, the carbon was burned, and the gold was smelted from the ashes—a comparatively expensive process. The first efforts at carbon desorption using concentrated NaOH and NaCN solutions required up to 48 h. Although this was better than smelting the gold from burned carbon ashes, it took too long. In the 1970s two methods were developed that greatly decreased the desorption time, pressure stripping and alkaline alcohol stripping.

In the pressure stripping method the loaded carbon is conditioned with caustic cyanide solution at 90°C (194°F) and eluted with water at 150°C (302°F). In the alkaline alcohol stripping method the caustic cyanide stripping solution contains 20% ethanol and the stripping solution is kept at 80°C (176°F). After stripping, any silver can be separated from gold by adding Na_2S and precipitating it as Ag_2S. However, if silver concentrations in the gold-bearing leach solutions are high (equal to gold concentrations), Ag_2S precipitation before carbon adsorption is preferable. Silver adsorption on carbon is less efficient than gold, thus requiring more carbon unless precipitated ahead of loading.

In some heap leaching operations the silver values far exceed the gold values. If silver concentration is too high (10–20 ppm), an alternative to activated carbon adsorption is used—Merrill-Crowe zinc dust cementation. Before the zinc metal goes into solution and causes the metallic gold and silver to form, the solution must be deaerated with vacuum pumps. Afterward, the nearly pure gold and silver is filtered from solution, remelted into ingots, and ultimately refined electrolytically.

Heap permeability. Another major innovation developed in the late 1970s involves a method of pelletizing or agglomerating the fines around the larger pieces of rock to enhance permeability. For heap leaching to be feasible, the ore must be porous and permeable to the leach solution. The ore must be crushed fine enough so that cyanide can diffuse through the host rock and the resulting gold cyanide complexes diffuse outward. However, the fines created during crushing can be washed into the voids by the percolating leaching solution and cause channeling and partial leaching. Any clay present swells and causes additional blocking of the voids in the heap, thus preventing solution flow.

Agglomeration consists of mixing the crushed ore with portland cement and lime, and mechanically tumbling so that fine particles adhere to the large ones. Moisture content is critical, and several hours of aging are needed for the cement to bond the particles. This simple pretreatment has increased the flow as much as 6000-fold, and has reduced the leaching cycle to days instead of weeks. The gold is recovered that much faster, making heap leaching even more economical. *See* SOLUTION MINING.

For background information *see* GOLD METALLURGY in the McGraw-Hill Encyclopedia of Science and Technology.

[KEITH A. PRISBREY]

Bibliography: P. D. Chamberlin, Heap leaching and pilot testing of gold and silver ores, *Min. Cong. J.*, pp. 47–51, April 1981; J. A. Eisele, A. F. Colombo, and G. E. McClelland, Recovery of gold and silver from ores by hydrometallurgical processing, *Separat. Sci. Technol.*, 18(12,13):1081–1094, 1983; G. M. Potter, Design factors for heap leaching, *Min. Eng.*, March 1981.

Heat pipes

Heat pipes are able to transfer heat efficiently between two locations by using the evaporation and condensation of a fluid contained therein. Originally developed for refrigeration and spacecraft thermionic generator applications, heat pipes (and their close relative, the thermosyphon) have become routine components in systems ranging from missiles to domestic tuner-amplifiers and furnaces. Their application in areas where reliable performance and low cost are of prime importance—for example, in electronics and heat exchangers—is the most important aspect of recent heat pipe progress.

Operating principle. The heat pipe, the idea of which was first suggested in 1942, is similar in many respects to the thermosyphon. A large proportion of applications for heat pipes do not use heat pipes as strictly defined below, but employ thermosyphons (Fig. 1a), sometimes known as gravity-assisted heat pipes. A small quantity of liquid is placed in a tube from which the air is then evacuated, and the tube sealed. The lower end of the tube is heated, causing liquid to vaporize and the vapor to move to the cooler end of the tube, where it condenses. The condensate is returned to the evaporator section by gravity. Since the latent heat of evaporation is generally high, considerable quantities of heat can be transported with a very small temperature difference between the two ends. Thus the structure has a high effective thermal conductance. The thermosyphon has been used for many years, and is also known as the Perkins tube. A wide variety of working fluids have been employed, ranging from helium to liquid metals.

One limitation of the basic thermosyphon is that

Methods of condensate return in heat pipes

Method	Device
Gravity	Standard thermosyphon
Capillary forces	Standard heat pipe
Centripetal forces	Rotating heat pipe
	Rotary heat pipe
Electrostatic volume forces	Electrohydrodynamic heat pipe
Magnetic volume forces	Magnetohydrodynamic heat pipe
Osmotic forces	Osmotic heat pipe
Vapor bubble pump	Inverse thermosyphon
Mechanical pump	Two-phase run-around coil

in order for the condensate to be returned by gravitational force to the evaporator region, the latter must be situated at the lowest point. The heat pipe is similar in construction to the thermosyphon, but in this case provision is made for returning the condensate against a gravity head. Most commonly a wick, constructed for example from a few layers of fine gauze, is used. This is fixed to the inside surface of the tube, and capillary forces return the condensate to the evaporator (Fig. 1b). Since the evaporator position is not restricted, the heat pipe may be used in any orientation. If, of course, the heat pipe evaporator happens to be in the lowest position, gravitational forces will assist the capillary force. Alternative techniques, including centripetal forces, osmosis, and others, may be used for returning the condensate to the evaporator (see table).

Capillary forces are by far the most common form of condensate return employed, but an increasing number of rotating heat pipes are used for cooling of electric motors and other rotating machinery. The use of electrohydrodynamic and magnetohydrodynamic forces has not achieved popularity, and even the osmostic heatpipe is still restricted to the research and development laboratory. A mechanical pump is used to return condensate in commercially available two-phase run-around coil heat recovery systems. While this may be regarded by some as a retrograde step, it is a much more effective method for condensate return than reliance on capillary forces.

Applications. The applications for which heat pipes are developed are related to five principal functions of the heat pipe: separation of heat source and sink; temperature flattening; heat flux transformation; temperature control; and acting as a thermal diode or switch. The two applications discussed below can involve all of these features. In the case of electronics cooling and temperature control, all features can be important. In heat exchangers employing heat pipes, the separation of heat source and sink, and the action as a thermal diode or switch, are most significant.

Cooling of electronic components. The use of heat pipes for cooling or temperature control of electronic components is the largest field of application, and a wide variety of heat pipe geometries may also be employed. The basic tubular heat pipe remains the

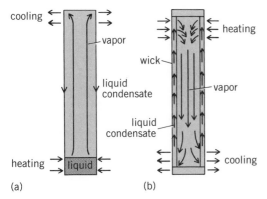

Fig. 1.   Heat transfer devices. (a) Thermosyphon. (b) Heat pipe. (Heat pipe can be in any position, not just vertical as shown.)

transistor

Fig. 2. Heat pipe "radiator" for transistor cooling. The transistor itself is also shown. (*Japan Radio Co. Ltd.*)

most common type, but flat heat pipes, direct-contact systems, and flexible heat pipes all play an increasingly important role. The most widely applied heat pipes in electronics cooling are units of conventional form that were developed for cooling of semiconductor devices and are manufactured using a low-cost production method for spirally grooved wicks. By using water as the working fluid in a copper container, heat transport capabilities in excess of 1 kW are achieved. The heat pipes, which normally operate horizontally, typically have a diameter

Fig. 3. High-frequency radio transmitter with exposed power amplifier cooled by heat pipes of type in Fig. 2. (*Japan Radio Co. Ltd.*)

of 0.6 in. (16 mm), and give a 30% improvement in heat dissipation and a 50% weight reduction when compared with a conventional heat sink.

A wide range of audio amplifiers employ heat-pipe heat sink systems of this type. As well as improving heat dissipation, the use of heat pipes also yields electrical benefits, including improvements in the distortion factor. The heat pipes are used to transfer heat to the periphery of the amplifier, where a finned heat sink can dissipate it to atmosphere. This prevents local overheating within the electronics enclosure.

A second major application is in power amplifiers. The heat pipe radiator (Fig. 2) is used as a heat sink for high-power transistors and thyristors. It contributes to low package sizes and general cost benefits. The overall heat sink resistance varies between 0.13 and 0.07 K W^{-1} (0.23 and 0.13°F W^{-1}) depending on air velocity, the thermal resistance of the heat pipe component being as low as 0.02 K W^{-1} (0.04°F W^{-1}) for a heat dissipation of 400 W. The heat sink size is 5.7 $\times$ 2.4 in. (145 $\times$ 60 mm). When the power amplifier is used in a general-purpose high-frequency radio communications transmitter of 1-kW peak envelope power, four heat pipe radiators each cool two power transistors (Fig. 3).

Heat exchangers. The use of bundles of multiple heat pipes as heat exchangers is second only to electronics thermal control in terms of number of units in operation. While heat-pipe heat exchangers in the form of air-air or air-liquid heat recovery systems are increasingly used in domestic and commercial buildings for heating, ventilating, and air-conditioning energy conservation, it is in the industrial field that the heat-pipe heat exchanger has attacted most interest, in terms of variety of uses and the number of heat exchanger concepts examined and, in some cases, installed. Gas-gas heat-pipe heat exchangers are used in a wide range of industrial processes, normally to recover heat from hot or humid exhaust streams to preheat make-up air. Heat may also be recovered for space heating.

The conventional heat-pipe heat exchanger is available in sizes covering heat transport duties ranging from a few tens of watts to several megawatts. The ability to separate source and sink conveniently has led to many derivatives which show the versatility of heat pipe and thermosyphon heat exchangers. These include systems for recovering heat from wastewater, condensers, and exchangers for improving the performance of domestic heat pumps; heat recovery in flue gas desulfurization; and heat transfer in spent nuclear fuel stores. Heat-pipe waste heat boilers (which use hot gases to generate steam for processes or space heating) have been used in situations where reactive fluids have to be separated, and the ability of heat-pipe heat exchangers (in common with the run-around coil, another heat recovery system) to accommodate different external surfaces on the evaporators and condensers benefits many energy-intensive processes.

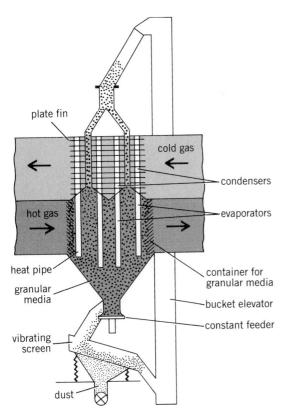

plate fin

cold gas

condensers

evaporators

hot gas

heat pipe

granular media

container for granular media

bucket elevator

constant feeder

vibrating screen

dust

Fig. 4.   Heat-pipe heat exchanger for use in heavily fouled gas streams.

An example is the development of heat pipe–based systems for use in heavily fouled environments. In such situations the use of unfinned evaporators, possibly immersed in a fluidized bed or another form of cleaning–heat transfer enhancement system, can be of benefit. Conventional extended surfaces may of course be retained on the condensers (which may be of different material than that of the evaporators), since the cold gas which is preheated by flowing over the condensers is normally much cleaner than the hot exhaust gas which flows over the evaporators. The concept (Fig. 4) has been developed primarily for use in the iron and steel industry. The heat pipes consist of tubes with longitudinal grooves on the inside surface, produced by hydrostatic extrusion. This form lends itself to the mass production of long heat pipes (normally gravity-assisted) of superior heat transport capability. The envelopes of the heat pipes consist of stainless or carbon steel, while the internal liners may consist of copper, so that compatibility with the working fluid can be ensured. In the evaporator sections, granular media held within a container having a perforated wall fill the space between the heat pipes. When the heat exchanger is in operation, the particles move slowly but continuously under gravity down through the container, preventing dust adhesion on the heat pipes and consequent clogging. At the bottom of the container, the dust-laden granules are cleaned by passing them along a vibrating

screen, prior to being recycled back to the top of the container via a bucket elevator.

For background information *see* HEAT EXCHANGER; HEAT TRANSFER in the McGraw-Hill Encyclopedia of Science and Technology.

[DAVID A. REAY]

Bibliography: P. D. Dunn and D. A. Reay, *Heat Pipes*, 3d ed., 1983; D. A. Reay, *Advances in Heat Pipe Technology*, 1982; L. L. Vasiliev, Low temperature heat pipes, *J. Heat Recovery Sys.*, 5:203–216, 1985.

High-pressure phase transitions

At pressures and temperatures characteristic of the Earth's mantle, minerals undergo transitions involving changes in their atomic packing to denser structure types. The high-pressure phase is the phase of lower volume so the volume of transition (ΔV) has a negative value. The entropy change (ΔS) of a high-pressure phase transition determines the pressure-temperature (*P-T*) slope of transition through the Clausius-Clapeyron equation, $dP/dT = \Delta S/\Delta V$. Experimental determination of dP/dT for pressures greater than 100 kilobars (10 gigapascals) can be difficult to attain. An alternate approach to determining ΔS involves modeling the vibrational contribution to the entropy by using infrared and Raman spectra to evaluate the change in the vibrational frequency spectrum as the crystal structure is changed. It has been suggested that positive entropies of transition or negative dP/dT slopes for high-pressure phase transitions can be expected to become more frequent for transitions occurring in the 150–1000-kilobar (15–100-GPa) region.

Kieffer's lattice vibrational model. Thermodynamic functions such as the heat capacity and entropy can be calculated from a crystal's vibrational density of states [$g(v)$], which is the phonon distribution in frequency space and is dependent on the crystal structure. Unfortunately, experimental techniques which determine a crystal's vibrational density of states are difficult to perform and are, at present, limited to simple structures. Therefore, various approximations of $g(v)$ have been used to predict and extrapolate thermodynamic functions of minerals.

One of the more recent models proposed for frequency distributions of lattice vibrations was developed by S. W. Kieffer and is shown schematically in Fig. 1*a*, where v_1, v_2, and v_3 represent frequencies of acoustic modes; v_l and v_u represent the lower and upper frequency limits of the optic continuum; and v_{E1} and v_{E2} represent the frequencies of simple Einstein functions known as Einstein oscillators. The primitive unit cell is taken as the fundamental vibrating unit of the crystal. Associated with this cell are 3*s* vibrational degrees of freedom, where *s* is the number of atoms in the primitive cell. Forsterite (Mg_2SiO_4), for example, crystallizes in space group *Pbnm* and has 28 atoms in its primitive unit cell and hence has 84 vibrational degrees of freedom. Three of these 3*s* degrees of freedom have fre-

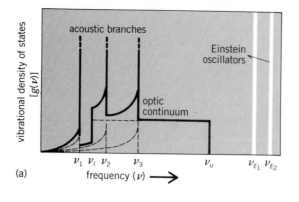

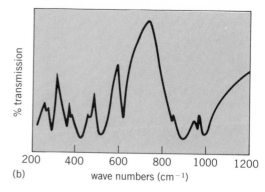

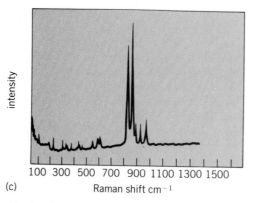

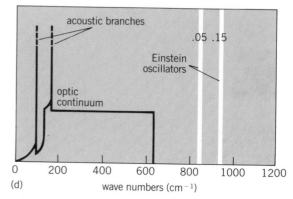

Fig. 1. Schematic representation of (a) the frequency distribution in Kieffer's lattice vibrational model; vibrational density of states is the distribution of vibrational frequen-

cies; (b) infrared spectra and (c) Raman spectra of forsterite, Mg_2SiO_4, used as input for (d) a Kieffer model of forsterite.

quencies of vibration which approach zero at long wavelengths of **k**, where **k** is a wave vector in reciprocal space. In this limit all particles in each unit move in parallel and with equal amplitudes. This motion is characteristic of elastic deformation under sound waves, and hence these three branches are termed acoustic branches.

In Kieffer's model, the acoustic branches are characterized at long wavelengths by directionally averaged sound velocities and are assumed to be dispersed sinusoidally across the Brillouin zone, which is the primitive unit cell in reciprocal space. In forsterite, directionally averaged acoustic velocities of 4.90, 4.96, and 8.56 km s^{-1} characterize acoustic branches that reach 89, 99, and 171 wave numbers (cm^{-1}) at the Brillouin zone boundary (Fig. 1d). The remaining $3s-3$ modes are optic modes, so named because modes of this type can be excited with light of an appropriate frequency in solids which are at least partly ionic. Infrared and Raman spectroscopy can be used to characterize optic modes in crystals.

In Kieffer's model, the optic modes are described in a simple way which satisfies spectroscopic observations of the minimum and maximum vibrational frequencies. If separate optic modes can be identified such as Si-O stretching modes, they are represented as weighted Einstein oscillators. High-frequency Si-O stretching modes range from 800 to 1000 wave numbers in the infrared and Raman spectra of forsterite (Fig. 1b and c) and are repre-

sented as separate Einstein oscillators (Fig. 1d), where the values 0.5 and 0.15 are the number of modes represented at the higher frequencies, the total number being normalized to 1.0. The remaining optic modes are assumed to be distributed uniformly between a range of frequencies whose lower and upper limits have been determined from spectroscopic data. This band of frequencies is called an optic continuum. The lower and upper limits of forsterite's optic continuum are 144 and 620 wave numbers determined from the far- and near-infrared and Raman spectra of forsterite (see Fig. 1d).

Vibrational contributions to entropies. With appropriate crystallographic, acoustic, and spectroscopic data, vibrational models of the low- and high-pressure phases involved in a transition can be formulated. Recent experiments using this type of approach have successfully constrained entropies of transition for the transitions of olivine to modified spinel to spinel in the Mg_2SiO_4 polymorphs, and for the phase transitions of wollastonite to garnet to perovskite in the $CaGeO_3$ polymorphs. Three factors determine which phase in the transition has the higher entropy. The first factor is the presence of low-frequency optic modes. If optic modes are present at low frequencies for a given phase, the vibrational heat capacity and entropy of that phase will be larger than those of a corresponding phase which does not have low-frequency optic modes (Fig. 2a). The second factor is the fraction of acoustic modes. Acoustic modes are present at low frequencies, and

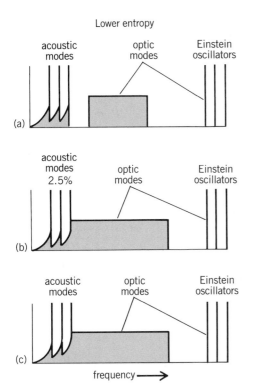

Fig. 2. Vibrational contributions to entropy from (*a*) the presence of low-frequency optic modes, (*b*) the percentage of acoustic modes, and (*c*) the overall distribution of optic modes.

a higher fraction of acoustic modes will increase a phase's heat capacity and entropy (Fig. 2*b*). The third factor is the overall distribution of optic modes. If the distribution of optic modes is concentrated at lower frequencies with no high-frequency modes present, the heat capacity and entropy of that phase will increase (Fig. 2*c*).

Entropies. The common mineral structures that exist at or near atmospheric pressure such as silica and its derivatives—feldspar, olivine, pyroxene, and amphibole—have a relatively loose packing of metal atoms and aluminosilicate units which have densities less than those of mixtures of corresponding oxides. Transitions among these phases involve, to a first approximation, a rearrangement of the packing of these isolated units with a decrease in the empty space between them in the higher-pressure polymorphs. Figure 3 shows a schematic portrayal of observed phase transitions in compounds with chemical formulas AB_2O_4 and ABO_3, where A represents a divalent cation such as Mg^{2+}, Ca^{2+}, or Fe^{2+}, and B represents a tetravalent cation such as Si^{4+} or Ge^{4+}. From 0 to 75 kilobars (0 to 7.5 GPa) the only changes in the cation coordination that occur are those for aluminum, which changes from 4-fold to 6-fold, and in the distorted polyhedra of large alkali and alkaline-earth cations. The former is important in stabilizing the garnet structure which has 8-, 6- and 4-coordinated sites. Stabilization by aluminum arises from the ability of aluminum to assume octahedral coordination at lower pressures than silicon. The entropies and volumes of transition

for transitions occurring in this pressure range have the same sign. In other words, phase transitions such as olivine to spinel (or modified spinel), pyroxenoid to garnet, and pyroxene to ilmenite (Fig. 3) have positive pressure-temperature slopes. The higher entropies of the low-pressure phases can be related to characteristics of their vibrational spectra. Low-pressure phases such as olivines, pyroxenes, and pyroxenoids have low-frequency modes below 200 wave numbers. The presence of optic modes at low frequencies contributes to the entropies of these phases so that transitions involving these phases generally have negative entropies of transition and positive pressure-temperature slopes.

At pressures above 75 kilobars (7.5 GPa), new phases with higher density are produced by increasing the coordination number of other cations. Silicon, for example, changes from tetrahedral coordination in structures such as olivine, pyroxene, and spinel to octahedral coordination in structures such as stishovite, ilmenite, and perovskite. Garnets and perovskites have structures in which divalent cations are 8- and 12-coordinated. The overall increase in density is achieved by a denser and more symmetrical packing of polyhedra, but is accompanied by an increase in nearest-neighbor distances. Therefore, with increasing coordination number and increasing bond length, the individual metal-oxygen bonds weaken.

The trend toward higher entropies for high-pressure phases with increasing coordination number can be related to systematic changes in the vibra-

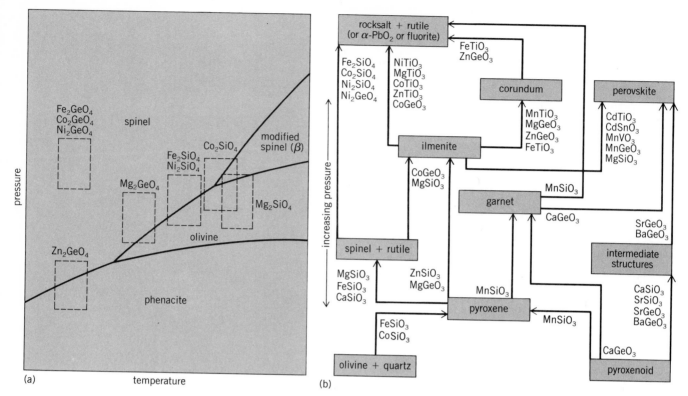

Fig. 3. Schematic representation of high-pressure phase transitions in (a) AB$_2$O$_4$ compounds (*after A. Navrotsky, Silicates and related minerals: Solid state chemistry and thermodynamics applied to geothermometry and geobarometry, Prog. Solid State Chem., 11:203–264, 1976*), and (b) ABO$_3$ compounds at approximately 1273 K or 1832°F (*after A. Navrotsky, Energetics of phase transitions in AX, ABO$_3$ and AB$_2$O$_4$ compounds, in M. O'Keeffe and A. Navrotsky (eds.), Structure and Bonding in Crystals II, Academic Press, pp. 71–93, 1981*).

tional spectra of these phases. The high-pressure phase often has higher symmetry and a smaller primitive unit cell than the low-pressure phase. Thus the low-frequency acoustic modes constitute a greater fraction of the total number of vibrational degrees of freedom, and the entropy increases. Also, high-frequency modes associated with tetrahedral Si-O stretching motions ranging from 800 to 1100 wave numbers are transformed into midfrequency lattice modes (400 to 800 cm^{-1}) associated with lattice vibrations involving bonds of [SiO$_6$] octahedra. Low-frequency modes associated with large distorted coordination polyhedra groups found in garnet and perovskite may occur. The overall effect is to compress the vibrational spectrum into a lower range of frequencies, leading to a generally increased entropy. Because of the interplay of complex factors, crossovers in heat capacities as functions of temperature are frequently observed when comparing phases of quite different structures.

Negative pressure-temperature slopes. The factors described above are responsible for the higher entropy of CaGeO$_3$ perovskite relative to CaGeO$_3$ garnet. The transition from CaGeO$_3$ garnet to perovskite is characterized by a positive entropy of transition and the pressure-temperature slope is negative. From a lattice vibrational point of view, MgSiO$_3$ perovskite is expected to behave analogously to CaGeO$_3$ perovskite and be a phase of high

entropy. The relative importance of acoustic modes, the existence of low-frequency optic modes, and the disappearance of the high-frequency optic modes together with the incorporation of modes below 800 wave numbers in the general region of the topic continuum are all changes that would contribute to a high entropy of MgSiO$_3$ perovskite. MgSiO$_3$ perovskite is believed to be an important constitutent of the Earth's lower mantle. The ilmenite-to-perovskite transition in MgSiO$_3$ may explain, in part, the abnormally high rate of increase in seismic velocities occurring at a depth of approximately 1040 mi (650 km) in the Earth. A negative pressure-temperature slope for the MgSiO$_3$ ilmenite to perovskite transition appears likely. A 1040-mi (650-km) discontinuity of a phase transition with a negative pressure-temperature slope has some interesting implications for mantle dynamics. That discontinuity would act as a barrier to mantlewide convection. A sinking lithosphere plate would meet a strong mechanical brake when passing through this depth.

For background information *see* BRILLOUIN ZONE; CRYSTALLOGRAPHY, LATTICE VIBRATIONS; SOLID-STATE CHEMISTRY; THERMODYNAMIC PRINCIPLES in the McGraw-Hill Encyclopedia of Science and Technology.

[NANCY L. ROSS]

Bibliography: M. Akaogi et al., The Mg$_2$SiO$_4$ polymorphs (olivine, modified spinel and spinel):

Thermodynamic properties from oxide melt solution calorimetry, phase relations and models of lattice vibrations, *Amer. Mineralog.*, 69:499–512, 1984; S. W. Kieffer, Thermodynamics and lattice vibrations of minerals: 3. Lattice dynamics and an approximation for minerals with application to simple substances and framework silicates, *Rev. Geophys. Space Phys.*, 17:35–58, 1979; A. Navrotsky, Energetics of phase transitions in AX, ABO_3 and AB_2O_4 compounds, in M. O'Keeffe and A. Navrotsky (eds.), *Structure and Bonding in Crystals II*, pp. 71–93, 1981; N. L. Ross et al., Phase transitions among the $CaGeO_3$ polymorphs (wollastonite, garnet and perovskite structures): Studies by high pressure synthesis, high temperature calorimetry and vibrational spectroscopy and calculation, *J. Geophys. Res.*, John Jamieson Memorial volume, in press.

Holography

Photographic recording of light pulses using two-photon fluorescence or ultrafast Kerr cells driven by laser pulses has been possible since 1975. Very fast optical phenomena such as refractive index changes in laser-produced plasmas have been recorded by holography using short illumination pulses. However, it was only in 1983 that the inherent properties of holography were first used directly to produce a frameless three-dimensional motion picture of ultrafast phenomena such as light in flight. Holography using a pulsed light source (or a continuous-wave light source with limited coherence length) represents a method of gated viewing. The hologram plate is only sensitive to information when it is illuminated by the reference beam. Thus, for example, a picosecond reference beam pulse corresponds to an extremely fast shutter, and if the reference pulse illuminates the hologram plate at an oblique angle, it is even possible to produce a picosecond movie. Such methods can be used for the measurement of space, time, and velocity.

Wavefront studies. An example of light-in-flight recording by holography is a study of the behavior of a spherical wavefront as it is focused by a positive lens. An opaque, diffusely reflecting, flat surface, painted white (the object screen) is illuminated from left by a divergent beam from a laser producing picosecond pulses. A cylindrical lens is fixed to the middle of the object screen with its axis normal to the surface so that it focuses the light which arrives almost parallel to the screen. In front of the object screen, at a distance of several feet (a few meters), a hologram plate is placed which records the screen as it is illuminated by the pulsed laser light.

The path length of the object beam traveling from the spatial filter in front of the laser to the middle of the object screen and from there to the middle of the hologram plate is equal to the path length of the reference beam traveling from the spatial filter to the middle of the hologram plate via two mirrors. The plate is illuminated from the left at an oblique angle by the reference beam which consists of a picosecond laser pulse. As this pulse passes along the hol-

ogram plate it first makes its left part sensitive to object light, and then the right part. Thus, what happens at the object screen is recorded like a motion picture along the hologram plate. The time span recorded along the plate is of the order of 800 picoseconds.

The ratio of intensity between reference and object beams is about 1:10, which is about one-hundredth of that of ordinary holograms. This is because the object light which arrives at the plate at those moments when it is not simultaneously illuminated by the reference beam has no influence on the recording of the hologram; it simply acts as an added incoherent illumination that darkens the plate.

The pulsed light source consists of a mode-locked argon laser with pumps a dye laser placed in a cavity of the same length as the argon laser. The whole setup is referred to as a synchronously pumped mode-locked dye laser. When used to expose the hologram, it produces pulses of about 10 ps (3 mm or 0.12 in. long) separated by twice the cavity length, or about 2 m (6 ft).

Figure 1 shows some of the results of this experiment. It is a composite of five photographs taken through different parts of a single hologram plate used for recording. The photograph taken furthest to the left shows the spherical wavefront before it reaches the lens. The next one shows how two parts of the wavefront are just creeping out of the lens surface, while the next three show the focused light as it moves further toward the focal point. The time separation between consecutive photographs is about 200 ps.

Since light travels more slowly through glass than through air, the light that has passed through the lens is delayed compared to the original wavefront. The lens being thickest at its center, the light that passes through this part is delayed the most. Thus

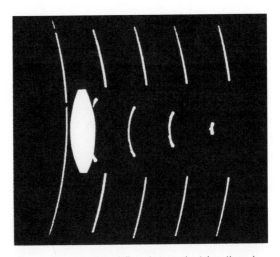

Fig. 1. A composite of five photographs taken through a single hologram plate that recorded a spherical wavefront that passed from left to right and was focused by a lens. An image of the lens has been included to make the picture clearer.

the originally convex wavefront is transformed into a concave shape. Since a wave travels perpendicular to its wavefront, the concave shape indicates that the light is traveling toward a focal point at the center of curvature of the wavefronts. In the original photograph the lens is not clearly seen, and an image of the lens has therefore been added to make the picture clearer.

Contouring. The above discussion shows how the shape of an unknown wavefront can be revealed by studying its intersection by a surface (the flat object screen). The method can just as well be reversed. The shape of an unknown three-dimensional object can be revealed by studying its intersection by a known wavefront (for example, a flat wavefront). Thus, light-in-flight recording by holography can also be used as a contouring method. One advantage of this method over other holographic contouring methods is that it is easier to evaluate automatically as it produces one single bright fringe that can be translated in depth by moving a detector along the hologram plate.

The possibilities of this technique are shown in the following demonstration. A stationary propeller from an ordinary fan is illuminated by a short pulse of laser light (using the same mode-locked laser as described above); the illumination and observation directions are approximately parallel to the shaft of the propeller.

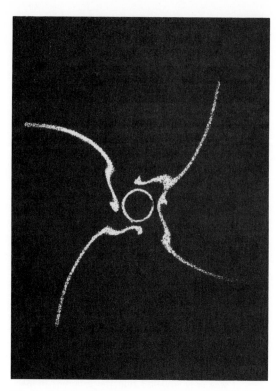

Fig. 3. Cross section of the propeller was made by photographing through the hologram plate at B_2 of Fig. 2, resulting in the cross section S_2.

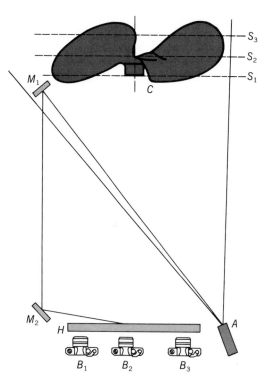

Fig. 2. Holographic setup used for measuring the three-dimensional shape of a propeller (C) which is illuminated by the divergent beam from a picosecond laser (A). The reference beam is reflected by the two mirrors (M_1 and M_2) toward the hologram plate (H). In the actual experiment the distance between H and C was much longer compared to the size of H and C than shown.

Reconstruction of the hologram discloses the propeller intersected by a thin sheet of light whose thickness is equal to half of the pulse length. The ideal case would be if this intersecting sheet of light had been flat, which would be the case if illumination and observation were collimated or made from infinite distances. In an actual experiment (Fig. 2) the intersecting surfaces were parts of ellipsoids, one focal point (A) being the point of illumination, the other (B) the point of observation at the hologram plate (H). As the object (C) and the separation between A and B were small compared with the distance from C to A and B, the intersecting sheet of light could be approximated by a section of a sphere (S) with almost zero curvature. For every distance from the left of the hologram plate, a corresponding cross section in depth exists so that by moving the observation point sideways behind the plate from B_1 to B_3 the cross section moves in depth from S_1 to S_3.

Figure 3 was made by photographing through the hologram plate at B_2 of Fig. 2 resulting in the cross section S_2. The hub has moved out of the light into darkness, and the central joint of the four blades is intersected. Its departure from flatness is easily detectable and measurable. The intersection of the four blades with S_2 reveals their inclinations and their symmetrical errors.

Relativistic effects. The above two examples show how light-in-flight recording can be used for measurement by studying the intersection between a wavefront and an object surface. However, the observed intersection is not identical to the "true" intersection. A flat wavefront passing by appears at first approximation to be rotated by 45°, because its most distant parts are seen at an earlier point in time. A closer study reveals that it is not only rotated but distorted into a paraboloid, its focal point being the eye of the observer. A spherical wavefront is distorted into an ellipsoid, the point source of illumination being one focal point (A in Fig. 2) while the other again is the eye of the observer *(B)*. If the distance separating A and B is caused solely by a high velocity of the holographer and the holographic equipment while they are making light-in-flight recordings of the stationary world, then the differences between ellipsoids and spheres causes apparent distortions identical to those usually referred to in the special theory of relativity.

Prospects. The technique of light-in-flight recording by holography represents four-dimensional recording, incorporating the three dimensions of ordinary holography plus time. The resolution in time and space depends on the pulse length. There has been a very rapid development of lasers that produce ultrashort pulses. The ultimate record will probably be a pulse of about three wavelengths, which for ordinary light represents about 2 micrometers. If the center of this pulse can be detected with an error of less than 10%, it will be possible to measure distance to an accuracy of less than 1 μm and time to an accuracy of less than 0.1 femtosecond. Although lasers that produce ultrashort pulses are now expensive and cumbersome, there is no technical reason why they should not become compact and inexpensive in the near future so that light-in-flight recordings can be used increasingly for measuring purposes by both the optical and the mechanical industry. Finally, the method can also be expected to become an important tool in fundamental science.

For background information *see* HOLOGRAPHY; LASER; RELATIVITY in the McGraw-Hill Encyclopedia of Science and Technology.　　[NILS ABRAMSON]

Bibliography: N. Abramson, Light-in-flight recording: High-speed holographic motion pictures of ultrafast phenomena, *Appl. Opt.*, 22:215–232, 1983; N. Abramson, Light-in-flight recording, 3: Compensation for optical relativistic effects, *Appl. Opt.*, 23:4007–4014, 1984; D. T. Attwood, L. W. Coleman, and D.W. Sweeney, Holographic microinterferometry of laser-produced plasmas with frequency-tripled probe pulses, *Appl. Phys. Lett.*, 26:616–618, 1975; M. A. Dugay, Light photographed in flight, *Amer. Sci.*, 59:551–556, 1971.

Homeo box

As recently as 1980, the idea that animals as different as flies and humans might specify early developmental patterns with common molecules would have prompted a derisive snort from most biologists. Yet today that idea is seriously discussed, and is even being tested by a variety of experiments. The discussion and experimentation are centered on a special class of genes found in the chromosomes of a variety of higher animals. All contain a common deoxyribonucleic acid (DNA) sequence called the homeo box. This name is derived from the sequences presence in homeotic (developmental control) genes from the fruit fly, *Drosophila melanogaster*.

The fruit fly has long been a favorite organism for genetic studies. It has a short generation time and is easy to rear in the laboratory, both of which are considerable advantages for the study of heritable traits. Some of these heritable traits profoundly alter the body plan of the developing fruit fly, resulting in the near-perfect duplication of a particular body part in the wrong position. For example, in a fly carrying a mutant copy of the homeotic gene *Antennapedia*, legs sprout from the head in a place that is usually occupied by a set of antennae. The *Antennapedia* gene is by no means the only gene in the fruit fly with the ability to confer such bizarre transformations. In fact, there appear to be at least two clusters of genes that can mutate to yield analogous developmental abnormalities.

Homeotic gene clusters. The pioneering work in defining these genes was done by E. B. Lewis in a continuing series of experiments begun in the mid-1950s. He and his coworkers defined the first known cluster of homeotic genes, which they called the Bithorax complex. The name of the complex comes from the duplications of thoracic structures that often arise from mutants in this cluster of genes. The duplicated structures often include entire body segments of the larval or adult fly. The body of both larval and adult flies is divided into easily visible metameres or segments: about six in the head region, three in the thoracic region, and eight in the abdominal region. The individual genes of the Bithorax complex act to specify the developmental fate of the segments in the posterior thoracic and abdominal regions.

A second, analogous cluster of developmental control genes has been recently defined by T. C. Kaufman and his colleagues, and named the Antennapedia complex. It includes the *Antennapedia* gene as well as other genes involved in controlling the development of the head and thoracic segments of the fruit fly. Thus, to a rough approximation, the individual genes of the Antennapedia complex seem to control the development of specific body structures in the anterior half of the fruit fly body plan, and the Bithorax genes to control development in the posterior half of the fly body (Fig. 1). Oddly enough, the gene order on the chromosome and the order of the segments whose development they control is the same. Though this may be a consequence of the evolutionary history of these two gene complexes, it really remains an unexplained, mysterious bit of symmetry in the arrangement of homeotic genes.

Lewis, who originally noted this colinearity between the Bithorax genes and the body segments they controlled, proposed that the cluster of genes

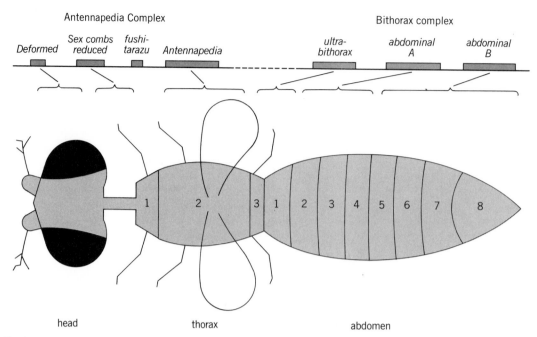

Fig. 1. Homeo box genes in *Drosophila*. A simplified map of the homeo box–containing genes in the Antennapedia and Bithorax complexes, showing which segments of the fruit fly body are controlled by a particular locus. Both gene complexes lie on the right arm of the third chromosome of *Drosophila*.

had arisen by the successive duplication of an ancestral gene. Once a duplication had taken place, one of the two copies would be able to assume new controlling functions, and establish new structures in some body segments. This process would eventually lead to a hypothetical organism in which each segment would have a different morphology under the control of a different (but related) homeotic gene. The primitive, relatively undifferentiated organism in which this evolutionary process began would presumably look similar to a present-day millipede, or even an earthworm. To the human eye, these animals, though having many body segments, have one much like another. According to the postulated process, a loss of one of the present-day genes from the Bithorax complex could return a segment to a more primitive morphology. In flies this might result in a mutant with millipedelike legs on abdominal segments that are normally lacking legs. Indeed, a mutation has been found.

Molecular genetic studies. Some of the results that have emerged from recent molecular genetic studies of these two clusters are in striking agreement with Lewis's ideas about the evolutionary history of the Bithorax genes. Analysis of cloned copies of the Bithorax and Antennapedia complex genes show them to be derived from a single ancestor; that is, the genes show DNA sequence homology at the molecular level, indicating a common evolutionary origin. This shared DNA sequence (the homeo box) seems to be exclusively associated in the *Drosophila* genome either with genes that control the development of specific body structures (such as *Antennapedia*), or with those genes that directly control the process of segmenting the fly body plan. Both sets of genes are involved in controlling the metameric pattern of the fly body plan—the latter group, segmentation genes, directly imposing it; the former group, the homeotics, assigning different structures to different elements of the metameric pattern.

Presence in higher animals. The most provocative result to emerge from the molecular genetic studies of these gene families is not that they share very similar copies of the homeo box, but that similar families of homeo box–containing genes are present in a variety of higher animal chromosomes. These animals form a very diverse group, ranging from earthworms to humans. The existence of this shared genetic information, which in *Drosophila* seems intimately involved in the metameric patterning of the body, suggests that a wide variety of animals may use similar genetic circuits to program embryonic development.

The reason that a common molecular controlling element like the homeo box has not been thought likely to occur on theoretical grounds is because early development in widely separated animal groups differs so drastically. Indeed the science of embryology is based on the comparative study of embryonic development, and one finds little to productively compare in the early development of fly and human. One of the few possible similarities between the body plans of insects and mammals is in their use of metamerism. Their bodies are, to a greater or lesser extent depending on the stage of development, composed of a series of serially repeated structures arrayed along the anteroposterior axis. These separate but integrated elements develop into different but related structures, based on their site of origin on the body axis. For example, legs and arms are serially related structures, closely related in their morphology, but developing from metameres at different locations on the body axis.

One of the problems that all animals face in embryonic development is the construction of ordered complexity from the simple fertilized egg. The construction of a metameric body plan may simplify this incredibly difficult process by allowing similar but nonidentical genetic programs to be activated in successive segments. This notion, though not expressed in modern-day genetic terminology, was proposed 90 years ago by William Bateson, who also coined the term homeotic. Bateson's idea was that much of the evolutionary variation that one observes in the animal kingdom could be explained by modifications of the common series of body segments that formed a fundamental part of the structure of most higher animals.

Though still only a tantalizing correlation, it is largely these segmented higher animals that appear to contain genes that are highly homologous to the homeo box sequence, while many nonsegmented animals clearly lack homeo boxes (Fig. 2). Thus, with this admittedly very incomplete evidence, the prospect appears that this conserved sequence may encode a developmental function that has been fixed during evolution to the process of metamerism. It also implies that the lesson learned about the development of fruit flies may also be directly applicable to aspects of development of a wide variety of higher animals, including humans.

Function. The detailed structure of the protein region that the homeo box sequence encodes supplies some hints as to its possible biochemical functions. It is very rich in positively charged amino acids, which is characteristic of proteins that must bind stably to negatively charged nucleic acids, such as DNA. It also exhibits weak vestiges of an amino acid organization that is characteristic of DNA-binding proteins from bacteria and bacterial viruses. These hints about the biochemical functions correspond nicely to previous speculations about the possible role of the products of homeotic genes. These speculations were most clearly and forcefully stated by A. Garcia-Bellido in a model referred to as the selector gene hypothesis. In this model the products of homeotic genes would be turned on in specific groups of cells along the fly body axis, and would accomplish their developmental function by activating large batteries of effector genes, whose products would differentiate one group of cells along the body axis from another. The homeotics would therefore act as molecular switches at key points in a hierarchy of developmental decisions, and not as specific building blocks themselves.

Use in study of development. One of the practical benefits of the discovery of the homeo box may be the most significant. The most powerful set of techniques in biology today are called recombinant DNA methods. They allow the detailed analysis and manipulation of specific genes by their cloning into bacterial or viral vectors. One of the drawbacks, however, is that the chromosomes of an animal con-

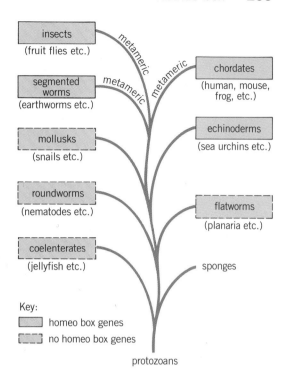

Fig. 2. Homeo box gene conservation in a proposed evolutionary tree, showing each group of organisms that contain detectable homeo box genes. The lineages whose members show clear metameric body organization are also marked.

tain tens or hundreds of thousands of genes, and the selection of a specific gene that controls an aspect of higher animal development has been an extremely difficult task. The use of the homeo box to pluck out these potential development-control genes promises to accelerate the analysis of these genes by an unprecedented degree. It has already allowed the easy isolation of many of the genes encoding molecular switches for *Drosophila* development, and promises to do the same for the analysis of other developmental systems, including those of mammals. New results on the genetic programs controlling animal development will probably lead to a much more sophisticated view of the molecular events controlling this process. If the promise of the homeo box is fulfilled, these new findings will lead to the understanding of the underlying similarities in what appears to be a bafflingly diverse set of developmental programs.

For background information *see* DEVELOPMENTAL BIOLOGY; EMBRYOLOGY; GENE ACTION; PROTEIN (EVOLUTION) in the McGraw-Hill Encyclopedia of Science and Technology.   [WILLIAM J. McGINNIS]

Bibliography: A. Laughon and M. P. Scott, Sequence of a *Drosophila* segmentation gene: Protein structure homology with DNA-binding proteins, *Nature*, 310:25–31, 1984; E. B. Lewis, A gene complex controlling segmentation in *Drosophila*, *Nature*, 276:565–570, 1978; W. McGinnis et al., A homologous protein-coding sequence in *Drosophila* homeotic genes and its conservation in other metazoans, *Cell*, 37:403–408, 1984.

Human ecology

Human ecology is the study of society in relation to environment. It is the science of human patterns and processes with the supporting landscape. A ma-

jor part of human ecology is concerned with the resource basis for humanity, now and in the future, that is, environmental economics.

Systems concepts are used for understanding the symbiosis of the human economy with environment. Figure 1 is a macroscopic minimodel showing that the basis of the human economy is the resource production of the Earth. A minimodel is one which is holistic in scope but simple in structure; it uses energy-language symbols to show kinetic and energetic relationships. The diagram is a simple way of showing a model that can be specified in more detail by the mathematical equations.

Earth-humanity model. Figure 1a shows the system supported by two main energy sources: the Sun which operates the atmosphere, oceans, and hydrologic cycles, and the heat deep in the Earth which drives the geologic mountain-building cycles. This heat came from radioactive decay processes and heat left over from the time when the solar system formed. Geologic and environmental production processes (producer symbol) use the energy sources to generate stores of environmental resources (storage symbol): minerals, soils, forests, fuels, and the like (resource reserves). The economy consumes (consumer symbol) these resources, while circulating money among people, which helps coordinate their services (broken-line circle labeled gross national product). The pathway from economic assets to the interaction symbol is a positive feedback that accelerates use.

The pathways of Fig. 1a show the flows and transformations of energy. As required by energy laws, all energy entering from sources is eventually accounted for, leaving the system in a degraded, no-longer-usable form (pathways going to heat sink symbol).

Figure 1b shows the essential circulation of materials. They are bound into the resource storages and dispersed again when the consumers release them as wastes. Some of the materials are scarce, but the economy is dependent on critical materials being available such as fertilizer for agriculture and metals for industry. Putting wastes in big dumps cannot work much longer. As long as there is energy available, critical materials can be reconcentrated as needed. As Fig. 1 shows, material recycling is the long-range solution. Some wastes can be reprocessed by the environmental systems; some need to be recycled more directly into reuse by the economy.

Figure 1 also shows the symbiosis of Earth and human economy, each stimulating the other. (Note the forward and feedback pathways.) The economy adapts to the available resources, and the environmental systems adapt to the actions of the human economy. It is a mutual organizing system in which each receives a reinforcing reward when the Earth and human economy are cooperatively coupled. A concept known as the maximum power principle holds that self-organizing systems of humanity and nature develop designs that maximize power—in

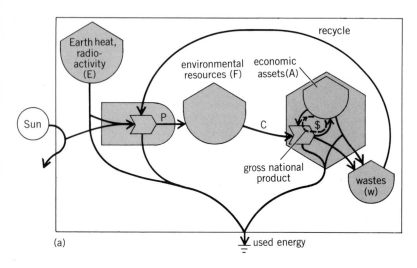

(a)

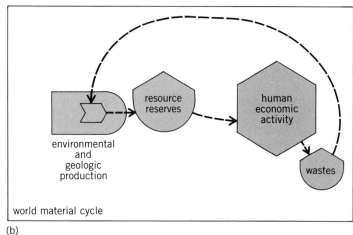

(b)

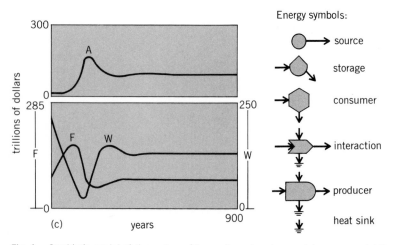

(c)

Fig. 1. Symbiotic model of the system of humanity and environmental resources. (a) Energy systems diagram of world energy flow. (b) Material cycle that is a part of the system in a. (c) Computer simulation of the model in a. A = economic assets; F = fuels, minerals, forests, and soils (environmental resources in storage); W = wastes.

other words, maximize the flow of useful energy. The cooperative circular design of Fig. 1 is an example; the producer stimulates the consumer, and vice versa.

Computer simulations provide scenarios of what would happen given the conditions put into the model—no more, no less. A computer simulation of the model is shown in Fig. 1c which starts with a large storage of Earth-produced resources. Some of these resources like coal, oil, and copper are produced by the Earth systems slowly and are known as nonrenewable resources (not appreciably renewed in human lifetimes); but other resources like wood and soil are known as renewable resources because they can be used as they are being produced.

The diagrams are a way of visualizing the relationships of a model that are expressed in mathematical equations when they are simulated with computer. The relationships in the equations for the computer simulation in Fig. 1c may be expressed in words as follows: The flow of production P is proportional to the product of the wastes and the available energy; the available energy is the unused sunlight and earth heat passing through the system. The flow of consumption (C) is proportional to the product of environmental resources and economic assets. The rate of change of environmental resources is the production minus the consumption (P − C). The rate of growth of economic assets is an inflow proportional to consumption minus depreciation. The rate of change of the wastes is proportional to a waste inflow that is proportional to consumption plus a waste inflow proportional to depreciation minus the recycle that is proportional to production. When equations are calibrated with world data, simulation generates the graph in Fig. 1c.

The simulation scenario (Fig. 1c) starts with a small human economy prior to the industrial revolution. Rapid development of the human economy in 300 years, in the industrial revolution, uses up some of the stored resources and then operates on renewable resources as they are produced. Its economic assets reach a maximum several hundred years after the start and then decline to a lower level which can be supported by renewable sources.

Although the world of humanity and nature is vastly more complex than a model can represent, simple models are useful to represent overview principles in a way that can be clearly understood and shared by those generating public policy. The model and scenario in Fig. 1 suggest that major public policy considerations should be addressed to preparing for the climax and decline of total assets, so as to decide what is really important to maintain for a vital economy on a lower resource budget.

Exponential growth model. Most people in the present world economy have a different model and scenario in their view of the world. This is drawn with energy symbols and simulation in Fig. 2a. Here economic assets are developed by a consumer

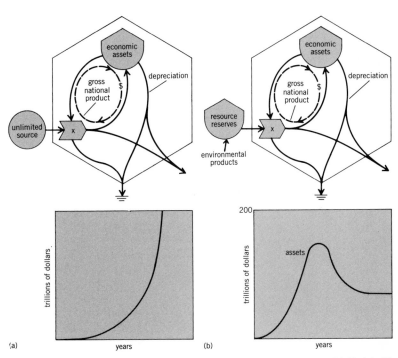

Fig. 2. Comparison of autocatalytic consumer growth on different sources. (a) Model with unlimited source producing exponential growth. (b) Model that is part of the system in Fig. 1. The source is initially large, part of which is being renewed with inflow of products. Rapid initial growth returns to lower steady state.

economy using unlimited resources. The majority of economists and political leaders believe that there will be technologies developed one after another so that a new resource can be substituted for each one as its storage runs out.

The mathematical relationships of the model in Fig. 2a may be expressed in words as follows: The rate of change of assets is the sum of an inflow that is the product of an unlimited source (constant availability) times the economic assets minus the depreciation outflow.

As Fig. 2a shows, unlimited resource availability supports steeply accelerating growth. The more assets that are developed, the more assets there are available to develop new assets. So long as there are new assets developed, new money can be added without its losing value. Therefore the money supply is proportional to assets. If money is added without any new assets, then more money would be circulating for the same amount of goods, services, assets, and so on. This is inflation. The model in Fig. 2a is a variation of the Domar-Harrod model of economics textbooks.

Figure 2b has the consumer part of Fig. 1 separated out for comparison with the unlimited growth model. The mathematical relationships of the model in Fig. 2b may be expressed in words as follows: The rate of change of resource reserves is the sum of the inflow of environmental products minus consumption (C). The rate of consumption is proportional to the product of the resource reserves and the economic assets. The rate of change of the economic assets is the inflow that is proportional to the consumption (C) minus the depreciation. Its growth starts out with steep acceleration just as in the case of the unlimited one, but it soon climaxes its growth and declines to a lower level that can be sustained on the inflow from Earth production processes (products).

There is international controversy over which of these two is correct. Those concerned with human ecology believe that large resources are required to maintain the educational levels and machinery for technology so that when nonrenewable energies and other resources are less, the economy which can be maintained must be smaller. To maintain high standards of living with a smaller economy would require a smaller population.

Hierarchy. The patterns of humanity and its resources are arranged in a hierarchy, with many small producers over the rural landscape converging their products to consumer centers, the cities. Figure 3 shows the pathways of products from rural producers converging to the hierarchical center. Figure 3 is a two-level hierarchy which is appropriate for the two-level system (Earth and humanity) in Fig. 1. Because the system has a symbiosis of the Earth producers and economic consumer societies, there are pathways of materials and services going from the consumer center back to help producer units.

Actually, the hierarchies on landscapes have many levels. For example, many grass blades are converged to support a cow; many cows are converged to support the farmer; many farmers converge products to a town; many towns converge products to a big city; and so on. The hierarchies of humans and those of the environmental systems are coupled. For example, the network of streams converging into larger streams and finally into great rivers is used by the human economy, with cities often developing at the convergence of large rivers, facilitating inputs of products and returns of materials to Earth cycles. The hierarchical centers of human economies show up strikingly as centers of bright lights at night as viewed from satellites.

The spatial organization within a city is hierarchical with roads and public transportation conveying people and raw commodities inward and wastes and products of urban work outward. A portion of the city's output goes to larger cities or overseas. Some of the difficulties experienced by existing cities can be traced to the hierarchical pattern being upset by poor planning. For example, many interstate highways were built through cities without considering the natural pattern.

Oscillation. Hierarchical systems of production coupled to consumption are observed to oscillate. The model in Fig. 1 with only slight changes in the pathway arrangement connecting production and consumption oscillates (adding a linear pathway from environmental resources to economic assets). Production builds up products; then consumption develops a surge of growth and later declines to a low value while production builds up products again.

One theory of the rise and fall of civilizations is based on the oscillation due to buildup of soils, forests, or farm assets, followed by a surge of consumption by new organization, population growth, conquests, culture, large building projects, and so on. Viewed over the world, the oscillating pulses at different places and different times are like a Christmas tree with flashing lights, each a surge of energy use at a different moment. That human ecological models are similar to and oscillate like those in chemistry, biology, and astronomy suggests that physical laws affect the adaptation of humanity to its resources.

Modern culture and economic development interacting intimately with the Earth have generated a symbiotic system of humans and environmental resources that is the basis for human existence. Perceptions of the way that this human ecological system works determine many public policy decisions by society and its leaders. With the development of systems concepts, scientific models and computer simulations are helping to clarify public policy overviews and decisions.

For background information *see* ECOSYSTEM; HUMAN ECOLOGY; SYSTEMS ECOLOGY; SYSTEMS ENGINEERING in the McGraw-Hill Encyclopedia of Science and Technology.

[HOWARD T. ODUM]

Bibliography: A. J. Bennet and R. J. Chorley, *Environment Systems,* 1978; B. J. L. Berry, E. C. Conkling, and C. M. Ray, *The Geography of Economic Systems,* 1976; Z. Naveh and A. S. Lieberman, *Landscape Ecology,* 1984; H. T. Odum, *Systems Ecology,* 1983.

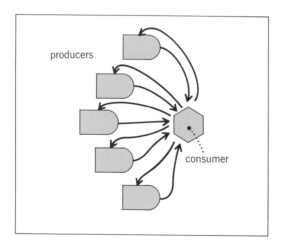

Fig. 3. Hierarchical spatial pattern of the system in Fig. 1. Small dispersed producers contribute to and receive from central consumers.

Human genetics

All humans belong to the same species, *Homo sapiens*; yet there are many populations, or races, that can be distinguished from each other in terms of morphological characters. The broad division into three major races, Caucasoid, Negroid, and Mongoloid, is generally accepted; many anthropologists add two more major races, Amerind (American Indians) and Australoid. The genetic differences between races are caused by the differences in gene frequencies at various loci. Therefore, by examining racial differences in gene frequencies, the genetic relationship and evolution of human races can be studied.

Gene frequencies are the relative frequencies of alleles at a locus in a population. For example, the ABO blood group locus has three major alleles, *A*, *B*, and *O*, and the frequencies of alleles *A*, *B*, and *O* in the English population are 25%, 6%, and 69%, respectively. These frequencies vary from population to population. In the early days of anthropological studies, the genetic relationship among different races was studied by examining the geographical distribution of gene frequencies at a few polymorphic loci such as the ABO and Rh blood group loci. It was later realized, however, that comparison of gene frequencies for one or two loci is not reliable, since each locus has a different distribution. Only when a large number of loci are examined does a genetic relationship become clear. This is partly because genetic variation between races is very small compared with the variation within races at the gene level. However, if a large number of loci are examined, even small differences can be detected with sufficient accuracy.

The genetic difference between a pair of populations is usually measured by a quantity called genetic distance, which is a function of gene frequencies. Once genetic distances are computed for a group of populations or races, their genetic relationships can be studied quantitatively. For this purpose, gene frequency data for loci that are detectable by electrophoresis or antigen-antibody reaction are usually used. In addition, some investigators have begun to use deoxyribonucleic acid (DNA) polymorphism data for this purpose.

Electrophoretic and antigenic loci. The results obtained from studies of electrophoretic and antigenic loci suggest that the Caucasoid and Mongoloid races may be genetically closer to each other than either is to the Negroid race (see illus.). However, this is not conclusive because the number of loci used is not sufficiently large. The genetic distances between populations within each of the three major races are generally much smaller than those between the major races. Particularly, European populations (such as English, French, Italian, Basque, and Finnish) are very closely related. In general, there is a tendency for geographically close populations to show small genetic distance. This is appar-

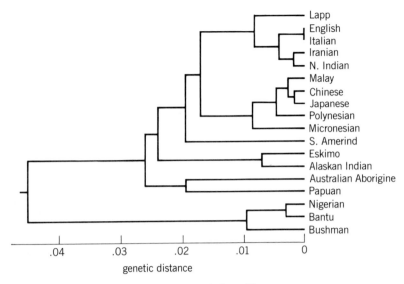

Genetic relationship of 18 representative populations of humans based on gene frequency data for 23 loci. (*After M. Nei and A. K. Roychoudhury, Genetic relationship and evolution of human races, Evol. Biol., 14:1–59, 1982*)

ently due to gene migration that has occurred in the past, or to recent splitting of the populations. Unlike the view generally held by anthropologists, the populations belonging to Amerind do not cluster with those belonging to Mongoloid. This is apparently due to genetic drift (random fixation of alleles) that occurred in the tribal Amerind populations. Amerind populations show large genetic distances even among each other. Similarly, the tribal populations belonging to Australoid show large genetic distances.

There are several factors that affect the genetic differentiation of human races, two of which are mutation and isolation. A well-defined race often has variant alleles that are unique to the race. For example, African Negroid people have many unique alleles such as F_y^4 at the Duffy locus and $PepC^0$ at the peptidase C locus. This suggests that the Negroid population has been isolated from other groups for a long time. Similar unique alleles have been found in several other populations, notably Australoid and South American Indian populations. Migration has an effect opposite to that of isolation. In the process of human evolution, gene migration seems to have occurred rather extensively in some groups of populations. For example, Europeans and Iranians are genetically quite close, though they live geographically far apart. Past migration can often be traced back by examining the distribution of rare alleles.

Language differences can be a barrier to interpopulational hybridization. Indeed, there is often a correlation between genetic distance and linguistic distance when a group of closely related populations are studied. However, the correlation between genetic distance and linguistic distance at the racial level is quite small. This is apparently because language is culturally inherited and can change rapidly under certain circumstances.

The correlation between genetic distance and morphological difference is also generally quite weak. For example, the Negritos and aboriginal Malays in Southeast Asia and Pygmies and Bushmen in Africa have a number of common morphological features such as short stature, dark skin, and frizzy hair. Because of these similarities, some anthropologists believed that they originated from the same common stock. Genetic distance data, however, have shown that the African and Southeast Asian populations are genetically quite different and are more closely related to their respective neighboring populations. Clearly, the evolutionary change of morphological characters is quite different from that of electrophoretic and antigenic loci. Morphological characters seem to be subject to stronger natural selection than average genes.

DNA polymorphism data. Data on DNA polymorphism are still quite limited, but available data indicate that the extent of DNA polymorphism is much higher than variation at the protein level. It has been estimated that two randomly chosen genes at a locus have, on the average, about four nucleotide differences, even if only the coding region of DNA is considered. However, variation between races is again much smaller than variation within races. Mitochondrial DNA (mtDNA) is known to have no recombination, so that a phylogenetic tree of polymorphic mtDNAs can be constructed. A number of authors have constructed such a tree, and the results obtained indicate that the divergence times of some pairs of polymorphic mtDNAs are very old, that is, much older than the time of the origin of *Homo sapiens* (about 200,000 years ago) or the time of divergence of the three major races. It has also been shown that although there is some tendency for the mtDNAs sampled from the same race to cluster, most of them are genealogically mixed with those from other races. Therefore, mtDNA data are not sufficient for discerning the evolutionary relationships of different races. To clarify these relationships, information on DNA polymorphism for many independently inherited nuclear genes seems to be necessary.

For background information *see* ANTHROPOLOGY; HUMAN GENETICS; POPULATION GENETICS; RECOMBINATION (GENETICS) in the McGraw-Hill Encyclopedia of Science and Technology.

[MASATOSHI NEI]

Bibliography: L. L. Cavalli-Sforza and W. F. Bodmer, *The Genetics of Human Populations*, 1971; A. E. Mourant, *Blood Relations*, 1983; M. Nei and A. K. Roychoudhury, Genetic relationship and evolution of human races, *Evol. Biol.*, 14:1–59, 1982.

Hydroelectricity

Small-scale hydro power is the generation of electricity by using hydraulic turbines in which the installed capacity of the plant lies within the range of 5 to 5000 kW. The term can be applied to high- and low-head schemes.

Since about 1900, many small-scale hydro power projects have been commissioned for industrial and domestic use, where suitable water sources are available in the locality. Early schemes were mainly built in Europe and the United States, but later in other countries to provide power for processing plants such as those required on tea and rubber plantations. Most installations were designed to meet a particular need and particular situation and were generally operated as independent units. These early plants had low overall efficiencies and required the presence of operating and maintenance personnel. In consequence, as more advances were made in electrical engineering, especially the transmission of power over longer distances, many of the installations became uneconomic and were abandoned.

The situation has now changed. The increased cost of fossil fuels, the design of more efficient small hydro plants, and the introduction of automatic controls and more appropriate technology have all contributed to making small-scale hydro power a more viable proposition. As a result, there has been an upsurge of interest during the 1980s.

The attractiveness of hydro power is that it is an environmentally clean renewable energy source, but it has a high initial capital cost compared with thermal power. This is of particular relevance to the economics of small-scale hydro, and attention has to be given to minimizing these costs by simplifying the design of both the plant and the civil works and by the introduction of standardized equipment.

Three main areas for development can be identified: (1) the refurbishment of old and disused plant; (2) the exploitation of existing untapped sources, such as can be found in river control schemes and water supply or irrigation networks; and (3) the installation of isolated plants in developing countries. Small schemes can be ideally suited to remote areas, provided appropriate technology is applied, including the use of local resources.

Turbines. All conventional types of turbines are suitable for small-scale hydro, the choice depending on the hydraulic conditions. For low- to medium-head applications, the turbines most commonly used are Francis or propeller types; for high heads, Pelton impulse turbines are adopted.

Francis turbines. These are of the reaction type. The shaft arrangement is generally vertical, but economies can be achieved on small-scale installations by adopting a horizontal axis (Fig. 1). These turbines are suitable for a head range of 30–1000 ft (10–300 m) and ratings greater than 100 kW. Where the head is less than 30 ft (10 m), it is more economical to install an open-flume-type Francis turbine. This arrangement avoids the need for a spiral casing and is simple to construct and maintain.

Propeller turbines. The most common form of the propeller-type turbine is the Kaplan. It has adjustable blades giving a high efficiency over a wide range of heads. It is normally set on a vertical axis; however, in small-scale hydro, considerable economies with civil and ancillary works can again be achieved by adopting a machine with a horizontal axis. Several manufacturers produce propeller-type machines with this configuration, such as bulb and axial-flow turbines.

In the bulb turbine (Fig. 2), the propeller drives a generator which is contained within a bulb, which in turn is located in the center of the water passage. In order to fit inside the bulb, the generator must be of small diameter and hence low inertia; it is therefore suitable only for use in conjunction with electrical systems of a size adequate to maintain electrical stability. A development of the bulb turbine which overcomes this problem is the straight-flow turbine, in which the generator is set around the periphery of the water passage rather than within the bulb itself. These types of machine are suitable for heads between 6 and 60 ft (2 and 20 m) and ratings greater than 300 kW. For lower ratings, down to 50 kW, within this head range the peripheral ring may simply comprise a pulley wheel with a belt drive to a separately mounted generator. Alternatively, an axial-flow, tube-type turbine may be used. In these machines the generator is located outside the waterway and is connected by a shaft to the turbine rotor. The generator can be easily maintained, and by incorporating a gear box between the rotor and generator cheaper high-speed generators can be used.

Impulse turbines. For high heads, above 600 ft (200 m), Pelton wheels are used, whereas for ratings up to about 1 MW in a head range of 6–600 ft (2–200 m) crossflow turbines are more appropriate (Fig. 3). The latter are low-speed radial- or impulse-type turbines of moderate efficiency. They have a simple blade geometry and, for low heads, blades can be made from cheap materials because the forces are small. Connection to a generator can be by direct coupling or, for small machines, by belt drive. Gear boxes can be provided to enable the use of economical generators. Overall the crossflow turbine is a satisfactory machine for the lower end of small-scale hydro power.

Generators and control. Except for very small systems, alternators of the brushless type equipped with automatic voltage control normally supply the output load. For small single isolated hydro plants, an unsophisticated turbine governor can be used to control the generator. This asynchronous system, although cheap, requires manual control, but can be used where there is a plentiful supply of free-flowing water and where the energy produced either can be immediately utilized or can go to waste. For larger schemes, where a number of machines are employed or where the generator is connected to an existing power supply network, more precise governing and sychronizing equipment is necessary.

Controls and associated equipment. The need to fit control gates depends upon the layout of the system. Simple control gates operated by screwed rods or hydraulic rams can be fitted, or butterfly valves

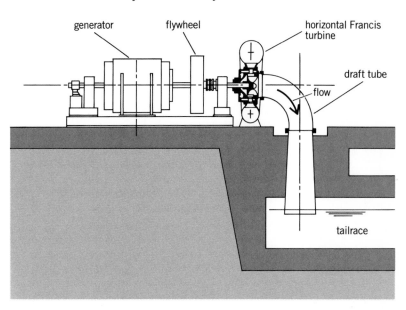

Fig. 1.   Horizontal Francis turbine installation.

for the tube-type turbines. Control gates and butterfly valves can be used for starting and stopping machines and for emergency closure.

Some form of screening facility is normal and, again depending upon the river and type and size of turbine, cleaning equipment can vary from simple hand raking to continuous automatic cleaning. Some facility, such as stop beams or gates, may also be necessary for completely shutting off the unit.

Appropriate technology. Associated with the appraisal and engineering of hydroelectric works are various aspects which are independent of the installed capacity. These include assessing the hydrology and geology of the site and the detailed design of the turbines, generators, gates, trash screen,

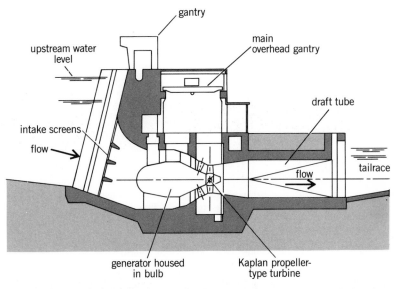

Fig. 2.   Longitudinal section through bulb turbine power station.

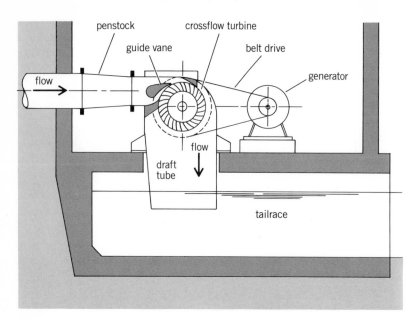

Fig. 3.   Crossflow turbine installation.

buildings, and other associated works. These costs are small when compared to the benefit from a large installation, but the same level of investigation and design, if applied to a small scheme, could in itself render the scheme uneconomic. Clearly, the smaller the scheme, the lower are the detailed design costs which can be absorbed by the available benefits.

This economic constraint has encouraged certain differences in approach between the designs of small and large schemes. The most obvious has perhaps been the trend toward supplying smaller machines from standard predesigned ranges rather than tailoring each machine for an individual site. In many cases, the associated generating and control equipment is supplied as part of the package. In some cases, this may also extend to the civil engineering works. Smaller installations have also tended to feature a high proportion of horizontal-axis machines. Often these offer economies in terms of reduced excavation and foundation requirements. Geotechnical assessments are also reduced for small installations.

It has been said that if more than minimal investigation work is considered to be required at a site then the site is not likely to be viable. Nevertheless, because of the safety aspects, it is essential that professional advice is engaged before any attempt is made to retain any significant volume of water.

A more recent development, which has significantly added to the expansion of small-capacity hydro power, is low-cost, automatic, microprocessor control technology. This can produce benefits in three distinct areas: (1) Regulation can be controlled continuously in response to changing load and inflow conditions. This produces a particularly efficient mode of operation. (2) In developed countries, the need for expensive full-time supervision is dispensed with. (3) In developing countries, automatic control enables the system to operate, under normal contingencies, in the absence of trained supervisory personnel.

Effect of location. A convenient way of subdividing installation and design is according to the requirements for developed and developing countries. In the first case, very small machines are generally installed as adjuncts to existing water control works, a water supply system, a lock, or an existing dam. Machines nearer the upper limit of the small-hydro capacity range may merit the construction of new barrages, particularly if the barrage is to include several such machines. In developed countries, skilled personnel and maintenance facilities are readily available, and the design emphasis tends to be on the efficiency of the plant and its operation as a means of maximizing benefits.

In developing countries, where small units are to be provided to supply remote communities, there is generally insufficient existing infrastructure to which hydro plants can be added. Provision of small hydro therefore entails the construction of complete works. For this to be done economically, the use of local materials and techniques is essential. Construction entails a high input of low-cost hand labor, building may be in local stone with hatches and covers in timber, and so forth. There are several instances of very small plants where hydraulic passages have been formed by using deformed oil drums encased in rubble mortar. In developing countries, where skilled personnel and spare parts tend to be scarce, overruling design constraints include low-cost construction, reliable operation, and general conceptual simplicity.

Where existing plants are to be refurbished, the amount of new work will clearly depend on the degree of dilapidation of the exisitng installation. In assessing the requirements for new work, however, the same general criteria as outlined above can be assumed to apply.

For background information *see* ELECTRIC POWER GENERATION; GENERATOR; HYDRAULIC TURBINE; VALVE; WATERPOWER in the McGraw-Hill Encyclopedia of Science and Technology.

[J. V. CORNEY]

Bibliography: J. S. Gulliver and A. Dotan, Cost estimates for hydropower at existing dams, Proceedings of the ASCE, *J. Energy Eng.*, 110(3):204–214, 1984; International Water Power and Dam Construction: Proceedings of the 1st European Conference on Small Hydro, Monte Carlo, December 6–8, 1982; International Water Power and Dam Construction: Proceedings of the 1st International Conference on Small Hydro, Singapore, February 13–16, 1984; Watt Committee on Energy Ltd., Small Scale Hydro-power Rep. 15, presented at 16th Consultative Council meeting, London, June 5, 1984.

Hydroponics

A mixed resin hydroponic system has been developed that provides excellent control of nutrient chemistry for hydroponic plant growth. This system is less costly and more flexible than earlier flow culture hydroponic systems.

Systems. Hydroponics is the technique for growing plants in nutrient solutions, either with or without inert rooting media. Long-term stability of hydroponic systems depends not only upon the homogeneity of the growth media but also upon the maintenance of nutrients. Hydroponic systems may be classified into three general categories—traditional, flow cultures, and resin-buffered systems—on the basis of how the balance of nutrients is maintained during plant growth.

Traditional hydroponic methods call for the systematic replacement of nutrient solution, and initial concentrations of nutrients much higher than those typically found in soil solutions. Routine replacement of the nutrient solution causes cyclic fluctuations in the concentrations of nutrients in solution that are often large, and the high initial concentrations of phosphate and other nutrients frequently exceed the threshold of toxicity to plants. When nutrient toxicity occurs, plant growth is reduced at least until the concentration in solution is reduced to subtoxic levels.

More advanced hydroponic systems provide nutrients at concentrations similar to those occurring in soil solutions. Flow cultures rely upon the continuous flow of nutrient solution past plant roots to displace partially depleted solution, and maintain the concentrations of nutrients. Displaced solution is sometimes discarded, but more frequently its composition is determined and chemical additions are made as it is recycled to large holding tanks. Tanks containing hundreds of gallons of hydroponic solution provide a mixing area and a reserve solution of mineral nutrients. Flow culture systems have been well tested, and remain a proven method for studying relationships between plant growth and nutrient concentrations. However, the large surface area of the solution distribution system, including plant containers, piping, tanks, and pumps, makes the system prone to contamination and limits its suitability for studies of nutrient essentiality. Flow cultures allow precise and accurate control of nutrients at concentrations typical of those that occur in soil solutions, but they are cumbersome and costly, and require an extensive nutrient distribution system for each experimental treatment.

Resin-buffered systems. Ion-exchange resins have been used since the 1940s as a source of nutrients for growing plants, though the early systems consisted of mixtures of sand or other plant support media with a single cation or an anion-exchange resin to buffer pH or supply a few nutrients. These early efforts resulted in fairly complex growth media with nutrient compositions that were difficult to define and control.

Mixed resin hydroponic systems have been developed recently that accurately define and control the activities of nutrient ions at concentrations similar to those that occur in soil solutions. These systems contain mixtures of synthetic ion-exchange and chelating resins that effectively buffer the activities of nutrient ions and pH by mimicking the ion-exchange reactions that typically control nutrient ion activities in soil solutions. Resins may be contained within water-permeable packets that are placed in well-stirred nutrient solution, or may be located in columns through which the solution is circulated. These systems can buffer the activities of macronutrient cations in hydroponic solution at virtually any preselected composition, and the levels of micronutrient cation activities may be selected accurately and independently.

The activity of any given micronutrient cation is maintained in a constant matrix of hydroponic solution even when the treatment levels of the micronutrient differ by several orders of magnitude. The activity of phosphate is continuously buffered at preselected levels that may be chosen to simulate those that occur in soil solutions. Other than a single macronutrient cation (that is chosen to be the primary reference ion), cation and phosphate activities may be selected without appreciably altering the pH or the activities of other nutrient cations in solution. Synthetic metal chelates such as ethylenediaminetetraacetic acid (EDTA) can be added to well-buffered systems to increase the concentrations of micronutrient metals without affecting the activities of the ions maintained by the resins. Thus the relationships between the activities of ions and the mineral nutrition of plants may be distinguished from effects due to concentration. Naturally occurring compounds exuded by plant roots may be allowed to accumulate because frequent replacement of the nutrient solution is unnecessary. Finally, trace-metal contaminants are scavenged from nutrient solution by the resins. This is a unique feature that is especially important when testing the function and essentiality of the various trace metals for the growth of higher plants.

Solution maintenance. Several types of resins with specific characteristics are required (see table) in order to maintain the solution composition. Strong acid resins with sulfonic acid functional groups provide the primary buffering of the activities of cal-

Resins used in the mixed resin hydroponic system

Type	Functional group	Nutrients buffered*
Strong acid	SO_3^-	Ca^{2+}, Mg^{2+}, K^+, Mn^{2+}
Chelating	$N(CH_2COO)_2^{2-}$	Cu^{2+}, Zn^{2+}, Ni^{2+}, Fe^{3+}, (Mn^{2+}, Ca^{2+})
Weak acid	COO^-	H^+, (Ca^{2+}, Mg^{2+}, K^+, Mn^{2+})
Polynuclear hydroxyaluminum surface†	$Al_6(OH)_{12}(H_2O)_{12}^{6+}$	$H_2PO_4^-$, (HPO_4^{2-}, PO_4^{3-})

*Ions in parentheses are buffered to a lesser extent.
†The final ratio of OH to Al is determined by pH adjustment.

cium, magnesium, potassium, and manganese. After selecting the macronutrient composition for the hydroponic solution, the strong acid resins are loaded with macronutrient cations in the ratio that will sustain the appropriate activities. Chelating resins containing iminodiacetate functional groups effectively buffer the activities of copper, zinc, nickel, iron [provided in the nutrient solution as iron-ethylenediaminedi-(o-hydroxyphenylacetic acid) FeEDDHA, a soluble chelate of iron], and to a lesser degree manganese and calcium. These resins exhibit a great affinity, capacity, and selectivity for micronutrient cations over macronutrients.

Chelating resins loaded with micronutrients are mixed in appropriate ratios to provide the preselected activities of micronutrient cations. The pH of the hydroponic solution is buffered by the carboxylic acid functional groups of a weak acid resin. This type of resin exhibits an affinity for hydrogen ions over most other cations. The pH sustained by the weak acid resin is adjusted to the preselected value by using souble pH buffers.

The activity of phosphate is controlled by a surface of polynuclear hydroxyaluminum that is adsorbed onto strong acid resin. Strong acid resins containing polynuclear hydroxyaluminum surfaces are not commercially available, so this surface must be prepared in the laboratory prior to preparing mixtures of resins. After adjusting the equilibrium pH of the hydroxyaluminum surfaces to the same value as that of the weak acid resin, phosphate is loaded onto the resin at a concentration corresponding to its preselected activity. The hydroxyaluminum surface has a great capacity and specificity for phosphate sorption. The resin and its hydroxyaluminum surface also contribute to buffering the activities of calcium, magnesium, potassium, manganese, and pH. Relatively small volumes of synthetic ion-exchange resins are usually required because the resins have large capacities for the exchange of nutrient ions.

The minimum amount of each type of resin, R, may be estimated from the equation below, where P

$$R = \frac{(P)(C)100}{(D)(L)}$$

is the maximum amount of plant tissue anticipated, C is the respective element concentrations of the plant species when grown at full nutrient sufficiency, D is the percent depletion that is deemed acceptable during the course of the experiment, and L is the concentration of the respective nutrients loaded on the resin primarily responsible for its buffering.

Not all nutrients are buffered by the mixtures of ion-exchange and chelating resins. Boron is supplied to the hydroponic solution as a trace constituent leached from spun glass. Nitrate and sulfate may be supplied continuously in very dilute nutrient solution that replaces the water transpired by the plants, or may be added at intervals with makeup water. Molybdate and chloride are required in very small amounts and may be present in sufficient

quantity even at their initial low concentrations. If not, they may be added at extremely low concentrations with the nitrate and sulfate. These supplements are necessary to counter the slow depletion of soluble salts during extended growth of plants, because anions other than phosphate are not buffered by the resins.

The ion-exchange and chelating resins buffer ions in solution by controlling only the activity ratios of phosphate to other anions and of the reference cation to other cations, not their absolute concentrations. Although the absolute activities of nutrients in solution depend in part upon the total concentration and the respective charges of the ions in solution, the ionic strength of the hydroponic solution may be maintained for weeks with a decrease of less than 10% through the additions of dilute nutrient solution from Mariotte bottles. Any potential for nutrient imbalances in solution is countered by the buffering power of the resins.

For background information *see* BUFFERS (CHEMISTRY); CHELATION; HYDROPONICS; ION EXCHANGE; pH in the McGraw-Hill Encyclopedia of Science and Technology.

[RONALD T. CHECKAI]

Bibliography: R. T. Checkai et al., A method for controlling the activities of free metal, hydrogen, and phosphate ions in hydroponic solutions using ion exchange and chelating resins. *Plant and Soil*, in press; R. T. Checkai, R. B. Corey, and P. A. Helmke, A method for controlling the ionic activities of nutrients in solution culture, *Agron. Abstr.*, p. 82, 1981; R. T. Checkai, W. A. Norvell, and R. M. Welch, Regulating the ionic activities of nutrients in the rhizosphere using ion exchange and chelating resins, *Agron. Abstr.*, p. 146, 1985; S. M. Combs, R. B. Corey, and R. L. Chaney, Effect of Cd^{2+}/Zn^{2+} activity ratios on elemental composition of tomato grown in a resin-buffered hydroponic system, *Agron. Abstr.*, p. 201, 1984.

Immunodiagnostics

Immunodiagnostic testing encompasses analytical procedures that utilize natural components of the immune system to achieve sensitive and specific measurements on a broad spectrum of substances of diagnostic importance in clinical medicine. These substances include proteins, hormones, therapeutic drugs, and infectious organisms. The immunoassay has become an established technique in the clinical laboratory, and constitutes one of the most rapidly growing segments of the commercial diagnostic market. This article reviews the various immunoassay techniques, with emphasis on newer developments. Some of the approaches described here are not commercially available but reflect the direction of immunoassay development.

Basic immunoassay. Immunoassays employ reagents that include antibodies and antigens. Antibodies are proteins produced by the immune systems of higher animals in response to infection or exposure to a foreign substance. These proteins are capable of binding a particular foreign substance or infective agent with high specificity and affinity; the substance to which a particular population of antibodies binds is called the antigen. In immunoassays, the antigen is usually the substance to be analyzed, the analyte. Antibodies that bind the analyte of interest are prepared by introducing the latter substance into the circulatory system of an animal, such as rabbit or goat. Antibodies subsequently produced by the animal's immune system are then collected in its serum. Alternatively, antibodies may be isolated from hybridoma cell cultures produced by fusing mouse myeloma cells with the antibody-producing cells of spleen. This technology is used to produce monoclonal antibodies in which all antibody molecules are identical. Antibodies isolated from serum contain many different antibody molecules which bind differing regions of one or several related antigens. Although the polyclonal antibodies are still acceptable or preferred in most applications, many of the newest immunoassay techniques employ monoclonal antibodies exclusively.

The fundamental reaction involved in immunoassays is the binding of antigen (Ag) by antibody (Ab) to form the antibody-antigen complex (Ab:Ag) as shown by reaction (1). Either antibody or antigen

$$Ab + Ag \rightleftarrows Ab{:}Ag \qquad (1)$$

may be assayed by supplying the binding partner to sample and determining the amount of Ab:Ag formed. In nonlabeled immunoassays the presence of the complex is monitored directly with techniques that are sensitive to the larger size of the complex relative to the individual antigen or antibody. To improve detection, antibody or antigen may also be attached to a second entity that possesses some easily measured property, such as radioactivity or fluorescence. Assays employing such labeled reagents are conveniently divided into two groups, heterogeneous and homogeneous. In heterogeneous assays it is generally necessary to separate the antibody-antigen complexes from the noncomplexed reactants prior to measuring the amount of label present. Homogeneous assays, on the other hand, do not employ a separation step but rely on a substantial change in properties of the label when the labeled reagent is part of an antibody-antigen complex. In general, heterogeneous immunoassays allow detection of the lowest concentrations of analyte, while unlabeled assays are simpler but less sensitive.

Nonlabeled immunoassays. Antibody-antigen complexes can often be detected visually when antigen is large enough to be bound by more than one antibody (such an antigen is called polyepitopic). Crosslinking of antigen by antibody may then occur, resulting in the production of precipitates or turbid solutions. Indeed, immunoprecipitation, discovered in 1897 by Rudolf Kraus, is the oldest immunodiagnostic technique. Immunoprecipitation has now spawned numerous techniques that use precipitation reactions in gel supports, including plate and radial immunodiffusion, and a series of electrophoretic procedures such as immunoelectrophoresis, rocket immunoelectrophoresis, and counterimmunoelectrophoresis. Although immunoprecipitation assays are widely used and have contributed much to the advancement of immunology, they are generally limited to the assay of serum proteins present at relatively high concentration. Increased sensitivity is afforded by the use of light-scattering measurements (nephelometry) to measure the presence of fine precipitates.

Certain antibody populations are capable of agglutination, that is, the crosslinking of cellular antigens. The aggregated cells may be detected visually by their sedimentation pattern, either under a microscope or with the unaided eye. Agglutination of red blood cells (hemagglutination) is a commonly used semiquantitative technique in hematology and serology. Other antigens may be assayed with this technique by chemically immobilizing antigen material to the surface of red blood cells or latex particles. This procedure has been used to assay urinary or plasma hormones and is the basis of the current home pregnancy tests.

Heterogeneous labeled immunoassays. High-sensitivity immunoassays became a reality in the 1960s as a result of the pioneering work of S. Berson and R. Yalow using radioisotope-labeled antigen. These radioimmunoassays provided a million-fold greater sensitivity relative to the existing nonlabeled techniques. Radioimmunoassays could be used to detect not only polyepitopic antigens but also small antigens such as drugs and certain metabolites to which only one antibody molecule could bind. The assay is performed by mixing the sample with radioisotope-labeled antigen and equilibrating this with a limited amount of antigen-specific antibody. Two competitive equilibria exist [reactions (2) and (3)], as antigen in the sample (Ag) and radiola-

$$Ab + Ag \rightleftarrows Ab{:}Ag \qquad (2)$$

$$Ab + Ag^* \rightleftarrows Ab{:}Ag^* \qquad (3)$$

beled antigen (Ag*) compete for binding to the antibody.

Due to the competition, samples containing a greater amount of the antigen will cause less radioactive antigen to be bound to antibody. Antibody antigen complexes are then physically separated from unbound components, the radioactivity of one or the other fraction is measured, and the sample antigen concentration is determined from a calibration curve. Separation of antibody-antigen complexes from free antigen is performed by several methods, including selective precipitation of antibody with a second antibody which binds the first. A particularly effective method uses antigen-specific antibody immobilized on a solid support, such as glass beads or polystyrene. After equilibration with sample and labeled antigen, the solid support is simply washed to effect separation.

Variations of the radioimmunoassay include the use of radioisotope-labeled antibody with either immobilized antigen or antibody. In the former case, sample antigen and immobilized antigen compete for binding to labeled antibody—the amount of label bound to the support is dependent, therefore, upon the sample antigen concentration. In the latter case, polyepitopic antigen is first bound by the immobilized antibody, followed by additional binding to labeled antibody. Consequently, radioactivity associated with the solid support is proportional to the amount of sample antigen present.

Radisotope-labeled assays are the most sensitive of the immunoassays, with detectability in the nmol/liter to pmol/liter range. New advances in this area include increased automation, application of centrifugal analyzers, the use of magnetizable solid supports to facilitate mixing and separation, and the introduction of multihead gamma counters to increase throughput of the radioactivity measurements. Radioisotope assays suffer from several disadvantages, however. A health risk exists for laboratory workers who prepare labeled material and perform the assays. Regulation of radioisotope use is a problem in many countries, and reagent stability does not allow extended storage of some of the more desirable radioactive products. For example, the commonly used isotope iodine-125 has a half-life of 60 days.

Alternative labels. Because of the disadvantages of radioimmunoassays, researchers have sought alternative labels that would provide sensitivities comparable to radioisotopes. Perhaps the most explored label alternative is protein enzymes. Many enzymes are stable for years and may be conveniently conjugated to antibodies and antigens. Enzymes also offer amplification of detectable material since each molecule of enzyme can convert a number of reactant molecules into detectable products. These products include highly colored compounds, chemiluminescent compounds, and fluorophores. The most commonly used enzyme labels are peroxidase, alkaline phosphatase, and β-galactosidase.

Enzyme-labeled immunoassays are performed in a manner analogous to radioisotope-labeled assays. Only the final detection step is different, as warranted by the measurable property of the enzyme product. At present, heterogenous enzyme immunoassays are not widely used in clinical laboratories. The enzyme assays require the additional enzyme-catalyzed reaction which introduces additional variability in the result. Many enzyme immunoassays, however, now report sensitivities comparable to that of radioimmunoassays, and the desirability of a nonisotopic assay should allow the displacement of some radioimmunoassays.

Fluorophores, notably fluorescein and rhodamine, are also being used for immunoassay labels. Although fluorescent compounds are detectable at picomolar concentrations, natural components of body fluids which fluoresce or scatter light often mask label fluorescence and cause fluorescence-based immunoassays to be an order of magnitude less sensitive than radioisotopic assays. Recent advances in fluorescence labeling include the use of highly fluorescent phycobiliproteins, isolated from algae,

which have a hundredfold greater absorbance than other common fluorophores and also possess nearly 100% fluorescence efficiency. Time-resolved fluorescence also has been introduced to discriminate against background fluorescence and scattered light. This technique uses fluorophores with very long excited-state lifetimes, such as chelated lanthanides. The fluorophores are excited with short pulses of light, usually by pulsed laser systems. The initial light detected following each excitation pulse contains the scattering and the unwanted fluorescence components and is not measured. Only emission persisting after this initial period, predominantly the label fluorescence, is measured. A commerical instrument based on this technology has recently become available to researchers.

Homogeneous labeled immunoassays. Homogeneous assays generally do not employ a solid phase or include separation steps. The omission of the separation procedures provides a shorter assay which is more amenable to automation. Homogeneous assays, in contrast to heterogeneous nonradioisotopic assays, have been welcomed into the clinical laboratories, due to their simplicity and speed, in situations where analyte is present in relatively high concentration (micromolar and greater). Particular application has been found in toxicology labs for therapeutic drug monitoring and assay of certain hormones and drugs of abuse. Enzymes are the most utilized labels at this time. In the homogeneous enzyme assay one reagent is composed of a small antigen chemically attached to enzyme. The labeled antigen reagent is added to the sample and competes with analyte antigen present in the sample for binding to a limited amount of antibody reagent. Binding of the large antibody (approximately 150,000 daltons) to the labeled reagent results in the inhibition of enzymatic activity, primarily due to steric hindrance. A high concentration of analyte (antigen), therefore, relieves the enzyme inhibition by preventing labeled antigen from binding to antibody. Examples of enzymes frequently employed are lysozyme, malate dehydrogenase, and β-galactosidase.

Homogeneous competitive assays may also be performed with fluorophore labels. The fluorescence quenching or enhancement assays take advantage of the sensitivity of fluorophores to their environment. Some fluorophores, when attached to antigen reagent, display an increased or decreased fluorescence efficiency when the reagent is bound to antibody. In a competitive assay the intensity of sample fluorescence will therefore be modulated by the amount of analyte antigen present. An alternative approach to the quenching assay uses two properly chosen fluorescent labels, one on antigen reagent and one on antibody reagent. When the two reagents are bound to one another, a process called dipole-coupled energy transfer occurs, which results in light energy absorbed by the first fluorophore label being transferred to the second label with a consequential reduction of emission from the first fluorophore. When analyte is present at high concentration, the labels are separated and emission from the first fluorophore is regenerated.

A fluorescence assay now in clinical use is based on fluorescence polarization. By exciting a fluorophore with polarized light and measuring the parallel and perpendicular polarized components of the emission separately, the rate of rotation of the fluorophore may be determined. Fluorophore attached to a small antigen (such as a drug or hormone) rotates relatively rapidly as compared to the same conjugate bound to antibody. Competitive binding with sample analyte, therefore, will cause polarization changes in the label fluorescence which are a measure of analyte concentration.

Conclusion. The multitude of immunoassay formats available has provided a realization of the utility of immunodiagnostic testing. Components of bodily fluids ranging from the relatively larger amounts of serum proteins to minute quantities of circulating hormones, may be analyzed routinely. Current trends in immunodiagnostics are toward simple, rapid, and inexpensive tests which maintain good sensitivity. This is evidenced by over-the-counter home test kits and test strip assays which may be performed in the doctor's office. While heterogeneous immunoassays provide greatest sensitivity, continued improvement in homogeneous approaches to assay technology should ultimately meet all three goals.

For background information *see* IMMUNOASSAY; IMMUNOFLUORESCENCE; RADIOIMMUNOASSAY in the McGraw-Hill Encyclopedia of Science and Technology.   [LARRY MORRISON]

Bibliography: R. C. Boguslaski and T. M. Li, Homogeneous immunoassays, *Appl. Biochem. Biotech.*, 7:401–414, 1982; R. P. Ekins and S. Dakubu, The development of high sensitivity pulsed light, time-resolved fluoroimmunoassays, *Pure Appl. Chem.*, 57:473–482, 1985; M. N. Kronick and P. D. Grossman, Immunoassay techniques with fluorescent phycobiliprotein conjugates, *Clin. Chem.*, 29:1582–1586, 1983; R. F. Schall, Jr., and A. S. Fraser, Enzyme immunoassays: A technical review, *Med. Device Diagnost. Ind.*, pp. 66–71, May 1983.

Immunogenetics

The genes coding for the alpha (α) and beta (β) chains of the T-cell antigen receptor have now been cloned and the corresponding proteins found to consist of a variable and a constant domain connected by a joining segment. A stretch of nonhydrophilic amino acids, located at the C terminus of these polypeptides, presumably represents transmembrane portions of the proteins. The extreme carboxyl ends are composed of short hydrophilic peptides, consisting of the intracytoplasmic ends. Thus, apparently the general structure of the T-cell receptor is very similar to the upper portion of the antibody molecules with the exception of an extension that anchors these structures on the cellular membranes. A schematic illustration of the structure of the T-cell receptor, a major histocompatibility complex (MHC) class II protein, and an immunoglobulin is given in Fig. 1.

Genes of the T-cell receptor. The α- and β-chain genes are organized in a manner similar to the immunoglobulin genes. They contain a number of noncontiguous variable (V) region and constant (C) region genes. Situated between these variable and constant genes are diversity (D) and joining (J) segments. The detailed arrangement of these genes in the germline deoxyribonucleic acid (DNA) are distinct from each other and those of the immunoglobulin (Ig) genes. A schematic diagram illustrating the organization of the α and β chains of these genes is given in Fig. 2.

Rearrangements of T-cell receptor genes. In the mature T cell, the germline deoxyribonucleic acid (DNA) pattern shown in Fig. 2 is altered in that a single V gene is joined with a single D and J gene. It is only after such rearrangement that it is possible to produce a mature T-cell receptor molecule. These rearrangement processes are accomplished by two separate steps. First, a single D-gene segment is brought into a position beside a single J-gene segment. Second, a V gene is transposed next to the combined DJ segment. Through the use of multiple V, D, and J segments, it is presumably possible to generate all the required diversity of the T-cell receptor repertoire. Somatic mutations, which are important factors for increasing immunoglobulin diversity, appear to play a less important role in the generation of diversity for T-cell receptor genes.

Rearrangement of these genes and the use of these receptors have been found in at least two subsets of functional T cells: the helper T cells and the cytotoxic T cells. The rearrangements of these genes may serve also as a marker of a malignant clone of T cells. The validity of T-cell receptor gene rearrangement as a marker of T-cell malignancy, and its use in diagnosis has been tested. Such studies have indicated that the T-cell receptor gene is rearranged in cloned T-cell lines and T-cell malignancies. The findings of rearrangement in T-cell malignancies confirmed the clonal nature of this disease. The rearrangements of these genes were not found in B-cell lines, B-cell malignancies, or other nonhemopoietic cells. Somewhat surprising is the finding that a large fraction of the cells from patients with common acute lymphoblastic leukemia (common ALL) have rearranged T-cell receptor genes. This casts doubt on the true nature of these cells and the previous hypothesis that common ALL belongs to the pre-B lineage. This former hypothesis was based on the finding that cells from common ALL have rearranged immunoglobulin genes. A small fraction of patients with acute myeloblastic leukemia also has rearranged T-cell receptor genes. This is also unexpected, as myeloid cells are not believed to use these immune recognition genes. The usefulness of T-cell receptor genes in the diagnosis of malignancies is exemplified by the finding that a high proportion of patients with lymphocytosis contain clon-

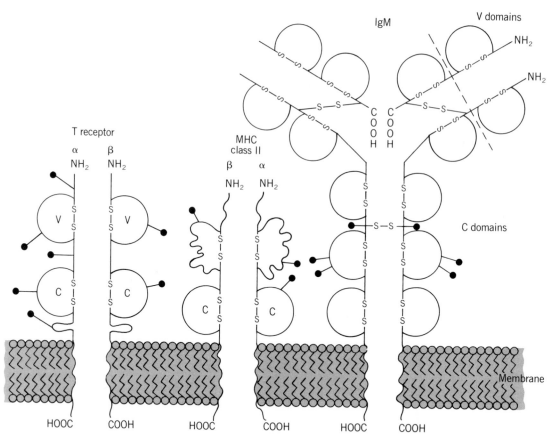

Fig. 1. Schematic diagram of the structure of the T-cell receptor, a class II gene from the major histocompatibility complex and an immunoglobulin gene. V = variable segment, C = constant-region gene, IgM = immunoglobulin macro, S-S = disulfide linkages. (*After A. Williams, News and Views, Nature, 308:108, March 8, 1984*)

ally rearranged T-cell receptor genes. These results clearly indicate that these patients with lymphocytosis may be already in a premalignant or malignant stage of disease. A summary of all these findings appears in the table. It seems that, like the immunoglobulin genes, the T-cell receptor genes are useful clinically in the diagnosis and possible management of patients with T-cell malignancies.

Chromosomal translocations. In some B-cell malignancies, activation of certain oncogenes can be achieved by translocation of these genes into the immunoglobulin gene structures. Evidence is now available that T-cell receptor genes are also involved in the translocation in cells from certain patients with T-cell malignancies. These translocations presumably also involve the activation of oncogenes

that may be involved in the initiation of certain T-cell neoplasias.

Future considerations. With the search for the T-cell antigen receptor genes completed, a new set of questions and goals arises. The phenomenon of dual recognition of antigen and MHC gene products by T-cell receptors can now be directly addressed. One powerful approach will be to genetically reconstitute a T-cell receptor by gene transfer of both the α and β chains of a T-cell antigen receptor with specific function. Another interesting problem would be to understand the mechanism for selection against T cells with receptors that are tolerant to self-antigens. The role of the thymus in the selection of these T-cell receptors will also be intensely investigated. Finally, the use of T-cell receptor genes in the in-

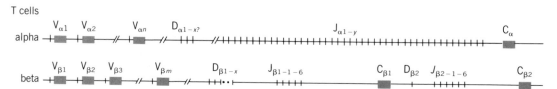

Fig. 2. Organization of the α- and β-chain genes of the T-cell antigen receptor. V = variable segment, D = diversity segment, J = joining segment, C = constant-region gene. Numerical subscripts indicate the position of the clusters and the number of segments.

Gene rearrangement of T-cell receptor and immunoglobulin in leukemia and related disorders

	Rearrangement, %	
	T-cell receptor	Immunoglobulin
T-cell leukemia	100	0
B-cell leukemia	0	100
Common acute lymphoblastic leukemia	30	100
Myeloblastic leukemia	10	5
Lymphocytosis	70	0

vestigation of certain clinical abnormalities such as autoimmune diseases and immunodeficiencies will also occupy researchers for years to come. *See* CELLULAR IMMUNOLOGY.

For background information *see* CELLULAR IMMUNOLOGY; IMMUNOGLOBULIN; LEUKEMIA in the McGraw-Hill Encyclopedia of Science and Technology.

[TAK W. MAK]

Bibliography: T. W. Mak and Y. Yanagi, Genes encoding the human T cell antigen receptors, *Immunol. Rev.*, 81:221–233, 1984; M. Minden et al., Somatic rearrangement of T cell antigen receptor in malignancies, *Proc. Nat. Acad. Sci. USA*, 82:1224–1227, 1985; P. Ohashi et al., Reconstitution of an active surface T3/T cell antigen receptor by DNA transfer, *Nature*, 316:606–609, 1985; Y. Yanagi et al., A human T cell-specific cDNA encodes a protein with partial homology to immunoglobulin chain, *Nature*, 308:145–149, 1984.

Information processing (psychology)

The development of modern technology has greatly changed the hierarchy of important human skills. Strength and motor performance have become less necessary for many tasks; so have perceptual skills, although these will always be needed. Intellectual skills, particularly those of judgment and decision making, have become the crucial elements of human performance.

Difficulties of decision making are typically blamed on the inadequacy of the available information; therefore, much technological sophistication has been mobilized to remedy this problem. Computers and other electronic devices supply the decision maker with an abundance of data. However, even the best attainable information often leaves many uncertainties. It has become evident that a key element in decision making is the ability to interpret and integrate multiple items of information, the reliability and validity of which are imperfect.

Information processing and decision making are studied by specialists from many disciplines. Their efforts center on two broad questions: What are decision makers doing? What should they be doing to make their decisions more effective and efficient?

Probabilistic reasoning. Because of the importance of probabilistic reasoning to decision making, a great deal of research has been devoted to describing how people assess and use the probabilities of uncertain events. For the most part, this research indicates that people have great difficulties judging probabilities, making predictions, or otherwise attempting to cope with probabilistic tasks. Frequently, these difficulties can be traced to the use of judgmental heuristics. These heuristics may be valid in some circumstances, but in others they lead to biases that are large, persistent, and serious in their implications for decision making. For example, the availability heuristic leads people to judge an event as likely or frequent if it is easy to imagine or recall relevant instances. In general, memorability and imaginability are appropriate cues for judging frequency and probability. However, because these cues can be influenced by factors unrelated to likelihood (such as recency, emotional impact, or familiarity), reliance on them sometimes leads to systematic misjudgment of frequencies and probabilities. For example, the frequencies of dramatic and highly memorable causes of death such as homicides and natural disasters are overestimated, whereas frequencies of death from less dramatic causes (such as asthma or diabetes) tend to be underestimated.

Two other important biases in judgment have to do with hindsight and overconfidence. Research on hindsight indicates that being told that some event has happened increases a subject's feelings that it was inevitable. In retrospect, a subject believes that people had a much better idea of what was going to happen than they actually did have. Research on confidence shows that, across a wide variety of tasks, people estimate much higher probabilities of success than are warranted. In a number of studies, people judged the probabilities that their answers to general-knowledge questions were correct (for example: Which magazine has the largest circulation, *Time* or *Newsweek*?). At 100% confidence, people were right about 70–80% of the time. At 80% confidence, they were right 60% of the time. At 60% confidence, they were right about 55% of the time.

Process-tracing methods. Cognitive psychology, with its emphasis on describing the processing of information, has had a profound effect on the study of decision making. Three process-tracing methods have been used to study the way in which a decision maker uses information. These methods involve verbal protocols, information monitoring, and eye-movement analysis. Verbal protocols require people to think out loud as they make decisions. Information-monitoring studies record people's selection of information items en route to making a decision. Eye-movement studies record where people fix their gaze as they perform a decision task. The results of these process-tracing methods are represented in a flow diagram or decision net, describing the sequences of thoughts and rules that a person employs while considering the various decision alternatives.

Process-tracing studies have made explicit the numerous elementary operations and rules that guide decision making. They have also shown how features of the decision task, such as the number of alternatives or the way that information about each alternative is displayed, govern the selection and use of these elementary rules. For example, as the number and complexity of the alternatives increase, people examine less information about each alternative.

Decisions are greatly influenced by the way that relevant information is presented. A recent study has shown that supermarket shoppers failed to use unit price information when it was displayed under each product on the shelf. However, if unit prices were listed in order of high to low in each section of the market, shoppers were much more likely to buy the more economical products.

A particularly dramatic example of presentation effects comes from another study in which people (including physicians) were asked to imagine that they had lung cancer and had to choose between two therapies, surgery or radiation. Surgery provides a longer life expectancy if one survives the operation (which has a 10% mortality risk). The two therapies were described in detail. Then, some people were presented with the probabilities of surviving for varying lengths of time after the treatment. Other people received the same probabilities framed in terms of dying rather than surviving (for example, instead of being told that 68% of those having surgery will have survived after 1 year, they were told that 32% will have died). Framing the statistics in terms of dying dropped the percentage of people choosing radiation therapy over surgery from 44% to

18%. The effect was as strong for physicians as for laypersons.

Decision aids. In light of the fallibility of unaided decisions, it is important to examine techniques for improving decision making. When decisions are repeatable (for example, selection or rejection of applicants for jobs or college admission), they can be handled quite effectively by a rule-based procedure. Often the rules can be derived by close observation or process-tracing studies of experts doing that task. The experts' decision processes can be programmed and executed by computers, a technique that is known as expert systems.

Another technique for enhancing decisions in a repeatable environment is simulation. Simulation places the decision maker in situations that are similar in certain important aspects to those likely to be encountered in the real world. Simulation has the advantage of exposing the decision maker to a rich variety of situations. Performance can be evaluated and immediate feedback provided. On the negative side, simulation must be carefully designed to present the critical aspects of the real decision if a proper generalization is to be obtained.

Decision analysis is a general technology for aiding unique decisions when the stakes are high and time is ample. Decision analysis assumes that all relevant considerations in a decision can be assigned to one of four components: options, possible consequences, the values of those consequences,

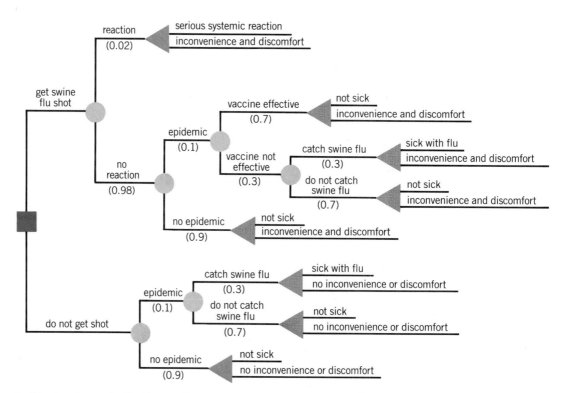

Decision tree for a swine flu dilemma. The square decision node indicates the choice branches. Circular nodes lead to outcomes having various probabilities of occurrence (indicated by numerical values between 0 and 1). Terminal outcomes are attached to triangular nodes. (*After R. D. Behn and J. W. Vaupel, Quick Analysis for Busy Decision Makers, Basic Books, 1982*)

and the uncertainties associated with their occurrence. An important tool is the decision tree, which diagrams the stream of uncertain consequences arising from a decision (see illus.). Decision analysis has been applied to numerous problems in business, medicine, space exploration, and military decision making.

For background information *see* ARTIFICIAL INTELLIGENCE; DECISION THEORY; INFORMATION PROCESSING; SIMULATION in the McGraw-Hill Encyclopedia of Science and Technology.

[PAUL SLOVIC]

Bibliography: D. Kahneman, P. Slovic, and A. Tversky, *Judgement under Uncertainty*: *Heuristics and Biases*, 1980; B. J. McNeil et al., On the elicitation of preferences for alternative therapies, *New Eng. J. Med.*, 306:1259–1262, 1982; H. Raiffa, *Decision Analysis*, 1968; J. E. Russo, The value of unit price information, *J. Market. Res.*, 14:193–201, 1977.

Infrared astronomy

In January 1983, *IRAS* (*Infrared Astronomy Satellite*) was launched by NASA on a mission to survey the sky at wavelengths between 10 and 100 micrometers. Once in its 500-mi-high (800-km) orbit circling the Earth's poles, *IRAS* began to map the universe, free of the warm terrestrial atmosphere, which both absorbs incoming infrared light and radiates so powerfully that only the strongest cosmic signals can be seen above its background. *IRAS* was equipped with a set of sensitive infrared detectors placed at the focus of a unique telescope, whose primary was chilled by liquid helium to a temperature of 10 K (18°F) above absolute zero in order to minimize the background infrared interference from the telescope itself. Until the liquid helium in its cryostat was exhausted in November 1983—well beyond its planned life—*IRAS* worked to perfection, providing the first sensitive "pictures" of the infrared sky. Its rich data base has continued to challenge and alter conceptions of the cool part of the universe.

Circumstellar disks. Early in the *IRAS* mission, one of the stars selected to calibrate the detectors, Vega, was found to be much brighter than expected at wavelengths greater than 25 μm. The color of this excess radiation suggests the presence of material emitting at a temperature of about 100 K (−280°F). Vega's proximity to the Earth made it possible to map the region responsible for the excess infrared radiation: a halo of dust grains orbiting around the star. Only large particles, bigger than 1 mm, could both account for the excess infrared luminosity and survive around Vega.

The *IRAS* science team quickly decided to devote precious observing time to measuring the infrared luminosity of several hundred nearby stars. At least eight of the Sun's neighbors are now thought to be surrounded by circumstellar dust clouds similar to Vega's. Guided by the *IRAS* discovery of excess infrared luminosity, astronomers were able to image

the emitting region around one of these stars, β Pictoris. Optical light, scattered in the Earth's direction in part by the same dust which glows in the infrared, reveals the geometry of the *IRAS*-discovered excess radiation: a thin, solar-system-like disk. The ease with which such disks have been found argues that they form with great frequency. It is not yet known whether they are the remains of a process similar to that which gave rise to the Earth and the other planets of the solar system. However, the discovery of the *IRAS* disks has provided new impetus both to searches for disks in much younger evolutionary phases and to searches for other planetary systems.

Objects of substellar mass. Various methods of "weighing" large cosmic aggregates—galaxies and clusters of galaxies—suggest that they are filled with objects which cannot be "seen" even with the most sophisticated tools. It is not known what unseen objects or particles account for the "missing mass" in the universe. *See* COSMOLOGY.

One possible form for the missing mass is a "brown dwarf," an object of mass less than 10% that of the Sun, and unable as a result to sustain the nuclear reactions which provide the energy source for most stars. In the first few hundred million years of its life, a brown dwarf "glows" at infrared wavelengths, as it converts the energy of gravitational contraction into radiation at a temperature of about 1000 K (1300°F). As a brown dwarf ages, it contracts further and cools. Because brown dwarfs are cool and small (and hence faint) for most of their lives, they have until recently eluded detection. Since 1984, however, ground-based infrared searches in the vicinity of nearby low-mass stars have located several brown dwarf candidates, companions to their more luminous stellar siblings and perhaps caught in their early, "hot" evolutionary phases. *See* STELLAR EVOLUTION.

IRAS's great sensitivity to infrared radiation from cool bodies makes it a powerful tool in the search for brown dwarfs. Several "false alarms" have disappointed the *IRAS* science team. However, astronomers believe that a careful search of the most sensitive subsets of the *IRAS* data base should make it possible to determine whether brown dwarfs are astrophysical oddities or a fundamental building block of the universe.

Observations of young stars. Newly formed stars are shrouded from view by dust grains in their parent molecular cloud complexes and by cocoons of circumstellar natal material. Moreover, they are cool during much of their early life. The combination— low temperature and high obscuration by grains— makes them impossible to observe optically, but ideal objects for *IRAS*. Several thousand infant stars have already been detected by *IRAS*.

The dark cloud designated Lynds 1641 (illus. *a*) contains a few optical "signposts" of recent star formation: stars that can be observed near the cloud's surface and whose luminosity and temperatures al-

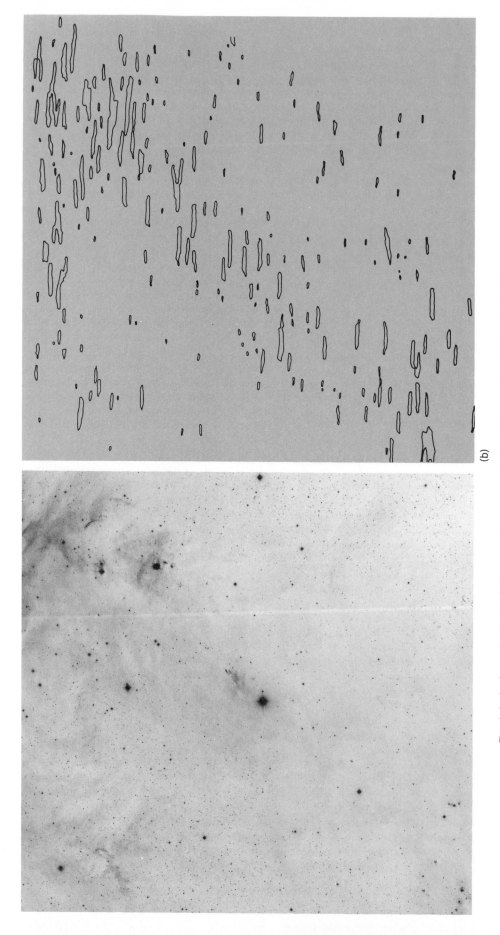

(b)

The dark cloud complex Lynds 1641. (a) Very sensitive optical image taken with the 48-in. (1.2-m) Schmidt telescope on Palomar Mountain, California. The image is printed in the negative (white sky and dark stars). The light "gap" running from the top right to the lower left is the Lynds 1641 cloud (Palomar Observatory Sky Survey; National Geographic Society). (b) Contour map of the same area reconstructed from IRAS survey data. The elongated images are actually pointlike stars; their unusual appearance results from a combination of the scan pattern chosen for the IRAS survey and the arrangement of individual detectors in the focal plane of the telescope.

low them to be classified as "young" (less than 1 million years of age). Although from this evidence L1641 appears to be an active stellar womb, the stellar birth process is largely hidden from view: almost no stars shine through its optically impenetrable haze of submicrometer-sized dust particles. However, the cloud contains nearly 10^5 solar masses of gas, largely hydrogen molecules, the fuel for forming new stars.

IRAS sees a very different and exciting picture (illus. *b*): a cauldron of star-forming activity with more than 400 luminous sources visible at 12 μm inside the optical boundary of Lynds 1641. The survey has revealed the brighter members of a populous, newly formed cluster of stars. Among its cluster members are several objects visible only at the longest wavelengths accessible to *IRAS*: 60 and 100 μm. These objects may be protostars—young stars still obscured from direct viewing at shorter wavelengths by placental clouds; only the radiation absorbed by dust in these clouds and reradiated in the infrared is detected by *IRAS*.

In other areas of the sky, *IRAS* finds bright far-infared sources located almost at the center of dense concentrations of natal molecular material (mapped by ground-based radio telescopes tuned to the millimeter-wavelength-emission characteristic of abundant molecules). Because stars wander with a characteristic "peculiar motion," the measured distance of each of these newly born stars from its birth center (at the molecular gas density maximum) can be used to estimate their maximum age. Several such stars discovered by *IRAS* can be no older than a few tens of thousands of years. Careful study of these stellar neonates should eventually provide unparalleled insight into the details of the stellar birth process.

Infrared galaxies. One of the surprises from the *IRAS* survey was the discovery of a class of galaxies which radiate predominantly in the infrared. Some of these infrared galaxies apparently house exotic "nuclear engines" at their centers—black holes accreting gas from the galactic disk or halo. In the most extreme cases, these galaxies radiate 100 times more energy in the infrared than at optical wavelengths, and are nearly 100 times as luminous as "normal," nearby bright galaxies. In such cases, the energy radiated by the nuclear engine must be absorbed by clouds of opaque dust, and then reradiated in the far infrared. Understanding these infrared-bright galaxies may turn out to be essential to unlocking the secrets of other galaxies with active nuclei such as the quasars.

In a majority of cases, infrared-bright galaxies appear to be undergoing "bursts" of star-forming activity. Clusters of stars, 1000 to 10,000 times as populous as the Lynds 1641 aggregate, may form in the disks or nuclear regions of galaxies. These star-forming events are so vigorous that, were the infrared galaxies to keep up their present rate of star-forming activity, the hydrogen "fuel" for future star-forming episodes would be consumed in a matter of a mere 100 million years (less than 1% of the nor-

mal lifetime of a galaxy). A clue to the origin of these mysterious and brief "star bursts" is provided by very sensitive television images taken with ground-based telescopes. These show that a good fraction of these galaxies are members of interacting pairs. Although the space between the galaxies is vast, the number of "close encounters" is not negligible over the $10–15 \times 10^9$ year lifetime of the universe. Gravitational interactions between passing galaxy pairs raise powerful tides, altering the motions of galactic gas and in some cases shocking and compressing the gas, thus triggering the formation of stellar clusters. These interacting pairs could turn out to be the laboratories for the discovery of the processes which trigger star formation and control the kinds of stars that are formed.

For background information *see* BLACK HOLE; GALAXY, EXTERNAL; INFRARED ASTRONOMY; STELLAR EVOLUTION in the McGraw-Hill Encyclopedia of Science and Technology.

[STEPHEN E. STROM]

Interferometry

Moiré interferometry is a rather new optical technique for high-sensitivity measurements of deformations of engineering materials, machine parts, and structural parts. Its introduction is timely: as theory and technology grow increasingly sophisticated, the need for physical data—experimental evidence of the behavior of real engineering bodies—becomes increasingly vital. Moiré interferometry satisfies this need in large measure by providing whole-field deformation data with subwavelength sensitivity.

A solid body subjected to external forces may be described by an orthogonal coordinate system in which the x and y axes lie on the surface of the body and the z axis lies perpendicular to the surface. The external forces deform the body such that each point suffers displacement components U, V, and W, which represent the movement of the point in the x, y, and z directions, respectively. Moiré interferometry yields contour maps of the U and V displacements throughout the body. It is especially significant that the in-plane displacements U and V are measured, since these are more closely related to strains and stresses in the body. This contrasts with classical and holographic interferometry, which are more useful for measuring out-of-plane displacements W.

Specimen and optical system. In moiré interferometry, a diffraction grating is applied to the surface of the specimen and it deforms together with the underlying specimen. This specimen grating is a reflection-type phase grating of frequency $f/2$ furrows per millimeter or lines per millimeter. It is formed on the specimen by the replication method illustrated in Fig. 1a. The mold is a crossed-line holographic grating, overcoated with a metallic film of evaporated aluminum. It is pried off after the adhesive (usually epoxy) has polymerized, but the metallic film remains bonded to the adhesive. The result is a very thin crossed-line diffraction grating on the

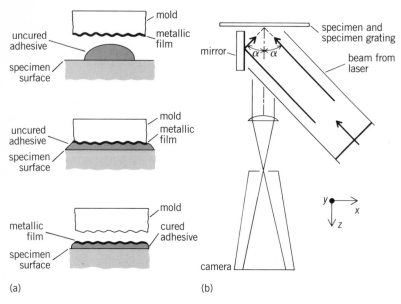

Fig. 1.   Experimental method for moiré interferometry. (*a*) Replication technique to form a high-frequency, high-reflectance grating on the specimen. (*b*) Optical arrangement.

specimen—a double grating with furrows, or lines, in both the *x* and *y* directions.

The specimen grating is viewed together with a superimposed reference grating of frequency *f*. The reference grating, however, is a virtual grating. It is formed by two beams of coherent light (Fig. 1*b*), one

of which reaches the specimen directly at angle $-\alpha$, while its companion beam is reflected by a plane mirror to reach the specimen at angle $+\alpha$. A steady-state pattern of optical interference is generated by the two intersecting coherent beams; lines of constructive and destructive interference, created by the presence and absence of light, are produced on the specimen surface. These lines constitute a virtual reference grating, whose grating pitch *g* and frequency *f* are given by Eq. (1), where λ is the

$$f = \frac{1}{g} = \frac{2}{\lambda} \sin \alpha \qquad (1)$$

wavelength of light employed.

The virtual reference grating interacts with the deformed specimen grating to form the pattern recorded in the camera. A rigorous explanation, however, considers diffraction of each of the two incident beams by the specimen grating. For the specified conditions, wherein the specimen grating frequency is (nominally) half the reference grating frequency, two first-order diffractions—one from each incident beam—emerge from the specimen grating. They emerge essentially perpendicular to the specimen and enter the camera. These two beams have wavefront warpages associated with the distortions of the specimen grating; they, too, combine by optical interference to create the two-beam interference pattern seen in the camera screen.

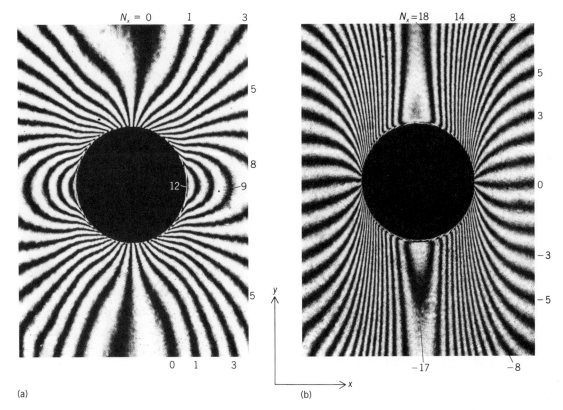

Fig. 2.   Moiré interferometry patterns depicting in-plane displacements *U* and *V*. The specimen is a unidirectional graphite-epoxy tension member with a central hole. (*a*) Fringe orders N_x, giving *U* displacements. (*b*) Fringe orders N_y, giving *V* displacements.

Sensitivity. Figure 2 is an example of such a pattern. The information therein is quantitative; each fringe, that is, each contour line, has an index number called the fringe order. When the reference grating lines are oriented perpendiular to the x axis, the pattern formed is assigned fringe orders N_x (Fig. 2a); when the reference grating lines are perpendicular to y, the N_y pattern is formed (Fig. 2b). The relationship between specimen displacements and fringe orders is given by Eq. (2). The factor of pro-

$$U = (1/f)N_x$$
$$V = (1/f)N_y \qquad (2)$$

portionality, $1/f$, is the sensitivity of the measurement. High sensitivity corresponds to small displacement per fringe order and therefore to high frequency of the reference grating. By Eq. (1), the theoretical upper limit of frequency is $2/\lambda$, which for visible light is slightly more than 4000 lines/mm. Fringes of excellent visibility were produced in an experimental demonstration conducted at 97.6% of the theoretical limit, using $f = 4000$ lines/mm. A frequency of $f = 2400$ lines/mm was used for the applications shown in Figs. 2 and 3. The corresponding sensitivity is 1/2400 mm, or 0.417 micrometer per fringe order. Where desired, fringes can be interpolated easily to 1/5 fringe order, providing a displacement resolution of at least 0.1 μm.

Input beams. Generally, both the U and V displacement fields are required in an analysis. In Fig. 1b, the lines of the virtual reference grating are perpendicular to the page, that is, perpendicular to the x direction. The pattern formed by the interaction of the reference and specimen gratings is the N_x pattern, depicting the U field. Since the specimen grating is a crossed-line diffraction grating, the specimen (or the optical apparatus) could be rotated 90° about the z axis to obtain the N_y or V pattern. An attractive alternative, however, is to create a separate virtual reference grating with its lines perpendicular to the y direction. This is accomplished by an optical apparatus that provides two additional input beams, with incidence angles $+\alpha$ and $-\alpha$ in the y-z plane. Then, the N_x and N_y patterns can be photographed separately by blocking the light of the x and y reference gratings in sequence.

The laser power required for this work depends upon the diffraction efficiency of the specimen grating and the magnification of the image in the camera. Laser power from 0.5 to 200 mW has been used for various applications.

Applications. Moiré interferometry has been applied to a diversity of problems. The two cases here are representative.

Figure 2 shows the N_x and N_y fringe patterns for a composite specimen, a tensile specimen with a 0.25-in.-diameter (6.3-mm) central hole. The composite material was graphite-epoxy with unidirectional fibers parallel to the y axis. Patterns were recorded for a series of load levels up to 90% of the failure load. The patterns of Fig. 2 represent a low

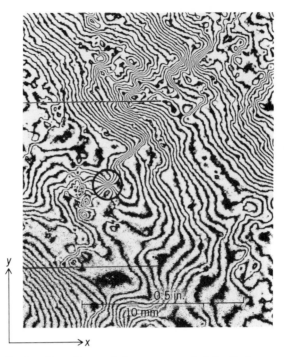

Fig. 3. Nonuniform or heterogeneous part of the V displacement field for a plastically deformed copper specimen. The two horizontal lines are index lines scribed on the specimen for the purpose of determining size or magnification from the image.

load, since those obtained for higher loads have very closely spaced fringes that cannot be printed clearly at this scale; the fringes were well resolved in the photographs for all load levels and enlargements were made for analysis.

The fringe orders are known for every point in the field of view, so displacements and strains can be determined throughout the field. An anomaly in performance was found in this work. The maximum shear strain occurs at a point near the hole where the fringe gradient in Fig. 2b is maximum. The shear strain at that point increased proportionally with the load at low load levels, but its rate of increase rises dramatically at higher levels. At 90% of the failure load, the shear-strain concentration factor was an order of magnitude greater than at 25% of the failure load.

A small lack of symmetry is noticeable in Fig. 2. This was caused in part by small variations in material properties along the specimen, and in part by use of imperfect gratings such that the fringe pattern for the zero load condition was not a true null field but exhibited initial fringes. When these initial fringe orders have significant effect, they can be subtracted from the subsequent fringe patterns by either whole-field optical methods or point-by-point data reduction methods. At high load levels, the initial field was insignificant compared to the load-induced fringes.

Figure 3 shows the N_y displacement field for a plastically deformed pure copper specimen at zero load. It shows only the nonuniform or heterogeneous part of the residual displacement field, since the uniform part was subtracted off by an optical technique. The uniform plastic strain was 0.85%, while the heterogeneous strain in the circled zone was 0.53%, to produce a local maximum of 1.38% permanent strain. Bands of plastic slip that occur at approximately ±45° to the direction of loading are seen through the field. The character of heterogeneous and nonlinear deformations of engineering bodies must be studied experimentally, and moiré interferometry offers a uniquely compatible technique.

For background information *see* INTERFEROMETRY; MOIRÉ PATTERN in the McGraw-Hill Encyclopedia of Science and Technology.   [DANIEL POST]

Bibliography: S. Kobayashi (ed.), *Handbook for Experimental Mechanics*; D. Post, Moiré interferometry at VPI & SU, *Exp. Mech.*, 23:203–210, 1983; D. Post, Moiré interferometry for deformation and strain studies, *Opt. Eng.*, 24:663–667, 1985; E. M. Weissman and D. Post, Moiré interferometry near the theoretical limit, *Appl. Opt.*, 21:1621–1623, 1982.

Ionosphere

The ability of high-power radio waves to modify the Earth's ionosphere has been recognized since the discovery of the Luxembourg effect in 1933. The program broadcast by the high-power Luxembourg transmitter was often heard on the low-power signals transmitted from Beromunster, Switzerland, on a completely different frequency. This cross modulation could be explained in terms of the changes in ionospheric electron temperature and hence in electron collision frequency induced by the high-power waves. These changed the absorption characteristics of the lower ionosphere and subsequently affected the signals from Beromunster passing through the disturbed region of the ionosphere. In the early 1970s it was realized that high-power radio waves could produce a wide range of instabilities in the ionospheric plasma in addition to the collision phenomena such as the Luxembourg effect. These instabilities have a range of spatial and temporal scales and can greatly influence the propagation characteristics of other radio waves traversing the modified volume. High-power heating facilities were constructed at Boulder, Colorado; Arecibo, Puerto Rico; Tromsø, Norway; and in the Soviet Union to study these new induced phenomena.

The heating experiments have enabled a number of ionospheric processes to be studied in a controlled manner for the first time. They have also led to the discovery of a wide range of plasma phenomena which are of importance in both geophysical and laboratory plasma physics. The scale sizes of some of these irregularities are such that they could not be reproduced in laboratory plasmas. Thus the heating technique has enabled the ionosphere to become a unique plasma physics laboratory where many of these phenomena can be studied and their relevance to practical radio and radar systems established.

Heating facilities. A typical heating installation consists of a transmitter capable of delivering about 2 MW of radio-frequency power in the range from 2 to 10 MHz. The transmitter feeds a large antenna array that normally forms a beam directed vertically into the ionosphere. Provision is made for radiating either ordinary (O), extraordinary (X), or linear polarization. The antenna gain will typically be about 20 dB. Thus, an effective radiated power of about 200 MW can be produced, yielding power fluxes of 150 μW m^{-2} and 65 μW m^{-2} at E-layer (70 mi or 110 km) and F-layer (150 mi or 250 km) heights respectively. The heating transmitter can be modulated or can radiate a continuous-wave signal.

The interaction of the high-power wave with the ionospheric plasma involves a number of complicated processes that can be subdivided into four general classes shown in Fig. 1 and discussed below: collisional interactions, small-scale F-region modification, large-scale F-region modification, and stimulated electromagnetic emissions.

Collisional interactions. The collisional absorption of heater wave energy in the lower ionosphere (D-region) produces a rapid increase in electron temperature that causes an increase in electron collision frequency and hence in the absorption coefficient. These changes will influence the propagation of other radio waves passing through the heated region, particularly at frequencies in the high-frequency (HF) band (3–30 MHz) and below. The time constants of these heating processes are very short (approximately 0.5 ms), and thus amplitude modulation of the heater signal can be transferred to other (low-power) radio waves passing through the disturbed region as in the classical Luxembourg effect.

The electron temperature modulations produced in the D- and E-regions by amplitude-modulated heating also produce electron density modulations through the electron-temperature-dependent recombination rates. Consequently, the ionospheric zero-frequency (dc) conductivities are modulated; and in the presence of a constant (dc) electric field, an alternating current is generated that radiates at the modulation frequency of the modifying heater (high-

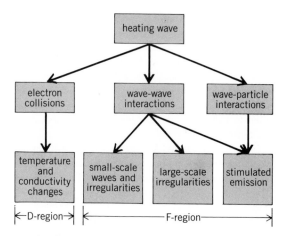

Fig. 1. Major modification processes induced in the ionosphere by a high-power radio wave.

frequency) wave. Significant signal levels in the ultralow-frequency (ULF), extremely low-frequency (ELF), and very low-frequency (VLF) bands (10^{-3} Hz–30 kHz) have been radiated by modulating the heater in this manner (Fig. 2). Low-frequency radiations have been stimulated from both auroral and equatorial electrojets, and this type of radiation generation could have new applications in long-distance and subsurface communications.

Small-scale F-region modification. When a radio wave propagates through the upper regions of the ionosphere (F-region), it encounters small irregularities in the electron density distribution and some of the incident wave energy is converted into plasma (Langmuir) waves. This wave-wave conversion process is insignificant for low-power waves, but as the heater wave power is increased so are the amplitudes of the Langmuir waves. As a consequence of the Langmuir wave growth, the electron density irregularities are enhanced and take the form of plasma striations elongated in the direction of the Earth's magnetic field. In this way the irregularities grow and more energy is absorbed from the heating wave until saturation occurs. This type of absorption is referred to as wideband or anomalous absorption. Theory indicates that striation growth should occur only for O-mode heating, and this is confirmed by experiment. The striations have scale sizes ranging from a few meters to a few kilometers, and their formation and decay time constants are typically about 0.1 and 10 s respectively.

Once the plasma striations are created, anomalous absorption will occur for any radio wave (even one of low power) traversing the modified region. A theoretical upper limit for this attenuation of 18 dB has been predicted, and this value has been confirmed experimentally. Anomalous absorption of the heating wave itself has also been measured. Figure 3 illustrates the variation of reflected signal strength of the heating wave as the radiated power is increased from zero to a maximum value and then reduced again to zero. The existence of a threshold for anomalous absorption and the extreme nonlinear behavior associated with the propagation of high-power waves through the ionosphere are clearly evident in these results. There is a marked asymmetry between the power-increasing and power-decreasing parts of the cycle.

High-frequency backscatter has been observed from the field-aligned striations for a wide range of frequencies (3 to 17 MHz). The range of scale sizes present ensures that the Bragg criterion is satisfied and maximum backscatter intensity is observed when the radar vector is perpendicular to the magnetic field (striation) direction. In addition to the normal radar backscatter, plasma line backscatter has also been observed even in the very high-frequency (VHF) and ultrahigh-frequency (UHF) bands (30 MHz–3 GHz). In this process the frequency of the scattered signal differs from the transmitted radar frequency by an amount equal to the heater wave frequency. This type of backscatter occurs from propagating waves rather than from field-

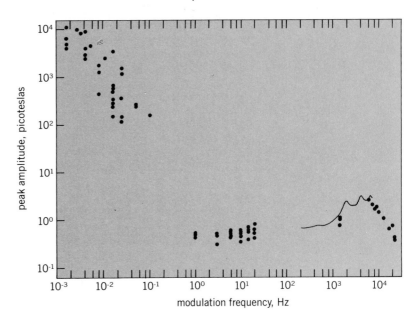

Fig. 2.   Relative amplitudes of low-frequency emissions from the auroral electrojet modulated by the Tromsø heater. Curve at lower right is data from a continuous frequency sweep. (*After P. Stubbe et al., Ionospheric modification experiments with the Tromsø heating facility, J. Atm. Terrest. Phys., 1985*)

aligned striations. These scattering waves are produced by other wave-wave processes.

The presence of medium-scale (kilometer) irregularities has been confirmed by monitoring the scintillation effects produced on signals from satellite beacons propagating through the disturbed region. Plasma striations of this magnitude also produce spread-F traces on ionosonde recordings and were regularly observed in the heating experiments at

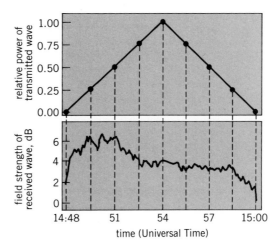

Fig. 3.   Variation of field strength of heater signal reflected from the ionosphere as transmitted power is increased from zero to a maximum effective radiated power of 260 MW and then decreased to zero. Marked nonlinear effects are evident, including a decrease in the reflected signal strength as radiated power is increased from about 0.5 to 0.75 of the full power. (*After P. Stubbe et al., Ionospheric modification experiments in northern Scandinavia, J. Atm. Terrest. Phys., 44:1025–1041, 1982*)

Boulder, Colorado. These spread-F features are far less in evidence at the high latitude (Tromsø, Norway) facility, emphasizing the importance of the field-aligned characteristics of the heater-induced irregularities.

Large-scale F-region modification. Part of the heater wave energy is expended through anomalous absorption in raising the electron temperature in the F-region throughout the heated volume. In the lower F-region, below heights of about 125 mi (200 km), the increase in temperature produces a decrease in the recombination rates and hence an increase in electron density. Above about 125 mi (200 km), the temperature increase enhances the diffusion rate and a depletion of the electron density occurs. These changes in electron density alter the refractive index of the medium within the heater beam. The possibility thus exists for the creation of converging or diverging lenses in the ionosphere. Such features have importance for high-frequency propagation since they can produce focusing and defocusing of the radio-wave energy.

Stimulated electromagnetic emissions. Measurement of the spectrum of the ionospherically reflected O-mode heating wave indicates that secondary electromagnetic waves are generated within the ionosphere. These are observed within ± 100 kHz of the primary heating frequency, and their spectra are highly structured. These emissions have been accounted for in terms of scattering processes involving Langmuir waves and low-frequency plasma density perturbations generated by the parametric decay instability. Heating at two independent frequencies not only generates stimulated emission around each frequency, but produces a new emission at a frequency equal to the arithmetic mean of the two heating frequencies. The processes involved in these stimulated emissions are complicated and not fully understood.

In addition to the enhanced electromagnetic-wave activity, superthermal electrons have been observed in the heated region due to wave-particle interactions. These energetic electrons can excite atoms and molecules, and optical emissions have been observed from the heated region as these return to their ground state. Particularly prominent are the oxygen green (557.7-nm) and red (630-nm) lines. The heating experiments at Tromsø, Norway, have been accompanied by very weak optical emissions in contrast to the relatively strong effects observed at other heating facilities.

For background information *see* Ionosphere; Plasma physics; Radio-wave propagation in the McGraw-Hill Encyclopedia of Science and Technology.   [T. B. Jones]

Bibliography: W. J. G. Beynon (ed.), Ionospheric modification, *J. Atm. Terrest. Phys.*, 44:1005–1171, 1982; J. A. Fejer, Ionospheric modification and parametric instabilities, *Rev. Geophys. Space Phys.*, 17:135–153, 1979; A. V. Gurevich, *Non-Linear Phenomena in the Ionosphere*, 1978; W. Utlaut (ed.), Ionospheric modification by high power transmitters, *Radio Sci.*, 9:881–1090, 1974.

Isotope (stable) separation

A new type of plasma centrifuge, developed since 1980, is distinguished by its source of plasma, which is a vacuum-arc discharge rather than a gas discharge. The vacuum-arc discharge produces a highly ionized plasma consisting of ions of the cathode material.

Plasma centrifuge principles. The plasma centrifuge is based on the concept of a cylinder of ionized matter (plasma), contained by a magnetic field and set into rotation by application of an electromagnetic body force. Centrifugal force causes heavier ions to move nearer the periphery of the rotating plasma, resulting in partial separation between the constituent isotopes. A typical gas plasma centrifuge (Fig. 1) consists of two stationary, coaxial electrodes between which there is a gas at some pressure, embedded in an externally produced magnetic field $\vec{B}$. When radial current $\vec{J}$ is discharged between the electrodes, the gas is partially ionized. The $\vec{J} \times \vec{B}$ Lorentz force causes the plasma to rotate in the azimuthal direction, and the confining magnetic field keeps the rotating plasma from contact with the walls of the container.

The balance of forces in the radial direction for a stably rotating gas centrifuge or a fully ionized plasma centrifuge leads to Eq. (1) for the ratio of

$$\left(\frac{n_2}{n_1}\right)_r = \left(\frac{n_2}{n_2}\right)_0 \exp\left[\frac{(M_2 - M_1)\omega^2 r^2}{2kT}\right] \quad (1)$$

abundances n_1 and n_2 of two isotopes in the radial direction, where M_2 and M_1 are the masses of the two isotopes, ω the rigid rotor angular frequency, r the radius (0 refers to the axis of the rotating fluid), T the temperature, and k Boltzmann's constant. Equation (1) shows that centrifugal separation in either a gas centrifuge or a plasma centrifuge depends exponentially on the ratio of the difference in ordered rotational kinetic energies of the isotopes to their random thermal energy. The rotational kinetic energy difference depends in turn on the square of the angular frequency ω. Gas centrifuges, which are based on high-speed rotating machinery, operate at low temperature but are limited to low values of ω. Thus for $\omega r \simeq 300$ m/s (1000 ft/s), $T = 300$ K (80°F), and $\Delta M = 3$ atomic mass units (for exam-

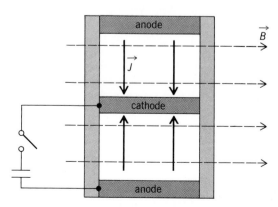

Fig. 1. Cross section of plasma centrifuge.

ple, in the case of uranium-238 and uranium-235), Eq. (2) holds, which implies roughly 6% enrich-

$$\left(\frac{n_2}{n_1}\right)_r \Big/ \left(\frac{n_2}{n_1}\right)_0 \simeq 1.06 \qquad (2)$$

ment per pass. In a plasma centrifuge, $\omega r \simeq 10^4$ m/s (6 mi/s), $T = 10^4$ K, and $\Delta M = 3$ atomic mass units gives Eq. (3), which implies about 500% en-

$$\left(\frac{n_2}{n_1}\right)_r \Big/ \left(\frac{n_2}{n_1}\right)_0 \simeq 6.0 \qquad (3)$$

richment per pass. However, these ideal estimates must be tempered by various losses in practice, as discussed below.

Plasma centrifuge types. Plasma centrifuges themselves fall into two categories: gas discharge and vacuum-arc discharge.

Gas discharge centrifuge. Gas discharge centrifuges are usually partially ionized. The neutral atoms in such centrifuges do not feel the electromag-

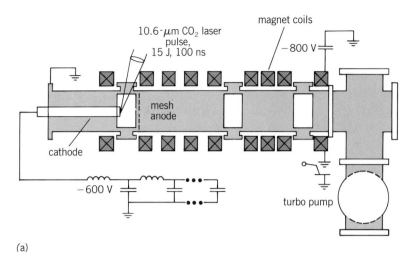

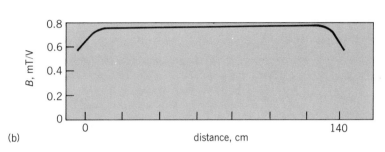

Fig. 2. Vacuum-arc centrifuge. (*a*) Schematic diagram. (*b*) Variation of magnetic field B, in milliteslas per volt across the magnet coils, with distance along the centrifuge.

netic forces directly and are carried along by collisions with the ionized particles. The drag due to neutral particles limits the rotation velocity, and viscous forces due to the neutrals lead to turbulence, mixing, and loss of separation. For these reasons, it was long believed that plasma centrifuges could not compete favorably with gas centrifuges as enrichment devices.

Vacuum-arc centrifuge. In the vacuum-arc centrifuge (Fig. 2), the source of plasma is a vacuum-arc discharge between a carbon or metal cathode and a gounded mesh anode. The plasma produced by such a discharge consists of ions of the cathode material and is highly ionized. Because the discharge is struck in a high vacuum, no deleterious effects due to neutral particles are present, and higher isotopic enrichments are possible. About 50% of the mass eroded from the cathode is transmitted through the anode mesh to produce a highly ionized, magnetized plasma column which rotates as a rigid rotor, with ω of the order of 10^5 rad/s. The plasma streams down the vacuum vessel axis as it rotates, with a gradual increase in radial centrifugal separation. Enriched material may be collected by placing an appropriate collector across the plasma column near the end of the vessel.

Detailed calculations show that a variety of isotopes useful in nuclear physics and nuclear medicine can be enriched by this device. The table shows calculations of the product yield and power requirements for enrichment of [203]Tl and [48]Ca.

Estimated parameters for isotope enrichment with vacuum-arc centrifuge

Parameters	Values	
Calculation of product yield:		
Ionization rate of feed	3×10^{-5} g/coulomb	
Peak discharge current	5 kA	
Duty factor	50% (on for 16 ms, off for 16 ms)	
Average discharge current	2500 A	
Plasma production rate	270 g/h (that is, 27 mol/day of uranium, 540 mol/day of carbon)	
Isotope	^{203}Tl	^{48}Ca
Natural abundance	29.5%	0.18%
Proposed number of stages	4	1
Proposed angular frequency (ω)*	5×10^5 rad/s	2.5×10^5 rad/s
Enriched abundance	95%	50%
Fraction of isotope in feed material collected	30%	10%
Fraction of total eroded mass collected	9%	0.018%
Product yield	24 g/h	0.05 g/h
Calculation of power requirements:		
Average discharge current	2500 A	
Discharge voltage	200 V	
Discharge average power	500 kW	
Vacuum system power	20 kW	
Confining magnetic field	0.1 tesla	
Magnet coil current	125 A	
Magnet coil resistance	0.8 ohm	
Magnet power	12 kW	

*Highest angular frequency measured is $\omega = 2.5 \times 10^5$ rad/s.

Application. The current world demand for isotopes may be divided into three categories: ton/year quantities of nuclear fuel materials; kilogram/year (pound/year) quantities of stable isotopes such as ^{203}Tl for nuclear medicine; and gram/year quantities of exotic isotopes for use in basic nuclear physics–nuclear chemistry research. The calutron electromagnetic enrichment process is well suited to the third of these categories, since any new technology would require prohibitive capital investment for a very small annual market.

Nuclear fuel enrichment has received considerable attention since the mid-1970s as cheaper methods have been sought to replace the costly gaseous diffusion process. A plasma process involving selective excitation of a plasma resonance, a molecular separation process, and an atomic vapor laser separation process (AVLIS) have been developed, with the AVLIS process in the lead as the most viable alternative. It would require scaling of the vacuum-arc centrifuge by many orders of magnitude in order to make it capable of the tons/year throughput required by the nuclear energy industry. It is not clear whether economies of scale will make this process competitive with the already developed laser separation schemes.

The second of the above categories may be the most appropriate area of application for the vacuum-arc centrifuge, since the throughputs of the order of 1 mol/day should match the required demand, and the total market value of the enriched product could justify the initial capital investment. In addition to the potential for a practical end use, the vacuum-arc centrifuge provides an excellent medium for basic arc discharge physics and plasma physics.

For background information *see* CENTRIFUGATION; ISOTOPE (STABLE) SEPARATION in the McGraw-Hill Encyclopedia of Science and Technology.

[MAHADEVAN KRISHNAN]

Bibliography: G. F. Brand, B. W. James, and C. J. Walsh, A metal seeded rotating plasma, *J. Phys.*, D12:1495, 1979; M. Geva, M. Krishnan, and J. L. Hirshfield, Element and isotope separation in a vacuum-arc centrifuge. *J. Appl. Phys.*, 56:1398–1413, 1984; M. Krishnan and R. R. Prasad, Parametric analysis of isotope enrichment in a vacuum-arc centrifuge. *J. Appl. Phys.*, 57:4973–4980, 1985; R. R. Prasad and M. Krishnan, A matched-impedance, low voltage, vacuum-arc centrifuge. *Rev. Sci. Instrum.*, 57:74–78, 1986.

Land reclamation

Indiscriminate disposal of chemical wastes and spills of hazardous substances leave residues that pose risks to humans and the environment. Recent federal and state legislation in the United States has encouraged research and development in new approaches to reduce these risks in the future through land reclamation efforts.

Contaminated sites. Land reclamation technology is designed to mitigate the hazards of chemical residues and, where possible, to restore the environment. The nature of contamination hazards is dependent upon both the environment in which the residues occur and the chemical and physical properties of the contaminants. Chemical residues which originate in the soil are transported by water and air, and through the food chain. Since surface waters and the atmosphere have relatively short residence times, they typically cleanse themselves without the implementation of reclamation efforts. Vegetation may cleanse itself rapidly, depending on the life cycle of the affected plants. On the other hand, groundwater, soil, and sediments often require some level of human intervention. Similarly, volatile or easily degraded contaminants will be reduced quickly below levels of concern in the environment, while stable, nondegradable substances will persist at hazardous levels. As a consequence, land reclamation technology is largely focused on persistent chemical residues in soils, groundwater, and sediment.

Remove and dispose. This is a term applied to alternatives based on the segregation of contaminants from the environment and subsequent use of conventional disposal technology—for example, treatment followed by placement in a secure landfill, ocean disposal, or underground injection. The disposal methods are derived from wastewater treatment and waste disposal technology, in that the material removed is often very dilute, large in volume, and accompanied by a significant mass of inert material such as soil. This can severely affect the cost of treatment; for example, incineration can be expensive.

Removal technology has been borrowed from other fields of endeavor. Soil removal is achieved with standard excavation techniques using scrapers for shallow contamination and backhoes for deeper contamination. Some experimentation has been conducted on solvent extraction and washing of soils. This can be accomplished in place or in disturbed soils, that is, soils that have been removed for processing. In-place approaches involve injection or release of solvents at the surface, followed by recovery in withdrawal wells. Solvent selection must balance selectivity for the contaminant of interest with toxicity and the risk of leaving solvent residuals. Soil washing is a related technology using water-soluble detergents to free contaminant from soil particles.

Where groundwater is contaminated, removal is achieved with a network of withdrawal wells. Slurry walls or grouting may be utilized to improve the degree to which a contaminant plume is captured. In some settings, invert wells can be constructed between the contaminated aquifer and a deeper layer. With this alternative the captured water is injected directly into the deeper receiving strata.

Sediment removal is accomplished by dredging. Devices developed specifically for contaminated sediment removal are designed to minimize resuspension and loss of sediments through vacuum action. Alternately, dredging operations can be conducted within a turbidity curtain, a physical barrier placed around the site to minimize off-site sediment transport.

Alternatives for removal and disposal are often very expensive because of the large volumes of material involved. They may also be politically difficult to implement in terms of finding a site for ultimate disposal. Choice of a distant disposal site will add significant secondary risks associated with long-distance hauling of large quantities of hazardous wastes.

Destroy-in-place. The most desirable approach to land reclamation is to destroy contaminants in place. This alternative is often contaminant-specific and is limited to a small number of contaminants. Chemical destruction can be achieved with neutralization or oxidation-reduction. Standard wastewater chemistry is applied by using the environment as the containment vessel: acids can be neutralized with lime or sodium bicarbonate; alkalies can be neutralized with sulfuric or hydrochloric acid; cyanide, sulfides, and other reduced species can be oxidized with chlorine or hydrogen peroxide; and chromium $(6+)$ and other oxidized species can be reduced with sulfite. Mixing and dosing pose the major problems. In the former case, it is difficult to ensure complete contact; in the latter, it is virtually impossible to select a dose that will eliminate all contaminant without leaving a new residual that is potentially just as hazardous.

An intriguing new approach for halogenated organics involves the addition of amines or other hydrogen donors to the contaminated media. These compounds enhance photodehalogenation by sunlight and thus make the contaminant more susceptible to biological degradation. This method works only for surfaces that receive or can be readily exposed to sunlight. In one case olive oil was employed on dioxin residues from the explosion of a chemical plant at Seveso, Italy.

Biological approaches are the most broadly applicable of the destroy-in-place alternatives. Indeed, biological degradation occurs when no action is taken. By definition, however, land reclamation is required for the more persistent chemicals that resist degradation. For these materials, special measures may be necessary, including (1) addition of nutrients and moisture to create an optimum growth environment for bacteria; (2) use of acclimated or specifically selected cultures; (3) use of genetically engineered strains; or (4) addition of enzymes capable of deactivating the contaminant.

Mixing is a major problem. One approach employs an activated sludge unit aboveground to which contaminated water is fed from withdrawal wells. Mixed liquors are then returned to the ground so that the bacterial culture establishes itself in the soil or aquifer. This approach, which combines in-place and remove-and-dispose techniques, helps circumvent problems from bacterial solids being filtered out by the soil. Dosing is not a problem since the bacterial population is regulated by the availability of food. There is a need to evaluate the impact of releasing new life forms, that is, mutations and genetically engineered strains. Recent work has focused on bacterial strains which can degrade organic insecticides and polychorinated biphenyls.

Isolation. Soil contamination is a problem only to the degree to which the contaminant can migrate into water or the atmosphere. One solution is to terminate the routes of migration by isolating the contaminant. Isolation techniques involve the major mechanisms of migration, those phenomena which provide the most significant routes for migration of contaminants—namely, infiltration of precipitation with subsequent leachate generation and percolation to the groundwater; vaporization of volatile chemicals with upward migration through the soil; contact of surface runoff with contaminants followed by dissolution of contaminants or suspension of soil particles to which contaminants are adsorbed; and direct contact of groundwater with contaminants resulting in dissolution and transport downflow.

Caps are applied to prevent infiltration that may generate leachate as well as to channel or retard vapor loss. Retardation is sufficient for a material such as radon gas from uranium mill tailings since it decays to a nonvolatile form; channeling of other gases such as vapors from chlorinated solvents facilitates collection and treatment. In this technique, gases are collected from a landfill area. Low-permeability layers are designed to channel or direct vapors as they move upward through the soil. Thus the gases can be concentrated at vents or collection points where they can be destroyed or captured for reuse. It is not economical to collect gas from the entire affected surface area. Channeling is employed to create a discrete point of collection. This approach was implemented as a part of the initial cap construction at Love Canal, New York.

Typically caps are constructed from several feet of clay with a coarse layer as a capillary barrier and a vegetated cover to enhance evaporative losses. Asphalt may also be used in caps as well as synthetic membrane liners, fabrics which have very low permeability and can be used to keep infiltration from coming in (functioning as a cap) and also to prevent leachate from flowing out (functioning as a liner). A new hygroscopic alum-based material has been demonstrated as a rerplacement for clays. Thin layers stop both moisture and vapor movement at a level comparable to that of much thicker clay layers.

Liners are made of the same material as caps, but are placed beneath contamination to prevent the movement of leachates. Liners are difficult to emplace without first excavating the wastes and, therefore, are more often associated with remove-and-dispose alternatives. A rough form-in-place clay liner can be installed by drilling holes on a grid pattern and using water pressure to form platelike cavities at the bottom of each hole. Clay slurries are then pumped into the holes and allowed to collect in the cavities, forming a series of overlapping clay disks. Hydrofracture, high-pressure water injection, has also been domonstrated to create wedge-shaped cavities which extend beneath contamination and can be filled with clay slurries. If the wedges intersect, they form a linerlike plane. However, these in-place techniques cannot consistently produce a liner of high intergrity.

Lateral movement of contaminants is controlled with cutoff walls. These may consist of sheet piling, a trench filled with a slurry of clay (slurry wall), or injected grouts. The completed wall is highly impermeable and can surround an area of contamination both to keep contaminants in and to divert clean waters. Grouts are used to form walls in fractured rock systems. The integrity of the resultant wall is highly dependent on the degree to which the fractures are characterized and interconnected.

In-place fixation may also be applied to isolate contamination by putting the entire block of affected soil in a state that resists leaching. Early attempts to achieve fixation with organic-based grouts proved expensive and difficult to control. In harbor sediments, greater success has been achieved with cement-based agents using a water setting formula in conjuction with vertical mixing devices. In terrestrial soils, fixation can be achieved with in-place vitrification; electrodes are placed in the ground and subjected to high voltage. A graphite layer on the soil allows current to flow and undergoes resistance heating. As the temperature rises, adjacent soils melt and the molten soil becomes the conductor and heating element, allowing the melt to proceed downward; organic contaminants are pyrolyzed during this process. When the current is stopped, the soil cools to an obsidianlike glass that encapsulates the contaminants.

Isolation techniques are often feasible, but do not constitute a final solution because the contamination remains even though migration is halted or slowed. Changes in environmental conditions, weathering, chemical action, or human intrusion may ultimately expose the contaminants for further migration.

Minimize exposure. In some cases, the cost of land reclamation cannot be justified on the basis of anticipated benefits; it may be best to prevent human exposure. This can be done by evacuating residents or, if only the water is affected, developing alternative water supplies. The latter can include bringing in water from a clean source or treating water at the well head by, for example, placing aerators in wells to strip out volatile solvents prior to use.

For background information *see* ENGINEERING GEOLOGY; GROUNDWATER HYDROLOGY; SEWAGE TREATMENT; WATER POLLUTION in the McGraw-Hill Encyclopedia of Science and Technology.

[GAYNOR W. DAWSON]

Bibliography: J. Ehrenfeld and J. Bass, *Handbook for Evaluating Remedial Action Technology Plans*, U. S. Environmental Protection Agency, EPA-600/2-83-076, August 1983; *Proceedings of the 5th National Conference on Management of Uncontrolled Hazardous Waste Sites*, Hazardous Materials Control Research Institute, Silver Spring, Maryland, November 7–9, 1984; H. L. Rishel et al., *Costs of Remedial Response Actions at Uncontrolled Hazardous Waste Sites*, U.S. Environmental Protection Agency, EPA-600/S2-82-035, March 1983.

Laser

Recent developments in laser technology include the fabrication of free-electron lasers and the development of coupled semiconductor laser arrays. Free-electron lasers offer the possibility of tunability over a wide range of wavelengths, excellent optical beam quality, and very high average power. Semiconductor laser arrays overcome the power limitation of approximately 100 mW on widely used semiconductor injection lasers.

FREE-ELECTRON LASER

A free-electron laser is a new type of laser in which the light is produced by electrons which are free, in the form of an electron beam in a vacuum, rather than bound in the atoms or molecules of a conventional gas, liquid, or solid laser medium. Invented in 1969, free-electron lasers show promise of becoming powerful sources of tunable laser radiation for a variety of applications.

Operating principle. A beam of electrons from an electron accelerator is directed through a set of magnets arranged so that the magnetic field points alternately up and down (Fig. 1). When the electrons travel through the magnetic field, they are deflected alternately left and right and follow a "wiggly" path. As they oscillate from side to side, the electrons emit electromagnetic radiation (light) at the oscillation frequency. In general, the electrons in free-electron lasers travel at nearly the speed of light, so the radiation appears principally in the forward direction. As a result, the wavelength of the light is Doppler-shifted to shorter wavelengths given by the equation below, in which λ_L is the wavelength of the

$$\lambda_L = \frac{1 + (eB\lambda_W/2\pi mc)^2}{2\gamma^2}\,\lambda_W$$

light, λ_W is the period of the wiggler magnet, γ is the energy of the electrons in units of their rest energy (0.511 MeV), B is the magnetic field in the wiggler magnet, m is the electron mass, and c is the velocity of light. When a laser beam is incident on the electron beam in the wiggler and satisfies the resonance condition given by the equation above, the electrons move in phase with the optical beam. The radiation from the electrons is then in phase with the incident beam and amplifies it. The radiation reflects back and forth between the mirrors of the optical resonator until it builds into an intense beam. Part of the beam is allowed to pass through the far mirror to form the output laser beam.

Typically, wigglers are designed so that the factor $eB\lambda_W/2\pi mc$ is of the order of unity; with available permanent magnet materials, this corresponds to a magnet period of about 1 in. (2–3 cm). Wiggler magnets have been built with overall lengths of from 3 to 15 ft (1 to 5 m). For an electron energy of 100

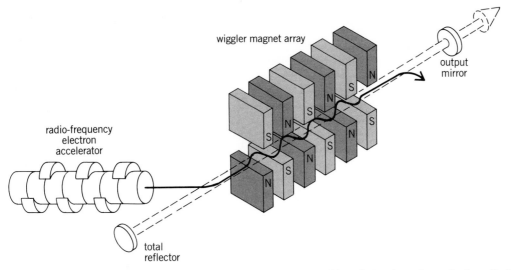

radio-frequency
electron
accelerator

wiggler magnet array

output
mirror

total
reflector

Fig. 1. Configuration of a free-electron laser. (*After C. A. Brau, Progress toward free-electron lasers for applications, Nucl. Instrum., B10/11:276–279, 1985*)

MeV (γ = 200) and a wiggler period of 0.8 in. (2 cm), the laser radiation appears at a wavelength of approximately 0.5 micrometer, in the green portion of the spectrum.

Experiments. The first experimental demonstration of a free-electron laser, in 1976, operated at a wavelength of 10 μm, in the infrared portion of the spectrum. Although the efficiency was very low, the results were in agreement with theoretical expectations. Soon afterward, it was proposed that the efficiency could be enhanced by designing the wiggler to compensate for the loss of resonance (given by the above equation) which occurs when the electrons transfer their energy to the laser beam. When this approach was tested at Los Alamos in 1981, nearly 4% of the electron energy was converted into light, in perfect agreement with theoretical predictions. The power produced by the laser was 6 kW, in pulses lasting 100 microseconds. By varying the electron energy from 20 down to 10 MeV, it was possible to tune the wavelength from 9 to 35 μm, in agreement with the above equation.

Other wavelengths have been addressed in experiments by groups at the University of California Santa Barbara, TRW, and the University of Paris, South. At Santa Barbara, a far-infrared device has been operated with a wavelength near 400 μm. The electron beam in this laser is accelerated to about 3 MeV by a Van de Graaff–type electrostatic accelerator. The TRW group used a radio-frequency linear accelerator to accelerate an electron beam to 65 MeV and produce laser radiation at wavelengths of 1.6 and 0.5 μm. The latter radiation corresponds to the third harmonic of the wiggler. The Paris group used an electron storage ring at 160 MeV to produce laser radiation at a wavelength of 0.6 μm, in the orange part of the visible spectrum. Although the radiation was weak, only about 100 microwatts, it lasted more than an hour, corresponding to the storage time of the electrons in the ring.

Advantages. Free-electron lasers offer several advantages compared to conventional lasers. The most obvious is tunability, which is achieved principally by varying the electron energy γ. A variety of accelerators, including linear induction accelerators, electrostatic accelerators, microtrons, radio-frequency linear accelerators, and storage rings, have been used to produce laser radiation at wavelengths from the microwave region through the infrared to the visible portion of the spectrum. In the future this range will be extended into the ultraviolet and probably beyond. Since accelerators can generally be operated over a range of electron energy, it is possible to tune the wavelength of a given device. The fourfold tuning range obtained in the Los Alamos experiments discussed above represents the broadest tuning range ever obtained by any laser of any type.

Excellent optical beam quality is another advantage of free-electron lasers. This is important when the beam must be focused to a very small, intense spot or when the beam must be transported to a target a great distance away. Experiments have demonstrated focal spots only 4% larger than the fundamental limit established by diffraction. This property of free-electron lasers results from the fact that the optical resonators of free-electron lasers tend to be long and slender, with the result that they filter the optical beam to remove nonfocusable components.

Finally, free-electron lasers offer the possibility of achieving very high average power. The principal obstacle to high power in conventional lasers is removal of the waste heat. The most powerful conventional lasers are gas lasers in which the waste heat is removed by flowing the gaseous laser medium out of the laser at high speed. In free-electron lasers, the waste heat resides in the electron beam, which is traveling at nearly the speed of light and exits the laser in a few billionths of a second.

Applications. A variety of applications is foreseen for free-electron lasers, most of which make use of their tunability. Free-electron lasers operating in the far-infrared portion of the spectrum (approximately 100 to 400 μm) will be useful for research in semiconductor physics, and the device at Santa Barbara, described above, is intended for this purpose. No other light sources of comparable intensity are available in this spectral region.

Powerful free-electron laser light sources in the near-infrared, visible, and ultraviolet portions of the spectrum will be useful for a variety of applications in chemical research and chemical processing; the precise tunability of free-electron lasers permits them to be adjusted to act only on specific molecules in very specific ways. Materials research in the far-ultraviolet portion of the spectrum will also benefit from the development of free-electron lasers since they promise to be orders of magnitude more powerful than synchrotron light sources now in use.

For even shorter wavelengths, harmonic radiation from free-electron lasers can be used. This radiation appears naturally at multiples of the fundamental frequency of the laser. Although it is ordinarily several orders of magnitude weaker than the fundamental radiation, the harmonics share the focusing properties of the fundamental.

In medicine, free-electron lasers may be tuned to different wavelengths which affect tissues in various ways. Alternatively, dyes may be placed in tissues or organs and used to absorb selectively laser light of specific wavelengths to achieve desired effects.

Finally, the possibility of achieving precisely focused beams of enormously powerful radiation from free-electron lasers makes them potentially useful for applications in strategic defense.

[CHARLES A. BRAU]

SEMICONDUCTOR LASER ARRAYS

The high reliability and efficiency of semiconductor injection lasers make them the preferred choice as the light sources in many applications (for example, optical communications and optical recording). However, due to power density limitations on the laser mirrors (to avoid catastrophic degradation) and limitations on the cross-sectional area of the lasing beam (to avoid undesirable multimode operation), the power output that is available from a conventional semiconductor injection laser in a reliable continuous-wave (cw) single-mode operation is limited to approximately 100 mW.

Some applications (for example, free-space optical communications, optical recording and printing, and pumping solid-state lasers) either require or could benefit from an increase in the available power levels. This can be accomplished through several possible approaches, most of which are based on combining the radiation emitted by several lasers. While there are many methods and configurations where power combining can be realized, there is a special incentive to implement them monolithically, that is, with all the individual lasers fabricated on the same substrate. By using this ap-

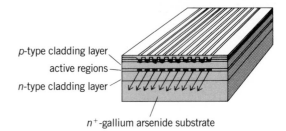

p-type cladding layer
active regions
n-type cladding layer
n⁺-gallium arsenide substrate

Fig. 2. Schematic configuration of a typical monolithic semiconductor laser array. (After D. R. Scifres et al., Lateral beam collimation of a phased array semiconductor laser, Appl. Phys. Lett., 41:614–616, 1982)

proach, the entire array of lasers resides on a single chip, and thus the advantage of having a single semiconductor device as the light source is retained.

Incoherent and phase-locked arrays. A typical monolithic array (Fig. 2) consists of some number N of parallel-stripe–geometry lasers. Such devices have two basic regimes of operation: incoherent or phase-locked. In the situation where there is virtually no coupling between adjacent lasers, the array operates as an incoherent ensemble of emitters. This situation can be obtained by either increasing the interelement spacing or by an appropriate design of the interelement regions. Since all the lasers operate incoherently (that is, without any fixed phase relationships), the far-field radiation pattern of the array is the same as that of a single laser (except for the obvious N-fold increase in intensity).

A more interesting—and potentially more useful—situation results when adjacent lasers interact with one another. When the interaction is sufficiently strong, the lasers can operate as one coherent ensemble, with fixed phase relationships between adjacent array elements. The monolithic device is then referred to as a phase-locked array. Alternatively, the monolithic phase-locked array can also be viewed as a broad-area laser whose typical filamentary operation has been stabilized by dividing it into several regions, each capable of sustaining exactly one filament. Most of the experimental work carried out so far has been on these types of arrays. Typical interelement center-to-center separation in this case is approximately 10 μm for gain-guided lasers, and 5 μm for real-index–guided devices, for which coupling between lasers is usually weaker. The main advantage of phase-locked arrays is the increased directivity of their output beams. Since their total radiation field is obtained by summation of field amplitudes (and not intensities, as in the incoherent case) of the array elements, the resulting far-field pattern can have an N^2-fold increase in the light intensity in the forward direction (on-axis), provided that all the lasers operate in the same phase.

This result is analogous to the operation of phased arrays of microwave antennae. Figure 3 compares the far-field radiation pattern in the junction plane of a single laser, an incoherent array of 10 lasers, and a phase-locked array of 10 lasers that operate in phase. Also shown is the far-field pattern for the case where adjacent lasers operate out of phase (180°) from one another. In this case, there are two main beams, each radiating in an off-axis direction; this is less desirable in most applications.

The actual details of the operation of phase-locked arrays are more complicated, and reasonably accurate modeling requires the self-consistent solution of the lasers' electromagnetic wave equations and quantum-mechanical rate equations. Even when each of the N array elements is a single-mode laser, the array structure can potentially support N eigenmodes, sometimes referred to as supermodes. Usually, the fundamental (lowest-order) supermode corresponds to the in-phase operation of all the array elements, and the highest-order supermode corresponds to out-of-phase operation between adjacent elements. The combination of eigenmodes that actually lases depends on their relative modal gains, which in turn are determined by the distribution of the pumping currents in the lasers.

Operating arrays. There have been extensive research efforts toward demonstrating actual operation of phase-locked monolithic arrays, understanding the underlying physical phenomena, and improving their characteristics. The first device, demonstrated in 1970, consisted of only one pair of lasers. In 1978 the first multielement arrays (with $N = 5$ and 10) were fabricated. Since then, more than a dozen array versions of the generic type shown in Fig. 2 have been constructed. The simplest version (and the first one to be implemented) consists of a uniform array of gain-guided lasers, where the individual laser stripes can be defined, for example, by proton implantation. The array with the largest number of elements demonstrated so far ($N = 40$) is of this type. This array also emitted the highest power level reported in continuous-wave operation, 2.6 W. A version of this type of array with 10 elements became commercially available in 1984. Projected lifetimes are currently about 3000 h at room temperature at the maximum specified power (200 mW in continuous-wave operation), and approximately 10 times longer at 100 mW output power.

Arrays consisting of lasers that are dominantly real-index–guided have also been demonstrated. Although such arrays are more difficult to fabricate, they may have superior performance in terms of their modal characteristics as compared to their gain-guided counterparts. For example, the highest power level emitted in continuous-wave single-longitudinal-mode operation (80 mW) was achieved in real-index–guided arrays.

Modal control. One of the basic problems of monolithic phase-locked arrays is that they usually do not achieve single-supermode operation. As a result, the far-field radiation pattern is degraded from the ideal (diffraction-limited) case, and unstable characteristics may also develop. One of the reasons for this situation is the fact that in most arrays current is supplied through a common electrode. When the current is thus applied, its distribution among the lasers is usually not consistent with the requirements for single-supermode operation. This problem

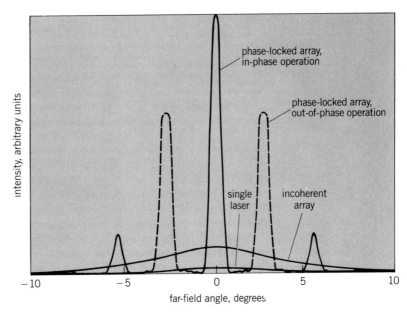

Fig. 3. Far-field radiation patterns of a single laser, an incoherent array of 10 lasers, a 10-element phase-locked array in the in-phase operation, and a 10-element phase-locked array in the out-of-phase operation.

can be rectified by providing the currents through more than one contact. An array with a separate contact to each laser was demonstrated in 1983, and by an appropriate tailoring of the currents in the array elements single-supermode operation was achieved. Control of the currents has also been demonstrated to allow wavelength tuning and beam scanning of the emitted radiation.

Even when the currents are controlled, undesirable supermodes (and in particular the highest-order one) can still be favored, or insufficiently discriminated against, in common array structures. Two main approaches have been attempted toward the solution of this problem. The first involves nonuniform arrays, where the laser elements have different widths (for example, linearly increasing widths from one side of the array to the other end). In such structures, the near-field envelopes of the resulting new supermodes are spatially separated to such a degree that with gain tailoring the fundamental supermode can be easily selected. The second approach involves breaking and mixing the laser stripes along the cavity length so that the conventional supermodes are no longer the eigenmodes of the array. An array combining both methods (nonuniform stripe widths and stripe interlacing along the array cavity) was demonstrated in 1985 and emitted more than 500 mW into a 2° single-lobed far-field pattern.

Technological issues. Although significant progress has been made in the area of monolithic phase-locked semiconductor laser arrays, more work is needed in order to bring these devices to a more technologically mature level from their present developmental stage. One issue that must be addressed is the packaging of the arrays. Since they operate at higher power levels, more heat is generated and hence low and uniform thermal impedance heat-sinking is critical. Higher output power levels also imply higher current levels—hundreds of milliamperes as compared to tens of milliamperes in single lasers; thus special care must be taken in considerations of bonding materials. Modal control is still an open issue, and more advanced device designs will probably be needed in applications requiring single-supermode operation.

Recent developments of lasers that emit their light perpendicular to the substrate plane (either surface-emitting lasers or regular lasers with specially etched mirrors) make it possible to consider the fabrication of two-dimensional monolithic arrays. While it seems feasible to implement incoherent arrays, prospects for on-chip two-dimensional phase-locked arrays (which can increase beam directivity in two dimensions) still seem very remote, mainly due to heat dissipation problems.

External cavity configuration. As mentioned above, there are alternatives to the monolithic implementations of phase-locked arrays. In one possible method, demonstrated in 1985 (Fig. 4), the discrete array elements have their front facet antireflection-coated (so that they operate only as gain elements, not as lasers), and are placed in a common laser cavity. A focusing lens is placed in front of the laser array, followed by a spatial filter in its focal plane, where the far-field pattern of the array is generated. The filter is designed to match the far-field pattern of the array when all the lasers are phase-

locked, and thus its effects are minimal in this case. However, when the array tries to operate incoherently, its far-field pattern is significantly different (Fig. 3), and thus the spatial filter presents large intracavity loss to that mode of operation. Implementation of phase-locked array operation in external cavities has several advantages: (1) two-dimensional array configurations can be easily realized; and (2) since the radiation from the lasers can be transferred into the common cavity through optical fibers, they do not necessarily have to be placed at close proximity to one another. Thus a larger number of elements can be included without generating excessive heating. Future high-power sources may include monolithic arrays as elements in external cavity arrays.

For background information *see* LASER; OPTICAL COMMUNICATIONS in the McGraw-Hill Encyclopedia of Science and Technology.

[JOSEPH KATZ]

Bibliography: C. Lin (ed.), *Optoelectronics for the Information Age: Lightwave Technology Based on Optical Fibers and Semiconductor Lasers*, 1987; T. C. Marshall, *Free-Electron Lasers*, 1984; D. R. Scifres et al., Phase-locked semiconductor laser array, *Appl. Phys. Lett.*, 33:1015–1017, 1978; D. F. Welch et al., High-power (575 mW) single-lobed emission from a phased array laser, *Electron. Lett.*, 21:603–605, 1985; A. Yariv, *Optical Electronics*, 3d ed., 1985.

Laser photobiology

Lasers have taken on a major role in medicine over the last few years. There are procedures where their use is preferred or complementary to conventional treatment, and new procedures using laser techniques have emerged. With more highly developed lasers and the ability to deliver laser radiation along flexible optical fibers, the scale of application in medicine and surgery has grown rapidly. At the same time a greater understanding has been built up of the mechanisms of the interaction of laser radiation with bodily tissue. The extensive range of current research and clinical activity of lasers in medicine includes further developments in ophthalmology and surgery and significant new applications in dermatology, endoscopy, cancer therapy, and cardiology. Also, laser techniques are valuable aids in diagnosis and medical research.

The intense radiation from lasers can be focused to very small spot sizes, a few times the wavelength of the radiation, to produce extremely high power levels at well-defined tissue sites. Radiation from lasers which operate in the wavelength region from 200 to 2000 nanometers can be transmitted in optical fibers; this includes all those in the table except the carbon dioxide (CO_2) laser. The ability to transmit radiation along flexible optical fibers of small diameter means that the radiation can be delivered to internal body sites with the use of endoscopes. Q-switched laser pulses have durations of about 20 nanoseconds, but only medium-energy Q-switched

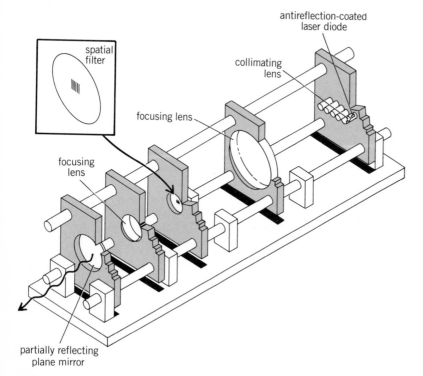

Fig. 4. Configuration of a phase-locked laser array in an external cavity configuration. The collimating lens is a short section of a special optical fiber attached to the laser diode. (*After R. H. Rediker, R. P. Schloss, and L. J. van Ruyven, Operation of individual diode lasers as a coherent ensemble controlled by a spatial filter within an external cavity, Appl. Phys. Lett., 46:133–135, 1985*)

Characteristics of lasers used in medicine and surgery

Laser medium	Spectral region and wavelength, nm	Operation	Power, W	Pulse energy, J	Pulse duration, s
Dye solution	Ultraviolet to infrared, tunable, 340–1100	Pulsed	0.5–20	0.1–5	10^{-6} to 10^{-5}
		Continuous	0.02–5	—	—
Argon	Visible, 488 and 515	Continuous	1–20	—	—
Neodymium–yttrium-aluminum-garnet (Nd-YAG)	Near-infrared, 1064	Pulsed	1–100	0.1–5	10^{-8} or 10^{-3}
		Continuous	1–50	—	—
Carbon dioxide (CO_2)	Mid-infrared, 10,600	Continuous	5–100	—	—
Excimer	Ultraviolet, 193–351	Pulsed	1–20	0.01–0.2	10^{-8} to 10^{-7}
Helium-neon (He-Ne)*	Visible, 633 or 543	Continuous	$1–5 \times 10^{-3}$	—	—

*Used mainly for alignment and target identification purposes.

pulses can now be transmitted in optical fibers without scatter loss or damage to the fiber core.

Tissue interaction. While the interaction mechanisms of laser radiation with bodily tissue are known in outline, a detailed understanding has not yet been developed. The dominant mechanism is a thermal process following absorption of the laser radiation. The mid-infrared 10.6-micrometer beam of the carbon dioxide laser is strongly absorbed by the high water content in cellular tissue, and as the water is heated and vaporized the tissue disrupts. This is characterized by an interaction distance of less than 1 mm for the carbon dioxide laser (see illus.), and typically 6 mm for the neodymium–yttrium-aluminum-garnet (YAG) laser since the absorption by water at the neodymium-YAG laser wavelength is low and the beam is mainly absorbed in pigmented tissue. The argon laser beam is absorbed by hemoglobin and pigmented tissue and has an intermediate characteristic absorption distance. The effect of the carbon dioxide laser is highly localized, and this leads to the use of this laser for surgical cutting.

The thermal contraction which initially occurs causes sealing of small blood vessels and coagulation with a reduction in hemorrhage. Coagulation with the carbon dioxide laser is mainly by conduction from the path of the laser beam, and the largest size of blood vessel which can be sealed is limited to about 1 mm. The larger interaction volume of the neodymium-YAG laser enables larger blood vessels to be sealed. To gain the advantages of good excision with the carbon dioxide laser and coagulation with the neodymium-YAG laser, recent investigations have been made on arrangements incorporating both lasers.

Tissue interaction with higher-power pulsed lasers is by various mechanisms. The output of a pulsed laser can be of durations 10^{-3}, 10^{-6}, or 10^{-8} s or even shorter. The high instantaneous power of the beam when focused on the tissue produces high electrical fields and electrical breakdown and sufficiently high temperatures to form a plasma, causing disruption of the surface. Excimer lasers produce ultraviolet radiation in the range 193–351 nm which is strongly absorbed by biological tissue, and with a laser pulse of 10^{-7} s the surface can be photoablated. Since the focal spot size of the excimer laser is a few micrometers, a very small and well-defined region can be selectively destroyed.

Laser surgery. The strong interaction of carbon dioxide laser radiation with tissue enables it to be used for tissue cutting. At the wavelength of the carbon dioxide laser, the beam is absorbed in the silica core of normal optical fibers, although more recently work has been undertaken on developing special hollow reflecting cores or infrared transmitting fiber optic systems. The usual method of transmitting the carbon dioxide laser beam has been to use an articulated arm with mirrors to steer the beam or to use the beam directly from the laser. The carbon dioxide laser can deliver high continuous powers up to 100 W or more and can be focused to spot sizes of about 0.1 mm. The laser can be used as a cutting tool for incision or to remove diseased tissue in a precise manner. It is now the standard treatment for removal of early cancers of the cervix where major surgery can be avoided so that treatment can be carried out without anesthetic. The carbon dioxide laser is increasingly used to excise polyps, to remove warts and minor skin lesions, and to open constricted regions. The precise cutting offered by the laser en-

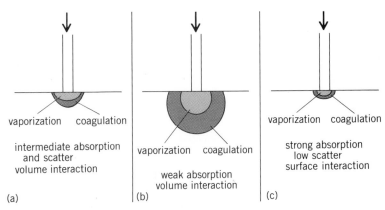

Interaction of lasers used in medicine with bodily tissue. (a) Visible lasers: tunable dye (0.340–1.1 μm) and argon (0.488, 0.515 μm). (b) Near-infrared laser: neodymium–yttrium-aluminum-garnet (Nd-YAG; 1.06 μm). (c) Mid-infrared laser: carbon dioxide (CO_2; 10.6 μm).

ables tissues in the brain and spinal cord to be removed without damaging neighboring sites. The use of an operating microscope in conjunction with the laser provides accurate viewing and manipulation of the focused beam; regions such as the trachea and larynx can be accessed and precisely treated.

Dermatology. A notable application of lasers in dermatology has been in the alleviation of the condition known as portwine stains. These can be extremely disfiguring and often occur in prominent areas such as the face. The stain is formed by an abnormally large number of small blood vessels just below the surface of the skin leading to red coloring of varying severity. Laser therapy can be used to seal the vessels while keeping the epidermis intact. The laser radiation is normally delivered by optical fiber to a hand-held probe situated about 0.4 in. (1 cm) above the skin surface.

Photodynamic cancer therapy. A highly promising but still experimental procedure in cancer therapy is the sensitizing of tissues with drugs and then selectively destroying the tissue by laser radiation. The special value of this procedure is that certain drugs are preferentially retained in malignant tissue rather than normal tissue. Irradiation into the appropriate absorption band excites the drug, which in turn converts tissue oxygen into a highly active form, singlet oxygen, which attacks and oxidizes the malignant tissue. Since more radiation is absorbed by the malignant tissue, the destruction is significantly greater than for normal tissue.

The system on which most research and clinical application has been carried out involves the drug hemataporphyrin derivative (hpd), extracted from blood, which is irradiated by a dye laser. Following intravenous injection, a period of 2 to 3 days is allowed to elapse, by which time the drug has been selectively taken up by the malignant tissue. The technique has been successfully applied to a wide variety of tumor sites.

Ophthalmology. One of the first medical applications of lasers was in ophthalmology, where a detached part of the retina could be reattached by laser heating. A further application has been in the treatment of diabetic retinopathy, in which the growth of vessels masking the retina causes progressive blindness; these may be removed by argon or tunable dye lasers.

In the condition of glaucoma, drainage of fluid from the anterior chamber of the eye is blocked, causing an increase in pressure in the eye which can impair vision. Internal drainage can be induced by forming small holes in the iris with laser radiation to provide alternative drainage channels. Recently investigation has begun into reshaping the cornea surface with excimer lasers by careful ablation of the surface; in this way the need for strong spectacle lenses to overcome major refractive defects may be avoided.

Endoscopy. Internal sites in the body can be accessed by the use of endoscopes, which may be rigid or flexible and can carry fiber-optic cables for illumination, viewing, and transmission of laser power. An early application of this technique was the treatment of hemorrhage from peptic ulcers where photocoagulation of vessels in the ulcer can be achieved. More recent work has involved treating various tumors in the gastrointestinal tract, lungs, and bladder. Palliative relief can be achieved in conditions which are too advanced for surgery to be considered; blocked airways at the entrance to the lungs or esophagus can be sufficiently cleared to alleviate the condition. Recent investigations have been made on the fragmentation of kidney stones with pulsed laser radiation. A possible mechanism for this involves the laser pulse driving an acoustic wave from the surface so that the mechanical disturbance breaks up the stone.

Diagnosis. In diagnosis, laser-induced fluorescence is used in the analysis and sorting of cells by the technique of flow cytometry. Cells to be sorted are passed singly in a narrow flow across a laser beam whose wavelength is chosen to excite the characteristic fluorescence from the cell, and this fluorescence is detected. Conventionally with a microscope up to 200 cells per minute may be observed to look for an abnormality; in the same time the laser flow cytometer can interrogate a thousand times as many. Fluorescent organic probes which attach to antibodies can be detected, and other applications are in detecting and diagnosing tumor cells and leukemia, studying genetic characteristics of the fetus, sorting and analyzing chromosomes for genetic information, and testing for viruses.

[T. A. KING]

Cardiovascular system. Laser application to the cardiovascular system shows great promise for the treatment of several cardiac and vascular abnormalities. The preliminary work over the last several years has clearly demonstrated the laser's potential. While each of the techniques discussed below is still in the investigational stage, basic and animal laboratory research is proceeding at a rapid pace, and clinical trials are beginning.

Recanalization of occluded arteries. Atherosclerotic heart disease, due to fatty material buildup in coronary arteries, is the leading cause of death in the United States. The atherosclerotic plaque can be efficiently vaporized by laser thermal energy, suggesting the possibility that partially or completely occluded arteries could be opened, reestablishing blood flow to the heart. A limitation of the process is that laser energy can easily burn through the wall of the artery, and the safety of the vaporization process is still in question. Several new techniques offer promise for improved safety.

The lasers which are commercially available, primarily the carbon dioxide, argon, and neodymium-YAG lasers, produce vaporization by a thermal mechanism. A recent refinement to the neodymium-YAG laser has, however, made it possible to deliver pulsed energy at low repetition rates. This eliminates the pathologic injury to surrounding healthy tissue which has been observed following angioplasty with traditional lasers. Newer lasers in the ultraviolet range of the electromagnetic spectrum, such as the excimer laser, remove the atherosclerotic material by a nonthermal mechanism. Lasers which eliminate the thermal mechanism may provide more precise control of the injury and limit the damage which occurs to the wall of the artery adjacent to or underlying the plaque.

Another promising technique is the use of photosensitizing materials. Examples of these materials are the hematoporphyrins, which have been successfully employed in tumor photosensitization for photoablation, as discussed above. These materials, when injected intravenously, accumulate in the plaque material to a greater degree than into the adjacent normal arterial wall. The photosensitizing material causes enhanced absorption of the laser energy. Low-level laser energy, insufficient to injure the normal, nonsensitized arterial wall, is absorbed by the photosensitizing agent to a degree sufficient to cause injury to the plaque.

It is probable that such laser recanalization can be accomplished by using flexible laser fibers and cardiac catheterization techniques, avoiding the surgical intervention required with coronary bypass procedures. The availability of very thin, flexible fiber-optic endoscopes may allow endovascular visualization and visual control of the laser injury, further enhancing the safety of the procedure.

Control of arrhythmias. Disturbances of the heart rhythm, arrhythmias, are occasionally resistant to medications. Laser injury to the endocardium of the heart is being evaluated as a method for destruction of the arrhythmia source. The current investigations utilize laser energy applied during open heart surgery and also through flexible laser fibers inserted into the heart via cardiac catheters. Preliminary studies indicate that laser injury can interrupt arrhythmias and may do so with less overall damage to the heart than that produced by other ablative procedures. This technique may be particularly useful for the control of resistant ventricular tachycardia, a potentially lethal rhythm disturbance. It has also been shown that laser energy can interrupt electrical conduction through the atrial-ventricular junction of the heart, a technique which may be useful in certain patients with intractable supraventricular arrhythmias.

Myocardial and valvular laser surgery. Laser energy vaporizes myocardial and valvular tissue as well. The device may be useful for the removal of excess or pathological myocardial tissue, such as in hypertrophic cardiomyopathy, or for relief of stenotic heart valves, either congenital in origin, or those caused by rheumatic fever. The advantage of the laser over currently used scalpel techniques may be greater control and precision.

Vascular welding and endarterectomy. Laser energy can be used to "weld" together the ends of cut vessels. This may offer an advantage over standard suture anastomosis in very small vessels and may avoid suture-induced foreign body reactions. The absence of suture at the anastomosis may reduce the thrombogenicity of the anastomosis, and in some hands vascular welding is faster than suture technique.

Endarterectomy of atherosclerotic-occluded vessels may be enhanced by performing the dissection by laser rather than by scalpel. The laser has the advantage of heat-sealing the wall during dissection, resulting in a smoother surface with less likelihood of an intimal flap or postdissection thrombus.

For background information *see* ARTERIOSCLEROSIS; CARDIOVASCULAR SYSTEM DISORDERS; HEART DISORDERS; LASER; LASER PHOTOBIOLOGY; OPTICAL FIBERS in the McGraw-Hill Encyclopedia of Science and Technology.

[G. MICHAEL VINCENT]

Bibliography: A. H. Andrews and T. G. Polyani (eds.), *Microscopic and Endoscopic Surgery with the CO₂ Laser*, 1984; L. I. Deckelbaum et al., Reduction of laser-induced tissue injury using pulsed energy delivery, *Amer. J. Cardiol.*, 56:662–667, 1985; J. A. Dixon (ed.), *Surgical Applications of Lasers*, 1983; J. M. Isner et al., Laser myoplasty for hypertrophic cardiomyopathy: In vitro experience in human postmortem hearts and in vitro experience in a canine model (transarterial) and human patient (intraoperative), *Amer. J. Cardiol.*, 53(11):1620–1625, 1984; S. N. Joffe, M. Muckerheide, and L. Goldman, *Neodymium YAG Laser in Medicine and Surgery*, 1983; Lasers in biology and medicine, special issue, *IEEE J. Quant. Elec.*, vol. QE-20, December 1984; J. R. Spears et al., Fluorescence of experimental atheromatous plaque with hematoporphyrin derivative, *J. Clin. Invest.*, 71:395–399, 1983; G. M. Vincent et al., Neodymium YAG laser ablation of simulated ventricular tachycardia in a canine model, *Lasers Surg. Med.*, 5(2):168, 1985

Lichens

Lichens, which are associations of algae and fungi, have long been of interest to biologists, both because of their frequent occurrence in a wide variety of habitats, and because of the unique nature of the symbiosis between the alga and the fungus. Recent advances in research on lichens have led to a broadening of the understanding of the general characteristics of the group, and to new insights into the physiological basis of the central symbiosis. Many scientists now believe that the fungal-algal relationship is not mutual, but rather the result of active parasitization of the alga by the fungus.

GENERAL CHARACTERISTICS

The fungal component (mycobiont) of a lichen is an ascomycete or, in a few tropical genera, a basidiomycete. These particular fungi do not occur free in nature, only in the lichenized state. The algal component (phycobiont) is most often the green unicell *Trebouxia* (Chlorococcales), but tropical lichens often contain filamentous *Trentepohlia*. The gelatinous lichens (Collemataceae) contain the cyanobacterium (blue-green alga) *Nostoc*. When both

Fig. 1. A foliose lichen, *Parmotrema austrosinense*.

alga and fungus grow together, they form a distinct, long-lived plant body (thallus), usually classified as one of three growth forms: crustose (closely attached to and often penetrating the substrate), foliose (leaf-like with an upper and lower surface; Fig. 1), and fruticose (cylindrical or shrubby, free-growing or basally attached; Fig. 2). Most lichens are light mineral gray or yellowish green, but a few are brilliantly pigmented orange or red.

The fungus makes up the vast bulk of the thallus and is differentiated into various kinds of tissues, including a dense upper cortex, a thick loosely arranged medulla, and in most foliose lichens, a lower cortex from which rhizines or other attachment organs project. The algae occur in a thin layer between the upper cortex and medulla. Recent ultrastructural studies have shown surprising complexity in the upper cortex of some major foliose groups. A thin palisade cortex is overlain by a pored epicortex about 0.6 micrometer thick with pores 15–30 micrometers in diameter (Fig. 3). This structure appears to assist in gas exchange through an otherwise impermeable upper cortex.

Lichens produce several unique vegetative structures. Soredia are microscopic bodies about 50 μm in diameter consisting of a few algal cells sur-

rounded by fungal hyphae. They are easily dislodged from the thallus and become airborne. Isidia are fingerlike protuberances of the upper cortex with inclusions of algal and fungal medullary tissue. A few families, especially Lobariaceae and Peltigeraceae, produce internal or external cephalodia, which are small vegetative bodies (1–2 mm wide) that contain blue-green algae different from the green algae of the host lichen. They have a role in nitrogen fixation.

Lichens produce nearly 300 different organic substances, the vast majority of which are unique phenolic secondary metabolites, which make up 1–5% of the dry weight of the thallus. The role of these secondary compounds is unknown, although some have proven antibiotic properties and discourage predation by invertebrates. The most common chemical groups are depsides and depsidones (aromatic esters with two benzene rings), but anthraquinones, aliphatic acids, dibenzofurans, and triterpenoids are also well represented. These substances are extremely important to taxonomists in interpreting the limits and evolution of species.

Growth and reproduction. Lichens are perennial plants with very slow growth rates. In temperate zones, foliose lichens may grow 2–6 mm/year and a well developed thallus may be 25–50 years old. Crustose lichens grow 0.1–2 mm/year and may be 100 years old. In arctic regions, some crustose lichen colonies are estimated to be as much as 9000 years old on the basis of extrapolation from size of colonies on dated moraines.

Very little is known about the reproduction of lichens. Lichen apothecia are essentially identical with those of related nonlichenized fungi. The large majority are discomycetes, but many pyrenomycetous types are found. Both ascohymenial and ascolocular fruiting structures and unitunicate and bitunicate ascal types are known in lichens. Sexual stages leading to formation of spores by meiosis in the asci are assumed to be similar to those in nonlichenized fungi, although very few studies have been made. In theory a new lichen could be formed when spores are released from the asci, germinate, and come into contact with a suitable symbiotic alga. It is not known how often this happens in nature, but for many crustose lichens it appears to be the only possible method of reproduction.

Vegetative reproduction is the main form of propagation among foliose and fruticose lichens. Soredia, isidia, and other diaspores that contain both fungal and algal elements function as vegetative propagules and are transported over fairly long distances by wind, water, and animals.

Classification. About 14,000 lichen species distributed in 600 genera are known. Lichens are now considered to be fungi, but they were long maintained as a class (Ascolichenes) separate from the closely related, nonlichenized fungi. Since the 1970s a number of different classification systems for lichens (and fungi) have been proposed which arrange lichen orders within the framework of fungal

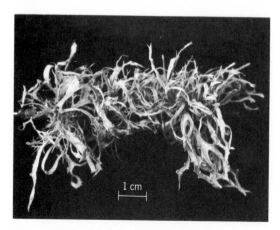

Fig. 2. A fruticose lichen, *Ramalina leptocarpha*.

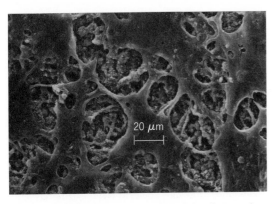

Fig. 3. Scanning electron view of the epicorticate surface of *Parmeliopsis aleurites*.

orders under class Ascomycotina and class Basidiomycotina as follows:

> Class Ascomycotina
> Order: Lecanorales
> Gyalectales
> Pertusariales
> Peltigerales
> Graphidales
> Lecanidiales
> Arthoniales
> Opegraphales
> Verrucariales
> Pyrenulales
> Dothidiales
> Caliciales
> Class Basidiomycotina
> Order: Aphyllophorales
> Agaricales

Over 50 families of lichens are recognized, but lichenologists are not in general agreement on their delimitation. Most research in lichen systematics is now directed toward revising the traditional genera into more natural, homogeneous groups, using ascal and hymenial characters, vegetative structures, and chemistry.

The origin and geological age of lichens are unknown since there is no fossil record. Continental distribution patterns suggest at least a Gondwanaland origin for many groups. The success of lichenization as a biological adaptation is attested by the fact that more than one-third of the species of ascomycete fungi are lichens.

Ecology and adaptations. Lichens are among the most widespread plant groups. There are several lichen families (Lichinaceae, Verrucariaceae) which occur in the intertidal zone on rocky shorelines. The reindeer mosses (*Cetraria*, *Cladina*, and *Cladonia*) carpet huge areas of the forest floor in subarctic regions. Boreal, temperate, and tropical forest have rich lichen communities from the base of the trunks to the canopy. In some northern coniferous forests the lichens with blue-green algal phycobionts actually contribute significant amounts of fixed nitro-

gen to the forest ecosystem. Because of their unique physiological traits, especially slow growth and low turnover of metabolites, lichens are well adapted to desert conditions, where they grow on rocks and soil and remain in a state of desiccation for long periods or utilize dew as a source of water. In Antarctica, endolithic lichens occupy a unique niche in sandstone rocks, a 1-cm-thick (0.4-in.) layer just under the rock surface, protected from the harsh environment where virtually no organisms can survive in the open.

Lichens are important as a food source for many insects and invertebrates, especially snails and slugs. Reindeer in Scandinavia feed on fruticose lichens (*Cladina*) as a standing crop in pastures, which must be carefully rotated to prevent overgrazing. In winter, caribou in arctic Canada subsist mostly on lichens. Humans' use of lichens as food is now largely historical, limited by the slow growth and regeneration, although rock tripe (*Umbilicaria*) is eaten locally in Japan as a delicacy.

Lichens have some use as sources of antibiotics in medicine, and usnic acid, derived from lichens, is used in salves for treatment of abrasions and burns. A few polysaccharides derived from lichens have shown antitumor activity.

The most important lichen products are the essential oils extracted from "oak mosses" (*Evernia* and *Pseudevernia* species) harvested in the Balkan countries. The oils are used for compounding in finished retail perfumery. Lichens were commonly used as dyestuffs before the commercial development of coal-tar dyes, but they are now of interest only to textile artists.

Growth of lichens on archeological ruins and other monuments is known to have detrimental effects. Damage is caused both by chelation by leached lichen substances and mechanical action by hyphae penetrating between rock crystals. Stone surfaces in ancient temple sites throughout the tropics, tombstones in northern countries, and even stained glass in church windows are affected. Various mild fungicides and algicides can be used as control measures.　　　　　　　[MASON E. HALE]

PHYSIOLOGY

The functional relationship between fungal and algal symbionts of lichens has been a subject of much speculation and study. A popular view of lichens is that they are examples of mutualism, a type of symbiosis in which both partners benefit. The long-lived nature of these associations and their ability to grow in extreme environments support this view. However, recent studies have resurrected earlier views that lichens are fungi that parasitize algae. The parasitism is gradual and controlled by the fungus and does not result in death of the lichen. Algal cells that are killed by the fungus are replaced by division of other cells; in this way a stable algal population is maintained.

Nutrient flow between symbionts. Inside a lichen, the fungus causes the algal cells, by means of

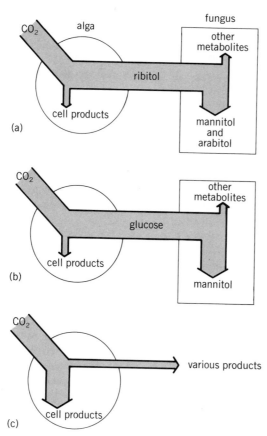

(a)

(b)

(c)

Fig. 4. Movement of photosynthetic products from photobiont to mycobiont (fungus), and in isolated photobionts. (a) *Trebouxia* (algal) photobiont. (b) *Nostoc* (blue-green algal, or cyanobacterial) photobiont. (c) Isolated *Trebouxia* or *Nostoc* photobiont. (*After D. C. Smith, What can lichens tell us about real fungi?, Mycologia, 70:915–934, 1978*)

a mechanism that is not understood, to release up to 90% of the carbon compounds they fix in photosynthesis. Green algal symbionts release polyols such as ribitol, erythritol, or sorbitol, while blue-green (cyanobacterial) photobionts (photosynthetic partners) release glucose. The compounds move rapidly from the alga to the fungus, in some cases within 90 s after carbon is fixed photosynthetically. When photobionts are removed from a thallus and cultured separately, they release only small amounts of compounds, none of which are polyols or glucose (Fig. 4).

Physiological buffering. Compounds released by the photobiont are absorbed by the fungus and slowly converted to polyols such as mannitol or arabitol, which are used for growth and development and for coping with environmental stresses. When dry lichens are wetted, they respire at much higher levels than normal for several hours. During this period of resaturation respiration, the lichen respires the stored polyols. According to some scientists, the polyols serve as a physiological buffer for proteins and other structural compounds whose concentrations remain constant while those of the polyols rise and fall according to the wetting and drying of the lichen.

Urease theory. The urease theory was developed to explain how the fungus regulates the flow of nutrients from the alga. Urease is a common enzyme in lichens. According to this theory, the fungus produces urease during periods of its active growth. The enzyme breaks down urea, which is normally present in the lichen, to carbon dioxide and ammonia. The carbon dixoide stimulates photosynthesis of the alga, while the ammonia increases algal respiration and carbohydrate breakdown and release; in this way, the fungus obtains the energy compounds it needs to grow. The fungus controls the release of compounds from the algae by synthesizing secondary compounds, such as usnic acid, which inactivate urease and thus slow the rate of photosynthesis.

External nutrients. How much and what kinds of nutrients a lichen obtains from its external environment is not known. The fungal partner has a remarkable ability to absorb minerals and organic compounds from rainwater that passes over the lichen thallus. Some of these compounds may be used by the algal symbiont as nutrients. This possibility is strengthened by laboratory studies which have shown that *Trebouxia*, the most common algal symbiont of lichens, grows much better in media that contain organic compounds. As with polyols, however, these nutrients are converted slowly into other compounds. The slow utilization of nutrients by lichens is consistent with their limited growth and development. The rapid absorption of compounds by lichens compensates for their generally short periods of time during which they are wet and, therefore, metabolically active. Lichens are similar to sponges in that they have little control over the water content of their thalli and dry quickly after being wetted.

Nitrogen fixation. Lichens which have blue-green bacteria as symbionts fix atmospheric nitrogen into ammonia. In a lichen, the fungus inhibits the nitrogen-assimilating enzymes of the photobiont, thereby causing most of the fixed ammonia to be released (Fig. 5). The ammonia is then taken up by the fungus and used to synthesize proteins and nucleic acids. Only about 8% of the known species of lichens contain nitrogen-fixing symbionts, but in some parts of the world, such as tropical rainforests and evergreen forests, these lichens contribute valuable nitrogen compounds to the soil when they decay.

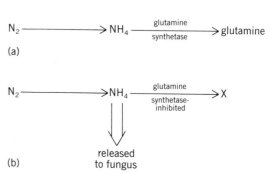

Fig. 5. Nitrogen assimilation pathways in the *Nostoc* lichen photobionts. (a) *Nostoc* in isolation. (b) *Nostoc* in lichen. (*After D. C. Smith, What can lichens tell us about real fungi?, Mycologia, 70:915–934, 1978*)

Physiological acclimation. Lichens adjust their basic physiological processes to the environmental conditions under which they are growing. This ability to adjust operates under different combinations of light, temperature, day length, and thallus water content. In effect, a lichen monitors the external conditions of its habitat and changes accordingly. For example, the lichen *Xanthoria fallax* growing in dry habitats assimilates CO_2 maximally when its thallus water content is about 40% that of saturation; in wet habitats it has optimal rates of CO_2 assimilation at about 50–70% of thallus saturation. Some woodland species of *Peltigera* adjust rapidly to different light intensities and can photosynthesize optimally in the presence or absence of the forest leaf canopy. The same lichens can also acclimate their photosynthetic rates to winter and summer temperatures.

Synthetic lichens. Several lichens, such as *Cladonia cristatella* (British soldiers) and *Usnea strigosa* (fog lichen), have been synthesized in the laboratory, starting with their separate fungal and algal symbionts. Hybrid forms of these lichens have also been created. Synthetic lichens have been produced under axenic conditions (that is, without the presence of other microorganisms). Thus, the often-mentioned possibility that bacteria may be a third symbiont in a lichen association seems not to be the case. Most importantly, laboratory synthesis of lichens has led to advances in the understanding of fungal transformation, and of the production of secondary metabolites.

Fungal transformation. The initial stages of lichen synthesis begin when fungal hyphae encircle and bind to algal cells (Fig. 6). Some investigators believe that a recognition process occurs at this stage and, as in other symbiotic systems, proteins called lectins are involved in binding together specific symbionts. After the initial contacts, the fungus produces tissuelike structures around the algal cells and forms a thallus. In foliose lichens the thallus is leaflike and the algal cells are spread out like the chloroplasts of a leaf. Other types of thalli are crustlike, upright, or filamentous.

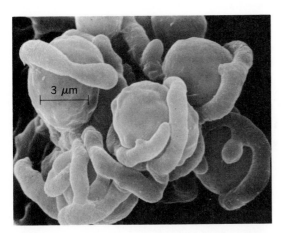

Fig. 6. Scanning electron micrograph of fungal hyphae enveloping algal cells in early stages of synthesis of the lichen *Cladonia cristatella*.

Although the physiological basis of lichen synthesis is not known, it is clear that the stimulus that causes the fungus to form a lichen thallus comes from the alga. Whether such a stimulus is the same for all lichens is not known. There may be different types of stimulatory factors for lichen fungi since one species of fungus responds to only a few kinds of algae. The possibility of gene transfer between lichen symbionts is being considered.

Secondary compounds. Lichens produce many unique chemical compounds, such as depsides, depsidones, and dibenzofurans. Some of these compounds are produced by the fungus, alone in culture, but most are products of the symbiosis. A current hypothesis is that the algae synthesizes a compound that inhibits the fungal enzyme orsellinic acid decarboxylase. This enzyme, in free-living fungi, is involved in the production of a variety of products, including quinones, which are toxic to algae. With the enzyme inhibited in lichens, however, different chemical compounds, generally called lichen acids, are produced, which are harmless to the algal symbionts. The enzyme inhibitor made by the algae appears to be a general one, since one fungus forms similar secondary compounds with a variety of unrelated algae.

A lichen fungus receives two unrelated messages from its algal symbiont. One message, perhaps a genetic one, stimulates the fungus to produce a particular type of thallus. This message is a specific one between a fungus and a particular type of alga. The other message is a more general one, and it triggers the formation of secondary compounds by the fungus. Finding the identity of these messages will be the object of future studies in lichen physiology.

For background information *see* ALGAE; CYANO-PHYCEAE; ECOLOGICAL INTERACTIONS; FUNGI; LI-CHENES in the McGraw-Hill Encyclopedia of Science and Technology. [VERNON AHMADJIAN]

Bibliography: V. Ahmadjian and M. E. Hale, *The Lichens*, 1973; L. J. Goff (ed.), *Algal Symbiosis*, 1983; M. E. Hale, *The Biology of Lichens*, 1983; J. D. Lawrey, *Biology of Lichenized Fungi*, 1985; H. F. Linskens and J. Heslop-Harrison (eds.), *Encyclopedia of Plant Physiology*, new series, vol. 17: *Cellular Interactions*, 1984; F. E. Round and D. J. Chapman (eds.), *Progress in Phycological Research*, vol. 1, 1982.

Liquid crystals

Liquid-crystalline polymers form a new class of materials which possess unique chemical and physical properties. Certain liquid-crystalline polymers can be highly oriented to produce materials with exceptional tensile properties and in some cases unique optical properties. Although many of the structural characterisitcs of liquid-crystalline polymers are identical to those found for small-molecule liquid-crystalline materials, there are several unique features. Especially important are the relatively high polymer thermal transitions which permit freezing a liquid-crystalline phase for application over a wide range of temperatures.

Liquid-crystalline polymer properties. The liquid-crystalline state is a unique condition in which long-range molecular orientational order persists in the absence of various types of short-range translational order. Liquid-crystalline phases can be formed by dissolution of liquid-crystalline material in a solvent (lyotropic behavior) or by melting a liquid-crystalline material (thermotropic behavior). These phases are divided into three broad categories: nematic, cholesteric, and smectic. Briefly, the nematic phase consists of parallel molecules with no positional order along the molecular long axis and poorly defined positional order perpendicular to that axis. The cholesteric phase is a twisted nematic in which the order parameter (pointed along a molecular axis) exhibits a helical twist with a pitch on the order of 0.4 micrometer to several micrometers. The smectic phase is a layered structure consisting of sheets of molecules with orientational order and translational order between the smectic layers. Many different types of smectics exist displaying varying degrees of order both within an individual smectic layer and between layers.

Specific polymers can be made which display each of the liquid-crystalline phases described above. In addition, polymers bring the concept of a glass transition to the field of liquid crystallinity. For polymers, the glass transition temperature is the point below which macromolecular motion is frozen. This has important implications for tensile and optical properties. By designing the primary structure of a polymer, a variety of desirable characteristics may be obtained for use over a preselected temperature range. The earliest and most easily attained of these characteristics is fiber tensile strength and stiffness.

The highly oriented nature of liquid-crystalline polymers produces highly anisotropic physical properties. For example, the tensile modulus (that is, stiffness in the direction of orientation) of the commercially available lyotropic Kevlar fiber is in the range of 9–17 Mpsi (million pounds per square inch; 60–120 gigapascals) depending on the process conditions, while that of conventional nylon or polyethylene terephthalate (PET) fiber is only about 0.9 and 1.8 Mpsi (6 and 12 GPa). Similarly, the tensile modulus of an injection-molded thermotropic polyester formed from hydroxybenzoic acid and hydroxynaphthoic acid is at least three times that of a conventional isotropic polyester such as PET (2.5 Mpsi versus 0.5 Mpsi, or 17 GPa versus 3.4 GPa), and the flexural strength of the former is twice that of the latter. Such impressive properties are due to the nearly rigid and linear units which compose the polymer chain. These units are also responsible for inducing the mesomorphic or liquid-crystalline phase, and are therefore termed mesogenic units or mesogens.

Liquid-crystalline polymer phases also possess attractive optical properties. For example, smectic liquid-crystalline polymers are able to form optically clear, uniformly aligned samples which are free of

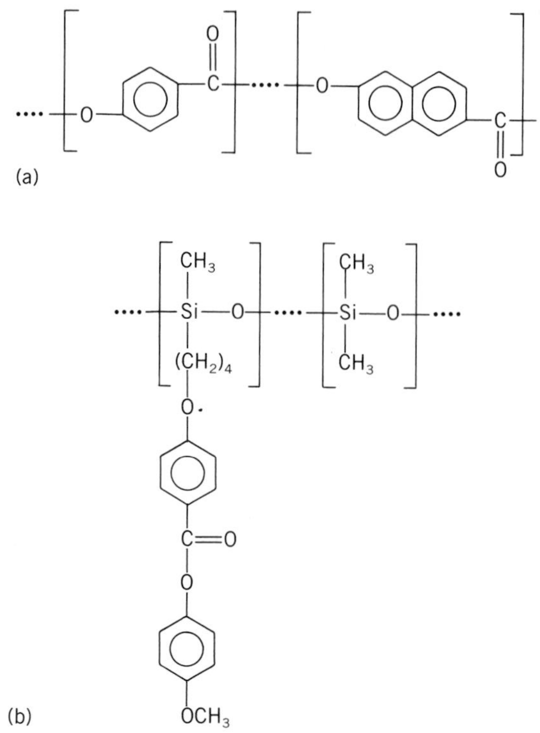

Fig. 1. Chemical structures of typical liquid-crystalline polymers. (a) Main-chain liquid-crystalline polymer. (b) Side-chain liquid-crystalline polymer.

defects. By selectively heating a region of the material with a laser beam, the material may be reversibly disordered to form an optically detectable turbid microdomain of low reflectivity. Such a material therefore has potential use for the optical storage of information.

Classification. In general, there are two broad classes of molecular architectures for liquid-crystalline polymers—namely, main-chain liquid-crystalline polymers and side-chain liquid-crystalline polymers. In the main-chain liquid-crystalline polymers the mesogenic or rigid segments form the backbone of the molecular chains, whereas in the side-chain liquid-crystalline polymers the mesogenic or rigid segments are pendant to a flexible polymer chain to form a comblike structure (Fig. 1). Lyotropic and thermotropic examples exhibiting nematic, smectic, or cholesteric phases are known for both architectures. Flexible spacers (for example, alkyl chains) between mesogenic groups in main-chain liquid-crystalline polymers or between the backbone and mesogenic units in side-chain liquid-crystalline polymers are sometimes necessary to form a particular liquid-crystalline phase. Mesogenic units are typically rigid aromatic structures with delocalized electrons.

Synthesis. Main-chain aromatic liquid-crystalline polyesters and polyamides are polymerized through condensation reactions. For a liquid-crystalline aromatic polyester, the aryl diol monomers usually are acetylated before polymerization to increase the efficiency of polymerization. Some transesterification polymerizations start with the phenyl esters of the diacids with aryl diols, or with acetylated aryl diols at relatively high temperatures. Suitable catalysts may sometimes be used. Polymerization is most conveniently carried out in the melt, that is, in the absence of any solvent, in a temperature range of 390–680°F (200–360°C). For a liquid-crystalline polyamide, such as Kevlar, polymerization proceeds by the reaction of *p*-phenylene diamine and terephthaloyl chloride at low temperature in an appropriate solvent. The degree of polymerization is increased with the presence of tertiary amine salts.

Two approaches have been often employed to synthesize side-chain liquid-crystalline polymers. One uses reactive vinyl groups, and the other employs reactive poly(hydromethylsiloxane), which has a reactive hydrogen atom. In the first approach, polymers are made through either a radical or an ionic reaction of vinyl monomers in solution. Reaction temperatures are generally much lower than those used for main-chain thermotropic liquid-crystalline polymers. Reaction initiators may be, for example, anionic such as butyl lithium or a free radical as

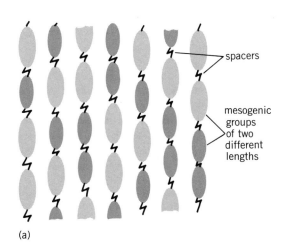

(a)

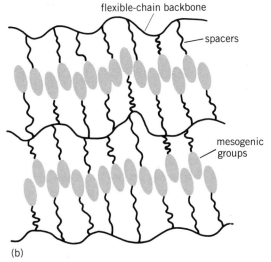

(b)

Fig. 2. Structures of liquid-crystalline polymers on a molecular length scale. (*a*) Nematic structure for main-chain liquid-crystalline polymers. (*b*) Smectic C structure for side-chain liquid-crystalline polymers.

obtained from organic peroxides or aliphatic azo compounds. The reaction of poly(hydrogen alkyl siloxanes) in solution with selected spacer units and mesogens is usually done with catalysts such as platinum and hexachloroplatinic acid.

Structure. On a micrometer length scale, the structures of liquid-crystalline polymers are the same as those found for small-molecule liquid crystals. All of the major phase types found for liquid crystals (nematic, cholesteric, and smectic) are also found for liquid-crystalline polymers. On a molecular length scale, liquid-crystalline polymers present some novel features.

For example, main-chain liquid-crystalline polymers often display a nematic structure in which long polymer molecules are packed in a parallel array (Fig. 2*a*). The individual chains are arranged on a two-dimensional lattice perpendicular to the molecular axis (for example, a hexagonal lattice with unit cell dimensions such that the interchain distance is approximately 0.45 nanometer). However, there is little or no axial registration between neighboring polymer molecules. (This is as if one held a bundle of pencils but did not align the ends.) If the molecules were axially registered, then the material would be an ordinary three-dimensionally crystalline solid. Typically polymers with a regular repetition of monomer (that is, a homopolymer) develop axial registration. However, a random copolymer made of monomers with unequal molecular lengths cannot develop true three-dimensional crystallinity. Such a random sequence is one method to promote a nematic phase. Other methods to reduce axial registration include adding bulky substituents to the chain.

The molecular architecture of side-chain liquid-crystalline polymers is not as clearly understood. The structural arrangement of side-chain mesogens clearly follows that found in a small-molecule liquid crystal of the same phase. For example, a smectic C polymer contains sheets of mesogens in which the mesogen long axis is tilted with respect to the sheet and for which there is no regular crystalline arrangement of mesogens within the layer or between layers (Fig. 2*b*). Thus, the liquid-crystalline polymer is envisaged as forming a comblike structure with neighboring molecules. The nature and degree of ordering of the chain axis to which the mesogenic units are attached is not presently understood.

Phase transitions and composition. In order to fully use the attractive properties that liquid-crystalline polymers offer, methods have been developed to minimize processing difficulties by lowering solid to liquid-crystal transition temperatures and improving thermal stability. One approach to develop tractable liquid-crystalline polymers is to form a rigid structure from two or more monomers of different lengths which occur in random sequence so that crystallization is inhibited. This method (Fig. 3) has been successfully used to control the solid-to-nematic transition temperature (that is, the processing temperature) for a random copolyester of *p*-

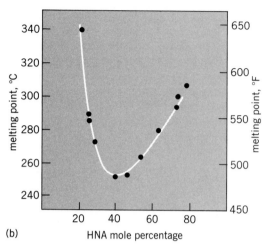

(a)

(b)

Fig. 3. Behavior of a random copolyester of *p*-hydroxybenzoic acid (HBA) and 6,2-hydroxynaphthoic acid (HNA). (*a*) Chemical structures of HBA and HNA. (*b*) Variation of the melting point (the solid-to-nematic transition temperature) with composition.

hydroxybenzoic acid (HBA) and 6,2-hydroxynaphthoic acid (HNA).

For background information *see* LIQUID CRYSTALS; POLYMER; POLYMERIZATION in the McGraw-Hill Encyclopedia of Science and Technology.

[GORDAN W. CALUNDANN; TAI-SHUNG CHUNG;
JAMES B. STAMATOFF]

Bibliography: G. W. Calundann and M. Jaffe, Anisotropic polymers, their synthesis and properties, *Proceedings of the Robert A. Welch Conference on Chemical Research, XXVI, Synthetic Polymers*, pp. 247–291, 1982; H. Finkelmann, H. Ringsdorf, and J. H. Wendorff, Model considerations and examples of enantiotropic liquid-crystalline polymers, *Makromol. Chem.*, 179: 273–276, 1978.

Magnetic field

Magnetic fields are an essential tool of experimental science. They play a particularly central role in solid-state physics, because high fields can drastically modify the motion of electrons in crystals, thereby probing their dynamics. The current understanding of the electronic properties of metals, semiconductors, and superconductors—understanding that is crucial to modern electronics—could not have been achieved without strong magnetic fields. Yet, until 1960, the highest steady fields available to all but a few scientists were those provided by iron-core electromagnets, about 3 teslas (30 kilogauss) in air gaps of a few centimeters. Much research has been performed in such magnets, but their fields are not high enough for many important experiments.

To generate fields higher than the saturation limit

of iron, it is advantageous to dispense with the iron and simply build electromagnets. The magnetic field is then directly generated, according to Maxwell's equations, by the current circulating in the winding. In a solenoid it is possible, in principle, to achieve a high field by passing a sufficiently large current through it. This approach was pioneered by Francis Bitter, who generated 10 T (100 kG) with high-power, water-cooled magnets in 1939. Practically speaking, the fields of electromagnets are limited by the extreme power required—in the megawatt range—and by the cooling system needed to prevent the solenoid from melting. Only large facilities can provide such powers. The Francis Bitter National Magnet Laboratory (NML), located on the MIT campus in Cambridge, Massachusetts, was established in 1960 to develop magnetic field facilities beyond 10 T (100 kG), and to use them for solid-state physics research.

The problem of high-magnetic-field generation was changed in the 1960s by the discovery of high-field superconductors. These materials, which have zero electrical resistance at low temperatures, maintain superconductivity to critical fields which can exceed 10 T (100 kG). Solenoids wound with superconducting niobium titanium (NbTi) wire, operated at liquid helium temperatures (4.2 K or $-452°F$), now routinely provide fields to 9 T (90 kG). Moreover, since 1975 superconducting magnet performance has been extended to 15 T (150 kG)—and in one instance to 17.5 T (175 kG)—as a result of advances in fabricating superconducting materials such as niobium tin (Nb_3Sn) and vanadium gallium (V_3Ga). Thus, in recent years high-magnetic-field laboratories have given priority to facilities and science in the field range above 15 T (150 kG).

Continuous fields. Nearly all high-field magnets for research are solenoids. They are classed as resistive, superconducting, or pulsed. The winding volumes of dc magnets are typically short, hollow cylinders, two to three bore diameters long, and have outside diameters two to three times as large as the bore. To generate high fields efficiently, this annular volume should be filled as completely as possible with current flowing circumferentially, in a conductor, about the axis. A typical 15-T (150-kG) magnet has a "current density" of 10,000 A/cm^2.

In resistive magnets of any size, such current densities require enormous amounts of power. For example, a 25-T (250-kG), 3-cm-bore (1.2-in.) electromagnet requires a 10-MW power supply, and 10 MW of cooling capacity. The solenoid winding is then approaching its thermal and mechanical limits, and an efficient cooling system is required. The coolant flows under pressure through channels in the coil which, at the same time, must be sufficiently strong to withstand the electromagnetic stresses imposed upon it. Designing to remove heat is a problem that feeds back upon itself because, as part of the winding is dedicated to cooling, the reduction in the filling factor necessitates even more power.

Bitter solenoid. The Bitter solenoid, which employs a stack of annular copper disks interleaved to form a continuous helix as its winding, is one of the most successful solutions to these conflicting design problems. This solenoid has proved itself through many years of service, and magnets based on it now regularly operate with high reliability. The magnets are rugged and easy to assemble, important considerations in a facility which serves many users and must regularly perform preventive maintenance on its magnet systems. The basic component of the Bitter solenoid, the annular copper disk including holes for coolant flow, is punched from copper sheet. Since no machining is involved, the process is inexpensive and can easily be standardized. However, the relative simplicity of these solenoids entails a penalty in electrical efficiency, because the current distribution in their windings is less than optimal.

Polyhelix. The Bitter construction was regarded as the ultimate until the thrust toward higher fields prompted an in-depth structural analysis, culminating in the polyhelix, a coil constructed of independent, nested helices. The main advantage of this design is structural, and it derives mainly from severing the radial continuity between regions within the winding. Freeing the inside from radial tension reduces the hoop stress, which is the dominant stress component. Dividing the winding volume into concentric shells, as the polyhelix does, also provides the magnet designer with an opportunity to control the radial distribution of current density. This possibility has been exploited in coils which are about 15% more efficient, in terms of field generated per unit power, than Bitter solenoids.

Bitter solenoids produce 20 T (200 kG) in a 2-in. (5-cm) bore, and 23 T (230 kG) in a 1-in. (2.5-cm) bore; the polyhelix generates 25 T (250 kG). These peak fields of resistive magnets are largely determined by the power supplies of the laboratories, about 10 MW in each case. However, even if more power were available, there would be severe thermal and mechanical problems in engineering an all-water-cooled magnet to make use of it.

Hybrid magnets. To achieve higher continuous fields, a hybrid concept was introduced, wherein a Bitter solenoid is operated in the background field provided by a large-bore superconducting magnet. A total field of 30.7 T (307 kG) has been achieved, and construction of a 33-T (330-kG) hybrid is planned in the near future.

Hybrid magnets, requiring both water-cooled and superconducting technologies, are more difficult to operate than Bitter solenoids, and are quite expensive. This expense can only be justified by the quality of the research done in the magnets. Fortunately, several unique and quite unexpected experiments have been performed in the hybrid.

Pulsed fields. Barring a major breakthrough in high-field superconductivity, continuous fields appear to be limited to the 30–40-T (300–400-kG) range. To provide higher fields, several laboratories have developed pulsed-magnet capabilities. The cooling problem that limits dc magnets can be avoided in pulsed systems, which store energy in the heat capacity of the winding. Energy is delivered to the magnet in bursts which are so short that nothing burns or melts. Between pulses, the heat must be removed and the energy source recharged.

Pulsed magnets can generate 45 T (450 kG) nondestructively in copper-wound coils and 70 T (700 kG) in superstrong coils of maraging steel (a high-strength, low-carbon iron-nickel alloy in which a martensitic structure is formed on cooling). These systems are driven by large, high-voltage capacitor banks that provide a few shots of millisecond-to-microsecond duration per hour. Many experimental programs have been completed in the pulsed-field laboratories. *See* ELECTRICAL POWER ENGINEERING.

Pulsed magnet systems require special instrumentation to acquire data in short times from a noisy electrical environment. These conditions complicate all experiments and render some impossible; given a choice, scientists almost invariably prefer dc magnet systems to pulsed ones. However, for fields exceeding 33 T (330 kG), there is presently no alternative. Laboratories dedicated to pulsed-field research have therefore gone to great lengths to develop unique pulsed-field instrumentation; their efforts have made possible a surprising variety of experiments.

A 45-T (450-kG) pulsed-field system is relatively simple and not overly expensive. The coils are wound with standard copper wire, the capacitor bank is modest in size, and the coolant is a container of liquid nitrogen per shot to reduce the resistivity of the copper. Beyond 45 T, the difficulty and expense of performing pulsed field research rapidly escalate. Copper coils fail there, torn apart by enormous electromagnetic forces. Higher fields require far stronger materials, and an exceedingly tough form of steel has been selected for such applications. Maraging steel coils are 6 times stronger than copper coils and have superior thermal properties. However, their resistivity is 30 times that of copper, necessitating a very large, fast, high-voltage condenser bank. Moreover, the operation of such a system entails severe safety problems since the energy released when a coil fails is comparable to that of a small artillery shell.

Although maraging steel coils generate 70 T (700 kG) and may ultimately reach 90 T (900 kG), this scale-up process is approaching its limit, since it can be demonstrated that the size and cost of a durable, pulsed-field coil, as well as the condenser bank which drives it, scale exponentially with peak field. It will be exceedingly difficult—perhaps impossible—to build a nondestructive, 100-T (1-MG) pulsed-field system. Beyond this limit, implosions, either electrical or explosive, can compress flux to generate fields to 1000 T (10 MG). The coil, and often much of the experimental apparatus, are destroyed in these implosions, however. Thus, though there has been considerable interest in such systems, it is still quite difficult to perform experiments with them. For the foreseeable future, most high-field research will be performed in dc magnets or nondestructive, pulsed-field systems.

For background information *see* MAGNETIC FIELD; SOLENOID (ELECTRICITY); SUPERCONDUCTING DEVICES in the McGraw-Hill Encyclopedia of Science and Technology.

[MATHIAS J. LEUPOLD; PETER A. WOLFF]
Bibliography: F. Herlach (ed.), *Strong and Ultrastrong Magnetic Fields*, Topics in Applied Physics, vol. 57, 1985.

Magnetic materials

Two of the principal methods of achieving better performance from electromagnetic devices are the development of new soft magnetic materials or the improvement of existing materials. In the last few years the development of amorphous metallic alloys, and, in particular, of amorphous ribbon materials, has made available new matrials which are suitable for devices used in the electrical engineering industry such as transformers and transducers. This innovation has also encouraged improvements to be made to the existing materials used in power transformers, namely 3% grain-oriented silicon-iron. Research on amorphous metallic alloys has also played a part in the discovery of a new class of high-performance permanent magnets (hard magnetic materials), based on combinations of neodymium, iron, and boron.

Amorphous ribbon materials. Amorphous metallic alloys, also known as metallic glasses, are of the basic composition $T_{80}M_{20}$, where T represents one or more of the transition metals—iron, cobalt, or nickel—and M represents one or more of the metalloid or glass-former elements—phosphorus, boron, carbon, or silicon. These alloys may be made by several methods such as electrodeposition and chemical deposition, but the preferred method of manufacture is rapid cooling from the melt because it is faster, is applicable to a wider range of composition, gives material of greater uniformity, and is adaptable to large-scale production.

Manufacture. In the basic manufacturing system an ingot of the required composition is melted in a crucible. On the application of an inert gas under pressure to the crucible, the melt is ejected through one or more orifices in the end of the crucible onto a rotating copper drum, where it is instantly quenched at a cooling rate typically of the order of 10^6 K/s (10^6 °F/s). The solidified melt is formed into a ribbon by the rotation of the drum, and the finished ribbon may be collected after it has left the drum. The crucial parameters of operation are melt temperature, gas pressure, dimensions and configuration of the orifices, angle at which the ejected melt impinges on the drum, and speed of the surface of the drum. Amorphous ribbons, made by this technique, are normallly 1 to 150 mm (0.04 to 6 in.) wide and 25 to 50 micrometers thick with lengths from a few meters (several feet) up to kilometers (miles). A typical production speed of approximately 2 km/min (1.2 mi/min) of ribbon is attainable.

Properties. A comparison of the properties of amorphous ribbons with conventional polycrystalline magnetic materials is shown in Table 1. The values shown have been optimized by magnetic field anneal treatment. It can be seen that amorphous materials can have properties similar to conventional nickel-iron alloys and hence have application as a replacement for the existing materials.

High-frequency applications. The use of amorphous alloys in high-frequency electronic ballasts, switched-mode power supplies, magnetic modulators and amplifiers, induction cores in linear accelerator systems, magnetic shields, and tape heads has been investigated. Their high resistivities, typically two to three times those of crystalline materials, encourage use at high frequencies. Amorphous alloys cannot compete with ferrite pot cores in low-power applications, but the lower losses and higher-saturation inductions make their use in compact toroidal transformer designs economic. It has been calculated that the capability of amorphous alloys in terms of maximum transferable power per transformer volume exceeds that of manganese-zinc (Mn-Zn) ferrites at 20 and 50 kHz and is similar to that of the more expensive 0.03-mm-thick permalloy.

Use in power transformers. Amorphous alloys have power losses, in the frequency range 50 to 400 Hz, that are much lower than those of the best silicon-

Table 1. Comparison of properties of amorphous ribbons with conventional polycrystalline magnetic materials

Material	Composition	Saturation magnetization, T	Coercive force, A/m	Saturation magnetostriction, μstrain	Curie temperature, °C (°F)	Resistivity, μΩcm	Density, g/cm³ (lb/ft³)
Amorphous ribbons							
Metglas* 2605CO	$Fe_{67}Co_{18}B_{14}Si_1$	1.74	4	35	415 (779)	130	7.56 (472)
Metglas* 2605SC	$Fe_{81}B_{13.5}Si_{3.5}C_2$	1.61	4.8	30	370 (698)	125	7.30 (456)
Metglas* 2826MB	$Fe_{40}Ni_{38}Mo_4B_{18}$	0.88	1.2	5	353 (667)	160	8.02 (500)
Conventional materials							
Mumetal	$Fe_{14}Ni_{77}Mo_4Cu_5$	0.8	1.0	0	350 (662)	60	8.8 (550)
Grain-oriented silicon-iron	$Fe_{97}Si_3$	2.0	24	4	730 (1346)	50	7.7 (480)

*Allied Corporation's registered trademark for amorphous alloys of metals.

iron alloys, typically by a factor of 2 to 5. At a sensible working flux, commercial transformer steel has total losses of approximately 1 W/kg at 60 Hz, while amorphous alloys have total losses in the range 0.2–0.5 W/kg. It is this property of lower power loss which has sustained and encouraged the development of amorphous alloys, particularly their application as core materials in power transformers working at 50, 60, and 400 Hz. The full-scale substitution of amorphous ribbons for the existing electrotechnical steels used in transformers in the electrical supply industry has the potential for saving two-thirds of the electrical energy wasted in the cores, typically 3×10^{10} kWh annually in the United States.

In 400-Hz transformers, which are used primarily in airborne and military applications, the critical parameters of volume or weight can be reduced because of the low loss of the amorphous ribbons. A 400-Hz transformer which uses an amorphous ribbon core can be designed with a higher copper loss than a transformer which uses a 0.1-mm-thick laminated silicon-iron core without exceeding a specified temperature rise because of the fivefold decrease in core loss. The higher copper loss enables the winding volume and hence the core size to be reduced because a smaller gage of wire can be used. It has also been shown, by considering the maximum power capability of similar toroidal cores with a restricted temperature rise, that an amorphous core has a 20 and 60% higher output power than cores made from 0.1- and 0.3-mm-thick laminated grain-oriented silicon-iron, respectively.

Distribution transformers which are used in the final voltage step-down stages of supplying electrical energy to consumers are continuously energized for 25 to 40 years. Therefore, highly efficient transformers are necessary because the costs of the losses are comparable to the initial purchase price of the transformer. It has been claimed that the use of amorphous ribbons offers the opportunity for a decrease in core losses by as much as 75% and the only possibility of producing transformers with an efficiency of greater than 99%.

Table 2 compares two 25-kVA distribution transformers, and shows that the use of the new materials has advantages. However, certain problems in their application in power transformers still exist, such as the thinness of the ribbon, postanneal brittleness, stress sensitivity, and an inferior stacking factor. However, a program of design, construction, and life-testing of distribution transformers with toroidal amorphous ribbon cores has been undertaken. Results from this program will enable distribution transformers using the new materials to be accurately appraised and compared to the existing designs using conventional electrotechnical steels. *See* TRANSFORMER.

Improved silicon-iron. Since 1980, attempts have been made to improve the properties of grain-oriented silicon-iron, particularly to decrease the power loss which is the primary criterion of assessment for transformer applications. One avenue of research has been to manufacture 6% silicon-iron with a sheet thickness of the order of 0.3 mm by the rapid-quenching technique used to make amorphous ribbons. This 6% silicon-iron material has zero magnetostriction, an improved resistivity and hence lower loss, but is microcrystalline and not amorphous due to the decreased cooling rate which accompanies the larger thickness. It is not possible to manufacture 6% silicon-iron by conventional steelmaking techniques because of brittleness, but the rapid-quenching technique does not introduce the optimum grain orientation and expensive heat treatments are necessary to relieve the stresses introduced by the quenching.

Basic research into the properties of 3% grain-oriented silicon-iron has shown that a minimum loss is obtained with materials with small grains which are all perfectly and consistently oriented by 2° out of the plane of the lamination. Therefore, attempts have been made to produce a very well-oriented material with a small magnetic domain size (which is the reason for the necessity for small grains). Highly oriented materials were introduced in the early 1970s, and the production of such materials with small domain size (produced by domain refinement) is now being studied. The methods being investigated for producing domain refinement in grain-oriented steels are laser surface scribing with pulsed and continuous waves, surface scratching, discharge arcing, and local cold deformation after the secondary recrystallization anneal.

The most advanced material is laser-scribed domain-refined 3% grain-oriented silicon-iron, and transformer manufacturers must now assess this material in working designs and determine the optimum lamination thickness. It has been suggested that the optimum thickness for nonlaser-scribed material is approximately 0.18 mm, but may be lower for the domain-refined material. The domain refinement process (typically the periodic introduction of local stress by scribing with laser) lowers the total loss by 10%. These recent improvements in electrotechnical steels make the wholesale introduction of a new technology of material and transformer production based upon amorphous ribbons even more

Table 2. Comparison of two 25-kVA distribution transformers

Material	Core loss, W	Coil loss, W	Total loss, W	Core weight, kg (lb)	Total weight, kg (lb)
Grain-oriented silicon-iron (M4)	85	240	325	65 (143)	182 (401)
Metglas* 26052-S2	16	235	251	77 (170)	164 (362)

*Allied Corporation's registered trademark for amorphous alloy of metals.

problematical, and only more research and development work will determine the materials to be used in future transformers.

[K. J. OVERSHOTT]

Nd-Fe-B permanent magnets. Since about 1900, permanent-magnet technology has progressed from the steel horseshoe magnet to high-performance samarium-cobalt–based materials. Although these latter magnets currently offer excellent properties, high cost has prevented their large-scale adaptation in, for example, the automotive industry. In recent years the search for less expensive, alternative materials has centered on the rare earth–iron alloys, in particular those containing the so-called light rare earths—lanthanum, cerium, praseodymium, and neodymium. Emphasis on these elements has been motivated by the fact that they comprise over 95% of the rare earths in a typical ore body and, hence, are the least expensive. Research on these materials has led to the discovery of a new class of high-performance magnets based on combinations of neodymium, iron, and boron (Nd-Fe-B). Energy values achieved for these materials are already substantially higher than those of the samarium-cobalt materials, while costs are expected to be substantially lower. Use of these new materials should allow significant reductions in the size and weight of a variety of consumer and industrial products which currently employ permanent magnets, as well as stimulate the use of permanent magnets in an entirely new range of applications.

Permanent magnets play a very important role in technology. Prominent consumer products which use

Table 3. Energy products of permanent magnet materials

Material	Energy product, $-(BH)_{max}$, MGOe (kJ/m^3)
Ferrite	4 (32)
Alnico	12 (95)
SmCo$_5$	18 (143)
Sm$_2$Co$_{17}$	27 (215)
Nd-Fe-B	35 (280)

permanent magnets include the telephone, television, speakers, computers, and printers. A good example of their widespread use is the automobile. Prominent among the various applications are a wide variety of small dc motors such as those used for the heater and air conditioner blowers, seat actuators, windshield wipers, window lifts and fuel pump. Other applications include a variety of actuators, gages and sensors. In all of these instances the use of a cost-effective, higher-performance magnet could allow increased operating efficiency, or a reduction in size and weight, or both, a significant consideration in the efforts of automobile manufacturers to increase fuel economy.

Performance. Permanent magnets have two important parameters which measure their performance. These are the induction or magnetic strength, and the coercive force, or resistance to demagnetization. These two parameters define the energy product, which can be expressed in units of megagauss-oersteds (MGOe) or kilojoules per cubic meter (kJ/m^3). This value, which is the figure of merit commonly used to compare permanent magnets, is inversely proportional to the volume of magnet needed to produce a given magnetic flux across a fixed air gap. Historically there have been three commercially important classes of permanent magnets: alnico, ferrite, and samarium-cobalt. The properties of these classes are compared with those of Nd-Fe-B in Fig. 1 and Table 3. Energy products of commercially available Nd-Fe-B material already significantly exceed those of the other classes, and even higher values up to 45 MGOe (360 kJ/m^3) have been achieved in the laboratory. Moreover, the theoretical energy product of this new material is as high as 64 MGOe (510 kJ/m^3), almost twice that of the rare earth–cobalt grades and 16 times that of ferrite magnets. The development of Nd-Fe-B magnets with energy products exceeding 50 MGOe (400 kJ/m^3) is expected in the very near future.

Discovery. The discovery of this new class of permanent magnets was aided by advances made in rapid-solidification technology and materials throughout the 1960s and 1970s. Notable among these rapidly solidified materials were the so-called metallic glasses. These are combinations of transition metals, primarily iron, and various metalloid elements, prominently boron, that are added to promote glass formation by increasing the tendency of the material to become amorphous or noncrystalline. The search for a light rare earth-iron–based perma-

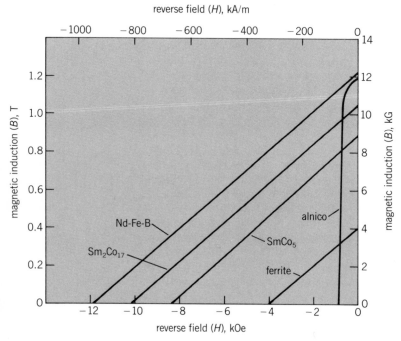

Fig. 1.   Magnetic induction B of various permanent-magnet materials versus reverse field H. The intercept along the reverse-field axis defines the coercive force of the material. The energy product is the maximum product of induction and field along these curves, $-(BH)_{max}$.

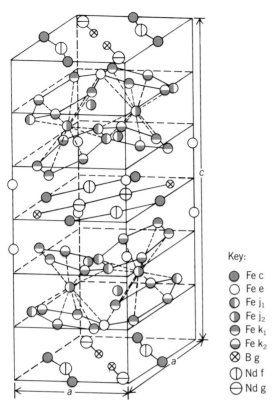

Key:

●	Fe c
○	Fe e
◐	Fe j₁
◑	Fe j₂
◒	Fe k₁
◒	Fe k₂
⊗	B g
◍	Nd f
⊖	Nd g

Fig. 2. Crystal structure of the Nd₂Fe₁₄B compound. Various symbols indicate specific crystallographic sites. The *c/a* ratio has been exaggerated to more clearly show the atomic arrangement. (*After J. F. Herbst et al., Relationships between crystal structure and magnetic properties in Nd₂Fe₁₄B, Phys. Rev., B29:4176–4178, 1984*)

nent magnet has been a longstanding one, dating back to the discovery of the rare earth–cobalt magnets in the mid-1960s. Development was severely hampered, however, by the fact that these materials formed no suitable intermetallic phases, a necessary ingredient for magnetic hardening. One method by which early researchers sought to overcome this problem was by applying rapid solidification techniques to form nonequilibrium or metastable microstructures. It was this work, starting in the late 1970s, which led to the discovery of the exceptional permanent-magnet properties of the Nd-Fe-B alloys.

Crystal structure. While boron was almost certainly added in an attempt to modify the rapid-quench behavior of these materials, it is now clear that its important role is to promote the formation of the equilibrium Nd₂Fe₁₄B intermetallic phase responsible for the permanent-magnet properties of these materials. The crystal structure of this compound has been established by neutron diffraction analysis (Fig. 2). The crystal structure is tetragonal and the magnetic ordering is ferromagnetic; the easy or preferred axis of magnetization lies along the *c* axis of the unit cell. The structure is highly anisotropic, resulting in strong coupling between the magnetic moments and crystal lattice. This strong coupling makes it difficult to rotate the moments out of their preferred direction, and is the origin of the

high coercive force in these materials.

Manufacturing processes. Nd-Fe-B magnets are currently manufactured by two competing processes. Although the roots to their discovery reside in rapid-solidification technology, it was subsequently found that they could be manufactured by the same powder metallurgy technology used to prepare samarium-cobalt magnets. Briefly, this involves grinding the Nd-Fe-B ingot in stages to a fine, approximately 3-micrometer-diameter powder, at which point each particle consists of individual single crystals which are susceptible to alignment by a large magnetic field. After alignment, the powder is compacted to approximately 70% density. The material is then sintered at roughly 1100°C (2000°F) to high density. Magnets prepared by this technique are frequently given a postsinter heat treatment to achieve optimum properties.

The competing process, in contrast, still involves a rapid-solidification, specifically a melt-spinning, operation. In this process a fine stream of molten alloy is quenched onto a cold rotating substrate to form directly a magnetically hard microstructure consisting of extremely fine (1-nanometer) crystallites. In order to manufacture a permanent magnet from this ribbonlike material, however, two significant problems had to be solved: first, how to consolidate the ribbon material into a bulk magnet and, second, how to crystallographically align the fine-grained material to maximize its energy product. Researchers discovered that both consolidation and alignment could be achieved by hot deformation, during which the individual crystallites within the rapidly quenched ribbons rotate such that their *c* or easy axis lies normal to the direction of plastic flow. Energy products as high as 40 MGOe (320 kJ/m³) have been achieved by this process. Advantages of this method include the fact that the ribbon material is inherently more stable than the finely ground, potentially explosive powder needed for the powder metal process. In addition, because the magnet is formed in a die, there is very little need for the extensive finishing operations frequently required for sintered materials.

Origin of high coercive force. Although the processing involved in these two competing manufacturing techniques is quite different, the end result is qualitatively the same, namely, a microstructure consisting of grains most of which are crystallographically aligned or oriented along the same direction. The theory behind the formation of such a microstructure is the so-called single-domain theory. In their unmagnetized state, all magnetic materials consist of magnetic domains, which are small regions within which all of the magnetic moments are aligned parallel. These regions are separated by entities called domain walls. Under the influence of a magnetic field, these walls move and are swept out of the magnet. The key to achieving high coercivity lies in preventing the reformation of these domain walls when a reverse magnetic field is applied. In theory, particles which are sufficiently small will resist the formation of a new domain wall because the cost in energy of producing the wall is proportionately higher than for a grain with larger volume. Hence the field needed to bring about magnetization reversal, that is, the coercive force, is higher in fine-grained materials. In practice this is an extremely simplified theory. Evidence for this is the fact that Nd-Fe-B magnets produced by the two different methods have particle sizes which vary by as much as an order of magnitude and yet display similar coercivities; this large difference in grain size remains one of the unanswered questions surrounding these materials. Nevertheless, this concept does provide a starting point for understanding the nature of the coercive force in these materials and its dependence on the particle size of the magnetic microstructure.

Limitations. There are several limitations to these new magnets. The Curie or magnetic ordering temperature is only 585 K (593°F); much lower than the approximately 1000 K (1340°F) value for samarium-cobalt materials. Since magnets lose their magnetic strength above this temperature, the properties of the Nd-Fe-B materials fall off more rapidly with temperature, thus limiting their use in some applications. In addition, because these are iron-based materials, the magnets are more susceptible to rusting or corrosion. Consequently, some form of protection will be required in humid or corrosive environments. As with the discovery of any new class of material, however, it is expected that further research will result in improvements in these areas. Moreover, the $Nd_2Fe_{14}B$ intermetallic phase may be only one of a number of similar or related compounds which may provide the basis for even more interesting and technologically significant permanent-magnet materials.

For background information *see* FERROMAGNETISM; MAGNETIC MATERIALS; METALLIC GLASSES; TRANSFORMER in the McGraw-Hill Encyclopedia of Science and Technology.

[JOHN J. CROAT]

Bibliography: F. E. Luborsky, *Amorphous Metallic Alloys*, 1983; *Proceedings of the 26th Conference on Magnetism and Magnetic Materials*, Dallas, 1981, *29th Conference*, Pittsburgh, 1983, *30th Conference*, San Diego, 1984; C. M. Srivastava (ed.), *Recent Advances in Materials Research*, 1984.

Marine engineering

Ships and offshore structures subjected to the severity of the seas must withstand extreme loads and motions in order to perform the designer's mission (Fig. 1). In modern ships and crafts, such as hydrofoils (Fig. 2), SWATH (small-waterplane-area twin-hull), and surface effect ships, designers try to take advantage of the unique aspects to improve seakeeping performance. The unpredictability or randomness of the ocean environment, however, prevents the naval architect or ocean engineer from precisely calculating normal operating and maximum vessel responses. Estimation of standard operating responses is required since good seakeeping characteristics frequently determine the economic feasibility of a proposed design. Alternatively, estimation of maximum responses is required to determine the ultimate survivability of the design. In harsh environments, minor malfunctions of sophisticated offshore systems may result in the loss of property and lives. *See* SHIP DESIGN; SWATH SHIP.

In an effort to more accurately describe the behavior of the vessels under these conditions, designers are increasingly relying upon experimental and analytical techniques such as model tests, complex computer simulations, and extreme value analysis. A variety of sea states are simulated, and the resulting responses are either measured on small-scale models or calculated by using large mainframe computers.

Recent casualties. In addition to the loss of the vessel and crew, the aftereffects of an accident at sea, such as pollution and environmental damage, can cost hundreds of millions of dollars.

The Great Lakes bulk carrier *Edmund Fitzgerald* sank on November 10, 1975, with a loss of 29 lives. An investigation stated that faulty hatch covers leaked and caused the ship to take on water and capsize, but this view is not necessarily shared by all.

The oil tanker *Amoco Cadiz* ran aground on March 16, 1978. The ship experienced a rudder failure, later attributed to improper maintenance of the steering gear. Although taken in tow, the tanker hit the rocks and broke up, resulting in an oil spill that covered over 90 mi (150 km) of French coastline.

The offshore drilling unit *Ocean Ranger* capsized and sank with all 84 persons on board on February 15, 1982. An investigation of the accident indicated that water coming through a broken portlight caused a malfunction in the unit's ballast control panel.

These casualties demonstrate how failures of local mechanical systems can lead to the complete loss of the vessel or offshore drilling unit. The sea states in each of these cases were severe but not outside the design envelope of the vessel.

Vessel responses in random waves. A small-amplitude, regular wave train with one characteristic frequency has a profile that closely follows a sinusoidal curve. However, ocean waves may be thought of as containing many sinusoidal waves with many different frequencies. The addition or superposition of these individual waves gives the irregular, random ocean surface frequently observed. Determining the statistical properties and extreme values of the water surface are some of the more challenging problems in oceanography. A designer of ocean structures, while also interested in wave statistics and extreme waves, is more concerned with vessel responses.

Rank estimator. The motion of an ocean vessel is a function of many variables, such as sea state, wind direction and severity, hull shape, vessel weight distribution, and vessel speed. While advances in theoretical and experimental hydrodyna-

Fig. 1.   Ship heading into long-crested seas off the west coast of the United States. The height of the waves is equal to or greater than the ship's bridge. (*San Francisco Examiner*)

mics have made it possible to more accurately describe the hydrodynamic aspects of seakeeping, consensus as to the precise definition of a good-seakeeping hull form is still lacking. As a result, the relative seaworthiness of various ships has been difficult to quantify.

Addressing this problem, the United States Navy has developed a set of specific seakeeping performance requirements for new ship designs. Coupled with these requirements is the determination of a relative seakeeping parameter, the seakeeping rank estimator R. This rank estimator assumes that the major contributing factors to vessel seaworthiness can be found in long-crested, head seas. Vessel responses in oblique and beam seas, such as roll, are assumed to be controlled by motion stabilizers (bilge keels, antirolling tanks, or active antiroll fins). An empirical formula for R was developed, based upon a small number of hull parameters. For destroyer-type hull shapes, this is given by Eq. (1). Here, C_{wf}

Fig. 2.   Hydrofoil PHM-1 *Pegasus* in rough water. (*From R. L. Johnston, ed., Hydrofoils, Nav. Eng. J., 97(2):142–199, February 1985*)

is the waterplane coefficient forward of midships,

$$R = 8.42 + 45.1\,C_{wf} + 10.1\,C_{wa} - 378\,T/L \\ + 1.27\,C/L - 23.5\,C_{vfp} - 15.9\,C_{vpa} \quad (1)$$

C_{wa} is the waterplane coefficient aft of midships, T/L is the draft-to-length ratio, C/L is the cut-up ratio where C is the distance from the forward perpendicular to the point where keel rises in the afterbody, and C_{vpf} and C_{vpa} are the vertical prismatic coefficients forward and aft of amidships respectively.

The rank estimator implies that good seakeeping characteristics can be achieved by increasing the vessel's length, waterplane coefficient, and cut-up point while decreasing the vessel's draft and vertical prismatic coefficient. This has the net effect of producing long and shallow draft vessels, with most of the hull displacement concentrated at or near the waterline. The vessel sections tend to resemble the letter V. A vessel with a value of R greater than 10

Fig. 3. Model tests. (a) Conventional hull form in rough water. (*Skibsteknisk Laboratorium, Copenhagen*). (b) Jackup offshore drilling platform in 100-knot (50-m/s) winds and 115-ft (35-m) seas. (*University of Michigan*)

is considered to have excellent seakeeping characteristics. An "ideal" seakeeping model hull was constructed and tested to demonstrate the concept of rank estimator. It had $R = 8.4$, which exceeds many existing ships. For example, the DD-963 destroyer class has $R = 2.5$, indicating that it is a relatively unseaworthy hull.

Extreme responses. The evaluation of the rank estimator rates hulls operating in typical ocean environments in a relative sense and does not quantify maximum responses. If the vessel is to survive the so-called design wave, a sea state representing a worst case, its responses in that environment must be determined.

The extreme wave height will not necessarily generate the extreme vessel response. Since the ocean structure is a dynamic system, different combinations of wave slopes and wave periods, while smaller than the extreme wave, may cause larger vessel responses. In either case, extreme responses or wave heights can only be estimated in a probabilistic sense. The random nature of the process or response implies that there is only a small probability that the calculated extreme will actually occur. Typically assumptions are made about the nature of the process, such as whether it is gaussian-distributed and narrow-banded. (A narrow-banded random process is one that has a characteristic frequency where most of its energy is concentrated.) With these assumptions, mathematical theories have been developed to estimate the extreme value that has a given low probability of exceedence. For example, the most probable maximum response above the mean in N oscillations for a narrow-banded process is given by Eq. (2), where RMS is the root mean

$$\text{Most probable maximum} = \text{RMS}\,\sqrt{2\ln(N)} \quad (2)$$

square of the process and $\ln(N)$ is the natural logarithm of N. The root mean square of the vessel's response may be calculated on a computer or measured in a model test.

Contrary to the classical theories, recent comparisons with measured data, both full-scale and model-scale, indicate that under some conditions these estimates do not represent the upper bound. In model tests, an oceangoing barge operating in scale 26-ft (8.5-m) seas produced extreme bow relative motions 60% greater than the empirically estimated "most probable maximum." This calculated underprediction may result from an incomplete understanding of the physical process or incorrect assumptions in the statistical theories. An example of the incomplete modeling of a process would be using a linear water wave theory in very steep nonlinear waves.

Model tests. Due to the uncertainties associated with the analytical methods, designers and builders of ocean vehicles have attempted to model the ocean environment in a confined experimental test basin. This type of laboratory can produce the relatively small-scale wind and waves which are necessary to evaluate the design's response to severe operating and survival conditions (Fig. 3). It is only recently that many of the relevant scaling laws used in model testing have been developed.

By using available oceanographic data, a representative sea state is selected for use in the model test. When normal operating conditions are being modeled, a moderate ocean environment may be used, and when survival conditions are being modeled, an extreme environment is picked. The sea state is described by various parameters such as the significant wave height, wave period, and wind velocity. These parameters are used in computer programs to generate a collection of actual time histories, representing the incident waves. The time histories are then used to electronically control the test basin's wave-making device.

As described above, a random sea state with the same general descriptive parameters will have many different time histories, and each time history has different extreme or largest wave. Futhermore, as discussed above, the largest wave does not necessarily produce the largest vessel response. The recommended procedure for testing in irregular seas is to expose the model vessel to a model sea state with a full-scale time period equivalent of 30 min. It is hoped that the largest response will occur in this time span. However, if the process being investigated is nonlinear, this method may not yield accurate estimates of extreme responses. Research is continuing in the areas of probabilistic determination of extreme hydrodynamic loads and experimental modeling of extreme vessel performance.

For background information *see* PROBABILITY; SHIP DESIGN; SHIP POWERING AND STEERING; SIMULATION; STOCHASTIC PROCESS; TOWING TANK in the McGraw-Hill Encyclopedia of Science and Technology.

[ARMIN W. TROESCH]

Bibliography. R. Johnson and P. Cojeen, An Investigation into the loss of the Mobil offshore drilling unit *Ocean Ranger, Mar. Tech.*, 22(2):109–125, April 1985; R. Keane and W. Sandberg, Naval architecture for combatants; A technology survey, *Nav. Eng. J.*, 96(5):47–64, September 1984; J. McVoy (ed.), Modern ships and crafts, special issue, *Nav. Eng. J.*, vol. 97, no. 2, February 1985; J. O'Brien (chairperson), *Extreme Loads Response Symposium*, Society of Naval Architects and Marine Engineers, 1981; A. Troesch, Design waves and extreme responses: A case study of extreme relative motions, *J. Energy Resources Technol., Trans. ASME*, publication pending.

Marine mining

Interest in minerals from the seabeds was highly publicized during the negotiations of the United Nations Conference on the Law of the Sea, and centered on the ubiquitous potatolike manganese nodules, distributed over most of the world's deep oceans at depths of 13,000–20,000 ft (4000–6000 m). From the early 1960s through 1978, several hundred million dollars was spent by six major international industrial consortia to characterize these deposits and to develop mining and metallurgical systems for commercial exploitation.

Three major at-sea tests to evaluate mining systems and to determine potential environmental impacts were completed in water depths exceeding 13,000 ft (4000 m). Major conclusions were that mining of the nodules presented no insurmountable technical problems with state-of-the-art technology, and that predictable environmental impacts of mining at sea appeared to be within acceptable and controllable limits. Several available on-land processing options to produce the manganese, cobalt, nickel, and copper from the nodules were reported.

Congress passed a Deep Seabed Hard Mineral Resources Act in 1980 to license United States industrial firms wishing to pursue mining developments pending ratification of the Law of the Sea Treaty. West Germany, the United Kingdom, Italy, France, and the Soviet Union followed suit. The United States refused to become a signatory to the final convention because of concern over certain mining issues. In 1984, under the United States law, four consortia applied for and were awarded exploration licenses for nodules in areas of the international seabed (Fig. 1), but poor metal markets have curtailed the planned activities of these groups. In 1983 President Reagan, following a precedent already set by over 50 other countries, proclaimed a 200-nautical-mile (370-km) Exclusive Economic Zone around the United States and its possessions and territories, nearly doubling the total area of the Earth's surface under United States jurisdiction and opening a vast new frontier to minerals development under United States laws.

Mining activities. Meanwhile, interest continued worldwide in the embryo marine mining industry, and efforts were concentrated on the development of five commodity groups, of which three represented traditional mining targets and two represented discoveries of previously unknown resources in deep water. Shallow-water exploration and mining of continental shelf deposits have continued for construction materials, including sand, gravel, aragonite, oyster shells, and coral in Japan, the Bahamas, the United States Gulf coast, and Pacific islands; for placer deposits of minerals containing chromium, tin, iron, titanium, thorium, gold, and diamonds in the west Pacific, Thailand, Indonesia, Japan, Australia, India, New Zealand, and Southwest Africa as well as the United States; and for phosphorites in the central Pacific, New Zealand, and eastern United States. Deep-water seabed investigations in the major oceans since 1978 have resulted in discoveries of iron manganese oxide encrustations carrying unexpectedly high quantities of cobalt and platinum; and hydrothermal sulfide deposits containing lead, zinc, copper, silver, and other metals. Of all of these activities, the three most significant appear to be the discoveries of very large resources of phosphorite off the eastern United States, high-cobalt manganese crusts in the Pacific, and hydrothermal sulfide deposits forming at active deep-

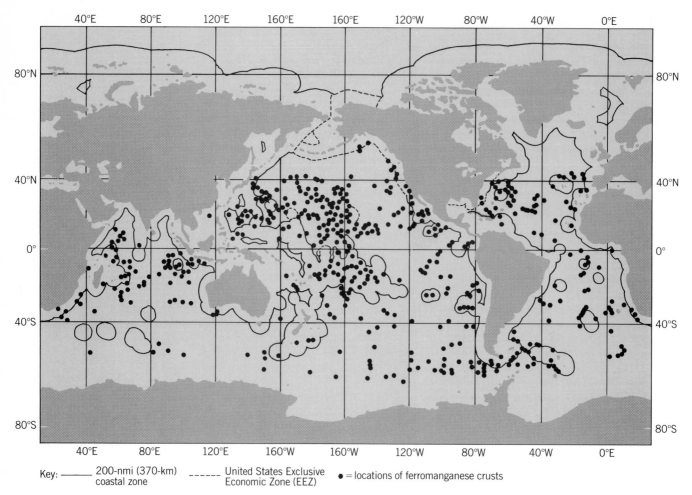

Fig. 1. Map showing worldwide distribution of known manganese crusts. (*After F. T. Manheim, Marine cobalt resources, Science, 1986*)

Key: ——— 200-nmi (370-km) coastal zone   - - - - - United States Exclusive Economic Zone (EEZ)   ● = locations of ferromanganese crusts

seabed spreading centers in the world's oceans and seas.

Bedded marine phosphorite. Reportedly vast discoveries of phosphate rock off North Carolina were made during recent research activities offshore funded by the U.S. National Science Foundation. Geologists verified phosphate deposits in an area running about 30 mi (50 km) parallel to the coast near Wilmington, North Carolina, and from nearshore to about 100 mi (160 km) out in water depths from 50 to 150 ft (15 to 40 m). Potential resources are estimated to be in the tens of billions of tons. If environmentally acceptable mining technology such as borehole mining can be economically applied, then the United States could retain its leadership in world phosphate production for many years to come.

High-cobalt, manganese oxide crusts. Rocks dredged from the slopes and tops of seamounts or submerged islands in the north-central Pacific in 1981–1982 were discovered to have manganese oxide crusts containing as much as 2.6% cobalt. Further investigations indicated that cobalt-rich crusts were widespread in the world's oceans. These crusts occurred most commonly between depths of 2600 and 8200 ft (800 and 2500 m) and in certain areas

contained mean values of about 1% cobalt as well as significant amounts of manganese, nickel, and platinum. Crust thickness varied from mere stains to over 4 in. (10 cm).

To explore the potential of a stable source of the strategic metals cobalt and platinum within the United States Exclusive Economic Zone, and to study the environmental implications of leasing activities, the Department of the Interior initiated additional investigations in the area of the Hawaiian and Johnston islands. Current studies indicate that the Hawaiian Exclusive Economic Zone might have as much as 360 million tons (320 million metric tons) of ore containing 0.4–1.0% cobalt. Platinum content in the Johnston Island Exclusive Economic Zone has been measured as high as 0.8 part per million. Further environmental and economic studies are being carried out in the region to characterize the deposits in detail before decisions are made to offer any of them for lease. Meanwhile, interest is being shown throughout the Pacific island areas in development of those deposits which could influence the economies of some of the island nations and the Trust Territory of the Pacific Islands. Table 1 gives the composition of some significant metalliferous oxide deposits.

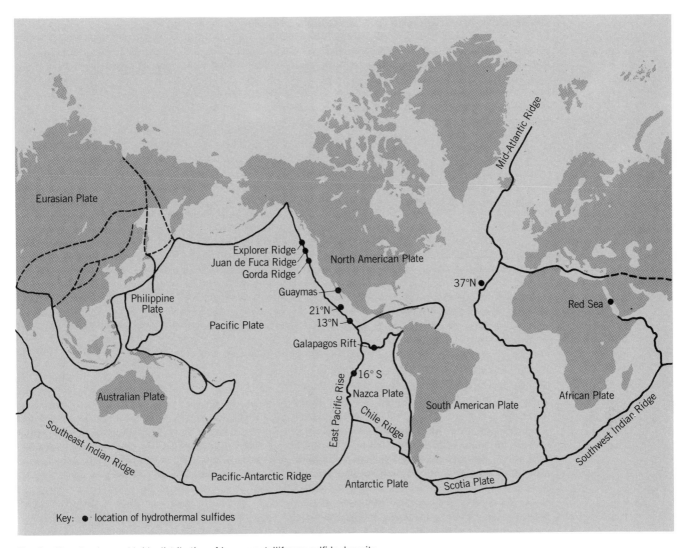

Fig. 2. Map showing worldwide distribution of known metalliferous sulfide deposits.

Metalliferous sulfide deposits. Deposits of metal-sulfide muds containing significant quantities of lead, zinc, vanadium, silver, and other metals were discovered in 1966, following earlier scientific investigations of hot brine pools in the Red Sea. In 1978, during geological and geophysical investiga-tions of tectonic activity around the axis of the East Pacific Rise at 21°N, a team of scientists from the United States, Mexico, and France, using the deep-submersible vessel *Cyana*, discovered massive sul-fide deposits, principally of zinc, copper, and iron. Since that time investigations have disclosed at least

Table 1. Selected data on metalliferous oxide deposits found on island and continental margins and deep seabeds

Metal or compound*	Metalliferous oxide crusts					Metalliferous oxide nodules	
	Blake Plateau†	Hawaii‡	Johnston Island‡	Indian Ocean	World	Clarion/ Clipperton fracture zones‡	World
P₂O₅	0.7						
Mn	17.2	17.0–33.0	18.4–33.2	18.01	20.54	25.4	17.4
Co	0.32	0.41–1.64	0.48–1.92	0.42	0.58	0.24	0.27
Ni	0.60	0.25–1.11	0.22–0.82	0.41	0.48	1.28	0.55
Cu	0.10	0.03–0.12	0.02–0.18	0.13	0.17	1.02	0.34
Zn	0.05			0.06	0.07	0.14	0.12
Pt	0.37		0.14–0.8				
Pb	0.06			0.17	0.15	0.04	0.09

*As reported in % except Pt (ppm).
†Off the Atlantic coast of the United States.
‡North Pacific Ocean.

Table 2. Selected data on hydrothermal sulfide deposits on the deep seabed

Area	Year of discovery	Discovered by	Reported metals					Volume of deposit, ft³ (m³)*
			Zn, %	Cu, %	Pb, %	Ag, ppm	Au, ppm	
Explorer Ridge‡	1984	Canada	9	8	0.32	112	0.8	7,000,000 (200,000)†
Juan de Fuca Ridge‡	1981	U.S.	54	0.22	0.25	260		3,500,000 (100,000)
Gorda Ridge‡	1981	U.S.	Not available					
Guaymas Basin§	1980	U.S./Mex.	30	1.0	0.1	300		42,000,000 (1,200,000)
E. Pac. Rise 21°N	1978	U.S./Mex./Fr.	41	0.61	0.05	380	Trace	572,000 (16,200)
E. Pac. Rise 13°N	1981	France	Not available					
Galapagos Rift¶	1981	U.S.	0.2	5.0	<10		0.05	280,000,000 (8,000,000)
E. Pac. Rise 16°S	1966	U.S./Fr.	Fe/Mn/Cu/Cr/Ni/Pb					Not available
Mid-Atlantic 37°N	1974	U.S./Fr.	Fe/Mn oxides					Not available
Red Sea, Atlantis II Deep	1948/1963	Sw./U.S./U.K.	3.4	0.5	—	41	0.5	700,000,000 (20,000,000)

*Based on estimates of dimensions of deposits observed by bottom photography and crewed submersible (density approximately 2.5 g/cm³).

†Forty individual deposits reported, ranging in size up to 660 × 330 × 33 ft (200 × 100 × 10 m) thick.

‡Off western United States–Canada.

§Northwest Mexico.

¶West of Ecuador.

10 other locations in the Pacific where massive sulfide deposits have been formed in association with active spreading centers (see Figs. 2–3 and Table 2).

Following the discovery of sulfide minerals in the Juan de Fuca–Gorda ridge system which lies partially within the United States Exclusive Economic Zone off Oregon and California, the federal government initiated a program to examine the ridge system in detail. The purpose of the program was to determine if it might be economically and environmentally feasible to lease the minerals for develop-

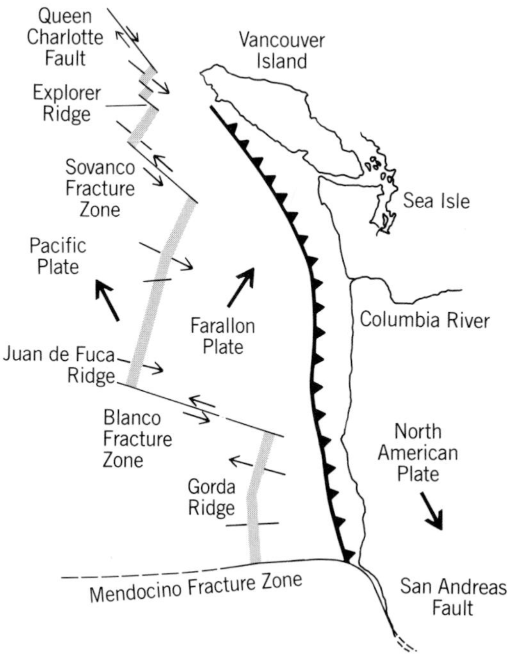

Fig. 3. Map showing Juan de Fuca study area.

ment. Meanwhile, discoveries by Canadian researchers in the Explorer Ridge in the northern part of the Juan de Fuca ridge system have indicated multiple deposits of up to 300,000 tons (270,000 metric tons) each. The deposits have been reported to contain average values of 9% zinc, 8% copper, 112 ppm silver, and 0.8 ppm gold. These discoveries, all of which have been made in water depths between 6600 and 10,000 ft (2000 and 3000 m), have increased scientific understanding of hydrothermal mineral deposition, and should lead to enhanced exploration capabilities for similar deposits on land. The realization that the deep seabeds which now form the flanks of spreading ridges through the world's oceans may be host to a prolific variety and quantity of mineral deposits should give impetus to continued exploration for deep-seabed minerals.

Exploration, mapping, and engineering. Tremendous advances have been made since 1975 in the development of satellite positioning, high-accuracy wide-swath bathymetric mapping, and wide-swath sidescan sonar techniques which produce images of the seabed similar to those resulting from high-altitude airborne radar mapping of the land. All these systems combine to produce an advance in seabed mapping techniques which is sorely needed to inventory the resources of the vast new areas claimed under the concept of Exclusive Economic Zones. Simlarly advances in engineering systems for oil production from deep and hazardous water areas and for the exploration of space will serve to bring nearer the time when the enormous resources of the sea and the seabeds will be utilized to supplement the ever-increasing minerals demand. *See* MARINE GEOLOGY; SATELLITE NAVIGATION SYSTEMS.

For background information *see* MARINE MINING; MARINE RESOURCES in the McGraw-Hill Encyclopedia of Science and Technology.

[MICHAEL J. CRUICKSHANK]

Bibliography: D. S. Cronan, *Underwater Minerals*, p. 362, 1980; J. Gardner, *Atlas of the Exclusive Economic Zone: Western Coterminous United States*, U.S. Geological Survey, I-1792, 1986; F. T. Manheim, Marine cobalt resources, *Science*, 1986.

Marine sediments

Subduction of marine sediments has been the focus of some recent research. As oceanic plates are consumed at convergent margins, marine sediments are carried into subduction zones. Of the sediment column entering the subduction zone, the upper portion is often removed from the descending plate and added to the base of the trench slope (accreted), whereas the underlying sediment section is transported landward beyond the toe of the slope (subducted). The thickness and percentage of sediment subducted vary markedly between trenches as well as along individual trenches. The surface between subducted and accreted sediments is a décollement, usually localized along a distinct stratigraphic horizon, and often rides up over subducting topographic features. The location of the decollement is controlled probably by contrasts in physical properties of the sediments above and below the décollement. Much of the subducted sediment carried beneath the toe of the trench slope is added probably to the base of the accretionary prism at greater depth by a poorly understood process termed underplating.

Subduction and accretion. Three dominant tectonic processes operate in subduction zones: sediment accretion, sediment subduction, and sediment underplating. Sediment accretion occurs where sediments are offscraped above thrust faults and added to the toe of the trench slope. When the thrust soles out within the sedimentary section, the sediments above the decollement are accreted while those below are subducted. The percentage, and the absolute thickness, of sediment that is accreted versus that subducted varies widely among trenches as well as along individual trenches. This selective subduction process occurs primarily in trenches where thick terrigenous sediment columns are being subducted. For example, Deep Sea Drilling Project and seismic reflection data at the toe of the trench slope off Barbados show that the upper 575–660 ft (175–200 m) of sediment on the descending oceanic plate is offscraped at the toe of the slope and the remaining 820–990 ft (250–275 m) of sediment is underthrust beyond the toe of the slope. Along the Nankai Trough, the trench sediment thickness ranges from 330 to 3300 ft (100 to 1000 m). The trench sediment and the upper part of the underlying Shikoku Basin sediment are offscraped, and the remainder of the basin sediment is subducted (Fig. 1). Although the thickness of accreted sediment thus varies along the trench, the thickness of subducted sediment remains nearly constant at approximately 1000 ft (300 m). Along the Makran continental margin of Iran and Pakistan, where a 4.3-mi-

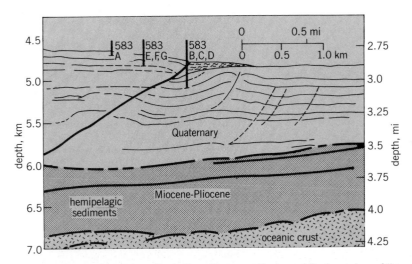

Fig. 1.   Enlargement of interpretive depth section across the base of the inner slope of the Nankai Trough. The hemipelagic sediments underlie sediments ponded in the Nankai Trough. The numbers and letters refer to locations of Deep Sea Drilling Project sites. (*After D. E. Karig et al., Varied responses to subduction in Nankai Trough and Japan Trench forearcs, Nature, 304:148–151, 1983*)

thick (7-km) sediment column is carried into the trench, the upper 1.5–1.8 mi (2.5–3 km) of sediment is accreted. Seismic reflection data from the central Aleutian Trench demonstrate that approximately the upper half of a 1.2–1.8-mi-thick (2–3-km) sedimentary section is stripped from the descending Pacific Plate and accreted.

The Middle America Trench off Guatemala, although characterized by no long-term net accretion of sediment, is the site of accretion of some trench sediment. Seismic reflection profiles and near-bottom observations demonstrate that the décollement surface is localized within trench sediments. Only the upper 330 ft (100 m) or so of trench sediment are offscraped and accreted to the toe of the slope, whereas the remainder of the trench sediments and the underlying pelagic section are subducted.

Décollement surface. The décollement surface generally is nearly planar along trenches where thick sedimentary sections are carried into the subduction zone. Along trenches where the incoming sedimentary section is thinner, basement topography controls, apparently, the shape of the décollement. Seismic reflection profiles across the toe of the Middle America Trench off Costa Rica show that a décollement separates subducting oceanic plate sediments from the continental slope (Fig. 2). Only the upper few tens of feet of sediment are offscraped, while the underlying sedimentary section is subducted. An asymmetric horst block, about 250–490 ft (75–150 m) high on its seaward side and up to 1000 ft (300 m) high on its landward side, is being subducted beneath the toe of the trench slope. Apparently, the décollement seeks to maintain a particular stratigraphic position. The décollement mimics the relief of the horst, climbing steeply from the graben to ride up over the horst.

The décollement surface in the central Aleutian accretionary wedge behaves differently along strike.

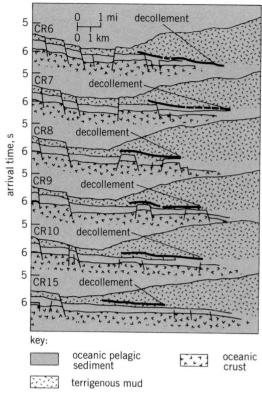

key:

▨ oceanic pelagic sediment

▨ terrigenous mud

▨ oceanic crust

Fig. 2. Line drawing interpretations of single-channel seismic reflection lines across the Middle America Trench off Costa Rica showing location of decollement that separates offscraped sediments from subducted sediments. Vertical exaggeration is about 2.5×. CR numbers identify seismic lines. (*After T. H. Shipley and G. F. Moore, Sediment offscraping and dewatering at the base of the trench slope: A three-dimensional view of the decollement, J. Geophys. Res., 1986*)

Some seismic reflection profiles show the décollement riding up over basement topography. Other seismic lines show the décollement cutting across gentle folds in the subducting sediment.

Deep Sea Drilling Project data indicate that the decollement location at the toe of the Barbados Trench slope is controlled by a boundary between smectitic and radiolarian mud in the descending oceanic plate sedimentary section. The associated increase in density and strength is believed to yield large fluid pressures that facilitate the underthrusting of the pelagic sediments beneath accreted sediments. The decollement thus appears to be controlled by the presence of zones within the sediment section containing local porosity maxima and shear strength minima.

Fate of subducted sediments. Although it has been demonstrated that variable thicknesses of marine sediments are subducted beyond the toe of the trench slope in many areas, it is not clear how far landward they are carried. It has been suggested that subducted sediments are carried down to great depths to be partially melted and to contribute to island arc volcanoes. Recent evidence, however, suggests that a considerable fraction of the subducted sediments are likely to be plastered to the bottom of the accretionary prism (underplated). Mass-balance calculations require that about half of the sediments subducted in the Middle America Trench off Mexico are underplated. The uplift history of the trench slope off Mexico has been interpreted to indicate that sediments are underplated within 9–18 mi (15–30 km) of the toe of the trench slope, causing uplift of the midslope region. Seismic reflection data from the Nankai Trough also suggest that some of the subducted sediments are underplated landward of the toe of the slope. Little is known about the scale of this process. Underplating has been suggested by some workers to be a viscous-flow process, although others have suggested that it occurs by large-scale imbrication of thrust-fault-bounded blocks.

For background information *see* FAULT AND FAULT STRUCTURES; MARINE SEDIMENTS; PLATE TECTONICS; STRUCTURAL GEOLOGY in the McGraw-Hill Encyclopedia of Science and Technology.

[G. F. MOORE; T. H. SHIPLEY]

Bibliography: D. E. Karig, H. Kagami, and DSDP Leg 87 Scientific Party, Varied responses to subduction in Nankai Trough and Japan Trench forearcs, *Nature*, 304:148–151, 1983; J. McCarthy and D. W. Scholl, Mechanisms of subduction accretion along the central Aleutian Trench, *Geol. Soc. Amer. Bull.*, 96:691–701, 1985; J. C. Moore et al., Geology and tectonic evolution of a juvenile accretionary terrane along a truncated convergent margin: Synthesis of results from Leg 66 of the Deep Sea Drilling Project, southern Mexico, *Geol. Soc. Amer. Bull.*, 93:847–861, 1982; J. C. Moore, B. Biju-Duval, and DSDP Leg 78 Scientific Party, Offscraping and underthrusting of sediment at the deformation front of the Barbados Ridge: Deep Sea Drilling Project Leg 78A, *Geol. Soc. Amer. Bull.*, 93:1065–1077, 1982; T. H. Shipley and G. F. Moore, Sediment offscraping and dewatering at the base of the trench slope: A three-dimensional view of the décollement, *J. Geophys. Res.*, 1986.

Massif

In the southern Appalachians, rocks of Grenville age [about 1.0–1.2 billion years ago (b.y.)] are exposed as massifs (terranes of older crystalline rocks) that form the basement upon which late Precambrian (0.75 b.y.) and younger rocks accumulated, both of which subsequently were deformed (about 0.30–0.45 b.y.) during the Paleozoic Era. Most of these Grenvillian massifs contain older layered (metavolcanic) gneisses which were intruded by slightly younger plutonic rocks. Still younger anorthosite (rock composed of 90 + % feldspar) diapirs cut not only these rock types but also a Grenvillian sequence of largely metasedimentary rocks in a geographically distinct terrane. During the peak of Grenville metamorphism these massifs were at depths in the crust equivalent to pressure (*P*) of 7–8 kilobars (700–800 megapascals) and subjected to temperatures (*T*) of about 750–950°C (1380–1740°F). The resulting mineral assemblages in the massifs, as reconstructed prior to faulting, indicate

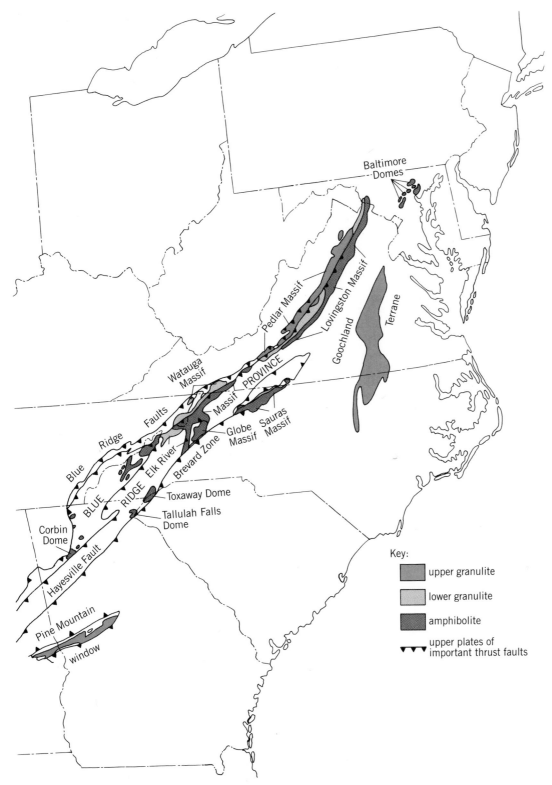

Fig. 1.   Location map showing distribution of massifs and dome containing Grenvillian rocks. Rocks in the Pedlar Massif, Goochland Terrane, and Pine Mountain window were metamorphosed at the highest *T* and *P* conditions (upper granulite facies), whereas the other massifs and domes were metamorphosed at lower *T* and *P* conditions (lower granulite and amphibolite facies).

that the intensity of Grenville metamorphism in the Blue Ridge geologic province (Fig. 1) decreased eastward from the upper (higher *T* and *P*) to the lower granulite facies to the amphibolite facies (lower *T* and *P*). Much farther to the east, upper granulite facies rocks are again exposed.

Although the existence of Grenville-age rocks in the southern Appalachians has been known for

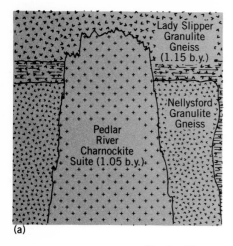 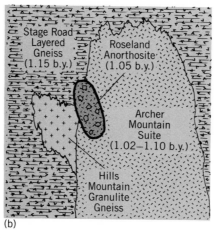 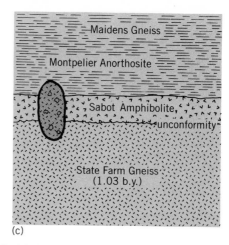

Fig. 2. Diagrammatic sketches showing relative ages as well as structural and stratigraphic relationships among rocks in the (a) Pedlar Massif, (b) Lovingston Massif, and (c) Goochland Terrane.

many years through radiometric age determinations and limited field studies, the last 10 years or so has seen increasing emphasis placed on studying these rocks. These Grenvillian rocks are the oldest known rocks in the Appalachian mountain chain and occur in three belts of massifs and domal cores exposed beneath late Precambrian (about 0.75 b.y.) and younger rocks of the Appalachian orogenic belt.

Along the western margin of the Blue Ridge mountains (Fig. 1) are the large Pedlar and Watauga massifs in Virginia and North Carolina with a train of smaller Grenvillian domes extending farther southwestward into Georgia. Juxtaposed about 0.35 b.y. over the top of the Pedlar and Watauga massifs are the Lovingston, Globe, and Elk River massifs along faults of the Fries Fault System. To the east of this belt are the Baltimore Gneiss Domes in Maryland and the Sauras Massif of North Carolina. Still farther east is the Goochland Terrane of Virginia and North Carolina, and to the southwest Grenvillian rocks are exposed in the Pine Mountain window of Georgia and Alabama.

Petrology. The Grenvillian massifs and domes of the southern Appalachians can be broadly lumped into three general types on the basis of stratigraphy and petrology: (1) deeper-seated crustal terranes characterized by charnockitic (hypersthene-bearing) plutonic rocks intruded into layered gneisses and then metamorphosed to the higher range of P and T that characterize the upper granulite metamorphic facies; (2) shallower-seated crustal terranes characterized by biotite dioritoid plutonic rocks intruded into layered gneisses and then metamorphosed to the lower range of P and T indicative of the lower granulite facies and amphibolite facies; and (3) predominantly metasedimentary and mafic metavolcanic rocks which unconformably overlie older biotite dioritoid rocks, but still are metamorphosed to the higher P and T range of the granulite facies.

The Pedlar Massif and Pine Mountain window (an erosional exposure of a lower thrust sheet beneath a higher one; Fig. 1) both contain rocks typical of type 1, although only plutonic rocks are present in the window. The Lovingston Massif as well as most of the other massifs and domes fall within type 2. The Goochland Terrane (Fig. 1) and possibly the exotic Border Gneiss associated with the Roseland Anorthosite diapir (in the central part of the Lovingston Massif) appear to be the only representatives of type 3. The Pedlar Massif, Goochland Terrane, and Lovingston Massif, discussed below, are used as representatives of each of the above types (Fig. 2).

Pedlar Massif. The rocks consist of the older, dark green, finely layered (centimeter-scale) Nellysford Granulite Gneiss apparently overlain by a younger sequence of layered (meter-scale) light gray to green gneisses that form the Lady Slipper Granulite Gneiss, which is about 1.15 billion years old. The layering in the Nellysford Granulite Gneiss is due to metamorphic segregation of quartz and feldspar versus mafic minerals into alternating centimeter-thick layers. The coarser layering in the Lady Slipper Gneiss probably reflects premetamorphic layers of differing composition. These granulite gneisses were intruded by the Pedlar River Charnockite Suite about 1.05 b.y. Subsequently (around 1.0 b.y.) all of these rocks were metamorphosed at temperatures and pressures of about 750°C (1380°F) and 7 kilobars (700 MPa), respectively. The resulting mineral assemblage typically consists of orthopyroxene + garnet + feldspar + quartz + biotite for these quartzofeldspathic rocks, and orthopyroxene + clinopyroxene + feldspar + spinel for the rarer mafic rocks. The layered gneisses and the massive charnockites are calc-alkaline and appear to have evolved from a common older (1.5 b.y.) deep crustal source—probably as igneous rocks derived by partial melting during a collision cycle of crustal plates.

Lovingston Massif. The oldest rocks of this massif are coarsely layered (meter-scale) biotite-rich gneisses, schist, and augen gneisses that make up the Stage Road Layered Gneiss, which is about 1.15 billion years old but contains detritus from 1.9-billion-year-old rocks. This layered gneiss was first intruded by small plutons of the Hills Mountain Granulite Gneiss and then (about 1.1 b.y.) by the biotite dioritoids, charnockites, and pegmatites of the Archer Mountain Suite. The diapiric emplacement of the Roseland Anorthosite and its associated Border Gneiss took place around 1.05 b.y. at a temperature of about 950°C (1740°F) and a pressure of about 8 kilobars (800 MPa). Grenville metamorphism culminated about 1.0 b.y. and produced mineral assemblages of biotite ± hornblende ± orthopyroxene ± feldspar + quartz + ilmenite in most rocks except those around the anorthosite where mineral assemblages are similar to those in the Pedlar Massif.

Goochland Terrane. The oldest known rocks in this area are part of the State Farm Gneiss which is similar to the 1.1-billion-year-old Archer Mountain Suite of the Lovingston Massif. Overlying this gneiss are the Sabot Amphibolite (mafic metavolcanic rock) with the Maidens Gneiss superjacent to the amphibolite. All of these rocks were cut by the Montpelier Anorthosite—a rock very similar to the Roseland Anorthosite that was emplaced about 1.05 b.y. Grenville metamorphism (about 1.0 b.y.) in the Goochland Terrane is preserved only as relict mineral assembages in the coarsely layered Maidens Gneiss. In the pelitic rocks the mineral assemblage typically was sillimanite + K-feldspar + muscovite, and in other rocks of the Maidens Gneiss the assemblages were orthopyroxene ± clinopyroxene ± garnet ± feldspar. These preserved assemblages suggest that Grenvillian metamorphic conditions here were similar to those in the Pedlar Massif.

Structure. Although much detailed structural work remains to be done on the massifs of the southern Appalachians, there is a sufficient amount of information to indicate complex deformation during the Grenville orogenic event similar to that which occurred in the Adirondack Mountains of New York—the classic area of Grenvillian rocks in the United States. The present-day structural relationships among the massifs have been used to reconstruct relative prefaulting geographic positions prior to Paleozoic deformation associated with the regional deformational orogenic events that formed the Appalachian orogen. This work (Fig. 1) indicates that more than 30 mi (50 km) of movement occurred along the Fries fault system when the Globe and Elk River massifs were thrust over the Watauga Massif. To the north the Lovingston Massif was juxtaposed over the Pedlar Massif (Fig. 3).

Normally, older (or deeper) rocks are thrust upward over younger (or shallower) rocks. In this case, however, rocks originally at shallower depths (Lovingston Massif) have been thrust over those originally at greater depths (Pedlar Massif) in the crust. This discrepancy can be accounted for by examining the palinspastic reconstruction that indicates significant uplift and erosion of the Pedlar Massif relative to the Lovingston Massif prior to the start (about 0.75 b.y.) of the depositional cycle of the Appalachian orogen. Perhaps the Fries Fault originated as an extensional fault (dropping the Lovingston Massif down) during the initial rifting that accompanied

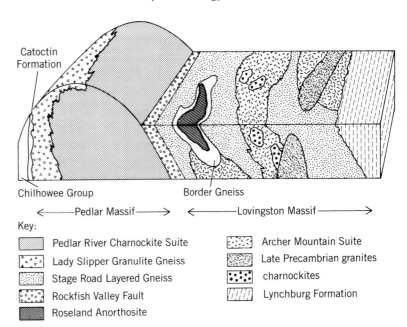

Fig. 3. Schematic block diagram showing the structural relationship between the Pedlar and Lovingston massifs in the Blue Ridge of central Virginia. The Rockfish Valley Fault is part of the Fries Fault System. (*After A. K. Shinha and M. J. Batholomew, Evolution of the Grenville Terrane in the central Virginia Appalachians, in M. J. Bartholomew et al., eds., The Grenville Event in the Appalachians and Related Events, Geol. Soc. Amer. Spec. Pap. 194, 1984*)

volcanism and plutonism about 0.75 b.y. If so, then the uplifted Pedlar Massif was eroded more, exposing rocks formed at a deeper crustal level. Subsequently, the Fries Fault was reactivated as a thrust, and the previously downdropped Lovingston Massif was then thrust over the now eroded Pedlar Massif. *See* BASEMENT ROCK.

For background information *see* FACIES (GEOLOGY); MASSIF; METAMORPHISM; MOUNTAIN SYSTEMS; OROGENY in the McGraw-Hill Encyclopedia of Science and Technology.

[M. J. BARTHOLOMEW]

Bibliography: M. J. Bartholomew, Palinspastic reconstruction of the Grenville terrane in the Blue Ridge Geologic Province, southern and central Appalachians, U.S.A., *Geol. J.* (England), 18(3):241–253, 1983; M. J. Bartholomew et al. (eds.), *The Grenville Event in the Appalachians and Related Topics*, Geol. Soc. Amer. Spec. Pap. 194, 1984; S. S. Farrar, Tectonic evolution of the easternmost Piedmont, North Carolina, *Geol. Soc. Amer. Bull.*, 96(3):362–380, 1985; H. S. Pettingill, A. K. Sinha, and M. Tatsumoto, Age and origin of anorthosites, charnockites, and granulites in the central Virginia Blue Ridge: Nd and Sr isotopic evidence, *Contrib. Mineral. Petrol.*, 85(3):279–291, 1984.

Medical parasitology

Parasites live in a unique environment in the mammalian host; as foreign organisms, they must develop evasive strategies to persist and reproduce in the presence of an efficient host immune response. Among the most intriguing of these strategies is antigenic variation, whereby the antigenic character of the parasite continually changes, thus evading clearance by the immune response. Antigenic variation finds its most elaborate expression in African trypanosomes. These flagellated protozoans cause sleeping sickness in humans and related diseases in animals; they are endemic to large areas of Africa, where a variety of wild animals serve as reservoir hosts. While the disease is usually chronic in wild animals, domestic cattle are more severely infected and suffer weight loss. Human sleeping sickness, which is usually fatal, also exists in acute and chronic forms caused by *Trypanosoma brucei rhodesiense* and *T. brucei gambiense* respectively.

Sleeping sickness has been recognized for centuries, but the infectious trypanosomes were first identified as the causative agent only in the early 1900s. Although the relapsing nature of the disease was first observed in infected patients by blood smear analysis, it was not until several decades later, in 1968, that Richard LePage demonstrated that each relapse population was composed of antigenically distinct parasites. During the course of the infection, successive populations of parasites proliferate, stimulating an effective immune response which results in fever and elimination of all parasites except those that have undergong an antigenic switch. This phenomenon of antigenic variation, repeated many times, results in the characteristic relapse pattern of disease.

The ability to change antigenic identity is inherent to the trypanosome: antigenic types are characteristic to each strain and variation is not caused by the host, since it has been shown to occur in laboratory cultures. Laboratory infections initiated with a single trypanosome have produced over 100 distinct antigenic variant types. Indeed, the upper limit of the antigenic repertoire may be closer to a thousand. Late in the infection, parasite-induced immunosuppression is evident, and frequently there is secondary infection of the respiratory system, which may result in death. Immune complex disease and neurological impairment may also result; in advanced disease the parasites invade the central nervous system, as shown by loss of sleep control and aberrant motor activity.

Variant surface antigen. The antigenic character of the trypanosome is determined primarily by its surface glycoprotein, which is organized in a monolayer external to the plasma membrane. This variant surface antigen provides a barrier against both nonspecific cytotoxic host factors and circulating antibodies which are present after the first immune response and are directed against antigen common to all variants. In each antigenic switch, the variant surface antigen is replaced by a new, antigenically distinct, glycoprotein layer. Analysis of isolated variant surface antiens has shown that while each has a unique amino acid sequence, the proteins do retain common structural characteristics consistent with necessary constraints on functionally similar proteins that must be exported and bound to the cell surface.

Comparisons between antigenically different glycoproteins reveal that the amino-terminal two-thirds of the protein is variable with respect to amino acid sequence while the carboxyl-terminal amino acids fall into three conserved homology groups. Two regions of the variant surface antigens are glycosylated: the carboxyl-terminal third of the protein (in which glycosylations vary in extent and position), and the carboxyl-terminal amino acid itself. Each variant surface antigen studied so far has amino- and carboxyl-terminal peptides not found in the mature protein; the amino terminals presumably function as signals in transmembrane transport, while the carboxyl terminals have an uncertain role. As might be expected, x-ray diffraction, DNA sequence, and immunological analyses all suggest that the more variable region of the antigen represents both the immunodominant domains and those that are exposed on the cell surface and presented to the host.

Mechanisms of antigenic variation. Mechanisms that result in an antigenic switch—the replacement of one particular variant surface glycoprotein with another—operate at the level of variant surface antigen gene expression. Trypanosomes contain hundreds of variant surface antigen genes, but only one of these in each variant type is an active template for messenger RNA synthesis. Most variant surface antigen genes are organized in tandem clusters in the genome, but many are located adjacent to chromosome ends or telomeres. The telomeric location is somehow significant since the variant surface antigen genes that are actually expressed are found only at telomeres. Although trypanosomes have a modest genetic complexity (about 4×10^7 base pairs), they have about 100 chromosomes, of which many are small (50–150-kilobase DNA), which means they have a high telomere-to-chromosome ratio compared with other organisms.

The molecular basis of antigen switching appears to require at least two types of genomic events. The first of these involves the replacement of one variant antigen coding sequence with a copy of another different antigen gene sequence. This additional gene copy invariably replaces a telomeric variant antigen gene sequence. Antigen switching by this mechanism occurs because a new, antigenically distinct, coding sequence replaces the expressed gene in a transcriptionally active telomeric site. Presumably the telomeric site remains active but now contains a new antigen gene template. The frequency of this gene conversion is quite low, since new variants occur only at a frequency of about 1 in 10^4 to 10^6 cells.

The second mechanism does not require duplication of the surface glycoprotein sequence. Rather, switching of transcription is accomplished by activation of a variant antigen coding sequence located in a previously silent telomeric site and the concomitant inactivation of another. The mechanisms by which telomeric copies of variant surface glycoprotein genes are exclusively and preferentially activated remain one of the research challenges for molecular biologists studying these processes. The gene duplication may provide a basis for changing the possible antigen repertoire by varying the telomeric candidates which can be activated by the mechanism for expression.

Trypanosome life cycle. African trypanosomes have a complex life cycle involving several developmental stages in both the insect (tsetse fly) and mammalian hosts. Antigenic variation and the production of variant surface glycoprotein are restricted to those life cycle stages that affect the mammalian host. These are the mammalian bloodstream stages, which are the proliferative extracellular forms capable of antigenic variation, and the terminal stage of insect form development, after which the trypanosome is ready for transmission to mammals. Shortly after the fly ingests parasite-infected blood, the bloodstream type of trypanosomes cease production of the surface antigen. The parasites, while reproducing in the fly, undergo several developmental transitions, during which time they migrate from the midgut to the tsetse salivary glands. The production of the variant surface antigen is somehow reactivated in the salivary gland when nondividing metacyclic forms, the terminal developmental stage in the insect, mature. When it takes its next blood meal, the fly transmits these mammalian infective metacyclic forms, and the cycle resumes.

A modest number of antigenic types occur in the metacyclic populations, but these do not include the variant types that were initially ingested by the fly. This suggests that an additional mechanism exists to ensure expression of the subset of antigenic types by metacyclic forms. The metacyclic variants appear to switch to other antigenic types at higher frequency than do variants arising during mammalian infection. The metacyclic antigen types do not appear to be either unique or stable; antigenic types that are indistinguishable from metacyclic types occur later in infections, and new antigenic types occur quite frequently in metacyclic parasites that have been repeatedly and sequentially transmitted through both insect and mammalian hosts.

Prospects for control. Control of African trypanosomiasis is a complicated problem. Substantial effort has been devoted to vector control, the development of diagnostics and chemotherapeutics, and assessment of the potential for vaccination. Vector control programs have enjoyed some success, but even successful cases have inevitably suffered reversal. Two classes of chemotherapeutics have been developed, which differ both in their ability to affect parasites in the central nervous system and in their toxicity; those that are effective against parasites in the central nervous system are highly toxic, while the other class of agents are potential carcinogens. The primary hope for chemotherapeutics is in new research that targets unique characteristics of the parasite for in-depth study. Glycolytic enzymes in trypanosomes in novel cytoplasmic organelles (glycosomes), novel metabolic processes, and novel transcriptional processes may each provide enzymatic targets with properties unique to the parasite, and are thus candidates for the development of new chemotherapeutic drugs. The hope that a vaccine directed against variant surface antigens might be forthcoming through modern biotechnological applications has diminished but has not been excluded. The basis for this pessimism comes from several observations: the variety of the repertoires exhibited by various natural populations of African trypanosomes; the lack of a fixed order of variant expression; and the ability to alter the metacyclic variant repertoire. Nevertheless, some natural populations do appear to have a limited variant repertoire, and the metacyclic repertoire is small (about 10). Hence, some limited but significant protection may be possible using complex vaccines. This may be especially true for the control of disease in cattle, where resistance to disease may eventually be genetically engineered or selected, and the exposure to diverse parasite populations minimized.

For background information *see* AFRICAN SLEEPING SICKNESS; ANTIGEN; MEDICAL PARASITOLOGY; TRYPANOSOMATIDAE in the McGraw-Hill Encyclopedia of Science and Technology.

[NINA AGABIAN; KENNETH STUART]

Bibliography: N. Agabian et al., Antigenic variation and the role of the spliced leader sequence in trypanosomatid gene expression, in *Genome Rearrangement*, Proceedings of the ICN-UCLA Symposium on Genome Rearrangement, pp. 153–172, 1985; J. E. Donelson and A. C. Rice-Ficht, Molecular biology of trypanosome antigenic variation, *Microbiol. Rev.*, 49(2):107–125, 1985; M. Parsons, R. G. Nelson, and N. Agabian, Antigenic variation in African trypanosomes: DNA rearrangements program immune evasion, *Immunol. Today*, 5(2):43–50, 1984.

Metallic glasses

Typically, metals possess crystalline structures in which atoms form ordered, repeating patterns. Metallic glass alloys differ from conventional metals in that they consist of atoms arranged in a near-random configuration. This novel, noncrystalline structure in a metallic system lends the material unique physical properties. Metallic glass alloys combine the strength and hardness characteristics of a silicate glass with the plasticity and toughness common to metals. Proper selection of composition gives metallic glass alloys corrosion resistance superior to that of stainless steel. Ferromagnetic metallic glass alloys magnetize and demagnetize more easily than any other known material.

In fact, ferromagnetism was once thought to be impossible in noncrystalline materials. It is ironic that the vast majority of commercial applications for metallic glass alloys are based on this phenomenon. The most notable ferromagnetic application involves the substitution of metallic glass for grain-oriented electrical steel for the core material in utility transformers. The resulting improvement in transformer efficiency offers the potential for saving over 3 × 10^{10} kWh of electrical energy in the United States alone.

History. In the 1930s, the first metallic alloys with noncrystalline structures were produced by vapor deposition. Some 20 years later, it was recognized that electrodeposited nickel-phosphorus, known for its excellent hardness and corrosion resistance, was also noncrystalline. Such structures were previously associated only with materials such as silicate glasses, which in effect were supercooled liquids.

Silicates are viscous in their liquid state with individual molecules (or atoms) having limited mobility. Crystallization proceeds slowly in these materials as substantial atomic motion is required to form a crystal structure. When liquid silicates are solidified, only modest cooling rates are necessary to suppress crystallization entirely.

In contrast, liquid metals are characterized by low viscosity. The constituent atoms move about quite freely, and crystallization occurs rapidly on cooling. In 1960, P. Duwez and coworkers first produced noncrystalline metallic structures directly from the liquid state. Their technique involved propelling a droplet of molten gold-silicon into a chilled copper surface. On impact, the droplet spread into a thin film and solidified. Cooling rates approaching 1,000,000°C/s (1,800,000°F/s) were achieved in this process, a technique known as splat quenching.

Process development. In the years following Duwez's pioneering works, it was demonstrated that a broad range of alloys of commercial interest, those based on iron, nickel, and cobalt, could be produced as metallic glasses by a variety of rapid solidification techniques. One feature common to these methods is that at least one dimension of the quenched material is small, typically less than 50 micrometers, to ensure that the bulk cooling rate is high. Nonetheless, rapid solidification technology has evolved so that metallic glasses are now produced in useful forms and sufficient volumes for practical applications.

Early process technology was directed at the refinement of splat quenching. Droplets of molten metal were quenched onto inclined surfaces, between two flat surfaces in a piston and anvil configuration or between twin rollers. At best, these techniques produced metallic glass samples that were slightly larger and longer. Their limitation was that they continued to process one droplet at a time.

The next generation of rapid solidification technology involved quenching a continuous jet of molten metal to produce metallic glass filaments. Experimental geometries for these jet casting processes included impinging the molten jet onto the inside or outside of a rotating drum or into the nip of twin rollers. Ribbons could be produced at linear speeds equivalent to the surface speed of the rotating drum. As these speeds reached 82 ft/s (25 m/s), the ability to cast large volumes seemed imminent. However, the dimensional stability of liquid jets restricted these techniques to the production of narrow-width (up to 0.2 in. or 5 mm) ribbon. Although jet casting could be performed in a continuous manner, the production rates associated with narrow ribbon are small.

The development of planar flow casting circumvented this barrier and made possible the continuous production of wide metallic glass strip. In this process (Fig. 1), molten metal is forced under pressure through a slotted nozzle in close proximity to a water-cooled rotating substrate. A melt puddle is formed at the exit port of the nozzle. Unlike the free molten jet, which is subject to deterioration through the forces of surface tension, the melt puddle in planar flow casting is constrained in a stable, rectangular shape by the nozzle lips and the substrate. Wide strip of uniform cross section is solidified from the puddle through the rotation of the substrate. Planar flow casting has been demonstrated in the production of 12-in.-wide (300-mm) metallic glass sheets at rates of 6600 lb/h (3000 kg/h).

Alloy development. Concurrent with the development of rapid solidification process technology came the identification of alloy systems capable of metallic glass formation. Most of these efforts have focused on transition metal–metalloid alloys, because of the relative ease of glass formation and relative stability of the glassy state in this system. These alloys consisted of mixtures of transition-metal elements such as iron, nickel, and cobalt and metalloid elements phosphorus, boron, carbon, aluminum, and silicon. The metallic content of the alloy typically has varied from about 70 to 90 at. %, depending on the specific elements involved and the desired material characteristics. Thousands of transition metal–metalloid metallic glass alloy compositions have been prepared and studied, optimizing a variety of mechanical, electrical, and magnetic properties.

Power magnetic applications for metallic glass alloys pose certain guidelines in choice of composition. Magnetic saturation induction (B_s) must be maximized to allow the transformation of large quan-

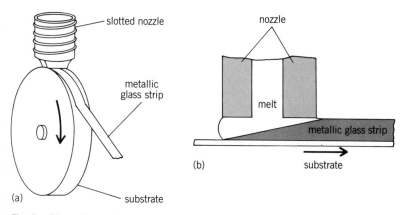

Fig. 1. Planar flow casting. (a) Schematic representation of the nozzle-substrate orientation. (b) Cross-section diagram of the nozzle-substrate, indicating melt flow and metallic glass strip formation.

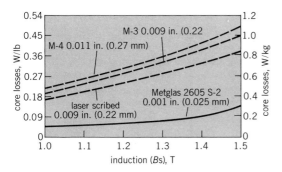

Fig. 2. Power loss values at 60 Hz for three grain-oriented electrical steels (identified as M-4, M-3, and laser-scribed) and a metallic glass alloy ($Fe_{78}B_{13}Si_9$). The measurements shown represent the thicknesses of the magnetic materials.

tities of energy within reasonable volumes. In devices such as utility transformers, power loss is minimized by aligning magnetic domains with the direction of magnetic flux. Given the absence of crystal structure and, hence, magnetocrystalline anisotropy, metallic glasses must be annealed in a magnetic field to achieve this domain alignment, and the metallic glass alloy must be stable with respect to this anneal. The annealed material must, in turn, be stable to a transformer core environment, typically oil immersion at temperatures up to 300°F (150°C), for the lifetime of the device. Finally, as the transformer core material in utility transformers represents the most significant cost component of the device, it is desirable that this be accomplished with iron, the most plentiful and least expensive metal, as the sole transition-metal component.

These varied requirements are best met with alloys of the iron-boron-silicon system such as $Fe_{78}B_{13}Si_9$. The atomic ratios of the components of this alloy were chosen to compromise competing forces. For example, increasing the iron content would increase the magnetic saturation induction, but at the expense of the thermal stability needed for annealing and life-cycle requirements. Although the magnetic saturation induction of this metallic glass alloy is slightly less than that of grain-oriented electrical steel (1.6 vs. 2.0 teslas), power loss at 60 Hz is approximately 25% that of electrical steel, as illustrated in Fig. 2.

Utility transformer applications. From recent estimates of electricity generation capacity and use in the United States it is calculated that over 5×10^{10} kWh of energy are dissipated annually in core loss. Over 60% of this total occurs in the cores of distribution transformers, which provide the final voltage step-down in the electric power distribution system. Given the efficiency of an $Fe_{78}B_{13}Si_9$ metallic glass alloy compared to conventional grain-oriented electrical steel, 75% of these core losses could be conserved.

Some barriers must be overcome in order to realize the energy-saving potential of metallic glasses. First, because the magnetic core material used in transformers represents a significant cost component

to the manufacturers of these devices, metallic glasses must be manufactured and sold at a price commensurate with their economic value in terms of energy saving. Advances in process and alloy development have made mass production of low-cost alloys possible. The last major obstacle to be overcome is the cost of ferroboron, one of the raw materials utilized in iron-boron-silicon alloy preparation. Recent developments in the efficiency of ferroboron production may lead to dramatic price reductions.

A second consideration is economical manufacture of transformers with metallic glass cores. The physical characteristics of metallic glasses, which are thin, hard, and brittle, present a formidable challenge to the designers and manufacturers of electrical devices. New design concepts, detailed engineering analysis and, probably, highly automated manufacturing technology are needed to realize the core-loss energy-savings potential of metallic glasses in core-coil assemblies that would effectively withstand and rigors of industrial usage.

The third key element is the willingness of end users, electric utilities, and industrial companies to pay today for the value of future energy savings. Without this incentive, there would be little motivation to develop the manufacturing technology for metallic glass and low-loss electrical devices. The leading electric utilities and manufacturing companies now include energy efficiency in the purchasing decisions for transformers. While there is considerable variation in the perceived value of future energy savings, sufficient incentive may be present to motivate the development of more efficient transformers with metallic glass cores.

For background information *see* METALLIC GLASSES; TRANSFORMER in the McGraw-Hill Encyclopedia of Science and Technology.

[NICHOLAS DECRISTOFARO; PATRICK CURRAN]

Bibliography: R. Hasegawa (ed.), *Glassy Metals: Magnetic, Chemical and Structural Properties,* 1983; F. E. Luborsky (ed.), *Amorphous Metallic Alloys,* 1983.

Metamorphism

One of the significant advances in recent years in the study of metamorphism is the recognition of the relationship between thermal evolution at convergent plate margins and metamorphic evolution. Geophysical heat-flow modeling of thickened crust, such as is found in orogenic belts, has provided guidelines for the interpretation of the pressure-temperature (*P-T*) evolution of metamorphic rocks. Conversely, it is now recognized that accurate *P-T* data from metamorphic rocks will provide the basic data needed to compute the thermal budget of orogeny, was well as provide important constraints on the tectonic processes that produced the metamorphism. As a result of these advances, the nature of research efforts in metamorphic petrology is shifting away from a study of the rocks as an end in itself and toward a study of the rocks as a means to understand the evolution of the continental crust.

Thermal modeling. The thermal models computed to date are one-dimensional models (that is, no lateral heat transport) that approximate the orogenic process as a simple crustal thickening event such as

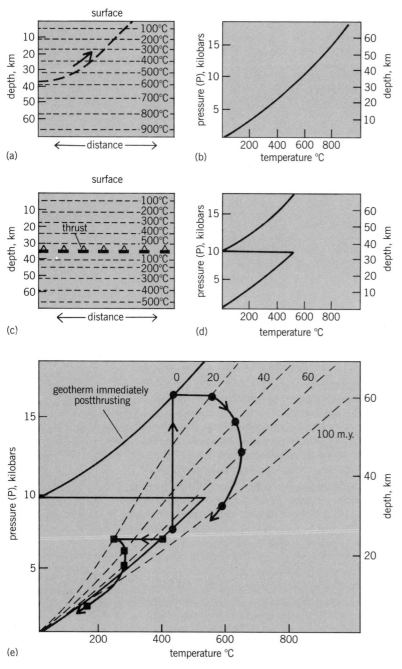

(a)

(b)

(c)

(d)

(e)

Fig. 1.   Thermal model of continental crust thickened by thrusting. (a) Prethrusting thermal structure of the crust; broken line is the incipient trace of the thrust fault that will thicken crust. (b) Prethrusting steady-state geotherm. (c) Immediately postthrusting thermal structure. The crust is doubled in thickness, and the thermal structure of the crust is repeated. (d) Perturbed postthrusting geotherm with a sawtooth appearance. (e) Evolving geotherms and resultant rock P-T paths. With time, the perturbed geotherm d evolves toward a steady-state value as shown by the broken lines. Numbers on geotherms are millions of years. Thermal relaxation is accompanied by erosion, so the rocks follow P-T paths as shown by the solid circles (lower plate) and solid squares (upper plate). Arrows indicate the direction of changing P and T. 1 km = 0.6 mi; 1 kilobar = 10^2 MPa. °F = (°C × 1.8) + 32. (After P. C. England and A. B. Thompson, Pressure-temperature-time paths of regional metamorphism, J. Petrol., 25:894–955, 1984)

overthrusting on a continental scale. The various factors that enter into the thermal models are discussed in the following paragraphs. The preorogenic crust is assumed to be in a steady state with a steady-state geothermal gradient (Fig. 1a and b). This geothermal gradient is a function of (1) the amount of heat flow from the underlying mantle, (2) the amount of heat production within the crust from the decay of radioactive elements, and (3) the thickness of the crust (thicker crust contains more heat-producing radioactive elements).

A large-scale tectonic disturbance is postulated such as subduction or continental overthrusting. One of the important results of the thermal modeling is the realization that tectonic processes take place on a time scale that is considerably shorter than thermal relaxation. Therefore, the thrusting event produces a large-scale perturbation in the thermal structure of the crust. For example, continental overthrusting, such as occurs today where the Tibetan Plateau is overthrusting the continent of India, results in an inversion of the thermal structures so that hot rocks are placed on cool rocks (Fig. 1c and d).

Relaxation of the perturbed thermal structure is accompanied by uplift and erosion of the thickened crust (Fig. 1e). With time the perturbed geotherm will decay by conduction toward a steady-state value that is determined by the mantle heat flow, the concentration of radioactive elements in the crust, and the crustal thickness. In addition, the thickened crust is isostatically unstable and will tend to rise and form mountains (such as the Himalayas). Because erosion rates are proportional to the height of mountains, the greater the uplift rate, the greater the erosion rate and consequent unloading of the crust. Crustal unloading decreases the pressure on the underlying rocks, as well as alters the steady-state geotherm configuration by changing the thickness of the crust.

The primary conclusion to be drawn from this modeling is that the geotherm within a region that is tectonically active will not be in steady state but will be constantly evolving. Therefore, a rock sitting within the crust will follow a path in pressure-temperature space (P-T path) that depends on the nature of the thermal disturbance, the erosion rate, and the thermal characteristics of the crust. For crustal thickening models, the paths show initial heating accompanied by decompression, followed by cooling with decompression. This result was rather surprising because most metamorphic petrologists had assumed that heating was accompanied by increased burial, not by exhumation. Figure 1e shows two idealized P-T paths for rocks from the upper and lower plates of a thrust fault derived from thermal modeling.

Metamorphic P-T paths. Metamorphic rocks are found in the cores of mountain belts, having been crystallized at elevated temperatures and pressures. The thermal models provide a framework for the interpretation of the P-T evolution of these rocks, but it is the metamorphic rocks themselves that record a history of their evolution.

There are several ways that metamorphic *P-T* paths can be determined, some of which are rather new. The classic approach is to investigate the texture of the rock in order to determine the sequence of mineral growth and the reaction history. These observations can then be related to theoretical or experimental calibrations to constrain the *P-T* history. For example, if andalusite (a low-pressure polymorph of Al_2SiO_5) is found to be replaced by kyanite (the high-pressure polymorph of Al_2SiO_5) in a rock, it can be concluded that the pressure of the rock must have increased, with the temperature fairly constant.

Geothermometry and geobarometry, which involve the solving of pressure- and temperature-sensitive equilibria for *P* and *T*, are used to obtain an instantaneous *P-T* point where a rock most nearly achieved chemical equilibrium. Many useful equilibria are available that can be applied to pressure and temperature estimation in metamorphic rocks of most bulk compositions and mineralogies. However, this approach yields only one point on the *P-T* path—generally the point where the temperature is maximum.

Additional information on the *P-T* path can be obtained from geothermometry and geobarometry on inclusions of one mineral within another if the included mineral has been chemically isolated from the surroundings since the time of inclusion. In this case it is assumed that the composition of the included mineral reflects the equilibrium composition at some earlier time in the rock's history.

Another method to obtain *P-T* paths from metamorphic rocks is to use chemically zoned minerals. If pressure and temperature conditions change as a rock recrystallizes, the minerals in the rock will change composition as they grow because the reactions producing the minerals are pressure- and temperature-sensitive. If intracrystalline diffusion is slow, the mineral does not homogenize as it grows, but will zone chemically. Garnet and plagioclase are two metamorphic minerals that commonly display zoning. If the reactions producing these minerals are known, the chemical zoning can be used as a type of chemical recording of the reaction history, and thus the *P-T* history.

Fluid inclusions also have the potential of providing *P-T* information. Fluid inclusions are small (1–50-micrometer) bubbles enclosed in metamorphic minerals (such as quartz) that contain various amounts of H_2O, CO_2, CH_4, NaCl, KCl, and other compounds. The density and composition of a fluid inclusion define a line of constant density in *P-T* space (isochore), which the rock must have passed through at some time in its history. Fluid inclusions are most useful for determining the late portion of a *P-T* path, because early-formed fluid inclusions generally decrepitate during the uplift cycle.

Metamorphic P-T paths and tectonic evolution. *P-T* paths have been determined for rocks from several tectonic environments, and these correspond

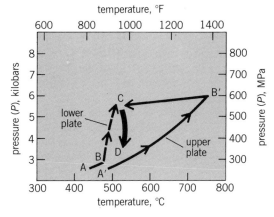

Fig. 2. Pressure-temperature paths for rocks that crop out near Bellows Falls, Vermont. The two plates of a thrust sheet are represented; arrows show the direction of change in pressure and temperature. Each plate underwent a different thermal history prior to thrusting and the same thermal history following thrusting. The lower plate (dashed line) experienced low *P-T* metamorphism prior to thrusting (A to B) and an isothermal increase in pressure during thrusting as the upper plate rode over it (B to C). The upper plate (solid line) experienced high *P-T* metamorphism just prior to thrusting (A′ to B′) and was cooled substantially during thrusting (B′ to C′). The thrust was fully emplaced at C, and both plates subsequently followed the same thermal history (C to D).

well with the paths predicted from thermal modeling.

As an example, Fig. 2 shows *P-T* paths from the upper and lower plates of a thrust fault in the vicinity of Bellows Falls, Vermont. The two rock units have different thermal histories prior to thrusting and show different effects as a result of the thrusting: the lower plate shows a near-isothermal increase in pressure whereas the upper plate shows near-isobaric cooling as the thrust is emplaced (B to C and B′ to C in Fig. 2). Following thrusting, the units are juxtaposed and follow the same *P-T* paths as the rocks move toward the surface. The implication of these results is that metamorphic *P-T* paths can be used to help map thrust faults where the geology is ambiguous or the upper plate is eroded away.

Time and metamorphic P-T paths. Metamorphic rocks endure at elevated temperatures and pressures for millions of years, and the age determined by most geochronologic techniques will not be the age at which a particular mineral grew, but rather the age when the system closes to diffusive homogenization. Because diffusion is thermally activated, closure generally occurs when the system has cooled sufficiently. The specific temperature at which closure occurs depends on the mineral system, so it is possible to obtain a set of closure ages, each for a different mineral system and each corresponding to the time when the rock passed through a specific temperature. This type of time information can only be used on the cooling portion of the cycle. Much future research will be directed toward determining pressure-temperature-time paths for orogenic terranes, and relating these to thermal and tectonic models for the evolution of the continental crust.

For background information *see* GEOCHRONOMETRY; GEOLOGIC THERMOMETRY; METAMORPHISM; OROGENY; PLATE TECTONICS in the McGraw-Hill Encyclopedia of Science and Technology.

[FRANK S. SPEAR]

Bibliography: P. C. England and A. B. Thompson, Pressure-temperature-time paths of regional metamorphism, *J. Petrol.*, 25:894–955, 1984; F. S. Spear et al., *P-T* paths from garnet zoning: A new technique for deciphering tectonic processes in crystalline terranes, *Geology*, 12:87–90, 1984; F. S. Spear and J. Selverstone, Quantitative *P-T* paths from zoned minerals: Theory and tectonic applications, *Contrib. Mineral. Petrol.*, 83:348–357, 1983.

Meteorological radar

Mesosphere-stratosphere-troposphere (MST) radar is an ultrasensitive radar system unaffected by precipitation or in cloud cover. It is used to probe the structure and dynamics of the atmosphere between approximately 1 and 100 km (0.6 and 60 mi), a region known as the middle atmosphere. While most MST radar systems operate at frequencies about 50 MHz, frequencies in the neighborhood of 400 MHz or higher are used for studies of the stratosphere and troposphere. Typical peak power ranges from 1 kW to 1 MW. Most systems use antenna arrays with dimensions from 1000 to 10,000 m² (11,000 to 110,000 ft²). A part of such an array is shown in the illustration. With their good time and range resolutions, MST radar systems find application as unique ground-based instruments for continuous remote sensing of the dynamic state of the middle atmosphere. They can observe and measure winds, waves, turbulence, frontal passage, atmospheric sta-

bility, and other meteorological phenomena.

Radar echoes. Echoes received by MST radars are caused by scattering and reflection from small-scale structures in the refractive index of the atmosphere which arise primarily from atmospheric turbulence. For frequencies used in MST radar operations, the refractive index of the atmosphere is given by Eq. (1), where e is the partial pressure of water

$$n = 1 + 3.7 \cdot 10^{-1} \cdot e/T^2 + 77.6 \cdot 10^{-6} \cdot p/T - 40.3 \cdot N_e/f_0^2 \quad (1)$$

vapor (humidity) in millibars, p the pressure in millibars, T the Kelvin temperature, N_e the number density of free electrons per cubic meter, and f_0 the radar operating frequency in hertz. Variations of e, p, T, and N_e are induced by the effects of turbulence. These give rise to the variations in the refractive index which become the targets for the radar. The radar echoes are caused mainly by scattering from variations with the scale of one-half the radar wavelength; pertinent scale sizes range from a fraction of a meter to a few meters (1 m = 3.3 ft). The radar reflectivity contributions for an MST radar operating at approximately 50 MHz varies with altitude. For the lowest few kilometers (1 km = 0.6 mi), turbulence-induced humidity fluctuations dominate. Above that, temperature fluctuations become the major contributors up to approximately 50 km (30 mi). In general, contributions due to pressure fluctuations are negligible. The decreasing turbulence strength in the upper stratosphere above 35 km (21 mi) or so makes the radar reflectivity so low that it becomes exceedingly difficult to obtain useful data. Above 50 km (30 mi), however, the presence of the free electrons in the atmosphere enhances the radar reflectivity. Free electrons become the major contributor in the mesosphere. This part of the reflectivity depends on electron density and the associated ionization gradient caused by turbulence. Since the density of free electrons in the mesosphere is directly proportional to the solar zenith angle, the radar reflectivity at mesospheric height is appreciable only during daytime. Nighttime mesosphere echoes for MST radars are sporadic. They can be supplemented by echoes arising from ionized meteor trails that occur during the day as well as the night. After signal processing of the received echoes, the power and the Doppler spectrum of the backscattered radar signal can be obtained for specified range and time resolutions. Atmospheric parameters can then be derived from these data.

Wind velocity measurement. As in any Doppler radar application, the mean Doppler frequency of the received signal contains information about the component of the velocity of the target along the radar beam direction, the so-called line-of-sight velocity. For the MST radar, the targets are the small-scale structures in humidity, temperature, or electron density induced by turbulence. Many exist in the scattering volume of the radar, defined by the beamwidth and a distance corresponding to one-half the pulse length of the radar. Since the structures

Yagi antennas of Chung Li radar in Taiwan.

are carried by the wind, a measure of their mean velocity yields the wind velocity. Indeed it has been established by comparison with winds measured by other techniques that the mean Doppler shift of the radar signal is an accurate measure of the mean radial wind in the scattering volume.

To measure three-dimensional vector wind, a minimum of three antenna beam-pointing directions are needed. If the pointing directions are chosen to observe orthogonal, uniform, horizontal wind components u and v with antenna elevation pointing angle θ_c and the vertical component w with vertically pointing antenna, then the radial Doppler velocities (V_1, V_2, and V_3) measured by the radar are related to the wind as shown in Eqs. (2). Typical MST radar

$$
\begin{aligned}
V_1 &= u \cos \theta_c + w \sin \theta_c \\
V_2 &= v \cos \theta_c + w \sin \theta_c \qquad (2) \\
V_3 &= w
\end{aligned}
$$

systems operate with elevation angle θ_c about 75°. Assuming horizontal uniformity of the wind, the three-dimensional wind vector (u,v,w) can be determined from the measured radial velocity V_1, V_2, and V_3. Horizontal gradients of the wind field cause error in determining the horizontal components of the wind. A five-beam or seven-beam system, operating on the assumption of a three-dimensional linearly varying wind field, can be applied to improve the accuracy for estimating locally varying winds, such as jet streams. In many forecasting applications the vertical velocity of the atmosphere is the single most important variable. The unique capability of MST [or stratosphere-troposphere (ST)] radar for direct continuous measurement of vertical velocity can provide valuable input to synoptic-scale analysis and forecasting.

In addition to the Doppler technique, another method known as the spaced antenna method has been used in MST radars to measure the horizontal wind fields. This technique measures the temporal and spatial variations of the radar echoes by three or more vertically beamed antennas spaced a few meters apart. A full correlation analysis among the signals received by the different antennas yields the horizontal wind vector.

Wind profilers. A class of MST radars known as wind profilers has been built and tested in the past few years. These are relatively low-cost radars that are capable of measuring vertical profiles of horizontal winds in the troposphere and lower stratosphere (approximately 1 to 18 km or 0.6 to 11 mi). They are specially designed to be operated continuously and unattended in nearly all weather conditions. Data from the remote radar sites are automatically transmitted to a central control station. The operating frequency of the existing wind profilers are about 50 MHz (very high-frequency, or VHF, band) and 400 MHz and 900 MHz (ultrahigh-frequency, or UHF, band). For the VHF system, height resolution of approximately 300 m (980 ft) can be achieved, while 100-m (330-ft) resolution can be obtained with UHF systems. Experiments have been planned, and

some carried out, to test the various possible applications of the wind profilers. Data from a network of wind profilers have been used in real-time for testing of improved weather forecast methods and for mesoscale research. Potential future applications of data from large networks of wind profilers include aviation weather forecasting and mesoscale and synoptic-scale investigations.

Middle atmosphere waves. In addition to measuring the gross features of the total mean wind field, MST radar can be applied to study smaller-scale, time-varying phenomena such as turbulence and wave motions in the atmosphere. Two features of MST radars make them well suited for this purpose: (1) fine temporal and spatial resolution, which can be as fine as a few seconds and a few tens of meters in regions where the echoes are strong; and (2) continuity of the data in time and in height. The good resolution makes it possible to study the propagation and evolution of internal gravity waves. The influence of small-scale motions such as gravity waves, turbulence, and convection of the general circulation of the middle atmosphere is appreciable; dissipation and momentum deposition by these waves in the mean flow can be followed.

The continuity of MST radar data can be applied to the study of atmospheric tides and planetary waves. Of particular interest are the long-period equatorial waves (Kelvin waves and mixed Rossby-gravity waves) which have a strong influence on the dynamics of the equatorial atmosphere. Individual wave events and their relation to the atmospheric dynamic states such as jet streams and convective activities can be diagnosed by the radar. The data can also be used to obtain the power spectral density of the atmospheric wind velocity fluctuations from a few minutes to many hours. This information can be applied to construct the height profiles of the average kinetic energy distribution in the atmosphere, which are very useful in studies of energy dissipation rates of wave processes.

Morphology and structure of turbulence. MST radar is used to measure turbulence strength by two methods. In the first the absolute power of the received echoes is measured and converted to turbulence strength. The second uses the Doppler spectral width of the received signal, which is related to the dissipation rate of turbulence energy. Turbulence in the middle atmosphere is intermittent in time and inhomogeneous in space. Regions of active turbulence are often confined to thin horizontal layers. Data from a network of MST radars will facilitate study of the morphology of the turbulence. In addition, by using high-spatial-resolution radars with narrow antenna beams, it is possible to study the space-time structure of atmospheric turbulence layers and estimate turbulence intensity and turbulent diffusion. It is possible to investigate the relation between turbulence and breakdowns of gravity wave and tides.

Determination of tropopause height. Hydrostatic stability in the atmosphere plays an important role in determining the dynamics of the atmosphere. Vertical motions and mixing are suppressed in highly stable atmospheric regions, making it possible for large vertical gradients in temperature, wind, humidity, and so on, to develop. These gradients cause Fresnel (specular) reflection of the radar signal. Early in the development of MST radar at VHF frequencies, it was discovered that considerably higher echo power is received when the antenna is pointed vertically than when it is obliquely pointed. This aspect sensitivity is the consequence of specular reflection of the radar signal from stable layers. Enhanced echo power and narrowed Doppler spectral width of the vertical returns have been shown to be directly related to the stability of the atmosphere. Therefore, power profiles of vertically pointed MST radar echoes can be used to indicate the stability of the atmosphere, especially at tropospheric and stratospheric heights, from which temperature gradients can be estimated.

This method of detecting stable atmospheric regions provides a direct and objective technique for determining the height of the tropopause. When the tropopause is well defined and consists of a pronounced discontinuity in atmospheric stability, its height is given by the height at which the power of the radar echo from the vertically pointed antenna starts to increase (above the altitude of the 500-millibar or 50-kilopascal level). Tropopause height can be determined from this technique to within 1 km (0.6 mi). This is far better than the resolution of current ground-based or satellite-based radiometric temperature sounders. Therefore, use of VHF radar in conjunction with radiometric sounders should offer an improved technique for determining atmospheric mean temperature profiles.

Observation of frontal passage. Since radar echoes are tightly related to vertical temperature structure of the atmosphere, VHF radars can be used to locate fronts. As with the tropopause, enhanced radar reflectivity appears when regions of strong temperature gradients associated with the fronts pass by the radar. In addition to locating the frontal structure, these radars can provide information about wind and vertical velocity associated with the fronts. For example, one experiment has shown that during the passage of a warm front, radar measured upward mean vertical velocity ahead of the front and downward motion behind the front. The vertical velocity measurements will improve understanding of how precipitation develops in association with the frontal structure. The good temporal and height resolution of the radars allows study of detailed structures, and it may help clarify the dynamics of the mixing process that tend to damp the frontal system. The radars may also provide a new tool for forecasting on shorter time scale.

Further applications. Additional applications of MST radar include the study of troposphere-stratosphere coupling, determination of the mean zonal and meridonal wind components in the mesosphere, investigation of mesocale convective systems, and transport of minor constituents and pollutants. There is also the potential for operational use for nowcasting and forecasting, especially when coupled with the radiometric temperature sounder.

For background information *see* DOPPLER RADAR; FRONT; RADAR METEOROLOGY; WEATHER FORECASTING AND PREDICTION in the McGraw-Hill Encyclopedia of Science and Technology.

[C.H. LIU]

Bibliography: R. J. Doviak and D. S. Zrnic, *Doppler Radar and Weather Observations*, 1984; E. E. Gossard and K.C. Yeh (eds.), Special issue on radar investigations of the clear air, *Radio Science*, March/April 1980; C. H. Liu and S. Kato (eds.), Special issue on MST radar, *Radio Science*, November/December 1985.

Microbial ecology

Many microbial habitats in nature, such as bacterial films at surfaces in aquatic environments, are characterized by steep gradients of chemical species produced or consumed by the metabolic activities of the densely packed microorganisms. Knowledge about the concentrations of the most important chemical species in these microenvironments is essential for the understanding of microbial processes in nature, such as decomposition and microbial photosynthesis. Until recently, most information about the microbial transformations in biofilms and similar substrates with dense microbial growth was obtained by measuring the exchange of chemical species with the surrounding macroenvironment. Such approaches are often unsatisfactory, however, since they give information about the net result of the metabolic processes but tell nothing about the spatial distribution of these processes within the substrate. Often very tight couplings exist between microbial transformations occurring in neighboring microlayers; the measurement of a net import to or export from the whole community may therefore be directly misleading.

One approach to the study of the chemical environment on a scale which is relevant to the microscopic world of bacteria and microalgae is based on the use of gas-selective and ion-selective microelectrodes. Such microelectrodes have been extensively used in human physiology since the 1950s, but their use in microbial ecology has only recently begun. Oxygen, dissolved sulfide, and pH can now be analyzed in microbial communities with a high spatial resolution. This article focuses on the microprofiles of oxygen, photosynthesis, and oxygen consumption which can be obtained from data gathered from oxygen microelectrodes.

Oxygen microelectrode. An oxygen microelectrode senses oxygen only at its very tip, which can be only 2 micrometers in diameter (Fig. 1). By using such an electrode, oxygen readings can be made with a spatial resolution of a few micrometers. The small size of the electrode makes it extremely fast in responding: a 90% response to changes in oxygen concentration is often achieved within 0.2 s. These

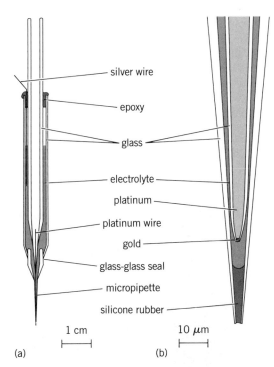

Fig. 1. Oxygen microelectrode used for measurement of oxygen in microbial communities. (*a*) Microelectrode elements. (*b*) Enlargement of tip. (*After N. P. Revsbech and D. M. Ward, Oxygen electrode that is insensitive to medium chemical composition: Use in an acid microbial mat dominated by Cyanidium caldarium, Appl. Environ. Microbiol., 45:755–759, 1983*)

features make it possible to obtain highly accurate readings of oxygen which are very well defined in space and time.

The work with tiny sensors such as the one in Fig. 1 requires access to specialized equipment such as micromanipulators and a sensitive ammeter to amplify the weak signal (10^{-12} to 10^{-10} ampere) produced by the microelectrode. The fragility of the electrodes, which may be the most apparent limitation, is, however, not a major problem. Microelectrode measurements have been performed deep into substrates such as sandy sediments and soils which at first glance may seem very badly suited for the introduction of a glass tip a few micrometers thick.

Oxygen in biofilm. Microelectrode measurements have revealed that oxygen levels in biofilms of many types are highly dynamic. For example, oxygen profiles in a biofilm rich in cyanobacteria (cyanobacterial mat) change radically after the light is removed (Fig. 2*a*). In the cyanobacterial mat, the oxygen penetration was only about 0.05 mm in the dark-incubated biofilm; and even in the light, when photosynthetic oxygen production by the cyanobacteria caused an oxygen concentration in the photic layer (the uppermost layer) of four times atmospheric saturation, oxygen penetrated only 0.9 mm into the biofilm. The effect of the light-dark shift on the oxygen concentrations is thus very dramatic. Other chemical species may display similar changes between day and night. This has been demonstrated

for sulfide and pH, but even parameters such as ammonium and phosphorus may show large diurnal oscillations. Oscillations in pH from 7.3 in the dark to 10.4 in the light have been measured in photosynthetically active biofilms collected at marine habitats.

The highly fluctuating chemical environment within photosynthetically active biofilms selects for either very versatile organisms or for microorganisms that can move. Diurnal migrations of mat microorganisms have often been described. For example, the surfaces of sulfide-rich microbial mats are often white in the dark, whereas they are green in the light. The white color is due to sulfide-oxidizing bacteria that move toward the oxygen-sulfide interface. At night this interface is found very close to the sediment surface, though it is found at some depth during the light hours.

Photosynthesis. The rate of photosynthesis in each layer of biofilm can be calculated from the rate of decrease in oxygen concentration just after turning off light. The spatial resolution of the determination of photosynthetic rate is about 0.1 mm when the rate of decrease in oxygen concentration 1 s after extinguishing light is used for the calculation. The vertical distribution of photosynthetic activity in a cyanobacterial mat as measured by microelectrodes is shown in Fig. 2*b*. The thickness of the photic zone was in this case only 0.4 mm, but extremely high rates of photosynthesis (up to 199 mmol O_2 dm^{-3} h^{-1}) were measured within this thin layer. The total photosynthetic rate, integrated over the 0.4-mm layer having photosynthetic activity, was 43 mmol O_2 m^{-2} h^{-1}, which corresponds to

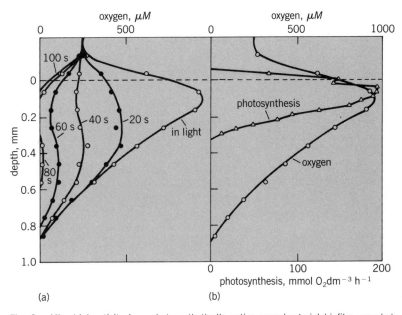

Fig. 2. Microbial activity in a photosynthetically active cyanobacterial biofilm sampled from the surface of an organic-rich, coastal sediment. (*a*) Oxygen profile in light (250 μEinst m^{-2} s^{-1}) and at 20-s intervals after darkening. (*b*) Steady-state oxygen profile in the light and profile of photosynthetic activity. Some photosynthetic activity was measured in the water immediately above the surface of the biofilm due to the only 0.1-mm spatial resolution of the photosynthesis determinations. All experiments were performed at the natural water temperature, 21°C (70°F).

about 0.4 g carbon $m^{-2} h^{-1}$. This photosynthetic rate is similar to the maximum rates measured in the water of extremely nutrient-rich lakes, where the photic zone is about 1000 times thicker. The data illustrated here are all from the same very active biofilm. Less rich photosynthetic communities, such as the benthic communities on tidal flats, often show considerably lower activities. The benthic photosynthetic activity, however, often exceeds the photosynthetic activity in the overlying water.

Respiration. The respiratory activity in biofilms can be very high. Several types of sewage treatment plants, for instance, are based on microbial degradation of dissolved organic matter in biofilms (trickling filters, activated sludge, and so on). It is often of considerable interest to know the vertical distribution of respiratory activity within a biofilm. Such information may be obtained by microelectrode analysis. Figure 3 shows a profile of oxygen consumption calculated from the data shown in Fig. 2. The calculation was based on computer treatment of a diffusion model, and knowledge of the diffusion coefficient of oxygen within the biofilm is therefore essential. Fortunately, diffusion coefficients for oxygen can be determined with an accuracy of about 1% by microelectrode measurement of oxygen diffusion in poisoned biofilm. The maximum of respiratory activity in the photic region is to be expected, as the high production of organic matter in this layer inevitably also leads to high rates of respiration. The maximum at the oxic-anoxic interface is due to oxidation of reduced inorganic and organic molecules and ions diffusing up from the anoxic region. Sulfide oxidation has been shown to account for most of the oxygen consumption at the oxic-anoxic interfaces in some microbial mats collected from marine sediments.

Application in other microbial communities. It has been shown above how biofilms can be analyzed by use of oxygen microelectrodes. Other microbial communities may be analyzed in a similar manner.

Analysis of marine sediments down to 20 m (66 ft) depth has been conducted, and analysis at 4000 m (2.5 mi) depth is planned. Algal mats in hot springs at 70°C (94°F) as well as arctic sediments at −1°C (30°F) have been analyzed. Microelectrode studies of symbiotic associations containing microalgae (such as foraminifera, corals, and salamander eggs) have also been conducted. These associations often have very high photosynthetic rates, and oxygen and pH may fluctuate during light-dark cycles as illustrated for oxygen in Fig. 2. Microelectrodes have also proven to be a valuable tool for the study of pure cultures of bacteria living in steep chemical gradients. The sulfide-oxidizing bacteria *Beggiatoa* spp. have been studied while growing in gradients of oxygen and sulfide. The measurements gave information about the microoxic niche occupied by these bacteria, about their taxis, and about their growth yield when grown with sulfide as the only energy substrate.

Electrodes can be made very sturdy for use in soils. Microelectrode measurements of oxygen in soil crumbs have been correlated with rates of denitrification, which was shown to occur only when the soil crumb contained an anoxic center.

For background information *see* FRESH-WATER ECOSYSTEM; MARINE ECOSYSTEM; MARINE MICROBIOLOGY; PHOTOSYNTHESIS.

[NIELS P. REVSBECH]

Bibliography: E. Gnaiger and H. Forstner (eds.), *Polarographic Oxygen Sensors: Aquatic and Physiological Applications*, 1983; B. B. Jørgensen and N. P. Revsbech, Colorless sulfur bacteria, *Beggiatoa* spp. and *Thiovolum* spp., in O_2 and H_2S microgradients, *Appl. Environ. Microbiol.*, 45:1261–1270, 1983; N. P. Revsbech et al., Microelectrode studies of photosynthesis and O_2, H_2S, and pH profiles of a microbial mat, *Limnol. Oceanogr.*, 28:1062–1074, 1983; N. P. Revsbech and D. M. Ward, Oxygen microelectrode that is insensitive to medium chemical composition: Use in an acid microbial mat dominated by *Cyanidium caldarium*, *Appl. Environ. Microbiol.*, 45:755–759, 1983.

Microbial geochemistry

Pollution of natural water supplies is a serious problem in modern society. Recent studies have revealed that various microorganisms can have both beneficial and deleterious effects in some portions of the hydrosphere. This article discusses the use of microorganisms to transform organic pollutants into harmless substances in the groundwater, and the role of some microorganisms in producing acid mine drainage.

Organic pollutants in groundwater. Groundwater pumped from aquifers provides water in many industrialized areas. Unfortunately these sources have frequently become polluted by organic chemicals that are released into aquifers from accidental spills, underground storage tanks, landfills of industrial or municipal wastes, septic tanks, industrial impoundments, agricultural pest control, and even

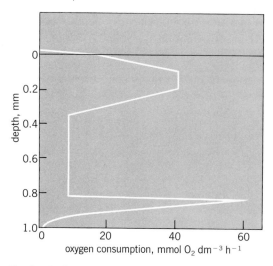

Fig. 3. Profile of oxygen consumption in the biofilm, calculated from the data shown in Fig. 2.

rainwater containing dissolved pollutants. While some of the microorganisms found in aquifers can transform a number of pollutants into harmless substances, occasionally new substances are formed that are even greater threats. Whether a particular class of organic contaminants will be degraded depends on the physiological action of the organisms, which in turn is influenced by the geochemical environment. Thus, degradation of most subsurface pollutants will be accomplished only where groundwaters possess certain geochemical properties.

Groundwater containing oxygen. Many water-table aquifers contain dissolved oxygen. These aquifers can support aerobic microorganisms that can degrade a wide variety of organic contaminants, including: benzene, toluene, the xylenes and other alkylbenzenes, acetone, methylethylketone, methanol, ethanol, and *t*-butanol released into groundwater from gasoline or solvent spills; naphthalene, the methylnaphthalenes, fluorene, acenaphthene, dibenzofuran, and a variety of other polynuclear aromatic hydrocarbons released from spilled diesel or heating oil; and many methylated phenols and heterocyclic organic compounds seen in certain industrial wastewaters. Synthetic organic compounds can also be degraded, including: mono-, di-, and trichlorobenzenes; the mono-, di-, and trichlorophenols; some of the soft detergents such as linear alkyl sulfonates and alkyl phenol ethoxylates; the detergent builder nitrilotriacetic acid (NTA); and some of the simpler chlorinated compounds such as methylene chloride (dichloromethane) and 1,2-dichloroethane.

The extent of biodegradation in groundwater depends on oxygen concentration. For the compounds listed above, approximately two parts by weight of oxygen are required to completely metabolize one part by weight of the organic compound. For example, microorganisms in a well-oxygenated groundwater containing 4 milligrams/liter of molecular oxygen can degrade only 2 mg/liter of benzene; the solubility of benzene in water, 1780 mg/liter, is much greater than the groundwater capacity for aerobic degradation. The prospects for aerobic metabolism of pollutants will depend on their concentration, as well as on the concentration of other degradable organic materials in the aquifer. Concentrated plumes of organic contaminants cannot be degraded aerobically until dispersion or other processes dilute the plume with oxygenated water.

Many commonly encountered organic pollutants in aquifers are synthetic organic solvents that do not ordinarily degrade in oxygenated waters. Examples include tetrachloroethylene (also known as perchloroethylene, PCE), trichloroethylene (TCE), *cis*- and *trans*-1,2-dichloroethylene, ethylene dichloride (1,2-dichloroethane), 1,1,1-trichloroethane (TCA), 1,1,2-trichloroethane, carbon tetrachloride, and chloroform.

Groundwaters producing methane. Methane-producing organisms (methanogens) find molecular oxygen highly toxic. These organisms are active only in highly reduced environments and produce methane by fermentation of several simple organic compounds such as acetate, formate, methanol, or methylamines. Molecular hydrogen can support a form of respiration in which the hydrogen is used to reduce inorganic carbonate to methane. Although the microorganisms that actually produce the methane use a very limited set of organic compounds, they can act in consort with the other microorganisms that break more complex organic compounds down to substances that the methanogenic organisms can use. These partnerships or consortia can totally degrade a surprising variety of natural and synthetic organic compounds.

The rates of reactions are usually slow and often require long lag periods before active transformation begins. Microbiologists, accustomed to microorganisms that grow to high densities in only a few days, rarely conduct experiments that last longer than a few weeks. However, the residence time of organic pollutants in aquifers is at least months to years, and is frequently decades to centuries. As a result, much of what was learned in earlier laboratory studies cannot be applied to the subsurface environment. Currently microbiologists are reexamining the potential for biodegradation of organic contamination in methane-yielding groundwaters and are finding many unexpected reactions.

Previously it was believed that molecular oxygen was required as a cosubstrate for the enzyme that initiates metabolism of compounds such as benzene, toluene, the xylenes, and other alkylbenzenes. Thus, their metabolism would not be expected in methanogenic environments. Recently the metabolism of these substances was demonstrated in methanogenic river alluvium that had been contaminated with landfill leachate. When radioactive toluene was added to this material, at least half the carbon was metabolized completely to carbon dioxide. The same material also metabolized several methyl- and chlorophenols. In this reaction pathway, chlorine atoms were removed first, then the phenol was consumed, rather than a direct attack occurring on the phenol nucleus while the chlorines were still attached.

Halogenated solvents that persist in oxygenated groundwater can be transformed in methanogenic groundwater (see illus.). Examples include trichloroethylene, tetrachloroethylene, the dichloroethylenes, 1,1,1-trichloroethane, carbon tetrachloride, chloroform, and ethylene dibromide. The chlorinated ethylenes undergo a sequential reductive dehalogenation from tetra- to trichloroethylene, then to the dichloroethylenes (primarily the cis isomer), and finally to vinyl chloride (monochloroethylene; illus. a). In some materials, appreciable quantities of vinyl chloride accumulate; this is undesirable because this compound is considerably more toxic and carcinogenic than its parent compound. In other materials the vinyl chloride is further metabolized. The factors that control the fate of vinyl chloride are at present entirely unknown. The chloroalkanes follow a similar pattern; carbon tetrachloride is converted to chloroform, then to methylene chloride, while 1,1,1-dichloroethane is converted to 1,1-trichloroethane, which in turn goes to ethyl chloride (illus. *b*).

These reductive dehalogenations resemble respirations. In aerobic respiration, molecular oxygen accepts an electron and is reduced to the hydrogenated compound, water. The chlorinated compounds accept electrons and are reduced to the corresponding hydrogenated compound, while the chlorine is released as a chloride ion. Whether these reductive dehalogenations benefit the microorganisms that carry them out is not known. However, the active microorganisms must have a source of hydrogen or

Biotransformations of chlorinated organic contaminants that may occur in groundwaters that harbor methane-producing bacteria. (*a*) Tetrachloroethylene, also known as perchloro-ethylene (PCE) and tetrachloroethene. (*b*) 1,1,1-Trichloroethane (TCA), also known as methylchloroform. (*c*) Carbon tetrachloride, also known as tetrachloromethane.

some other organic compound to provide the electrons for the reduction of the chlorinated compounds. The source of electrons can be a co-occurring contaminant, such as volatile fatty acids in landfill leachate, or it can be a geologic material. Reductive dechlorination of trichloroethylene has been associated with flooded surface soil, buried soils in glaciated area, buried layers of peat, and coal seams.

Groundwaters reducing sulfate or nitrate. Once oxygen is depleted, certain classes of organic compounds can be degraded by bacteria that respire nitrate or sulfate. Groundwaters recharged through soils that support intensive agriculture often have high concentrations of nitrate, and those with appreciable concentrations of sulfate are widespread, particularly in arid regions. Microorganisms respiring nitrate can degrade a number of phenols and cresols (methylphenols). Recently, it has been shown that nitrate-respiring organisms in river alluvium could degrade all three xylenes (dimethylbenzenes). However, the microorganisms could not degrade *para*-dichlorobenzene. Nitrate-respiring microorganisms can also degrade carbon tetrachloride and a variety of brominated methanes, but they have not been shown to degrade chloroform or those chlorinated ethylenes or ethanes which are also stable in oxygenated groundwater.

Like the methanogens, the sulfate-respiring bacteria can participate in consortia that degrade a wide variety of natural organic compounds. In contrast to the behavior of methanogenic subsurface material, chlorinated derivatives of naturally occurring aromatic compounds were not degraded in river alluvium containing appreciable sulfate concentrations (200 mg/liter) and exhibiting active sulfate respiration. However, tetrachloroethylene and trichloroethylene underwent reductive dehalogenations, as they did in methanogenic material.

[JOHN T. WILSON]

Acid mine drainage has a deleterious impact on receiving streams and lakes by causing a marked reduction of the normal plant, animal, and microflora components, thus representing a serious environmental problem in regions where high sulfur coal is mined. In the 11-state Appalachian coal mining region of the United States, it has been estimated that 10,500 mi (16,800 km) of streams have been polluted by acid drainage, of which about half emanates from inactive abandoned mine sites.

Many naturally occurring metallic ore deposits are metal sulfide minerals. Some metallic sulfides, notably iron pyrite and marcasite (both FeS_2), are frequently found in intimate association with coal deposits. Microscopic lenses of pyritic mineral are often dispersed throughout the coal and usually represent the major sulfur component in coal which has sulfur content in excess of 1%.

When coal is mined, the chemically reduced pyritic minerals, along with the coal, shale, and other minerals, become exposed to moisture and the oxygen in air and tend to become oxidized. This chem-

ical oxidation of iron pyrite and other pyritic minerals, producing sulfuric acid plus oxidized metal sulfate, is the source of mine acid. The nonmarketable minerals which are extracted with the coal are frequently left in piles on the surface when the coal is cleaned and taken to market. In the case of surface (strip) mining, the refuse material may be buried in an open pit after the marketable coal has been removed, and may be exposed to groundwater flow.

Whether on the surface or in a mine shaft, pyrite exposed to oxygen and water becomes oxidized to water-soluble by-products including sulfuric acid [reactions (1)]. The acid and soluble ions are then

$$4FeS_2 + 15O_2 + 8H_2O \rightarrow 8H_2SO_4 + 2Fe_2O_3 \qquad (1)$$

leached away with groundwater flow or surface runoff into receiving streams.

The ferric iron (Fe^{3+}) by-product via the oxidation of the ferrous iron component of iron pyrite will subsequently hydrolyze with water to produce a yellow-brown or red-brown precipitate of ferric hydroxide and ferric hydroxysulfate, depending upon a number of environmental conditions [reactions (2) and (3)].

$$Fe_2O_3 + 3H_2O \longrightarrow Fe(OH)_3 \qquad (2)$$

$$Fe(OH)_3 + H_2SO_4 \longrightarrow Fe(OH)(SO_4) + 2H_2O \qquad (3)$$

Acid mine drainage is in effect a dilute solution of sulfuric acid which has the capability of dissolving away many minerals with which it comes in contact. Acid mine water can have a pH as low as 1.8 and a dissolved mineral content of up to 10% by weight.

A key consideration in the formation of acid mine drainage is the presence of acid-tolerant and acidophilic microorganisms, some of which live nutritionally by deriving energy via the oxidation of iron pyrite. One group of bacteria known as acidophilic thiobacilli are characterized by their ability to utilize reduced sulfur, including the S^{2-} of pyrite, as their exclusive source of energy. These organisms also use that energy to reduce carbon dioxide as their only source of cellular carbon. Since they can therefore grow entirely on nonorganic substances, they are referred to as chemosynthetic autotrophs and their life-style can be compared to the photosynthetic autotrophs (that is, algae and green plants).

One bacterial species, *Thiobacillus ferrooxidans*, is capable of using the oxidation of both the ferrous iron and sulfide components of iron pyrite nutritionally in the presence of air and water. These bacteria increase the reaction rate shown in reaction (1) by more than 1 million times the abiotic chemical rate, and thereby exert a significant influence on the formation of acid in active and abandoned mines and in storage piles of high-sulfur coal as well as in high-sulfur coal refuse.

Many variations on the nutritional theme described above are present in the populations of microorganisms that make up the consortium that lives in and produces mine acid. It is known that heterotrophic bacteria which derive energy from organic substances utilize the organic metabolic by-products of the chemoautotrophs and promote a synergistic relationship in the microbial community, thereby further increasing the production of mine acid from pyrite.

For background information *see* AQUIFER; BACTERIAL METABOLISM; GROUNDWATER HYDROLOGY; METHANOGENESIS (BACTERIA) in the McGraw-Hill Encyclopedia of Science and Technology.

[PATRICK R. DUGAN]

Bibliography: M. L. Apel, Leachability and revegetation of solid waste from mining, *Project Summary U. S. Environmental Protection Agency*, Rep. EPA-600/S2-82-093, 1982; G. Bitton and C. P. Gerba (eds.), *Groundwater Pollution Microbiology*, 1984; P. R. Dugan, Bacterial ecology of strip mine areas and its relationship to the production of acid mine drainage, *Ohio J. Sci.*, 75(6):266–279, 1975; P. R. Dugan, Biochemistry of acid mine drainage, in *Biochemical Ecology of Water Pollution*, pp. 123–137, 1972; J. Libicki, S. Wassersug, and R. Hill, Impact of coal refuse disposal in groundwater, *Project Summary U.S. Environmental Protection Agency*, Rep. EPA-600/S2-83-028, 1983; W. van Duijvenbooden, P. Glasbergen, and H. van Lelyveld (eds.), *Quality of Groundwater, Studies in Environmental Science 17*, 1981; C. H. Ward, W. Giger, and P. L. McCarty (eds.), *Ground Water Quality*, 1985.

Monoclonal antibodies

The hybridoma method of making monoclonal antibodies was first described in the mid-1970s. Since that time, these have played an increasingly important role in many aspects of biology and medicine. One of the most common uses of monoclonals is in diagnostic assays, particularly for hormones. They have increased the sensitivity of such tests resulting in, for example, earlier diagnosis of pregnancy. Further, tests for pathogens can be carried out much more rapidly with monoclonal antibodies than in the several days necessary for traditional diagnosis by culturing. Monoclonal antibodies are also widely used in purification procedures, their extreme specificity allowing them to separate one substance from a mixture of many others. For example, they are used in the immunopurification of various types of interferons.

However, monoclonals produced by traditional hybridoma technology are still mostly of rodent origin, and the technique can only be used to immortalize the cells producing those types of antibodies normally expressed by rats and mice. Recent advances in recombinant deoxyribonucleic acid (DNA) technology, coupled with the development of techniques for introducing DNA into lymphocytes, has opened the way to making a new generation of monoclonal antibodies. The genes that encode monoclonal antibodies can be isolated from hybridoma cells and manipulated before being reintroduced into lymphoid cells. With this approach, it is possible to make completely new antibodies and antibody-related proteins.

Such second-generation monoclonal antibodies will be of value in a number of ways. Perhaps the most obvious is the ability to produce truncated antibody molecules containing only the portion that includes the antigen-binding domains. Second, chimeric antibodies that contain domains originating from two different species could be produced. This would allow, for example, the antigen-binding variable region domains of an antibody raised in a mouse or rat to be combined with human constant regions, thus giving a human effector function. Third, recombinant antibodies can be assembled with completely novel effector functions; for example, the heavy-chain constant regions could be replaced by an enzyme. This may be of value in two ways: First, it could provide a convenient method of purification of the enzyme. Alternatively, it could yield a molecule that both binds to a particular antigen and catalyzes a certain reaction. The latter would be valuable in, for example, immunoassays.

Techniques and expression systems. A number of research groups reported the introduction of DNA into lymphoid cells in 1983, thereby opening the way to production of second-generation monoclonal antibodies. Mouse myelomas are plasma cell tumors, plasma cells being the cells responsible for the synthesis and secretion of antibodies in mammals. These are ideal for expressing transfected antibody genes, because they recognize immunoglobulin gene transcription signals, they are equipped for protein secretion, and they carry out the posttranslational modifications necessary for the expression of functional antibodies. Experiments have been carried out with both bacteria and yeast as the host cells for immunoglobulin gene transfection, but myeloma cells at present provide the most attractive expression system. Many myeloma-derived cell lines have already been shown to be capable of yielding transfected antibodies.

Once an immunoglobulin gene has been cloned, it can be manipulated to replace parts of it with corresponding parts from a different type of immunoglobulin, perhaps from a different species, or with a gene for an altogether different type of protein, such as an enzyme. These recombinant genes can then be inserted into a vector suitable for expression in myeloma cells. These vectors usually also contain a selectable marker to allow selection of stably transfected cells. The selectable markers most commonly used are *gpt* and *neo*. The *gpt* gene encodes a guanine-xanthine phosphoribosyltransferase activity that confers resistance to mycophenolic acid, whereas *neo* allows transfected cells to survive in the presence of the antibiotic G418.

Once such a recombinant plasmid has been constructed, it can be introduced into myeloma cells in a variety of ways, including calcium phosphate coprecipitation, diethylaminoethyl (DEAE) dextran–

mediated DNA uptake, electroporation, and spheroplast fusion. In any of these techniques, only a small proportion of the cells will both take up the plasmid DNA and incorporate it into their genome; these few cells continue to grow when the selective agent is applied to the medium after 1 or 2 days. Such stable transfectants may then be cultured and used for antibody production.

An example of the use of this technology is the construction of a chimeric antibody that contains the variable region of a mouse antibody directed against the hapten 4-hydroxy-3-nitrophenacetyl (NP) and the human C_ϵ constant region (see illus.). In this case, the starting point is a mouse IgM heavy-chain gene, which encodes specificity against the NP hapten, and which has been cloned into a suitable plasmid vector. When this heavy-chain gene is transfected into the cell line J558L, which produces only light chains, functional antibodies (which contain both heavy and light chains) are secreted that are specific for the hapten NP. Before the genes are introduced into J558L, the constant regions of the mouse IgM gene are removed, leaving only the variable-region gene in the vector. Next, the human IgE constant-region gene is inserted in place of the original mouse one. When this plasmid construct is transfected into J558L by spheroplast fusion, chimeric antibodies are secreted that retain both the NP-binding activity (from the mouse variable region) and the authentic human effector function of triggering histamine release from basophils (from the human constant region).

By using similar techniques, chimeric IgM and IgG antibodies have also been made. In each case, their known antigen specificity allows a facile assay system for testing for clones that produce functional antibody. This hapten specificity also forms the basis for a simple one-step method of purifying antibodies to homogeneity in which culture supernatants are passed over a column on which the appropriate hapten is immobilized. The antibody is retained on the column, and can be eluted with an excess of free hapten.

Applications. Many research groups have attempted to produce human monoclonal antibodies, both by hybridoma technology and by establishing continuously growing human B-cell lines that have been immortalized by using Epstein-Barr virus. However, problems have been encountered in obtaining cell lines that stably secrete large quantities of human antibodies. Further, traditional hybridoma technology requires that a rodent be injected with the required antigen, and once an immune response has been elicited, the spleen is removed for fusion. It is obviously not possible to do this with humans, and difficulties have been encountered in developing a suitable procedure. One way in which these problems may be partly overcome is by the production of rodent-human chimeric antibodies of the type described above.

Because of the difficulties encountered in routinely producing human monoclonal antibodies,

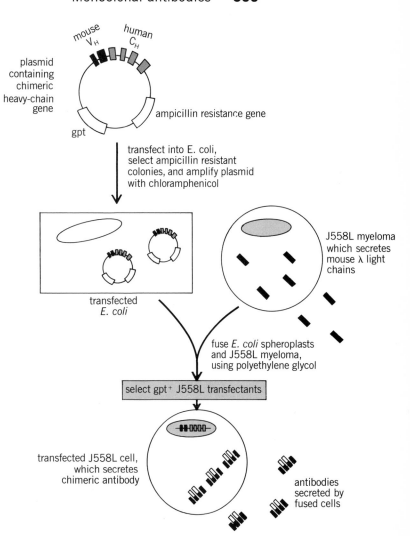

Introduction, by protoplast fusion, of a plasmid encoding a chimeric heavy chain into the mouse myeloma J558L. Amp R is the gene for ampicillin resistance, and is necessary to select transfected bacterial cells harboring the recombinant plasmid.

monoclonals used in therapy are mostly of rodent origin. These are easily raised against specific antigens, and can be produced in reasonable quantities. The therapeutic and diagnostic uses to which mouse monoclonals have been put include tumor imaging, in which a monoclonal specific for a tumor antigen is radiolabeled and is administered to a patient, and the tumor is visualized by radioimmunoscintigraphy. Mouse monoclonals have also been used to remove alloreactive T cells (responsible for graft-versus-host disease) in bone marrow transplantation. However, administering repeated doses of mouse monoclonals to humans is liable to elicit an antiantibody response, reducing the efficacy of the therapy. This problem could potentially be overcome by using chimeric antibodies that incorporate a rat or mouse variable region with a human constant region. These would retain the advantage of easy production against a specific antigen and, since only the variable region is foreign, may elicit a reduced antiantibody response.

Furthermore, by transfecting cells with the genes for only a part of an antibody, the antigen-binding portion (or F_{ab}) of the antibody can be produced on its own, independently of the rest of the antibody molecule. These may be of value in therapy; for example, it is possible that in tumor imaging, their relatively small size may allow them a greater degree of access to tumors than is possible for complete antibody molecules.

By using these methods to establish a set of cell lines that secrete monoclonal antibodies differing only in their effector functions, it will be possible to carry out a detailed comparison of the properties of the human immunoglobulin classes and subclasses. Moreover, by applying the techniques of site-directed mutagenesis to modify specific regions of the antibody, much valuable information will be obtained about the nature of the interactions between antibodies and their antigens as well as the modes of action of human antibody effector function regions.

Novel recombinant antibodies. One hope for monoclonal antibodies was that they could be used in the treatment of malignant disease by constructing immunotoxins. These consist of antibodies directed against a tumor conjugated with a toxin such as ricin, abrin, or diptheria toxin. When administered to a live organism, the antibody part directs the immunotoxin to the tumor and the toxic agent kills it. The use of such immunotoxins may be of considerable value in cancer therapy. Since the genes coding for many of these toxins have been cloned, it is possible to link those genes to the ones for antibodies directed against specific tumors in order to construct a recombinant antibody which is, effectively, an immunotoxin.

The principle of producing recombinant antibodies with novel effector functions using this approach has been demonstrated recently. Myeloma cell lines have been established that secrete recombinant IgG hapten-specific antibodies, in which two of the heavy-chain constant-region domains have been replaced by the nuclease of *Staphylococcus aureus*. This antibody was successfully obtained from a transfected cell line, and retained both its antigen-binding specificity and the nucleic acid–degrading activity contained within its novel nuclease moiety.

One of the most common uses of monoclonal antibodies is in enzyme-linked immunosorbent assays (ELISA), which are used to detect very small quantities of certain compounds, for example, to measure hormone levels in blood or urine. At present, a specific antibody preparation is coupled to an enzyme chemically. Recombinant antibody technology can be used to produce antibodies with enzymic effector functions, ensuring reproducibility of coupling efficiency, and alleviating the possibility of uncoupled antibody competing for binding sites. *See* IMMUNO-DIAGNOSTICS.

Another potentially valuable application of the second-generation monoclonal antibody technology involves using a recombinant antibody-enzyme fu-

sion as a way of producing and purifying enzymes. In this case, the F_{ab} region of the recombinant antibody provides a tag to ensure that the novel product is secreted, and provides, as described above, a simple method of purification. Theoretically, this could provide a valuable general method of producing, and easily purifying, foreign proteins in mammalian cells. It is not clear, however, how far the approach of fusing antibodies with enzymes can be extended. Problems may be encountered with enzymes that are not active as monomers, and the function of some enzymes may be altered by the close proximity of the F_{ab} moiety that composes the antigen specificity.

For background information *see* ANTIBODY; GENETIC ENGINEERING; IMMUNOGLOBULIN; IMMUNOLOGY; SOMATIC CELL GENETICS in the McGraw-Hill Encyclopedia of Science and Technology.

[JOHN O. MASON; GARETH T. WILLIAMS]

Bibliography: G. Köhler and C. Milstein, Continuous culture of fused cells secreting antibody of predefined specificity, *Nature*, 256:495–497, 1975; M. S. Neuberger, Making novel antibodies by expressing transfected immunoglobulin genes, *Trends Biochem. Sci.*, 10:347–350, 1985; M. S. Neuberger, G. T. Williams, and R. O. Fox, Recombinant antibodies possessing novel effector functions, *Nature*, 312:604–608, 1984; V. T. Oi et al., Immunoglobulin gene expression in transformed lymphoid cells, *Proc. Nat. Acad. Sci. USA*, 80:825–829, 1983.

Muon spin relaxation

An atom, nucleus, or elementary particle that has an intrinsic spin angular momentum interacts with a magnetic field in such a way as to align its spin vector along the field axis. In addition, the spin experiences a torque that causes the spin vector to precess about the direction of the field with a frequency proportional to the field strength. Consequently, changes in the field can be tracked by monitoring the spin direction and precession frequency. This is the principle behind a magnetic resonance experiment. Conventionally, a large number of spins in a sample material are aligned with an applied magnetic field, creating a net spin polarization. The decay of this polarization is then monitored to provide detailed information about the fields internal to the sample. Recently, physicists and chemists have been using muons to probe internal magnetic fields in a similar but unique way. The technique, denoted by μSR (for muon spin relaxation, rotation, or resonance), is used to study many phenomena in condensed-matter science, including superconductivity, magnetism, diffusion and trapping, impurity states in insulators and semiconductors, and chemical reactions of muonium ($\mu^+ e^-$) atoms.

Muon properties. A muon is a point-like elementary particle with a spin of ½ (in units of Planck's constant), a charge of ± 1 (in units of the electron charge), and a magnetic moment about three times that of the proton. The mass ratio for an electron:muon:proton is given approximately by 1:200:1800. Polarized muons can be produced when high-energy protons interact in a target and can be conveniently focused onto a sample of interest by using magnetic lenses. Muons decay via the weak interaction with a half-life τ_μ of 2.2 microseconds into an energetic positron or electron ($e^\pm$) and two neutrinos (ν_μ and ν_e) via the reactions $\mu^+ \rightarrow e^+ \bar{\nu}_\mu \nu_e$ and $\mu^- \rightarrow e^- \nu_\mu \bar{\nu}_e$. The $e^\pm$ are emitted preferentially along the axis of the muon spin vector in a classic example of parity violation. The loss of polarization in the sample is monitored by observing the spatial anisotropy of this decay.

In a crystal lattice, energetic positive muons are slowed down by ionization until they reach thermal energies and reside at interstitial or vacancy sites. In a metal the μ^+ retains essentially all of its polarization in the slowing-down process. In semiconductors or insulators the μ^+ may bind an electron in the final stages of its thermalization, producing a hydrogenlike atomic state referred to as muonium ($\mu^+ e^-$). This also may occur in liquids or gases. Like the μ^+, the muonium atom may diffuse through the medium or, particularly at low temperatures, may be trapped. Sometimes muonium may chemically bond to the molecules or atoms of the medium, as discussed below.

After thermalization, negative muons are captured by the atoms of the sample material and wind up in low-lying orbits far inside the atomic electrons because of the large muon mass. Approximately 80% or more of the initial polarization is lost in this atomic capture process. Nevertheless, muon spin relaxation is still useful, particularly if the nuclei in the material under study have no spin angular momentum themselves, making nuclear magnetic resonance experiments impossible. In addition, the negative muon can easily sample the nuclear magnetism itself, because part of the μ^- wave function actually penetrates the nucleus. Indeed, for nuclei heavier than carbon, the μ^- are captured by the nucleus before they decay naturally, thus reducing their lifetime.

Technique. In a muon spin relaxation spectrometer (Fig. 1), an incident muon triggers a plastic-scintillator detector in front of the sample, thus registering its arrival time. The decay positron or electron triggers another detector pair when the muon decays. Approximately 1 to 10 million such events are collected, the difference between each arrival and decay time being histogrammed into time bins by a clock, giving rise to the idealized spectra shown in Fig. 1. A typical time resolution is a few nanoseconds, but a resolution of about 250 picose-

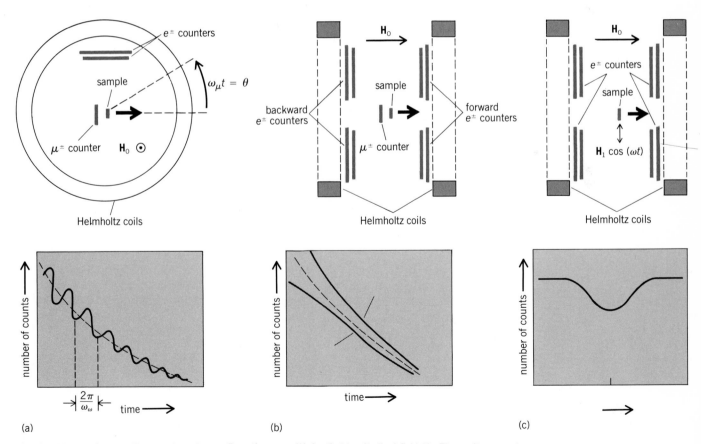

Fig. 1.   Muon spin relaxation spectrometer configurations (above) and corresponding time or frequency spectra (below). (a) Applied field H_0 transverse to muon spin (heavy arrow). The scintillation counters for the muons ($\mu^\pm$) and positrons or electrons ($e^\pm$) are indicated, along with the sample. (b) Applied longitudinal field H_0. The positron counters are located in the forward and backward directions. (c) Muon spin resonance. (*After T. Yamazaki, ed., Collected Papers on Muon Spin Research, Japan Society for the Promotion of Science, 1979*)

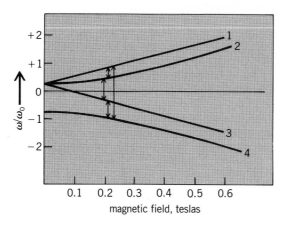

Fig. 2. Energy-level diagram for muonium arising from the dipolar coupling of two spin-½ particles, μ^+ and e^-, in a transverse magnetic field.

conds has been achieved in practice. The $e^{\pm}$ counting rate is given by Eq. (1), where $G(t)$ is a function

$$N(\theta,t) = [1/(4\pi\tau_\mu)]\exp(-t/\tau_\mu)[1 + A\,G(t)\cos\theta] \quad (1)$$

giving the depolarization and θ is the angle between the $e^{\pm}$ velocity and $\mu^{\pm}$ spin vectors. The asymmetry A is determined experimentally and is typically 0.2–0.3.

In a field H_0 transverse to the muon spin (Fig. 1a), the muon precesses at its Larmor frequency ω_μ given by Eq. (2), where γ_μ is the muon gyromag-

$$\omega_\mu = \gamma_\mu\,H_0 \quad (2)$$

netic ratio, equal to 8.5×10^8 radians/(second tesla). Here θ equals $\omega_\mu t$. The measured muon spin relaxation frequency gives the time-averaged magnetic field at the muon site, which may be due to an applied field, to internal fields (as in a ferromagnet), or both. Depolarization occurs from two processes: (1) a distribution of static fields can cause the muons at different sites to precess at different rates, producing a dephasing of the ensemble; or (2) a fluctuating transverse field whose frequency is on resonance can cause a muon to flip its spin from up to down, or vice versa. The resonance frequency is equal to the muon's Zeeman frequency ω_μ in its local longitudinal field. In an applied longitudinal field (along the muon spin direction) no precession occurs and θ equals 0° or 180° for forward and backward $e^{\pm}$ detectors (Fig. 1b), so depolarization is due only to the second mechanism. Muon spin resonance experiments (Fig. 1c) have also been performed by flipping the μ^+ with an applied resonant field. The muon spin is flipped when the oscillating transverse field $H_1 \cos(\omega t)$ has the resonance condition $\omega = \omega_\mu$, where ω_μ is again given by Eq. (2). The resonance frequency can be used to determine the magnitude of internal fields that are too large to be determined by the precession method because of limitations in the time resolution.

Several advantages of the muon spin relaxation technique over conventional nuclear magnetic resonance experiments are: (1) There is no need for an applied magnetic field. (2) Almost any material can be studied. (3) There are no radio-frequency field penetration problems. (4) The μ^+ has no quadrupole moment and has no core atomic electrons that complicate the interpretation of other nuclear or atomic probes. (5) The small mass of the μ^+ makes it a unique probe of quantum tunneling mechanisms. The technique also has disadvantages, such as the short muon lifetime, and the muon's charge and magnetic field, which can perturb its environment making it an impurity probe.

Research. Areas of active muon spin relaxation research include muonium chemistry, the study of muon diffusion, and the probing of magnetism in solids.

Muonium. The muonium atom in vaccum is the same size and has the same energy levels as hydrogen, but with about one-ninth the mass. The energy levels of muonium in a transverse magnetic field arise from the dipolar interaction of the two spin-½ particles (μ^+ and e^-) and are displayed in the Breit-Rabi diagram of Fig. 2. In muon spin relaxation experiments, muon precession frequencies are observed corresponding to the magnetic dipole transitions labeled by arrows in Fig. 2, namely ω_{12}, ω_{23}, ω_{34}, and ω_{14}. The $\mu^+ - e^-$ hyperfine frequency ω_0 equals $2\pi(4.4 \times 10^9 \text{ radians/s})$.

Most muonium chemistry studies are concerned with two areas: muonium reaction rates and muonic free radicals. Because the muonium atom exhibits quantum tunneling, new phenomena unavailable to hydrogen reaction rate studies can be measured. In addition, muonium studies generally yield more precise data than do their hydrogen counterparts. Finally, because there are no second-order effects due to muonium-muonium reactions (since only one muon exits at a time), the experiments are relatively easy to interpret. Muonium-substituted free radicals are derived by muonium addition to unsaturated carbon or oxygen bonds (as in the case for benzene), a

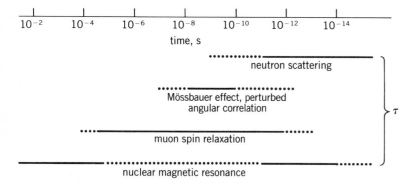

Fig. 3. Ranges of electronic spin fluctuation times τ measurable with various techniques. The solid lines indicate times usually measurable, and the dotted lines indicate times measurable in favorable cases.

process which produces an unpaired electron spin. The resultant muonium frequencies allow identification of the site of the muon, which is often away from the reaction center of the molecule. This fact enables the μ^+ to act as a unique tracer with which the chemical reaction of the radical can be observed.

In the elemental semiconductors silicon, germanium, and diamond, muonium forms in two types of states that are particularly interesting because no hydrogen electron paramagnetic resonance signals have ever been seen in these materials. Muonium is also observed in insulators and compound semiconductors.

Diffusion. In copper, for example, positive muons are self-trapped at temperatures between 10 and 80 K (-441 and $-315°F$), and the depolarization function $G(t)$ [introduced in Eq. (1)] is gaussian in time, with a width characteristic of the distribution of nuclear dipole fields at octahedral sites. At higher temperatures, this line width narrows as the muon hops to many different sites in its lifetime. The temperature dependence of the linewidth determines the muon diffusion rate, which can be compared to heavier interstitial particles such as hydrogen. As the temperature is lowered below 5 K ($-450°F$), the μ^+ again becomes delocalized, probably due to the onset of quantum tunneling processes. Quantum tunneling of the μ^+ has been observed in several materials, notably high-purity iron, copper, and aluminum.

Magnetism. The muon spin relaxation technique has been used to probe magnetism in two basic areas: the study of hyperfine fields and spin densities at interstitial sites using μ^+, and the study of magnetic ion dynamics in ordered and disordered magnets using $\mu^\pm$. In the first category, information is obtained that is important for an understanding of the electronic structure of hydrogen in metals and its associated properties, such as diffusion, localization, solubility, and chemical bonding. Very little proton nuclear magnetic resonance data are available. In the second category the muon spin relaxation technique can be used to measure magnetic ion fluctuation times in a regime that is complementary to other probes (Fig. 3). Fluctuation times in the range 10^{-11} to 10^{-4} s are easily measured in zero-field muon spin relaxation. This information can be used to test dynamical models of magnetic materials.

For background information *see* LEPTON; MUON-IUM; NUCLEAR MAGNETIC RESONANCE (NMR) in the McGraw-Hill Encyclopedia of Science and Technology.

[ROBERT H. HEFFNER]

Bibliography: R. H. Heffner and D. G. Fleming, Muon spin relaxation, *Phys. Today*, 37(12):38–46, 1984; A. Schenk, *Muon Spin Rotation Spectroscopy*, 1985; D. C. Walker, *Muon and Muonium Chemistry*, 1983; T. Yamazaki and K. Nagamine (eds.), *Hyperfine Interactions*, vols. 17–19: *Muon Spin Rotation and Associated Problems*, 1984.

Nobel prizes

For 1985 nine recipients of the Nobel prizes were announced by the Swedish Royal Academy.

Medicine or physiology. Michael S. Brown and Joseph L. Goldstein, molecular geneticists working at the University of Texas Health Center, Dallas, received this award for discovering low-density lipoprotein (LDL) receptors and their function in cholesterol metabolism. These receptor molecules on the surface of cells remove lipoproteins (cholesterol-rich particles) from the bloodstream, and the lipoproteins are subsequently metabolized by the cells. Excessive amounts of cholesterol can clog arteries and lead to atherosclerosis, heart attacks, and strokes. Through research into familial hypercholesterolemia (inherited high blood cholesterol) the geneticists found that an individual's susceptibility to high blood cholesterol was directly linked to deficiency of LDL receptors. When the cells accumulate enough cholesterol, they briefly discontinue making the receptors, resulting in an increased amount of cholesterol in the blood. Current research promises to develop a cholesterol-reducing resin that will inhibit the synthesis of cholesterol and also increase the number of LDL receptors.

Physics. Klaus von Klitzing, a computer expert and director of the Max Planck Institute for Solid State Research, Stuttgart, West Germany, received this award for discovering the quantum Hall effect. He measured the Hall resistance across metal oxide semiconductor field-effect transistors in very strong magnetic fields and at temperatures near absolute zero. He found that as the magnetic field is increased the Hall resistance exhibits plateau regions in which it is constant, independent of the magnetic field strength, and that in these plateau regions the Hall resistance is given quite accurately by a universal relation involving only fundamental constants of nature.

Chemistry. Herbert A. Hauptman, director of the Medical Foundation of Buffalo, New York, and Jerome Karle, of the Naval Research Laboratory, Washington, D.C., received this award. With the aid of computers they developed mathematical techniques for use in x-ray crystallography to determine the three-dimensional structures of molecules. Over 45,000 molecules have been studied with these methods, enabling the discovery of new antibiotics and vaccines.

Economics. Franco Modigliani, a professor at the Massachusetts Institute of Technology, was awarded this prize for his "life-cycle" savings theory and his theory on the financial market's evaluation of stocks. According to the life-cycle savings theory, people will put away money for their own old age or retirement years, not for their descendants. He found that people will save less if they are assured of retirement income. He also argued that financial markets assess a company's stock value by its expected future earnings. This principle has become a standard in corporate finance.

Literature. Prize-winner Claude Simon, a French novelist in the *nouveau* trandition, has written 15 novels. His highly praised novel *Flanders Road* (1961) gives an account of French POWs during World War II.

Peace. The peace prize was given to the founders of the International Physicians for Prevention of Nuclear War: Bernard Lown of the United States and Yevgeny I. Chazov of the Soviet Union. Both are cardiologists; Lown is a professor at the Harvard School of Public Health, Boston, and Chazov has been a personal physician to top Soviet officials, and Deputy Minister of Health. The international organization has 135,000 members from 41 countries; 20,000 are from the United States, 60,000 from the Soviet Union. The Nobel committee praised the organization for performing "a considerable service to mankind by spreading authoritative information and by creating an awareness of the catastrophic consequences of atomic warfare."

Nuclear physics

Recent advances in nuclear physics include (1) the development of detailed computer simulations of the complex phenomena that occur in energetic collisions of nuclei; (2) understanding of the general features of nuclear orbiting, a phenomenon in which two colliding nuclei temporarily form a rotating complex that carries out at least a significant fraction of a full rotation; and (3) the observation of pion emission in relatively low-energy nucleus-nucleus collisions and the development of a model for this process that is analogous to bremsstrahlung.

SIMULATION OF HEAVY-ION COLLISIONS

For most of its history, research in nuclear physics has been concerned with probing the structure of the atomic nucleus in its ground state, or at small excitation energy. During this time many unique properties of the nucleus have been established. In particular, it is now known that the atomic nucleus represents the only strongly interacting quantum-mechanical many-particle system whose structure is governed by the strong interactions which exhibits a shell structure similar to electrons in atoms and which exhibits superfluid properties similar to liquid helium.

The period since the mid-1960s, has seen dramatic changes in both the scope and emphasis of this work. New machines have been developed which allow scientists to accelerate entire nuclei collisions between two full-size nuclei. Such collisions have proved very useful in establishing conditions of matter flow and energy deposition that subsequently permit scientists to study the elementary properties of nuclear matter.

Since the mid-1970s, high-speed computers have been used to simulate the details of these complicated nuclear reactions. The task of constructing computer models of these reactions has presented a variety of challenges to nuclear researchers. A typical simulation of the collision between two nuclei entails following the motion of several hundred neutrons and protons together with the forces that bind them into nuclei, as well as following the collective excitations of the nuclei themselves for time periods which are comparable to the collision time of the two nuclei. Because the forces in a nucleus are quite strong, examination of the state of each proton and neutron at hundreds of thousands of times during the collision is required. Assuming that the computer simulations can actually be performed, there still remains a difficulty in extracting useful information from the vast amount of numerical information generated in the simulation.

Coulomb barrier. The collision between two nuclei (usually called heavy ions) results in an exceedingly complex set of phenomena which is best organized in terms of the total mass of the reaction, and in terms of the total bombarding energy of the projectile nucleus. This organization principally reflects the interplay of the long-range electromagnetic forces and the short-range nuclear forces during the collision. For example, during the collision of two oxygen nuclei at bombarding energies less than about 25 MeV, the repulsive Coulomb forces among the protons in the respective nuclei prevent the two ions from approaching sufficiently close that the nuclear interactions can play a role. In this way the long-range electromagnetic force acts as a barrier, called the Coulomb barrier, to prevent the heavy ions from interacting.

The bombarding energy needed to overcome the Coulomb barrier varies throughout the periodic table as the combination of projectile and target varies; for collisions of oxygen on oxygen this energy is about 25 MeV, while for collisions of uranium on uranium it is about 1400 MeV. Alternately, the bombarding energy can be divided by the number of protons and neutrons in the projectile nucleus to obtain an energy per particle: about 1.6 MeV per particle for oxygen on oxygen and about 5.8 MeV per particle for uranium on uranium. For collisions of identical nuclei having the same number of neutrons and protons, the bombarding energy per particle, E_c, needed to overcome the Coulomb barrier is given by Eq. (1), where A is the mass number of projectile

$$E_c \sim 0.25\, A^{2/3} \quad (\text{MeV}/A) \qquad (1)$$

nucleus. This expression gives a qualitative representation of the Coulomb barrier for heavy-ion collisions insofar as it is approximately correct for light nuclei but overestimates the barrier energy by about a factor of two for collisions as heavy as uranium on uranium.

Pauli principle and nuclear forces. Thus collisions of heavy ions which probe the strong short-range forces between nucleons are those whose energies are greater than the Coulomb barrier energy. The dynamics of these collisions are governed by nonrelativistic quantum mechanics until the bombarding energy per particle is a sizable fraction of the nucleon mass, about 1000 MeV per particle.

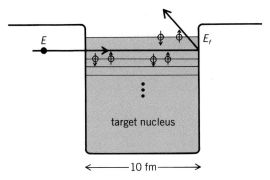

Fig. 1. Action of the Pauli exclusion principle during a heavy-ion collision.

However, in general, the complexity of the quantum mechanics of these strongly interacting nuclear systems make them difficult to simulate with present computer technology. But at lower energies, about 35 MeV per particle, the dynamics of these collisions become much simpler. This reflects two rather subtle properties of quantum mechanics: the Pauli exclusion principle and the fact that the nucleons which build up nuclei obey Fermi-Dirac statistics. The Pauli principle postulates that two identical nucleons cannot be placed at the same point in space in exactly the same state, while the Fermi-Dirac statistics require that a collection of nucleons must have a totally antisymmetric wave function under the process of exchanging two nucleons. Thus the wave functions identifying the state of identical nucleons

in two different heavy ions must rearrange themselves so that these nucleons do not come close together. During a heavy-ion collision this process of exclusion and exchange can occur with no energy loss or gain. Hence nucleons in different nuclei do not interact directly with one another, but only through the potentials that bind the nucleons into nuclei. This effect was first deduced in the 1930s by P. A. M. Dirac within the context of atomic structure, and only recently was its application to heavy-ion collisions discovered.

In Fig. 1 these processes are illustrated graphically. The nucleons in the target nucleus, denoted by the open circles, fill the allowed spaces of the nucleus up to a maximum energy, E_f, given by Eq. (2), the Fermi energy of the nucleus. A nucleon

$$E_f \approx 35.0 \text{ MeV} \qquad (2)$$

from a projectile nucleus having an energy per particle of E, denoted by the solid circle, passes through the target nucleus without colliding with the nucleons in the target due to the exclusion principle. However, it can interact with the mean field of the target, as indicated by the potential wall of the nucleus, at which time it scatters in a new direction. If the bombarding energy E is greater than E_f, the heavy ions exhibit a very different kind of reaction than if E is less than E_f.

Computer simulations. Simulations of heavy-ion collisions have been carried out at energies of between 2 and 20 MeV per particle by using the time-dependent Hartree-Fock (TDHF) theory, in which

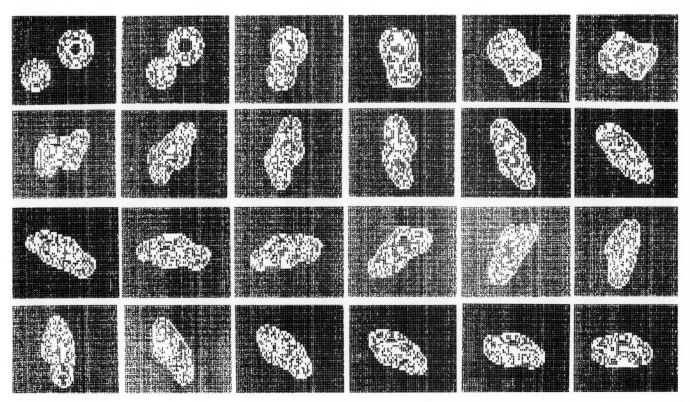

Fig. 2. Time frames of a computer simulation of the matter density in a collision between oxygen and calcium nuclei.

the nucleons in the heavy ions interact via a mean field as proposed by Dirac. At energies of a few MeV per particle above the Coulomb barrier, heavy-ion collision phenomena organize simply in terms of the total mass of the collision. Two prominent phenomena in such collisions are fusion and deep inelastic scattering.

Heavy-ion fusion. For light-mass collisions in which the total mass number of the system is less than about 100, the probable result is heavy-ion fusion. A computer simulation of the fusion process between oxygen and calcium nuclei at a bombarding energy of 12 MeV per particle is shown in Fig. 2.

The resulting fused system is an isotope of the nickel nucleus. Each frame in Fig. 2 shows the matter density of the nuclei in a segment of the collision plane comprising about 40×40 femtometers, and at time intervals of 10^{-22} s. The sequence begins in the upper left-hand corner and proceeds by rows across the figure. In this collision all of the bombarding energy is converted into rotational and vibrational excitation energy. The process of vibration and rotation continues for about 10^4 frames until the energy is uniformaly distributed among the various modes of the system, at which time secondary processes begin to play an important role. These computer simulations were carried out at Lawrence Livermore National Laboratory on the Cray 1s computer and required about 120 min of central processing unit time for the 24 frames shown in the figure.

Deep inelastic scattering. For heavy-ion collisions in which the combined mass number is 200 or greater, the probable outcome of a collision in the energy range greater than the Coulomb barrier and less than about 20 MeV per particle is a process called deep inelastic scattering. In this process, discovered about 1967, the two ions come together and exchange large amounts of energy and mass, as in the case of fusion. However, the lifetime of the rotating and vibrating complex is very short, about 10^{-21} to 10^{-20} s.

Figure 3 shows six time frames from the simulation of the matter density in a typical deep-inelastic scattering event, the collision of a krypton nucleus with a lanthanum nucleus at a bombarding energy of 8 MeV per particle. The collision plane is specified in terms of cylindrical polar coordinates and is completely symmetric upon rotations about the z axis (the horizontal axis) so that only half of the collision plane is shown in the figure. The collision is localized to a region of space of about 10×50 fm. Each frame is labeled by the elapsed time T during the simulation in units of 10^{-21} s. The computations were performed at the Daresbury laboratory, in England, on a Cray 1s computer. Although only six time frames are shown in the figure, the entire collision comprises nearly 10^4 frames and required nearly 10^3 min of central processing unit time.

Figure 3a shows the two nuclei about to collide; in Fig. 3b the two nuclei display considerable overlap. By this stage of the collision, they have begun the process of exchanging mass and energy. In Fig. 3c the two nuclei are trying to separate from each other. However, because the matter is now highly excited, it flows together to form a neck between the two nuclei. In Figs. 3d–f the nuclei continue to separate from one another, and in the process the neck fractures leaving a small amount of matter behind. After Fig. 3f the two nuclei continue to separate from one another with an energy corresponding to about 2.5 MeV per particle.   [MICHAEL R. STRAYER]

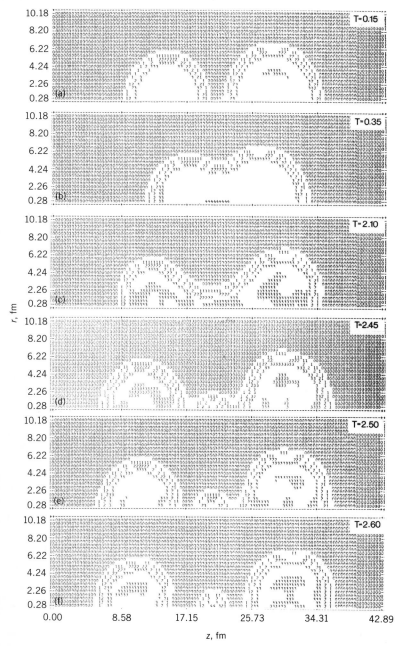

Fig. 3.   Six time frames *a–f* of a computer simulation of the matter density of a collision between krypton and lanthanum nuclei. The collision plane is specified in terms of cylindrical polar coordinates in which the z axis (horizontal axis) is the axis of symmetry, and r is the perpendicular distance to this axis. Elapsed time *T* in the frames, in units of 10^{-21} s, is indicated.

NUCLEAR ORBITING

When atomic nuclei collide with one another, they sometimes form a transient, rotating dinuclear complex. Under special circumstances the system remains in this configuration long enough to carry out a significant fraction—and perhaps considerably more than that—of a full rotation. Because of the analogy with the motion of the planets about the Sun, this phenomenon has been termed orbiting. Orbiting in nuclear reactions is analogous in several respects to behavior observed in atomic molecules.

Whether in a planetary, atomic, or nuclear system, orbiting requires for its existence a delicate balance between the various attractive and repulsive forces that come into play. As a consequence, orbiting research in nuclear physics is a unique and sensitive probe of the dynamics of nucleus-nucleus collisions. A detailed understanding of nuclear orbiting has not yet been achieved, but a semiquantitative picture is beginning to emerge.

Conditions for orbiting. Nuclei interact with each other primarily by means of the Coulomb and nuclear forces, with the gravitational and weak forces being unimportant in determining the overall evolution of a nucleus-nucleus collision. The delicate balance required for orbiting can be understood in terms of the potential energy functions plotted in Fig. 4, where a collision is illustrated schematically (Fig. 4a). These potentials are defined as the work performed against a given force in bringing the nuclei to a distance d starting from infinite separation. They are approximated here as simple functions of distance alone, in an essentially phenomenological treatment. A more rigorous theoretical discussion of orbiting can be given in terms of the time-dependent Hartree-Fock method, discussed above.

The Coulomb force between the positively charged protons in the two nuclei is both repulsive and long-range, and this is reflected by the way that the Coulomb potential energy increases as the nuclei approach each other (Fig. 1b). An additional repulsive force, associated with angular momentum in the system, comes into play at short distances; the additional energy required to overcome this "centrifugal" force is represented by a potential energy function that rises steeply at short distances (Fig. 1c). Taken together, these repulsive forces tend to prevent the nuclei from undergoing close collisions. The nuclear force between the colliding nuclei is the sum total of the numerous pairwise strong interactions between nucleons in the two nuclei. It is attractive, short-ranged, and multidimensional, but for many purposes it can be modeled by a simple function of nucleus-nucleus separation (Fig. 1d).

The total potential energy of the combined Coulomb-centrifugal repulsion and nuclear attraction (Fig. 1e) reaches its maximum value when the mass densities of the diffuse nuclear surfaces just barely overlap. If the colliding nuclei approach each other with an initial kinetic energy equal to this maximum value, their radial velocity of relative motion falls to zero at the top of the repulsive barrier. At this point the system consists, momentarily, of two touching nuclei with a fixed distance between their respective centers of mass. Angular momentum tends to induce rotations of the system, and thus if the separation

distance between the two nuclei were frozen in time the resulting motion would be that of a rigid rotor.

In the absence of additional factors, the configu-

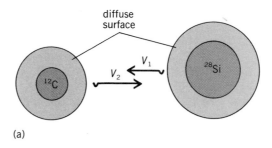

(a)

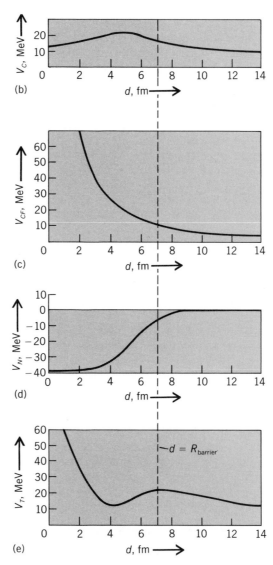

(b)

(c)

(d)

(e)

Fig. 4. Interactions between colliding nuclei. (a) Nuclei approaching with velocities V_1 and V_2 in the center-of-mass system. Diffuse nuclear surfaces have from 10 to 90% of the central density. (b) Potential energy function associated with the electrostatic Coulomb force V_C, plotted versus nucleus-nucleus separation d. (c) Potential energy function associated with the centrifugal force V_{CF}. (d) Potential energy function associated with the nuclear force V_N. (e) Total potential energy function V_T, showing the location of the Coulomb barrier.

ration consisting of two touching nuclei, rotating about their common center of mass, is intrinsically unstable. Any fluctuation removes the system from this point of unstable equilibrium. At bombarding energies near the Coulomb barrier, the dinuclear complex generally coalesces rapidly into a compact, compound nucleus in a fusion process, either because the initial kinetic energy is higher than the repulsive barrier at the classical turning point or because quantum-mechanical tunneling through the barrier occurs at lower energies. Sometimes the dinucleus complex separates immediately into two or more fragments that fly apart, in a direct reaction.

Contributing factors. The fact that orbiting occurs, under appropriate circumstances, indicates that special mechanisms work to maintain the dinuclear complex at its point of otherwise unstable equilibrium. Insight into different aspects of these mechanisms has been gained by studying orbiting under a variety of circumstances, differing in bombarding energy and in the masses of the nuclei involved.

Nuclear quasimolecules. For example, nuclear quasimolecules are formed in the collisions of light nuclei at bombarding energies near the Coulomb barrier. These molecules exhibit fully developed rotational and vibrational behavior, in the form of resonances in the reaction yields and, as such, they represent a particular form of orbiting. Such nuclear molecular phenomena point to the role of nuclear deformations and the properties of special nuclear states in establishing a quasistable dinuclear complex.

Orbiting in deep-inelastic scattering. Orbiting nuclear behavior in nuclear collisions also takes place at higher bombarding energies, where it is accompanied by the phenomenon of deep inelastic scattering, discussed above. In this process, the fragments emerging from the nuclear reaction have the relatively low kinetic energies characteristic of the Coulomb barrier, just as for the nuclear molecules mentioned above, even though the incident energies are typically several times higher than the Coulomb barrier. A very large fraction of the kinetic energy associated with the approach of the colliding nuclei is dissipated, or converted to another form of energy. Experiments have shown that the missing kinetic energy reappears in the form of internal energy of the reaction products. In parallel with the dissipation of kinetic energy, incident orbital angular momentum is converted to internal angular momentum, or spin. Not all of the initial orbital angular momentum can be dissipated, however: the remainder induces rotations of the dinuclear system which now has very little radial velocity of relative motion because of the kinetic energy dissipation.

Characteristic features. One signature of orbiting at the higher bombarding energies is the observation of reaction cross-section contours similar to those shown in Fig. 5. With orbiting, distinct ridges appear in the parameter space defined by the reaction angle θ and reaction product kinetic energy *E*. Nor-

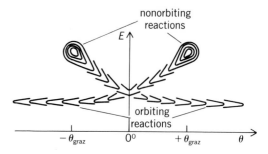

Fig. 5. Contours of constant cross section in a nuclear collision as a function of reaction angle θ and kinetic energy of the nuclei after the collision *E*, showing characteristics of orbiting. (*After J. Wilczynsky, Nuclear moments and nuclear friction, Phys. Lett., 47B:484–486, 1973*)

mal, nonorbiting, nuclear reaction yields are clustered in the vicinity of the bombarding energy and grazing angle θ$_{graz}$, and the corresponding cross-section magnitudes (which reach their maximum here) depend strongly on reaction angle. Following a period of energy dissipation, orbiting appears as an enhanced cross section for deflection to smaller angles, and past zero degrees to so-called negative angles. In contrast to the nonorbiting case, when orbiting is fully developed the magnitudes of the orbiting yields remain undiminished over a wide range of observation angles, and the kinetic energies of fragments emitted from the orbiting system are independent of angle (and significantly less than the bombarding energy). The first observation implies that the system has a relatively long lifetime, while the second indicates that the dissipative degrees of freedom of the rotating complex have equilibrated over a time interval that is short compared to the rotation interval. These consequences of orbiting appear not only in the two nuclei that initiated the reaction but also in a range of neighboring nuclei, reflecting the transfer of charge and mass between the orbiting partners.

Evidence for clutching. The final and perhaps most characteristic feature of orbiting is the existence of a "clutching" or "sticking" mechanism that somehow holds the two nuclei together as they rotate. Quantitative evidence for this mechanism was discovered in a series of experiments at Oak Ridge National Laboratory that showed how orbiting depends on bombarding energy. The kinetic energies of the reaction products of orbiting are independent of angle, but they increase linearly with bombarding energy. The slope of the latter dependence implies the formation of an intermediate complex, consisting of a rotating system with the moment of inertia of two touching nuclei. The energy of rotation, which reappears in the final stage of the reaction as kinetic energy of the fragments, is essentially that of a rigid, electrically charged rotator. The portion of the initial kinetic energy and angular momentum not tied up in orbiting rotational motion goes into internal excitations of the reaction product nuclei in a way that preserves the conditions necessary for orbiting. The Oak Ridge experiments also showed

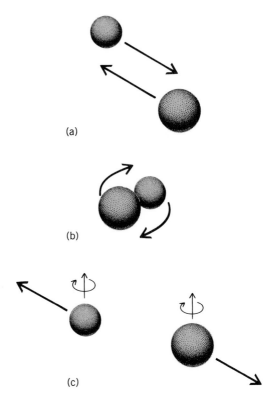

Fig. 6. Picture of orbiting. (a) The nuclei approach with nonzero impact parameter. (b) Equatorial sticking or clutching induces a revolution of the two nuclei about their mutual sticking point. This also induces a spinning of each nucleus. (c) A fraction of the initial kinetic energy has been converted into internal rotational energy of the parting nuclei.

that, for collisions of relatively light nuclei, fragments are thrown from the rotating complex with equal probability in all directions. Such behavior implies complete rotations of the dinuclear system. Taken together, these findings lead to the picture of orbiting sketched in Fig. 6.

Open questions. The special mechanisms that make orbiting possible—and, in particular, their dependence on nuclear deformation, special nuclear states, and nucleon exchange, and on energy and angular momentum dissipation in general—are only partly understood and remain topics of current research. They cannot be explained in terms of the one-dimensional potential energy functions of Fig. 4. Some aspects of orbiting emerge naturally in the self-consistent mean field approach of the time-dependent Hartree-Fock method. Other theoretical approaches, ranging from the use of quantum hydrodynamics to a meld of classical friction with the Schrödinger equation, also have met with partial success. These studies continue to enrich the theoretical understanding of the nuclear many-body problem. On the experimental side, it will be important to test suggestions that collisions of the heaviest nuclei lead to the formation of long-lived orbiting complexes in which very exotic atomic physics processes can occur.

[KARL A. ERB; DAN SHAPIRA]

PION EMISSION IN NUCLEUS-NUCLEUS COLLISIONS

The emission of pions is observed in relatively low-energy collisions of heavy nuclei. Such pions must result from a process in which the individual nucleons of a nucleus act together in a collective manner. A model has been developed in which this pion emission is analogous to bremsstrahlung, the radiation of photons observed when charged particles are decelerated. This pion bermsstrahlung model has been highly successful in explaining a number of experimental observations and provides an important tool for investigating the properties on nuclear matter.

Pion emission. A pion is a particle with a rest-mass energy of about 140 MeV. Until the advent of modern particle physics, it was considered to be an elementary particle, that is, a particle without internal structure, but it is now known to consist of a quark and an antiquark. It has zero spin and exists in three species, π^+, π^0, π^-; that is, it can either be neutral (π^0) or carry a positive or negative elementary charge.

Pions are usually produced in high-energy proton-nucleon collisions. In such collisions, the energy of the incoming proton must, of course, be larger than the rest energy of the pion in the center-of-mass frame. Otherwise, energy conservation forbids the formation of the pion.

In nucleus-nucleus collisions, however, pions are observed at nucleon energies much lower than this threshold energy. If for example a uranium nucleus is bombarded by another uranium nucleus, with approximately 240 nucleons, pions are observed at bombarding energies as low as 20 MeV per nucleon. The energy of an individual nucleon-nucleon en-

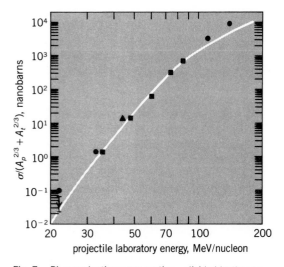

Fig. 7. Pion production cross section σ divided by the geometrical factor $A_p^{2/3} + A_t^{2/3}$ (where A_p and A_t are the mass numbers of the projectile and target nuclei), as a function of projectile energy in the laboratory frame. The curve indicates the prediction of the pion bremsstrahlung model, and the data points indicate experimental results. 1 nanobarn = 10^{-37} m^2.

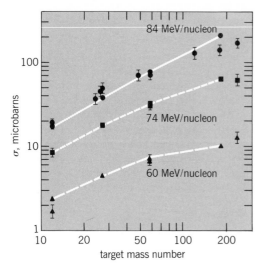

Fig. 8. Pion production cross section σ for ^{12}C + X at energies of 60, 74, and 84 MeV per nucleon as a function of the mass number of the target nucleus X. The lines indicate that predictions of the pion bremsstrahlung model, and the data points indicate experimental results. 1 microbarn = 10^{-34} m^2.

fore called a coherent or collective or sometimes cooperative phenomenon.

Pion bremsstrahlung model. Such a process was predicted in 1980. It was proposed that the emission of pions is analogous to the radiation of photons that occurs when a charged particle is accelerated or decelerated. It is well known from classical electrodynamics that an accelerated charge radiates; in particular, if a charged particle is rapidly decelerated, its photon cloud is unable to follow the slowing-down motion, and hence a part of the virtual photon cloud is stripped off, a phenomenon known as bremsstrahlung. The pionic cloud around a nucleus behaves in a similar manner. If the nucleus is decelerated rapidly enough, the pionic cloud around it cannot follow the slowing-down process and is partly stripped off. Clearly, this is a coherent process for pion production. All the nucleons jointly produce the pion cloud; they are also jointly accelerated in heavy-ion accelerators and are jointly decelerated upon impact on a target nucleus. During the rapid deceleration of the projectile nucleus, the pions are radiated off.

Comparison with experiment. The pion bremsstrahlung model has been strikingly successful in explaining a number of experimental observations. One of its impressive successes is in predicting the excitation function for pion production at low energies per nucleon (Fig. 7). Obviously, the experimental observations are very well reproduced by the theoretical bremsstrahlung model, except at bombarding energies higher than 100 MeV per nucleon. At these higher energies, an additional process sets in. Because of nuclear compression, nuclear matter is heated up. The high-temperature nuclear matter emits pions in analogy to the emission of photons by a hot gas. At low bombarding energies, this process is negligibly small. At higher bombarding energies (above 100 MeV per nucleon), it dominates over the bremsstrahlung process.

The dependence of pion production on the masses of the projectile and target nuclei is also well understood (Fig. 8). The fact that the observed cross sections are somewhat smaller than those calculated for very heavy target nuclei may indicate the reabsorption of the pion while it is passing through the heavy nucleus.

The most important comparison of experiment and theory involves the angular distribution of the emitted pions relative to the beam axis (Fig. 9). The data still contain large margins of error, but they clearly show a dip at an angle θ of 90° (cos θ = 0), indicating that there are less pions emitted perpendicular to the collision axis. Again, this phenomenon is easily explained by the bremsstrahlung model: The emission amplitudes from the two oppositely decelerated nuclei in the center-of-mass frame are proportional to the deceleration vectors. Hence, they interfere destructively at 90°.

Applications. The new pion production mechanism in itself is very interesting, since it constitutes the first cooperative phenomenon observed in elemen-

counter is lower than the pion energy threshold by a factor of 14. Nevertheless, the law of conservation of energy is not violated in such a uranium + uranium collision because the total energy available is 240 × 20 MeV = 4.8 GeV, which is so large that nearly 20 pions could be produced. This is obviously a process where the individual nucleons of a nucleus act together in order to produce pions. At such low energies, an individual nucleon can never create a pion, but coherent, collective action of all the nucleons should be able to produce pions with no difficulty. This type of pion production is there-

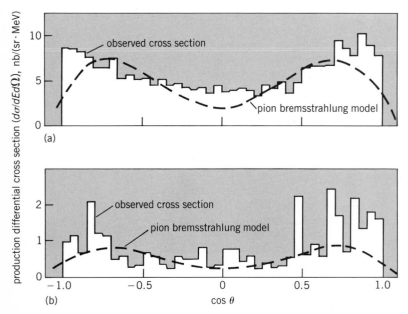

Fig. 9. Angular distributions of pions in the energy windows (a) 50–100 MeV and (b) 100–150 MeV, relative to the beam axis. Abscissa is cos θ, where θ is angle of pion emission relative to beam axis. 1 nanobarn = 10^{-37} m^2.

tary particle physics. Such phenomena should also exist for other bosons (γ-quanta, e^+e^--pairs, other mesons, and so forth).

Pion bremsstrahlung also gives a very sensitive test of the nuclear deceleration profile, so far otherwise unattainable. It appears that nuclei are stopped at a very short distance of interpenetration, in good agreement with the hydrodynamical model. This behavior was certainly not expected by some nuclear physicists, but is in agreement with the discovery of nuclear shock waves at higher energy.

Also, calculations of pion bremsstrahlung for various friction models of heavy-ion collisions have shown that the standard nuclear friction proportional to the relative velocity (Stokes' friction) describes the situation at higher velocities quite well, but that at low velocities the friction must be of higher power in the velocity (Newton's friction). Hence, very important and interesting properties of nuclear matter can be investigated with pion bremsstrahlung as a tool.

For background information *see* BREMSSTRAHLUNG; DEEP INELASTIC COLLISIONS; ELEMENTARY PARTICLE; EXCLUSION PRINCIPLE; FERMI-DIRAC STATISTICS; NUCLEAR MOLECULE; NUCLEAR REACTION; NUCLEAR STRUCTURE; POTENTIAL BARRIER; SCATTERING EXPERIMENTS (NUCLEI) in the McGraw-Hill Encyclopedia of Science and Technology.

[WALTER GREINER]

Bibliography: R. Bass, *Nuclear Reactions with Heavy Ions*, 1980; B. Braun-Munzinger et al., Pion production in heavy ion collisions at $E_{lab}/A = 35$ MeV, *Phys. Rev. Lett.*, 52:255–259, 1984; D. A. Bromley (ed.), *Treatise on Heavy Ion Science*, vol. 1, 1984, vol. 3, 1985; H. Noll et al., Cooperative effects observed in the π^0 production from nucleus-nucleus collisions, *Phys. Rev. Lett.*, 52:1284–1288, 1984; D. Shapira et al., Deep-inelastic back-angle yields and orbiting in ^{28}Si + ^{12}C, *Phys. Lett.*, 114B:111–114, 1982; M. Uhlig et al., Spin-isospin density fluctuations in nuclear collisions, *Z. Phys.*, A319:97–106, 1984; D. Vasak et al., Pionic bremsstrahlung in heavy ion collisions, *Nucl. Phys.*, A428:291–304, 1984; D. Vasak et al., Pion radiation from fast heavy ions, *Phys. Scripta*, 22:25–36, 1980; J. Wilczynski, Nuclear molecules and nuclear friction, *Phys. Lett.*, 47B:484–486, 1973.

Nuclear structure

Recent studies of nuclei with mass number A around 220 have shown the need to postulate new types of collective motion in which reflection symmetry may be broken. While this symmetry breaking may occur, the nuclear structure is expected to be invariant under the combined operation of rotation and parity; the eigenvalue of this operator is called the simplex quantum number. This quantum number may be contrasted with the signature quantum number of reflection symmetric systems.

Collective model. The structure of medium- and heavy-mass atomic nuclei, with A greater than 100, and away from the closed shells, can be understood in terms of a collective model in which the motion of the nucleons are correlated and the nuclear excitations at low energies are dominated by the collective motion. The classic example of collective motion is the rotational band structure of many nuclei in the rare-earth ($A \approx 150$–190) and actinide ($A \geq 230$) regions of the periodic table. The rotation occurs about an axis perpendicular to the symmetry axis of the nucleus, in other words a minor axis. The lowest energies as a function of angular momentum follow to good approximation Eq. (1), where E_x is

$$E_x(I) = \frac{\hbar^2}{2\mathcal{I}}\{I(I+1) - K^2\} \qquad (1)$$

the total energy, $\mathcal{I}$ is the effective moment of inertia of the nucleus I is the total angular momentum, and K is the projection of the total angular momentum on the symmetry axis.

Signature in axially symmetric nuclei. For an axially symmetric deformation, a prolate or oblate spheroid, the nuclear degrees of freedom are invariant under rotation about an axis perpendicular to the symmetry axis, the operation $\mathbf{R}_x$. Rotation by 180° restores the nuclear shape (Fig. 1). Every operator in quantum mechanics has an eigenvalue; the eigenvalue of $\mathbf{R}_x$ is designated r, which is frequently defined in terms of the signature α by Eq. (2). In an

$$r = e^{-i\pi\alpha} \qquad (2)$$

even-mass nucleus the states have integral intrinsic angular momentum or spin and, therefore, $\mathbf{R}_x^2 = 1$, and $r = \pm 1$ and $\alpha = 0$ or 1 according to Eqs. (3). For an odd-mass nucleus the states have half-

$$r = +1, \alpha = 0 \quad \text{for } I = 0, 2, 4, 6, \ldots \qquad (3a)$$
$$r = -1, \alpha = 1 \quad \text{for } I = 1, 3, 5, 7, \ldots \qquad (3b)$$

integral spin and, therefore, $\mathbf{R}_x^2 = -1$, and $r = \pm i$ and $\alpha = \pm \frac{1}{2}$, according to Eqs. (4).

$$r = -i, \alpha = +\tfrac{1}{2} \quad \text{for } I = \tfrac{1}{2}, \tfrac{5}{2}, \tfrac{9}{2}, \ldots \qquad (4a)$$
$$r = +i, \alpha = -\tfrac{1}{2} \quad \text{for } I = \tfrac{3}{2}, \tfrac{7}{2}, \tfrac{11}{2}, \ldots \qquad (4b)$$

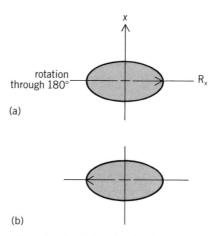

Fig. 1. Operation that defines the signature quantum number in a nucleus with a prolate (axially symmetric) deformation. (*a*) Rotation through 180° about the *x* axis (perpendicular to the symmetry axis) [the operation $\mathbf{R}_x$] results in (*b*) restoration of the original shape.

In other words, the states in a rotational band in an axially deformed nucleus can be labeled in terms of the total angular momentum I, parity $p = \pm$, projection of angular momentum along the symmetry axis K, and signature α. For example, the $K^p = \frac{5}{2}^+$ band would have $\alpha = +\frac{1}{2}$ members $\frac{5}{2}^+$, $\frac{9}{2}^+$, $\frac{13}{2}^+$, . . ., and $\alpha = -\frac{1}{2}$ members $\frac{7}{2}^+$, $\frac{11}{2}^+$, $\frac{15}{2}^+$, ... In a nucleus with quadrupole, prolate or oblate deformations, enhanced electromagnetic transitions of electric quadrupole character connect states which differ by two units of angular momentum. Therefore, the $\alpha = +\frac{1}{2}$ members of the $K^p = \frac{5}{2}^+$ band would be connected by enhanced intraband electric quadrupole transitions, and the $\alpha = -\frac{1}{2}$ members of the $K^p = \frac{5}{2}^+$ band would also be connected by enhanced intraband electric quadrupole transitions, while transitions between $\alpha = +\frac{1}{2}$ and $\alpha = -\frac{1}{2}$ states differing by one unit of angular momentum, for example, the $\frac{9}{2}^+ \rightarrow \frac{7}{2}^+$ transition, would in general be a mixture of electric quadrupole and magnetic dipole transitions. In the reference frame defined by the nucleus, properties such as excitation energies and electromagnetic transition probabilities are independent of the signature for a system that is invariant under $\mathbf{R}_x$.

Excitation energies, for example, are measured in the laboratory reference frame. For a rapidly rotating nucleus the different signatures in the traditional rotational band will not follow Eq. (1) in the laboratory. Excitation energies are transferred into the rotating frame of the nucleus by Eq. (5), where ω is the rotational frequency and I_x the projection of the

$$E_x'(I) = E_x(I) - \omega(I)I_x \qquad (5)$$

angular momentum on the rotation axis. There will be no difference in the behavior of the excitation energies in the rotating frame as a function of rotational frequency for the two signatures of a rotational band for an axially symmetric nucleus that is invariant unde rotation $\mathbf{R}_x$.

The observation several years ago in light-mass rare-earth nuclei of signature-dependent excitation energies in the rotating frame and signature-dependent electromagnetic transition probabilities indicated that a new degree of freedom was operative in the collective excitation of these nuclei. Much of the observed phenomena could be understood in terms of the breaking of the axial symmetry, thereby introducing a triaxial shape.

Simplex quantum number in octupole nuclei. Recently, there has been evidence that quadrupole deformation, that is, a prolate or oblate spheroidal shape, may not be sufficient to account for nuclear structure. The first experimental evidence that octupole deformations are needed came when calculations of nuclear binding energies in the $A = 220$ region could not reproduce the data unless a finite octupole deformation was introduced.

A nucleus with the symmetries of an octupole shape (Fig. 2) is no longer invariant under $\mathbf{R}_x$; that is, the reflection symmetry is broken. Operating next with the parity operator $\mathbf{P}$, which inverts right

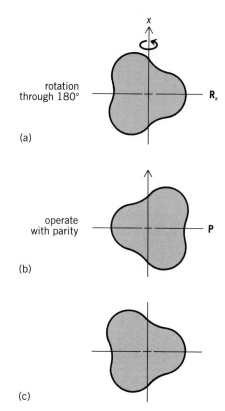

Fig. 2. Operation that defines the simplex quantum number in a nucleus with an octupole shape. (a) Rotation through 180° about the x axis [the operation $\mathbf{R}_x$], followed by (b) the parity operation $\mathbf{P}$, results in (c) restoration of the original shape.

and left, restores the nuclear shape. In other words, an octupole shape is invariant under the operation $\mathbf{S}$ defined by $\mathbf{S} = \mathbf{P}\mathbf{R}_x^{-1}$. The operation $\mathbf{S}$ is analogous to reflection about the symmetry axis. The eigenvalue of $\mathbf{S}$ is the simplex quantum number s: $sr = p$, where p is the parity of the state and r is the eigenvalue of $\mathbf{R}$ defined above. For an even-mass nucleus, $\mathbf{S}^2 = 1$ and $s = \pm 1$, according to Eqs. (6). For an odd-mass nucleus, $\mathbf{S}^2 = -1$ and $s = \pm i$, according to Eqs. (7).

$$s = +1 \qquad \text{for } I = 0^+, 1^-, 2^+, 3^-, \ldots \quad (6a)$$
$$s = -1 \qquad \text{for } I = 0^-, 1^+, 2^-, 3^+, \ldots \quad (6b)$$

$$s = +i \qquad \text{for } I = \frac{1}{2}^+, \frac{3}{2}^-, \frac{5}{2}^+, \frac{7}{2}^-, \ldots \quad (7a)$$
$$s = -i \qquad \text{for } I = \frac{1}{2}^-, \frac{3}{2}^+, \frac{5}{2}^-, \frac{7}{2}^+, \ldots \quad (7b)$$

For a shape in which the reflection symmetry is broken, the states can be labeled by the total angular momentum I, parity p, projection of angular momentum along the symmetry axis K, and simplex s. Also, because the shape is no longer invariant under parity $\mathbf{P}$, a $K = \frac{5}{2}$ "band," for example, would actually consist of a series of doublets of levels differing in parity; parity "doublet" structure has been observed in several odd-mass nuclei with $A \approx 220$. In a system with broken reflection symmetry, the members of a $K = \frac{5}{2}^\pm$ band can be labeled in terms of (s,p) values, and levels with the same (s,p)

values will be connected by enhanced electric quadrupole transitions. That is, there will be four "bands," which can be labeled in terms of (s,p) values:

$$(+i, +): {}^{5/2^+}, {}^{9/2^+}, {}^{13/2^+}, \ldots$$
$$(-i, +): {}^{7/2^+}, {}^{11/2^+}, {}^{15/2^+}, \ldots$$
$$(+i, -): {}^{7/2^-}, {}^{11/2^-}, {}^{15/2^-}, \ldots$$
$$(-i, -): {}^{5/2^-}, {}^{9/2^-}, {}^{13/2^-}, \ldots$$

Electromagnetic transitions between states with the same simplex, but differing parities, would proceed by enhanced electric dipole transitions; transitions between states with the same parity, but differing simplex, would proceed by mixed electric quadrupole and magnetic dipole transitions.

In the laboratory, deviations from Eq. (1) could be observed for different simplices and signatures. However, examination of excitation energies in the rotating frame should indicate that simplex- and signature-dependent properties do not exist for nuclear structures with axial quadrupole and octupole shape.

To date, relatively few searches for simplex-dependent effects in nuclei as a function of angular momentum have been undertaken. The only example has been in ^{219}Ac with 89 protons and 130 neutrons. The level scheme of this nucleus has four bandlike structures of opposite parity. The excitation energies in the rotating frame as a function of rotational frequency indicate considerable simplex dependent effects. The origin of these may be similar to the triaxiality observed in $A \approx 160$ nuclei, but simplex-dependent effects do indicate that degrees of freedom in addition to axial quadrupole and octupole collectivity are needed.

For background information *see* EIGENVALUE (QUANTUM MECHANICS); NONRELATIVISTIC QUANTUM THEORY; NUCLEAR STRUCTURE; PARITY (QUANTUM MECHANICS); SYMMETRY LAWS (PHYSICS) in the McGraw-Hill Encyclopedia of Science and Technology.

[JOLIE A. CIZEWSKI]

Bibliography: M. W. Drigert and J. A. Cizewski, Observation of parity doublets in ^{219}Ac, *Phys. Rev.*, C31:1977–1979, 1985; M. Gai et al., Molecular alpha-particle clustering in ^{218}Ra: Dipole collectivity in the vicinity of nuclear shell closures, *Phys. Rev. Lett.*, 51:646–649, 1983; G. B. Hagemann et al., Evidence for signature-dependent transition rates at large rotational frequencies, *Phys. Rev.*, C25:3224–3227, 1982; G. B. Hagemann et al., Signature dependent M1 and E2 transition probabilities in ^{155}Ho and ^{157}Ho, *Nuc. Phys.*, A424:365–382, 1984; R. K. Sheline et al., Evidence for near-stable octupole deformation in ^{225}Ra, *Phys. Lett.*, 113B:13–16, 1983.

Ocean-atmosphere relations

A net transfer of heat from the tropics to the poles by the atmospheric-oceanic envelope of the planet is required to maintain the present climate of the Earth. In the tropics the ocean transfers more heat poleward than does the atmosphere, while the atmosphere is dominant at higher latitudes. In the Northern Hemisphere the Atlantic Ocean apparently transfers more heat poleward than does the Pacific Ocean. The net transfer, also called transport or flux, of heat through a vertical plane extending across the Atlantic from west to east depends on the temperature of the various current features flowing north and south through the plane. At latitudes between 20° and 30°N, the net poleward heat transport in the North Atlantic Ocean is strongly dependent on the temperature of the Gulf Stream in the Straits of Florida; in this region the Gulf Stream is also called the Florida Current.

Global energy balance. Recent satellite observations have confirmed that at the top of the atmosphere the Earth experiences a net surplus of radiation in tropical regions and experiences a net deficit of radiation at polar latitudes. Yet other observations demonstrate that over long time periods neither are the tropics continuously warming nor are the poles continuously cooling. Thus, a mechanism to maintain the thermal equilibrium of the planet (a component of the Earth's climate) is required. For many years it was believed that the atmosphere was the primary vehicle that transported heat from the tropics to the poles. The ocean was thought to be primarily a reservoir that exchanged heat locally with the atmosphere. Recent information indicates that the atmospheric transport is indeed dominant at latitudes north of about 30°N, but that oceanic transport is responsible for most of the transfer of heat from the tropics (Fig. 1).

The processes that cause the motions in the at-

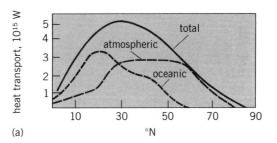

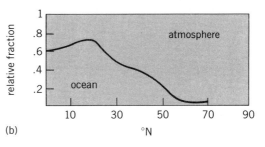

Fig. 1. Global heat transport. (*a*) Magnitude of the total oceanic and atmospheric heat transports in the Northern Hemisphere. (*b*) Fraction of the total poleward transport accounted for by the atmosphere and by the ocean. (*After H. Vonder Haar and A. H. Oort, New estimate of annual poleward energy transport by northern hemisphere oceans, J. Phys. Oceanog., 3:169–172, 1973*)

mosphere and the ocean, also called the circulation of the fluids, can be expressed as terms in a set of mathematical equations known as a model. Only in very simple models can analytical solutions be obtained; usually an approximate solution must be found by using computers. This approximation is called a numerical model. Numerical general-circulation models of the coupled ocean-atmosphere system are being developed to understand and eventually to predict the impact of the redistribution of heat on global climate. These numerical models will ultimately include all the relevant physical processes needed for these predictions and will be run on the new generation of very fast computers. However, even at this early stage in the evolution of general-circulation models as the important processes are being defined, it is clear that poleward transport of heat by ocean currents plays a major role in determining climate. For example, a simplified general-circulation model was tested with and without ocean circulation. The results from the model that included ocean circulation are characterized by increased surface temperatures at high latitudes, reduced snow and sea-ice coverage, and reduced sensitivity of the atmospheric climate generated by the model to changes in atmospheric carbon dioxide (an important result for studies of the effect of humans on climate). The results represent long-term aver-

ages; it is expected that variability in the amount of heat transported by ocean currents will induce variability in the atmospheric climate.

Ocean heat transport. There is uncertainty concerning where, how, and how much heat is transported poleward by the oceans. In fact, there is some evidence that the South Atlantic Ocean may work backward in the sense that heat is transported equatorward by the oceans. The large uncertainties in heat transport estimates are due to the lack of direct observations of the variables needed to compute heat transport.

The direct computation of net oceanic heat transport through a plane extending from the surface to the bottom of the ocean and crossing the ocean from west to east requires many observations of velocity and temperature. If the net flow through the plane is zero, heat transport is proportional to the integral of temperature times the northward component of velocity evaluated over the area of the plane. That is, heat transport is effected by differences in temperature between compensating northward and southward flows. Ideally, direct observations of temperature and velocity would be available at sufficient density both in time and in space to provide an estimate of the heat transport over some time period. However, both the temperature pattern and currents of the ocean are extremely complicated, and sufficient data (particularly velocity observations, which are more difficult and expensive to obtain than temperature observations) to compute accurate estimates of heat transport are not available, nor are they likely to become available in the immediate future.

Historically, therefore, investigators have decomposed the total velocity field into components, which through various simplifications can be evaluated from available data. For each component, the transport of the northward flow and the compensating southward flow must be equal. The net heat flux is then proportional to the difference in temperature between the two flows multiplied by the transport. In a convenient but somewhat arbitrary approach, the total oceanic velocity and temperature fields are decomposed into depth-averaged and depth-dependent components. The depth-averaged component can be related to the circulation in a horizontal plane, the subtropical gyres of the ocean which are characterized by northward flow of warmer water on the western boundaries and southward flow of colder water in the interior. The depth-dependent component can be related to the circulation in a vertical plane, which is characterized by poleward flow of warmer water at the surface and equatorward flow at depth.

In the case of the depth-averaged component of the Northern hemisphere, considerable data are available to describe the northward-flowing currents on the western boundaries of the ocean. Assuming an equal southward flow in the midbasin, net heat flux is proportional to the difference in temperature between the two flows times the transport of the cur-

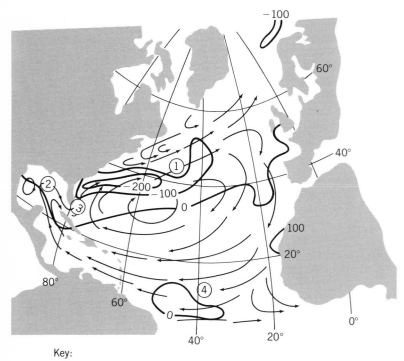

Key:

1 = Gulf Stream
2 = Loop Current
3 = Florida Current
4 = North Equatorial Current

Fig. 2. Schematic diagram of the subtropical gyre of the North Atlantic Ocean and mean annual heat gain in watts per square meter. Negative heat gains indicate transfer of heat from the ocean to the atmosphere. (*After A. F. Bunker, Computation of surface energy flux and annual air-sea interaction cycles of the North Atlantic Ocean, Mon. Weather Rev., 104:1122–1139, 1976*)

rent on the western boundary. In the case of the depth-dependent component, currents can be estimated from density and surface wind distributions by simplifying the equations used to model these currents.

Heat transport in North Atlantic. At the latitude of Miami, the Florida Current is the western boundary current of the North Atlantic subtropical gyre. Many direct observations of the velocity and temperature structure of the Florida Current are available. Recent measurements confirm previous findings demonstrating the importance of the Florida current in transporting heat poleward at the latitude of Miami. The volume transport of the Florida Current averaged over 1 year was 1.07×10^8 ft^3 (3.05×10^7 m^3) per second (as a point of reference, the average discharge of the Mississippi River is only about 700,000 ft^3 or 20,000 m^3 per second). The average temperature of the Florida Current during this period was 63.3°F (17.4°C). Historical data collected in the Atlantic Ocean indicate that the depth-averaged temperature of the southward flow in the midbasin is 41.7°F (5.4°C). Thus, the net northward heat transport effected by the horizontal circulation of the subtropical gyre of the North Atlantic was 1.47×10^{15} W during 1982–1983.

The contribution to heat transport from currents in the vertical plane can also be estimated from data collected in the Straits of Florida during 1982–1983 and historical data collected in the midbasin. The net contribution to heat transport from this mode at the latitude of Miami was -2×10^{13} W (that is, to the south). Thus, the total northward heat transport through this section was 1.2×10^{15} W. This estimate of heat flux is very close to other estimates of heat flux calculated from independent data sets using different methods.

It has been demonstrated that the Gulf Stream loses some of its surface layer heat content further downstream from the Straits (Fig. 2). The region of large heat loss off the east coast of the United States is coincident with regions of large variability in meteorological variables, suggesting a possible causal relationship. Further studies are required to quantify the role of oceanic heat transport on these regional features as well as on the global climate.

For background information *see* GULF STREAM; HEAT BALANCE, TERRESTRIAL ATMOSPHERIC; MARINE INFLUENCE ON WEATHER AND CLIMATE; OCEAN-ATMOSPHERE RELATIONS; OCEAN CURRENTS in the McGraw-Hill Encyclopedia of Science and Technology.

[ROBERT L. MOLINARI]

Bibliography: A. F. Bunker, Computation of surface energy flux and annual air-sea interaction cycles of the North Atlantic Ocean, *Mon. Weather Rev.*, 104:1122–1139, 1976; R. L. Molinari et al., Subtropical Atlantic climate studies: Introduction, *Science*, 227:292–295, 1985; M. J. Spelman and S. Manabe, Influence of oceanic heat transport upon the sensitivity of a model climate, *J. Geophys. Res.* 89:571–586, 1984; T. H. Vonder Haar and A. H. Oort, New estimates of annual poleward energy transport by Northern Hemisphere oceans, *J. Phys. Oceanog.*, 3:169–172, 1973.

Optical detectors

Interest in photoconductors for use in optical receivers for light-wave communications has increased due to recently reported sensitivity measurements made using devices with indium gallium arsenide (more precisely, $In_{0.53}Ga_{0.47}As$, that is, material with 53 atoms of indium and 47 atoms of gallium for every 100 atoms of arsenic) as the photoconductive material. This material is particularly well suited for photoconductor applications since it can be used to fabricate photoconductors which have high gain-bandwidth products and are sensitive to wavelengths λ less than 1.65 micrometers, thus encompassing the entire spectral range of interest for long-wavelength communication systems.

In this article the theoretical noise limitations of long-wavelength photoconductor receivers are discussed, and are compared to experimental results. In addition, the sensitivities of photoconductive receivers are compared with receivers employing either pin photodiodes or avalanche photodiodetectors. The discussion presented can easily be generalized to consider the sensitivity of photoconductors used in other spectral ranges. Of particular interest is the so-called short-wavelength region between 0.8 and 0.9 μm.

Photoconductor receiver operation. A typical $In_{0.53}Ga_{0.47}As$ photoconductor (Fig. 1) consists of a slab of absorbing material on an insulating substrate, such as iron-doped indium phosphide (InP), with ohmic contacts made to the absorbing layer. In operation, a potential is applied between the cathode and anode contacts, and the current flowing through the device in the absence of light is deter-

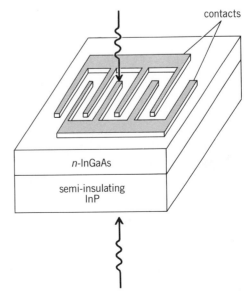

Fig. 1.   Cross section of a prototypical $In_{0.53}Ga_{0.47}As/InP$ photoconductive detector.

mined by the dark resistance R_D of the channel. When light is incident on the photoconductor, additional carriers are generated, thereby reducing the channel resistance and increasing the current flowing in the external circuit. Photoconductors have gain G equal to the ratio of the majority carrier lifetime τ to the transit time t, and can be as high as 1000. Values of G between 10 and 50 are more typical of $In_{0.53}Ga_{0.47}As$ photoconductors. Since the response time of the device is equal to the carrier lifetime, the gain-bandwidth product is simply given by Eq. (1). Thus, to increase the intrinsic device speed, the distance between cathode and anode

$$G/2\pi\tau = 1/2\pi t \tag{1}$$

must be minimized. This is accomplished by using the interdigitated electrode configuration shown in Fig. 1. The gain-bandwidth limit implies that gain can be obtained only with a concomitant loss in detector speed, as shown in Eq. (1).

Receiver noise. The various contributions to photoconductor receiver noise are Johnson noise, generation and recombination noise of photogenerated carriers, $1/f$ noise, field-effect transistor (FET) channel noise, gate leakage shot noise, and noise arising from intersymbol interference due to slow detector response to high-bit-rate data streams. Johnson noise is the dominant source of noise in photoconductors and is due to thermal fluctuations of the channel dark resistance, R_D. Assuming an impedance-matched front end, that is, assuming that the amplifier front-end resistance R_F is equal to R_D, then the total root-mean-square (rms) Johnson noise current (per unit bandwidth) is given by Eq. (2).

$$\langle \overline{i_j^2} \rangle = 8\,kT/R_D \tag{2}$$

Here, k is Boltzmann's constant and T is the thermodynamic temperature.

Receiver sensitivity. In Fig. 2 the receiver sensitivity is plotted as a function of digital bit rate B

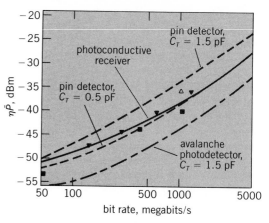

Fig. 3. Photoconductive receiver sensitivity $\eta\overline{P}$ as a function of bit rate B for a channel dark resistance $R_D = 400\ \Omega$ (assuming a receiver with other characteristics similar to those used in Fig. 2), compared with the sensitivity of pin detector receivers with front-end capacitances of $C_T = 0.5$ picofarad and 1.5 pF, and an avalanche photodetector receiver with $C_T = 1.5$ pF. Curves indicate calculated values. Data points refer to experimental results as follows: photoconductors (open triangles); pins with $C_T \approx 0.5$ pF (closed triangles); avalanche photodetectors (squares).

over the range 50 megabits/s $< B <$ 5 gigabits/s and at $\lambda = 1.3\ \mu m$ for several values of channel resistance R_D. This range of bit rates is typical of what is presently being considered for use in optical transmission systems. Also a gain G of 40 and a recombination time τ of 2 nanoseconds are assumed. At low bit rates, receiver sensitivity increases as the square root of the channel resistance, as expected for Johnson noise–limited response. The degradation in sensitivity at high bit rates is a result of the slow response (2 ns) of the photoconductor.

For comparison, in Fig. 3 the $R_D = 400\ \Omega$ photoconductor receiver sensitivity is plotted along with sensitivities typical of $In_{0.53}Ga_{0.47}As$ pin and avalanche detector receivers. It is apparent that photoconductors can result in an improvement in sensitivity of several decibels over pin detectors with capacitances C_T greater than 1 picofarad. However, the sensitivities of photoconductor receivers are found to be somewhat worse than those of the best pin receivers studied, for which a front-end capacitance of only 0.5 pF has been reported. Although difficult to achieve with present hybrid pin–field-effect transistor circuits, front end capacitances with C_T of about 0.5 pF may be routinely attainable in the near future by using monolithically integrated pin–field-effect transistor circuits. In contrast, high-performance avalanche photodiodes can give 5 to 7 dB higher sensitivity than the best photoconductor receivers.

Also shown in Fig. 3 are data points corresponding to experimental results reported for these several device types. The data point corresponding to the photoconductor receiver lies roughly 2 dB above the theoretical sensitivity value. This device had R_D, G, and τ approximately equal to those values used in generating the solid curve shown. A sensitivity pen-

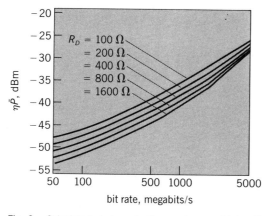

Fig. 2. Calculated photoconductive receiver sensitivity $\eta\overline{P}$ as a function of bit rate B for several values of channel dark resistance R_D, assuming a photoconductive receiver with characteristics determined by the parameters in the table, and gain G of 40 and recombination time τ of 2 nanoseconds. Sensitivity is measured in decibels per milliwatt (dBm), where 1 mW = 0 dBm.

alty of approximately 1.5 dB above the theoretically predicted sensitivity was incurred with this particular device due to an optical "filling factor" of less than unity. The small filling factor results from the incomplete illumination of the channel due to obstruction of the incident light by the electrodes (Fig. 1). The difference between the avalanche photodiode results, and the calculation is probably due to the large primary dark current and device capacitances compared to those used in the analysis.

In conclusion, ideal long-wavelength photoconductors with gains of 40 and channel resistances R_D of 400 Ω can have sensitivities as high as the best pin photodetector receivers obtained so far. The comparison between photoconductors and pin receivers becomes less favorable, however, if the inherently inferior coupling between the photoconductors and the incident light beam is considered. In addition, avalanche photodiode receivers are more sensitive than either photoconductors or pin receivers over the entire bit rate range of 50 megabits/s to 5 gigabits/s considered in this analysis.

For background information *see* ELECTRICAL NOISE; MICROWAVE SOLID-STATE DEVICES; OPTICAL DETECTORS; PHOTOCONDUCTIVITY in the McGraw-Hill Encyclopedia of Science and Technology.

[S. R. FORREST]

Bibliography: C. Y. Chen, B. L. Kasper, and H. M. Cox, In$_{0.53}$Ga$_{0.47}$As/InP photoconductive detector with high receiver sensitivities, *7th Topical Meeting on Integrated and Guided Wave Optics, Technical Digest*, Pap. THA4, Orlando, Florida, April 24–26, 1984; S. R. Forrest, The sensitivity of photoconductor receivers for long-wavelength optical communications, *IEEE J. Lightwave Technol.*, LT-3:347–360, 1985; H. Kressel (ed.), *Semiconductor Devices for Optical Communication*, Topics in Applied Physics, vol. 39, ch. 4, 1982.

Optical microscope

Microfabrication and micromovement technology has developed to the point where it is about to be applied in all areas of science. This opens the possibility of using the optical near-field in a variety of applications with great technological significance. Near-field scanning optical microscopy (NSOM) is an example of this development in microfabrication and micromovement combined with readily available microcomputers, lasers, and extremely low-light-level detection systems. This technique is a form of superresolution light microscopy, and has all the advantages of light microscopy, such as operation in ambient environments without the use of ionizing radiation, combined with the high resolution available with scanning electron microscopes.

Optical near-field. When light emanates from an illuminated spot on an object or is passed through an aperture or a slit in an opaque screen, it goes through three distinct regimes. In Fig. 1 an aperture is illuminated with plane waves. In the near-field region, immediately after transmission through the aperture, the radiation is highly collimated to the

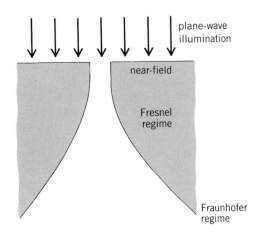

Fig. 1. Radiation field after it has passed through a sub-wavelength aperture in an opaque screen.

size of the aperture. Subsequently, the radiation field begins to spread out in two stages. In the first stage, called the Fresnel regime, the opening angle of the spreading radiation field changes with distance from the aperture. This is followed by the Fraunhofer or far-field region where the opening angle is constant as a function of distance from the screen. It is in the far-field that all of the conventional notions of geometrical optics are operative.

To achieve a somewhat deeper understanding of the near-field where the radiation is highly collimated and not diffracted, the behavior of the radiation field of light of wavelength λ = 500 nanometers when it passes through a slit 50 nm (λ/10) wide in an 180-nm-thick opaque screen, was calculated exactly. The results of this calculation are shown in Fig. 2 as a three-dimensional plot of the Poynting

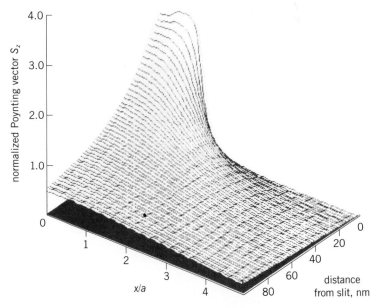

Fig. 2. Graphical representation of the Poynting vector of the radiation flux emanating from a slit one-tenth the wavelength of the radiation in terms of distance from the slit and distance parallel to the aperture plane expressed as *x/a*, where *x* = 0 is the center of the slit and *x* = *a* is the edge of the slit.

vector of the radiation flux, as a function of both distance from the aperture plane and distance parallel to the aperture plane. Certain essential features of the near-field are dramatically illustrated. First, it is collimated to a distance of at least the radius of the aperture which is 25 nm in this case, and experiments and other calculations have shown that there are various ways to extend this regime of collimation. Second, at the aperture plane there is a dramatic increase in the intensity of the radiation field which is due to the presence of the aperture. This important intensity increase falls off exponentially in a manner similar to the fall-off in the collimation.

Aperture-object distance control. Thus, in order to use to full advantage the intense, collimated radiation in the near-field, the aperture must be close to the illuminated object. One obvious way of achieving this is for the aperture and object to be in contact, but the best way is to have some feedback loop that can accurately control (within ± 3 nm) the aperture-object distance. Luckily, several methods exist for constructing such a feedback loop. One method involves measuring a tunneling current between a conducting object and the metal aperture. This current is a sensitive monitor of the aperture-object separation. Other methods that can be used to monitor this separation are measurements of other electrical properties such as capacitance between the aperture and object or even a measurement of the rapid fall-off in the near-field intensity. The near-field intensity fall-off could be of special importance in cellular imaging, especially in view of the fact that cell membranes can readily be labeled with fluorescent molecules, and fluorescence has been detected with high sensitivity through submicrometer apertures (Fig. 3). Fluorescent molecules

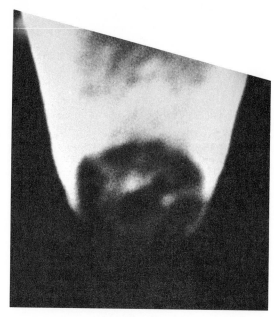

Fig. 4. Scanning electron micrograph of a submicrometer aperture at the tip of a pipette.

in a cell membrane could be used both as near-field intensity markers and for contrast enhancement in cell membrane imaging.

Pipette apertures. Planar aperture arrays as seen in Fig. 3 could be very important for a variety of near-field applications. However, for imaging with near-field scanning optical microscopy of real surfaces, which are rough, such apertures have limited potential. To alleviate this problem, 50-nm apertures have been constructed at the tip of metallized glass pipettes. Scanning electron micrographs of these apertures (Fig. 4) make it possible to accurately characterize their dimensionality and geometry. This characterization is important in order to understand both the mechanism of light transmission and the resolution achieved with a particular aperture. Pipette apertures are cheap and easy to construct and, most important, allow probing, with near-field scanning optical microscopy, of objects with rough surfaces.

Test structures. The availability of pipette apertures is an important step in the general applicability of near-field scanning optical microscopy. However, to demonstrate this technique it is also important to have well-characterized test structures. Such test patterns have been constructed with grating masks fabricated by using electron beam lithography at the National Resource Research Facility for Submicron Structures at Cornell University. The masks are used for contact-printing the resulting grating of aluminum on glass.

Near-field scanning optical microscope. The near-field scanning optical microscope (Fig. 5) consists of an ordinary tungsten light source, filtered to transmit 500-nm light, which is focused through a pipette with an aperture at its tip. For all the initial

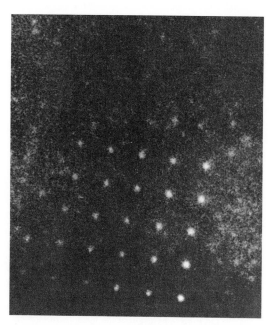

Fig. 3. Fluorescent light at a wavelength of 600 nm transmitted through an array of 120-nm apertures.

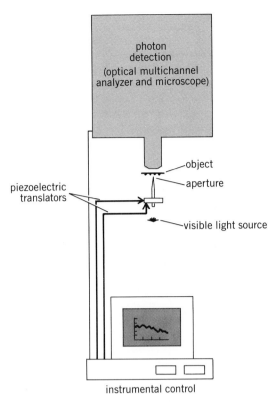

photon
detection
(optical multichannel
analyzer and microscope)

object

aperture

piezoelectric
translators

visible light source

instrumental control

Fig. 5. Initial version of a near-field scanning optical microscope.

measurements with this instrument, the aperture was in contact with the test grating as each data point was recorded. It was then moved with the piezoelectric translators shown in Fig. 5 to the next position. As the aperture moved along a line across the grating, the transmitted light through the grating was detected with an optical multichannel analyzer yielding intensity variations in the light, and this variation was plotted as a function of the pipette position. To build up a two-dimensional image of the grating, many such horizontal line scans at different vertical positions could be combined. From this line scan of a well-characterized test object, a resolution of less than 150 nm was estimated, even though the aperture diameter was 500 nm. The second-generation instrument that is currently being constructed is expected to achieve the goal of a resolution better than 50 nm.

Applications. A light microscope with such a resolution when combined with the use of fluorescent cellular probes for contrast enhancement should extend the applicability of near-field scanning optical microscopy from applications in advanced technology to those in biology and medicine. Thus, near-field scanning optical microscopy should become the chief tool in areas from microelectronics to chemistry, biology, and medicine where nondestructive, high-resolution imaging is required without the use of high vacuums or ionizing radiation. In addition, since rapid scanning is possible, it is entirely feasible to conceive of measuring dynamic phenomena

with high spatial resolution. Finally, the concept of near-field imaging opens new avenues for extremely high-density (1 gigabyte/cm^2) optical storage of information.

For background information *see* DIFFRACTION; OPTICAL MICROSCOPE in the McGraw-Hill Encyclopedia of Science and Technology.

[AARON LEWIS; MICHAEL I. ISAACSON; ERIC BETZIG; ALEC HAROOTUNIAN]

Bibliography: E. Betzig et al., Near-field scanning optical microscopy (NSOM): Development and biophysical applications, *Biophys. J.*, 49:269–280, 1986.

Optical telescope

During the past 50 years astronomy has been revolutionized by the development of new spectral windows. Radio astronomy and x-ray astronomy, for example, starting from the crudest beginnings, have matured to the point where advanced instruments have enormously broadened the view of the universe. Yet during this period the size of telescopes used to study optical wavelengths, in many ways still the most valuable window, has not in general advanced, nor has the technology used to make them.

The Hubble space telescope, operating at optical and ultraviolet wavelengths, will indeed represent a major breakthrough. It is designed to achieve diffraction-limited images 10 times sharper than the best to be seen through the atmosphere. However, with an aperture of 45 ft^2 (4 m^2) in area it does not compete with even existing ground-based telescopes for problems where large light grasp is essential, or for deep surveys which require mapping a large area of sky.

Multiple-mirror telescope. A new generation of large telescopes for these problems is now being developed. A breakthrough in technology was made with the multiple-mirror telescope (MMT), put into operation in 1979. Its light-gathering power is achieved by six telescopes, each with a 28 ft^2 (2.6 m^2) collecting area, mounted rigidly in one frame and all gathering light from the same object, to give a total area of 170 ft^2 (16 m^2). Thus relatively small mirrors give power similar to the 215 ft^2 (20 m^2) of the single 200-in. (5-m) mirror at Palomar. While not in itself of unusual size, the multiple-mirror telescope has proven that the very accurate controls needed to combine light from different mirrors can be achieved in practice, opening for the first time the way to telescopes larger than any single reflector.

Segmented-mirror design. Following in this direction, a telescope has been designed that will use 36 mirrors, each 24 ft^2 (2.2 m^2) in area, to give a total area of 850 ft^2 (79 m^2), nearly four times the light grasp of Palomar. With so many mirrors it is not practical to configure each one as a separate telescope, as in the multiple mirror telescope, but all will be figured as segments of one parent paraboloidal surface. This step requires new technology,

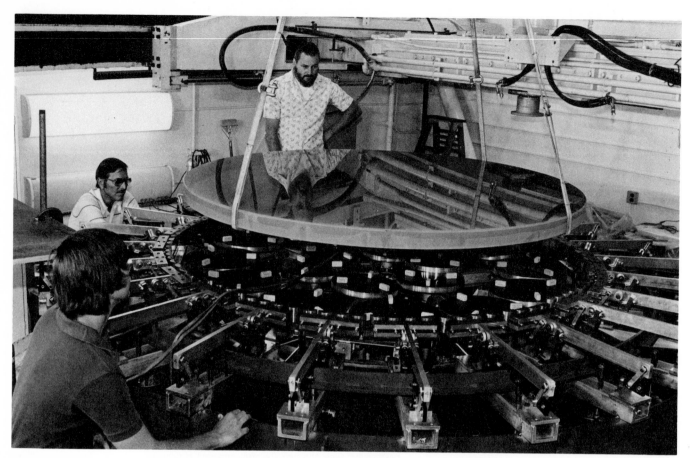

Fig. 1.   A 6.5-ft-diameter (2-m) glass disk being fitted into a stressing mechanism that will keep it bent during polishing. When released, it will spring into the desired aspheric shape, part of a large paraboloidal parent surface. (*Kitt Peak National Observatory*)

since existing methods fail in the polishing of non-axisymmetric surfaces, particularly those requiring an accurate surface right to the mirror's edge. A new method for making such surfaces was devised, and a full-size prototype of a single segment has been constructed. A relatively thin disk of glass is bent at the edges in such a way that if a spherical surface is polished in while it is bent, the disk will spring into the desired shape when released (Fig. 1). It remains then to fit and hold together all the pieces. There must be only the thinnest gaps between them, or the telescope will not work well at infrared wavelengths. The complete surface will be made as a mosaic of hexagons, with each hexagon cut from one of the polished disks.

In the multiple-mirror telescope, exact coalignment of the six telescopes is maintained by periodic computer analysis of the final image with all the light combined. The segmented-mirror design will also use image analysis to achieve exact overlapping of images from the 36 mirrors. However, a new method has been devised to stabilize the alignment, analogous to soldiers lining up on parade. Each mirror will sense and adjust its position relative to its nearest neighbors. Readings from 168 sensors between adjacent edges will be relayed to a central computer, which will then calculate the correction

to be made to the position actuators, three per mirror. The placement of each one must be accurate to within a small fraction of the wavelength of light, or about a millionth of an inch (25 nanometers).

Multiple-mirror design. A different approach to constructing large telescopes is to make each mirror element as large as possible, so that only a few elements need be used. These would be configured as independent telescopes, all pointing in tandem as in the multiple-mirror telescope, but generally operated with separate instruments on each telescope. Only when necessary would the light be brought to a common focus. This would be done, for example, to study relatively bright objects with even sharper detail than the Hubble space telescope, by comparing how the crests and troughs of light waves arrive at the different telescopes, a technique known as Michelson interferometry.

Production of large mirrors. The problem of making the largest possible telescope mirrors from a single piece of glass has been the subject of an intensive research program. A technique has been developed for casting glass in a complex mold to form a one-piece stiff but lightweight honeycomb structure. It is not unlike the method used to make the Palomar blank, when liquid glass was ladled into a circular mold. An array of blocks secured to the mold bottom left a ribbed backing to the disk. In the new method the blocks are raised on short columns, so that the glass running under the blocks forms a back plate. This forms the glass into the most rigid of lightweight structures, honeycomb ribs sandwiched between two facesheets (Fig. 2).

Construction has begun on a furnace to cast such blanks, 26 ft (8 m) in diameter and weighing about 15 tons. The furnace is mounted on a turntable, so that it can be rotated while the glass is liquid, to form the required parabolic dish shape to the upper face. Formed this way, the surface is not of optical quality, but is close enough to short-cut many of the usual tedious and expensive steps of preparing a large blank for precise grinding and polishing.

Mirrors 26 ft (8 m) in diameter, with an area of 540 ft² (50 m²), could be polished by the same methods used currently for 13-ft (4-m) mirrors. However, these methods are not adequate to polish very deep parabolas with focal length no larger than the mirror diameter. For very large telescopes such deep mirrors would be especially advantageous to keep the telescope mount and enclosure down to manageable proportions and reasonable cost. A new method for polishing, specifically aimed at these strongly aspheric surfaces, makes use of a polishing tool whose shape can be changed under computer control as it is moved over the mirror. The tool curvature is changed to match the changes in shape of the desired aspherical surface at different positions. In this way the polishing process will tend to produce the aspherical surface rather than the usual sphere.

Thermal design. Another advance of a more general nature is the recognition of the need for great care in the thermal design of telescopes and their enclosures. Even slight temperature differences cause convection, and the resulting shimmer may blur an otherwise excellent image. Large structures, including large mirrors, are especially susceptible to this problem, because they respond only slowly to changes in ambient temperature. It is thus desirable to build telescopes at high or ocean sites where changes in the nighttime temperature are minimized. Also, the mirror should be thermally responsive, as is the honeycomb mirror ventilated with air at ambient temperature. Other parts of the telescope also must be at the air temperature, and this is done in the multiple-mirror telescope by insulating heavy structural members and by wrapping the tube structure in aluminized tape to prevent overcooling by radiation on clear dry nights.

Plans for large telescopes. As these new technologies are being advanced, so are plans for building a new generation of large ground-based telescopes. The California Institute of Technology announced in 1984 a gift from the Keck Foundation to construct a 33-ft (10-m) segmented-mirror telescope on Mauna Kea in Hawaii. It will be built in conjunction with the group at Lawrence Berkeley Laboratory that developed the segmented-mirror design discussed above. The University of Arizona and

Fig 2.   A 6-ft-diameter (1.8-m) honeycomb sandwich glass mirror, a prototype for 26-ft-diameter (8-m) mirrors. The surface was formed directly as a paraboloid by spinning liquid glass. Its very fast surface (*f*/1) will be finished by new grinding and polishing methods. (*Steward Observatory, University of Arizona; photograph by Dan Watson*)

Ohio State University have also announced their intention to construct a telescope, making use of one or more of the short-focus 26-ft (8-m) honeycomb mirrors, on the 10,700-ft (3260-m) peak of Mount Graham in Arizona. Another similar mirror is to be used in a 26-ft (8-m) telescope in Chile by the Carnegie Institute of Washington. Designs for multiple-mirror and segmented-mirror telescopes have been developed at the National Optical Astronomy Observatory in conjunction with the Arizona and Berkeley groups, and a decision has been taken to pursue a design that calls for four 25-ft (7.5-m) honeycomb mirrors in a multiple-mirror configuration.

These telescopes should be in operation by 1995, perhaps with others planned in Japan and Europe. A very active period in astronomy can be expected as these instruments work in conjunction with radio telescopes and new observatories in space.

For background information *see* INTERFEROMETRY; OPTICAL TELESCOPE in the McGraw-Hill Encyclopedia of Science and Technology.   [ROGER ANGEL]

Bibliography: J. R. P. Angel and N. J. Woolf, 15 m MMT design study, *Astrophys. J.*, in press; L. D. Barr and B. Mack (eds.), Advanced technology optical telescopes II, *Proceedings of International Society of Optical Engineering*, vol. 444, 1984; M.-H. Ulrich and K. Kjar (eds.), *Very Large Telescopes, Their Instrumentation and Programs*, 1985.

Ore and mineral deposits

Scientists have been studying the origin of platinum-group elements in mafic and ultramafic rocks for nearly 200 years. Concentration of these elements in placer, hydrothermal, or nickel-copper deposits is generally well understood. However, the concentration of platinum-group elements in large mafic and ultramafic layered intrusions is just beginning to be understood. Recently it has been documented that at least two magmas were involved in the formation of known rich platinum-group element-bearing ultramafic and mafic layered intrusions. Two models that involve mixing of two magmas have been proposed to account for the concentration of such elements in these layered intrusions. The role played in concentrating platinum-group elements in layered intrusions by migrating liquids, chlorine-rich hydrothermal fluids, and platinum-iron alloys is uncertain but possibly important. Field, theoretical, and experimental investigations are in progress to further the understanding of the processes that concentrate platinum-group elements in this type of remarkable deposit.

Early studies. Scientists studying platinum-group mineral occurrences have found that these deposits are associated with or derived from mafic or ultramafic rocks that had their origin in the Earth's mantle. This relationship is thought to result from the platinum-group elements' tendency to alloy with metallic iron. Thus, when the primitive Earth was in a molten state, these elements were concentrated in a dense iron-rich liquid that sank toward the Earth's center. The platinum-group elements in this iron-rich liquid were later partitioned between the core and mantle of the Earth. In turn, throughout geologic time there has been an exchange of material between the mantle and crust of the Earth, and this exchange has given rise to deposits of platinum-group minerals. These deposits can be classified as (1) placer, (2) hydrothermal, (3) platinum-group mineral-bearing nickel-copper deposits hosted in mafic and ultramafic rocks, and (4) large mafic and ultramafic layered intrusion deposits.

Later studies. Placer deposits were the sole source of all platinum-group elements during the late 1800s and very early 1900s. It was found that these placer deposits were associated with large blocks of mantle material that had been thrust into or onto the crust as well as intrusive and extrusive ultramafic and mafic rocks. These rocks contained anomalous but minor amounts of platinum-group elements. Although these rocks could not be profitably mined, they could break down chemically and physically, releasing small fragments or particles of platinum-group minerals or alloys that might be concentrated in stream or river gravels to form high-grade platinum-group element-bearing deposits.

In 1919 the great Sudbury nickel-copper platinum-group element-bearing deposit of Canada was discovered. Because mafic or ultramafic magmas mostly originate in the mantle, they contain anomalous but small amounts of platinum-group elements (usually less than 100 parts per billion combined platinum-group elements). If these magmas become saturated with sulfur, small drops of molten sulfide will form. These molten sulfide drops are able to concentrate platinum-group elements from several to several hundred times that of the host magma. Thus, nickel-copper ore bodies enriched in platinum-

group elements, such as Sudbury, may form. These types of ore deposits produce nickel and copper, with the platinum-group elements being an important by-product.

Significant but as yet unimportant concentrations of platinum-group elements have been reported in mafic rocks that have been extensively altered by hydrothermal fluids. Thus it appears (1) that hydrothermal solutions are able to scavenge platinum-group elements as they pass through mafic rocks and concentrate them at select locations or (2) that platinum-group elements are not concentrated in silicate minerals but are concentrated in late-stage hydrothermal solutions.

The discovery of the Merensky Reef in 1924 in the Bushveld Complex of South Africa revealed a large body of mafic to ultramafic rock with a concentration of platinum-group elements rich enough to be mined solely for platinum-group minerals. Until the early 1970s this occurrence was thought to be one of a kind, and explanations were proposed concerning the concentration of platinum-group elements in this remarkable mineral deposit.

Platinum-group minerals in the Merensky Reef are associated with sulfide and chromite grains. Because of this relationship, it was suggested that sulfide or chromite grains could act to concentrate these elements. However, such grains present at other locations in the Bushveld Complex are not associated with ore-grade concentrations of platinum-group elements. Therefore the concentration must require something more than the presence of sulfide or chromite grains.

Recent studies. The discovery of the J-M Reef in the Stillwater Complex of Montana in the early 1970s sparked renewed theoretical, experimental, and field investigations in the Bushveld and Stillwater Complexes. The Bushveld was no longer the single known occurrence of this kind of concentration of platinum-group elements.

Platinum-iron alloy model. It was proposed that platinum-iron alloy precipitated directly from the Bushveld magma and that some of the base-metal sulfide-platinum-iron alloy intergrowths may have formed when sulfur reacted with the original platinum-iron alloy grains.

During studies of the Columbia River basalts, textures and compositions were found that suggest sulfide liquid and platinum-iron alloy can coexist at magmatic temperatures and pressures. Furthermore, it was found that platinum concentrations in sulfide liquid coexisting with platinum-iron alloy are equal to or less than the platinum content of sulfide found in certain deposits such as the J-M Reef of the Stillwater Complex and UG-2 Reef of the Bushveld Complex. Perhaps even more important is that the platinum-to-sulfide ratios in the J-M and UG-2 Reefs suggest that some of the platinum was precipitated as a second phase that could possibly be a platinum-iron alloy. Thus it appears that precipitation of a second phase, possibly platinum-iron alloy, could play an important role in forming these deposits.

Migrating liquid model. A model has been proposed in which liquid initially occupying the space between mineral grains becomes concentrated in platinum-group elements as it migrates upward through a thick accumulation of mineral grains. This liquid enriched with platinum-group elements is combined with the overlying magma, or possibly sulfide grains precipitated from the magma, to form a high-grade platinum-group mineral-bearing layer. The migration of the liquid initially occupying the space between mineral grains should produce chemical discontinuities that are separated from the various silicate layers of the intrusions. This relationship has not yet been demonstrated for the platinum-group mineral-bearing layers of the Bushveld or Stillwater Complexes.

Two-magma model. Based on field and chemical relationships, evidence was offered indicating that the Stillwater Complex is the product of two magmas. In addition, the data have suggested that the first magma has ultramafic characteristics and that the second had anorthositic affinities. In this model the J-M Reef formed in response to mixing of these two magmas, with most of the platinum and palladium being derived from the first magma and most of the sulfur from the second. Rare-earth-element data have been presented that support and strengthen the concept of the Stillwater Complex being the product of two magmas.

The results of analyses of sills thought to be feeders for the Bushveld Complex documented that the Bushveld Complex was also a product of two magma series and their hybrids. Two magma types were recognized, termed U-type and A-type, to reflect their ultramafic and anorthositic products. The U-type magma was found to be exceptionally rich in platinum and palladium and thus differed from ordinary basalt.

Plume model. The plume model was developed to explain the mixing and resultant concentration of platinum-group elements in the Stillwater and Bushveld Complexes. In this model, a new pulse of mafic magma, slightly lighter than the host magma, enters the magma chamber as a plume or jet and spreads laterally to form a density-stratified hybrid layer at some intermediate level in the magma chamber. Sulfide liquid and olivine grains form during the ascent of the plume and formation of the hybrid layer in response to mixing of the magmas and resultant cooling of the hybrid layer. The sulfide liquid scavenges platinum-group elements from a large volume of magma during mixing of the two magmas, eventually settling to the floor of the magma chamber along with the olivine grains to form an ore deposit rich in platinum-group minerals.

Double-diffusive convective magma mixing model. This model was developed to account for the mixing and resultant concentration of platinum-group elements in the Stillwater and Bushveld magmas. Double-diffusive convection has two cases, a diffusive case and a finger case. In the diffusive case, convection occurs in layered liquids where two or more components of the liquids have different diffusivities. Each pair of liquids is separated by a thin interface, and all heat and chemical transfer occurs through this interface by diffusion. Two liquid layers should mix by advection of the diffusive interface during precipitation of minerals. The diffusive case prevails where a relatively hot, dense liquid underlies a cooler, less dense one. The finger case prevails where a relatively hot, low-density liquid overlies a liquid that is denser only because it is cooler. In this case, liquid layers are mixed by mass flow.

In the double-diffusive convective model, two parent magmas, the first having ultramafic characteristics and the second anorthositic affinities, differ sufficiently in density to form separate layers in a magma chamber. The crystallization and mixing of these magmas is controlled by double-diffusive convection, with sequences of rock layers forming concurrently by downdip accretion from a column of liquid layers that are separated by diffusive interfaces. The sulfide liquid and resultant concentration of platinum-group elements are believed to have formed during the downdip accretion process in response to mixing effects and temperature changes.

Hydrothermal model. It has recently been recognized that chlorine-rich biotite, apatite, graphite, and amphibole are associated with the J-M and Merensky Reefs. The chlorine-rich nature of these minerals strongly indicates that chlorine-rich fluids may have been associated with late-stage processes and that such chlorine-rich fluids may have been important in transporting platinum-group elements during or after the formation of the reefs.

For background information *see* ORE AND MINERAL DEPOSITS; PLATINUM in the McGraw-Hill Encyclopedia of Science and Technology.

[S. G. TODD]

Organic chemical synthesis

Cyclopentanoids are compounds whose key structural unit consists of five carbon atoms arranged in a cyclic array or ring. Their importance stems from the properties that such substances can possess. For example, the pyrethrins are an early example of what is now known as an allelochemical, a naturally occurring insecticide employed as a defense mechanism, in this case by pyrethrum flowers. This environmentally safe pesticide derives by combining a cyclopentanoid known as a rethrolone (the five-member portion of a pyrethrin) with another structural fragment possessing a three-member ring (cyclopropanoid) which is either chrysanthemic acid or pyrethric acid in the natural pyrethrins. A particularly important family of cyclopentanoids consists of the prostaglandins, human hormones that are involved in many biological control functions in extremely small amounts. Sometimes, several rings are fused together to create di- or tricyclopentanoids (for example, coriolin, an antibiotic and antitumor agent) which possess two or three such rings. Some of these compounds have antibiotic or anticancer properties. Elucidating the biological function of

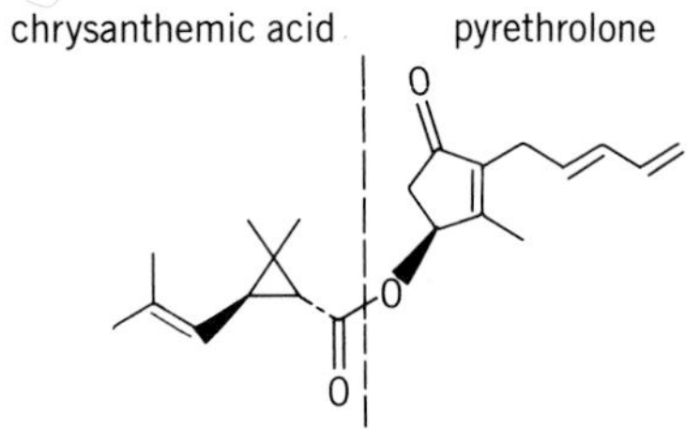

chrysanthemic acid   pyrethrolone

Pyrethrin I

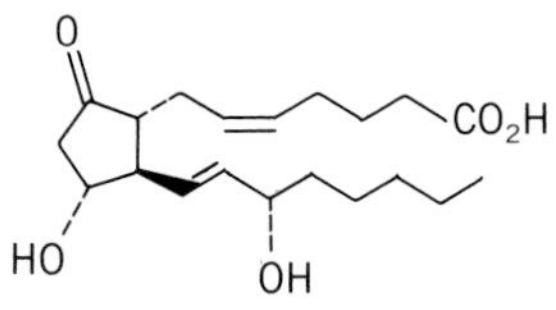

Prostaglandin E$_2$ 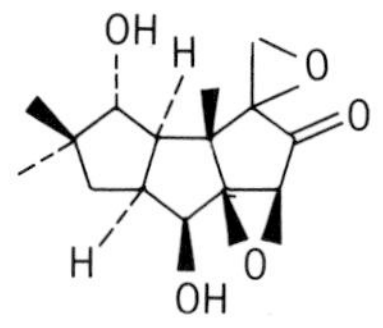   Coriolin

1,4-dicarbonyl compound

Aldol unit

bond formed by aldol condensation

cis-Cinerolone   (2)

some compounds, as in the case of the prostaglandins, requires chemical synthesis to provide adequate amounts. Optimizing the properties of any compound for a particular end use, for example, as an insecticide or a pharmaceutical, requires fine-tuning the structure—a task that also requires chemical synthesis.

Modernization of classic approaches. A classic approach to making ring compounds involves the aldol condensation, first reported in 1838. While it is tempting to believe that little new can be found in a reaction that has been known for 150 years, modern thought regarding chemical reactivity and stereochemistry has led to a revolution in the application of this reaction in the synthesis of complex natural products. The first chemical synthesis of one of the prostaglandins employed this reaction as the key five-member ring-forming reaction, as have most of the subsequent syntheses.

An important development has been the expanding base of starting materials to make the requisite substrate that undergoes the aldol condensation, a 1,4-dicarbonyl compound [(I) in reaction (1)]. Fur-

Furfural

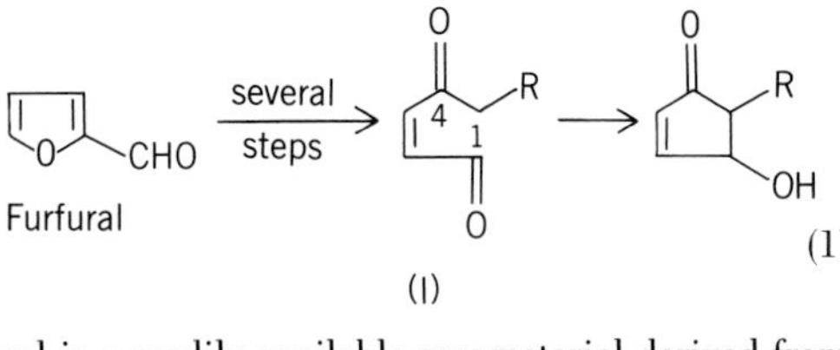

several steps

(I)   (1)

fural is a readily available raw material derived from various agricultural products such as corn. Very little use exists for this abundant substance at present; however, the demand for this chemical may increase as a result of the recently developed process to manipulate furfural chemoselectively to form the aldol substrate. This simple and short route to five-member ring compounds has served as a practical method for producing the prostaglandins.

Recent synthesis of the rethrolone fragment of the pyrethrin insecticides uses a strategy centered on the aldol condensation. In one such approach, only three synthetic operations were required. The last one [reaction (2)] embodies the aldol condensation

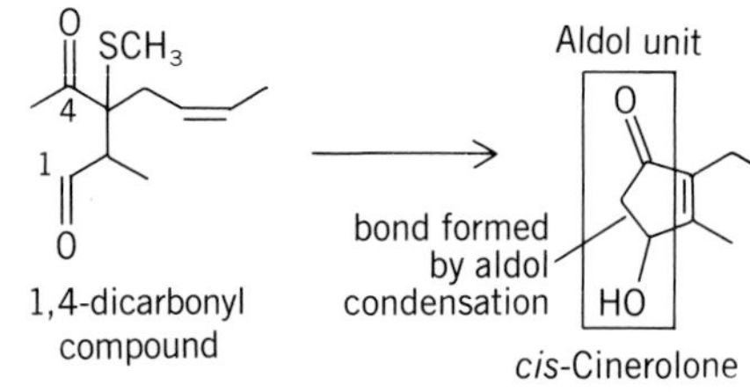

and the simultaneous elimination. Such extraordinary efficiency creates an opportunity to improve the properties of the pyrethrins by making available synthetic analogs.

In many instances, the initial aldol products split out the elements of water to produce a multiple bond between two carbon atoms (also known as an olefinic linkage). Such a process (olefination) can be controlled by using a variant introduced by Georg Wittig. In this version, an organophosphorus reagent is employed as shown in the synthesis of the bicyclopentanoid in reactions (3), which was a key inter-

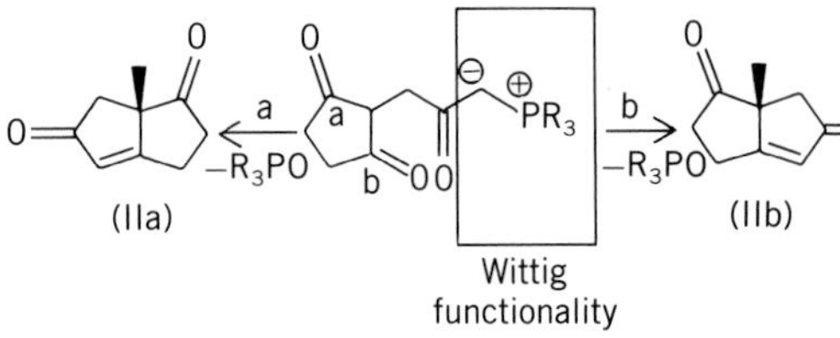

(IIa)

Wittig functionality

(IIb)   (3)

mediate in the synthesis of the antibiotic and antitumor agent coriolin. In this Wittig olefination, condensation at the carbonyl group labeled "a" in the compound containing the Wittig functionality (an inner salt or zwitterion in which the negative and positive charges are on adjacent atoms in which the positive charge resides on phosphorus) generates (IIa), whereas condensation at carbonyl group "b" generates (IIb). These compounds represent mirror-image isomers, a type of differentiation that is very difficult to achieve outside of enzyme-catalyzed reactions. By properly choosing the phosphine portion [the functional group derived from phosphine (PH$_3$)] of the molecule, such a discrimination is possible in this reaction.

Because of the importance of cyclopentanoids, virtually every type of carbon-to-carbon bond-forming reaction has been applied to their synthesis. An example starts with tartaric acid, which is widely distributed in nature. In one modern synthesis it serves as a building block which is optically pure, that is, it is available as a pure mirror-image isomer (enantiomer) from nature. Tartaric acid represents a rare case in which both mirror-image isomers occur naturally. Reactions (4) illustrate how D-tartaric acid can be converted into a four-carbon building block that contains iodine at each terminus. Such a compound is an alkylating agent since it is able to transfer its carbon fragment or alkyl group to an electron-rich center (nucleophile). The nucleophile in this case is a carbon anion that is stabilized by placing two sulfur substituents on the carbon. Repeating this process but within the same molecule (intramo-

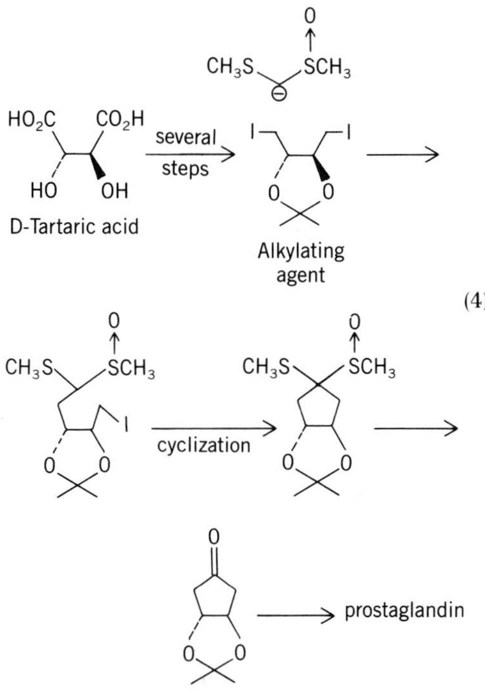

D-Tartaric acid

Alkylating
agent

(4)

cyclization

(IV)

→ prostaglandin

lecularly) creates the ring—a process termed cyclization. The particular type of nucleophile employed in reactions (4) is easily reacted with water to replace the C—S bonds by C—O bonds, thus forming a carbonyl group. The structural pair (III*a*) and (III*b*) symbolizes the structural change accomplished

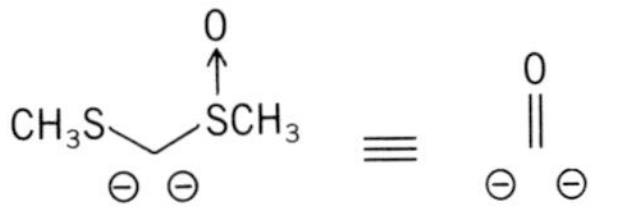

(IIIa)   (IIIb)

by reactions (4). The short arrows representing S—O bonds in reactions (4) are a type of donor bond in which the bonding electrons both come from S. Classic synthetic strategy normally created compounds like the optically pure carbonyl compound (IV) by using the normal reactivity pattern of a carbonyl group as a center of electron deficiency. The strategy of reactions (4) inverts the normal reactivity pattern—a phenomenon referred to as umpolung or polarity reversal.

Thermal reorganizations. Reactions like the aldol condensation and alkylation apply to a broad array of cyclic compounds in addition to five-member rings. Some reactions are unique to the synthesis of five-member rings. A special type of ring enlargement of a three-member ring by two atoms to form five-member rings, illustrated in reaction (5), is

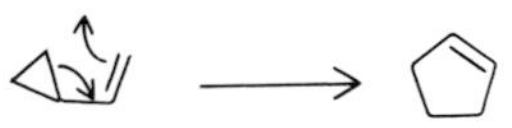

(5)

known as a vinylcyclopropane rearrangement. Accommodating a cyclic array of only three carbon atoms demands a severe distortion of the normal geometry around carbon and thereby creates strain.

Conversion of the three-member ring to a five relieves this strain. Such a process is particularly useful in annulating a five-member ring onto an existing cyclic system. Aphidicolin is an antiviral and antimitotic substance isolated from *Cephalosporium aphidicola*. The high biological activity for a substance that is so devoid of functionality generates a need to tinker with the structure. The five-member ring of aphidicolin may be created by the [3 + 2] type of annulation depicted in structure (V). In a

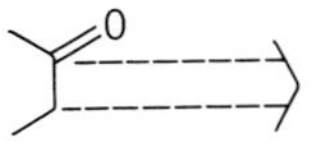

(V)

[3 + 2] type of annulation, two reaction partners, one of which contributes three atoms and the other two atoms, combine to form a five-member ring. Translating this pictorial representation of the structural change desired into reality takes advantage of the vinylcyclopropane rearrangement as outlined in reactions (6).

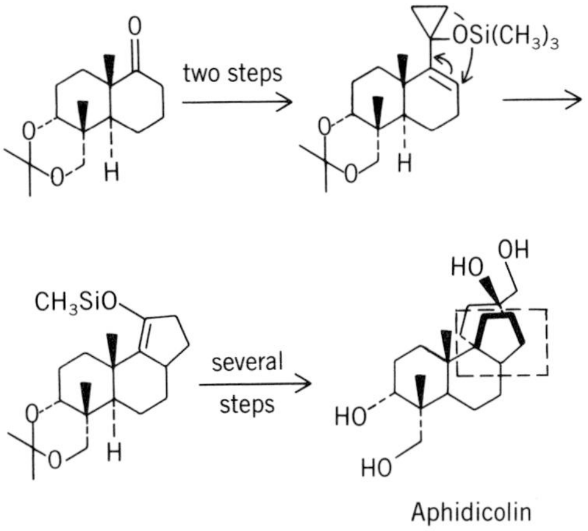

(6)

Aphidicolin

Catalysis. A particularly important ring construction approach is the direct condensation of two molecules in a cyclic array. A most important illustration of this type of construction is the [4 + 2] cycloaddition, also known as the Diels-Alder reaction. A [4 + 2] cycloaddition involves a ring-forming process in which two reaction partners condense to form a six-member ring, one partner contributing four atoms and the other two atoms. The importance of this reaction for the synthesis of complex natural products has only recently been appreciated and demonstrated. The synthesis of five-member ring compounds containing carbon as well as other kinds of atoms (heterocycles) can also be accomplished by an electronically similar process known as 1,3-dipolar cycloaddition. Antitumor agents based on heterocycles containing nitrogen known as pyrrolizidines have been synthesized with this approach; an example of this strategy is shown in reactions (7). A process resembling the 1,3-dipolar cycloaddition but making cyclopentanoids requires the introduction of a catalyst, a template that affects the reactivity of two reaction partners but is unchanged itself. Such a cycloaddition which has been used to make an antitumor agent known as brefeldin A is illustrated in reaction (8). In this process, the structural equivalent is the 1,3-dipole (VI) adding to a two-

carbon reaction partner to give the five-member ring.

Rational design of catalysts offers the synthetic chemist an extraordinary opportunity to invent reactions of high selectivity. Up to the recent past, the degree of synthetic selectivity achieved by enzymes was only an unreachable goal. However, recent successes show that the extraordinary molecular complexity of enzymes is unnecessary to effect selective chemical transformations. The development of the palladium catalyst to permit the cycloaddition shown in reaction (8) is one such example. Another illus-

tration is the ability of a rhodium-based catalyst to effect a cyclization to form a five-member ring as in reactions (9). Of the 17 hydrogen atoms available for insertion in forming the new carbon-carbon bond, the rhodium chooses the one labeled H_R preferentially. R* in reactions (9) is a homochiral auxiliary, a fragment possessing an optically pure asymmetric center.

Future prospects. The increasing understanding of materials of biological and nonbiological importance continues to reveal the importance of organic molecules built around cyclopentanoids. As new structures are revealed, they suggest additional avenues to pursue. Improved efficiency in synthesizing these molecules becomes a major task for progress. Catalysts will undoubtedly play a key role. By understanding the basic principles employed by enzymes to achieve selectivity, chemists will be able to improve techniques for rationally designing simple catalysts.

For background information *see* CONDENSATION REACTION; DIELS-ALDER REACTION; FURFURAL; MOLECULAR ISOMERISM; ORGANIC CHEMICAL SYNTHESIS; STEREOCHEMISTRY in the McGraw-Hill Encyclopedia of Science and Technology.   [BARRY M. TROST]

Bibliography: A. Mitra, *The Synthesis of Prostaglandins*, 1977; K. Nauman, *Chemistry of Plant Protective Agents and Pesticides*, vol. 7, *Chemistry of Synthetic Pyrethroid Insecticides*, 1981; L. A. Paquette, Recent synthetic developments in polyquinane chemistry, *Topics in Current Chemistry*, vol. 119, 1984; B. M. Trost, Cyclopentanoids: A challenge for new methodology, *Chem. Soc. Rev.*, 11:1419, 1982.

Organic geochemistry

Major advances in the understanding of the biological origins and geological fate of organic compounds in sediments have stemmed from, and been facilitated by, recent improvements in analytical techniques and instrumentation for the determination of complex mixtures. These developments have led to significant increases in the number and variety of compounds identified, and in the ability to detect and characterize them at low concentration levels (10 parts per billion).

Organic constituents of sediments. A small proportion of sedimentary organic matter can be extracted with organic solvents, yielding a highly complex mixture of compounds that have been termed chemical or molecular fossils, or biological markers. The insoluble polymeric material, or kerogen, has a macromolecular structure formed by various condensation processes from biological debris, notably carbohydrate and amino acid residues, during sediment deposition and consolidation. Recently a number of reagents have been shown to provide a useful means of selectively releasing molecular fragments, including intact biological markers, from kerogens. For example, the specific cleavage of ether bonds enables the determination of constituents of archaebacterial membranes that possess such structural features.

Combining the techniques of mass spectrometry and gas chromatography permits rapid discrimination and identification of individual components within complex mixtures of organic compounds. Since individual families of compounds are structurally related, they often possess common, distinctive ions in their spectra. Such a homologous series of compounds can be profiled from the investigation of the response with analysis time of their characteristic ions—a technique known as mass fragmentography or mass chromatography. Here, the selective monitoring of only a few chosen ions rather than the scanning of an entire spectrum can significantly enhance the sensitivity of detection of key compound types. It is also now possible to determine the responses from components of a particular carbon number with a complex mass chromatogram.

Such developments in analytical methodology have led to the growth in the nature and range of organic compounds recognized in geological materials. For instance, well over 200 pentacyclic triterpenoids belonging to the hopanoid structural family (Fig. 1) have now been characterized in sediments and petroleums, in contrast to the 20 or so that have been recognized in living organisms. Molecular organic geochemistry can be considered an extension of natural product organic chemistry in a geological framework, where the already complex distribution of biosynthetic compounds becomes even more complicated by geological processes.

Biological origins of sedimentary organic matter. There are major differences between the contributions from disparate types of organisms in different depositional environments that affect both the precise nature and the relative proportions of organic matter derived from various sources. Such features are reflected in the extractable fractions and also in the properties of the bulk organic matter. For example, kerogens containing predominantly algal or mixed algal-terrigenous organic matter tend to be oil-prone, whereas those formed principally from land plants, especially lignites and coals, tend to be gas-prone.

In molecular terms, the types and amounts of specific organic compounds produced by a given species of organism are governed by its biosynthetic processes, which are an evolutionary inheritance. They depend, in part, on the habitat of the organism which is reflected in the contributions from biota to sediments. They may also be affected by the aquatic food chain; zooplankton feeding on marine algae assimilate those compounds that fulfill their dietary requirements and excrete those superfluous to their needs in fecal pellets. These pellets are believed to be a major mechanism for the transport of organic matter from surface waters to bottom sediments. Laboratory cultures and, preferably, collections of natural populations of appropriate species are investigated, together with feeding experiments to assess the influence of food web processes. Also, the evaluation of well-defined environments helps to establish links between the composition of contributing organisms and those found in present-day sedi-

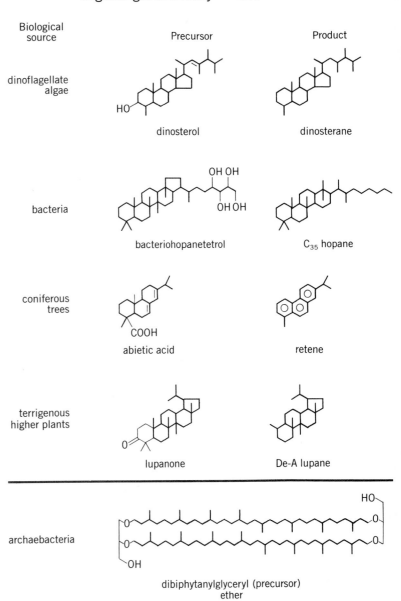

Fig. 1.   Chemical structures of a selection of precursor-product relationships. Each organic compound occurs in the specific organisms stated; none of the products are biosynthesized directly, but all have been identified in sediments. The structural differences between each of the precursor-product pairs are the presence of various functional groups in the former, whereas the latter are all hydrocarbons. Importantly, the structural skeleton of each precursor (heavy bonds) is preserved in its product. (*After A. S. Mackenzie et al., Chemical fossils, the geological fate of steroids, Science, 217:491–504, 1982*)

ments. Such experiments cannot, however, address these relationships in the geological past.

From the specificity of their chemical structures, many compounds are diagnostic of sediment inputs from particular organisms. Thus, marker compounds exist for algae, notably dinoflagellates, bacteria (especially archaebacteria), and terrigenous higher plants such as conifers (Fig. 1). The ability to recognize archaebacteria stems largely from the observation that their membranes differ chemically from those of other organisms, being composed of ether-linked isoprenoid moieties rather than ester-linked compounds. Similarly, while sterols act as rigidifiers in the cell membranes of most organisms, except bacteria, 4-methylsterol constituents appear to be specifically biosynthesized by dinoflagellates and therefore can denote their input to sediments.

A major development in this general area of research is the use of the relative proportions of di- and triunsaturated long-chain alkenones, that is, unsaturated methyl and ethyl ketones, in sediments as a measure of water paleotemperatures. This new tool for climatic assessment stems from the recognition that the coccolithophorids that synthesize these compounds generate greater amounts of the triunsaturated alkenones under colder growth temperatures. Thus, the determination of their fluctuations in the recent sedimentary record traces glacial-interglacial cycles, paralleling the oxygen isotope record of planktonic foraminifera.

Maturity changes in organic matter. Few of the compounds synthesized by living organisms are stable under geological conditions. Therefore, during sediment deposition and subsequent burial, their structures are altered chemically in various ways by the processes of diagenesis and catagenesis. The reactions involved, in which specific structural modifications occur, are referred to as precursor-product transformations. The stepwise nature of these transformations means that the product of one reaction may become the precursor for the next. Initially, in shallow sediments these changes are principally ef-

fected by microbial processes, involving the selective uptake and degradation of components by microorganisms. The modifications first affect the functionalities of the organic compounds, such as hydroxyl, carboxyl, and carbonyl groups, leading to the formation of saturated, unsaturated, and aromatic hydrocarbons. Subsequently, these hydrocarbons may undergo structural changes effected by thermal processes.

These molecular transformations occur systematically and can be used to follow the progression of diagenetic and catagenetic alteration of organic matter. Apparently the functionalized sterols derived directly from living organisms are transformed to either steroidal alkanes or aromatic steroid hydrocarbons via sterene (steroidal alkene) intermediates (Fig. 2). Among these changes is the alteration of the molecular shape of steroids. The "flat" form of sterols required for membrane structures is inherited by the early-formed steranes (steroidal alkanes), but it is subsequently modified through changes in stereochemistry into a more spherical shape of greater thermal stability. Aromatization also modifies the molecular shape of steroids and can occur independently in any of the ring cycles of the steroid nucleus, giving rise to A-, B-, or C-ring aromatic steroids. Further aromatization produces triaromatic steroids in which all three rings are aromatic.

Steroids are cited here as examples, but virtually all biological markers undergo a similar sequence of transformations (Fig. 1). Unlike most changes in mineral species, the alterations observed for organic

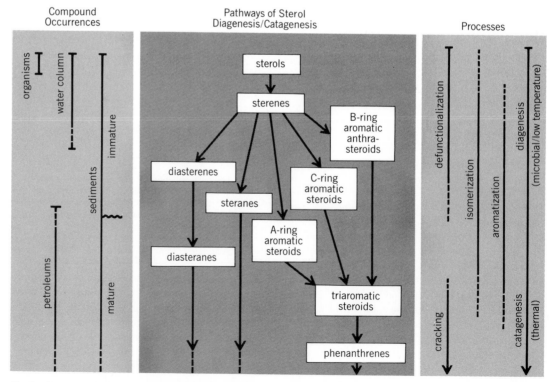

Fig. 2. Summary of the presumed geological fate of sterols. The biological and geological ranges of occurrence of steroid types and the main processes that effect their structural modifications are shown. The broken lines indicate that the reactions and compound occurrences do not cease abruptly, but gradually disappear. (*After S. C. Brassell, Molecular changes in sediment lipids as indicators of systematic early diagenesis, Phil. Trans. Roy. Soc. Lond. A, 315:57–75, 1985*)

compounds are nonreversible. Thus, the distributions of these components reflect their maximum depth of burial and temperature history. A number of precursor-product transformations that occur over discrete zones of diagenesis-catagenesis can therefore be used to assess the maturity of sediments and petroleums. These various reactions permit comparisons of the thermal histories of different sedimentary basins.

Thermal alteration of extractable organic compounds is not, however, the only process that affects their overall composition. After migration and accumulation in a reservoir, rock petroleums can experience a progression of changes effected by bacteria introduced by meteoric water circulation. These sequential changes have been characterized and documented, so that the extent of microbial alteration of a crude oil can be evaluated from its molecular composition.

Applications in petroleum exploration. First, the molecular composition of sediments and petroleums provides a useful assessment of their environments of deposition. For example, marine and terrestrial source rocks differ in terms of specific compounds and their overall distributions and relative proportions, for example, of steranes and hopanes. Second, biological markers are uniquely valuable in the determination of thermal history—a vital factor in deciding whether a particular sediment is sufficiently mature to be a petroleum source rock.

In such applications, gas chromatography–mass spectrometry provides the means of "fingerprinting" sediment extracts and petroleums from their mass chromatograms. For example, when the distributions of diagnostic biological marker compounds, as seen in mass chromatograms, for an oil and a proposed source rock are similar, it may suggest a correlation between them. Alternatively, discrepancies in the distribution patterns of mass chromatograms may be attributed to differences in the maturity of the oil and its proposed source rock, which might prompt the search for a more deeply buried source rock elsewhere in the basin. Significant differences between the compound distributions of various samples can indicate that they are unrelated to each other. Such techniques provide a ready test of correlations between oils and source rocks and between different oils. These procedures can also be applied when oils are biodegraded, since biological markers survive in all but the most extreme cases. Additional uses of this approach lie in environmental studies for the characterization and monitoring of oil spillages and pollution.

For background information *see* GAS CHROMATOGRAPHY; MASS SPECTROMETRY; ORGANIC GEOCHEMISTRY; ORGANIC SEDIMENTS (GEOCHEMISTRY) in the McGraw-Hill Encyclopedia of Science and Technology.

[SIMON C. BRASSELL]

Bibliography: A. S. Mackenzie, Applications of biological markers in petroleum geochemistry, in J. Brooks and D. H. Welte (eds.), *Advances in Petroleum Geochemistry*, vol. 1 116–214, 1984; A. S. Mackenzie et al., Chemical fossils: The geological fate of steroids, *Science*, 217:491–504, 1982; G. Ourisson, P. Albrecht, and M. Rohmer, The microbial origin of fossil fuels, *Sci. Amer.*, 251(2):44–51, 1984; B. P. Tissot and D. H. Welte, *Petroleum Formation and Occurrence*, 1984.

Paleobotany

This article discusses current theory on the origin of seed plants (as gymnosperms) and the subsequent evolution of flowering plants (angiosperms) from among the gymnosperms. Included are new fossil materials and modern techniques of data analysis.

Origin of flowering plants. One of the most perplexing questions of science remains the origin of the angiosperms. This question is currently being studied by several paleobotanists. Reinvestigations of the data already known, as well as new collections of fossil plants yielding additional data, have produced evidence sufficient to lead to the proposal of new hypotheses about the origin of flowering plants.

When did the flowering plants originate, where did they evolve, from what basic ancestral stock did they evolve, what was the nature of the first flowering plants, and what features of flowering plants uniquely adapted them for their extensive radiation into the most diverse plant group ever to live on the Earth. Important new data have become available from Lower Cretaceous plant-bearing sediments of eastern North America, from offshore cores from eastern Africa, and from mid-Cretaceous plant-bearing sediments in midcontinental North America.

Fossil angiosperms. The earliest clearly accepted record of fossil angiosperms is from Lower Cretaceous Barremian sediments (125–119 million years before present, or mybp) in England, eastern Africa, the Middle East, and the Potomac Group of eastern North America. These fossils are infrequently occurring monosulcate columellate pollen grains. The earliest angiosperm pollen reported is found near the Equator; this pollen becomes common in younger sediments in the temperate areas of both the Northern and Southern hemispheres. Some have suggested that angiosperm origin and initial radiation were centered near the Equator along the margins of the rift valley system as the ancient land masses of Gondwanaland and Laurasia split apart into the several continental units recognized today. Several widespread primitive families occur in both the northern (Laurasia) and southern (Gondwanaland) primordial land masses. This fact is consistent with the notion that the flowering plants may have originated when these two great land masses were much closer together. Thus this theory of the place in time of the origin of angiosperms is supported by the Northern and Southern hemisphere distribution of several modern angiosperm families.

Recent research on isolated pollen grains from upper Lower Cretaceous sediments (Albian, 113–97.5 mybp) has revealed pollen that occurs in tetrads and has aperture types and sculpturing similar to that of some extant pollen in the Winteraceae. Thus the Winteraceae is the oldest recognized mod-

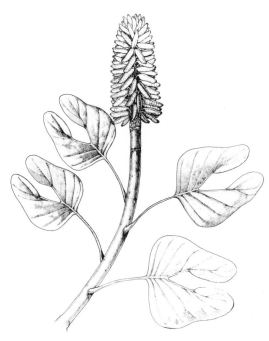

Fig. 1. Reconstruction of *Archaeanthus linnenbergeri* in fruit. This reconstruction is based upon fossils collected in the Dakota Formation, Cenomanian age (100 mybp), found in west-central Kansas. The scars where other floral organs were attached are seen below the fruits. Several features of the ancient plant are similar to those found in various modern families in the Magnoliidae. (*Drawing by Megan Rohn; from D. L. Dilcher and P. R. Crane, Archaeanthus: An early angiosperm from the Cenomanian of the western interior of North America, in D. L. Dilcher and W. L. Crepet, eds., Historical Perspectives of Angiosperm Evolution, Ann. Missouri Bot. Gard., 71:344–630, 1984*)

ern angiosperm family. The remains of leaves, flower parts, and fruiting axes, shown in Figs. 3 and 4, are from slightly younger sediments and are similar to those of other families in the Magnoliidae. The Hamamelidae (Platanaceae-like) are also found in sediments of upper Lower Cretaceous and mid-Cretaceous age. These two subclasses are recognized as the basic stem groups of the class Magnoliopsida (commonly known as dicotyledons). Although the dicotyledons show an earlier and more abundant fossil record, there is a good fossil record of the monocotyledons (class Liliopsida) by mid-Cretaceous times. One of the long-standing views of the flowering plants was that the monocots were "more primitive" than the dicots, or at least such was the sequence presented in the great compendia of A. Engler and K. Prantl. In the early 1900s it became an article of faith among phylogenetic botanists that the primitive flowering plants were "magnolioid" in structure, and this has generally been supported by subsequent research in angiosperm biosystematics. However, it has recently been speculated that the monocotyledons may be the older and more primitive of the two classes of flowering plants.

While a single ancestor of flowering plants has not been discovered, analyses of the available fossils combined with detailed comparisons with the characters of living plants have led to the proposal that particular groups of plants may be part of a common

ancestral complex. Some extant plants are placed together in the Gnetales, which is a sister (closely related) group to the angiosperms but is not a flowering-plant group. This means they are both derived from the same ancestral complex. The ancestral stock for early angiosperms and the Gnetales is thought to be from the extinct early Mesozoic seed ferns, the pteridosperms. Many characters are shared between the angiosperms and some Mesozoic pteridosperms. Such characters include net venation of the leaves, the association of reproductive organs with leaves, clustering of the reproductive organs into compact collections of ovule- or pollen-producing organs, and the enfolding of leaflike structures (cupules) around unfertilized ovules while they develop into seeds. (Such cupules are not equivalent to structures referred to as cupules in "Origin of gymnosperms" below.)

Many characters are also shared between the angiosperms and the living order, Gnetales. The Gnetales are seed plants which lack flowers but share more similar features with the angiosperms than with any other group of living plants. The fossil record of pollen thought to be from the Gnetales goes back to the Triassic (245–208 mybp, early Mesozoic), predating the generally accepted earliest record of angiosperm pollen by about 100 million years.

Characteristics. Some research is now concerned with sorting out the specific characters which make angiosperms unique and noting the occurrence of similar characters in living and fossil nonangiosper-

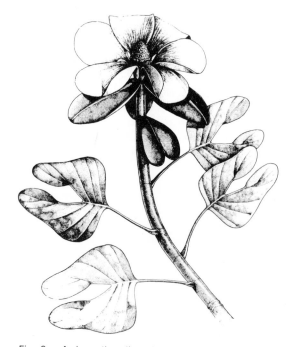

Fig. 2. *Archaeanthus linnenbergeri*. Reconstructed as a leafy twig bearing a flower. All organs associated together in this reconstruction were found associated in the same sediments and were filled with similar resinous bodies. (*Drawing by Megan Rohn; from D. L. Dilcher and P. R. Crane, Archaeanthus: An early angiosperm from the Cenomanian of the western interior of North America, in D. L. Dilcher and W. L. Crepet, eds., Historical Perspectives of Angiosperm Evolution, Ann. Missouri Bot. Gard., 71:344–630, 1984*)

mous plants in an attempt to gain a clearer understanding of their early phylogeny. Such character state analyses, called cladistic analyses, result in cladograms (a graphic representation of a cladistic relationship) which illustrate the similarities of groups of living and fossil plants based on an order ranking of particular characters. These analyses, when combined with information from other detailed research work on fossil plants, are now providing some useful data from which to work out the nature of the elusive first flowering plant.

Flowering plants or angiosperms (Magnoliophyta) have a unique set of characters. The origins of the distinctive characters that they share have long been the subject of investigation by botanists who work with living plants as well as those that work with fossil plants. Not all of the features that make up the set of characters unique to the angiosperms are present in primitive angiosperms, and several are sometimes found in plants that are not angiosperms. One such feature is the collection of reproductive structures into flowerlike aggregations. In flowering plants these organs are the closed carpel which encloses one or more ovules and matures upon fertilization into a fruit containing one or more seeds, and one or more stamens which bear the pollen or male gametophyte. The carpels and stamens collectively are often surrounded by the sterile organs such as petals and sepals. Because the seeds of angiosperms are enclosed in a carpel, the pollen (male gametophyte) must grow a tube through the carpel (sporophyte) tissue in order for the sperm to reach the egg in the immature ovule. This has made possible the development of a special situation in which the proteins of the carpel may interact with the proteins of the pollen tube to halt its growth or with those of the pollen grain to arrest pollen germination. Angiosperms are unique in this ability to chemically control which male gametes are allowed

to reach the unfertilized egg. Evidence of this incompatibility system is present early in the history of the evolution of the flowering plants. This early development permitted the evolution of flowers that contained both female (ovule-bearing) and male (pollen-bearing) organs in close proximity for ease in insect pollination (Fig. 5) and with a barrier against self-pollination. Insect pollination allowed for the exchange of pollen between the flowers of different plants and encouraged an increase in plant-animal specializations in their reproductive biology. This coadaptive evolution of plants and animal pollinators is very important in the reproductive biology of flowering plants and has been the principal reason that flowering plants, with over 250,000 species, are the most diverse plant group to have lived in the world. Some ancient angiosperms, such as the Platanaceae, seem to contain early wind-pollinated flowers. The occurrence of wind pollination in various modern families and genera suggests that there were several independent origins for this mode of pollination. [DAVID DILCHER]

Origin of gymnosperms. The evolution of seed plants occurred over 350 mybp (during the Late Devonian) and was among the most significant events in the development of modern vegetation. Prior to this time, all land plants were pteridophytes that reproduced by means of naked spores, and were heavily dependent upon favorable environmental conditions for successful completion of sexual reproduction. The first seed plants were gymnosperms (the so-called naked-seed plants), and their seeds were borne on leaves or leaflike structures. Later, during the upper Carboniferous (about 325–280 mybp) more sophisticated gymnosperms, including conifers, apeared in the fossil record. Since their origin, seed plants have increased in prominence until they now constitute more than 90% of the land flora. Flowering plants make up the majority of living species, but conifers (the largest group of living gymnosperms) are the dominant vegetation over large areas of mountainous and high-latitude terrain.

In the fossil record, the evolution of gymnosperms is marked by the earliest appearance of seeds or ovules (unfertilized seeds). These occur in deltaic deposits that accumulated at the margins of continents in equatorial regions during the Famennian Stage of the Devonian Period. Some seeds are dispersed in the rock matrix, while others are found in an enclosing system of forking axes termed a cupule (Fig. 1). Small fragments of foliage, clusters of sporangia, and stems with three-armed steles are found in the same deposits. Fossils of these types probably also represent parts of the earliest seed plants, but until there is documented evidence for this interpretation, studies of gymnosperm origins will focus primarily on the ovules and ovule-bearing cupules.

The evolution of seed plants is considered to have resulted from three significant modifications to the life cycle of a free-sporing heterosporous pteridophyte: reduction in the number of functional megaspores to one per megasporangium; enclosure of the megasporangium (= nucellus) in the integument (=

Fig. 3. Reconstruction of pentamerous flower showing whorls of sepals, petals, stamens, and carpels. This reconstruction is based upon fossils collected in the Dakota Formation, Cenomanian age (100 mybp), found in south-central Nebraska. This is the oldest, most completely known flower in the world. It shares characters of more than one modern angiosperm order in the Rosidae. (*Drawing by Bronwyn Elkuss; courtesy of D. L. Dilcher*)

seed coat); and a functional change from megasporangial dehiscence to indehiscence of the megasporangium. The last change permanently retained the functional megaspore (with its developing megagametophyte and embryo) within the nucellus, and facilitated a dramatic alteration in reproductive biology.

Reduction in the number of megaspores per megasporangium, and either enclosure of the megasporangium or elaboration of the megasporangial wall occurred independently in several groups of Paleozoic vascular plants. Included are lycopsids (such as *Lepidocarpon* and *Achlamydocarpon*), sphenopsids (*Calamocarpon*), and the fernlike plant *Stauropteris*. However, none of these developed the megasporangial indehiscence, pollination, and postpollination biology that characterizes reproduction in gymnosperms and angiosperms. Also, none of the other plants with seedlike structures have given rise to living descendants. Therefore, they will not be considered further.

Evolution of integument. For many years, studies of gymnosperm origins concentrated on interpreting homologies of the tissue that surrounds the seed megaspore (the nucellus), and on the evolution of the seed coat (the integument). Most botanists now interpret the nucellus to be a modified megasporangium with one functional megaspore, and there is good evidence for the evolution of the integument. During the past 25 years a large number of Devonian and lower Carboniferous ovules that show a wide spectrum of integumentary features have been described. The simplest (and presumably most primi-

tive) of these have a more or less naked megasporangium that is encircled loosely by a whorl of narrow processes (telomes). The others show a complete intergradation from specimens of this type to specimens with the nucellus completely enclosed by the integument, as it is in the seeds of living gymnosperms. Together, the primitive ovules provide strong evidence that the seed coat originated by coalescence of a ring of telomes around the megasporangium. Three ovules are shown with partly coalesced integumentary lobes attached to the cupule in Fig. 1.

Evolution of reproduction. Now that evolution of the integument is well documented, investigations are focusing on the changes in reproductive biology that led to the origin of gymnosperms. As a result, a heightened appreciation of the evolutionarily important functional facets of seed plant biology is developing.

Differences in genetic potential as a result of a change from reproduction by free-sporing heterospory to reproduction by primitive gymnospermous biology probably are negligible. In both instances the level of heterozygosity is potentially greater than that expected of a homosporous pteridophyte, because the gametes have independent meiotic origins (sperm and egg cannot come from the same gametophyte). However, there are dramatic differences between the heterosporous pteridophytes and primitive gymnosperms both in fertilization biology and in dissemination of the new sporophyte generation. The origin of these latter features now forms the central focus of many evolutionary studies.

Heterosporous pteridophytes shed both their microspores and megaspores. Therefore, fertilization occurs away from the parental sporophyte plant. It is successful only if the megaspores and microspores land close enough together so that the sperm can swim through a thin film of water to the female gametophyte, and then down the neck of the archegonium to the egg. Also, it is the megaspore (containing the megagametophyte with archegonia, eggs, and ultimately the embryo) that acts as the unit of propagation. For terrestrial pteridophytes, the place where the megaspore lands is the place where the new sporophyte must grow or die.

In contrast, the functional megaspore of a seed plant is not shed from the megasporangium. Rather, a pollen chamber for the reception of microgametophytes develops at the tip of the megasporangium. Pollination eliminates the association of megagametophytes and microgametophytes by chance, and occurs at a developmental stage when the megasporangium is still attached to (and nourished by) the parental sporophyte.

Following pollination, the ovule is sealed to produce an internal environment conducive to further development of the gametophytes, fertilization, and embryo development. As a result, delicate reproductive processes are not dependent upon conditions in the external environment being favorable. Microorganisms with the potential to infect or parasitize the developing ovule (or seed) also are ex-

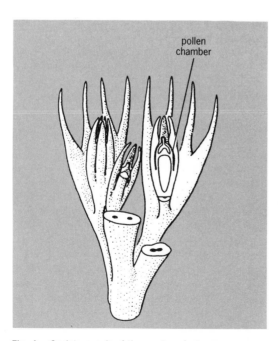

Fig. 4. Ovulate cupule of the most ancient gymnosperms. Front part has been cut away to expose three ovules. Ovule at right is shown in longitudinal section to reveal the pollen chamber. Note that the integument of each ovule consists of several separate lobes in the apical region. (*After W. H. Gillespie, G. W. Rothwell, and S. E. Scheckler, The earliest seeds, Nature, 293:462–464, 1981*)

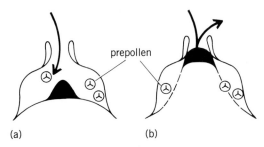

prepollen

Fig. 5. Pollen chamber of the most primitive gymnosperm ovules. (a) Pollination stage of development, with opening at apex. Arrow indicates access to the pollen chamber. (b) Postpollination stage with the central column pressed into the narrow pollen-chamber tip to seal the ovule. Arrow indicates lack of access to the pollen chamber as the result of the postpollination sealing mechanism.

cluded by this process.

In living gymnosperms the integument seals the ovule, but in the most primitive gymnosperms the integumentary lobes were incompletely fused (Fig. 1), and could not perform this function. Instead, there was a mound of tissue (the central column) on the pollen chamber floor (Fig. 2a) that sealed the ovule. After pollination, the central column was pushed upward into the narrow tip of the pollen chamber by the growing megagametophyte (Fig. 2b). In this way the ovule was sealed in a fashion similar to corking a bottle from the inside.

Because the megasporangia of gymnosperms are indehiscent, megaspores are not shed from the megasporangium. Rather, the seed or ovule becomes detached by abscission and acts as the unit of propagation. By this mechanism, the new sporophyte phase of the life cycle is protected from destructive environmental influences and nourished by the food stored in the megagametophyte.

Because the most significant changes associated with the origin of seed plants resulted from alterations in the developmental and reproductive features of the megasporangium, the evolution of gymnosperms was a dramatic biological event. It included several significant and functional changes. These may be summarized as follows: (1) Dehiscent megasporangia are replaced with indehiscent megasporangia that have a pollen chamber. The pollen chamber brings together the mega- and microgametophytes and seals the ovule from the external environment. (2) Sperm swim to the egg within the enclosed environment of the sealed ovule, rather than through a thin film of water on the outside of the megagametophyte. (3) The seed (or ovule) replaces the naked megaspore as the propagule, and the embryo or seedling is not exposed to the external environment until germination.

Modern gymnospermous reproduction incorporates numerous additional changes in the structures that bear the seeds (primarily cones), in the structure and function of the integument, and in the microgametophytes. In the seeds of extant gymnosperms there is no central column, and the ovule is sealed by other tissues (the integument in most). Of

perhaps greatest significance, changes in the microgametophytes of advanced seed plants allow fertilization to be accomplished more efficiently with the aid of a pollen tube. In cycads and the maidenhair tree (*Ginkgo biloba*) the pollen tube functions only to absorb nutrients into the microgametophyte, but in all other living seed plants it delivers the sperm directly to the egg.

Sexual reproduction in gymnosperms is neither as subject to limitations of a desiccating environment nor as susceptible to infection as that of free-sporing plants. As a result, the evolution of gymnosperms set the stage for colonization of large areas of land surface that previously had been unavailable or only marginally available to vascular plants. New zones of ecological space were created, and previously underutilized zones were more intensely inhabited. Thus, through a more efficient mode of sexual reproduction and an enhanced potential to survive under a wider range of ecological conditions, the gymnosperms became the most dominant form of terrestrial vegetation by Mesozoic time.

For background information *see* Fossil seeds and fruits; Gneticae; Liliopsida; Magnoliophyta; Paleobotany; Palynology; Pinophyta; Pollination in the McGraw-Hill Encyclopedia of Science and Technology.

[GAR W. ROTHWELL]

Bibliography: H. N. Andrews, Early seed plants, *Science*, 142:925–931, 1963; J. F. Basinger and D. L. Dilcher, Ancient bisexual flowers, *Science*, 224:511–513, 1984; D. L. Dilcher and W. L. Crepet (eds.), Historical perspectives of angiosperm evolution, *Ann. Missouri Bot. Gard.*, 71:344–630, 1984; W. H. Gillespie, G. W. Rothwell, and S. E. Scheckler, The earliest seeds, *Nature*, 293:462–464, 1981; L. J. Hickey and J. A. Doyle, Early Cretaceous evidence for angiosperm evolution, *Bot. Rev.*, 43:3–104, 1977; G. W. Rothwell, Classifying the earliest gymnosperms, in B. A. Thomas and R. A. Spicer (eds.), *Systematic and Taxonomic Approaches in Palaeobotany*, Systematics Association Special Volumes, Linnean Society of London, in press; W. N. Stewart, *Paleobotany and the Evolution of Plants*, 1983; T. N. Taylor and M. A. Millay, Pollination biology and reproduction in early seed plants, *Rev. Palaeobot. Palynol.*, 27:329–355, 1979; J. W. Walker, G. J. Brenner, and A. G. Walker, Winteraceous pollen in the Lower Cretaceous of Israel: Early evidence of a magnolialean angiosperm family, *Science*, 220:1273–1275, 1983; M. S. Zavada, The relation between pollen exine sculpturing and self-incompatibility mechanisms, *Plant Syst. Evol.*, 147:63–78, 1984.

Paleoclimatology

Past climatic conditions on the surface of the Earth are related to changes in solar insolation, to plate tectonic processes that affect topography and sea level, and to geochemical cycles that include the composition of the atmosphere. During the Phanerozoic, the last 650 million years of Earth's history for which abundant fossils provide evidence for paleoclimatic conditions, the Earth has alternated between shorter intervals with climatic oscillations from glacial to interglacials (Pennsylvanian-Permian; Pleistocene) and longer intervals with warm equable climates and ice-free poles. Changes in solar insolation resulting from variations in the Earth's orbital elements are moderated by a variable greenhouse effect resulting from changes in the CO_2 content of the atmosphere and the dynamics of the ice-earth system to produce the glacial-interglacial cycles. A significantly greater CO_2 content of the atmosphere results in a greenhouse effect which elevates the mean annual temperature of the Earth by increasing polar temperatures, resulting in warm equable climates over the entire globe. Local climate variability depends on topographic relief and the distribution of land and sea. The recent advances in paleoclimatology are largely a result of the development of relevant plate tectonic, climatic, and geochemical models.

Glacial-interglacial oscillations. A relation between changes in the Earth's climate and variations in insolation resulting from changes in the Earth's orbital elements was suggested by the Yugoslav astronomer Milutin Milankovitch in the 1920s. The precession of the equinoxes, namely, the position within the orbit when the Earth is at perhelion, has a cycle of approximately 25,500 years. The obliquity, namely, the tilt of the Earth's axis with respect to the orbital plane, ranges from 22°0 to 24°5 on a 41,000-year cycle. The eccentricity (ellipticity) of the Earth's orbit varies with a mean periodicity of 95,000 years. Each of these orbital elements, but especially the eccentricity, can be perturbed by the gravitational attraction of other bodies in the solar system, primarily Jupiter and Saturn. The solar insolation at a given latitude is a result of the combination of these orbital variations. The mean annual global insolation remains almost constant, but at a given latitude a seasonal contrast may differ and at high latitudes insolation may vary significantly.

The ratio of the stable isotopes of oxygen, $^{16}O/^{18}O$, usually expressed as the change in relative abundance of ^{18}O, $\delta^{18}O$, has been used to examine the glacial-interglacial oscillations quantitatively. The oxygen isotope ratio in sea water is incorporated into the $CaCO_3$ shells of marine organisms through biologic processes which may involve some isotopic fractionation, termed the vital effect, but by examining the shells of a single species deposited over an interval of time, the temporal variations of the oxygen isotope ratios in sea water can be determined. Since ^{16}O is lighter than ^{18}O, it tends to escape from the sea surface more readily as evaporation occurs, so that rain and snowfall are depleted in ^{18}O. As glaciers build up on land, the ocean becomes enriched in ^{18}O. The isotope ratio also responds to the ambient temperature at the time of shell formation, cooler temperatures resulting in shell material enriched in ^{18}O. Thus both growth of glaciers and cooler temperature result in more positive $\delta^{18}O$. The record of fluctuation of $^{18}O/^{16}O$ in the calcite tests (shells) of *Globigerinoides sacculifer*,

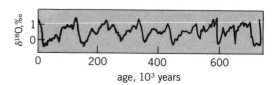

Fig. 1. Changes in $\delta^{18}O$ in shells of *Globigerinoides sacculifer* from a Caribbean deep-ocean core. (*After E. J. Barron, Ancient climates: Investigation with climate models, Rep. Prog. Phys., 47(12):1563–1599, 1984*)

a species of surface-dwelling planktonic foraminifer, over the past 750,000 years, expressed as $\delta^{18}O$, is shown in Fig. 1. Spectral analysis of this record shows that it is made up of four major frequencies, as shown in Fig. 2. These frequencies correspond closely to the Milankovitch orbital variations, the 19,000- and 23,000-year peaks being related to the 25,500-year precession cycle.

Although the general correspondence between waxing and waning of glaciers and the Milankovitch orbital variations seems well established, the actual mechanism of climatic change remains elusive. The dominant 100,000-year glacial-interglacial cycle corresponds to the change in eccentricity of the orbit, which produces only small insolation variations; some of the eccentricity insolation minima show the same absolute values as other eccentricity insolation maxima. Further, the glacial-interglacial responses lag substantially behind the eccentricity maxima and minima, and variations in the orbital parameters alone do not produce the sawtooth shape of the observed glacial-interglacial curve recorded by oxygen isotopes (Fig. 1).

Modeling of ice sheet dynamics suggests that as ice builds up, it both spreads horizontally and depresses the Earth's crust through isostatic response to its weight. This causes a change in the elevation

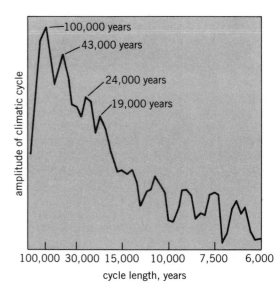

Fig. 2. Spectral plot showing the amplitude and frequency of the sine waves whose sum best approximates the curve derived from ratios of oxygen isotopes. (*After C. Covey, The earth's orbit and the ice ages, Sci. Amer., 250(2):58–66, 1984*)

of the surface of the ice above the snow line, above which snow cover is permanent; eventually the top of the ice cap sinks below the snow line and rapid deglaciation occurs. Modeling studies incorporating a 10,000-year isostatic adjustment time scale along with the orbital variations can duplicate the 100,000-year sawtooth cycle, but are still not fully in agreement with observations. Changes in meltwater supply to the ocean, the extent of sea ice, and the rate of production of oceanic deep water may also be involved in amplifying the effects of changing insolation. Evidence from glacial-age atmospheric samples trapped in Greenland ice suggests that variations of atmospheric CO_2 of up to 50% may also be important in glacial-interglacial climate change.

Warm equable global climates. During most of its history, the Earth lacked polar ice caps and had no alternation of glacials and interglacials. The most intensively investigated of these periods of warm equable global climate is the Cretaceous. Mean annual temperatures in the tropics are thought to have been similar to the present, about 305 K (90°F), but mean annual polar temperatures could not have been less than the freezing point of fresh water, 273 K (32°F), or more than 30 K (54°F) higher than today.

Recent climate model experiments have taken into account the different distribution of the continents and higher global sea level; they indicate an equator-to-pole climatic gradient less extreme than that of today, but suggest that polar ice should be present if land occurs at the pole. However, if the atmosphere contained significantly more CO_2 than at present, a greenhouse effect would have been produced, changing tropical temperatures only slightly, but resulting in much warmer polar temperatures. A model with four times present CO_2 levels indicates mean annual polar temperatures of about 278 K (41°F).

The idea of greater atmospheric CO_2 as modulator of the Cretaceous climate was suggested by another research group, which attributed the larger CO_2 content of the atmosphere to greater volcanic activity. In other recent studies a quantitative geochemical model for atmospheric CO_2 content in the context of the carbonate-silicate geochemical cycle was developed; this was based on the consideration of plate tectonics as the driving mechanism that forces geochemical cycling. In this model the subduction of ocean carbonate sediments releases CO_2, producing a warmer, more equable climate and resulting in greater chemical weathering of rocks on land. The CO_2 returned through volcanoes is eventually consumed by the weathering of silicate rocks. Disequilibrium in the system, resulting in increased atmospheric CO_2, is produced by more rapid sea-floor spreading such as occurred during the mid-Cretaceous, with consequent increased subduction and volcanic activity. In the simplest version of the model, it is expected that the globally averaged mean annual temperature at the Earth's surface is a function of the rate of sea-floor spreading. To a re-

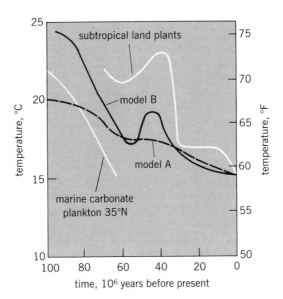

Fig. 3. Global mean annual surface temperature versus time, comparing two model predictions with midlatitude estimates derived from carbonate shells of oceanic planktonic organisms and land plants from Japan, Europe, and North America. Model A assumes linear decline in sea-floor spreading rate since the mid-Cretaceous; model B is based on the variable spreading rates proposed by J. R. Southam and W. W. Hay. (*After R. A. Berner, A. C. Lasaga, and R. M. Garrels, The carbonate-silicate geochemical cycle and its effect on atmospheric carbon dioxide over the past 100 million years, Amer. J. Sci., 283:641–683, 1983*)

markable extent, this seems to be true, as shown in Fig. 3. The geochemical models suggest Cretaceous atmospheric CO_2 levels of three to a hundred times present, with the most likely levels being a few times the present.

Another factor influencing Cretaceous climates was the reversed circulation of the deep ocean. At present the thermohaline circulation of the oceans is driven by plumes of dense cold saline water formed near the polar regions, primarily in the Weddell Sea and in the Norwegian-Greenland Sea. In the Cretaceous, the vertical circulation of the ocean was driven by plumes of dense warm hypersaline water formed in marginal seas in the subtropical arid zones. Thus, instead of oceanic deep circulation being dominated by the equatorward flow of deep water formed at high latitudes, the deep Cretaceous oceans were dominated by polar flow of deep waters formed at low latitudes; this reversal of the deep circulation of the ocean could affect global heat transport significantly.

Plate tectonics and local climate. Plate tectonic processes, particularly continental rifting, subduction, and continental collisions, may produce topographic relief strongly influencing climate. The present-day Himalayan-Tibetan uplift causes dislocation of the zonal climatic belts. Subduction of hot young ocean crust, such as is going on beneath western North America, causes large-scale regional uplift with attendant downwind climate modification. The continental rifting process involves regional uplift and graben formation, resulting in long-lived large fresh-water lakes which may contain significant accumulations of organic carbon, as is the case along the East African Rift. As continental separation occurs, a narrow semienclosed seaway with marginal uplifts is formed; this becomes a preferred site for evaporate deposition, and as much as half of the world evaporite deposits have formed in this setting. As the seaway becomes more openly connected with the ocean, it may become a site of intensive upwelling and removal of both opaline silica and organic carbon, as in the case of the present Gulf of California.

For background information *see* CLIMATOLOGY; GREENHOUSE EFFECT, TERRESTRIAL; PALEOCLIMATOLOGY; PLATE TECTONICS in the McGraw-Hill Encyclopedia of Science and Technology.

[WILLIAM W. HAY]

Bibliography: E. J. Barron, Ancient climates: Investigation with climate models, *Rep. Prog. Phys.*, 47:1563–1599, 1984; W. H. Berger and J. C. Crowell (eds.), *Climate in Earth History*, 1982; R. A. Berner, A. C. Lasaga, and R. M. Garrels, The carbonate-silicate geochemical cycle and its effect on atmospheric carbon dioxide over the past 100 million years, *Amer. J. Sci.*, 283:641–683, 1983; H. D. Holland, *The Chemical Evolution of the Atmosphere and Oceans*, 1984.

Parasitic plants

Parasitic flowering plants are distinguished primarily by the development of an intrusive organ, the haustorium. This specialized structure invades the host plant's vascular tissue and forms a physiological union between parasite and host. The haustorial connection is usually to the host's xylem; water and nutrients are drawn by osmotic factors and by the transpiration stream of the parasite. Many species are obligate parasites, that is, they cannot survive without a host; others are facultative parasites, that is, they can grow without a host, but development is less vigorous. Though physiologically capable of growing alone, in nature the facultative parasites always are found with attachments to other plants.

Prominent obligate species are the stem parasites *Arceuthobium* (dwarf mistletoe) and *Cuscuta* (dodder), and root parasites *Striga* (witchweed) and *Orobanche* (broomrape). Host recognition is important for survival of obligate parasites; seeds germinate only when close to a host root. Chemical cues are probably essential for host recognition.

Seed germination. For the obligate parasitic plants, germination is an irreversible commitment to find a suitable host. Large food reserves in seeds can prolong the prehost stage; germination requirements for such large-seeded plants (for example, *Arceuthobium*, *Cuscuta*, and *Cassytha*) are unspecialized. In *Striga* and *Orobanche*, however, seeds are small (0.2–0.3 mm) with little stored food, and germination requirements are complex—mature seeds require a water soak or conditioning of about 10 days. Germination inhibitors have been reported in some seeds, while in others it has been suggested that water conditioning is essential for the accumulation of a germination stimulant to threshold levels.

A host germination stimulant is required for many parasitic species; compounds that promote germination include auxins, cytokinins, gibberellins, ethylene, coumarin, thiourea, xyloketose, and sodium or calcium hypochlorite. Natural host germination stimulants are difficult to identify because of low concentrations in host root exudate, possible synergism of several components, and structural instability of the active fraction as it is purified or subjected to spectral analysis. The most significant finding has been the isolation of a natural product from cotton (not a host of *Striga*). This compound, called strigol (structure I), and several analogs promote germina-

(I)

tion in *Striga* and *Orobanche* at very low concentrations.

Substances such as strigol could provide family or genus recognition for parasites that require a specific host. Regulation of germination in *Striga* is less precise. Under field conditions, for example, suicidal germination is significantly promoted by injection of ethylene into the soil, or by the planting of trap crops that produce active germination stimulants but are nonhosts.

Following exposure to host stimulant, germination can be very rapid. In *S. asiatica*, radicle emergence occurs in less than 12 h at 81°F (27°C). For many species, mobilization of food reserves is directed exclusively to radicle growth at the expense of plumule development; this maximizes opportunity for host contact. Recent studies on *Alectra vogelii* and *S. gesnerioides* show little deoxyribonucleic acid (DNA) synthesis during prehost contact. Cell divisions cease while the cell elongation responsible for growth continues, but cell divisions are renewed when plants are grown on a nutrient medium. There is little information available on biochemical changes during germination.

Host contact. The salient feature of parasitic plants is the haustorium, which establishes contact and invades the host vascular tissue. In some species, haustorial initiation requires an exogenous chemical signal from the host. Receipt of the signal results in many rapid cellular changes in the radicle, including reduced elongation, enlargement of cortex cells, establishment of densely protoplasmic cells within the haustorial complex, and initiation of haustorial hairs (see illus.). The initiation process is the same for primary haustoria, formed at the radicular apex, and secondary ones, located at lateral positions proximal to the radicle meristem.

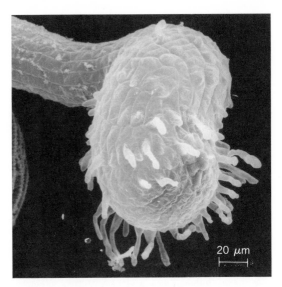

Scanning electron micrograph of *Striga asiatica* 12 h after induction of the primary haustorium.

Several substances promote haustoria growth in the laboratory, including cytokinins, a triterpene, several quinones and numerous phenolics (including caffeic acid), ferulic acid, sinapic acid, *p*-hydroxybenzoic acid, vanillic acid, vanillin, and syringic acid. In a recent study of *S. hermonthica*, syringic acid was the most active compound. In *S. asiatica* the most active inducers are xenognosin A (structure II), a flavanoid, and 2,6-dimethoxy-*p*-benzoquinone

(II)

(structure III), a quinone, isolated from *Sorghum* roots.

(III)

Promotion requires considerable molecular specificity, and this has been shown for haustorial induction in *Agalinis purpurea*. Two promoters, xenognosin A and xenognosin B (structure IV), were isolated

(IV)

from gum tragacanth. Formononetin (structure V), which is also contained in gum tragacanth (a foliar exudate) and differs from xenognosin B only in lacking the 2′ hydroxyl, produces no haustoria at any

(V)

concentration. The specificity of xenognosin A has been probed with analogs, and two structural features are required for activity: a meta relationship of hydroxyl and methoxyl groups, and an alkyl branching ortho to the methoxyl substituent.

The relevance of these substances to field conditions remains uncertain and is under active study. Active promoters have been successfully isolated from gum tragacanth and root extracts, but not from root exudates. The concentration of promoters in root exudate may be extremely low, or molecules may be bound to the host root surface.

Haustorial attachment is a nondiscriminatory event, even for several obligate species. Haustoria will adhere to a variety of biological and nonbiological substrates. Adhesion is enhanced by the specialized haustorial hairs that have a surface covered with a papillate network composed of hemicellulose material. In *A. purpurea* the proportion of papillate hairs parallels attachment competency.

Intrusive growth of the haustorium is thought to be facilitated by mechanical pressure and enzymatic lysis of host cells. Studies show the presence of acid phosphatase and other lysosomal enzymes at the host-parasite tissue interface that may disrupt host cell membranes. A recent study of *Cuscuta reflexa* reports higher pectin esterase, polygalacturonase, xylanase, and a glucosidase in the haustorial region; these enzymes may degrade host cell walls. Factors important in sustaining the development of the intrusive stage of the haustorium or directing its growth to specific cells of the host vascular tissue remain unknown.

Other host-parasite interactions. Dependence on the host varies among parasite species. True holoparasites, such as *Orobanche*, contain no chlorophyll, while other species, such as *Arceuthobium* and *Striga*, are photosynthetic. Some chlorophyll synthesis occurs in *Cuscuta* when separated from its host. *Striga asiatica*, *S. hermonthica*, and *S. gesnerioides* can be grown on simple inorganic media—plants are diminutive and photosynthetic competence may be diminished, but plants do flower and set seed. *Striga* species have lower amounts of chlorophyll and CO_2 exchange rates than nonparasites.

The movement of nitrogen in a suitable form from the host may be extremely important for normal development of the parasite. In some field studies, nitrogen applications reduced host damage by *Striga*, but other studies contradict this finding. Host influences on nitrogen composition of the parasite are significant. The nitrogen in xylem exudates of *Odontites verna* attached to barley was less than 10% nitrate, but up to 80% when parasitic on *Stellaria*. Amino acids and amides may be alternative nitrogen sources to nitrates.

The movement of nonnitrogenous organic substances from host to parasite is well documented, although identification of the substances translocated requires further study. In some instances, sucrose is the main carbohydrate transferred to the parasite.

Loss of water and nutrients can produce serious consequences for the host, including altered development, reduced size, and poor yield in crops. Altered growth of the host may be the result of host hormonal imbalances. *Sorghum* parasitized by *Striga hermonthica* is reported to have as much as a 90% reduction in cytokinins and up to an 80% reduction in gibberellins present in the xylem exudate. In dwarf mistletoe infections of black spruce (*Picea marians*) there are increases in the host of zeatin riboside and probably other cytokinins, increases in indoleacetic acid (IAA), and decreases in abscisic acid (ABA) concentrations.

Host range. The biological basis for host suitability, and also host resistance, remains obscure. Host specificity is absent in most mistletoes, but there are exceptions, such as *Viscum cruciatum*, which is found only on olive trees. Some conifers exhibit resistance to dwarf mistletoe, and the parasite shows definite host preferences. Yet two-thirds of the dwarf mistletoes parasitize species other than their principal hosts. Climatic factors may be important. In *Cuscuta* a few species have narrow host ranges, but most parasitize a wide variety of plant families. Even a single plant can attach to several different hosts simultaneously.

Root parasites generally have broad host ranges. Species of *Orobanche* usually have specific family preferences, and do not parasitize monocots, ferns, or gymnosperms. *Orobanche crenata* has the narrowest host range and parasitizes only the legumes. For *Striga* most hosts are grasses. A notable exception is *S. gesnerioides*, which parasitizes cowpea (*Vigna unguiculata*) and other dicots.

For background information *see* ECOLOGICAL INTERACTIONS; PLANT PATHOLOGY; PLANT PHYSIOLOGY in the McGraw-Hill Encyclopedia of Science and Technology.

[JAMES L. RIOPEL]

Bibliography: J. Kuijt, *The Biology of Parasitic Flowering Plants*, University of California Press, 1969; J. Kuijt, Haustoria of phanerogamic parasites, *Annu. Rev. Phytopathol.*, 17:91–118, 1977; L. Musselman, The biology of *Striga, Orobanche,* and other root-parasitic weeds, *Annu. Rev. Phytopathol.*, 18:463–489, 1980; J. Steffens et al., Molecular specificity of haustorial induction in *Agalinis purpurea*, *Ann. Bot.*, 50:1–7, 1982.

Particle accelerator

High-energy accelerators capable of generating ultrahigh-current electron beams are an active area of research. The motivation for developing these devices lies in the potential applications of high-current electron beams that include the development of high-power coherent radiation sources by the free-electron laser mechanism, x-ray radiography, and national defense. *See* LASER.

Among the various accelerating schemes that have the potential to produce ultrahigh-power electron beams, induction accelerators appear to be the most promising. Induction accelerators are inherently low-impendance devices and thus are ideally suited to drive high-current beams. In these devices the acceleration process is based on the inductive electric field produced by a time-varying magnetic field. The electric field can be either continuous or localized along the acceleration path.

Substantial progress has been achieved in the development of ultrahigh-current induction accelerators. The Advanced Test Accelerator (ATA) at Lawrence Livermore National Laboratory has met its design specifications, producing 10-kA 50-MeV electron beam pulses, and a 40-kA 9-MeV electron beam has been generated by the RADLAC-II Module (RIIM) at Sandia National Laboratory. In addition to these linear accelerators, a cyclic "table top" device at the University of California, Irvine, has produced electron beams with about 200 A circulating current at 1 MeV; and the Naval Research Laboratory modified betatron, a cyclic device that has been designed to generate 10-kA 50-MeV electron beams, has successfully completed its testing phase.

Principle of operation. The electric field that is responsible for the acceleration of charged particles can be written as the sum of an electrostatic and an inductive component. The electrostatic component is generated by electric charges; in mathematical terms its divergence is proportional to the charge density, while its curl is zero. The inductive electric field, on the other hand, is generated by a time-varying magnetic field. The curl of the inductive component of the electric field is proportional to the rate of change of magnetic field, while its divergence is zero. It is the inductive field that is responsible for the acceleration of particles in induction accelerators.

The energy gained by a charged particle in the presence of an electric field can be found by integrating the energy rate equation (that is, by summing the equation over infinitesimal intervals along the particle's path). This equation states that the time rate of change of the kinetic energy of a charged particle is proportional to the total electric field acting on the particle. If the energy rate equation is integrated from position 1 to position 2, two positions along the particle's path, and if it is assumed that the fields vary slowly in time, then the energy of the particle is found to change from W_1 to W_2 between these two positions. For an electrostatic field this energy change is given by Eq. (1), where

$$W_2 - W_1 = \frac{|e|}{mc^2}[\phi(2) - \phi(1)] \qquad (1)$$

$\phi(1)$ and $\phi(2)$ are the values of the electrostatic potential at positions 1 and 2, e and m are the particle's charge and mass, and c is the speed of light.

For an inductive field the energy change is given by Eq. (2), where Φ is the magnetic flux through the

$$W_2 - W_1 = \frac{|e|}{mc^3} \frac{\partial \Phi}{\partial t} \simeq \frac{\Phi(2) - \Phi(1)}{\tau} \quad (2)$$

particle orbit, $\Phi(1)$ and $\Phi(2)$ are the value of Φ at the time the particle reaches positions 1 and 2, and τ is the transit time of the particle. Therefore, in the presence of an inductive field the particle gains energy, provided that the magnetic flux through the particle orbit increases with time.

Only modest acceleration can be achieved with electrostatic fields because the maximum potential difference is limited by insulator flashover. However, this is not the case with inductive fields, since the voltage is induced only in the circuit threading the flux and the voltages of a sequence of circuits successively add to the energy of the accelerated particle. The striking difference between the two fields becomes apparent when the orbit of the particle is circular. After a complete revolution, $W_2 - W_2 \neq 0$ for an inductive field, if the magnetic flux through the particle orbit changes with time, while $W_2 - W_1 = 0$ for an electrostatic field.

Types of induction accelerators. Induction accelerators are naturally divided into linear and cyclic types. The linear devices are in turn divided into the Astron type, the Radlac type, and autoaccelerators. In the Astron type, ferromagnetic induction cores are used to generate the accelerating field, while "air core" cavities are used in the Radlac type. In the autoaccelerator the air core cavities are excited by the beam's self-fields rather than external fields. Similarly, cyclic accelerators can be divided into two categories. In the first category belong those cyclic accelerators that use weak focusing to confine the electron beam as in the conventional and the modified betatron. In the second category, beam confinement is achieved with strong focusing fields. This latter category includes the stellatron, racetrack, and rebatron accelerators, all of which are currently in a rather preliminary state of development.

The table gives the parameters of some representative high-current induction accelerators. Most of these accelerators are presently in operation. To gain a better insight into the operation of induction accelerators, the modified betatron, a cyclic, compact device that is presently under development, will be described.

Modified betatron. The conventional betatron accelerator consists of two transverse magnetic field components (the betatron field) B_z and B_r that vary with time. The toroidal accelerating electric field is generated by the time-varying magnetic field. During acceleration the radius of the electron remains constant provided the flux rule is satisfied; that is, the rate of change of the average magnetic field inside the orbit is equal to twice the rate of change of the field at the orbit.

Near the beam orbit, the magnetic field B_z varies with the distance r from the axis of the ring in proportion to $1/r^n$, where n is called the external field index. To achieve both axial and radial confinement in a betatron, it is necessary that the field index be between zero and one. Magnetic configurations that have such a field index can be easily obtained by suitably shaping the pole faces at the gap of the ferromagnetic core. However, the desired field index can also be obtained with air cores, although not as easily.

The beam current in a conventional betatron is limited by the space-charge orbital instability, a tendency of the beam to expand due to the mutual repulsion of the like charges of which it is composed. Orbital stability requires the self-field index n_s to be less than ½, a condition can also be expressed as Eq. (3), where I_{cb} is the maximum elec-

$$I_{cb} = 4.2(r_b/r_0)^2\gamma^3\beta^3 \text{ [kA]} \quad (3)$$

tron current that can be stably confined, r_0 is the major ring radius, r_b is the minor ring radius, $\beta = v/c$, and $\gamma^2 = 1/(1 - \beta^2)$, where v is the beam velocity. The relativistic factor γ is also equal to the ratio of the particle's mass to its rest mass.

The stability properties of the conventional beta-

Representative high-electron-current induction accelerators

Device	Laboratory or country	Year	Energy, MeV	Current, kA	Duration, ns	Repetition rate, Hz
Astron type						
FXR	Lawrence Livermore National Lab	1982	20	4	75	0.1
NBS	Naval Research Lab	1973	0.8	1	2000	Single pulse
ETA	Lawrence Livermore National Lab	1979	4.5	10	30	2
ATA	Lawrence Livermore National Lab	1984	50	10	50	1–10
Radlac type						
LIU-10	Soviet Union	1977	13.5	50	20 or 40	—
RIIM	Sandia National Lab	1985	9	40	40	Single pulse
MABE	Sandia National Lab	1983	8	80	40	Single pulse
Autoaccelerator	Naval Research Lab	1962	7.3	70	10	Single pulse
Cyclic type						
Betatron	Soviet Union	1964	100	0.1	*	Single pulse
Modified betatron	University of California, Irvine	1984	1	0.2	*	Single pulse
Modified betatron†	Naval Research Lab	1985	(50)	(5–10)	*	Single pulse

*Depends on extraction scheme.
†Parentheses indicate design value.

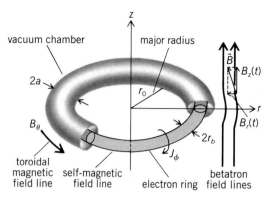

Modified betatron configuration, in which a strong toroidal magnetic field is added to a conventional betatron.

tron can be substantially improved by adding a strong toroidal magnetic field B_θ (see illus.). In contrast to the betatron or vertical magnetic field which affects mainly the major radius of the ring, the toroidal magnetic field affects primarily its minor radius. The force responsible for the control of the minor radius of the beam is proportional to $J_\phi B_\theta$, where J_ϕ is the poloidal ring current. As a result of this force, the modified betatron can confine rings with very large emittance or transverse velocities.

In the absence of walls, orbital stability in a modified betatron configuration requires the electron current I_{mb} to be less than the value given by Eq. (4). This is larger than the value given in Eq. (3) for

$$I_{mb} = 2.1(r_b/r_0)^2\gamma^3\beta^3(B_\theta/B_z)^2 \text{ [kA]} \qquad (4)$$

the maximum current I_{cb} in a conventional betatron by a factor $(B_\theta/B_z)^2/2$. Because B_θ/B_z is much greater than 1, the current that can be stably confined in a modified betatron is very large, even for moderate values of γ.

In the pressure of resistive metal walls, the electron current is limited by the drag instability. This instability is due to the finite resistivity of the vacuum chamber wall, and when the electron current exceeds an upper limit, the ring spirals outward and hits the surrounding wall. For $n = \frac{1}{2}$, the maximum electron current is given by Eq. (5), where a

$$I_{mb} = 4.2\beta^2\gamma^3a^2/r_0^2 \text{ [kA]} \qquad (5)$$

is the minor radius of the torus surrounding the beam. Since this limit is independent of B_θ, it cannot be improved by increasing the toroidal field. Furthermore, since the phase velocity of the drag instability is very high, the velocity spread along the direction of propagation does not affect its growth rate because the Landau damping of the unstable wave is negligible. Therefore, the limit imposed by Eq. (5) cannot be circumvented with thermal beams, that is, with beams that are not monoenergetic but whose velocity varies around an average value.

For relatively low values of axial energy spread, that is, a few percent, the electron current is limited

by the negative mass instability. Other, less important instabilities are the wall-resistive, orbital-resonance, ion-resonance, and streaming instabilities. The growth rate of all these instabilities is reduced with increasing toroidal magnetic field. Also, a moderate energy spread will help to stabilize these instabilities. Unlike the conventional betatron, the electron ring of an ultrahigh-current modified betatron can tolerate a substantial energy spread without significant expansion of its minor radius.

Although the addition of the toroidal magnetic field substantially improves the current-carrying capabilities of the conventional betatron, the beam injection and extraction from the modified betatron are more involved than in the conventional betatron. Several schemes have been developed for injecting the beam into a modified betatron accelerator, and extraction schemes are under development. The modified betatron at the Naval Research Laboratory will provide the test bed for these schemes, and the outcome from these experiments will be crucial to the fate of the concept.

Applications. It is expected that the technology of ultrahigh-current accelerators will impact many areas of the civilian economy. As an example, consider the application of ultrahigh-current accelerators to tunnel drilling. A 100-kA 1-GeV 100-ns-duration electron beam pulse contains 10 megajoules, which is approximately equivalent to 5 lb (2 kg) of TNT. Pulsed once every second, the energy delivered by such an accelerator over an 8-h period is equivalent to 144,000 lb (65,000 kg) of TNT. In addition to the large amounts of energy, the peak power of such an electron pulse is 10^{14} W, that is, approximately 100 times the combined power of all the electric utilities in the United States. If the 10-MJ beam strikes a 10-cm^2 (1.5-in.^{2}) area, the incident energy flux will be 1 MJ/cm^2 (6.5 MJ/in.^{2}). The thermal shock pressure within the material has been estimated to be 5×10^4 kpsi (350 megapascals), well above the tensile strength of any metal. Therefore, the shock wave generated by such a powerful pulse will shatter the most demanding materials present in the Earth's crust.

For background information *see* BETATRON; ELECTROMAGNETIC INDUCTION; PARTICLE ACCELERATOR in the McGraw-Hill Encyclopedia of Science and Technology.

[C. A. KAPETANAKOS]

Bibliography: J. Hecht, *Beam Weapons*, 1984; C. A. Kapetanakos et al., Equilibrium of a high-current electron ring in a modified betatron accelerator, *Phys. Fluids*, 26:1634–1648, 1983; C. A. Kapetanakos and P. Sprangle, Ultra-high-current electron induction accelerators, *Phys. Today*, 38(2):58–69, February 1985; P. Sprangle and T. Coffey, New sources of high-power coherent radiation, *Phys. Today*, 37(12):44–51, December 1983; P. Sprangle and C. A. Kapetanakos, Constant radius magnetic acceleration of a strong non-neutral proton ring, *J. Appl. Phys.*, 49:1–6, 1978.

Particle detector

The development of new ways to detect and identify subnuclear particles has long been an important aspect of modern experimental physics. Such particles are invisible and usually are moving at speeds close to the speed of light, so ingenious methods must be employed to detect them. This article discusses (1) large detectors, designed to study the results of interactions at ultrahigh-energy colliders; (2) the time projection chamber, which is one of the detectors used at these colliders and also has a wide range of other applications; (3) bismuth germanite radiation detectors, whose use in conjunction with high-resolution germanium detectors has had a dramatic influence on gamma-ray spectroscopy; and (4) bolometric neutrino detection, a proposed technique for detecting low-energy neutrinos, involving the measurement of small temperature changes and sound waves in ultracold materials.

DETECTORS FOR ULTRAHIGH-ENERGY COLLIDERS

High-energy colliders are a new class of particle accelerators used to provide collisions between elementary particles, such as electrons and positrons or protons and antiprotons, at very high energies. In colliding-beam accelerators (called colliders), two beams of high-energy particles are allowed to collide head-on. In this way, much higher interaction energies can be obtained. The present generation of ultrahigh-energy colliders, which either have recently come into operation or are now under construction, have interaction energies ranging from 100 GeV to thousands of GeV. In circular colliders, both beams are stored in large circular rings of magnets and are brought into collision repeatedly at several interaction points. In linear colliders, the two beams are accelerated by linear accelerators and brought into collision at a single interaction point. Colliders are classified according to the particles they accelerate and collide: electron-positron (e^+e^-) colliders, electron-proton (ep) colliders, and proton-proton or proton-antiproton (pp or $p\bar{p}$) colliders.

Collider detectors are large devices used to study the results of the particle interactions at these colliders. Their function is to detect and identify the particles produced in the collisions and to measure their energies and production angles. The particles produced in these interactions that can be detected are photons, electrons, muons, and various hadrons such as pions, kaons, neutrons, and protons. These particles have very different properties, and thus the detectors consist of several elements optimized to detect each of these different particles. Since these particles are generally produced in all directions, the detector elements surround the interaction point in consecutive shells, like the layers of an onion.

The major detectors in operation or under construction at the ultrahigh-energy colliders are listed in Table 1. All of these are very large devices, varying in diameter from 20 to 50 ft (6 to 16 m), with lengths typically similar to the diameter. Their design, construction, and experimental utilization are carried out by large collaborations of between 100 and 400 scientists each. These detector systems, once completed, become part of the overall facility at which they are located, and may well be used actively over periods of many years by many groups of experimentalists, each addressing completely different problems.

Detector elements. Figure 1 shows a cutaway view of a typical detector, indicating the main elements and their location in the assembled detector. There are large variations from detector to detector; however, most of them contain the following elements.

Magnetic field. The trajectories of charged particles are bent into circular arcs by a large magnetic field. A measurement of the radius of these trajectories gives the momentum and sign of the electric charge of each particle. The magnetic field configuration most commonly used is that of an axial field induced by a large solenoidal coil, with the axis of the coil, and therefore the field, along the axes of the incident colliding beams. Some detectors, however, use a dipole magnet with field perpendicular to the beams. The fields are typically between 0.5 and 1.5 teslas (5 and 15 kilogauss). Some of the detectors use superconducting coils, while others use conventional aluminum or copper conductors; such coils are typically 10 to 20 ft (3 to 6 m) in diameter and 10 to 23 ft (3 to 7 m) long.

Central tracking system. This is located inside the magnetic field, centered on and surrounding the interaction point. Its main function is to measure points along the paths of charged particles coming

Table 1. Ultrahigh-energy colliders and major detectors

Collider	Particles collided	Location	Maximum energy, GeV	Expected turn-on date	Major detectors
SppS	$p\bar{p}$	CERN, Geneva, Switzerland	640	1983	UAI, UAII
Tevatron I	$p\bar{p}$	Fermilab, Chicago, Illinois	2000	1986	CDF, D0
SLC	e^+e^-	SLAC, Stanford, California	100	1987	Mark II, SLD
LEP	e^+e^-	CERN	200	1989	ALEPH, DELPHI, L3, OPAL
HERA	ep	DESY, Hamburg, Germany	300	1990	H1, ZEUS
SSC*	pp	?	40,000	?	Not yet known

*Proposed but not yet approved.

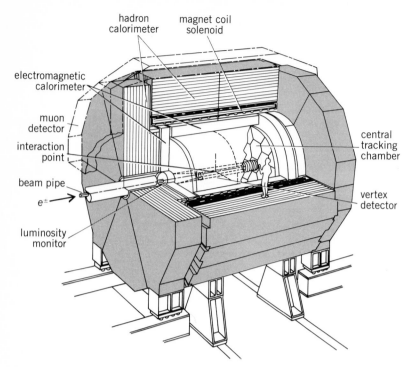

hadron calorimeter

magnet coil solenoid

electromagnetic calorimeter

muon detector

interaction point

beam pipe

$e^{\pm}$

luminosity monitor

central tracking chamber

vertex detector

Fig. 1. Cutaway view of ALEPH detector for the LEP electron-positron (e^+e^-) collider at CERN.

from the interactions, providing a measurement of the direction and radius of curvature (and thus the momentum or energy) of each particle. The tracking chambers are typically drift chambers or time projection chambers (discussed below), and consist of wires at a high voltage in a gaseous atmosphere. The times of arrival at these wires of ionization electrons produced by the passage of charged particles through the gas provide a measurement of the position of the trajectories in each layer of the tracking chamber. These devices are typically cylindrical in shape, 6 to 13 ft (2 to 4 m) in diameter and 6 to 13 ft (2 to 4 m) long.

Vertex detectors. These detector elements provide high-precision (typically 5 to 20 micrometers) measurements of points along the trajectories of charged particles very close to the interaction point (typically 0.4 to 4 in. or 1 to 10 cm). This information is used to reconstruct the decay points of very short-lived particles emanating from the interactions. The vertex detectors are the smallest components, typically 4 to 12 in. (10 to 30 cm) in diameter, and are those placed closest to the interaction point. They consist of high-precision drift chambers or of charge-sensitive silicon strips or chips (charge-coupled devices).

Particle identification system. The function of this detector element is to distinguish among the different types of charged particles, that is, to identify them as an electron, pion, kaon, or proton. In some detectors, these are major separate components located between the central tracking system and the magnet coil, and consist of Cerenkov counters or transition radiation detectors. In other detectors,

this function is performed by measuring the relative amount of ionization induced by the particle in the central tracking system.

Calorimeters. The function of these components is to measure the energy of both neutral and charged particles by means of total absorption. The previously described components are designed to be as light as possible, that is, to have as little material in them as possible, so that particles can pass through them undisturbed to the outer detector elements, such as the calorimeters. In contrast, the calorimeters are very heavy so that both neutral and charged particles interact in them and their total energy can be absorbed. The calorimeters consist of metal sheets, typically of iron, lead, or uranium, interleaved with layers of some sampling medium, such as scintillator, liquid argon, or gas. The optical or electric signals from the sampling medium are proportional to the total energy absorbed and, when properly calibrated, they provide a measurement of the total energy of the particle. Calorimeters are often divided into two parts; the electromagnetic component is optimized to measure the energies of photons (gamma rays) and electrons, and is followed by a hadronic component, designed to measure the energies of hadrons, such as neutrons, protons, pions, and kaons.

Muon detectors. The function of these components is to identify muons. Muons are the only known charged particles that can penetrate large amounts of material, such as several meters of steel, without being absorbed. Thus, charged particles penetrating the hadron calorimeters can be identified as muons by the muon detectors, located outside of the calorimeters, and typically surrounding the entire detector.

Luminosity monitors. These devices are located on the inside of the detector near the two entering beams and give a signal proportional to the total number of collisions that occur at the interaction point. This information is used to calculate the cross section (that is, the probability) of any particular reaction by comparing the number of events of that reaction to the total number of interactions.

Location of components. The relative location of the detector components is determined by their function and properties as described above. Closest to the interaction point is the vertex detector, followed by the central tracking device, which in turn is surrounded by the particle identification system. These three are always inside the magnet coil. The calorimeters are next, either inside or outside of the magnet coil. Frequently the electromagnetic calorimeter is inside the coil, and the hadron calorimeter is outside, with the iron plates of the hadron calorimeter also serving as the flux return iron of the magnet. The outermost layer is usually the muon detector.   [CHARLES BALTAY]

TIME PROJECTION CHAMBERS

The time projection chamber is a sophisticated detector used to detect, measure, and identify subnuclear particles produced in physics experiments at high-energy accelerators. The appeal of this new detector stems from its ability not only to reconstruct the trajectories of charged particles in three dimensions, but also to identify particles by measuring the ionization energy that they deposit along their tracks. The time-projection chamber can make these measurements over a large solid angle and in very crowded environments where many particles are created at the same time. From versions the size of a grapefruit for constructing decay vertices, to versions weighing 10,000 tons or more to detect proton decay, the time projection chamber in its current and proposed uses shows a large range of application.

Configuration. The idea of the time projection chamber was conceived in 1974 by D. R. Nygren. After the construction and testing of a small prototype model, a team of 30 physicists at Lawrence Berkeley Laboratory designed, constructed, and put into operation in 1982 a large 6-ft-long (2-m), 6-ft-diameter (2-m) time projection chamber at the electron-positron colliding beam accelerator (PEP) at the Stanford Linear Accelerator Center. It is essentially a large cylinder (Fig. 2) in which a very uniform electric field has been created by connecting the center midplane to a large negative voltage (-100 kV) and the endplanes to a more positive potential, and by placing metal equipotential rings along the surface of the cylinder to act as precision voltage dividers. The cylinder, filled with a mixture of gas of 80% argon and 20% methane at a pressure of 8.5 atm (850 kilopascals) is placed around the beam line of PEP in such a way that particles created in the collisions of the electrons and positrons (the electron's antiparticle) fly out through it. A powerful magnet surrounds the cylinder, with its field direction parallel with the electric field in the chamber. As discussed above, the main purpose of the magnetic field is to make the particles curve as they pass through the chamber.

Ionization trail. When a particle passes through the time projection chamber, it encounters atoms of

the argon-methane gas and knocks off electrons, creating a track, or pattern, called an ionization trail, that somewhat resembles a string of beads. The entire string appears to be formed instantaneously because the particle is moving so fast. The clusters of ionization electrons in the string, being electrically charged, feel the electric field in the chamber and are pulled toward the more positive pole at the end. The string drifts across the chamber with its shape intact, though the drifting electron clusters tend to broaden in space due to diffusion. What is new in the time projection chamber is the idea of drifting the particle tracks through a very large space in which the electric and magnetic fields are parallel to each other rather than at right angles. This design provides a unique benefit: the magnetic field suppresses unwanted diffusion of the electrons by causing them to follow magnetic field lines as they drift, and thus holds the shape of the "string of beads" intact, even over very long drift distances.

Spatial information. When the drifting clusters of ionization electrons reach the endcaps, they are detected by an array of 183 radially spaced proportional sense wires which amplify and collect the electrons. The positive voltage of the sense wires is set so that the signal out of the wire is proportional to the ionization drifting down. The radial position of a cluster of ionization electrons is determined simply by noting which radially spaced sense wire detected the cluster. The ground cathode plane under some of the sense wires is locally segmented into square pads, and the process of proportional amplification at the sense wire induces a signal on the nearest cathode pad. Thus, noting which pad has a signal gives the position of the ionization cluster along the sense wire. Thus the sense wire and cathode pad signals provide two coordinates in the plane perpendicular to the axis of the cylinder. Since the axial magnetic field bends the tracks in the same plane by an amount which depends on the particule's momentum, this information enables the momentum of the particle to be calculated.

The drift time of the ionization electrons from the point on the particle's trajectory where they were liberated to the sense wires at the endcap is measured; by using the known drift velocity (5 cm/microsecond), the axial coordinates of up to 183 points on the trajectory can be measured. Thus the time projection chamber provides three-dimensional information: radial (from the pattern of sense wires hit), azimuthal (from the signals on the cathode pads), and axially along the drift direction (from the time at which the information appears on the wires). It does not face the serious problem of unscrambling tracking ambiguities encountered in conventional systems of wire chamber detectors. The novel use of drift time to complete the three-dimensional information led to the coining of the name time projection chamber.

Particle identification. The clusters of ionization electrons are also used to measure the energy loss of the particle in the gas. Since this loss depends on

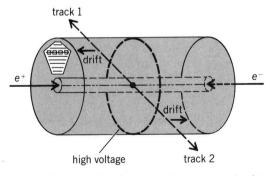

Fig. 2. Diagram of the time projection chamber, showing two tracks of ionization left by the passage of charged particles through the gas. The ionization electrons drift along axial electric and magnetic fields to the endcaps, where the signals from sense wires and cathode pads give spatial and time coordinates of the tracks, and also provide information about the identity of the original particles.

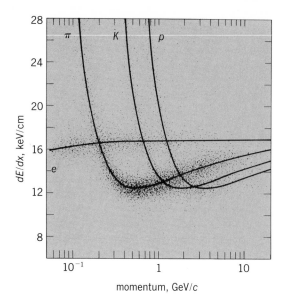

Fig. 3.   Scatter plot of ionization energy loss per unit distance traversed (dE/dx) versus momentum, for tracks in a sample of multihadronic events from electron-positron annihilation, as measured by the time projection chamber. The curves represent the predictions of theory for various particle types.

the particle velocity, the measurement can be used, together with the momentum measurement, to determine the mass and thus the identity of the particle. The amount of energy loss of a particle in a thin layer of material has enormous fluctuations. These fluctuations make a high-resolution measurement of energy loss impossible with a single sample, and the key to achieving high resolution is to take many independent samples. This is achieved in the time projection chamber because each of the 183 wires in the endcaps is used to sample the electrons from each track. The gas is at high pressure (8.5 atm or 850 kPa) to ensure that each sample has an adequate number of ionization electrons. With the excellent resolution of the time projection chamber, individual electrons, pions, kaons, and protons over a momentum range from 100 MeV/c (where c is the speed of light) to greater than 15 GeV/c can be identified. This can be done for all the particles in complex, multiparticle events by using the pulse height information from each wire.

Figure 3 shows the distribution of ionization energy loss per unit distance traversed, dE/dx, plotted against momentum measured in the time projection chamber for tracks in multiparticle events from electron-positron annihilation. The figure also shows the dependence predicted by theory for various particle types. In the region of low momentum, the pion, kaon, and proton bands are well separated. When the momentum is close to and above that which creates the minimum ionization (the lowest points on the curves in Fig. 3), a maximum-likelihood statistical method separates the various particle types. Figure 4 shows a typical multihadron event in electron-positron annihilation, as recorded by the time projection chamber at PEP. The measurements of ionization energy loss identify unambiguously all but one of the particles.

Thus the time projection chamber provides both three-dimensional spatial information about a particle's trajectory in space, and energy loss information about a particle's identity, all over a large solid angle in a compact system. This is what makes the time projection chamber so useful as a particle detector, enabling it to surpass most other conventional devices used to detect subnuclear particles.

Particle detection system. The time projection chamber built at Lawrence Berkeley Laboratory (like other collider detectors, as discussed above) is the heart of a massive particle detection system operating at the PEP colliding beam accelerator at Stanford. In addition to the time projection chamber, this system includes a large superconducting magnet to provide the 1.3-tesla (13-kilogauss) magnetic field, cylindrical drift chambers at the inner and outer radii of the time projection chamber to provide a trigger to the readout electronics when particles of interest pass through the detector, electromagnetic calorimeters to provide detection of photons and electrons, and a muon detection system of thick iron absorbers to provide a means of distinguishing muons from pions.

New chambers. Building on the experience of the Berkeley group with the first time projection chamber, other groups of physicists are using or planning to use time projection chambers in their research. In Europe at CERN two of the largest detector projects ever built, DELPHI and ALEPH (Fig. 1), will

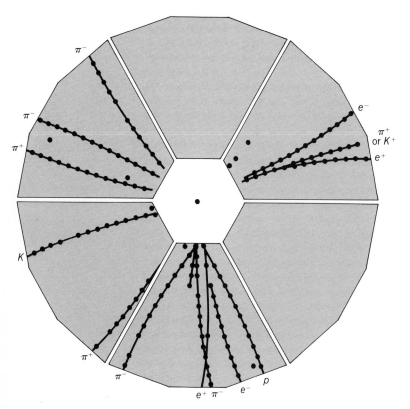

Fig. 4.   Typical multihadron event produced by electron-positron annihilation, as seen by the time projection chamber. This view along the axis of the chamber shows the signals from the cathode pads and the corresponding track fits.

use time projection chambers as their main detectors. Similarly, the TOPAZ detector being built in Japan for the new TRISTAN electron-positron collider has a time projection chamber at its core. Specialized time projection chambers have been built and operated successfully: the recoil spectrometer built at Fermilab to study high-energy diffraction dissociation, the detector built at TRIUMF in Canada to look for violation of lepton number conservation in muon capture, and a detector built at Los Alamos, New Mexico, to measure with unprecedented accuracy the positron spectrum from muon decay. New applications of time projection chambers are under development: Physicists have built a time projection chamber to study neutrinoless double beta decay, and efforts are under way to develop liquid-argon time projection chambers for massive experiments to study nucleon decay. At the other end of the spectrum, particle physicists are studying very small time projection chambers for vertex detectors to observe and tag short-lived particles produced very near primary collisions of subnuclear particles.   [RONALD J. MADARAS]

BISMUTH GERMANATE RADIATION DETECTORS

The field of gamma-ray spectroscopy has been dramatically influenced by the use of bismuth germanate (BGO) scintillation detectors in conjunction with high-resolution germanium (Ge) detectors. The success of a multidetector array (called TESSA) on the tandem accelerator at the Daresbury Laboratory in the United Kingdom has led to various new designs in Europe and the United States. The instruments measure the gamma radiation emitted following a nuclear reaction with the bismuth germanate detectors utilized in two distinct roles. One role is to measure the total radiation, and the part of the instrument that carries out this function is called a total energy ball. The second role is to improve the response of the germanium detector by rejecting events in which gamma rays do not deposit their full energy within the germanium crystal. Bismuth germinate detectors operating in this mode are termed suppression shields.

Origin of a gamma-ray shower. A heavy-ion accelerator, for example, produces a beam of a particular nuclide, such as calcium-48, with sufficient energy to overcome the Coulomb repulsion of nuclei in a target, such as cadmium-114, to take examples essentially at random from the periodic table. The beam and target nuclei, in this example, then fuse, creating a hot rotating ebrium-162 nucleus which rapidly cools by boiling off neutrons. Then more slowly, over a period of 10^{-15} to 10^{-12} s, 20 to 30 gamma rays are emitted to complete the cooling process and dissipate the large amount of angular momentum brought in by the heavy projectile. Multidetector arrays are designed to detect at least two of the members of this shower of gamma rays and typically do so by using germanium detectors with a high resolution of 0.2%. As many as possible of the remainder of the shower photons are then detected by higher-efficiency, lower-resolution detectors

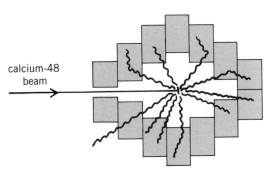

calcium-48 beam

Fig. 5. Schematic view of a gamma-ray shower being detected by a bismuth germanate total energy ball.

which completely surround the target (Fig. 5). There is an enormous number of different combinations of gamma ray energies which can make up the shower, and so it is essential to record data from approximately 10^7 to 10^8 showers to establish information on the structure of the nucleus. Because of the large number of gamma rays emitted in a shower and the necessity to ensure that only one gamma ray is detected in each high-resolution germanium detector, they must individually subtend a small solid angle at the large target. Therefore, a considerable number of detector systems are essential to obtain sufficient data in a typical experiment lasting 3 or 4 days. Each germanium detector is surrounded by its suppression shield, and the size of these shields directly affects the number of systems which can be placed around a target. The ball of low-resolution, high-efficiency detectors is situated between the target and the suppression shields. It must be compact; otherwise the germanium detectors will be far from the target and the data rates will become too low to obtain useful statistics.

Advantage of bismuth germanate. The scintillation material in general use since the 1950s for detecting gamma radiation has been crystals of sodium iodide activated with thallium, NaI(Tl). They have been used for suppression shields and total energy balls, but the resulting instruments are very large. For example, the total energy ball at the Max Planck Institute at Heidelberg, West Germany, has an outer radius of 18 in. (45 cm), whereas germanium detectors need to be not more than 10 in. (25 cm) from the target. The major advantage of bismuth germanate over thallium-activated sodium iodide is that its gamma-ray absorption coeffecient is 2.5 times greater (Table 2). This reflects both its higher density and the large atomic number of bismuth ($Z = 83$). Thus, there is a reduction in size of 2.5 in a linear dimension and up to a 16-fold volume reduction. Bismuth germanate is also nonhygroscopic and therefore easier to handle. It is an intrinsic scintillator and so free of the problems associated with the nonuniform dopant distribution of thallium in sodium iodide crystals. A disadvantage is that its light output is only 12–14% relative to thallium-activated sodium iodide crystals. This leads to resolutions, expressed as the full width at half the maximum

Table 2. Properties of sodium iodide and bismuth germanate detector materials

Property	Sodium iodide	Bismuth germanate
Chemical composition	NaI(Tl)	$Bi_4Ge_3O_{12}$
Density, g cm^{-3}	3.67	7.13
Linear absorption coefficient at 500 keV (cm^{-1})	0.4	1.0
Decay constant, μs	0.23	0.30
Hygroscopic	Yes	No

(FWHM) of the energy peak, of around 15% for 662-keV gamma rays from a cesium-137 source compared with 8% for thallium-activated sodium iodide. However, this poorer resolution is not crucial in many applications such as total energy balls and suppression shields where compactness is the critical factor.

Total energy balls. Two different geometric structures have been utilized as the basis for the total energy balls. A spherically symmetric polyhedral solution has been used for the large sodium iodide systems at Heidelberg, West Germany, and Oak Ridge, Tennessee, and for the proposed bismuth germanate ball at Chalk River in Canada. It involves a combination of pentagon and hexagon faces such that they subtend equal solid angles at the center of the ball which is the target position. The Oak Ridge and Chalk River arrays are composed of 72 separate detectors, while the Heidelberg ball has 162. The other structure in use has cylindrical symmetry with the individual detectors of hexagonal cross section, making a honeycomb arrangement with an inner hole where the target is placed. The advantage of this approach is that the photomultiplier tubes and associated cabling are confined to the ends of the cylinder, leaving the space outside the surface of the cylinder free for the suppression shields. It is more difficult to obtain both high efficiency and equal effective solid angles for each detector, but this has been achieved in the TESSA arrays by combining detectors in the outer rings.

The two primary quantities measured by the ball are the total energy deposited by the shower and the number of its component detectors in which this energy is deposited. This latter number is termed the fold, and is a measure of the number of gamma rays in the shower. The radius of the ball is directly related to the gamma-ray absorption coefficient of the scintillator, and thus a bismuth germanate ball can be extremely compact. The TESSA ball has an outer radius of only 4.7 in. (12 cm). Reflecting the statistical probability of the interaction of gamma rays with matter and the dominance of the Compton scattering mechanism over much of the relevant energy regime, any single gamma-ray photon may be detected by zero, one, two, three, or even more detectors. The fold resolution of the instrument can be improved by reducing the multiple detection probability. This gives a further advantage for bismuth germanate over thallium-activated sodium iodide, as the higher atomic number of bismuth greatly increases the fraction of interactions in which the gamma ray is totally absorbed due to the photoelectric effect compared to those interactions scattered through the Compton process.

Compton suppression shields. A germanium detector typically has the excellent resolution of 2 keV for a 1200-keV gamma ray, but only 15% of the observed intensity is in the full energy peak. The remaining 85% constitutes the Compton background which is associated with a lower-energy gamma ray being scattered out of the germanium crystal. A suppression shield consists of a 1.4-in.-thick (3.5-cm) cylindrical shell of bismuth germanate around the germanium detector to detect this scattered radiation (Fig. 6). Often there is a cone of thallium-activated sodium iodide at the front where the scattered gamma-ray energy is much lower and 1.2 in. (3 cm) of the crystal is sufficient to stop most of the radiation. The increased light output from the thallium-activated sodium iodide is then valuable in reducing the minimum gamma-ray energy which can be detected. The scintillators in the suppression spectrometer are shielded from direct radiation originating in the target by a few inches (several centimeters) of tungsten or lead. The mode of operation of the device is to accept gamma-ray events in the germanium detector only if there is no coincident detection of gamma radiation in the shield. This increases the fraction in the full energy peak to over 55%. The improvement is even more dramatic in experiments requiring coincidences between two germanium spectrometers as the fraction of both recording full-energy events increases from 2% to 30%.

Multidetector arrays. The suppression shield spectrometers can be used in a flexible manner and built into various array designs. For example, at the Daresbury Laboratory one array consists of the bismuth germanate ball with 12 suppression shields. A second structure is a polyhedral soccer ball structure enabling 30 suppression spectrometers to completely surround the target (Fig. 7). The data rates from this instrument are 1000 times greater than

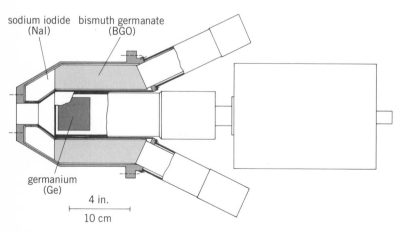

sodium iodide   bismuth germanate
(NaI)   (BGO)

germanium
(Ge)

4 in.

10 cm

Fig 6.   Suppression shield surrounding a high-resolution germanium detector.

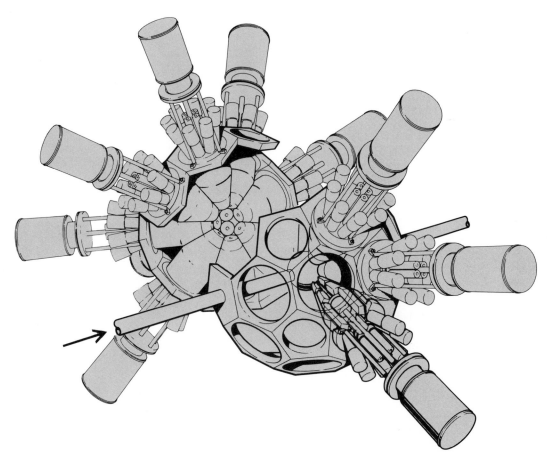

Fig. 7.   Thirty-suppression-shield spectrometer array at Daresbury Laboratory. Arrow indicates direction of gamma rays. (*After Daresbury Laboratory*, Annual Report 1984/85)

from a two-detector system for double coincidences and thus open up new avenues for research in gamma-ray spectroscopy.

[P. J. TWIN]

BOLOMETRIC NEUTRINO DETECTION

Neutrinos, because of their extremely weak interactions with matter, are the most difficult among the known stable or long-lived elementary particles to detect. The spectrum of neutrinos emitted by the Sun, in spite of its considerable scientific importance, has not yet been measured. Traditionally, detectors for low-energy neutrinos have relied upon the nuclear transmutations induced by them. However, such processes are often suppressed compared to the simple elastic scattering of neutrinos off nuclei and electrons. In addition, they occur only above certain neutrino energy thresholds, while elastic scattering occurs for neutrinos of even very low energies. Thus, it would by very useful to have a technique for detecting the small energy transfers involved in these scatterings. While many detectors are capable of measuring very small energies (less than 1 keV), such measurements can only be made on very small volumes of material. However, low-energy neutrinos typically require up to tons of target material before interaction rates become measur-

able. Thus the problem of measuring the elastic scattering of neutrinos off matter becomes one of devising a detector which not only is sensitive to small energy transfers, but also can monitor large amounts of material. A new technique, involving measuring small temperature changes and the production of phonons (sound waves) in ultracold materials, has been proposed which promises such features.

Neutrino interactions and bolometry. Low-energy (less than 1 MeV) neutrinos scatter elastically off matter in principally two ways: elastic scattering off electrons and coherent elastic scattering off nuclei. These two processes can be independently probed because of the very different kinematics involved. The ratio of energy that a low-energy neutrino can transfer to an electron compared to that transferred to a nucleus is roughly $m_e/m_{nucleus} \approx 2000$. Thus a given neutrino energy distribution will lead to a two-pronged recoil energy distribution for elastic scattering: a low-energy peak due to scattering off nuclei and a high-energy peak due to neutrino-electron scattering (see Fig. 8 and Table 3).

There is another distinction between neutrino-electron and neutrino-nucleus scattering. The first process depends strongly on which type of incident neutrino is doing the scattering, while the latter is independent of neutrino type. This implies the pos-

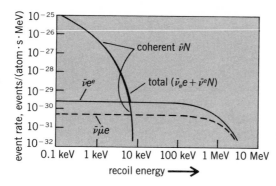

Fig. 8. Expected recoil energy spectrum from elastic scattering of antineutrinos off electrons ($\bar{\nu}_e e$ and $\bar{\nu}_\mu e$) and coherent elastic scattering of antineutrinos off nuclei ($\bar{\nu}N$), based on a typical incident antineutrino spectrum from a nuclear reactor. (*After B. Cabrera, L. M. Krauss, and F. Wilczek, Bolometric detection of neutrinos, Phys. Rev. Lett., 55:25–29, 1985*)

sibility of doing experiments to probe for the possibility of neutrino "oscillations."

Silicon bolometry principles Remarkably, for silicon (or similar materials) at 0.01–0.001 kelvin, even the energy deposited in the elastic scattering of a single low-energy neutrino is sufficient to raise the temperature of a macroscopic sample by a measurable amount. This is because silicon has a high Debye temperature (636 K or 705°F), and hence one of the lowest heat capacities of all materials at low temperatures. An energy deposit of 0.1 MeV will raise the temperature of 2.2 lb (1 kg) of silicon initially at about 1 millikelvin to about 5 millikelvins. This fact, which could allow large volumes to be monitored for small energy deposits, led to the proposal of detectors using large quantities of such materials of ultracold temperatures for neutrino bolometry.

After a neutrino scattering event, recoil electrons below 1 MeV will lose energy by ionization with a range of 1 mm, while recoil nuclei, with energies less than 10 keV, will lose energy in much smaller distances by directly producing lattice vibrations, called phonons. Eventually even the energy lost by electrons will also be transferred to such lattice vibrations. The main goal of the bolometric technique is to detect these vibrational phonons, either once they have scattered many times and distributed thermal energy uniformly or before they have scattered

and all the energy is in the form of one or several so-called energetic ballistic phonons.

Thermal phonon detection. By doping semiconductors appropriately, their resistance can be made to vary exponentially in the temperature range of interest. A small chip placed on the surface of a silicon block can then function as a thermometer.

Such resistive techniques are not likely to prove satisfactory at temperatures in the millikelvin region, because of the large specific heat of the thermometer compared to silicon. One approach would be to use superconducting thin films as thermometers. For such films the superconducting transition is expected to be fairly broad, with a linear temperature-versus-resistance region, allowing a potential extension of resistance techniques to lower temperatures. If heat leaks prove prohibitive, the thin films can be inductively monitored with a SQUID magnetometer to measure the change between superconducting and normal states. The time that the films remained normal would be related, after calibration, to the energy deposition. These techniques could allow energy resolution of the order of a keV and time resolution of the order of milliseconds.

Ballistic phonometry. Another possibility involves measuring directly the energetic phonons accompanying the initial energy deposition before they have scattered and thermalized. Superconducting tunnel junctions might be used to measure the energy of a ballistic phonon by its destruction of Cooper pairs. Timing resolution of the order of microseconds may be possible.

Backgrounds. Since such detectors will measure events in real time (a significant advantage over most other proposed detectors), vetoing of many unwanted backgrounds will be possible. The worst of these include cosmic-ray events from outside the detector, and radioactive decays inside. The former can be distinguished in a segmented detector, since neutrino-induced events will be localized while cosmic-ray events will not. Radioactive events are harder to veto in this way. However, silicon is about the purest substance available, with standard impurity levels of 1 part in 10^{15} easily available. Moreover, energy resolution should allow many radioactivity-related events to be vetoed. Backgrounds for reactor-size experiments should be containable at rates below the expected neutrino signal, but it is not known whether radioactivity can be controlled at

Table 3. Weighted total neutrino cross sections and event rates*

Source of neutrino	Process	Total cross section (σ), cm^2	Maximum recoil energy, MeV	Event rate, ton^{-1} day^{-1}
Solar pp	$\nu_e e$	8.1×10^{-46}	0.26	1.2
Solar ^{7}Be	$\nu_e e$	5.3×10^{-45}	0.66	0.75
Solar pp	$\nu_\mu e$	2.1×10^{-46}	0.26	0.3
Solar ^{7}Be	$\nu_\mu e$	1.0×10^{-45}	0.66	0.13
Solar ^{8}B	νN	3.3×10^{-41}	0.014	0.16
Reactor	$\bar{\nu}_e e$	5.3×10^{-45}	9.6	4500
Reactor	$\bar{\nu}_\mu e$	1.9×10^{-45}	9.6	1600
Reactor	$\bar{\nu} N$	2.5×10^{-45}	0.007	161,000

*After B. Cabrera, L. M. Krauss, and F. Wilczek, Bolometric detection of neutrinos, *Phys. Rev. Lett.*, 55:25–29, 1985.

the levels needed for a solar neutrino experiment. However, if ionization can be measured at the same time as thermal photons, it may be possible to obtain unambiguous signatures for neutrino-induced events, regardless of other backgrounds.

Applications. If it proves practical, based on prototype experiments now under way, neutrino bolometry could significantly advance understanding of nuclear and particle physics as well as astrophysics. Measurement of the entire solar neutrino spectrum would (besides its great significance for understanding the processes which power the Sun) resolve the controversy over whether the anomalously small high-energy solar neutrino flux may be due to neutrino oscillations. (Oscillations can also be probed in reactor-size experiments.) In addition, if the dark matter which is known to surround galaxies is made up of heavy weakly interacting elementary particles, then they can scatter off matter at rates and with energy deposits for which neutrino bolometers will be sensitive. *See* COSMOLOGY.

The possibility of quickly and accurately obtaining the spectrum of neutrinos emitted by nuclear reactors also promises a variety of technological applications. At present, reactor energy measurements are made by using thermal neutron fluxes which propagate out from the core with characteristic long thermal diffusion times. Neutrinos, on the other hand, propagate at or near the speed of light out from the core and, moreover, have a spectrum which is directly related to the energy-producing processes in a reactor. Thus, neutrino bolometers could provide significant new monitors for power reactors, allowing operation at higher power levels. Another possible use of bolometers is as a tool for weapons test verification in any arms control treaty.

Besides being used to detect neutrinos, large-scale bolometric devices capable of measuring small and infrequent energy deposits are likely to have wide-ranging and significant technological application, from environmental testing to new types of quality control for semiconductors.

For background information *see* GAMMA-RAY DE-TECTORS; GAMMA RAYS; NEUTRINO; PARTICLE ACCEL-ERATOR; PARTICLE DETECTOR; SOLAR NEUTRINOS in the McGraw-Hill Encyclopedia of Science and Technology.

[LAWRENCE M. KRAUSS]

Bibliography: B. Cabrera, L. M. Krauss, and F. Wilczek, Bolometric detection of neutrinos, *Phys. Rev. Lett.*, 55:25–29, 1985; E. Commins and P. Bucksbaum, *Weak Interactions of Leptons and Quarks*, 1983; R. Donaldson and J. G. Morfin (eds.), *Proceedings of the 1984 Snowmass Summer Study*, 1984; C. W. Fabjan and T. Ludlam, Calorimetry in high-energy physics, *Annu. Rev. Nucl. Part. Sci.*, 32:335–389, 1982; C. N. Holmes (ed.), *Proceedings of International Workshop on Bismuth Germanate*, Princeton University, 1982; Johns Hopkins University et al., *Proposal for a PEP Facility Based on the Time Projection Chamber*, 1976; S. C. Loken and P. Nemethy (eds.), *Proceedings of the 1983 Division of Particles and Fields Workshop on Collider Detectors*, 1983; R. J. Madaras and P. J. Oddone, Time-projection chambers, *Phys. Today*, 37(8):36–47, August 1984; J. N. Marx and D. R. Nygren, The time projection chamber, *Phys. Today*, 31(10):46–53, October 1978; Proceedings of the 1982 IEEE Nuclear Science Symposium, *IEEE Trans. Nucl. Sci.*, vol. NS-30, February 1983; E. Segré, *Nuclei and Particles*, 2d ed., 1977; P. J. Twin, Developments in 4π arrays for gamma-ray detection, in P. Blasi and R. A. Ricci (eds.), *Proceedings of International Conference on Nuclear Physics*, Florence, pp. 527–549, 1983.

Pegmatite

Pegmatites are coarse-grained igneous rocks that represent the ultimate fractionation products of silicic magmas. The residual liquids that are derived from crystallizing granitic plutons to form pegmatites are highly enriched in the lithophile metals lithium, beryllium, rubidium, cesium, tin, tungsten, niobium, and tantalum, as well as important volatile constituents such as boron, phosphorus, halogens, and water. Fractionation within a pegmatite dike system produces further enrichment in these lithophile rare elements to the point that their concentrations in lithium-rich, rare-element pegmatites are four to six orders of magnitude above the average values in typical (source) granitic plutons. Rare-element pegmatites, therefore, hold a wealth of information on the processes by which petrogenetically and economically significant lithophile elements are concentrated and ultimately precipitated in plutonic silicic magmas. The conditions of formation for rare-element pegmatites in terms of pressure, temperature, and fluid compositions have been poorly understood. Recent experimental and theoretical studies of lithium mineral stabilities, however, now provide a means of evaluating the pressure-temperature conditions of pegmatite crystallization. In turn, lithium mineral stabilities constitute a pressure-temperature frame of reference by which other important pegmatitic phenomena may be assessed.

Of the rare metals cited above, lithium is particularly useful in petrogenetic studies of pegmatite evolution. It is an abundant and common component of pegmatite systems; estimates of the lithium content of rare-element pegmatites range from about 1.5 to 2.0 wt % Li_2O. The small ionic radius (0.068 nanometer) precludes extensive incorporation of lithium into the other common alkali aluminosilicates (especially feldspars); this incompatible behavior leads to the generation of a discrete suite of lithium minerals (see table). Even in these minerals, the small lithium cation does not provide entirely satisfactory charge-size compensation for most octahedral sites. As a result, lithium minerals commonly undergo both subsolidus breakdown and metasomatic replacement; thus they serve as monitors of changing pressure, temperature, and fluid compositions.

Lithium aluminosilicates. Among the common lithium minerals in pegmatites, the lithium aluminosilicates petalite ($LiAlSi_4O_{10}$), spodumene (α-

LiAlSi$_2$O$_6$), and eucryptite (α-LiAlSiO$_4$) are especially useful in petrogenetic studies. As a group, these are the most common and abundant lithium minerals, so that the details of their stability relations are applicable to a large number of pegmatite deposits. Their compositions in pegmatites are near the ideal end members, and phase equilibria in the simple binary system LiAlSiO$_4$–SiO$_2$ are directly applicable to natural systems. In addition, all lithium-rich, rare-element pegmatites are quartz-saturated; thus lithium aluminosilicate stability relations are functions of pressure and temperature only.

The experimentally calibrated lithium aluminosilicate phase diagram constitutes a petrogenetic grid for lithium-rich pegmatites (see Fig. 1). The stability fields of the natural lithium aluminosilicates petalite and spodumene are bounded at high temperature by reactions that produce tetragonal β-spodumene and hexagonal virgilite, neither of which occurs in pegmatites; thus, lithium aluminosilicate saturation in pegmatites takes place at temperatures below about 700°C (1290°F) at any pressure. Below the β-spodumene and virgilite reaction boundaries, only spodumene and petalite are stable at magmatic

conditions, thus explaining why these are the only primary lithium aluminosilicates in pegmatites. Whether spodumene or petalite crystallizes as the primary lithium aluminosilicate phase probably depends more on depth of emplacement than on the temperature at which lithium aluminosilicate saturation occurs. The stability field of eucryptite + quartz is restricted to low pressures and temperatures. The presence of this assemblage reflects subsolidus reaction conditions and comparatively low sodium and potassium activities in coexisting aqueous fluids, because the assemblage eucryptite + quartz is particularly susceptible to replacement by albite or micas. The rare lithium zeolite bikitaite, LiAlSi$_2$O$_6$ · H$_2$O, apparently forms by replacement of eucryptite + quartz assemblages in response to decreasing pressure-temperature conditions and perhaps with increasing activity of H$_2$O in residual pegmatitic fluids.

At some of the largest and most economic rare-metal pegmatites, two or all three of the common pegmatitic lithium aluminosilicates are found together with quartz. In these important cases, the lithium aluminosilicate phase diagram can be used

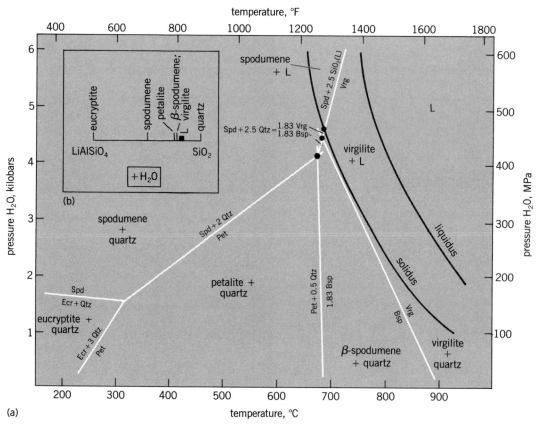

Fig. 1.   Petrogenic grid for lithium-aluminosilicate–bearing pegmatites. (a) Experimental pressure-temperature phase diagram for the bulk composition 20 mole % eucryptite +80 mole % quartz in the system LiAlSiO$_4$–SiO$_2$–H$_2$O. The symbol L represents hydrous silicate liquid. Stable portions of univariant reactions are represented by white lines. The solidus and liquidus indicated are for the bulk composition of this study. (b) Compositions of phases encountered in this system. The composition of the liquid (L) is projected from H$_2$O; the liquid composition varies between the eutectic composition 16 mole % eucryptite +84 mole % quartz and the bulk composition of 20 mole % eucryptite +80 mole % quartz. Bsp = β-spodumene; Ecr = eucryptite; Pet = Petalite; Qtz = quartz; Spd = spodumene; Vrg = virgilite. (*After D. London, Experimental phase equilibria in the system LiAlSiO$_4$–SiO$_2$–H$_2$O: A petrogenetic grid for lithium-rich pegmatites, Amer. Mineralog., 69:995–1004, 1984*)

to define pegmatite cooling paths (see Fig. 2). A case in point is the large and economically important Tanco deposit in southeastern Manitoba, where primary petalite was replaced pseudomorphously by an intergrowth of spodumene + quartz, and spodumene + relict petalite were subsequently altered to minor amounts of eucryptite + quartz. The cooling path of the Tanco pegmatite is thus constrained to have passed through the stability fields of all three lithium aluminosilicates and above but near the invariant point that contains these three minerals plus quartz.

Internal evolution of Tanco pegmatite. The combination of lithium aluminosilicate reaction relationships with fluid inclusion studies at the Tanco pegmatite has yielded the most detailed assessment yet of pressure-temperature conditions and fluid evolution within a highly fractionated pegmatite. The results indicate that the pressure at pegmatite consolidation was approximately 3 kilobars (300 megapascals; corresponding to depths of about 5.5 mi or 9 km). The crystallization of internal units, and the pseudomorphic replacement of petalite by spodumene + quartz, took place in the presence of a dense, hydrous, alkali borosilicate fluid. Aluminosilicate melt was fluxed by $Li_2B_4O_7$, which constituted approximately 14 wt % of the fluid. The high concentration of this alkali borate component served to increase silicate liquid-H_2O miscibility and to facilitate rare-metal concentration through extensive depolymerization of the hydrous melt. Between 420 and 470°C (790 and 880°F), and 2.6 and 2.9 kbar (260 and 290 MPa), the borate component of this fluid was removed by the crystallization of tourmaline, resulting in the deposition of albite, quartz, micas, and ore minerals (for example, microlite, beryl, and pollucite) and consequent evolution of a comparatively low-density, solute-poor, CO_2-bearing aqueous fluid. Reactions over this pressure-temperature interval mark the transition from dominantly magmatic to subsolidus hydrothermal conditions. At such high alkali borate contents, the transition from magmatic to hydrothermal conditions is in principle continuous (that is, supercritical).

Although exceptional, the Tanco deposit is not unique. The general characteristics of mineralogy and zonation are representative of lithium-rich, rare-element pegmatites around the world. The steps in the internal evolution of Tanco as outlined above, therefore, are probably applicable to many deposits of this type, and possibly to geochemically similar, highly fractionated tourmaline-rich granitoids (for example, the stanniferous granites of Cornwall, England).

Miarolitic pegmatites. A recent study of gem-bearing miarolitic pegmatites from several districts around the world has shown that these bodies crystallize at approximately the same pressure-temperature conditions as the compositionally similar massive pegmatites. Whether pockets form or not may be controlled in part by the timing and continuity of tourmaline crystallization, which liberates large mo-

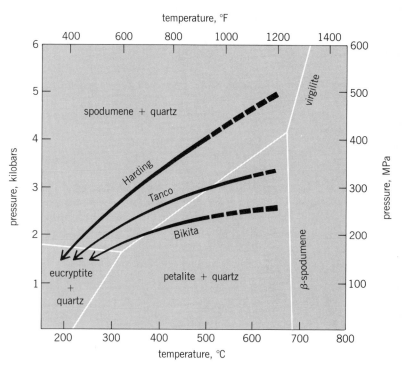

Fig. 2. Inferred pressure-temperature crystallization paths for the Harding, New Mexico; Tanco, Manitoba; and Bikita, Zimbabwe, lithium-rich rare-element pegmatites, as defined by lithium aluminosilicate + quartz assemblages at these deposits. For simplicity, the spodumene = β-spodumene and virgilite = β-spodumene reactions have been omitted. (*After D. London, Experimental phase equilibria in the system LiAlSiO₄–SiO₂–H₂O: A petrogenetic grid for lithium-rich pegmatites, Amer. Mineralog., 69:995–1004, 1984*)

lecular quantitites of water. Early and continuous crystallization of tourmaline would result in steady volatile loss, but the late appearance of abundant tourmaline would serve to create miarolitic pockets by the comparatively sudden release of water from residual melt. Above all, these studies have indicated that the halide concentrations of typical massive and miarolitic rare-element pegmatites are low (exsolved magmatic aqueous fluid contains less than 3 equivalent wt % NaCl), and that the processes leading to extreme mineralogical zonation and rare-element concentration and precipitation may be governed by other components, especially alkali borate and fluorine.

Metasomatic replacement. As noted above, the instability of lithium minerals and their susceptibility to metasomatic replacement make them sensitive monitors of changing fluid compositions and pressure-temperature conditions. The lithium aluminosilicates and lithium phosphates (amblygonite-montebrasite and lithiophilite-triphylite) respond differently to changes in fluid acidity or basicity and solute species, so that the combined paragenetic sequences of metasomatic replacement in each host phase provide a coherent picture of subsolidus fluid evolution. In general, incipient autometasomatism of lithium aluminosilicates produces secondary albite by Na ⇌ Li exchange, whereas phosphate alteration yields calcium-rich phases (for example, apatite) by Ca ⇌ 2Li and Ca ⇌ (Fe,Mn) exchange. This stage

Lithium minerals in granitic pegmatites

Name	Ideal composition
Silicates	
Petalite	$LiAlSi_4O_{10}$
Spodumene	$\alpha\text{-}LiAlSi_2O_6$
Eucryptite	$\alpha\text{-}LiAlSiO_4$
Bikitaite	$LiAlSi_2O_6 \cdot H_2O$
Lepidolite	$K(Li,Al)_3(Si,Al)_4O_{10}(F,OH)_2$
Zinnwaldite	$KLiFe^{2+}Al(AlSi_3)O_{10}(F,OH)_2$
Cookeite	$LiAl_4(AlSi_3)O_{10}(OH)_8$
Manandonite	$LiAl_4Si_3BO_{10}(OH)_8$
Bityite	$CaLiAl_2(AlBeSi_2)O_{10}(OH)_2$
Holmquistite	$Li_2(Mg,Fe^{2+})_3Al_2Si_8O_{22}(OH)_2$
Elbaite	$Na(Li,Al)_3Al_6(BO_3)_3Si_6O_{18}(OH)_4$
Liddicoatite	$Ca(Li,Al)_3Al_6(BO_3)_3Si_6O_{18}(OH)_4$
Brannockite	$KSn_2Li_3Si_{12}O_{30}$
Phosphates	
Amblygonite	$LiAlPO_4F$
Montebrasite	$LiAlPO_4(OH)$
Natromontebrasite	$(Na,Li)AlPO_4(OH,F)$
Tavorite	$LiFe^{3+}PO_4(OH)$
Lithiophilite	$LiMnPO_4$
Triphylite	$LiFePO_4$
Sicklerite	$Li(Mn^{2+},Fe^{3+})PO_4$
Lithiophosphate	Li_3PO_4
Tancoite	$HLiNa_2Al(PO_4)_2(OH)$
Bertossaite	$(Li,Na)_2CaAl_4(PO_4)_4(OH,F)_4$
Palermoite	$(Sr,Ca)(Li,Na)_2Al_4(PO_4)_4(OH)_4$

of replacement is followed by weakly acidic (K + H) metasomatism with the generation of abundant secondary micas. The sequence of alkali metasomatism followed by sericitic alteration is comparable to hydrothermal reaction schemes in other magmatic-hydrothermal deposits (for example, porphyry systems).

Other lithium minerals. Petrogenetic studies of other lithium minerals have the potential to yield additional information on fluid chemistry of pegmatite systems. For example, a growing number of analyses of amblygonite-montebrasite and lithium-rich micas and tourmalines leads to the conclusion that the fluorine contents of typical rare-element pegmatites are low [log f_{HF} = 10^1 bar at 600°C (1110°F), 2 kbar (200 MPa)]. One difficulty in extracting meaningful information from other lithium-rich minerals such as micas and tourmalines is that they are compositionally far more complex than the lithium aluminosilicates and the simple lithium phosphates, and hence their stabilities are governed by the activites of several volatile and nonvolatile components, as well as by pressure and temperature.

Lithium halos in wallrocks. Lithium exomorphism around pegmatites may be a valuable guide to economic rare-metal deposits. Rare-element pegmatites that intrude mafic to ultramafic hosts commonly produce metasomatic aureoles of lithium-rich biotite and tourmaline. Around very highly fractionated and rare-element-rich pegmatites, however, hornblende in the wallrocks is converted to the lithium amphibole holmquistite. Morever, the pegmatites that generate holmquistite are usually large and include the highest-grade pegmatitic deposits of

lithium, beryllium, and tantalum in the world. The presence of holmquistite in pegmatite wallrock may be especially important in pegmatite exploration, because the outer zones of lithium-rich pegmatites are mineralogically similar to those of barren pegmatites and may conceal ore-bearing internal units.

The lithium aluminosilicate phase diagram and theoretical analysis of lithium mineral stabilities provide the much-needed framework for understanding the internal evolution of rare-element pegmatite systems. The knowledge gained from the study of such pegmatites will be important ultimately in understanding the processes that control element concentration and zonation in silicic magmas.

For background information *see* MAGMA; PEGMATITE; PETROLOGY in the McGraw-Hill Encyclopedia of Science and Technology.

[DAVID LONDON]

Bibliography: D. London, Experimental phase equilibria in the system $LiAlSiO_4$–SiO_2–H_2O: A petrogenetic grid for lithium-rich pegmatites, *Amer. Mineralog.*, 69:995–1004, 1984; D. London, The magmatic-hydrothermal transition in the Tanco rare-element pegmatite: Evidence from fluid inclusions and phase equilibrium experiments, *Amer. Mineralog.*, R. H. Jahns Memorial Issue, in press; D. London, Holmquistite as a guide to pegmatitic rare-element deposits, *Econ. Geol.*, in press; D. London and D. M. Burt, Chemical models for lithium aluminosilicate stabilities in pegmatites and granites: *Amer. Mineralog.*, 67:494–509, 1982.

Petroleum

Gas, gasoline, kerosine, jet fuel, and part of diesel fuel are composed of petroleum hydrocarbons containing from 1 to 14 carbon atoms in their molecules. These hydrocarbons originate from the decomposition of organic matter in marine and lacustrine sediments of the Earth's crust. The process occurs in three stages (Fig. 1). During diagenesis (<50°C or 122°F) methanogenic bacteria form methane from substrates in the sediments. Traces of C_2 through C_{14} hydrocarbons and about 15% of the C_{15} through C_{40} hydrocarbon range in crude oil also are formed from low-temperature biological and chemical reactions. During catagenesis (Fig. 1; 50–200°C or 122–392°F) about 85% of the oil and 75% of the gas is formed from the cracking of the organic molecules (kerogen) in the sediment. Some asphalt also is formed as either a by-product or intermediate in the reactions. In the last stage, metagenesis or metamorphism, only gas is formed in appreciable quantities. An unknown number of intermediate reactions are involved with the starting and ending products approximately as shown in the overall reaction below.

$$2C_{10}H_{14} \rightarrow C_{10}H_{18} + 2CH_4 + 2C_4H$$

| Kerogen | Oil | Gas | Pyrobitumen |

Graphitic pyrobitumen residues have been found in many sedimentary rocks. If the oil is subjected to higher temperatures with greater burial, most of it is

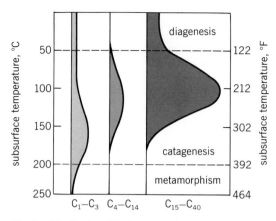

Fig. 1. Distribution of hydrocarbons in sedimentary rocks. Molecular sizes are indicated by the number of carbon atoms: gas, C_1 through C_3; gasoline, kerosine, and jet fuel, C_4 through C_{14}; heavy gas oil, lubricating oil, and residuum, C_{15} through C_{40}.

converted to gas. The oil floor in sedimentary basins, the depth below which there are no economic oil accumulations, varies between about 10,000 and 23,000 ft (3 and 7 km), depending on the geothermal gradient. The gas floor is unknown, but the concentrations of gaseous hydrocarbons such as methane decrease to very low values in deep, high-temperature ($>200°C$ or $392°F$) sediments. In the Aquitaine Basin of France, it was found that the generation of methane peaks at a subsurface temperature of around $150°C$ ($302°F$), after which the methane yields decreased sharply. The analyses of many deep well cuttings by geochemical service laboratories in the United States have shown that methane concentrations in the fine-grained sediments decrease to low values in the deepest ($>25,000$ ft or 7.6 km) high-temperature formations. To date, no commercial gas production exists below 27,000 ft (8.2 km). After generation, the hydrocarbons migrate from source to reservoir rocks in sedimentary basins by diffusion, by solution, and in an oil-gas phase by buoyancy with the latter mechanism dominant.

Generation. There are three major sources for methane: (1) methanogenic bacteria utilizing substrates such as carbon dioxide and hydrogen in near-surface sediments, (2) thermal degradation of organic matter disseminated in deep subsurface sediments, and (3) thermal degradation of coal and oil formed at depth. The biological origin of methane can occur from near the surface to sediment depths of about 1000 ft (305 m) or more in both lacustrine and marine environments. Thermogenic methane is formed mainly in the subsurface temperature range between 100 and $200°C$ (212 and $392°F$). It forms within and below the oil generation zone. It tends to disappear in deep sediments due to nongeneration when the ratio of hydrogen to carbon in the organic matter drops below 0.25. Generation stops apparently due to a lack of readily available hydrogen to form methane.

Hydrocarbons containing 2 to 14 carbon atoms per molecule are formed in trace amounts near the surface with branched hydrocarbon chains as the dominant products. With increasing burial the higher temperatures form straight-chain hydrocarbons, causing a shift in the ratio of iso- to normal (branched to straight-chain) hydrocarbons at the depth where intense hydrocarbon generation begins. Figure 2 shows the concentrations of total C_6-C_7 hydrocarbons in the fine-grained shales obtained from a well drilled offshore of Texas in the Gulf of Mexico east of South Padre Island. Only trace amounts of these hydrocarbons are found above 8000 ft (2.4 km). In the 8000–10,000-ft (2.4–3-km) interval, equivalent to a subsurface temperature of 100–110°C (212–230°F), there is an exponential increase in the concentration of these hydrocarbons due to thermal cracking of the organic matter (kerogen) in the shales. The concentrations decrease at depths beyond 14,000 ft (4.3 km) due to three factors: decreased generation, migration of the hydrocarbons out of the source rock, and cracking of the hydrocarbons to smaller molecules, mainly methane through propane. Within the 12,000–14,0000-ft-depth (3.7–4.3-km) range, not all samples show high concentrations due to variations in the generating capability of different kerogens within this interval and to early migration along more permeable bedding planes.

The solid lines in Fig. 2 are best-fit lines for the

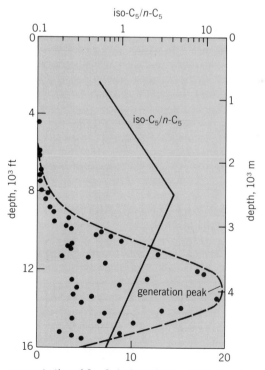

Fig. 2. Distribution of hexanes and heptanes (solid circles) compared to the ratio of isopentane to *n*-pentane in sediments from cuttings from a well drilled off the shore of South Padre Island, Texas. The broken-line curve represents an envelope around the data points.

ratio of methylbutane (isopentane) to *n*-pentane. The peak is just about at the threshold of increased hydrocarbon generation. This ratio continues to decrease with depth because the high-temperature thermal cracking reactions form predominantly straight-chain hydrocarbons. A similar peak is observed in the ratios of isobutane to *n*-butane and 2-methylpentane to *n*-hexane in Gulf Coast wells.

The generation peak in Fig. 2 represents the sum of 14 hydrocarbon structures that contain 6 or 7 carbon atoms per molecule. Distribution patterns for individual hydrocarbon molecules indicate that hydrocarbon structures with quaternary carbon atoms are generated deeper than those with tertiary carbon atoms. For example, the threshold of intense generation for 2,2-dimethylbutane and 2,2-dimethylpentane, both containing quaternary carbon atoms, was about 3000 ft (914 m) deeper than for the corresponding 2,3- homologs in the South Padre well. Increased generation of 1,1-dimethylcyclopentane also occurs about 2000 ft (610 m) deeper than 1,2-dimethylcyclopentane. It is believed that the hydrocarbons with tertiary carbon atoms form earlier due to greater stability of the intermediate tertiary carbonium ion or free radical. These observations are not confined to Gulf Coast sediments. On the North Slope of Alaska, shale cuttings from the Inigok No. 1 well showed the threshold of intense generation of isopentane to be at about 9000 ft (2.7 km), whereas that of neopentane with a quaternary carbon was at 11,000 ft (3.4 km). The Kugrua No. 1 well showed generation of 2,2-dimethylbutane and 3,3-dimethylpentane to continue about 2000 ft (610 m) deeper than its corresponding C_6 and C_7 homologs.

Migration. Recent studies have found that the concentration of C_2 through C_7 alkanes show a large decrease across a siltstone (source)–sandstone (reservoir) boundary for the C_4 through C_7 molecules but a relatively small decrease for C_2 and C_3. This difference has been attributed to diffusive transport which is faster for the smaller molecules. There also

may be a solubility effect because in these studies the aromatic hydrocarbons, benzene and toluene, had anomalously high migration rates. These aromatics are more soluble in sediment pore waters than the C_2 through C_7 alkanes. Although the combination of diffusion and solution mechanisms can cause hydrocarbon migration across sand-shale boundaries, the evidence is less convincing that extensive vertical migration occurs by these mechanisms within fine-grained shales except for methane and ethane. The more likely mechanism of vertical migration is as an oil or gas phase.

Figure 3 shows the distribution of four C_7 hydrocarbons compared to methane and ethane in the Inigok No. 1 well. A major source of the giant oil accumulations on the North Slope of Alaska is the Kingak shale of Jurassic age which extends from 9260 to 12,210 ft (2822 to 3722 m) in this well. It is overlain by the Pebble Shale and Torok Formations and underlain by the Shublik Formation, both of which may be minor contributors to the oil accumulations. The present-day temperatures of the Kingak Shale range from about 95°C (203°F) at the top to 110°C (230°F) at the bottom. The highest concentrations of the C_7 hydrocarbons, *n*-heptane, methylcyclohexane, 1,1-dimethylcyclopentane, and toluene, are in the Kingak shale source rock. The threshold of intense generation for these hydrocarbons starts at about 8000 ft (2.4 km) in the overlying Torok Formation. There is no evidence of downward migration of these C_7 hydrocarbons and only a suggestion of upward migration of the most water-soluble hydrocarbon, toluene. Likewise, ethane shows only small increases in concentration outside of the Kingak shale source. Yet, it is known that these hydrocarbons do migrate from source to reservoir rocks in sedimentary basins. This suggests that the major migration pathways are not through the fine-grained rock, but along more permeable avenues such as faults, fractures, bedding planes, unconformities, sheet sands, and other large pore openings that permit migration as an oil or gas phase. Migration by this last mechanism can occur even in microfractures with openings of a few hundred nanometers. Such microfractures have been observed frequently in source rocks buried within the oil generation range.

For background information *see* Hydrocarbon; Kerogen; Petroleum; Petroleum, origin of; Petroleum geology in the McGraw-Hill Encyclopedia of Science and Technology.

[JOHN M. HUNT]

Bibliography: J. M. Hunt, Generation and migration of light hydrocarbons, *Science*, 226:1265–1270, 1984; J. M. Hunt, *Petroleum Geochemistry and Geology*, 1979; K. LeTran, J. Connan, and B. Van Der Weide, Problems relating to hydrocarbon formation in the Aquitaine Basin, in B. Tissot and F. Bienner (eds.), *Advances in Organic Geochemistry*, pp. 761–789, 1974; D. Leythaeuser, R. G. Schaefer, and H. Pooch, Diffusion of light hydrocarbons in subsurface sedimentary rocks, *AAPG Bull.*, 67:889–895, 1983.

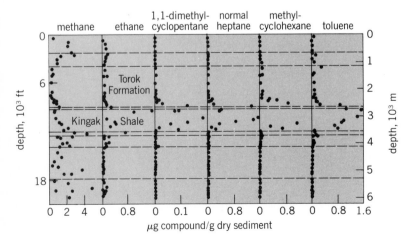

Fig. 3. Distribution of four C_7 hydrocarbons, 1,1-dimethylcyclopentane, normal heptane, methylcyclohexane, and toluene, compared to that of methane and ethane in the Inigok No. 1 well drilled in the Ikpikpuk Basin, North Slope of Alaska.

Physiological ecology (plant)

Plants that live on land have evolved a waxy cuticle that covers all aerial organs (Fig. 1) to prevent desiccation in air. A continuous cuticle, however, would prevent plants from absorbing carbon dioxide (CO_2) from the atmosphere for photosynthetic production of sugars in the mesophyll. Neither modern chemistry nor natural selection has been able to develop a material that restricts passage of water vapor but not of CO_2. This fundamental dilemma of plant life is resolved by development of stomata, which provide regulated pathways for diffusion of CO_2, water vapor, and other gases through this impermeable cuticle. Opening and closing of stomata are tightly regulated by guard cells, which respond to many aspects of the environment. Recent advances in the study of guard cell responses to light, humidity, and the phytohormone abscisic acid (ABA) allow partial characterization of the mechanism and ecological consequences of stomatal regulation.

The stomatal pore is defined by two guard cells (Fig. 1) with specialized cell wall thickening. The result of these wall properties is to cause guard cells to change shape, opening the pore between them, when their turgor increases due to osmotic adjustment. The ease with which a substance diffuses along its concentration gradient, the concentration gradient being the driving force which causes diffusion to occur, is known as conductance. The pore diameter determines the conductance for diffusion of gases through each pore. There are many such pores in the lower, and often upper, surface of leaves.

Conductance must be high when photosynthetic rate is high to allow entry of sufficient CO_2. It must be low when photosynthetic rate is low (for example,

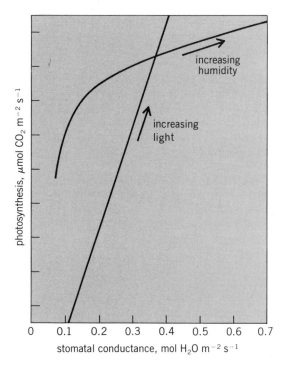

Fig. 2. Relationship of photosynthetic rate and stomatal conductance in a soybean leaf as light or humidity is increased. Light directly affects both guard cells and mesophyll cells so the two processes are well coupled. Humidity affects only the guard cells so that this coupling is disrupted, although photosynthesis increases somewhat with conductance because the supply of intercellular CO_2 increases.

due to darkness or plant stress) to minimize water loss that is not associated with carbon fixation. The coordination of conductance and photosynthesis is achieved through independent responses to the same environmental signals by the guard cells which control conductance and by the mesophyll cells which carry out photosynthesis.

Responses to light. The primary stimulus for stomatal opening is light. Both photosynthesis and stomatal conductance respond to photosynthetically active light (400–700 nanometers), resulting in functional coupling of the two processes as light increases (Fig. 2). In the field this results in characteristic daily time-courses, in which photosynthesis and conductance increase together in the morning and decline together with the setting sun.

Common energy conservation reactions in guard cells and mesophyll cells provide a mechanism for this coupling. Guard cell chloroplasts convert light energy into chemical energy by using the same pigments and the same light-driven electron transport and adenosinetriphosphate (ATP) synthesis reactions as mesophyll chloroplasts. In mesophyll cells this energy is used to fix atmospheric CO_2, while in guard cells it is used for osmotic adjustment to produce turgor changes and stomatal movement.

Guard cells have an additional, high-sensitivity light response that is specific for blue light (400–470 nm). It is not mediated by chloroplasts, and is not matched by a parallel response in the mesophyll

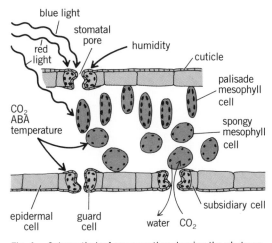

Fig. 1. Schematic leaf cross section showing the photosynthetic mesophyll, the epidermis, the impermeable cuticle, and the stomatal pore which allows exchange of gases (light arrows) between leaf and air. The pore is defined by two guard cells, surrounded by specialized subsidiary cells and less specialized epidermal cells. Photosynthetic reactions occur in chloroplasts in guard cells and mesophyll cells. Stomatal conductance and photosynthesis are differentially affected by various environmental parameters (heavy arrows). The solid circles in the two types of mesophyll cells and the guard cells represent the chloroplasts.

(Fig. 1). As a result, conductance and photosynthesis can be experimentally uncoupled by irradiation of a leaf with pure blue light. This response is activated by very low light intensities. It has been suggested that such light stimulates stomatal opening at dawn, when the diffuse light is enriched with blue wavelengths. Since conductance responds more slowly than photosynthesis, this could enhance carbon fixation in the early morning. Delayed stomatal opening could otherwise inhibit photosynthesis by restricting the uptake of CO_2.

Responses to humidity. Humidity, like blue light, directly affects conductance but not photosynthesis. The resulting lack of coupling (Fig. 2) provides a mechanism for suppressing conductance, even though photosynthetic capacity is high, during periods of low humidity and high potential water loss. During the afternoon, elevated temperatures cause the relative humidity to decrease and the evaporative demand (water vapor pressure difference between leaf and air) to increase. Conductance is often suppressed under these conditions in spite of the opening stimulus of high light. This may restrict entry of CO_2 and cause a midday depression of photosynthesis, even though mesophyll photosynthetic capacity is not directly affected. Under natural conditions the daily time-course of conductance may often be very accurately predicted by using only values of light and evaporative demand.

The mechanism of the humidity response is not well understood. Changing humidity is apparently perceived as a change in water loss and water content of the leaf epidermis. Recent kinetic analyses of stomatal responses to step changes in humidity suggest that turgor changes caused by passive water loss from guard and subsidiary cells do not directly cause the observed stomatal responses (the resulting small conductance changes are often in the wrong direction), but rather serve as the stimulus which induces guard cell osmotic adjustment. Aside from the initial stimulus, the stomatal response to humidity is apparently mediated by the same metabolic processes and osmotic adjustment that mediate stomatal responses to light and other parameters. This area is a subject of considerable current research interest.

Responses to temperature. The stomatal responses to humidity and to temperature are difficult to separate because humidity and evaporative demand changes with temperature. When the two factors are perturbed independently, the response to humidity usually predominates and the direct response to temperature is relatively insignificant.

The most important stomatal responses to temperature involve temperature extremes. Conductance is suppressed during the day following an injurious chilling night, as is photosynthetic capacity. Thus the two processes remain well coupled. On the other hand, extremely high leaf temperatures cause conductance to increase to very high levels. It is not known whether this represents a regulated stomatal response which results in increased evaporative

cooling, or whether it represents high-temperature damage and a complete loss of stomatal control. Exposure of isolated epidermis to such elevated temperatures often results in guard cell death.

Responses to CO_2. Stomatal conductance increases as CO_2 inside the leaf is depleted. Stomata of C_3 plants (which fix CO_2 via the reductive pentose pathway) exhibit such strong responses to light that the response to intercellular CO_2 is almost completely obscured under natural conditions. In C_4 plants (which fix CO_2 via the C_4-dicarboxylic acid pathway and maintain a lower concentration of CO_2 inside the leaf), separate responses to light and CO_2 are observed. Stomatal responses to intercellular CO_2 may enhance the coupling of conductance and photosynthesis, since photosynthesis without adequate stomatal opening would deplete intercellular CO_2, inducing further opening.

Stomatal sensitivity to CO_2 varies not only between species but between individuals exposed to different conditions. Water deficits, chilling injury, and other stresses sensitize stomata to CO_2. These changes in sensitivity are often associated with stress-induced synthesis of abscisic acid, and can be induced experimentally by application of synthetic ABA.

Role of abscisic acid. It has recently been shown that guard cells contain the enzymes necessary to metabolize ABA (Fig. 3). Guard cells are known to accumulate ABA when it is presented to intact

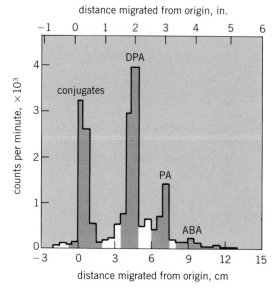

Fig. 3. Metabolism of abscisic acid (ABA) in isolated guard cells of *Commelina communis*. Cells were incubated in radioactive (^{3}H) ABA for 12 h and were extracted with methanol. Radioactive metabolites were separated by using thin-layer chromatography. Phaseic acid (PA) and dihydrophaseic acid (DPA) are the major short-term metabolites of ABA. Conjugates are biologically inactive forms of ABA, PA, or DPA bound to glucose. Metabolism of ABA was sufficiently vigorous that little unmetabolized ABA remained in the guard cells, even though excess ABA was present in the external medium. (*From D. A. Grantz et al., Metabolism of abscisic acid in guard cells of Vicia faba L. and Commelina communis L., Plant Physiol., 78:51–56, 1985*)

leaves or isolated epidermis. The outside surface of the guard cell plasma membrane contains a specific binding site for ABA. It is not known whether ABA exerts its influence on conductance from inside or outside the guard cells, nor is it known how ABA causes loss of guard cell turgor and stomatal closure.

Recent evidence suggests that ABA suppresses both conductance and photosynthesis. Since synthesis of ABA is induced by plant stress, this could account for the coupling observed as both functions are inhibited by chilling or water deficits. In addition to a likely role in stress responses, there is some speculation that ABA may have a role in other stomatal responses. The humidity response may be triggered by transient changes in water status of the epidermis. Localized synthesis and catabolism of ABA in guard cells could modulate guard cell osmotic adjustment and turgor and thus regulate stomatal conductance.

Conclusions. Current concepts of the coupling of conductance and photosynthesis invoke independent but parallel environmental responses that occur simultaneously in the epidermis and the mesophyll. The ecological significance of this coordination is that water loss through open stromatal pores is tightly coupled to fixation of CO_2, so that over a day, plants lose minimal water per unit of CO_2 fixed into sugars. The uncoupling of conductance and assimilation that can be demonstrated by manipulating blue light or humidity reflects guard cell responses that are not matched by parallel responses in the mesophyll. This regulatory scheme suggests that stomatal responses may be susceptible to genetic manipulation independently of photosynthetic responses, with the potential for improved adaptation of crop plants to specific environments.

For background information see ABSCISIC ACID; EPIDERMIS (PLANT); LEAF; PHOTOSYNTHESIS; PHYSIOLOGICAL ECOLOGY (PLANT); PLANT-WATER RELATIONS in the McGraw Hill Encyclopedia of Science and Technology.

[DAVID A. GRANTZ]

Bibliography: D. A. Grantz et al., Chloroplast function in guard cells of *Vicia faba* L.: Measurement of the electrochromic absorbance change at 518 nm, *Plant Physiol.*, 77:956–962, 1985; D. A. Grantz et al., Metabolism of abscisic acid in guard cells of *Vicia faba* L. and *Commelina communis* L., *Plant Physiol.*, 78:51–56, 1985; C. Hornberg and E. W. Weiler, High affinity binding sites for abscisic acid on the plasmalemma of *Vicia faba* guard cells, *Nature*, 310:321–324, 1984; D. Laffray et al., Moist air effects on stomatal movements and related ion content in dark conditions: Study on *Pelargonium* × *hortorum* and *Vicia faba*, *Physiol. Veg.*, 22:29–36, 1984.

Placer mining

Recent developments in beach and offshore mining operations have been concerned mainly with achieving greater productivity from conventional dredger types through faster digging capabilities, improved recoveries of heavy minerals in the finer-particle sizings, and the more efficient utilization of power, floor space, and height. Benefits from research carried out in other fields include the increased precision of positioning offshore and the introduction of automatic control systems to regulate dredging according to predetermined programs. Environmental considerations have led to an overall reduction in noise level and in better restoration practice.

Dredging. The near depletion of easily won material in sheltered locations has forced miners to extend offshore operations further into the open sea. Two large alluvial tin dredgers were constructed for offshore dredging in Indonesian waters around the mid-1970s. One of these was the largest offshore tin dredger ever built, with hull dimensions of 360 × 98 × 21 ft (110 × 30 × 6.5 m) for dredging to 148 ft (45 m) below the waterline. It was equipped with large (225 gallons or 0.85 m^3) buckets and was provided with a hydro pneumatic compensating system in the bucket ladder suspension to achieve a high degree of availability in the prevailing wind and sea conditions. The other, with hull dimensions of 354 × 105 × 15 ft (108 × 32 × 4.75 m), was designed for dredging to 164 ft (50 m) below the waterline. This model favored simplicity of design and smaller (165 gallons or 0.652 m^3) buckets for a lower nominal dredging rate. A special feature of its design was provision to obtain maximum bucket fill over a wide range of angles at the digging face.

Recent studies based upon the performance of these dredgers have suggested that 164–197 ft (50–60 m) below waterline may be the economic limit for digging with conventional bucketline dredgers. Onshore, a similar deep-digging dredger was completely rebuilt to incorporate an extended bow gantry, a longer digging ladder, new bucketline, and new electrical system. A new hydraulic control system and other modifications make it one of the deepest-digging (148 ft or 45 m below pond level) gold dredger in the world. Preliminary engineering studies have begun for reconstructing another dredger to dredge to around 190 ft (58 m) below pond level, but it is thought that at least some departure from conventional designs will be needed.

Overall, there is little doubt that a new generation of dredgers is needed for depths greater than 104 ft (50 m). The discovery of vast quantities of polymetallic modules on the deep-ocean floor has provided the incentive for multinational groups to engage in developing other methods of offshore mining. Research studies have considered a variety of design possibilities, and pilot-scale exercises have demonstrated the technical feasibility of several alternatives. Of these, the three most practicable for offshore placer mining are the continuous dragline bucket system (Fig. 1a); mining by remote control from submersible units operating from the sea floor to elevate feed to a floating treatment plant (Fig. 1b); and systematized clamshell dredgers (Fig. 1c). Crawler-type platform dredgers (Fig. 2) may find ap-

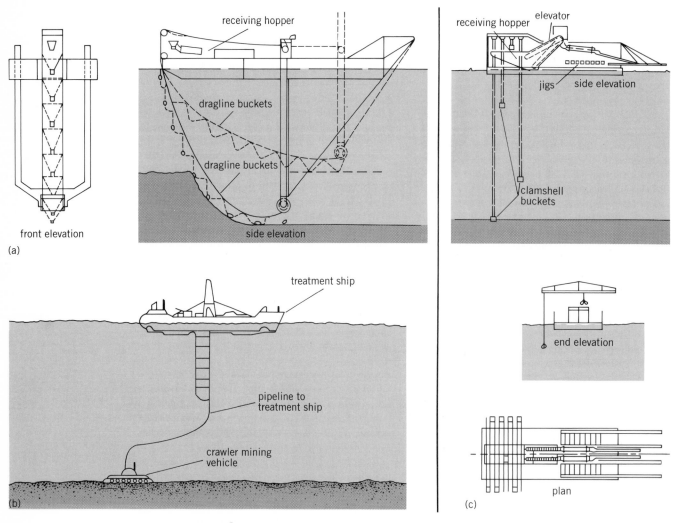

Fig. 1. Dredging innovations for offshore placer mining. (*a*) Arrangement of continuous dragline dredge; side elevation shows two digging positions of dragline. (*b*) Submersible dredging arrangement, remotely controlled. (*c*) Arrangement of projected multigrab dredge; end elevation shows one clamshell drag; plan shows four grabs on either side of vessel, one emptying.

plication for mining in some shallow waters where the seas are too rough for presently available dredgers to operate.

Noise pollution has also been studied, and modern dredgers are constructed with totally enclosed pressure-lubricated reduction units and sealed bucket pins to reduce noise levels. In addition, other means of reducing noise levels are now applied to the older dredgers.

Hydraulic suction dredgers. Modern beach mining operations are conducted largely by suction dredging using either the cutter-head or bucket-wheel systems. The two systems compete directly in most applications, and recent innovations reflect the progress made in overcoming their respective inadequacies.

Cutter-head systems. Cutter suction dredgers are large-capacity units originally designed for reclamation work and surface stripping. They were unique in their ability to mine and transport sediments in a slurry form over considerable distances.

In mineral sands mining, however, they suffered the disadvantages of high power costs because of fluctuating and generally low slurry densities, and excessive pond losses of valuable heavy minerals. However, recent design has borrowed from bucket-wheel technology, and a high-speed cutter wheel design has been developed that force-feeds the slurry into a suction duct mounted directly between cutter blades located on either side of a revolving drum. The cutter action pulls the dredge into the face, thus eliminating the need for costly spudding (anchoring pivot) systems. The high slurry density thus maintained reduces pond losses because the cutters clean up at basement equally on both forward and reverse swings.

Again, digging depths appear to be limited by economic considerations to about 164 ft (50 m) below the waterline, and although various proposals have been made to increase the capabilities of conventional dredges, none has been tested fully. Presently, the most favored concept is to use sectionally

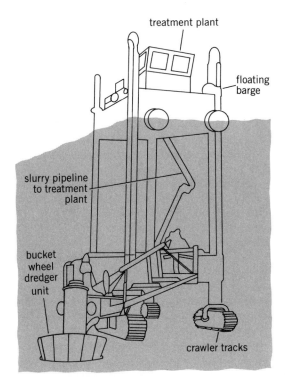

treatment plant

floating barge

slurry pipeline to treatment plant

bucket wheel dredger unit

crawler tracks

Fig. 2. Mobile crawler platform dredger.

articulated ladders in place of standard one-piece ladders. Because of reduced bending moments, such ladders could provide equal strength by using smaller steel sections, thus reducing their weight and allowing for expansion.

Bucketwheel systems. Improvements to bucket-wheel dredger design relate mainly to improving cost effectiveness by the use of modular construction, increased maneuverability, and flexibility of control. The system is comparatively new, and designers are still expanding its field of application.

One innovation, a cutter module, is a complete unit mounted on a short ladder section that can be bolted to any desired length of ladder without concern for alignment. Ladder sections are added, removed, or angled to suit any particular digging depth. This spud carriage system provides a close control of cutting advances and positioning at the face. Other new designs provide a variety of bucket-wheel configurations to suit the type of material being dredged, with special attention addressed to improving accessibility for maintenance and flexibility of control.

Positioning. Recent offshore activites in such fields as hydraulic engineering, hydrographic surveys, and oil exploration have resulted in the development of electronic position-fixing systems that are accurate, reliable, and simple to operate. A variety of systems is now available to offshore placer operators based upon various measurements between the mobile station (dredger or survey vessel) and one or more fixed stations at known positions onshore. In a project involving offshore exploration for tin and heavy minerals, an accuracy of ±3.3 ft (±1 m) was

achieved in the Andaman Sea, off Thailand, using a microwave range positioning system and three onshore transponders set up on permanent marks.

Treatment. The recovery efficiency of conventional gravity concentrating devices falls off sharply with particles smaller than 100 micrometers in size and, until recently, such particles were thought to be largely irrecoverable in gravity treatment practice. However, operators now expect high recoveries of gold and cassiterite down to 70 μm, and significant recoveries down to 20–30 μm for gold and 50 μm for cassiterite. It has been suggested that established practices elsewhere in the mineral industry may physically recover gold down to 5 μm in size and could find application for fine alluvial gold.

Of the newer machines developed for recovering fine gold, centrifugal concentrators have achieved 80–90% recoveries of fine gold down to 40 μm in ideal conditions. However, disadvantages such as high maintenance costs, intermittent operations, large clean water requirements, and low upgrading ratios, outweigh their advantages for the most part.

The most important advances in fine-particle recovery systems resulted from the development of spiral concentrators in a wide range of configurations with separating characteristics to suit individual applications. Space savings have also been significant, and the use of polyurethane-lined fiberglass and multiple starts around a common column has resulted in major reductions in spiral plant size and in the amount of structural support required. The effectiveness of the new generation of spirals in the recovery of fine tin is illustrated in the table for a recently developed spiral series.

Microcomputers and automatic control systems. The impact of computerization has been felt at all levels in the placer mining industry, and computerized models have been developed for all aspects of evaluation, mining, processing, and financial analysis. Microcomputers are now commonplace, and manufacturers have attained such a level of sophistication that a selected computer can be expected to satisfy operational requirements for the full life of an operation.

In modern mine planning, computer modeling is performed on encoded drill hole information, including both physical and chemical relationships. A

Distribution of tin in package plants concentrate*

Size, μm	Sn distribution, %	
	Concentrates	Tailings
+850	20	80
600	95	5
425	97	3
300	96	4
150	95	5
75	95	5
−75	63	37
Total distribution:	93	7

*Comparative jig recoveries are made only in the +150-μm size range.

conceptual dredging path is formulated, then digitized, and superimposed on the geological model. In-place grades and quantities of ore and overburden are calculated for each strip or block and transferred to mine scheduling routines to determine the most efficient mining sequence. The modeling process incorporates facilities for reflecting geomorphological interpretations and boundary conditions; the financial subsystem then evaluates the likely returns on investment for the various alternatives.

Thus, provided the data are sufficiently detailed and accurate, dredgers can now be equipped with control systems and instrumentation to execute any dredging operation according to a predetermined model. The control systems work in either semiautomatic or automatic mode to minimize losses and inefficiences arising out of human error. Geophysics as a tool in both exploration and mine planning is being used increasingly to supplement drilling data and reduce the cost of mine sampling programs.

For background information *see* DREDGE; GEOPHYSICAL EXPLORATION; MINING OPERATING FACILITIES; PLACER MINING; POSITION FIXING in the McGraw-Hill Encyclopedia of Science and Technology.

[EOIN H. MACDONALD]

Bibliography: A. G. Fricker, Metallurgical efficiency in the recovery of alluvial gold, *Proc. Aust. Inst. Min. Met.*, no. 289, February 1984; J. A. Hewitt, Present and future trends in capabilities and design of sea-going tin dredges, *Oceanology, International*, 1978; E. J. McLean and R. H. Goodman, *Modular Processing Plants to Improve Mineral Recovery at Indonesian Tin Mines*, Vickers Australia Ltd., Wyong, 1985.

Plant mineral nutrition

This article discusses new findings in evaluation of soil nutrient bioavailability and the functions of the chloride anion in plant nutrition and osmosis.

Soil nutrient bioavailability. A new method for evaluating soil nutrient bioavailability is based on measurement of the mechanisms and kinetics that are involved in moving nutrients from the soil into the plant root, where they can be used by the plant. Before they are available for uptake by the root, most nutrients must move through the soil to the root surface by mass flow in the convective flow of water and by diffusion. The kinetics of these processes together with the kinetics of nutrient uptake as related to concentration are combined with root size and rate of root growth in a model that can be used to calculate the amount of nutrient taken up by the plant. This method provides sensitivity measurements to determine which mechanism has the greatest effect on nutrient uptake.

Usually soil chemists have studied soil nutrient supply characteristics in order to measure the quantity of nutrients present in various forms in the soil. Nutrient uptake by plants, on the other hand, has been studied by plant physiologists, frequently using excised roots to measure the effect of nutrient concentration in solution on the rate of uptake per

unit surface or influx. Nutrient uptake by plants involves processes occurring in both the soil and the plant. It also is affected by the morphology and growth rate of the root system which provides the sink for nutrient uptake from the soil.

Soil nutrient supply. Plant roots absorb nutrients that are in soil solution, so that the initial concentration in the soil solution, C_{li}, is an important parameter for determining rate of supply. As nutrients are removed from solution, the concentration of nutrients absorbed on the solid phase, C_{si}, that are in equilibrium with C_{si} will move into solution. The capacity of the solid phase to supply nutrients to the solution is expressed mathematically as $\partial C_{si}/\partial C_{li}$ or b, the buffer power, a soil property which gives the quantity of nutrient supply when multiplied by C_{li}.

The amount of nutrient reaching the root by mass flow is determined by the rate of water absorption and the concentration of nutrients in the water. If mass flow does not supply a nutrient as fast as it is absorbed by the root, the concentration in the solution at the root surface will decrease and a concentration gradient extending out from the root will develop. The nutrient will diffuse along the concentration gradient toward the root. The effective diffusion coefficient, D_e, describes the rate at which the ion will diffuse to the root in a particular soil. It is calculated as though the soil were a uniform medium, and is much smaller than the value for diffusion in water for several reasons: (1) the ion only diffuses through the water phase, which occupies from 10 to 30% of the soil volume; (2) the diffusion path is tortuous, reducing the rate of diffusion (D_e), and (3) D_e is reduced by the buffer power of the soil. When evaluating soil nutrient supply by describing mathematically the processes and reactions occurring when nutrients are absorbed from soil by plant roots (a mechanistic approach), the soil parameters to measure are C_{li}, b, and D_e.

Root absorption of nutrients. Nutrient uptake from the soil is influenced by the size and morphology of the root system and by the relation between uptake rate per unit root surface, influx, and concentration at the root surface. The relation between phosphorus influx by corn (*Zea mays*) and the phosphorus concentration in solution can be described by the "influx" equation below, where I_n is net influx, I_{max} is

$$I_n = \frac{I_{max}(C_l - C_{min})}{K_m + C_l - C_{min}}$$

maximal influx, K_m is the concentration where net influx is one-half I_{max}, C_l is the concentration in the solution, and C_{min} is the concentration in solution where I_n is 0.

Uptake increases as the amount of roots increases. Root growth is usually exponential with time during early growth and later becomes linear. The maximum amount of roots of annual species are present when the plant changes from vegetative to reproductive growth.

Modeling uptake. When roots absorb phosphorus, which primarily diffuses to the root, a concentration

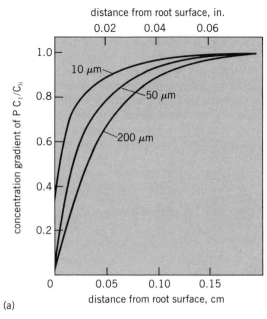

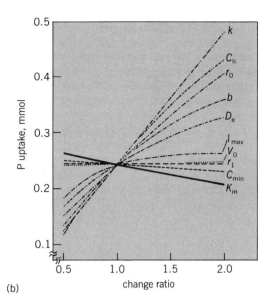

(a)

(b)

Fig. 1.   Evaluation of soil nutrient bioavailability. (a) P concentration gradient perpendicular to root after 15 days' uptake by corn roots. Values on curves are radii of roots or root hairs. Effective diffusion coefficient $D_e = 3.29 \times 10^{-9}$ cm^2 s^{-1}. Buffer power $b = 130$. Initial concentration in soil solution $C_{li} = 14$ μM. (b) Sensitivity analysis of model parameters for P uptake by corn roots.

gradient perpendicular to the root develops. This gradient is affected by the soil and root properties. Figure 1a shows the calculated gradient after 15 days' uptake of phosphorus (P) by corn roots or root hairs with radii of 10, 50, and 200 micrometers. Radial geometry for diffusion causes root radius r_0 to have an effect on the gradient. Mathematical models calculate this gradient, and from the concentration in solution, C_1, at the root surface and the relation of influx to C_1 they calculate uptake. The models have been verified for plants grown both in controlled climate chambers and in the field.

Sensitivity analysis. It is useful to know which properties of the uptake system have the greatest effect on uptake. A sensitivity analysis, where each parameter is changed independently while the others remain at a constant value, conducted for P uptake gave the results shown in Fig. 1b, where k is the root growth rate constant and had the greatest effect, as would be expected. Root radius, r_0, which affected root surface area, was a sensitive parameter also, as were the soil supply parameters of C_{li}, b, and D_e. However, the parameters describing uptake—I_{max}, K_m, and C_{min}—were less sensitive. Two other parameters—rate of water uptake, v_0, and half-distance between root axes, r_1—also had little effect; mass flow supplies very little phosphorus to the root, and the rate of diffusion was so slow for phosphorus that roots did not compete for phosphorus.

An understanding of the principal mechanisms governing nutrient uptake should make it easier to solve nutrient supply problems of crops growing in soil.   [STANLEY A. BARBER]

Chloride.   Chlorine is an essential element required in minute quantities for plant growth. Found almost universally in soils and plants, it occurs primarily as the chloride anion (Cl$^-$). Despite its presence in plants, Cl$^-$ was not considered essential for growth until 1954 when it was demonstrated that tomato species required Cl$^-$. Later nine other species were shown to have this requirement. Cl$^-$ is readily absorbed from the soil by plants and, depending upon its concentration in the soil, plants usually accumulate from 50 to 500 mmoles Cl$^-$/kg dry weight without detrimental effects. Unlike other micronutrients, Cl$^+$ is relatively nontoxic. However, some plants growing on saline soils accumulate much higher concentrations, which cause injury and even death. Chloride anion deficiencies are rare in plants, but they can occur if Cl$^-$ concentrations in the plant tissue fall below 20 mmoles/kg dry weight.

Functions.   Chloride has two major functions in plants: nutritional and osmotic. As a micronutrient, Cl$^-$ is essential as a cofactor in the oxidation of water in photosynthesis and as an activator of enzymes. Photosynthesis is the process by which green plants utilize light energy, CO$_2$, and water to synthesize organic constituents. The overall process involves two reactions: the oxidation of H$_2$O to form O$_2$ and H$^+$, and the reduction of CO$_2$ to produce carbohydrates. The chloride anion is apparently required in the splitting of the H$_2$O molecule. The actual mechanism is still unknown, but recent evidence indicates that binding of Cl$^-$ to chloroplast membranes is necessary to activate the O$_2$-evolving enzyme. These membranes contain the photosynthetic pigments of the chloroplast as well as the enzymes required for the primary light-dependent reactions. At least three other plant enzymes appear to require Cl$^-$ for optimal activity: α-amylase, asparagine synthetase, and adenosinetriphosphatase.

The second major role of Cl$^-$ is as a counterion for cation transport and as an osmotic solute. As a counterion, Cl$^-$ maintains electrical charge balance for the uptake of essential nutrient cations. In addition, it contributes to the osmotic balance of the plants. Water molecules move with Cl$^-$ and other solutes accumulated by the cells. Because Cl$^-$ readily moves into and out of cells and organelles, for example chloroplasts, and is not metabolized, it helps to control the amount of water and the water pressure in organelles, cells, and tissues. The pressure of water against the cell wall is called turgor pressure. Water continues to flow into the cell until the water potential in the cell is equal to that outside the cell. If the external solution becomes saline, water flows out of the cells until the protoplasm shrinks and pulls away from the cell wall. These cells are plasmolyzed and will die unless they take up water. When readily absorbable ions like potassium cation (K$^+$) and Cl$^-$ are available, the cells rapidly take up the ions and water and the cells become turgid again. This process is known as osmotic adjustment.

One important process that depends upon Cl$^-$ is the opening and closing of leaf stomata which regulate transpiration. Stomata open when water enters the guard cells causing them to swell apart, and they close when water moves out. Water flows in and out with the accumulation and loss of Cl$^-$, K$^+$, and in some cases malate, an organic acid ion. As far as is known, all species utilize some Cl$^-$ as a counterion for K$^+$. Some species, like onion, have an absolute requirement for Cl$^-$; others, like corn, utilize both Cl$^-$ and malate. Unlike K$^+$ and Cl$^-$ which must be transported into the cell, malate is an organic solute that is synthesized within the cell.

The rhythmic leaf movements of a number of legumes are caused by similar changes in cell turgor. K$^+$ and Cl$^+$ fluxes into and out of special cells known as motor cells provide the osmotic gradient for the uptake or loss of water. The swelling and shrinking of these cells cause the leaves to move.

Toxicity. Although plants can tolerate relatively high levels of Cl$^-$, some species may accumulate toxic concentrations in the tissue when grown on saline soils. Most nonwoody crops tolerate excessive levels of Cl$^-$ reasonably well. One exception is soybean, which includes some varieties that are sensitive. The difference between sensitive and tolerant varieties is that tolerant varieties restrict or moderate Cl$^-$ transport to the shoots. In sensitive varieties, Cl$^-$ accumulates in leaf cells to toxic concentrations. Initial symptoms of Cl$^-$ toxicity appear at the tip and along the margins of leaves where the tissue begins to dry up and die. Many woody plant species, such as fruit trees, shrubs, and vines, are susceptible to Cl$^-$ toxicity. However, susceptibility varies among species and even among varieties. Like soybean varieties, rootstocks also differ in their ability to prevent or retard Cl$^-$ transport to the shoots. This characteristic has been exploited with

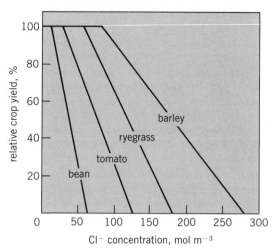

Fig. 2.   Relative yields of four crops as a function of chloride concentration in the soil solution.

tree crops by grafting sensitive scions onto rootstocks that exclude Cl$^-$ from the shoot.

While many crops are relatively resistant to Cl$^-$ toxicity, all crops are susceptible to stunting because of excess salt concentrations in the soil. This effect is caused by the osmotic properties of the various ions in the soil solution which decrease its total water potential. Since Cl$^-$ is usually prevalent in saline soils, it contributes to the osmotic effects. As mentioned above, plants adjust to osmotic stress by absorbing ions from the solution or by synthesizing organic solutes internally. Both processes consume metabolic energy. It is now believed that growth suppression occurs because energy normally utilized for growth is required for osmotic adjustment.

Salinity tolerance. Plants vary considerably in their ability to adjust to osmotic stress and, consequently, the amount of growth reduction caused by Cl$^-$ salts also varies. Approximately 70 different crops have been tested for their tolerance to salinity. Most of the available data were obtained in field plots that were salinized with Cl$^-$ salts of sodium and calcium, so the data can be converted to express tolerances in terms of Cl$^-$ concentration. Typically, salt tolerance data are given in terms of the percent growth reduction expected per unit increase in soil salinity. If growth or yield is plotted as a function of the Cl$^-$ concentration in the soil solution, two-piece linear response lines like those presented for some selected crops in Fig. 2 are obtained. These response lines show that plants tolerate Cl$^-$ up to a threshold level above which growth or yield decreases as a linear function of Cl$^-$ concentration.

For background information *see* OSMOREGULATORY MECHANISMS; PLANT METABOLISM; PLANT MINERAL NUTRITION; ROOT; SOIL CHEMISTRY in the McGraw-Hill Encyclopedia of Science and Technology.

[EUGENE V. MAAS]

Bibliography: S. A. Barber, *Soil Nutrient Bioavailability: A Mechanistic Approach*, 1984; T. C.

Broyer et al., Chlorine: A micronutrient element for higher plants, *Plant Physiol.*, 29:526–532, 1954; N. Claassen and S. A. Barber, Simulation model for nutrient uptake from soil by a growing plant root system, *Agron. J.*, 68:961–964, 1976; J. H. Cushman, Nutrient transport inside and outside the root rhizosphere: Generalized model, *Soil Sci.*, 138:164–171, 1984; E. V. Maas, Salt tolerance of plants, *Appl. Agr. Res.*, 1:12–26, 1986; E. Zeiger, A. J. Bloom, and P. K. Hepler, Ion transport in stomatal guard cells: A chemiosmotic hypothesis, *What's New Plant Physiol.*, 9(8):29–32, 1978.

Plant pathology

Virus infection of plants may lead to a systemic disease in which the virus multiplies and spreads throughout the entire plant, or to a localized reaction during which the virus is confined to the immediate vicinity of its initial point of entry. In several virus–host plant combinations, resistance takes the form of a hypersensitive reaction, in which necrotic local lesions develop on the inoculated leaves. Even though the virus is restricted to these lesions, its effects extend to other plant parts. When, following the appearance of lesions induced by a first inoculation, other leaves of the plant are challenge-inoculated with the same virus (an inoculation carried out to determine if resistance has been induced), smaller or fewer lesions develop than in leaves of comparable plants that were not inoculated previously. This phenomenon has been termed acquired or induced resistance. Strictly speaking, plants were already resistant before being induced, because upon inoculation the infection remained localized and loss of tissue was limited to a small necrotized area. In tissues with induced resistance, however, further restriction of the virus takes place.

Interest in induced resistance has grown during recent years. It has become clear that such resistance can be acquired after infection with different pathogens and perhaps treatment with chemicals. It can be effective not only against various viruses but often also against pathogenic fungi, bacteria, and even insects. The physiological and biochemical mechanisms underlying this nonspecific form of protection are being actively investigated in order to explore the possibilities of increasing general resistance to pathogens in crop plants under field conditions.

Physiological aspects. A systemic effect of localized virus infection was first reported in 1952. Resistance to carnation mosaic virus was discovered in upper leaves of a clone of carnation following inoculation of lower leaves with a local-lesion-inducing strain of the virus. Fewer lesions developed in these upper leaves than in upper leaves of control plants, the lower leaves of which had not been inoculated. Further work established a similar phenomenon in tobacco, bean, cowpea, and several other plant species. For ease in testing, both the inducing and the challenging pathogen (often the same) were chosen so that they produced local lesions which are easy to count and measure. Two types of induced resistance were distinguished: localized and systemic acquired resistance. In tobacco, a high degree of resistance developed in a 0.04–0.08-in. (1–2-mm) zone surrounding primary lesions. No lesions, or at most a few tiny ones, developed in this area following challenge inoculation. This localized acquired resistance with its effect primarily on lesion number appeared to contrast with systemic acquired resistance, in which primarily lesion size was reduced. Inoculation of half-leaves or of lower leaves on a plant induced a high level of resistance in the opposite half-leaves and in upper leaves, respectively. Upon challenge inoculation, lesions were consistently only one-fifth to one-third as large as were lesions in noninduced leaves, and they were usually, but less consistently, fewer in number.

In other experiments, systemic acquired resistance was reported mainly in terms of lesion numbers. However, since very small lesions are difficult to discern, apparent reductions in lesion numbers can be a consequence of lesions not progressing far enough to result in visibly detectable necrosis. Although reductions in lesion size and number may not be correlated, it is commonly assumed that all reports are dealing with the same general type of induced resistance and that the effects of localized and systemic acquired resistance are quantitatively rather than qualitatively different. Induced resistance can thus be defined as an increase in the capacity of the plant to limit pathogenic attack, as manifested by a reduction in the number or size of lesions developing upon challenge inoculation. This induced resistance is different from cross-protection, in which tissues infected with one strain of a virus are immune to superinfection with related strains of the same virus.

Physical injury does not induce acquired resistance, but induced resistance is not virus-specific and may develop as a response to viruses, fungi, or bacteria. Once induced, it is effective against most, though not all, viruses inducing local lesions, although when challenge inoculations are made with the same virus as the one used for induction, resistance is stronger than when inducing and challenging pathogen are different. Even when the challenging virus does not produce a localized reaction but invades the plant systemically, induced resistance can be manifested through a reduction in the rate of virus multiplication or the extent of symptom development.

Almost any virus, fungus, or bacterium that induces necrosis in tobacco induces systemic effects that are virtually indistinguishable. This indicates that induced resistance is part of the general, nonspecific defense reaction of plants to pathogens that induce localized infections. Substantial resistance may also be induced by viruses causing mosaic symptoms, yellow chlorotic lesions, or symptomless starch-containing lesions. Necrosis is not a prerequisite for the development of induced resistance, as also evidenced by experiments in which resistance was induced chemically without visibly harmful side effects. Various polyanions when injected into leaves reduce lesion numbers and sometimes also lesion size upon challenge inoculation. Benzoic acid derivatives, notably salicylic acid (aspirin), induce a resistance in tobacco that is characterized primarily by reduced lesion size rather than number. Upon injection with these chemicals, the effects are confined to the treated leaves, but protection is systemic when plants are watered repeatedly with the compounds.

In Samsun NN tobacco reacting hypersensitively to tobacco mosaic virus, induced resistance typically develops in 2–3 days, concomitant with the formation of the necrotic lesions. These primary lesions slowly enlarge up to 7–10 days, and the extent of acquired resistance increases until lesion growth ceases. The conditions under which the plants are grown do not affect induction as long as the resistance mechanism remains activated. Once induced, acquired resistance often persists for the life of the plant. Inoculations additional to the primary one barely increase resistance further. However, whereas plants grown from seed derived from once-inoculated plants do not express acquired resistance, seed transmission of induced resistance has been reported under conditions when parent plants had been repeatedly induction-inoculated.

Removal of the primarily infected leaf 2–3 days after inoculation does not greatly reduce the level of acquired resistance attained, nor does removal of leaves above the inoculated ones after approximately the same lag time interfere with the development of acquired resistance in the excised leaves. The signal responsible for the induction of acquired resistance is transported in the phloem: in tobacco, cutting the midvein of an upper leaf prevents the development of resistance distal to the cut. Likewise, killing a section of the petiole of each inoculated lower leaf with boiling water prevents development of resistance in upper leaves. The resistance-inducing substance moves both up and down the stem, although the extent of resistance induced is stronger in upper than in lower leaves. The substance can be transmitted by grafting. By using different tobacco cultivars as rootstock and scion, resistance was induced in scions when rootstocks were inoculated with a virus causing a hypersensitive reaction. An amphidiploid of *Nicotiana glutinosa* and *N. debneyi* appears to possess induced resistance constitutively. Upon inoculation with tobacco mosaic virus or tobacco necrosis virus, it produces very few, small local lesions. When uninfected amphidiploids are used as rootstock, scions of tobacco become systemically resistant. The amphidiploid thus possesses a nonspecies-specific resistance-inducing substance. The nature of this effective principle has not been elucidated, and it is not known whether it is the same as the one occurring in *N. tabacum*.

Biochemical aspects. Various treatments that affect lesion size after a primary inoculation influence lesion enlargement similarly upon challenge inoculation of induced tissues. On these grounds, it is taken that the same mechanism of virus localization operates under both conditions. In Samsun NN tobacco infected with tobacco mosaic virus, lesions developing on systemically resistant leaves enlarge more slowly and stop spreading earlier than lesions on noninduced leaves. Virus content is correspondingly reduced. This indicates a stimulation or enhancement of the mechanisms normally functioning in the hypersensitive reaction to limit lesion size. In induced leaves, the localizing mechanism becomes operative earlier than in noninduced leaves or operates at a higher level. However, in genetically different tobacco cultivars viruses causing localized necrosis may fail to induce systemic resistance; or when lesion spread is reduced upon challenge inoculation, virus content is apparently not. It has been suggested that in the latter case necrotization is inhibited rather than virus multiplication. However, virus may not be present farther outside the necrotized area than in primary lesions.

The virus localization mechanisms in developing lesions are not known. The expression of resistance is undoubtedly complex with many metabolic alterations acting in concert for the defense reaction to be effective. During the stage of slow lesion expansion before final lesion limitation occurs, cells in advance of infection become activated and an intense metabolic activity develops. Of the various biochemical changes occurring in the tissues surrounding primary lesions, several extend to the noninoculated parts showing systemic acquired resistance, as follows.

1. There is an increased capacity to synthesize the plant hormone ethylene from its precursor 1-aminocyclopropane-1-carboxylic acid (ACC). At high doses the commercially used growth regulator ethephon, from which ethylene is released in plant tissues, induces resistance, but ACC is only moderately active. In contrast to the effect of other chemicals, the inducing action of ethephon is systemic, and it is suggested that increased ethylene evolution is a determining factor in systemic induction.

2. There is increased peroxidase activity. This enzyme catalyzes the oxidation of various phenolic compounds by hydrogen peroxide. During lesion development, the activities of the enzymes involved in the synthesis of phenolic compounds are greatly stimulated and phenols accumulate before being oxidized. Upon challenge inoculation, enzyme activities are increased to a greater extent, and phenolic compounds are consumed earlier. Products of peroxidase action include ligninlike compounds which can form structural barriers to pathogen spread.

3. Pathogenesis-related proteins are present. These proteins do not occur in healthy plants but accumulate under virtually all conditions in which acquired resistance is induced. In spite of numerous investigations, no function for these proteins has been established. Their presence appears to be a marker reflecting the severe and prolonged stress

accompanying the induction of the resistant state.

4. Callose (β-1,3-linked glucose polymers) accumulates along cell walls. Cell wall thickening due to callose deposition does not appear to impede virus spread but is a marker of wound reactions. Like the presence of pathogenesis-related proteins, it is indicative of the stress history of the plant.

5. There is an increase in content of the hormone abscisic acid. Abscisic acid accumulates likewise under stress conditions and alters the water relations of the plant, thereby affecting susceptibility to infection. No effect on the rate of virus spread has been found so far.

Upon challenge inoculation of leaves with acquired resistance, all normal processes accompanying virus localization in primary lesions are accelerated and increased. The mechanism was already fully operative in the zone of localized acquired resistance because of the reaction of the plant to the primary infection. The ability to react is strongly enhanced in leaves with systemic acquired resistance. Thus acquired resistance manifests itself as an increased capacity to make use of the existing resistance mechanisms.

Agricultural implications. Induced resistance appears to be one of the mechanisms evolved by plants to cope with environmental stresses. It resembles the hardening experienced by plants exposed to gradually lowering temperatures, which enable them to tolerate subfreezing conditions. Similar adaptations to heat, drought, salt, mechanical, or wind stresses are well known to occur whenever plants are exposed to moderately adverse environments. Abscisic acid accumulation commonly occurs under stress conditions and has been implicated in adaptation. Increases in ethylene production and peroxidase activity likewise occur. Some of these stresses can reduce pathogen spread, but most do not. Induced resistance to viruses or to pathogens in general appears to constitute a largely different response, although its adaptive value may be rather similar. Severe stresses usually cause a reduction in growth and reproductive capacity. Systemically resistant plants that have been repeatedly inoculated to induce a maximum level of systemic resistance (induction inoculation) likewise show retarded development of young tissue and accelerated senescence of older parts, indicative of the costs that induced resistance may entail. Such costs may be caused by the divergence of assimilatory products to pathways not utilized in healthy plants: the synthesis of new proteins and enzymes and of abundant secondary metabolites.

However, substantial systemic resistance may be induced by only a few lesions on a single leaf, indicating that severe stress on only a tiny portion of a plant may be sufficient to induce systemic protective effects. If such treatments can be applied to major crop plants, induced resistance can be an attractive means of enhancing field resistance against many pathogens. Since acquired resistance depends on mechanisms already present in plants, it may be considered as natural and as safe for humans and the environment as is intrinsic disease resistance in plants.

Apart from using a pathogen to induce resistance, with its inherent danger of accidental disease, chemical treatments may also be considered. Under laboratory conditions, it has been possible to induce acquired resistance by different chemicals. However, the doses required are relatively high; and treatments have to be applied repeatedly, as the effect of the compounds is confined to those parts of the plant in which the chemical is present. Nontoxic fungal and microbial metabolites have recently been found effective in inducing nonspecific resistance against pathogenic fungi and bacteria as well as viruses, and may constitute more natural types of inducers. Preliminary field trials are in progress. When the biochemical basis of virus localization is more clearly understood, more directed measures may be taken to enhance specifically those processes that limit the effectiveness of the resistance mechanisms.

For background information *see* PLANT PATHOLOGY; PLANT VIRUSES AND VIROIDS in the McGraw-Hill Encyclopedia of Science and Technology.

[L. C. VAN LOON]

Bibliography: J. A. Bailey and B. J. Deverall (eds.), *The Dynamics of Host Defence*, 1983; A. B. R. Beemster and J. Dijkstra (eds.), *Viruses of Plants*, 1966; R. S. S. Fraser (ed.), *Mechanisms of Resistance to Plant Diseases*, 1985; J. Kuc, Induced immunity to plant disease, *Bioscience*, 32:854–860, 1982.

Plant physiology

This article discusses recent advances concerning the roles of calcium physiology and nitrogen metabolism in plant growth.

CALCIUM PHYSIOLOGY

Numerous physiological processes in plants appear to be regulated by calcium. As in animals, calcium plays a pivotal role in the regulation of growth and development throughout the life of the plant. The importance of calcium was independently recognized and documented by J. Sachs and W. Knop around 1860. Initially calcium's importance in plant growth centered on studies of nutrient deficiency. While specifics of the regulation at the physiological level became available during the 1950s, it is only now that the biochemistry of calcium is being unraveled.

Distribution and transport. Multicellular plants, all of which possess cell walls, may be divided into two generally discrete regions. The first of these is the symplast, which encompasses the cellular compartment containing the cytoplasm and associated organelles bounded by the plasma membrane. The second region is the apoplast or intercellular free space, which includes the cell wall and extracellular space outside the plasma membrane. Ionic or free calcium is present in both regions.

Calcium is initially absorbed by the roots and transported by both the apoplast and symplast, ultimately reaching the stele (xylem and phloem) for transport to all parts of the plant. Throughout the root, except at the tip, the apoplast transport is blocked just prior to the vascular tissue by a group of heavily suberized cells (Casparian strip) which ring the vascular tissue in the root. This prevents the passive transport of calcium into and through the cortical cells to the xylem. Thus the movement of calcium is closely regulated as a result of passing through living cells. Calcium transport in the xylem is primarily from the roots to meristematic regions. Movement within the xylem results from exchange with competing cations that are ionically bound to the negatively charged carboxyl groups of the lignins and pectins within the xylem vessels. Movement out of the xylem is primarily through the apoplast.

Symplastic movement into a cell requires active transport across the plasma membrane. Symplastic transport from cell to cell is accomplished by movement through the plasmodesmata. Plasmodesmata are small tubular connections between the plasma membranes of plant cells and appear to be present between all adjacent cells.

Most of the calcium in plants is not present in the free (ionic) form. Localization studies have shown that calcium at millimolar concentrations is bound within and to the components of the cell wall, presumably via the uronic acid residues of the pectin fraction. Bound calcium is believed to contribute to the strength and rigidity of the cell wall. High concentrations of bound calcium are also associated with certain cytoplasmic organelles, notably the plastids, vacuole, and mitochondrion. Calcium in somewhat lower concentrations is also associated with the chloroplast. It is present both in the stroma and on the outer surface of the thylakoid membranes, and it appears to play an important role in normal stacking of the thylakoid membranes.

Cytoplasmic calcium concentrations are in the nanomolar to micromolar range. Active accumulation by the mitochondria and active efflux across the plasma membrane keep the calcium levels within this range. Mitochondrial adenosinetriphosphatases (ATPases) involved in calcium accumulation appear to require inorganic phosphate. Although a plasma membrane Ca^{2+}/Mg^{2+} ATPase has been identified kinetically, this enzyme has not been isolated or purified. Accumulation of calcium by both plastids and the vacuole appears to be dependent upon Mg^{2+}-dependent ATPases, although these enzymes have not been specifically identified.

Plant growth. Plant physiologists define growth as irreversible enlargement. Application of millimolar calcium to hypocotyls (stems) or roots rapidly inhibits growth. In roots, inhibition is partially reversed with time, apparently resulting from an adaptation to the higher extracellular calcium. Inhibition can also be reversed by removing the applied calcium. Although the mechanism of inhibition is unknown, measurable growth is inhibited to a greater extent than is synthesis of new cell wall material. Submillimolar calcium has been reported to stimulate growth. Removal of calcium with chelating agents increases the growth rate in some plants.

Application of calcium undoubtedly raises the apoplastic calcium level, which is in dynamic equilibrium across the plasma membrane with the cytoplasmic calcium. This rapid rise results in a greater-than-normal concentration gradient across the plasma membrane and disruption of the transmembrane equilibrium. Since movement of calcium into the cell is passive, the cytoplasmic calcium concentration would be expected to rise rapidly from the normal micromolar level, leading to inhibition of calcium-dependent enzymes associated with the growth process.

From a mechanochemical viewpoint, much of the cell wall rigidity is attributed to the presence of calcium in the wall. Increased calcium in the cell wall may rigidify the wall to a greater-than-normal extent. This would decrease the extensibility and increase the yield stress of the wall so that the turgor pressure exerted during growth would be insufficient to expand the cell.

Regulation of hormone action. In most cases, exogenous calcium inhibits hormone action. However, cytokinin retardation of leaf senescence has been shown to be enhanced by calcium, while gibberellin-stimulated synthesis of α-amylase apparently requires calcium. One of the most thoroughly investigated plant hormones is indole-3-acetic acid (IAA), an auxin. Calcium is essential for the normal movement of auxin from the source of synthesis to the site of action (polar transport). Cell-to-cell transport of auxin apparently results from passive entry of IAA, in the ionic form, at the apical end of the cell, followed by active transport from the base of the cell. This concept has been elaborated as the chemiosmotic theory of polar auxin transport. The absolute requirement of polar transport for calcium implicates the presence of calcium-binding proteins associated with the basal auxin transport receptors.

Auxin-stimulated growth is also inhibited by calcium when millimolar concentrations are used. By contrast, the growth of maize seedling roots, which is normally inhibited by low auxin concentrations, loses auxin sensitivity if depleted of calcium prior to treatment with auxin. Adding calcium to calcium-depleted roots induces inhibition of growth by auxin. The interactions of calcium with the plant hormones, and with auxin in particular, indicate that calcium regulates hormone action both at the membrane level and at the protein level. Membrane-level regulation probably involves alteration of membrane integrity, while protein-level interactions undoubtedly involve specific binding to calcium-dependent enzymes and calcium-modulated proteins. In both cases, calcium appears to function as a second messenger in a manner similar to the messenger system identified in animals.

Modulation of proteins. Only a few proteins in plants are known to be modulated by calcium. Many of the processes regulated by the protein pigment phytochrome appear to involve changes in calcium binding and sequestering. These changes result from light-induced alterations in the equilibrium between the two forms of phytochrome, Pr and Pfr. Protein kinases which regulate enzyme activity by phosphorylating specific target proteins have been shown to be stimulated by calcium and by calcium in combination with calmodulin, a calcium-binding protein. Calcium uptake by microsomes is inhibited by calcium channel blockers (diltiazem, lidoflazine) which are known from animal studies to specifically block calcium-transport ATPases. Quinate: nicotinamide adenine dinucleotide (NAD^+) oxido reductase, which functions in the shikimic acid pathway, is regulated by reversible phosphorylation controlled by calcium. One calcium-regulated enzyme has been isolated and purified. Because of its presence in the chloroplast, this enzyme is unique to plants. NAD^+ kinase of the chloroplast is stimulated severalfold by calcium, while cytoplasmic NAD^+ kinase exhibits no such stimulation. The effect of calcium upon NAD^+ kinase is regulated by the ubiquitous protein calmodulin. Changes in calcium concentration alter calmodulin's stimulation of the kinase.

Calmodulin appears to be involved in many growth responses. Most studies of calmodulin in plants have employed calmodulin inhibitors, usually derivatives of phenothiazine. Auxin-stimulated acidification, which is a key component of the acid growth hypothesis for auxin-stimulated growth, is inhibited by the phenothiazines. Calmodulin has been isolated and identified from several plants, including spinach, maize, and zucchini. There is at least one report in which calmodulin has been isolated from the apoplast, and at least two reports that calmodulin is present in large amounts within this region. The presence of calmodulin at the outer surface of the plasma membrane offers an intriguing explanation for the regulation of both calcium efflux across the membrane and auxin-stimulated, calcium-dependent acidification. Calmodulin has also been implicated in the regulation of gravitropic curvature.

Gravitropism. When plant organs are placed perpendicular to a gravitational field, they will curve (reorient) in the direction of the field, a result of asymmetric growth rates of the lower and upper surfaces of the plant. This change in growth rates is believed to be caused by a change in the auxin level of the two surfaces. The time required to trigger the curvature (presentation time) is about 2 min, the curvature occurring even after the plant is returned to the normal vertical position. Such a short presentation time requires a rapid sensory and transduction system. Calcium transport occurs within 1–2 min, and redistribution, with accumulation on the slower-growing side, has been demonstrated in both sunflower hypocotyls and maize root tips. Indeed, asymmetric application of calcium at the tip of a maize root mimics the gravitropic responses. Cal-

modulin inhibitors block both this mimicry and normal gravitropic calcium redistribution. Although the sensory mechanism is not known, it is known that calmodulin-dependent asymmetric calcium redistribution is required. Calmodulin appears to play a pivotal role in the regulation of calcium-mediated responses in plants and may represent a major key to understanding hormone-regulated growth in plants.   [KONRAD KUZMANOFF]

NITROGEN METABOLISM

Animals ordinarily obtain an excess of nitrogen from their food, and reduced nitrogenous compounds are a major portion of animal excretory products. In contrast, plants are very conservative of nitrogen since it is frequently available to them only in limited quantities, and even then its incorporation often requires the expenditure of a significant amount of energy. In nitrogen chemistry, there is a spectrum of compounds running from ammonia, which is fully reduced and therefore possesses a lot of free energy, to nitrate which is fully oxidized and is very low on the free-energy scale.

Ammonia is the only form of nitrogen that can be biochemically incorporated into the macromolecules of living organisms. It is made available either by fixation (reduction) of atmospheric nitrogen or by recycling of decaying organic waste. In the first instance, the ability to fix nitrogen is limited to certain prokaryotic species (bacteria and blue-green algae) which possesses the enzyme nitrogenase; these organisms occur either free in soil and water, in loose associations with plant roots, or in highly specialized symbiotic associations where bacteria are enclosed within cells of root tissue. This could be thought of as a primary source of nitrogen from an ecological point of view. However, the majority of plants rely on recycled sources of nitrogen, and these secondary sources are the focus of the following discussion.

In soil, large nitrogen-containing organic molecules are slowly degraded by a wide range of organisms to release ammonia. This ammonia may be taken up directly by plants or by microorganisms (for example, *Nitrosomonas*) which can use it as a source of metabolic energy. The oxidation of ammonia (called nitrification) eventually produces nitrate. Except for the fact that it is very soluble and subject to leaching, nitrate is a relatively stable soil component. Most plants readily take up nitrate and, in well-aerated soils, it is probably the most universal nitrogen source.

Under conditions of limited oxygen availability, nitrate becomes an alternate agent for accepting electrons from metabolic processes. Certain bacteria (*Micrococcus, Thiobacillus*), by transferring electrons to nitrate, reduce it to N_2 gas and, in turn, extract energy from their food sources. This process, known as denitrification, undoes the work of the nitrogenase enzyme in the nitrogen-fixing bacteria. While this is seen as a loss by those interested in improved soil fertility, it is environmentally essential in order to maintain the overall balance between free nitrogen in the atmosphere and the reduced forms found in soil and water.

Nitrogen transport. After being absorbed by roots, nitrate either is transported directly to the shoot by the xylem transpiration stream or is reduced to ammonia in the roots. In the latter case, most of the ammonia is incorporated by the roots into organic molecules such as amino acids, amides, and ureides which are then translocated to the shoot. In general, the nitrate-processing and -transport behavior of a plant depends on the species and to some degree environmental conditions. For example, the cocklebur (*Xanthium*) transports nitrate almost exclusively, barley (*Hordeum*) transports a mixture of nitrate, amino acids, and amides, and some legumes [for example, the pea (*Pisum*) and bean (*Phaseolus*)] transport ureides as well as the other components. If roots are cold, their metabolism is inhibited, and the proportion of nitrate transported increases. While the shoot can use the various forms of reduced nitrogen that it receives from the roots, it is probably more efficient for the plant as a whole to reduce the nitrate directly in the leaves simply because energy from light is available.

Nitrate reduction. Nitrate can be reduced to ammonia in a two-stage process which involves two widely distributed enzymes, nitrate reductase and nitrite reductase (see illus.). These enzymes are found in roots and shoots of higher plants as well as in fungi, algae, and prokaryotic microorganisms.

Nitrate reductase. This is an inducible enzyme, and it is synthesized de novo when nitrate is present in a tissue. Factors which control nitrate reductase synthesis and destruction have important effects on the activity of the enzyme. In particular, light promotes nitrate reductase activity, possibly because it stimulates synthesis of the enzyme and also because it causes increased transpiration, which in turn brings new substrate into the vicinity of the enzyme. Moreover, light drives the reactions which reduce the nitrite produced by nitrate reductase; this nitrite, if allowed to accumulate, would inhibit nitrate reductase activity. Nitrate reductase activity appears to exhibit diurnal cycles, being high during the light and low in the dark. However, the cyclic nature of nitrate reductase activity persists even in constant conditions of light or dark and apparently follows a circadian rhythm. In addition, water deficits in plants also have significant inhibitory effects on nitrate reductase activity.

Nitrate-to-nitrite reduction is the slower of the two reduction steps so that nitrate reductase activity is considered to be the rate-limiting step in the production of reduced nitrogen in plants. Thus nitrate reductase has been studied very extensively. Early procedures for measuring nitrate reductase activity involved grinding tissue to make a crude enzyme extract which was then tested for its ability to reduce nitrate to nitrite in the presence of an electron donor. More recently, methods have been developed for measuring nitrate reductase activity in intact tissue. Briefly, pieces of a plant (leaves, roots, or stem) are incubated in the dark in a solution containing a buffer, nitrate, and usually a wetting agent. Since nitrite reductase is inactive in the dark, nitrite accumulates in the tissue, and it can be recovered and measured when it subsequently leaks into the surrounding incubation fluid. Initially, the intact-tissue technique was criticized because it gave lower activities than those obtained from ground-tissue procedures. However, data on long-term nitrogen incorporation by plants have been much more consistent with the values predicted by short-term measurements in intact tissue than with those from ground-tissue studies.

Nitrite reductase. This enzyme catalyzes the conversion of nitrite directly to ammonia (see illus.). Despite requiring the incorporation of six electrons, this appears to occur in a single step; there is currently no evidence of any intermediate molecular forms, and the reduction is completed entirely by the one enzyme. Nitrite reductase in leaves is located on the membranes of the chloroplast, where it has direct access to the high-energy products of the light reactions of photosynthesis. In particular, reduced ferredoxin, which provides the electrons needed by nitrite reductase, is available in a nearly

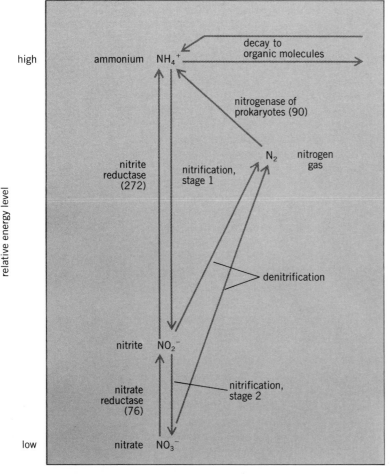

Diagram of interconversions of inorganic nitrogen molecules involved in plant nutrition and their relative free-energy levels. Numbers in parentheses indicate minimum energy required in kilojoules per mole.

limitless supply as long as there is light. Nitrite reductase also occurs in plastids in roots, but here its energy source must be generated by metabolism of carbohydrate.

Like nitrate reductase, nitrite reductase is thought to be induced by nitrate but, once synthesized, it is a somewhat more stable enzyme. Less is known about the behavior of nitrite reductase than of nitrate reductase, but recent approaches to studying the enzyme promise to reveal more about its behavior. This research involves the effects of physical factors such as temperature, light intensity, pH, nitrite concentration, wetting agents, and bicarbonate on nitrite reductase activity in intact tissue. In addition, the effects of herbicides, particularly those which affect photosynthesis, have been monitored in a variety of species. It has been found, for instance, that under optimum conditions nitrite reductase in intact leaf tissue is from four to ten times as active as nitrate reductase under similar conditions, and ensures that nitrite, which is considered to have long-term toxic effects, will not accumulate to any extent in green tissue.

The optimum temperature for nitrite reductase activity is relatively high [in the 35–40°C (95–104°F) range], and concentrations of 20 millimolar of nitrite or more are needed to saturate the enzyme. Light intensity effects indicate that the enzyme is adequately supplied with energy when light levels reach approximately 10–20% of full sunlight. The nitrite-reducing system would be expected to compete with the carbon-fixing system for products of the light reactions. It has been observed that the activity of bean leaf nitrite reductase, which is approximately 20 micromoles of nitrite used per gram of fresh weight per hour in the absence of carbon, drops to a rate of 6–8 micromoles per gram of fresh weight per hour when the incubation medium contains 0.5% sodium bicarbonate as a carbon source. Under normal growth conditions, it is likely that both the nitrite reduction and carbon fixation systems would take reducing agents, as required, from the pool produced by the light reactions.

Comparison of the relative amounts of fixed carbon and nitrogen compounds which appear in growing plants suggests that the demands of the nitrogen reduction system are relatively small. Over a period of 6 days, the amount of energy required to fix the nitrogen that accumulated in rapidly growing 24- to 30-day-old corn plants, supplied with nitrate, was about 5% of that required to fix carbon. While this would be a real demand on the energy supply of the plant, it is reasonable in that approximately 3–5% of the dry matter of a plant is nitrogen. It seems unlikely that nitrogen incorporation, even from sources such as nitrate, which is extremely low on the energy scale, would cause a significant reduction in carbon fixation.

A microorganism that converts ammonia to nitrate is theoretically tapping a source of energy equivalent to approximately 350 kilojoules/mole. Inefficiency in the process will mean that the microorganism does not obtain all the energy, but nitrate, because of its highly oxidized status, ends up at an energy level much below that of ammonia and, for that matter, below atmospheric nitrogen as well. It requires an input of about 76 kJ/mole to reduce nitrate to nitrite and an additional 272 kJ to reduce nitrite to ammonia. Thus, strictly from the point of view of an energy budget, nitrate is the most costly source of nitrogen available to plants. The advantage of nitrate is that practically all green plants as well as microorganisms and fungi have the necessary enzymes to convert it to ammonia. In contrast, the amount of energy required to produce a mole of ammonia from N_2 gas is approximately 90 kJ. Any plant that forms an association with microorganisms to obtain nitrogen from the atmosphere is getting a bargain in terms of the energy it must spend to get ammonia (the microorganism will use energy from carbohydrate supplied by the plant). In addition, the atmosphere is an inexhaustible source of nitrogen.

For background information *see* CELL WALLS (PLANT); NITROGEN CYCLE; PLANT CELL; PLANT GROWTH; PLANT HORMONES; PLANT METABOLISM; PLANT MINERAL NUTRITION; PLANT MOVEMENTS; PLANT PHYSIOLOGY in the McGraw-Hill Encyclopedia of Science and Technology.

[DAVID R. PEIRSON]

Bibliography: D. E. Metzler, *Biochemistry: The Chemical Reactions of Living Cells*, 1977; J. S. Pate, Transport and partitioning of nitrogenous solutes, in W. R. Briggs (ed.), *Annual Review of Plant Physiology*, vol. 31, pp. 313–340, 1980; D. R. Peirson and J. R. Elliott, In vivo nitrite reduction in leaf tissue of *Phaseolus vulgaris* L., *Plant Physiol.*, 68:1068–1072, 1981; G. Refano et al., Modulation of quinate: NAD^+ oxidoreductase activity through reversible phosphorylation in carrot cell suspension, *Planta*, 154:193–198, 1982; S. J. Roux and R. D. Slocum, Role of calcium in mediating cellular function important for growth and development in higher plants, in W. Y. Cheung (ed.), *Calcium and Cell Function*, vol. 3, 1982.

Plant tissue systems

The tissue systems in plants arise by the fundamental biological process of cell differentiation. The vascular tissues (procambium, primary xylem and phloem) constitute a model system for the study of the control of differentiation, because they are so distinctive from their surrounding tissues. By means of culture and wound regeneration systems, indole-3-acetic acid (IAA), which is in the general class of plant hormones called auxins, has been implicated in xylem and phloem differentiation. In these systems, vascular tissues differentiate from mature parenchyma in relatively old regions of the stem without first passing through a provascular stage.

Recent research has involved examination of the role of auxin in vascular differentiation in the intact plant *Coleus blumei*, which bears a distinctively patterned vasculature. Two collateral corner leaf traces, a collateral side bundle, and a number of phloem bundles, which may develop xylem conduct-

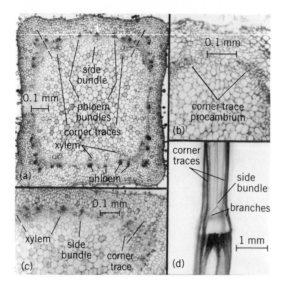

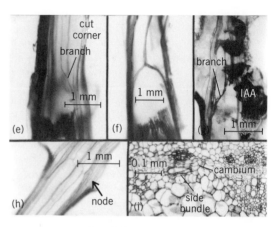

Vascular pattern of *Coleus*. (*a*) Configuration of vascular bundles in untreated stem transverse section. The bundles between the broken lines enter one leaf above. (*b*) Transverse section of internode beneath leaf excised 4 weeks before. (*c*) Transverse section of internode beneath leaf excised 4 weeks before and replaced with IAA. The bundles between the broken lines diverged from the stem into leaf site. (*d*) Cleared specimen through node of untreated stem showing branching of side bundle. (*e*) Cleared specimen in which right corner was severed above, and side bundle branched only toward that corner. (*f*) Cleared specimen in which right corner was severed, and IAA was applied into the incision. Side bundle branching was normal. (*g*) Cleared specimen in which right corner was severed, and the concentration of IAA (applied where indicated) was increased (*from D. K. Bruck and D. J. Paolillo, Jr., The control of vascular branching in Coleus, 1. The side bundle, Ann. Bot., 53:727–736, 1984*). (*h*) Cleared specimen showing restriction of xylem strand branching to nodal regions. (*i*) Transverse section through nodal region containing vascular cambium. Side bundle is just beginning to split, and the xylem branches are composed of secondary elements only.

ing cells in older regions, enter the stem from each leaf (illus. *a*).

Inductive substance. Auxin is synthesized by growing leaves and transported to the stem. The removal of young leaf primordia prohibited further vascular differentiation in the first internode beneath that leaf position. Procambium remained in the corners, where it was present during leaf initiation, but was not initiated in the side or phloem bundle positions (illus. *b*). When leaf primordia were replaced with IAA, the auxin-induced vascular strands nearly duplicated leaf-induced strands in position, morphology, rate of development, and branching pattern (illus. *c*). Procambium was induced in the side and phloem bundle positions. The substitution of IAA for the leaf primordia is consistent with auxin being an essential, if not the sole, factor supplied by the leaf for differentiation of xylem, phloem, and procambium in the subjacent stem of *Coleus*.

Three peculiarities were induced by the exogenous application of auxin. High levels of IAA caused xylem to form in the phloem bundle locations (illus. *c*). With low levels of IAA, the side bundle failed to develop xylem and was indistinguishable from phloem bundles. Corner-trace xylem connections occurred at the node of IAA application. These differences are discussed below.

The vascular tissues form in an intricate pattern of anastomosing strands of different types. The pattern is three-dimensional, with species-specific differences arising developmentally in each dimension by coordinated cell differentiation. Thus, analysis of the pattern is complicated by the occurrence of dif-

ferences over both space and time. It can be viewed as a hierarchical system of superimposed levels of complexity.

Differentiation sequences. The first level of the hierarchy is the sequence of differentiation of tissues within a vascular bundle. In any single bundle in *Coleus*, phloem differentiates before xylem. Xylemless phloem bundles occur, whereas phloemless xylem bundles do not. Increasing the auxin supply to the stem resulted in earlier and greater xylogenesis in phloem bundle locations. Decreasing the auxin supply resulted in an absence of side bundle xylogenesis. These results indicate that xylem differentiation requires a higher auxin level than phloem differentiation. This conclusion suggests that phloem differentiation precedes xylem differentiation because of a low level of auxin produced by young leaf primordia and transported through the procambial strand during early vascular development. A phloem bundle is then recognized as an early, auxin-deficient stage in collateral bundle formation. Phloemless xylem bundles do not occur because when the auxin level required for xylem differentiation is achieved, that for phloem differentiation has already been surpassed.

The second level of the hierarchy is the sequence of formation of entire vascular strands, which is manifested in the number and the nonsynchronous development of strands. In *Coleus*, the two corner traces are initiated nonsimultaneously. Both differentiate before the side bundle, which in turn precedes the phloem bundles. When IAA was applied in place of leaf primordia, corner xylem connections occurred at the node of application. Connections are formed by the lateral differentiation of xylem between strands in the interfascicular region. IAA was applied to the cut leaf base over a surface greater than the cross-sectional area of the leaf's vascular tissue. Presumably, the applied auxin had access to the stem via the ground tissue, and ground tissue transport increased the auxin availability between vascular bundles in the stem to promote atypical corner-trace connections. This argument leads to an explanation for the sequential increase in the number of vascular strands. The amount of auxin transported to the stem increases over time with increasing leaf age. Enlargement of the leaf base and stem produces a greater distance between existing vascular bundles. The rising auxin load exceeds the drainage capacity of the existing traces so that "leakage" occurs laterally into interfascicular regions and other strands differentiate.

Branching patterns. The third level of the vascular hierarchy is the branching (or connecting) of strands. Xylem strands in *Coleus* branch only in association with phloem strands on the same radii, whereas phloem strands can branch independent of xylem. The fact that a greater auxin level is necessary for the differentiation of xylem than for phloem can also explain this structural characteristic. Whenever the level of auxin transported laterally is adequate for xylem differentiation and thus branching, it is also adequate for phloem differentiation, but not vice versa.

The branching pattern is also established by the direction and location in which individual vascular strands branch. Branching appears to be due to lateral divergence of auxin from a newly differentiating strand to a preexisting strand. A developing strand makes a choice of connections between two preexisting strands. According to a recently proposed model, the choice is based on the balance of auxin fluxes in the preexisting strands. A steeper auxin gradient exists in the direction of the strand with the lower flux, so that auxin transport and vascular differentiation occur toward that strand to form a connection.

This model was tested in the developing vascular system of *Coleus*. The side bundle splits, and each branch joins a corner trace (illus. *d*). Incisions were made in one of the corner traces to isolate it from distal auxin sources and reduce its auxin flux. The side bundle xylem then developed only in the branch in the direction of the severed corner trace (illus. *e*). Restoring the original auxin balance by applying IAA to the severed corner trace restored normal side bundle branching (illus. *f*). Increasing the amount of auxin applied to the severed corner trace produced a more favorable auxin gradient toward the intact corner. Side bundle xylem differentiation again followed the gradient, connecting only with the intact corner trace (illus. *g*). Comparable experiments designed to divert developing corner traces gave similar outcomes.

These results are explicable solely in terms of auxin relations. However, additional evidence indicates that while auxin relations can control branch-

ing, they may not actually do so in the intact plant. First, the direction of xylem branching was governed by the auxin imbalance, but the location was restricted to nodal regions, as it is in intact plants (illus. *h*). Thus, auxin relations are not able to override intrinsic constraints imposed by the overall morphology. The morphological constraint in *Coleus* is thought to involve the nodal arrest of shoot elongation and the concomitant initiation of isolated nodal vascular cambia (illus. *i*). Because all xylem branches are composed entirely of cambial derivatives, the cambium may be required for xylem branching by providing a lateral auxin transport pathway. Second, the magnitudes of the auxin imbalances necessary to redirect branching were large. It is unclear whether the correct polarity of the imbalances or the requisite magnitude are ever realized in the intact plant. Third, while the xylem was responsive to the auxin manipulations, the procambium and phloem were unaltered. Because collateral bundles branch as a unit in intact plants, the discrepancy here casts doubt on whether auxin relations account for normal branching in the untreated plant.

Patterns established by differential cell differentiation are common in biological systems. The vascular pattern in *Coleus* is an example of one investigated in depth. Still, while auxin relations as discussed account for many of the phenomena addressed, other control mechanisms undoubtedly come into play. In addition, the levels and locations of auxin must be identified to confirm these hypotheses.

For background information *see* AUXIN; PLANT GROWTH; PLANT HORMONES; PLANT TISSUE SYSTEMS; XYLEM in the McGraw-Hill Encyclopedia of Science and Technology. [DAVID K. BRUCK]

Bibliography: D. K. Bruck and D. J. Paolillo, Jr., Anatomy of nodes vs. internodes in *Coleus*: The nodal cambium, *Amer. J. Bot.*, 71:142–150, 1984; D. K. Bruck and D. J. Paolillo, Jr., The control of vascular branching in *Coleus*, 1. The side bundle, *Ann. Bot.*, 53:727–736, 1984; D. K. Bruck and D. J. Paolillo, Jr., Replacement of leaf primordia with IAA in the induction of vascular differentiation in the stem of *Coleus*, *New Phytol.*, 96:353–370, 1984; T. Sachs, The control of patterned differentiation of vascular tissues, *Adv. Bot. Res.*, 9:151–262, 1981.

Plant-water relations

Many plants live continuously under flooded conditions or commonly experience intermittent floods. Flooding causes severe changes in soil as well as physiological changes in plants; this article summarizes sequential changes induced by flooding on both soil and plants. Emphasis is given to injury, growth responses of plants, and the morphological and physiological changes that precede changes in growth. Adaptations of plants to flooding with fresh and salt water are common, and are also characterized.

Effect of flooding on soils. The oxygen in the pores of well-drained soils is depleted progressively by respiration of roots and microorganisms, but is readily replaced by diffusion from the atmosphere along gradients of concentration or partial pressure. By comparison, in flooded soils water occupies the previously gas-filled pores, and gas exchange between the soil and air is drastically reduced. Furthermore, all of the residual oxygen in the water and soil is rapidly consumed by microorganisms. Hence, the amount of oxygen in a flooded soil is negligible except in the very thin surface layer that is in contact with oxygenated water. Not only is soil oxygen depleted by flooding, but also carbon dioxide, nitrogen, methane, and hydrogen increase appreciably. Following soil inundation, the aerobic soil organisms are first replaced by facultative anaerobes, which are followed by strict anaerobes, primarily bacteria. These anaerobic forms are involved in denitrification; reduction of manganese, iron, and sulfur; formation of methane; and nitrogen fixation.

Rapid decomposition of organic matter in unflooded soils is associated with release of large amounts of energy by aerobic organisms, including actinomycetes, fungi, and numerous bacteria. In flooded soils, however, decomposition of organic matter is slow and restricted to activities of anaerobic bacteria, which release little energy. Also, many compounds, some of which are phytotoxic, are produced in flooded soils. These include a variety of gases, hydrocarbons, alcohols, carbonyls, volatile fatty acids, nonvolatile acids, phenolic acids, and volatile sulfur compounds. Flooding also leads to destruction of soil structure by reduced cohesion, deflocculation of clay, presence of trapped air, uneven swelling, and destruction of cementing agents,

all of which lead to the disruption of soil aggregates. Inundation of soil also decreases the soil redox potential, raises the pH of acid soils (primarily because of change of iron ion from Fe^{3+} to Fe^{2+}), and lowers the pH of alkaline soils (mainly because of carbon dioxide accumulation).

Responses of plants to flooding. Anaerobic conditions that accompany flooding of soil cause changes that injure plants and reduce their vegetative and reproductive growth (Fig. 1). The specific mechanisms by which flooding injures plants are complex and involve oxygen deficiency, excess carbon dioxide, action of toxic compounds, and hormonal imbalances in plants. Completely anaerobic soil conditions eventually kill most plants, but longevity of different species varies widely. Most woody plants die within 3–4 years of continuous flooding (Fig. 2), including species usually rated as very tolerant to flooding. The nature and rapidity of plant response to flooding vary with species and cultivar, age of plants, duration and season of flooding, and condition of the flood water.

A few herbaceous plants (such as rice) and woody plants [such as *Taxodium distichum* (bald cypress), *Nyssa aquatica* (water tupelo), and *Salix nigra* (black willow)] thrive in flooded soils, but most plants do not. Wide variations often occur in flood tolerance of closely related plants, with *Betula nigra* (river birch) much more tolerant than *B. papyrifera* (white birch), *Eucalypatus camaldulensis* (red gum) more tolerant than *E. globulus* (southern blue gum), and *E. grandis* (rose gum) more tolerant than *E. robusta* (swamp mahogany) or *E. saligna* (blue gum). The flood tolerance of fruit trees also varies appre-

Fig. 1.   Effects of flooding on growth of seedlings of two trees. (*a*) Paper birch (*Betula papyrifera*) seedlings after 60 days of flooding. (*b*) Pine (*Pinus banksiana*) seedlings after 45 days of flooding. The unflooded seedling of each species is on the left. (*From T. T. Kozlowski, ed., Flooding and Plant Growth, Academic Press, 1984*)

Fig. 2. Trees killed by flooding as a result of dam construction. (*U.S. Forest Service*)

ciably, with quince and pear being much more tolerant than apricot, peach, or almond. However, such differences often are modified by variations among rootstocks on which scions of fruit trees are grown.

Flooding during the dormant season is relatively unimportant whereas flooding during the growing season is very harmful; this reflects the high oxygen requirements of growing roots. Growth of plants is reduced much more by flooding with standing water than with moving water. Even the most flood-tolerant woody plants (such as *Salix nigra*, *Taxodium distichum*, and *Nyssa aquatica*) are injured or their growth is reduced when flooded with stagnant water. By comparison, a short period of flooding with moving water sometimes stimulates plant growth in these species.

Growth. Flooding of soil decreases the rate of shoot elongation, inhibits leaf initiation and expansion, and induces leaf epinasty, senescence, and abscission in most plants. Cambial growth of most woody plants is reduced following prolonged flooding. However, such plants sometimes show an initial increase in stem thickening followed by a decrease. Some of the early increase in stem diameter of flooded plants is caused by swelling of rehydrated tissues and by stem hypertrophy (such as proliferation and expansion of cortical cells).

Root growth usually is reduced much more by flooding than is shoot growth. Not only are root initiation and growth suppressed but many of the physiologically active fine roots decay, as a result of activity of *Phytophthora* fungi which can tolerate low oxygen levels. The action of these fungi is stimu-

lated by the low vigor of the host plant and the attraction of zoospores to such root exudates as sugars, amino acids, and ethanol. If the flood waters drain away, the previously flooded plants often are very susceptible to drought because their now small root systems cannot absorb water fast enough to keep up with water loss by transpiration from the leaves.

Physiological responses. Growth reduction in flooded plants is mediated by altered food, water, mineral, and hormone relations. The rate of photosynthesis is reduced, often within a few hours after the soil is flooded. Initial reduction in photosynthesis is largely the result of closing of stomatal pores, which impedes diffusion of carbon dioxide from the atmosphere to the photosynthetically active mesophyll cells of the leaf. Stomatal closure of flooded plants may or may not be accompanied by overall dehydration of leaves. In the longer term, photosynthesis is reduced by changes in activity of carboxylating enzymes, lowered chlorophyll content of leaves, and a reduced amount of photosynthesizing tissue. The loss of photosynthesizing tissue is a result of leaf abscission and injury (necrosis), as well as inhibition of leaf formation and expansion. Soil anaerobiosis also reduces translocation of photosynthetic products from leaves to meristematic and storage tissues.

Following soil inundation, the synthesis, destruction, and transport of hormonal growth regulators are altered, leading to changes in amounts and balances of these compounds. The amounts of auxins, ethylene, and abscisic acid increase, while the levels of gibberellins in stems and roots usually decrease.

The energy released by aerobic respiration of roots is necessary for active uptake of mineral nutrients from the soil. In the absence of soil oxygen, however, the anaerobic respiration of roots releases too little energy to sustain active uptake of required amounts of mineral nutrients. Soil anaerobiosis also affects the differential permeability of root cells and causes loss of ions by leaching. Absorption of mineral nutrients is also inhibited because the normally aerobic mycorrhizal fungi, which supplement mineral uptake, are suppressed by anaerobiosis.

In flood-intolerant plants both the concentrations and total amounts of macronutrients (such as nitrogen, phosphorus, and potassium) absorbed by roots are lowered by flooding. The decrease in nitrogen concentration is partly the result of conversion of nitrate to nitrous oxide or nitrogen and loss from the soil. Absorption of calcium and magnesium is also reduced by flooding, but less than absorption of nitrogen, phosphorus, and potassium. By comparison, flood-tolerant plants such as rice absorb more mineral nutrients in response to flooding than do well-watered but not flooded plants.

Adaptations of plants to flooding. Many flood-tolerant plants are adapted to absorb oxygen from the atmosphere by leaves and transport it through the stem to the roots. Such oxygen then diffuses out of the roots, resulting in oxidation of reduced toxic compounds (such as ferrous and manganous ions), which reduces their phytotoxicity. Among the important morphological adaptations that facilitate oxygen movement in plants are aerenchyma tissues, hypertrophied lenticels, and root regeneration.

Aerenchyma tissues form in stems and roots of flood-tolerant plants by dissolution of entire cells or by separation of cell walls, either of which gives rise to gas-filled lacunae (Fig. 3). Such aerenchyma tissues assist in avoiding anoxia by facilitating oxygen movement between the well-aerated shoots and the roots. In many aquatic plants aerenchyma forms a system that is continuous from leaves to roots.

A number of woody plants respond to flooding by producing hypertrophied lenticels on the submerged stem and roots. Because they have large intercellular spaces, such lenticels are more pervious to gases than are the lenticels of nonflooded plants. The flood-induced lenticels not only assist in aeration of the stem and roots but also serve as openings through which toxic compounds associated with anaerobiosis are released from plants.

Whereas flood-intolerant plants do not grow new roots when flooded, many flood-tolerant plants initiate new roots on the original root system and on submerged portions of stems. Such replacement roots compensate for loss of function of the original roots by supplementing the absorption of water and mineral nutrients. They also play an important role in oxidizing the rhizosphere and detoxifying soil toxins.

Morphological adaptations of flood-tolerant plants often have been linked to ethylene production by flooded plants. Although leaf epinasty, senescence, and abscission; formation of aerenchyma tissues; and formation of hypertrophied lenticels and adventitious roots can be induced by exposing plants to ethylene, several other compounds, including carbohydrates, auxins, cofactors, and various enzymes, are also involved in such adaptations.

In addition to morphological adaptations, some plants have various biochemical mechanisms that make survival possible under prolonged soil anoxia or hypoxia. Foremost among these are: (1) maintenance of a high-energy charge (see equation below),

$$\text{Energy charge} = \frac{[\text{ATP}] + \frac{1}{2}[\text{ADP}]}{[\text{ATP}] + [\text{ADP}] + [\text{AMP}]}$$

which is the extent to which the ATP-ADP-AMP (adenosine triphosphate, adenosine diphosphate,

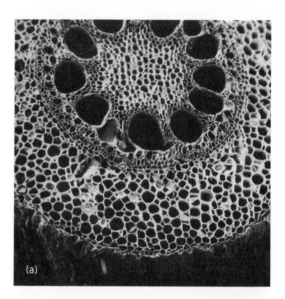

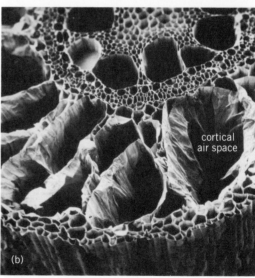

Fig. 3. Cross sections of roots of corn (*Zea mays*) showing effects of aeration on formation of aerenchyma tissue. (*a*) Root grown in well-aerated solution. (*b*) Root grown in unaerated solution. Note large cortical air spaces. (*From M. C. Drew, M. B. Jackson, and S. Gittard, Ethylene-promoted adventitious rooting and development of cortical air spaces (aerenchyma) in roots may be adaptive responses to flooding in Zea mays L., Planta, 147:83–88, 1979*)

and adenosine monophosphate) system is filled with high-energy phosphate groups; (2) maintenance of a supply of glucose and adjustment of carbon metabolism to avoid accumulation of toxic compounds; and (3) adjustment of metabolism so as to maintain a low energy charge and low rate of metabolism.

Salt water adaptations. Growth and survival of plants in brackish or saline water are associated with one or more adapations to the stresses of both anaerobiosis and salinity. Salt-tolerant plants include the halophytes and salt-tolerant glycophytes. Whereas true halophytes can complete their entire life cycle in saline environments, the glycophytes cannot. Salt tolerance varies greatly both among and within species.

Salt-tolerant glycophytes include: (1) species of nonsaline habitats which evolved salt-tolerant physiological races or ecotypes when subjected to salinity, and (2) species possessing various degrees of salt tolerance or with strains variously sensitive to salinity that have not evolved salt tolerance when subjected to a saline environment. In the latter group, selection of tolerance to some other stress factor (such as drought) often has imposed some degree of salt tolerance.

Adaptations to salinity may reflect capacity for tolerating salinity, or for avoiding it by passively excluding salt, actively extruding salt, or diluting salt as it enters plants. Some salt-tolerant woody plants, such as certain mangroves, have developed mechanisms that assist in gas exchange of their submerged roots. *Rhizophora mangle* has stilt roots which bear lenticels above the water level through which oxygen is absorbed and transported to tissues of the submerged roots. The numerous vertical air roots of *Avicennia nitida*, which protrude above the mud in which these plants are rooted, also have lenticels through which oxygen is absorbed from the atmosphere.

Other halophytes eliminate salts through epidermal glands. Extrusion of salt occurs by an active mechanism localized within such glands as indicated by secretion of brine from glands on detached leaves or on isolated leaf disks. Some mangroves absorb copious amounts of water and dilute the accumulated salt, and others shed leaves which contain large amounts of salt. Root membranes of still other mangroves exclude salt by separating fresh water from salt water.

For background information *see* PHYSIOLOGICAL ECOLOGY (PLANT); PLANT MORPHOGENESIS; PLANT-WATER RELATIONS; PLANTS, SALINE ENVIRONMENTS OF in the McGraw-Hill Encyclopedia of Science and Technology.

[T. T. KOZLOWSKI]

Bibliography: D. D. Davies (ed.), *The Biochemistry of Plants*, vol. 2: *Metabolism and Respiration*, 1980; T. T. Kozlowski (ed.), *Flooding and Plant Growth*, 1984; T. T. Kozlowski, Plant responses to flooding of soil, *BioScience*, 34:162–167, 1980; P. J. Kramer and T. T. Kozlowski, *Physiology of Woody Plants*, 1979.

Positronium

Positronium, first observed in 1951, is an atomiclike system consisting of an electron and a positron. Just as in the hydrogen atom, the energy levels of positronium are quantized, with the deepest levels bound by about 6.8 eV. The electron and positron spins can be aligned in the same direction (singlet states) or in opposite directions (triplet states). Annihilation of the positron and electron destroys the lowest-energy singlet state in about 10^{-10} s, but the lowest triplet state survives longer, about 10^{-7} s. Given modern spectroscopic techniques, this allows sufficient time for precise measurement of the energy levels of triplet states. Because of the absence of nuclei in positronium, these measurements provide an accurate test of theories of the electromagnetic force without interference from the strong force.

Since the formation of positronium requires the close approach of a positron and an electron, beams of slow positrons can be used as probes of the electron density in gases, in insulating solids, or near surfaces. Since the singlet and triplet forms of positronium have very different lifetimes and transitions between the two states can be induced by neighboring electrons, study of the decay of positronium can also be used to give information about electron densities on a microscopic scale. This is especially useful in the study of density fluctuations in gases near the critical point for condensation into liquids or solids.

Technology. The first requirement for positronium experiments is an intense beam of positrons. Portable sources are provided by long-lived radioactive isotopes such as ^{22}Na, ^{58}Co, or ^{68}Ge, which have half-lives between 2 months and 3 years. Greater intensities can be obtained by creating short-lived species, which live less than 1 day and thus must be made in the laboratory. Carbon-11 can be produced by bombarding ^{11}B with protons from a Van de Graaff generator, and ^{64}Cu can be created through collisions of ^{63}Cu with neutrons from a nuclear reactor. Another powerful, but expensive, source is obtained by allowing very high-energy electrons from a linear accelerator to strike a heavy-metal target, such as tungsten or tantalum.

Each of these sources produces fast positrons, which are too energetic for many applications. Considerable progress has been achieved in the development of moderators to slow down the positrons. The first successful approach involved the use of powders, such as the oxides of magnesium, aluminum, and silicon, often in the form of coatings. Surfaces of materials such as boron, gold, or tungsten are also efficient in absorbing fast positrons and emitting slow ones. These surfaces can be used even with relatively poor vacuum conditions, but better results can be obtained with clean surfaces of single crystals of aluminum, copper, nickel, or tungsten, or by coating such crystals with cobalt or sulfur.

Positronium is formed through the capture by slow

positrons of electrons from atoms or molecules in gases or nonmetallic solids. Most early experiments used gases, such as nitrogen, both to slow down the positrons and to provide the electrons for positronium formation. However, the required gas density is often so high that precise spectroscopic or collision studies become difficult. Many of the solid surfaces that are successful moderators also lead to the production of slow positronium atoms and can be used in the high-vacuum conditions needed for many experiments.

Spectroscopy. The study of the energy levels of positronium has long been regarded as a prime test of quantum electrodynamics, which is the basis of the theory of all electromagnetic phenomena. The fact that this theory is in practice highly accurate despite serious fundamental problems in its derivation has led physicians to strive toward more and more stringent tests of the theory. Motivation for further development of this theory has been provided by the similarity between quantum electrodynamics and quantum chromodynamics, which is a theory of the strong interactions that are dominant within atomic nuclei.

Hyperfine structure. The difference in energy between the triplet and singlet states of positronium is often referred to as hyperfine splitting. Recent experiments have established that for the ground state this separation corresponds to a frequency of $203,389 \pm 1$ MHz. The disagreement with theory is less than 20 MHz, which is within the uncertainty of current computational efforts.

Considerable progress in the interpretation of this result accrued from the realization that the effects of electron-positron annihilation must be taken into account both in the calculation of the energy levels and in the analysis of the experiment, since some of the positronium atoms disappear as their energy is being measured. It is in the computation of the annihilation corrections that the major uncertainty in the quantum electrodynamic currently lies.

Optical excitation. A new form of positronium spectroscopy has been made possible through the development of tunable dye lasers. The 5.2-eV interval between the 1^3S_1 and 2^3S_1 levels is accessible to a pair of photons of visible light. Although the simultaneous absorption of two photons is possible only with intense light beams, the capture of two photons of the same frequency traveling in exactly opposite directions reduces the frequency spread caused by the Doppler effect and leads to very precise measurements of the energy-level difference. This has been achieved by the use of two counterpropagating pulses in a 486-nanometer ring laser. Positrons from a cobalt source were gathered into bunches and collimated into a narrow beam by means of electromagnetic lenses. This beam was directed onto a clean aluminum surface with its crystal plane carefully oriented. The positronium emitted in the ground states from the surface was exposed to the laser beams. The absorption of two photons leads to excitation of the $n = 2$ level when the lasers are tuned exactly on resonance with the transition frequency. The excited atom is rapidly ionized by a third laser photon, and the freed positron can be easily detected.

The frequency of the 1^3S_1-2^3S_1 transition was measured as $1,233,607,185 \pm 15$ MHz. Although this measurement provides an accuracy of 12 parts per billion, the absolute magnitude of the uncertainty is greater than that of the hyperfine structure measurements. Further improvements to this pioneering experiment are feasible and should provide valuable tests of theory.

Decay rates. The ground singlet state, 1^1S_0, of positronium decays primarily through the emission of two gamma-ray photons, in a time which is too short for precise direct measurements. Decay of the lowest triplet state, 1^3S_1, produces at least three photons and is slow enough to facilitate accurate observations.

When positrons with energy of a few hundred elecronvolts strike a surface, positronium formation is accompanied by the emission of secondary electrons. This emission can be detected and used to start an electronic clock. The clock can be stopped when the observation of a gamma ray signals the annihilation of the positronium. The observed lifetimes follow an exponential distribution from which the decay rate can be deduced. The triplet decay rate measured by this technique is 7.050×10^6 s^{-1}. The corresponding calculated rate is 7.039×10^6 s^{-1}. This difference is larger than early estimates of the error in the calculation. It could be due either to unidentified systematic errors in the experiment, to a failure of quantum electrodynamics or, most likely, to an unexpectedly large contribution from higher-order corrections to the theory that have not yet been computed.

When positronium is formed in the presence of a magnetic field, the singlet and triplet states are weakly mixed. This mixing increases the decay rate of the singlet state by an amount that has been measured by using positronium created in a gas of isobutane. From calculations of the strength of the mixing, the singlet-state decay rate has been estimated as $(7.994 \pm 0.011) \times 10^9$ s^{-1}. This value is consistent with theoretical calculations.

Spin-dependent formation rates. The availability of spin-polarized beams of positrons, in which the positron spins are not randomly oriented, provides a new technique to measure the polarization of electrons. This is achieved through the observation of a change in the rate of formation of triplet positronium (in which the electron and positron spins must be parallel) when the polarization of the positron beam is reversed. A change in the formation rate indicates that the electrons must also be polarized. This method has been used successfully to study surface magnetism, and is being employed in a search for electron helicity in optically active molecules of biological interest. This latter investigation is designed to probe a possible causal connection between the spin polarization of electrons emitted in beta decay and the almost complete dominance of L-amino acids.

For background information *see* AMINO ACIDS; LASER SPECTROSCOPY; POSITRONIUM; QUANTUM ELECTRODYNAMICS in the McGraw-Hill Encyclopedia of Science and Technology.

[J. N. BARDSLEY]

Bibliography: J. W. Humberstone and M. R. C. McDowell (eds.), *Positron Scattering in Gases*, 1983; A. Rich, Recent experimental advances in positronium research, *Rev. Mod. Phys.*, 53:127–166, 1981.

Precambrian

The Precambrian Period extends from the formation of the Earth, about 4.6 billion years ago, to about 600 million years ago. Hence, the Precambrian Period covered more than 85% of geological time. Accordingly, most of the geological and biological milestones in the history of the Earth occurred during the Precambrian. Over the last few years, new ideas have emerged concerning the chemical composition and evolution of the atmosphere and the interaction between the atmosphere and surface during the Precambrian.

It is now believed that the composition of the atmosphere early in the Precambrian consisted of molecular nitrogen (N_2), carbon dioxide (CO_2), and water vapor (H_2O), with trace levels of molecular hydrogen (H_2) and carbon monoxide (CO), resulting from the release of these gases originally trapped in the solid Earth during the planetary formation or accretion phase. The older view held that the early Precambrian atmosphere was composed of methane (CH_4), ammonia (NH_3), and molecular hydrogen, not unlike the present atmospheric composition of Jupiter. However, recent theoretical calculations based on the photochemistry probable in such an atmosphere indicate that such a mixture would have been chemically unstable and, hence, very short-lived, if indeed such a mixture ever existed.

Outgassing and atmosphere formation. It is generally believed that the atmosphere formed via the release of gases originally trapped in the solid interior of the Earth during its formation. The gases (volatiles) were released probably from the solid Earth during the final stages of the planetary accretion stage. During planetary accretion, chunks of material composed of stone and metal, called planetesimals, accreted or coalesced within the primordial solar nebula (the interstellar gas cloud that condensed to form the solar system) to form the terrestrial planets, Earth, Venus, Mars, and Mercury. Modern versions of the planetary accretion hypothesis suggest that the Earth formed as a hot, molten, and geologically differentiated object: iron had already migrated to the core, leaving an iron-deficient mantle, surrounding the iron core. The Earth forming as a hot, molten body would explain the absence of rocks older than 3.7 billion years at the surface, because the surface would have been too hot for rocks to solidify prior to that time. Earlier accretion theories hypothesized that the Earth formed originally as a cold, geologically undifferentiated object; that after several hundred million years it heated up due to radiogenic heating, became molten, and then underwent geological differentiation.

Geochemical and geological considerations indicate that the oxidation-reduction state, and hence the chemical composition of the outgassed volatiles, was determined by the composition and structure of the Earth's interior. Volatiles released from a geologically differentiated Earth, that is, an iron-deficient mantle, should consist of oxidized rather than reduced gases. These volatiles include water vapor, carbon dioxide, and nitrogen, the very chemical composition of present-day volcanic emissions. Chemical analyses of present-day volcanic emissions indicate that their approximate composition by volume is: water vapor 80%, carbon dioxide 12%, sulfur dioxide (SO_2) 6%, and molecular nitrogen 1%. Because it is now believed that the Earth's interior was differentiated geologically prior to the beginning of volatile outgassing, the composition of outgassed volatiles in the early Precambrian was probably very similar to that of present-day volcanic emissions.

Ocean and carbonate formation. The next consideration is the fate of the volatiles released in the early Precambrian.

Water. The overwhelming outgassed component (80% by volume), water vapor is only a minor gas in the present atmosphere, at a concentration ranging from less than 1% to only several percent. On Earth, water is a unique compound, because it can exist in all three phases: as gaseous water vapor, as a liquid in cloud and rain droplets and in the ocean, and as a solid in the form of ice and snow. The amount of water vapor that may exist in the atmosphere is controlled by the saturation vapor pressure, which is solely dependent on the temperature of the atmosphere. Once atmospheric water vapor reaches its saturation vapor pressure, any additional outgassed water vapor would have condensed out of the atmosphere in the form of liquid droplets. As more water vapor was added to the atmosphere via volatile outgassing, the liquid droplets would have precipitated out of the atmosphere. The vast oceans of the Earth were formed as a result of this condensation and precipitation of outgassed water vapor. If the surface of the Earth were perfectly smooth, so that no mountains or valleys existed, the entire surface of the Earth would be covered with about 2 mi (3 km) of liquid water.

Carbon dioxide. The second most abundant outgassed volatile (12% by volume), carbon dioxide (like water vapor) is considerably deficient in the present atmosphere at a concentration of only about 0.03% by volume. The presence of liquid water on the Earth's surface had a significant impact on the fate of the outgassed carbon dioxide. Atmospheric carbon dioxide is very water-soluble, and rapidly dissolved into the newly formed oceans. For every molecule of carbon dioxide in the atmosphere today, there are about 50 carbon dioxide molecules dissolved in the oceans. Once dissolved in the oceans, carbon dioxide reacted chemically with calcium and magnesium ions and precipitated out of the ocean in the form of carbonates, such as limestone. Calcium and magnesium ions were first supplied to the Precambrian ocean by weathering of the ocean floor by flowing water. Once the continents first appeared and began growing (about 250 million years after the Earth formed), these ions also were provided by the weathering of continents by precipitation. For every molecule of carbon dioxide in the atmosphere today, approximately 30,000 carbon dioxide molecules are incorporated in the form of carbonates. Hence, the overwhelming amount of outgassed carbon dioxide is no longer in the atmosphere, but is in carbonates. Prior to the incorporation of atmospheric carbon dioxide into carbonates, the early Precambrian atmosphere may have contained 100 to 1000 times or more carbon dioxide than it presently contains.

Sulfur dioxide. The third most abundant outgassed volatile (6% by volume), sulfur dioxide is transformed readily to sulfuric acid (H_2SO_4) by various atmospheric photochemical and chemical reactions. Water-soluble sulfuric acid precipitates out of the atmosphere and is a source of sulfates to the surface. Due to the rapid transformation of sulfur dioxide to sulfuric acid, its atmospheric lifetime is very short (measured in days), and hence it is only a minor trace gas in the present-day atmosphere and was probably only a trace gas in the Precambrian atmosphere.

Molecular nitrogen. This is the fourth most abundant outgassed volatile (1% by volume). Molecular nitrogen is noncondensable, insoluble, and chemically inert. For these reasons, molecular nitrogen did not condense out of the atmosphere (as did water vapor), did not dissolve in the oceans (as did carbon dioxide), and did not undergo photochemical and chemical transformation (as did sulfur dioxide). Therefore, molecular nitrogen, although only a minor component of the outgassed volatiles, accumulated in the atmosphere to become the major constituent of the present atmosphere (78% by volume).

Chemical evolution. The time period over which volatile outgassing occurred and the atmosphere, oceans, and carbonate rocks were formed is not known precisely. Estimates for the time of outgassing range from as short as 100 million years to as long as 1 billion years. In any case, it appears that the atmosphere and ocean began forming very early in the Precambrian. The early Precambrian atmosphere was composed probably of carbon dioxide (initially, at levels of perhaps thousands of times greater than today, but decreasing with time, as more and more atmospheric carbon dioxide was incorporated into carbonates), water vapor (at about present-day levels), and molecular nitrogen (at levels which increased with time as volatile outgassing continued).

Recent theoretical photochemical calculations and laboratory experiments indicate that a Precambrian atmosphere composed of molecular nitrogen, carbon dioxide, and water vapor with small amounts of molecular hydrogen and carbon monoxide energized by solar ultraviolet radiation or atmospheric lightning would have formed more complex molecules, such as formaldehyde (H_2CO) and hydrogen cyanide (HCN). These molecules precipitating out of the atmosphere into the Precambrian ocean would have produced organic molecules of increasing complexity that eventually led to the first living cells. The abiotic synthesis of organic molecules of increasing complexity, the precursors of living systems, is known as chemical evolution.

Methane and ammonia. The new view of the Precambrian atmosphere composed of nitrogen, carbon dioxide, and water vapor is considerably different from the earlier idea which suggested that the Precambrian atmosphere was composed of methane, ammonia, and molecular hydrogen. However, recent photochemical and chemical calculations have indicated that both atmospheric methane and ammonia are destroyed rapidly and, hence, possess an atmospheric lifetime of less than a hundred years, a very short time interval on the geological time scale. Furthermore, a continuous source to replace the rapidly destroyed atmospheric methane and ammonia has yet to be identified. In the present atmosphere, both methane and ammonia are found at trace levels resulting from biological activity. However, a biological origin for these gases is clearly precluded in the early Precambrian. Due to their lack of sources, coupled with their rapid atmospheric destruction, it is hard to imagine significant quantities of methane and ammonia lasting for significant periods of time in the early Precambrian.

Oxygen, ozone, and solar ultraviolet radiation. Not only can the outgassing scenario outlined above shed light on the composition of the Precambrian atmosphere, but it can also explain the bulk composition of the present atmosphere, with one notable exception—molecular oxygen (O_2), the second most abundant constituent of the present atmosphere at 21% by volume. Molecular oxygen was produced as a by-product of photosynthetic activity, once photosynthetic organisms evolved (perhaps as early as 3 billion years ago). Photosynthetic organisms possess the ability to transform atmospheric water vapor and carbon dioxide to carbohydrates, used by the organisms for food. In this biochemical transformation, oxygen is given off as a by-product. Over geological time, oxygen accumulated to become the second most abundant constituent of the present-day atmosphere. The third most abundant atmospheric constituent (0.9% by volume), argon (Ar) is the radiogenic decay product of the potassium-40 found in the crust.

The buildup of photosynthetic oxygen had very important implications for the evolution of life on Earth. As oxygen accumulated in the atmosphere, respiration replaced fermentation as the energy-producing metabolic process in living systems. Respiration is considerably more efficient than fermentation in energy production. Accompanying the buildup of oxygen in the Precambrian atmosphere was the evolution of ozone (O_3), which is photochemically produced from oxygen. Ozone, while only a trace constituent, has the ability to absorb solar ultraviolet radiation between 200 and 300 nanometers (solar ultraviolet radiation shorter than 200 nanometers is absorbed by several gases, including molecular nitrogen, carbon dioxide, water vapor, and molecular oxygen). Ultraviolet radiation is biologically harmful to living systems. Once ozone evolved to about half of its present atmospheric level, it could absorb effectively solar ultraviolet radiation and shield the surface from this lethal solar radiation. Once this shielding occurred, life could leave the safety of the ocean and go ashore for the first time. Prior to the evolution of ozone, life was confined most probably to several meters below the ocean surface, where this depth of ocean water could absorb solar ultraviolet radiation and protect life. Photochemical calculations indicate that a sufficient level of ozone existed in the atmosphere to shield the surface from solar ultraviolet radiation when atmospheric oxygen evolved to about 10% of its present level. Atmospheric oxygen may have reached this level by the end of the Precambrian. Once ozone evolved and the continents were opened as a biological niche, life exploded in both numbers and diversity.

For background information *see* ATMOSPHERE, EVOLUTION OF; LIFE, ORIGIN OF; PHOTOCHEMISTRY; PHOTOSYNTHESIS; PRECAMBRIAN in the McGraw Hill Encyclopedia of Science and Technology.

[JOEL S. LEVINE]

Bibliography: P. Cloud, *Cosmos, Earth, and Man: A Short History of the Universe*, 1978; H. D. Holland, *The Chemical Evolution of the Atmosphere and Oceans*, 1984; J. S. Levine, The photochemistry of the early atmosphere, in J. S. Levine (ed.), *The Photochemistry of Atmospheres: Earth, the Other Planets, and Comets*, pp. 3–38, 1985; J. C. G. Walker, *Evolution of the Atmosphere*, 1977.

Printing

Register is an important attribute of four-color printing, so that the pictures will look sharp. If the colors are misregistered, the color balance is wrong. Register adjustments are necessary on printing presses because color pictures are made up of different colored inks laid down on the paper one at a time. The basic ink color combination is magenta, cyan, yellow, and black.

Color printing achieves photographic quality by modulating the size of dots of the four primary subtractive colors. (The dots result from prior processing of the artwork or photograph into four halftones based on color separation.) To achieve the full tonal range, the size of the dots must go from 0% to 100% coverage. Thus the dot patterns vary from distinctly separate dots to totally overlapping dots. When the dots are separate, the colors are additive. When they overlap, the overprinted color serves as a filter to block the color below, giving subtractive color combination. If the positions of one color's dot vary, causing distinct dots at some positions and overlapping dots at others, the hues will change. Correct printing practice requires that these individual ink dot positions be highly regulated with respect to each other; this is known as register.

Registration problems. Traditional printing operations involve two kinds of registration problems. One is the presetting or initial set-up, and the other is continuing adjustment while the press is running.

Initial set-up. The overall press-to-paper alignment must be made correctly at the initial set-up. It is often different for each job. One reason for a register problem in set-up is that artwork is often prepared by hand, and the individual color separations are manually aligned. This process takes place both on the film and on the printing plate that will be used on the press. Great precision is needed, with many possibilities for error during alignment of these mechanically. It is anticipated that total electronic makeup for the entire area of a printing plate will become possible. This will eliminate many of the register errors of traditional artwork layout. However, if this page makeup equipment remains expensive, it will be used only by the largest printing companies.

Preparation for press can present other mechanical alignment problems. One potential source is the bending of the plates, that is, putting in the creases needed to lock them onto the printing press. This problem can be overcome by use of steel pins that register holes in the plates at time of exposure to the film; later, pins are also used to hold the plate while bending. Then an error is possible when the plates are mounted on the press. Plate mounting is done by hand. Typically the lateral (side) misalignment due to mounting will be 0.025 in. (0.064 cm); circumferential error will be about 0.010 in. (0.025 cm) and is governed by the plate bending crease.

The circumferential register (around-the-cylinder register) is subject to many more variations on the press than lateral register. Thus the circumferential errors will be larger than the lateral errors. Since the circumferential register is in the direction that the main drive power is applied to the press, it is subject to slippage at clutch points. The clutches are designed to slip during rapid speed changes; therefore, after an emergency stop, the phasing of the printing units often has to be realigned.

Continuing adjustment. The most significant reason for misalignment on a web press is caused by paper following the blanket cylinders because of tacky ink. Depending on the amount of ink coverage, the paper can cling to the upper or the lower blanket on a blanket-to-blanket web press, or can alternate and flutter as the paper attempts to follow the tacky ink. This phenomenon can produce circumferential register errors as high as 0.060 in. (0.15 cm) from the expected settings based on geometry alone. It also can lead to dynamic impression-to-impression register errors (revolution-to-revolution) of 0.010 in. (0.025 cm). The paper is traveling 10 to 20 mi (16 to 32 km) an hour, and the typical distance between the printing units is 7 ft (2 m). In order to maintain register, the tension in the web must be held very high, typically 5 lb per inch of width (0.9 kg/cm). The only indication of excessive tension for good register is when the web breaks become excessive. The high tension helps remove some of the stretchiness of the paper by moving the paper close to its elastic limit. The nonuniformity of the modulus of elasticity (the stretch versus distance) causes variation even though tension is held constant. This variation in paper stretch from unit to unit will cause register variations.

The cost of poor register is significant. Major components of register waste are costs of paper, ink, and machinery. Since the press can print 10 to 30 impressions per second, delays in obtaining register can be very expensive in terms of wasted material.

Press register mechanisms. The adjustments of register on a web press have been governed by the types of mechanism used to bring the images into alignment. The lateral alignment mechanisms are usually lead screws that slide the plate cylinders sideways. There are very few side forces, so it has not been difficult to develop simple mechanisms to move and hold lateral register. In contrast, the circumferential register has been an intriguing problem for press designers. Two traditional approaches involve either a differential gear or sliding helical gears. The helical gear mechanism is simpler, and since the helical gears are also transmitting the power, they are the best gears in terms of tolerance and strength. Any looseness or slip in the gear train will be reflected to the images as register jitter.

In a differential gear design, the register motor drives one member of a differential input at a high reduction ratio, thus making a small change in the cylinder angle with respect to other cylinders.

The first major change in register control electronics was the introduction in the early 1970s of a digital register. This consisted of a series of thumbwheel switches that were preset by the press operator to the amount the register should move. Electrical pulses were applied simultaneously to the register correction motors and the digital switches. Thus, as the motors were driven toward the correct positions, the thumbwheel switches were driven toward zero.

After the invention of the microprocessor, the first computerized register was developed. It brought two major advances to the industry: the use of microprocessors in the pressroom, and the direct keypad entry of register moves. The second advance permitted register corrections to be entered as fast as the operator could type on a keypad. The computerized register also accepted small incremental corrections as multiple pushes of direction buttons.

Computer-assisted registration closed the loop around the press mechanism. It applied position sensors to the printing cylinders directly. When the press operator entered the correction, the cylinder was actually moved that amount, despite backlash and play in the mechanical components. Use of the microprocessor in registration led to several other breakthroughs, such as the introduction of memory so that plate registration could be recalled to a previous running position for a repeat job. For a completely new job, all the cylinders could be zeroed to a known point.

Another innovation was the introduction of calibrated register targets. It is important for the operators to know the amount of register error in order to enter the correction. These targets became widely used on all types of presses and on all types of calibrated register adjustments.

Two other types of register controls affected designs for total closed-loop controls. The first was cut-off controls which were used to control the register between printed images and the cutting of continuous material into sheets. The second was gravure printing controls. Gravure presses heat the paper web to dry the ink after each unit, causing changes in the paper dimensions. To maintain register, the gravure process requires continuous monitoring of each printed color with respect to the previous one printed on the web. Manufacturers of traditional gravure register systems have adapted the closed-loop computer-assisted register to commercial presses. The gravure technology works with marks either across or along the web of paper, giving significant flexibility.

This closed-loop register uses a triangular color mark printed on the web in an area of white space for each color printing unit. A photodetector and lamp measure the distance between marks to detect circumferential errors. They measure the distance between the slanted edges of the triangle to detect

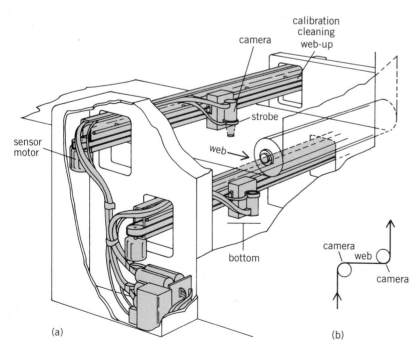

(a)

(b)

A sensor installation for register maintenance by closed-loop controls. (a) Components. (b) Schematic diagram of movement of the web through the unit.

lateral errors. The distance is recorded by an encoder on the press main drive shaft. However, the marks can incur substantial cost in ink and paper for any job that cannot accommodate the mark as part of the printing product design. Advances are being made to overcome this problem. One approach is to use cut-off controls that can maintain a mechanism in synchronization with an individual mark on the web. By considering each printing unit as a mechanism and printing only one mark on the page (or by using patterns which are part of the printing), the total register can be maintained circumferentially, and with certain software processing techniques laterally as well.

The most recently developed system maintains register by using closed-loop controls. A color video camera and stroboscopes stop and freeze the image under high magnification (see illus.). For setting up initial register, round marks are printed in the image area configured in a unique pattern whose 0.010-in. (0.025-cm) dots are virtually undetectable by the average viewer. Artificial-intelligence pattern recognition software searches out the dot patterns and locates the amount and direction of misregistration. The error-correction signals are sent to the individual press unit correction motors for total closed-loop register control.

For background information *see* PRINTING in the McGraw-Hill Encyclopedia of Science and Technology. [BUCKY CROWLEY]

Pseudomonas aeruginosa

Pseudomonas aeruginosa is a bacterium that has emerged as an important pathogen in recent years, especially among hospitalized patients. It was first isolated in 1882 and first recognized as a pathogen in 1890. *Pseudomonas aeruginosa* infections occurred infrequently until after the introduction of antimicrobial agents. At the present time, the organism causes less than 5% of community-acquired infections, but causes 10–15% of hospital-acquired infections.

Pseudomonas aeruginosa is an obligate aerobic, motile, gram-negative bacillus. It produces fluorescent pigments which may lend a characteristic greenish color to sites of infection. The organism can be isolated from soil and water, and has been found as a contaminant of fresh vegetables before delivery to the hospital. It thrives in moist environments, can survive for 300 days in water, and can proliferate in distilled water. For this reason, it has caused epidemics by contaminating solutions such as ophthalmic preparations, humidifiers, aerosols, and soap solutions, and equipment such as respirators, faucets, sink drains, water pitchers, hydrotherapy tanks, and shower stalls. Recognition of these potential sources of contamination has led to more careful attention to sterilization procedures and has reduced their importance as major vehicles for spread of infections. Only 4–12% of healthy people are carriers of *P. aeruginosa*, whereas 40–50% of hospitalized patients may become carriers.

The carrier rate is a function of the nature of the patient's illness, duration of hospitalization, and antibiotic administration.

Toxicity. *Pseudomonas aeruginosa* has a complex cellular structure that protects it from destruction and enhances its virulence. The outermost polysaccharide or slime layer protects the organism from phagocytosis by the host's cellular defenses. The organism contains pili that enable it to adhere to host cells. In addition *P. aeruginosa* produces pyocyanins, toxic substances that inhibit the growth of other organisms in the environment. The most toxic substance to the host is exotoxin A, a protein with activity identical to the diphtheria toxin. When injected into animals, it causes acidosis, shock, leukopenia, liver necrosis, renal tubular necrosis, and pulmonary hemorrhage. The fatality rate from *Pseudomonas* septicemia in humans was 38% for exotoxin-producing strains compared to 11% for non-endotoxin-producing strains. Furthermore, patients who survived *Pseudomonas* septicemia produced higher antibody titers against this exotoxin.

Pseudomonas aeruginosa also produces extracellular proteolytic enzymes, known as proteases, that dissolve elastin and fibrin and destroy collagen. When injected into the skin of animals, they cause necrosis and hemorrhage. Phospholipase is another enzyme secreted by *P. aeruginosa* that destroys pulmonary surfactant and may play an important role in pulmonary infection. The glycolipid produced by *P. aeruginosa* appears to enhance the activity of the phospholipase by enacting as a detergent. *Pseudomonas aeruginosa* likely also has endotoxic effects similar to other gram-negative bacilli, although the exotoxin appears to be a more important virulence factor.

Host defense. Both humoral and cellular host defense mechanisms protect against *Pseudomonas* infection. In one study, patients capable of producing high antibody titers against type-specific polysaccharide had a survival rate of 85%, whereas patients who produced low titers had a survival rate of only 48%. Such observations have led to the development of *Pseudomonas* vaccines and antisera for the prevention and treatment of infections. *Pseudomonas* antisera have been considered useful therapeutic adjuvants for the treatment of burn wound infections. Early studies of *Pseudomonas* vaccines for prophylaxis in highly susceptible populations proved to be discouraging because adequate antibody titers were not maintained for prolonged periods of time and there was a high frequency of local reactions at injection sites. Despite these early disappointing results, research in this area is continuing.

The role of cell-mediated immunity to *P. aeruginosa* has been less extensively studied. Phagocytosis by neutrophils is a critical host defense mechanism since *Pseudomonas* infections are especially prevalent in neutropenic patients. The virulence of *P. aeruginosa* is enhanced in animals infected with cytomegalovirus which depresses lymphocyte-mediated immunity. Patients with cystic fibrosis who are highly susceptible to *Pseudomonas* infection have depressed lymphocyte proliferative response and abnormal phagocytosis and killing of *P. aeruginosa* by phagocytic cells.

Characteristic infections. *Pseudomonas aeruginosa* causes numerous characteristic minor and serious infections of various body organs (see table). The organism flourishes on moist skin; hence, superficial skin lesions are especially common in tropical climates or in individuals exposed to moist environments. The "green nail" syndrome, an infection of the nailbed, occurs in individuals whose hands are frequently submerged in water. Toe web infection occurs in tropical climates or among soldiers on maneuvers in swampy land where the feet remain moist. Infection of the hair follicles has been described in individuals using hot tubs or indoor heated swimming pools because it is difficult to maintain adequate chlorination under these circumstances. These types of infection usually resolve by avoiding the moist environment.

Several types of skin lesions have been associated with *Pseudomonas* infection of the bloodstream. The most characteristic lesion is an ecthyma gangrenosum that usually arises in the groin, perianal area, or axilla. The typical lesion consists of a bluish to blackish central area of necrosis surrounded by an erythematous halo that is due to the organisms invading the walls of blood vessels, causing a vasculitis. Other skin lesions that may be associated with *Pseudomonas* bacteremia include clusters of painful vesicles, pinkish maculopapular plaques, and painful subcutaneous nodules.

Pseudomonas aeruginosa has a propensity for colonizing and infecting burn wounds. This may result in septicemia and death. Burned individuals are also at risk of developing *Pseudomonas* pneumonia, eye infection, and urinary tract infection. The prophylactic use of topical antipseudomonal agents has reduced the frequency of these complications substantially.

Pseudomonas aeruginosa can cause a variety of infections of the head and neck. Among these infections are conjunctivitis, infected corneal ulcer, orbital cellulitis, otitis externa, mastoiditis, and oro-

pharyngeal ulceration. Eye infections are of special concern because they can evolve into panophthalmitis and thus endanger the entire eye if not treated correctly. Serious sinus infection due to *P. aeruginosa* may develop in patients in intensive care units due to insertion of nasogastric tubes. Malignant otitis externa occurs predominantly in the elderly or diabetics; in which case the infection fails to respond to topical antibiotics and may progress to involve the parotid gland, temporomandibular joint, and cranial nerves. If not treated promptly with systemic antibiotics, the infection could be fatal. *Pseudomonas* meningitis occurs infrequently, and is difficult to treat successfully.

Other major organ infections of importance include endocarditis, pneumonia, osteoarthritis, typhlitis, and urinary tract infection. *Pseudomonas* endocarditis occurs predominantly in intravenous drug abusers but also may follow open heart surgery. Typhlitis is a necrotizing infection, usually localized to the cecum but sometimes involving extensive portions of the intestine. Typhlitis occurs primarily in neutropenic patients, and its cause is unknown, but it is often associated with *Pseudomonas* bacteremia.

Pseudomonas aeruginosa causes only 15% of nosocomial pneumonias, but the fatality rate is substantially higher than for other gram-negative bacillary pneumonias. Often patients have associated pharyngitis, otitis, or tracheitis. The usual x-ray pattern is of a bilateral diffuse bronchopneumonia with areas suggestive of abscess formation. *Pseudomonas* pulmonary infection is especially prevalent among patients with cystic fibrosis, especially after they have had extensive exposure to mists, aerosols, antibiotics, and frequent hospitalizations. Mucoid strains predominate as the disease progresses. Antibiotic therapy alleviates the symptoms of infection but fails to eradicate the organism.

Pseudomonas urinary tract infections are most likely to occur in patients with chronic urinary catheters who have received multiple courses of antibiotic therapy. Osteoarthritis is a special problem among intravenous drug addicts and usually involves the vertebral column, pelvis, and sternoclavicular joint. Osteoarthritis also may complicate puncture wounds of the foot in children. Unless systemic antibiotic therapy is instituted promptly, permanent damage to the foot may ensue.

Only 1% of cases of septicemia occurring before 1950 were caused by *P. aeruginosa*. Currently, it causes 7–12% of gram-negative bacillary septicemias. The frequency and prognosis are determined by the patients' underlying diseases, but the fatality rate is usually higher than for other gram-negative bacillary septicemias. The infection is especially likely to terminate fatally if appropriate antibiotic therapy is not instituted promptly.

Antibiotics. The first antipseudomonal antibiotics were polymyxin B and colistin. Although they had substantial activity in laboratory preparations, they were proved to be unreliable for the treatment of

Conditions associated with a high frequency of *Pseudomonas* infections

Condition	Type of infection
Diabetes mellitus	Malignant otitis media
Drug addiction	Endocarditis, osteoarthritis
Acute leukemia, neutropenia	Septicemia, typhlitis
Burn wound	Cellulitis, septicemia
Cystic fibrosis	Lower respiratory infection
Head and neck surgery	Meningitis
Neonates	Diarrhea
Corneal ulcer	Conjunctivitis, panophthalmitis
Vascular catheter	Suppurative thrombophlebitis
Urinary catheter	Urinary tract infection
Nasogastric intubation	Malignant sinusitis
Moist skin	Dermatitis
Puncion wounds of foot	Osteoarthritis

serious infections. The aminoglycosides (gentamicin, tobramycin, amikacin, netilmicin) have been used extensively, but they have potential renal and auditory toxicity and limited activity in neutropenic patients. Penicillins (carbenicillin, ticarcillin, mezlocillin, piperacillin, azlocillin), cephalosporins (cefoperazone, moxalactam, ceftazidime, cefsulodin), and other β-lactam antibiotics (aztreonam, imipenem) are effective antibiotics, but occasionally resistance emerges during therapy. Laboratory and animal studies indicate that aminoglycosides interact synergistically with these other antibiotics against many strains of *P. aeruginosa*. Combination therapy is believed to be most effective for human infections.

Pseudomonas aeruginosa is capable of developing antibiotic resistance by several mechanisms. Some strains produce enzymes that destroy the antibiotic. Other strains alter the characteristics of their cell wall so that the antibiotic cannot enter the cell. The organism may also change the site in the cell where the antibiotic exerts its effect so that it can no longer be effective. These mechanisms of resistance are modulated by the genetic material of the cell. Resistance may be the result of changes in the chromosome itself or may be due to the acquisition of plasmids, which contain extrachromosomal genetic material that can be transferred between organisms, even of differing species. Often these plasmids carry genetic material that conveys resistance to multiple antibiotics simultaneously.

For background information *see* HOSPITAL INFECTIONS; INFECTION; MEDICAL BACTERIOLOGY in the McGraw-Hill Encyclopedia of Science and Technology.   [GERALD BODEY]

Bibliography: G. P. Bodey et al., Infections caused by *Pseudomonas aeruginosa*, *Rev. Infect. Dis.*, 5:279–313, 1983; G. P. Bodey and V. Rodriguez, Advances in the management of *Pseudomonas aeruginosa* infections in cancer patients, *Europ. J. Cancer*, 9:435–441, 1973.

Pyrochlore

The pyrochlore structure type has long been recognized in a variety of naturally occurring cubic oxides with the general formula $A_{1-2}B_2O_6(O,OH,F)_{0-1}$ pH_2O. The atomic arrangement of the pyrochlore structure is similar to that of fluorite, but rather than the AX_2 stoichiometry of fluorite, it has one-eighth fewer anions and two kinds of cation sites. Thus, the pyrochlore stoichiometry can be generalized as $A_2B_2X_7$, where A and B are cations and the X's are anions. The fundamental portion of the pyrochlore structure is a B_2X_6 framework of corner-linked octahedra, whose (111) octahedral layers have an atomic arrangement like the (001) layers in the hexagonal bronzes. Characterization of pyrochlore group minerals has been hampered by their complicated chemistry, metamictization (radiation damage due to alpha decay of uranium and thorium), and alteration.

Synthetic pyrochlores exhibit even greater chemical diversity than minerals, owing to facile chemical substitution at the cation and anion sites. The structure also tolerates vacancies at the A-site and one of the anion sites. Among the over 500 synthetic pyrochlores, many exhibit potentially useful properties, which include catalytic activity, ferroelectricity, ferromagnetism, luminescence, ionic conductivity, and high thermal stability. In addition, pyrochlore and related structures are important constituents of polyphase, crystalline, nuclear waste forms which have been proposed for the long-term isolation of radionuclides. An important concern is the effect of alpha-decay events on the crystal structure and on properties such as fracture toughness and leachability.

Structure and chemistry. The structural formula for pyrochlore is ideally $^{VIII}A_2{}^{VI}B_2{}^{IV}X_6{}^{IV}Y$, where A and B are metal cations and X and Y are anions (see table); the roman numeral superscripts denote coordination numbers. The structure is cubic, $Fd3m$, $Z = 8$, $a = 0.9$–1.2 nanometers, and is commonly described as an anion-deficient derivative of the fluorite structure (Fig. 1). The coordinations of the ions in the ideal case are as follows: $\underline{A}X_6Y_2$, $\underline{B}X_6$, $\underline{X}A_2B_2$, and $\underline{Y}A_4$ (underscored terms are ions whose coordination is being presented). Empirical constraints on the cation sizes for the formation of oxide pyrochlores ($X = O^{2-}$) are $r_A = 0.085$–0.155 nm, $r_B = 0.040$–0.078 nm, and $r_A/r_B = 1.29$–2.30. The lower extremes are for high-pressure pyrochlores. As the A- and B-site cations become more nearly equal in size, a disordered defect fluorite structure becomes equally stable. For many pyrochlore compositions, this transition to defect fluorite occurs at high temperatures [for example, $1530°C$ ($2786°F$) for $Gd_2Zr_2O_7$] and is quenchable. For an ideal pyrochlore, the electrostatic neutrality principle restricts the cation charge combination to $A^{4+}B^{3+}$; however, this valence combination is not observed because cations of the appropriate relative size do not exist. Therefore, the common valence combinations observed, $A^{2+}B^{5+}$ and $A^{3+}B^{4+}$, cause the X anion to be slightly overbonded and the Y anion to be underbonded. Other cation valence combinations are also possible by having two or

Reported ionic substitutions for ideal ($A_2B_2X_6Y$) and defect ($A_{1-2}B_2X_6Y_{0-1}$) pyrochlores

Ion site	Examples
A	$\underline{Ag}^+$, $\underline{Ba}^{2+}$, $\underline{Bi}^{3+}$, Cd^{2+}, $\underline{Cu}^{2+}$, $\underline{Cs}^+$, Fe^{2+}, $\underline{Ga}^{3+}$, H^+, Hg^{2+}, In^{3+}, $\underline{K}^+$, $\underline{Lanthanides}^{3+}$, $\underline{Mn}^{2+}$, $\underline{Na}^+$, NH_4^+, $\underline{Pb}^{2+}$, Rb^+, $\underline{Sb}^{3+}$, Sc^{3+}, $\underline{Sn}^{2+}$, Sn^{3+}, $\underline{Sr}^{2+}$, $\underline{Th}^{4+}$, Tl^+, $\underline{U}^{4+}$, $\underline{U}^{6+}$, $\underline{Y}^{3+}$, $\underline{Zn}^{2+}$
B	Cd^{5+}, Cr^{3+}, $\underline{Fe}^{3+}$, Ga^{4+}, Ge^{4+}, Hf^{4+}, Ir^{4+}, Mo^{4+}, $\underline{Nb}^{5+}$, Nb^{4+}, Pb^{4+}, Pd^{4+}, Pt^{4+}, Re^{5+}, Ru^{4+}, $\underline{Sb}^{5+}$, Si^{4+}, $\underline{Sn}^{4+}$, $\underline{Ta}^{5+}$, Tc^{4+}, $\underline{Ti}^{4+}$, U^{5+}, V^{4+}, $\underline{W}^{6+}$, $\underline{Zr}^{4+}$
X	F^-, $\underline{O}^{-2}$
Y	$\underline{F}^-$, $\underline{O}^{2-}$, $\underline{OH}^-$, S^{2-}

*Underscore denotes those important in pyrochlore group minerals. Molecular H_2O is also a common constituent, particularly in minerals.

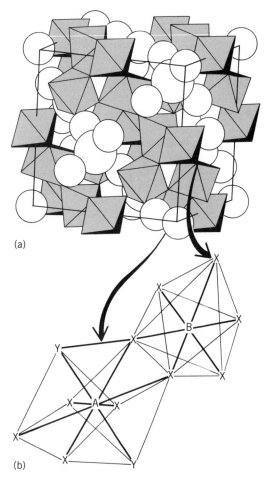

Fig. 1. Pyrochlore structure. (*a*) Polyhedral model (one unit cell), emphasing the corner-linked B_2X_6 octahedral framework with the A- and Y-site ions shown as small and large spheres, respectively. (*b*) The coordination geometries of adjacent A- and B-site cations, labeled according to the structural formula $A_2B_2X_6Y$. The arrows indicate the location in *b* of the portion of the structure enlarged in *a*.

more cations randomly distributed on each of the A- and B-sites. By removing combinations of A and Y ions, a host of defect structures (for example, $^{VI}A_2{}^{VI}B_2{}^{IV}X_6$, $^{VI}A^{VI}B_2{}^{III}X_6$) result, which fit the general formula $A_{1-2}B_2X_6Y_{0-1}$. Another structural variant, $\square^{VI}B_2{}^{III}X_6{}^{VI}M$, referred to as inverse pyrochlore, has the same B_2X_6 framework, but with the A cation positions vacant, symbolized by $\square$, and the Y anions replaced by large monovalent cations, M. The X anion in this case is typically either monovalent or divalent. In the open B_2X_6 framework of AB_2X_6 defect pyrochlores, the A cation can be in one of several crystallographic sites.

All the atoms in the ideal pyrochlore cell occupy special positions, so that the atomic arrangement is completely specified except for a single positional coordinate of the X anions and the cell edge. Variations of this coordinate changes the shapes of the A-site and B-site polyhedra (Fig. 1). For a value of this fractional coordinate of 0.3750 the A- and B-site polyhedra are regular cubes and trigonally flattened octahedra, respectively. Fractional coordi-

nates are fractions of the unit cell dimensions, to specify the atom positions within the unit cell. When the coordinate increases to 0.4375, the B-site becomes a regular octahedron, and the cubic A-site distorts into a trigonal scalenohedron. For oxide and fluoride pyrochlores, acceptable values of the anion coordinate are in the range 0.3750–0.4375; otherwise unreasonably short anion separations result. The range of anion coordinates observed for 27 ideal oxide pyrochlore subjected to crystal structure analysis is 0.3950–0.4400, with a mean value of 0.422 (±0.010). Because the single variable anion coordinate is a function of the cationic bond length ratio, algebraic equations can be used to calculate the anion positional parameter and cell edge.

Applications. The potential applications of pyrochlore materials are extremely varied owing to their extensive chemistry and defect structural variations. Their electronic behavior varies from insulating through semiconducting to metallic. Hence applications in electronic devices are prominent, and include high-permittivity ceramics ($Cd_2Nb_2O_7$, $Ln_2Ti_2O_7$, $Cd_2Nb_2O_6S$), thermistors ($Bi_2Ru_2O_7$, $Cd_2Nb_2O_7$, Bi_2CrNbO_7, Bi_2CrTaO_7, $Ln_2Fe_{4/3}W_{2/3}O_7$, $CdLnFeWO_7$, $CdLnCrWO_7$, $CdLnMnWO_7$), thick-film resistors and materials for screen printing (Pb and Bi precious-metal oxide pyrochlores), and switching elements ($Cd_2Os_2O_7$, $Ca_2Os_2O_7$, $Tl_2Ru_2O_7$). Oxide pyrochlores may also be useful as electrodes in magnetohydrodynamic (MHD) power generation ($Pr_2Zr_2O_7$ with 10% In_2O_3), heating elements in furnaces, oxygen electrodes (Pb and Bi precious-metal defect pyrochlores), semiconductor electrodes for solar energy conversion and solid electrolytes [$Tl_xNb_{2+x}O_6F_{1-x}$ or $A_{1+x}B_{1+x}W_{1-x}O_6$ (A = K, Tl; B = Nb, Ta) substitute electrodes for β-Al_2O_3 in the Na-S battery].

Radiation damage. The primary cause of radiation damage in pyrochlore group minerals and that of concern to the long-term stability of crystalline nuclear waste forms is alpha decay of constituent radionuclides (for example, in minerals, ^{235}U, ^{238}U, and ^{232}Th). The alpha-decay event consists of an alpha particle (helium nucleus, He^{2+}) and its recoil nucleus. The range of the energetic ($3–6 \times 10^8$ kilojoules/mole) alpha particle is approximately 10,000 nm, but along its trajectory, energy is dissipated mainly by electronic excitation, ionization, and, at most, several hundred atomic displacements. The alpha particle rapidly captures and exchanges electrons many times along its collision path, with its electronic charge decreasing as it slows, and finally stops with zero charge as a He atom. Launched in the opposite direction, the recoil nucleus ($6–10 \times 10^6$ kJ/mole) has a range of 100 nm, but because of its greater mass, it produces several thousand atomic displacements. The helium may collect to form bubbles, which may explain microvoids observed by electron microscopy. In reprocessed nuclear wastes, the cumulative alpha dose is 10^{18} alpha particles per cubic centimeter during the first 100 years, and in pyrochlore group minerals

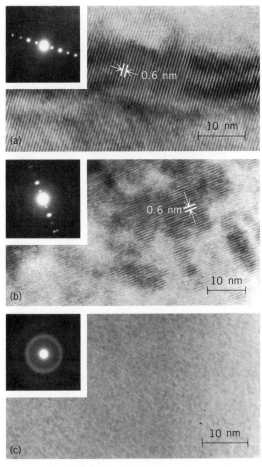

Fig. 2. Lattice-fringe images using the (111) reflection and electron diffraction patterns (insets) for the pyrochlore group mineral microlite, illustrating the effect of increasing alpha-event dose: (a) 10^{14}, (b) 10^{15}, and (c) 10^{16} alpha particles per milligram. (*From G. R. Lumpkin, B. C. Chakoumakos, and R. C. Ewing, Mineralogy and radiation effects of micro-lite from the Harding pegmatite, Taos County, New Mexico, Amer. Mineralog., vol. 71, in press*)

cumulative alpha doses of greater than 10^{16} alpha particles per milligram have been calculated for samples 1.3×10^9 years old. The accumulation and overlap of recoil tracks can eventually render a pyrochlore x-ray and electron diffraction amorphous. This metamict state has properties similar to a glass. The primary coordinations of the cations and the local short-range order of the crystalline material are preserved in the metamict state, but the long-range periodicity is disrupted. Macroscopic swelling and microfracturing occur, due to the increased volume of the metamict state relative to the crystalline state. In partially metamict material, the residual crystalline regions expand several percent by volume, which may be due to accumulating defects induced by the alpha particles alone.

The metamict state can be restored to a crystalline state by thermal annealing. Consequently the preservation of radiation damage in pyrochlores depends on the competition between the damage in growth rate due to recoil nuclei and the activation energy and rate of annealing. That pyrochlores of great age (greater than 10^9 years) still preserve radiation damage implies that natural annealing is not always efficient. In the laboratory, metamict pyrochlores recrystallize over the temperature range 400–700°C (752°–1292°F), depending on the chemical composition.

The transition from the crystalline to the metamict state has been followed in natural pyrochlore group minerals by electron microscopy (Fig. 2) and powder x-ray diffraction. The progressive structural change with increasing alpha dose is one of isolated defect aggregates (that is, individual alpha recoil tracks) up through doses of 10^{14} alpha particles per milligram with no effect on the material's ability to diffract x-rays or electrons; continued damage and overlap of these defect aggregates yield coexisting regions of amorphous and crystalline domains at 10^{15} alpha particles per milligram, with the final damage saturation value reached at doses greater than 10^{16} alpha particles per milligram.

Observations on the transition from the crystalline to metamict state for natural pyrochlores over long periods of time are being compared to synthetic pyrochlores which have been doped with short-lived actinides (such as ^{238}Pu or ^{244}Cm).

For background information *see* ALPHA RAYS; CRYSTAL DEFECTS; FERROELECTRICS; METAMICT STATE; NONSTOICHIOMETRIC COMPOUNDS; RADIOACTIVE MINERALS; RADIOACTIVE WASTE MANAGEMENT in the McGraw-Hill Encyclopedia of Science and Technology.   [B. C. CHAKOUMAKOS]

Bibliography: B. C. Chakoumakos, Systematics of the pyrochlore structure type, ideal $A_2B_2X_6Y$, *J. Solid State Chem.*, 53:120–129, 1984; J. Grins, Studies on some pyrochlore type solid electrolytes, *Chem. Commun.* (Stockholm), 8:1–70, 1980; D. D. Hogarth, Classification and nomenclature of the pyrochlore group, *Amer. Mineralog.*, 62:403–410, 1977; M. A. Subramanian, G. Aravamudan, and G. V. Subba Rao, Oxide pyrochlores: A review, *Prog. Solid State Chem.*, 15:55–143, 1983.

Quartz

The mechanical properties of quartz are strongly influenced by chemical interactions with water. Slow crack growth and frictional sliding depend upon the presence of water on crystal surfaces, and the plasticity of quartz depends upon trace amounts of water within the crystal's interior. Recent advances have been made in understanding the role of water in quartz deformation as a result of mechanical testing in controlled H_2O-atmospheres and characterization of crystalline defects related to water. Although the specific mechanisms associated with crack extension, frictional sliding, and plastic flow differ, water appears to aid them all by reducing their respective activation energies.

At the crystal surface, water is adsorbed in the form of a hydroxyl layer which reduces the surface free energy γ. In a two-stage process, water first dissociates by reaction (1) and then recombines with

$$H_2O \rightleftharpoons OH^- + H^+ \qquad (1)$$

dangling Si and O bonds at the surface to form silanol (Si-OH) and hydroxyl (O-H) groups. The adsorption of water onto the surfaces of a slowly propagating microcrack thus changes the energetics of crack growth. In addition to adsorption processes, the dissolution and growth of quartz in water affects stress corrosion cracking and crack healing. Changes in surface properties due to adsorbed water and dissolution influence the frictional behavior of quartz by decreasing adhesion at microscopic points of contact and aiding near-surface fracture.

Within the crystal, small concentrations of water interact with crystalline line defects known as dislocations. By a process similar to that at the surface, water dissociates and recombines to satisfy dangling Si and O bonds at dislocation cores and hydrolyzes neighboring Si—O bonds. The plastic deformation of quartz depends upon the movement or glide of these dislocations. Thus, the plastic yield strength of quartz is reduced as the activation energies for dislocation nucleation and glide are reduced.

Crack growth. Tensile crack growth in quartz occurs when the local stress (in units of force per area) at the crack tip exceeds its local strength (Fig. 1*a*). The local state of stress depends not only upon the externally applied force F, but also upon the orientation and length of the crack. A local crack extension force G is therefore defined, which can be expressed as a function of F for simple geometries. Crack growth occurs when $G > G_c$, where G_c is the critical crack extension force of quartz, and arrests when $G \leq G_c$.

The crack extension force G is otherwise known as the strain energy release rate (expressed in units of mechanical energy per unit length of crack extension). In the absence of energy losses by acoustic emissions, plasticity, and other irreversible losses, the mechanical energy released during crack growth depends upon the surface energy γ of quartz, given by Eq. (2). Adsorption of water reduces the fracture

$$G_c = 2\gamma \tag{2}$$

surface energy of quartz by an order of magnitude, from $\gamma = 3.5$–4.8 joules/m^2 measured in dry air to $\gamma^* = 0.3$–0.4 joule/m^2, the surface energy measured in moist air. Thus, the critical crack extension force G_c is reduced from 2γ to $2\gamma^*$.

In a dry, inert environment, crack growth occurs unstably when $G > 2\gamma$ at rates approaching acoustic velocities (Fig. 2). If the crack tip is exposed to moisture, however, slow crack growth occurs over the range $2\gamma^* < G < 2\gamma$. The rates of slow, quas-

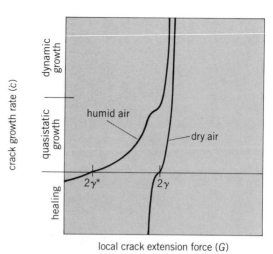

local crack extension force (*G*)

Fig. 2.   Graph showing crack growth rates $\dot{c}$ as a function of the local crack extension force *G*. (*After J. R. Rice, Thermodynamics of the quasi-static growth of Griffith cracks, J. Mech. Phys. Solids, 26:61–78, 1978*)

istatic crack growth are controlled by the kinetics of chemical interactions at the crack tip. Hydrolysis reaction (3) replaces strong Si—O bonds of quartz

$$(\text{Si}{-}\text{O}{-}\text{Si}) + \text{H}_2\text{O} \rightleftharpoons (\text{Si}{-}\text{OH}{:}\text{HO}{-}\text{Si}) \tag{3}$$

at the crack tip with hydrogen-bonded silanol groups and leads to a direct dependence upon the partial pressure of water. Thus, the rates of quasistatic crack growth depend upon relative humidity. Additional rate dependencies upon fluid chemistry arise from the dissolution of quartz and other chemical reactions at the crack tip related to stress corrosion cracking. Unlike quasistatic crack growth, however, the rapid, dynamic extension of cracks is insensitive to the chemical environment and follows more closely the behavior of dry cracks.

Frictional sliding. Frictional sliding between two quartz surfaces (Fig. 1*b*) occurs when the shear stress σ_s exceeds the product of the normal stress σ_n and the friction coefficient μ, that is, $\sigma_s > \mu\sigma_n$. The shear stress σ_s is the force applied parallel to the sliding surface divided by its area, and the normal stress σ_n is the force applied perpendicular to the surface divided by its area.

The friction coefficient μ depends upon the surface properties of quartz at the scale of its surface roughness. Adhesion and shear failure are important processes at load-bearing contacts that are affected by the presence of water. Adhesion at asperity contacts is reduced by the formation of a hydroxylated surface layer, as described in reaction (1). Shear failure of surface irregularities is aided by water by the same mechanisms that affect larger-scale crack growth and plastic deformation. As a consequence, friction coefficients under dry conditions (μ) range from 0.85 to 1.00, whereas the friction coefficients in moist air (μ^*) range between 0.55 and 0.65.

The presence of water also leads to effects of time and sliding rate on friction as a result of changes in the real area and age of adhesive contacts. These

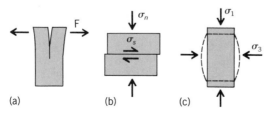

Fig. 1.   Mechanical behavior of quartz. (*a*) Tensile crack growth. (*b*) Frictional sliding. (*c*) Crystal plasticity.

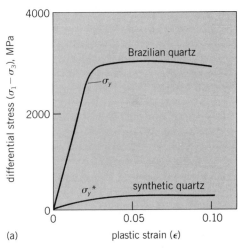

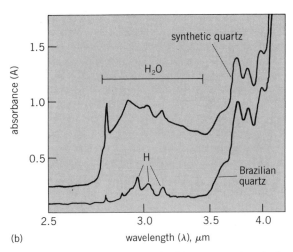

Fig. 3. Plasticity and water weakening. (a) Stress-strain curves for Brazilian quartz and synthetic quartz crystals. (b) Infrared absorption spectra of Brazilian quartz and synthetic quartz crystals.

time and velocity effects are important to unstable frictional sliding and lead to stick-slip behavior.

Plasticity. Plastic deformation of quartz occurs when the differential stress, defined as the difference between the stress terms σ_1 and σ_3 (Fig. 1c), exceeds its yield strength σ_y, that is, $(\sigma_1 - \sigma_3) > \sigma_y$. The onset of plastic strain ϵ, or the nonrecoverable change in length divided by the original sample length, occurs by the nucleation and movement of dislocations. Continued deformation is then controlled by the rates of dislocation nucleation, glide on inclined planes, and recovery by climb. Glide of dislocations involves breaking and recombining Si—O bonds along crystallographically controlled slip planes, and results in shearing within the quartz crystal. Climb of dislocations out of these slip planes occurs by diffusion of Si and O, and aids continued glide by reducing interferences with other defects.

Small concentrations of molecular water (H_2O) incorporated during crystal growth affect the yield strength of quartz by reducing the activation energies of dislocation glide and climb. Deformed under the same conditions, natural Brazilian quartz and synthetic quartz crystals (Fig. 3a) yield at $\sigma_y = 3000$ and $\sigma_y{}^* = 200$ megapascals, respectively, corresponding to hydrogen contents of 30 and 400 ppm (30 and 400 atoms H per 10^6 atoms Si; the term $\sigma_y{}^*$ represents the stress for quartz containing a small concentration of molecular water). Hydrogen impurities in Brazilian quartz are revealed by small, sharp absorption bands due to O-H stretching modes in its infrared spectrum (Fig. 3b). Molecular water impurities, on the other hand, give rise to a large, broad absorption band in the infrared spectrum of synthetic quartz characteristic of its rapid crystal growth. By reactions similar to those shown in reactions (1) and (3), water hydrolyzes Si—O bonds within the quartz crystal and satisfies dangling bonds at the cores of dislocations. Thus, the activation energies for the nucleation and glide of dislocations is reduced. Water also aids in the diffu-

sion of Si and O through the quartz structure, promoting dislocation climb and recovery.

For background information *see* BRITTLENESS; CRYSTAL DEFECTS; CRYSTAL GROWTH; INFRARED SPECTROSCOPY; QUARTZ; STRESS AND STRAIN in the McGraw-Hill Encyclopedia of Science and Technology.

[ANDREAS K. KRONENBERG]

Bibliography: R. D. Aines, S. H. Kirby, and G. R. Rossman, Hydrogen speciation in synthetic quartz, *Phys. Chem. Mineral.*, 11:204–212, 1984; B. K. Atkinson, A fracture mechanics study of subcritical tensile cracking of quartz in wet environments, *Pure Appl. Geophys.*, 117:1011–1024, 1979; J. H. Dieterich and G. Conrad, Effect of humidity on time- and velocity-dependent friction of rocks, *J. Geophys. Res.*, 89:4196–4202, 1984; D. T. Griggs and J. D. Blacic, Quartz: Anomalous weakness of synthetic crystals, *Science*, 147:292–295, 1965.

Quasicrystals

For decades, physicists have believed that the atomic structure of a pure solid must be either crystalline or glassy. A crystalline structure is highly ordered, with atoms arranged in structural units called unit cells repeated periodically (at equal intervals) throughout the solid. A glassy structure, usually formed by rapidly cooling a liquid far below its freezing temperature, is highly disordered, with atoms arranged in a dense but random array. This basic tenet of solid-state physics has been shattered with the discovery of a new solid that is neither crystalline nor glassy. The solid appears to be an example of a new class of ordered structures, called quasicrystals, which have surprising symmetries and physical properties that distinguish them from all previously known materials.

New solids. The new solid, discovered by D. Shechtman and colleagues, is an alloy formed by rapidly cooling a liquid mixture of aluminum and manganese. The alloy forms a solid composed of micrometer-size grains characterized by feathery (den-

Fig. 1. A feathery grain of the icosahedral alloy of aluminum and manganese. (*From D. Shechtman and I. A. Blech, The microstructure of rapidly solidified Al$_6$Mn, Met. Trans., 16A:1005–1012, 1985*)

dritic) arms (Fig. 1). The solid is metastable: heating it to high temperatures induces a transition to a more stable crystal state. When electrons are scattered from the solid, they diffract to form a set of sharp spots arranged in a pattern with icosahedral symmetry (Fig 2a). As in a crystal, the sharp spots mean that the structure has long-range translational order—that is, given the positions of a few atoms in the structure, the positions of the remaining atoms can be reliably predicted. The symmetry of the spot pattern means that the structure also has icosahedral orientational order—that is, the orientation of neighboring atoms or clusters is constrained (on average) to be along the symmetry axes of an icosahedron. The icosahedron is a regular polyhedron with 12 vertices and 20 identical triangular faces. (The centers of the black pentagons on the surface of a standard soccer ball lie at the vertices of an icosahedron).

The result is startling because, according to a rigorous theorem of crystallography, icosahedral symmetry is disallowed for ordinary crystals. The icosahedron has six fivefold symmetry axes, and the theorem states that fivefold symmetry is not possible for a periodic structure. A more familiar example of the theorem is the fact that a plane (or wall) cannot be tiled with just pentagons, even though it can be tiled with just squares or hexagons. A possible explanation for diffraction patterns that are disallowed crystallographically involves multiple scattering from tiny microcrystallites with different orientations. However, other evidence from electron micrographs, x-ray scattering, and other imaging techniques does not support this model.

Model. The leading explanation is that the solid is a quasicrystal, a theory proposed by D. Levine and P. Steinhardt. Quasicrystals represent a new

phase of solid matter that exhibits long-range orientational order and translational order, like a crystal. However, quasicrystals are quasiperiodic, rather than periodic—the atoms or clusters repeat in a sequence defined by a sum of periodic functions whose periods are in an irrational ratio. Clusters of unit cells appear in recurring motifs on every length scale, but there is no regular repeating pattern. Examples of quasiperiodic structures with the symme-

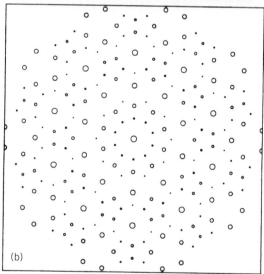

Fig. 2. Quasicrystal diffraction patterns. (*a*) Diffraction pattern along a fivefold symmetry plane of the icosahedral alloy of aluminum and manganese, showing concentric rings of 10 spots (*from D. Shechtman and I. A. Blech, The microstructure of rapidly solidified Al$_6$Mn, Met. Trans., 16A:1005–1012, 1985*). (*b*) Computed diffraction pattern for the fivefold symmetry plane of an icosahedral quasicrystal. The peaks, whose intensities are indicated by the circle size, lie along quasiperiodically spaced lines, which may be observed by holding the figure at a grazing angle (*from D. Levine and P. J. Steinhardt, Quasicrystals: A new class of ordered structures, Phys. Rev. Lett., 53:2477–2480, 1984*)

Fig. 3. Example of a quasicrystal lattice, formed from a square and a rhombus, with a disallowed (eightfold) symmetry. A set of rhombi with common orientation is shaded. By holding the figure at a grazing angle and sighting along the rows of shaded rhombi, the quasiperiodic sequence of spacings may be observed.

tries of periodic crystals, the so-called incommensurate crystals, were already known; but it had not been previously realized that structures with quasiperiodic rather than periodic translational order can have arbitrary orientational symmetry—including an infinite set of possibilities that are totally disallowed for crystals.

The inspiration for quasicrystals derived from a study of a set of nonperiodic two-dimensional tilings developed by R. Penrose. The tilings are composed of two fundamental shapes arranged in a pattern with fivefold orientational order, which is disallowed for crystals. An analogous tiling with a disallowed eightfold symmetry is shown in Fig. 3. This example is formed from a square and a rhombus so that all edges are oriented normal to the symmetry axes of an octagon, resulting in eightfold centers.

The intriguing tilings attracted the attention of numerous mathematicians and physicists who considered the possibility of atomic structures analogous to Penrose tilings and attempted to construct icosahedral analogs. However, the full symmetries of the tilings, especially the quasiperiodicity, were not identified. Once the full symmetries of the tilings were determined, methods of constructing analogs with other symmetries in other dimensions could be developed. In particular, a three-dimensional analog with icosahedral symmetry has been constructed (Fig 4). In general, quasicrystal lattices with arbitrary orientational symmetry have been proven possible.

It has also been shown that the diffraction pattern of an ideal quasicrystal consists of a pattern of true Bragg peaks (point spots) arranged in a pattern that reflects the orientational symmetry (Fig. 2b). Owing to the presence of relatively irrational periodicities in the structure, the spots densely fill the pattern (between any two spots are yet more spots), although

most spots are not bright enough to be observed experimentally. The match between the computed diffraction pattern for an ideal icosahedral quasicrystal and the experimentally observed pattern (Fig. 2) led to the proposal that the new alloy is a quasicrystal.

In a useful alternative description, several groups of physicists have shown that quasicrystal lattices can be generated by projections from higher dimensions; the icosahedral lattice is generated by projecting a six-dimensional hypercubic periodic lattice onto three dimensions. This approach allows a straightforward calculation of Bragg peak intensities. In another description, based on a standard approach to analyzing phase transitions in solids, the atomic structure is described as a set of mass den-

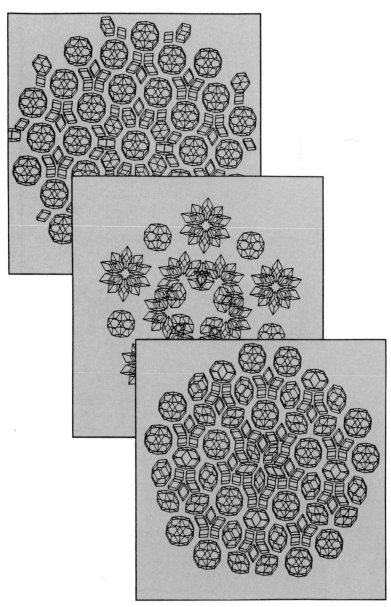

Fig. 4. Three overlapping layers of a three-dimensional icosahedral quasicrystal. In this case, the structural units are four different polyhedra which, in order of decreasing volume, are a rhombic triacontahedron, a rhombic icosahedron, a rhombic dodecahedron, and a rhombohedron. The four units are connected in the structure but have been shrunk in the figure so that they can be distinguished. (*From J. Socolar and P. Steinhardt, Quasicrystals II: Unit cell configurations, Phys. Rev. B, vol. 34, 1986*)

sity waves with adjustable magnitudes and phases. This method lends itself naturally to calculations of the structural stability, thermodynamics, elasticity, and occurrence of defects in the quasicrystal phase.

Nature of new alloy. The evidence that the new alloy is a quasicrystal is not yet totally conclusive. Direct measurements of the translational order indicate that it extends over a length of only a few tens of nanometers. This is most likely due to strain or defects in the samples, but there remains the possibility that the alloy is not quite a quasicrystal but a different phase with long-range orientational order but only short-range translational order. It is important to produce alloys with larger and more perfect grains both to resolve the issue and to accurately measure physical properties. Efforts in this direction have led to a growing list of candidate quasicrystal alloys, including reports of alloys with other symmetries disallowed for crystals. It is also important to develop the mathematical tools necessary to analyze the physical properties of quasicrystals. From the symmetries alone, it is clear that quasicrystals must have subtle physical properties different from other crystalline or glassy solids. For example, quasicrystals are much harder to (plastically) deform than ordinary solids; also, electronic and vibrational properties are, roughly speaking, intermediate between those of crystals and glasses with both extended and localized states.

Whether or not the alloy proves to be a quasicrystal, its properties have raised a fundamental question: Whether a quasicrystal atomic structure can, in fact, exist. This will not be resolved until a quasicrystal structure is positively identified or a physical principle is found to explain why nature cannot take advantage of this possibility.

For background information see CRYSTAL STRUCTURE; CRYSTALLOGRAPHY; POLYHEDRON; X-RAY DIFFRACTION in the McGraw-Hill Encyclopedia of Science and Technology.

[PAUL J. STEINHARDT]

Bibliography: P. A. Heiney, Quasi-crystals: Respectable icosahedral symmetry, *Nature*, 315:178, 1985; D. Levine and P. J. Steinhardt, Quasicrystals: A new class of ordered structures, *Phys. Rev. Lett.*, 53:2477–2480, 1984; B. M. Schwarzschild, Forbidden fivefold symmetry may indicate quasicrystal phase, *Phys. Today*, 38(2):17–19, February 1985; D. S. Shechtman et al., A metallic phase with long-ranged orientational order and no translational symmetry, *Phys. Rev. Lett.*, 53:1951–1954, 1984.

Remote sensing

Recent major advances in remote sensing have provided substantial improvements both in identification and mapping of geologic units and in mineral exploration. These advances result from the development of new instruments and the accompanying development of new methods of interpretation.

GEOLOGIC IDENTIFICATION AND MAPPING

Landsat Multispectral Scanner (MSS) data have been available since 1972 (see table). These data

enable geologists to discriminate remotely among rock types and also to identify uniquely the presence of limonitic rocks (iron oxides) at the Earth's surface. No other unique rock identifications, however, are possible with these broadband multispectral data. Laboratory and field studies such as the NASA/Geosat Test Case Project demonstrate that the geologic utility of multispectral satellite data could be greatly improved if spatial and spectral resolution was increased and spectral coverage was extended past 1.1 micrometer, where Multispectral Scanner coverage ended.

Instruments. The *Landsat* Thematic Mapper (TM) and NASA experimental aircraft instruments such as the Airborne Imaging Spectrometer (AIS) and the Thermal Infrared Multispectral Scanner (TIMS) are sensor systems available since 1982 (see table). These sensors span regions of the electromagnetic spectrum that contain diagnostic spectral features for characterizing many geologic materials. These new sensors also show higher spatial resolution for improved photogeologic mapping compared to Multispectral Scanner data. Used independently or in combination, data from these sensors allow not only improved delineation of geologic units but also determination of mineralogy based on spectral properties.

Thematic Mapper. Data from the Thematic Mapper can be used effectively for lithologic mapping and also for identification of certain rock types. Channel number 7 (2.2 μm) was included on the Thematic Mapper sensor primarily for geologic reasons. Laboratory and field spectral reflectance studies showed that this channel, used with channel number 5 (1.6 μm), could detect the presence of hydroxyl-bearing (clays) and carbonate materials. Limonitic (iron oxide) materials possess diagnostic absorption features in the 0.45–0.85-μm wavelength range, sampled by Thematic Mapper channels 1 through 4. No other specific mineral identifications have been demonstrated with Thematic Mapper data alone, although the additional spectra channels do allow improved discrimination of rock types compared to Multispectral Scanner data.

Photogeologic interpretations of Thematic Mapper images can also be used, in combination with topographic information, for stratigraphic studies, including the construction of "spectral" stratigraphic columns. The 100-ft (30-m) spatial resolution and cartographic fidelity of Thematic Mapper data are sufficient to allow images to be enlarged to 1:24,000 scale to match 7½′ (minutes; for example, 7½ minutes = 1:24,000 scale) topographic maps without any rectification. Thus, standard photogeologic methods can be employed to calculate strike and dip of stratigraphic units and determine stratigraphic thickness. Such columns portray outcrop resistance, true stratigraphic thickness, sequence, and spectral characteristics. Correlation with conventional stratigraphic columns and other spectral stratigraphic columns is also possible. Thus, inferences regarding rock types, lithologic sequences, and lateral variations in rock types can be

Characteristics of sensor systems[a]

Sensor (date available)	MSS (1972)	TM (1982)	AIS (1982)	TIMS (1982)
Platform	*Landsat 1–5*	*Landsat 4–5*	C-130 aircraft	Lear Jet/C-130 aircraft
Altitude	900 km[b]/700 km[c]	700 km	5 km[d]	5–10 km[d]
Swath width	185 km	185 km	290 m[d]	4 km[d]
Wavelength	(4) 0.50–0.60 μm	(1) 0.45–0.52 μm	1.2–2.4 μm	(1) 8.2–8.6 μm
	(5) 0.60–0.70 μm	(2) 0.52–0.60 μm	(128 bands)	(2) 8.6–9.0 μm
	(6) 0.70–0.80 μm	(3) 0.63–0.69 μm		(3) 9.0–9.4 μm
	(7) 0.80–1.1 μm	(4) 0.76–0.90 μm		(4) 9.4–10.2 μm
		(5) 1.55–1.75 μm		(5) 10.2–11.2 μm
		(7) 2.0–2.36 μm		(6) 11.2–12.2 μm
		(6) 10.4–12.5 μm		
Pixel size	80 m	30 m (0.45–2.36 μm)[e]	9 m[d]	12–25 m[d]
		120 m (10.4–12.5 μm)[e]		

[a]1 km = 0.6 mi; 1 m = 3.28 ft. Numbers (1) through (7) represent wavelength channels: 1–5 and 7 are the visible and the near-infrared, and 6 is the thermal infrared.
[b]*Landsat 1, 2,* and *3.*
[c]*Landsat 4* and *5.*
[d]Typical.
[e]Values in parentheses are the wavelengths for these resolutions.

made directly from the Thematic Mapper data.

Thematic Mapper images are useful for discriminating among a variety of lithologic units, and also for limonite-bearing rocks; however, the data lack specific spectral information for unambiguous identification of other minerals. This is due mainly to the relatively broad bandpasses of each Thematic Mapper channel. For example, carbonate minerals show a spectral absorption feature at 2.33 μm, within the range of Thematic Mapper channel 7. Thus, carbonate-bearing rocks can be confused with hydroxyl-bearing rocks (2.2-μm absorption feature) in the same Thematic Mapper image. Narrower bandpasses or some additional data are needed in order to accomplish this separation.

Airborne Imaging Spectrometer. The Airborne Imaging Spectrometer was designed to make remote identification of surface materials possible. The 128 Airborne Imaging Spectrometer channels are contiguous, and are each approximately 9 nm wide (compared to several hundred nanometers for Thematic Mapper channels). This sampling of the 1.2–2.4-μm wavelength region resolves most diagnostic absorption features associated with rock-forming minerals. This is especially true for materials containing OH (clays), CO_3 (carbonates), SO_4 (evaporites), and H_2O radicals and molecules.

Standard image-processing techniques are not useful for analysis of Airborne Imaging Spectrometer data. One of the most effective means of data analysis is to sample individual picture elements (pixels) and construct spectral reflectance curves. Thus, direct identification of surface materials is possible by comparing image spectra to laboratory or field spectra of well-characterized materials.

Thermal Infrared Multispectral Scanner. The six-channel Thermal Infrared Multispectral Scanner sensor measures spectral radiance or brightness temperature of the Earth's surface in the 8–12-μm wavelength region, in six channels. Spectral emittance information derived from these measurements contains diagnostic spectral features for many Earth materials. These features are particularly useful for detecting the abundance of silica in rocks. Bulk thermal properties, such as thermal inertia, thermal conductivity, thermal diffusivity, and density, may also be derived from ground temperatures acquired from Thermal Infrared Multispectral Scanner data. Such data are correlated very highly from one channel to the next because of a dominance of ground temperature. For this reason Thermal Infrared Multispectral Scanner data have been processed by using a modified principal-components technique called decorrelation stretching which displays spectral emittance information as image color, and temperature information as intensities. Silica-rich rocks are portrayed in red to red-orange image colors, clay-rich rocks in blue-red to purple, carbonate rocks in blue to blue-green, and sulfate materials (mainly evaporites) in yellow.

Applications. The Silver Bell porphyry copper deposit, located northwest of Tucson, Arizona, was studied as part of the NASA/Geosat Test Case Project to evaluate the utility of remote sensing data for mapping and detection of hydrothermal alteration zones associated with base- and precious-metal deposits. A west-northwest-trending fault zone localized the intrusion of shallow level stocks and sills into country rocks consisting of limestones, volcanics, and older intrusives. Hydrothermal alteration of the host rocks to mineral assemblages dominated by CH-bearing minerals accompanied mineralization. In this study a Thermal Mapper simulator image and an image-derived lithologic interpretation map were produced and compared with a color air photograph and a geologic map. It is possible to identify every previously mapped geologic unit on the Thermal Mapper simulator image. In addition, several subdivisions of the mapped units can be made which correspond to various alteration types. For example, dacite porphyry is displayed in several different image colors: purple where it is propylitically altered,

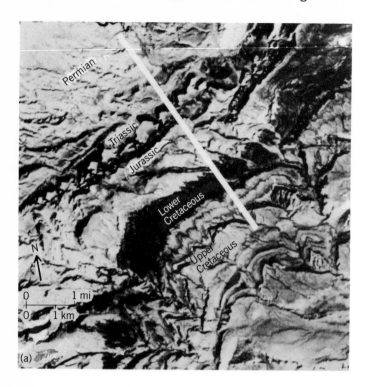

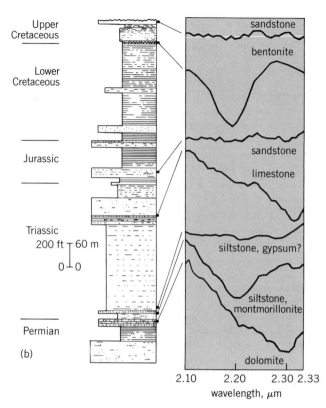

Fig. 1. Deadman Butte area of the Casper Arch, Eastern Wind River Basin, Wyoming. (a) Composite of Thermal Mapper data, 512 × 512 pixel image. (b) Stratigraphic column illustrating the various lithologies and Airborne Imaging Spectrometer–derived spectra of the units. The Airborne Im-aging Spectrometer groundtrack is indicated on the Thermal Mapper image. (H. Lang, Principal Investigator, Multispectral Analysis of Sedimentary Basins Project, Jet Propulsion Laboratory, California Institute of Technology)

green-blue where it is unaltered, bright blue where it is phyllically altered, and orange in areas of potassic alteration.

Figure 1 shows some results of a study of the Wind River Basin in Wyoming. Figure 1a is a Thematic Mapper image of the Deadman Butte area located on the west side of the Casper Arch, eastern Wind River Basin, Wyoming. This 512 × 512 pixel image covers an area of approximately 9 × 9 mi (15 × 15 km). The area shows excellent exposures of sandstones, shales, limestones, dolostones, and evaporites ranging in age from Permian to Upper Cretaceous. Airborne Imaging Spectrometer data were acquired along a flightline across strike to sample these strata. The Airborne Imaging Spectrometer groundtrack is indicated on the Thermal Mapper image. Airborne Imaging Spectrometer pixels were sampled to construct spectral reflectance curves. The stratigraphic column (Fig. 1b) illustrates the lithologies and Airborne Imaging Spectrometer-derived spectra for selected beds. Montmorillonite, calcite, dolomite, and gypsum were identified with these Airborne Imaging Spectrometer spectra as basis. Sandstone (orthoquartzite) is spectrally flat in the wavelength interval. In contrast, no direct mineral identifications could be made by using the Thematic Mapper data alone, although the data discriminated the stratigraphic units quite effectively.

The images shown in Fig. 2a and b cover part of the Death Valley National Monument in California. (The original images are in colors.) A geologic map of the area (Fig. 2c) is shown for comparison. The area includes the eastern Panamint Mountains (left side of images), alluvial fans extending from the mountains to the valley floor, and the Death Valley Salt Pan (right side of images). Bedrock units in the area consist of mainly Precambrian to Ordovician dolostones, shales, limestones, quartzite, and Tertiary tuffs and basalts. The Thermal Mapper image (Fig. 2a) was processed by using principal-components transformation. The Thermal Infrared Multispectral Scanner image (Fig. 2b) was processed by using a decorrelation stretch. Although no mineralogical identifications can be made with the Thermal Mapper data, more lithologic detail and delineation of alluvial fan units are evident as compared to the geologic map. Thermal Infrared Multispectral Scanner data, on the other hand, display compositional information; for example, the appearance of Trail Canyon and Tucki Wash fans are red in the original image, suggesting abundant silica. Their source terrain is high in the Panamint Mountains and consists of more quartzite and shales than do the small fans between them (appearing blue-green on the original image) which have local carbonate rock sources. Spectral contrast within fan units is related mainly to variability in surface weathering and varnishing.

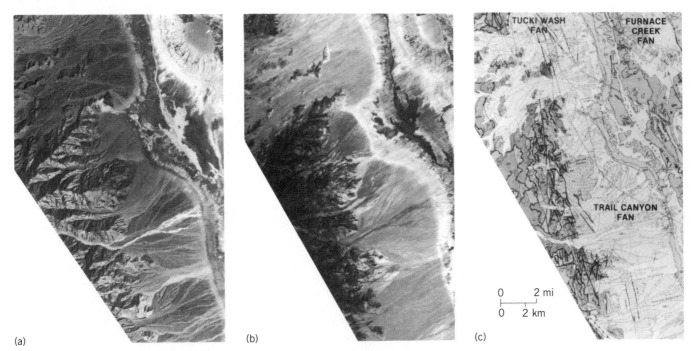

(a)   (b)   (c)

Fig. 2. Death Valley, California. (a) Thermal Mapper principal-components image of the area (*from E. D. Paylor et al., Performance evaluation and geologic utility of Landsat-4 thematic mapper data, Jet Propul. Lab. Publ. 85–86, 1985*). (b) Thermal Infrared Multispectral Scanner channels 1, 3, and 5 decorrelation-stretched image of the area (*from A. B.*

Kahle and A. F. H. Goetz, Mineralogic information from a new airborne thermal infrared multispectral scanner, Science, 222(4619):24–27, 1983). (c) Geologic map of part of the Death Valley National Monument (*from C. B. Hunt and D. R. Mabey, Stratigraphy and structure, Death Valley, California: U. S. Geological Survey Prof. Pap. 494-A, 1966*)

The salt pan is composed of salt, gypsum, and caliche, and intermixed with argillaceous material. These materials are separable on both images, but only the Thermal Infrared Multispectral Scanner image portrays composition directly; for example, the sulfate zone around the toe of the fans is identified by a yellow color in the original Thermal Infrared Multispectral Scanner image.

[ERNEST D. PAYLOR, II]

Mineral exploration. The use of remote sensing in mineral exploration is not new. Aerial photographs have been used since the 1920s to locate prospects through structural interpretation, and with the advent of *Landsat* in 1972, the further advantage of the synoptic view has been utilized to develop regional structural interpretations and models. New developments in spectral remote sensing, that is, the acquisition of images in two or more spectral bands simultaneously, have greatly increased the usefulness and value of remote sensing for mineral exploration. The new sensor developments fall into three groups: (1) the Thematic Mapper flown on *Landsat 4* and 5, an instrument that acquires data with higher spatial resolution and in seven spectral bands as opposed to four on the earlier *Landsat* Multispectral Scanner; (2) thermal infrared Multispectral Scanners that divide the region between 8 and 12 μm into six channels; and (3) imaging spectrometers that can provide a laboratory-type spectrum for each picture element (pixel) by simultaneously acquiring images in as many as 224 contiguous spectral bands.

Landsant Multispectral Scanner. This was the first space-borne imaging system to produce data in digital form. These multispectral data could then be manipulated by computer processing to produce new image products such as color ratio composites that highlight subtle spectral differences. With the Multispectral Scanner data it was possible to map alteration halos or gossans based on the absorption characteristics of ferric iron in limonite and jarosite. However, the Multispectral Scanner data had two major shortcomings. The spatial resolution was poor (260-ft or 80-m picture element size) and the wavelength coverage extended from 0.5 to only 1.0 μm. Within that wavelength region only limonite could be directly identified. Other important minerals such as clays, alunite, or carbonates were not detectable because the diagnostic overtone bending-stretching vibrations lie in the region 2.0–2.5 μm.

Thematic Mapper. The Thematic Mapper was launched aboard *Landsat 4* and 5 in 1982 and 1984, respectively. Thermal Mapper images are now the fundamental remote sensing data set used in resource exploration. The images cover an area of 115 × 115 mi (185 × 185 km) with 98-ft (30-m) picture elements and coverage in seven spectral bands. Two of the bands are centered at 1.6 and 2.2 μm. These two bands in the short-wavelength infrared make possible the detection of the presence of hydroxyl-bearing and carbonate minerals (Fig. 3) because the shape of the reflectance curve is defined by the wings of the fundamental hydroxyl and car-

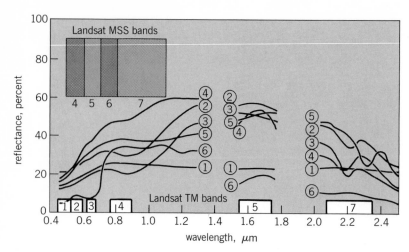

Fig. 3. Field-acquired reflectance spectra: 1, unaltered tuff fragments and soil; 2, argillized andesite fragments; 3, silicified dacite; 4, opaline tuff; 5, tan marble; 6, ponderosa pine. The gaps at 1.4 and 1.9 μm are the result of atmospheric water absorption. (*After A. F. H. Goetz and L. C. Rowan, Geologic remote sensing, Science, 211:781–791, 1981*)

bonate absorption features near 2.7 μm. The presence of more than approximately 30% vegetation cover limits the ability to detect hydroxyl-bearing minerals because of the similar ratio between the reflectances at 1.6 and 2.2 μm. Higher spectral resolution imaging can overcome this limitation. Mapping of quartz or silicified rocks is important in some kinds of mineral exploration. Silicates such as quartz and plagioclase, fundamental building blocks of igneous rocks, do not have diagnostic features in the wavelength region 0.4–2.5 μm; therefore, they cannot be mapped directly with Thematic Mapper data.

Thermal Infrared Multispectral Scanner. An airborne Thermal Infrared Multispectral Scanner has been developed which divides the region 8–12 μm into six spectral bands. This scanner has a sensitivity or noise equivalent change in temperature ($NE\Delta T$) of 0.18°F at 81°F (0.1°C at 27°C). This sensitivity makes it possible to detect subtle changes in spectral emissivity that are diagnostic of silicate minerals. Regions of lower spectral emissivity called reststrahlen bands are caused by metalliclike reflectance associated with the fundamental stretching vibrations of the silicon-oxygen tetrahedron. The wavelength position of the reststrahlen bands is dependent on the degree of sharing among the oxygen atoms. For instance, quartz (SiO_2), a mineral in which all the oxygens are shared among the tetrahedra, has a reststrahlen band at shorter wavelengths than olivine [$(Fe,Mg)_2SiO_4$], a mineral containing isolated SiO_4 tetrahedra. In natural settings, quartz has a deeper reststrahlen band and therefore produces a higher-contrast spectral signature. For this reason, the presence of free quartz in rocks is more readily mapped than any of the other silicate minerals. For instance, a distinction can be made between monzonite and quartz monzonite, important for porphyry copper exploration. However, if significant vegetation cover is present, subtle spectral

emissivity features are no longer detectable and other methods must be used.

Imaging spectrometer. Seventy percent of the land surface of the Earth is covered by vegetation. Until recently, image interpretation in vegetated terrain was confined to deriving structural information from photographs or radar images that showed the surface morphology in spite of vegetation cover. However, indications of chemical anomalies in the soil beneath the vegetation, potentially useful for mineral exploration, can now be identified by use of multispectral remote sensing data.

Thematic Mapper data can be used to identify water stress, degree of vigor, and amount of vegetation cover. However, more subtle forms of stress caused by anomalous concentrations of metals in the soil require high-spectral-resolution reflectance measurements for detection. The spectral region most affected by stress lies between 0.55 and 0.75 μm. The effect seen is a shift to shorter wavelengths in the red edge (sharp reflectance rise) of the 0.68-μm chlorophyll absorption band. The shift is on the order of 10 nm and is called the blueshift. Detection of the blueshift requires continuous sampling of the reflectance spectrum from approximately 0.5 to 0.8 μm at intervals of at least 10 nm. Anomalies associated with increased copper in the soil have been identified in conifers as well as deciduous trees by using an airborne spectroradiometer. Spectroradiometers can produce profiles of data only along an overflight ground track (Fig. 4). In order to accomplish continuous mapping, imaging spectrometers are required.

The very recent technological developments leading to the implementation of imaging spectrometry represent a major advancement in the effectiveness

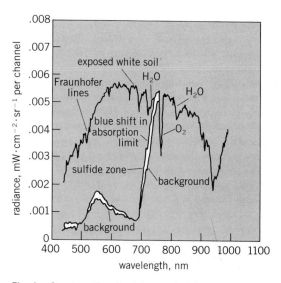

Fig. 4. Spectroradiometer data acquired from aircraft over a conifer stand in an area of sulfide mineralization. The blueshift in the long-wavelength edge of the 0.68-μm chlorophyll absorption band is seen. (*After S. H. Chang and W. Collins, Confirmation of the airborne biogeophysical mineral exploration technique using laboratory methods, Econ. Geol., 78:723–736, 1983*)

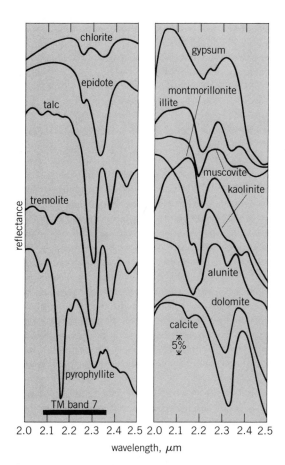

Fig. 5. Selected laboratory spectra of minerals containing overtone vibrational absorption features Al-OH (2.16 and 2.22 μm), and Mg-OH (2.3 to 2.35 μm), and CO_3 (2.3 to 2.35 μm). The band 7 bandwidth of the *Landsat* Thematic Mapper is also shown. (*After A. F. H. Goetz et al., Imaging spectrometry for earth remote sensing, Science, 228:1147–1153, 1985*)

of remote sensing for mineral exploration. Imaging spectrometers that are now being developed can sample the spectrum from 0.4 to 2.5 μm in 10-nm intervals throughout.

Imaging spectrometry will allow not only the detection of vegetation stress, but also the direct detection and identification of many OH-, CO_3-, and SO_4-bearing minerals. Figure 5 shows examples of laboratory spectra of some of these minerals in the 2.0–2.5-μm region. Minerals associated with hydrothermal deposits and skarn deposits an be mapped directly.

An instrument known as the Airborne Imaging Spectrometer is now flying aboard the NASA C-130 aircraft. Flights over an ancient hot spring system at Cuprite, Nevada, have shown that it is possible to identify directly the major minerals present on the surface: kaolinite, alunite, and opalite. In addition, a rare mineral, buddingtonite, an ammonium feldspar, was identified for the first time by imaging spectrometry. Subsequent field work showed that it was not possible to identify buddingtonite by eye, and only with difficulty by using laboratory x-ray diffraction techniques. The experience with bud-

dingtonite has shown that some minerals can only be identified in the field by remote-sensing spectral techniques.

In 1987 the next-generation imaging spectrometer called the Airborne Visible and Infrared Imaging Spectrometer was to be flown aboard the NASA U-2 aircraft. This instrument covers a swath 6.8 mi (11 km) wide with a pixel size of 66 ft (20 m) in 224 spectral bands simultaneously within the 0.4–2.5-μm region. With this sensor it will be possible to extract all the theoretically available information in the reflected signal from the Earth's surface. In 1991 NASA plans to orbit an imaging spectrometer, the Shuttle Imaging Spectrometer Experiment, that will make possible measurements of a selected areas on a global basis. A further development of SISEX will be the High Resolution Imaging Spectrometer Experiment planned for the Space Platform in the mid-1990s. An instrument of this type, incorporating a 30-mi (50-km) swath, 18-mi (30-m) picture element, and 192 spectral bands, will make possible worldwide coverage. The potential rewards from the use of imaging spectrometry and multispectral thermal imaging in mineral exploration are just beginning to be understood.

For background information see AERIAL PHOTOGRAPH; PROSPECTING; REMOTE SENSING in the McGraw-Hill Encyclopedia of Science and Technology.

[ALEXANDER F. H. GOETZ]

Bibliography: M. J. Abrams et al., Remote sensing for porphyry copper deposits in southern Arizona, *Econ. Geol.*, 78:591–604, 1983; M. J. Abrams, J. E. Conel, and H. R. Lang, *Joint NASA/Geosat Test Case Project Final Report*, American Association of Petroleum Geologists, 2 pts., 2 vols., 1985; W. Collins et al., Airborne biogeochemical mapping of hidden mineral deposits, *Econ. Geol.*, 78:737–749, 1983; A. F. H. Goetz et al., Imaging spectrometry for earth remote sensing, *Science*, 228:1147–1153, 1985; C. B. Hunt and D. R. Mabey, *Stratigraphy and Structure, Death Valley, California*, USGS Prof. Pap. 494-A, 1966; A. B. Kahle and A. F. H. Goetz, Mineralogic information from a new airborne thermal infrared multispectral scanner, *Science*, 222:24–27, 1983; E. D. Paylor et al., *Performance Evaluation and Geologic Utility of Landsat-4 Thematic Mapper Data*, Jet Propul. Lab. Publ. 85–86, 1985.

Respiratory system (invertebrate)

The gills of decapod crustaceans are ventilated by the rhythmic beating of the scaphognathites ("gillbalers") which are flattened, bladelike structures on the second maxillae, one of the pairs of modified appendages which form the mouthparts. The scaphognathites are situated in narrow channels just anterior to the branchial chambers, and their beating activity creates a current of water (or of air in the case of air-breathing decapods) through the branchial chambers to enable gas exchange to take place at the respiratory surfaces.

Ventilatory pauses. During activity or following disturbance, the beating of the two scaphognathites is often closely coordinated so that the branchial chambers are ventilated continuously. In quiescent animals under normal oxygen conditions, however, ventilation rates are reduced, and the ventilatory activity may exhibit intermittent pauses in which either only one scaphognathite pumps at a time (unilateral ventilation), or in which both scaphognathites stop beating, resulting in a complete cessation of branchial ventilation (bilateral apnea). Periods of apnea are frequently accompanied by a reduction in heart rate of varying severity. During brief ventilatory pauses, the rate reduction may be only slight, but during longer periods of bilateral pausing the animals frequently undergo periods of complete cardiac arrest, the onset and cessation of which are closely synchronized with that of ventilatory activity (Fig. 1).

The close relationship between scaphognathite activity and heart rate in decapods not only is seen during pausing, but also occurs during normal pumping and could possibly serve to maximize the efficiency of gas exchange at the gills. This coordination between cardiac and scaphognathite activity appears to be a characteristic and unique feature of decapod respiratory physiology, and is probably due to interaction between respiratory and cardiac motor neurons within the central nervous system.

Pause duration and frequency. The duration and frequency of ventilatory pauses varies greatly between species. In many prawns, for example, although such pauses may occur frequently (at approximately minute intervals in some species) the pauses are brief (perhaps each only a few seconds' duration) and represent little more than short interruptions of branchial ventilation. Nevertheless, in species such as *Palaemon elegans* and *Crangon crangon*, pausing behavior may be very erratic and can give rise to considerable minute-to-minute variation in ventilation rates. Among the crabs, however, the average duration of a bilateral pause can be somewhat greater, between 5 and 20 min.

Of all species studied, the duration of ventilatory pauses is greatest among those species of crab that burrow into soft marine sediments may thus remain inactive for long periods. In species such as *Atelecyclus rotundatus*, *Corystes cassivelaunus*, and *Ebalia tuberosa*, for example, the average duration of the ventilatory pauses may exceed 40 min.

Ventilatory pausing in quiescent animals is often characterized by a well-developed rhythmicity of alternating periods of pausing and of active ventilation. The onset of such pausing behavior occurs spontaneously in inactive animals and may last for periods of up to several hours. It is therefore important to distinguish this behavior from the sudden cessation of ventilatory and cardiac activity that may be recorded in decapods in response to some disturbing stimulus. In the latter case, the arrest period is generally briefer, lasting perhaps only a few seconds. It has been suggested that these disturbance-stimulated pauses have a defensive function, since the cessation of ventilatory and cardiac activity in buried or camouflaged crabs may help reduce the likelihood of detection by a potential predator. In contrast, true pausing behavior is recorded only in quiescent animals and cannot be correlated with any external disturbance. Indeed, it has been observed that any disturbance or stimulation of the quiescent animal during a period of rhythmic pausing behavior usually results in an abrupt return to uninterrupted ventilatory activity.

Unilateral and bilateral pausing. Unilateral and bilateral ventilatory pausing, although first observed in marine decapods, is not restricted to this group, but occurs also in fresh-water decapods and in semiterrestrial, air-breathing crabs. The widespread oc-

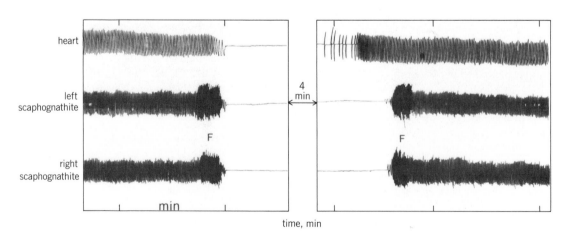

Fig. 1. Part of a continuous recording of the activity of the heart and of both scaphognathites in the crab *Atelecyclus rotundatus*, illustrating a respiratory pause. The onset and cessation of cardiac and scaphognathite pausing is closely synchronized, although in this species cardiac activity re- sumes a few seconds before scaphognathite beating. (*From A. C. Taylor, Branchial ventilation in the burrowing crab, Atelecyclus rotundatus, J. Mar. Biol. Ass. U.K., 64:7–20, 1984*)

currence of this behavior among the decapods has prompted a number of investigations of its functional significance. The consensus is that these animals do not need to ventilate the gills continuously in order to meet their metabolic oxygen demand during periods of inactivity. Instead, they can rely on either unilateral pumping or on periodic ventilatory activity alternating with periods of bilateral apnea. By avoiding continuous bilateral ventilation, these animals are able to make a significant reduction in the overall energy costs of ventilation. A detailed study of scaphognathite beating in the shore crab, *Carcinus maenas*, has indicated that the energy cost of sustained ventilation could be as high as 30% of the total oxygen consumption. If this value is also true for other species, then the animals have much to gain by adopting a pattern of intermittent pausing during periods of inactivity. It is significant, that both when active and when exposed to low levels of oxygen, the animals resume continuous bilateral pumping in order to maintain the supply of oxygen to the respiratory surfaces.

Oxygen activity. Previous studies on marine crabs have shown that during ventilatory pauses oxygen consumption can be reduced by 30–50%. These figures are somewhat higher than the estimated energy cost of scaphognathite pumping, but it should be recalled that cardiac arrest frequently accompanies ventilatory pausing and so a further energy saving may be gained in the cost of circulating the blood. In the semiterrestrial crabs the situation may be slightly different, for although the scaphognathites are still used to ventilate the branchial chambers, the energy cost of pumping air is likely to be considerably less than that of pumping water. Thus the potential energy saving will be much smaller in these animals than in their aquatic relatives. The explanation for the retention of this behavior by semiterrestrial crabs may therefore lie elsewhere. Since water conservation is a major problem facing semiterrestrial crabs, it is likely that pausing behavior which helps to minimize convective water loss from the branchial chambers may be far more bene-

ficial on land than any small saving in the energy cost of respiratory ventilation.

What controls the frequency and duration of ventilatory pausing in aquatic crabs? In inactive animals, nearly all the oxygen demand of the tissues is supplied by the oxygen carried in solution in the blood; very little comes from the oxygen carried by the respiratory pigment. The oxygen bound to the respiratory pigment acts as a store of oxygen which is available to the animal only if its oxygen demand suddenly increases, perhaps as a result of increased locomotor activity. This store of oxygen is also utilized during periods of apnea as shown by the gradual decline in the oxygen content of the blood during a ventilatory pause. The oxygen present in the water within the branchial chamber is taken up by the gills and tissues lining the chambers; but it is probably used locally since the cessation of blood circulation during pausing must prevent effective distribution to other areas.

Information available from studies limited to only one or two species seems to indicate that there is almost no increase in the L-lactate content of the blood and thus anaerobic metabolism does not make a significant contribution to energy production, and that aerobic metabolism is maintained at a basal rate throughout the ventilatory pause. The maximum duration of a pause will therefore depend both on the size of the oxygen store, which in turn will depend on the blood volume and the oxygen-carrying capacity of the blood, and on the rate of oxygen consumption during the pause. If this interpretation is correct, the variation in pause duration which may sometimes be observed during long-term recordings in some species must result from variations in the rates of oxygen consumption affecting the rates of depletion of the oxygen store. It is perhaps significant that pause duration often increases during such recordings as the animals become more quiescent and oxygen consumption rates are reduced.

There is evidence that both peripheral and central receptors may be involved in monitoring oxygen in

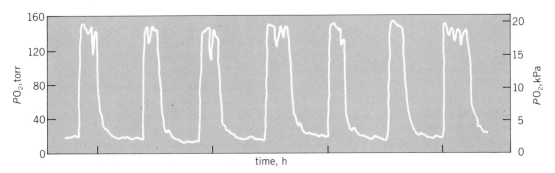

Fig. 2.   Recording of the P_{O_2} (oxygen content expressed as the partial pressure of the oxygen gas in water) of the water in one of the branchial chambers of the crab *Atelecyclus rotundatus* when buried in sand. The recording shows the gradual reduction in the P_{O_2} during ventilatory pausing and the sudden increase when ventilation recommences. The peaks represent periods of active ventilation, and the troughs represent periods of apnea. (*After A. C. Taylor, Branchial ventilation in the burrowing crab, Atelecyclus rotundatus, J. Mar. Biol. Ass. U.K., 64:7–20, 1984*)

decapod crustaceans, but recent studies which have looked at pausing behavior in decapods have indicated that it is unlikely that any oxygen receptors situated in the branchial chambers are directly involved in controlling the recommencement of ventilatory and cardiac activity. In some species, the oxygen content of the water within the branchial chambers gradually declines to an approximately constant level (1.33–2.67 kilopascals or 10–20 torr) during successive pauses (Fig. 2). The oxygen content remains at this level for several minutes before ventilation recommences, and it would appear unlikely that the resumption of both ventilatory and cardiac activity is a simple response to the oxygen in the branchial chambers reaching a certain level. Further evidence supporting this interpretation is experimental. Water having either a reduced or an enhanced oxygen content carefully introduced into the branchial chambers via a catheter has no significant effect on pause duration. It is much more likely that pause duration is controlled by receptors monitoring the oxygen (or carbon dioxide) content of the blood. Unfortunately, it has proved impossible so far to monitor blood oxygen content during pausing without disturbing the animals and causing the cessation of pausing behavior.

For background information *see* DECAPODA (CRUSTACEA); RESPIRATORY SYSTEM in the McGraw-Hill Encyclopedia of Science and Technology.

[A. C. TAYLOR]

Bibliography: L. E. Burnett and C. R. Bridges, The physiological properties and functions of ventilatory pauses in the crab, *Cancer pagurus*, *J. Comp. Physiol.*, 145:81–88, 1981; B. R. McMahon and J. L. Wilkens, Ventilation, perfusion, and oxygen uptake, in L. H. Mantel (ed.), *The Biology of Crustacea*, vol. 5, pp. 289–372, 1983; A. C. Taylor, Branchial ventilation in the burrowing crab, *Atelecyclus rotundatus*, *J. Mar. Biol. Ass. U.K.*, 64:7–20, 1984; J. L. Wilkens, P. R. H. Wilkes, and J. Evans, Analysis of the scaphognathite ventilatory pump in the shore crab *Carcinus maenas*, II. Pumping efficiency and metabolic cost, *J. Exp. Biol.*, 113:69–81, 1984.

Satellite astronomy

The *Solar Maximum Mission* is an Earth-orbiting solar observatory, operated through the National Aeronautics and Space Administration, that obtained data on solar activity during the sunspot maximum of 1980 and the decline to minimum in 1984–1985. The instruments aboard studied high-energy radiation from flares, dynamics of the corona, and variations of the total solar energy ouput. Analysis of *Solar Maximum Mission* data involved the efforts of over 400 solar scientists from 17 countries. Comprehensive observations led to a better understanding of the buildup, release, and propagation of flare energy. Observations of coronal activity led to an appreciation of the role of large-scale magnetic fields in major solar activity. High-precision monitoring of solar irradiance revealed temporal fluctuations

caused by the transit of sunspots across the solar disk, and a small but definite trend toward lower solar luminosity. The *Solar Maximum Mission* also was the first spacecraft to be repaired in orbit (April 1984) and operate successfully thereafter.

Instruments. Details of the instruments aboard the *Solar Maximum Mission* are given in the table. Five of the seven instruments observe the Sun in spectral ranges completely absorbed by the Earth's atmosphere. The gamma-ray experiment detects high-energy radiation from specific nuclear reactions. The hard x-ray burst spectrometer measures radiation from the interaction of energetic flare electrons with the ambient solar atmosphere. The hard x-ray imaging spectrometer produces pictures of the Sun in the light of hard x-rays. The x-ray polychromator uses two Bragg reflection spectrometers to measure soft (lower-energy) x-ray spectral line emission from gas heated by the flare energy release. The ultraviolet spectrometer polarimeter measures spectral line and continuum radiation from both flares and the quiet Sun; it could also measure solar magnetic fields. The coronagraph/polarimeter uses a television-type camera to take artificial eclipse pictures of the solar corona. The active cavity radiometer irradiance monitor measures the total power of the Sun directed at the Earth over a spectral band extending from the ultraviolet to the far-infrared.

The *Solar Maximum Mission* was launched on February 14, 1980, and operated for 9½ months. In December 1980, the satellite's fine-pointing system failed; only those instruments with full-sun fields of view (gamma-ray experiment, hard x-ray burst spectrometer, and active cavity radiometer irradiance monitor) could continue observations. The coronagraph/polarimeter and hard x-ray imaging spectrometer electronic control systems also failed. In April 1984, the space shuttle *Challenger* retrieved and repaired the satellite, returning it to active service.

Flares. Solar flares are explosive releases of energy (average total energy 10^{25} joules) beginning within an area of a few hundred miles; the size of a flare often increases a hundredfold within minutes. Hard x-ray imaging spectrometer, hard x-ray burst spectrometer, and x-ray polychromator observations showed that flares characteristically have two distinct temporal phases (Fig. 1). In the first, or impulsive, phase, radiation from the flare rises to a maximum in a few seconds or minutes, and can then vary rapidly in bursts with rise times as short as 10 ms. These x-ray bursts usually continue for several minutes with decreasing amplitude. Radiation from the second, gradual or thermal, phase appears soon after the beginning of the impulsive phase, grows in intensity as the impulsive bursts wane, and lasts up to several hours; this radiation is less energetic (softer) than that of the impulsive phase. Impulsive-phase radiation arises from the initial heating of the solar atmosphere; the separate bursts imply that several energy releases can occur in rapid succes-

Solar Maximum Mission instruments

Instrument		Field of view, arc-minutes	Spatial resolution, arc-seconds	Temporal resolution, s	Spectral range
Gamma-ray experiment (GRE)		Full sun	None	2	0.3–100 MeV
Hard x-ray burst spectrometer (HXRBS)		Full sun	None	0.128	20–490 keV
Hard x-ray imaging	fine field of view	2.8	8	0.5	3.5–30 keV
spectrometer (HXIS)	coarse field of view	6.4	32	0.5	3.5–30 keV
X-ray	flat crystal spectrometer (FCS)	2	15	70	0.185–1.897 nm
polychromator (XRP)	bent crystal spectrometer (BCS)	6	None	11.26	0.177–0.323 nm
Ultraviolet spectrometer polarimeter (UVSP)		4	3	0.1	115–360 nm
Coronagraph/polarimeter (C/P)		88	6	120	446.5–658.5 nm
Active cavity radiometer irradiance monitor (ACRIM)		Full sun	None	120	Ultraviolet to infrared

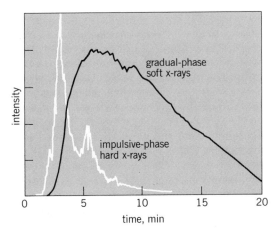

Fig. 1.   Intensity of x-rays from a flare as a function of time. Hard x-rays of the impulsive phase were observed by the hard x-ray burst spectrometer (HXRBS). Lower-energy (soft) x-rays of the gradual or thermal phase were observed by the x-ray polychromator (XRP).

sion. Gradual phase radiation arises from radiative cooling of gas heated to tens of millions of degrees during the impulsive phase.

Skylab x-ray telescope images showed (as did hard x-ray imaging spectrometer and x-ray polychromator images) that flares occur in magnetic loops filled with high-temperature, low-density gas. These coronal loops (Fig. 2) span tens of thousands of miles and rise a comparable distance above the photosphere. Gravity causes the gas density at the loop apex to be much smaller than at the photospheric footprints. Hard x-ray imaging spectrometer images of the impulsive phase showed that the most energetic x-rays appear almost simultaneously, within seconds, at the widely separated footpoints of a magnetic loop. This heating creates a cloud of high-energy electrons (constrained within the loop by the magnetic field) that passes through the low-density loop to deposit its energy in the high-density chromospheric footpoints. Timing experiments by the ultraviolet spectrometer polarimeter and hard x-ray burst spectrometer found that the beam thus formed generates a simultaneous flash of x-rays and ultraviolet line radiation on impact with the lower solar atmosphere. If the loop density is somewhat higher,

the beam may lose energy throughout its path, causing the loop to light up everywhere simultaneously, but still may deposit most of its energy at the footprints. In very energetic flares, gamma rays from nuclear reactions appear within a few seconds of each hard x-ray burst. The impact of the electron beam rapidly heats the dense gas at the footpoints, and causes it to expand up the loop; the radiation of the gradual phase originates in this hot gas. The x-ray polychromator observed that dissipation of the electrons' energy results in x-ray line emission at characteristic temperatures above 10^7 K. A blue-shifted copy of the spectrum also appears during the beginning of the flare, showing that part of this gas is moving upward with a velocity of several hundred miles per second. This higher-density gas fills the loop within a few tens of minutes, cools slowly by radiation and conduction, and eventually drains back down to the photosphere.

The flare energy release and particle acceleration mechanisms are still not understood in detail; indeed, the beam scenario itself is still controversial. Only the magnetic fields defining the loops contain

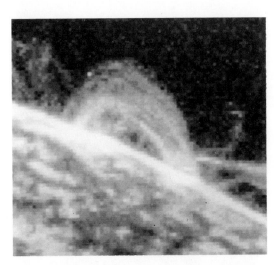

Fig. 2.   Coronal loop seen projecting above the edge of the Sun; the picture was made in ultraviolet light by the ultraviolet spectrometer polarimeter (UVSP). The Sun's disk extends from the bottom of the picture to the bright diagonal line.

Fig. 3. Coronal mass ejection of April 14, 1980. The black circular area at lower right covers the Sun's disk, and has a radius on the image of 6×10^5 mi (10^6 km).

enough energy to power the flare. Gamma-ray experiment and hard x-ray burst spectrometer observations of the repetitive onset of radiation simultaneously over the range from 40 keV to greater than 50 MeV severely constrain any model of flares. In a likely model, the magnetic fields of different loops

Fig. 4. *Solar Maximum Mission* secured in the cargo bay of the space shuttle *Challenger*. The observatory is about 20 ft (6 m) tall.

interact to produce strong electric fields and energy dissipation, much as in a short circuit produced by the touching of current-carrying wires. Flares have often been observed in regions where loop systems appear to interact; the bursts of the impulsive phase may be the result of repeated interactions between adjacent loops.

Corona. The solar corona, the region of hot (2×10^6 K or 4×10^6 °F) tenuous gas beginning about 6×10^3 mi (10^4 km) above the photosphere, reflects the structure of large-scale solar magnetic fields. The coronagraph/polarimeter continued studies of the large eruptions of material (coronal mass ejections) first studied by the *Skylab* coronagraph in 1973. An average coronal mass ejection contains 9×10^{12} lb (4×10^{12} kg) of gas moving at 280 mi/s (450 km/s), and has a spatial scale of several solar radii. In addition, the coronagraph/polarimeter observed almost continuous motion of all coronal structures.

Coronal mass ejections often have a distinctive three-part structure (Fig. 3): a bright filamentary core surrounded by a relatively dark space, both surrounded by a smooth, bright loop. From its structure and from the presence of hydrogen line emission, the core can be identified as an erupting prominence. The dark space is a high-magnetic-field, low-density region that surrounds the prominence. The leading loop is ionized coronal material set into motion by the rise of the strong magnetic field region initially below it. Another type of ejection may be seen as the expulsion of a large closed bubble of magnetic field and ionized gas. These eruptions, together with prominence eruptions and flares, may signify changes in the global solar magnetic field.

In 1980, coronal mass ejections occurred with a frequency of 0.9 per day, only 20% above the rate seen during the previous solar minimum, considerably lower than that expected from correlations with sunspot number. The rate near minimum in 1984-1985 was, however, significantly lower. More ejections occurred at high solar latitude during maximum than at minimum. Coronal mass ejections occur most often in association with prominence eruptions. They also appear following many flares, but a causal link between the two is less clear.

Solar variability. The active cavity radiometer irradiance monitor detected transient decreases of up to 0.3% in the total solar power emitted in the direction of the Earth; these variations were clearly related to the transit of dark sunspots across the disk of the Sun. A sunspot is dark because its strong magnetic field reduces the efficiency of gas convection, drastically inhibiting the flow of energy from the solar interior through the spot. The observations imply that, rather than flowing around the sunspot, the blocked energy must be stored for long times beneath the visible surface of the Sun. The active cavity radiometer irradiance monitor also observed a diminution of 0.1% in the average level of solar irradiance over a period of several years.

In-orbit repair. Because the *Solar Maximum Mission* was the first satellite equipped for retrieval by the space shuttle, in 1981 NASA decided to repair it in orbit. The attitude control system was designed to be replaced easily. The coronagraph/polarimeter main electronics box, mounted on an outside panel of the instrument housing, was easily accessible; the hard x-ray imaging spectrometer electronics, however, could not be repaired without endangering the other instruments. The space shuttle *Challenger* was launched into orbit on April 6, 1984, and approached the *Solar Maximum Mission* early on April 8. Several attempts to capture the satellite failed, causing the *Solar Maximum Mission* to tumble wildly. Since the solar power panels could not point to the Sun, the satellite's batteries drained rapidly. The tumble was stopped by using a set of on-board electromagnets to produce compensating torques by reaction against the Earth's magnetic field. On April 10, *Challenger* again approached the *Solar Maximum Mission,* and captured it with the manipulator arm (Fig. 4). The following day, two astronauts repaired the attitude control system and coronagraph/polarimeter in a 7-h spacewalk. *Solar Maximum Mission* was released at 308 mi (495 km) altitude in operational condition on April 12. Over the next month, all satellite and solar instrument systems were tested and put into full operation. Solar observations resumed in June 1984 and continued for over a year.

For background information *see* SATELLITE ASTRONOMY; SPACE SHUTTLE; SUN; X-RAY ASTRONOMY in the McGraw-Hill Encyclopedia of Science and Technology.

[RAINER M. E. ILLING]

Bibliography: A. Chaikin, Solar Max: Back from the edge, *Sky Telesc.,* 67(6):494–497, 1984; S. P. Maran and B. E. Woodgate, A second chance for Solar Max, *Sky Telesc.,* 67(6):498–500, 1984; Solar Maximum Mission science team, Descriptions of the satellite and scientific instruments, *Solar Phys.,* 65(1):5–109, 1980; R. C. Willson, H. Hudson, and M. Woodard, The inconstant solar constant, *Sky Telesc.,* 67(6):501–503.

Satellite navigation systems

The first use of the NAVSTAR satellite Global Positioning System (GPS) initiated a new era in positioning and navigation. The system includes a constellation of 18 satellites in 12-h orbits (19,650 km altitude) inclined at 60° from the Equator. Originally designed principally for military navigation, GPS can also be used as a revolutionary tool for geodetic and other nonmilitary positioning.

GPS signal. Each NAVSTAR satellite transmits GPS signals at two carrier frequencies, designated as L1 (1.6 GHz) and L2 (1.3 GHz). The most precise (and accurate) GPS techniques record carrier phases at specific points in time, providing measurements with a precision of a few millimeters, which is several percent of the 19-cm (L1) and 24-cm (L2) wavelengths. Both L1 and L2 are modulated at 10 MHz by a precision code (p-code) with a 29-m wavelength. This code carries satellite time and orbit information. The p-code is classified, and is designed for use in military navigation and positioning. The L1 carrier (but not L2) is also modulated at 1 MHz by a clear/acquisition code (c/a-code) with a 290-m wavelength which is available for civilian use. The wavelengths of the carriers and their modulations can be visualized as basic measuring units, ranging from 0.19 to 290 m. The highest accuracies can be derived from carrier phase measurements, followed by p-code and c/a-code measurements.

Point and relative positioning. Point positioning determines the coordinates of a single, stationary or moving, GPS receiver in an Earth-center-fixed coordinate system. GPS was designed primarily for real-time, dynamic, point positioning to an accuracy of several meters. The use of either the p-code or c/a-code to obtain the coded GPS message is required for point positioning.

Even before the development of GPS, the very long-baseline interferometry (VLBI) technique was in use. VLBI, using observations of remote radio sources called quasars, can determine the relative positions of radio telescopes separated by thousands of kilometers, to an accuracy better than 10 cm (4 in.). Quasars, which are located in the outer reaches of the cosmos 15 billion light-years away, are point sources of radio noise. In VLBI, radio noise originating from a number of quasars is recorded along with a precise time index from a hydrogen maser, the most accurate clock in existence. The noise recordings from each quasar are compared to find the exact time delay between the two sites for which the noise is correlated. This time delay multiplied by the speed of light, for quasars at a variety of viewing angles, allows an accurate measurement of the vector distance, or baseline, between the two VLBI sites.

The VLBI technique can also be applied to GPS signals, in which case relative carrier phase measurements (or relative phases of either of the modulated signals) can be obtained without knowledge of either the p-code or the c/a-code. The signal is treated as noise, and recordings from receivers with an accurate time index are compared for coherency. GPS receivers using this codeless approach cannot be used for point positioning. Of course, GPS receivers can also take advantage of the coded information for determining point or relative positions. Such receivers possess the disadvantage that they are useless if the codes are not available, as a result of intentional scrambling by military authorities or for any other reason.

Error correction. Major sources of error in GPS positioning and navigation stem from NAVSTAR orbit uncertainties, as well as signal refraction in the ionosphere and atmosphere. Errors from these sources tend to cancel for relative positioning, particularly over baselines shorter than 100 km (60 mi), and are therefore considerably smaller than point position errors. However, it is possible to correct for these major error sources, resulting in a significant improvement in either point or relative positioning accuracy.

The NAVSTAR ephemeris, or orbit, broadcast in the GPS coded message is a prediction based on the laws of physics and also on p-code ranging data from five tracking stations that are widely located globally. Corrections to this prediction are uploaded to each satellite twice per day. Although uncertainties in the broadcast ephemeris are estimated to be about 100 m (330 ft), it is possible to determine improved orbits by using carrier phase observations by GPS receivers located at sites with well-known coordinates, perhaps established by other space positioning techniques such as VLBI, satellite laser ranging, or lunar ranging. Such corrections, if they are carefully obtained, can improve the orbit accuracy by a factor of 100 or more.

Ionospheric refraction (delay) errors depend upon the total electron content of the ionosphere, which is quite variable on a diurnal and seasonal basis, and is strongly linked to solar activity. The average ionospheric delay during the night is 3 m (10 ft), and during the day 15 m (50 ft). However, this error can be corrected if both the L1 and L2 frequencies are observed, because it is frequency-dependent.

Atmospheric refraction errors depend upon wet (water vapor) and dry (other constituents) parts of the atmosphere. A barometric pressure measurement at a GPS receiver site is usually sufficient to correct the roughly 3-m (10-ft) dry part to 1 cm (0.4 in.) or so. The wet part, although smaller, varies from several centimeters to 1 m (3.3 ft) or more. It is also much more variable, and is difficult to determine by humidity or other local measurements at the receiver site, particularly during disturbed weather conditions. However, an instrument known as a water vapor radiometer, looking passively at microwave energy emitted by atmospheric water molecules along the line of sight to the observed NAVSTAR satellites, can obtain a good correction for wet refraction, again to 1 cm (0.4 in.) or so.

High-accuracy test. In the spring of 1985, eight days of GPS observations were carried out from 10 sites ranging across North America by researchers interested in improving the accuracy of GPS relative positioning measurements. Coded and codeless GPS receivers were used, and water vapor radiometers were colocated with receivers at three California sites. Orbit, ionospheric, and atmospheric corrections are possible by using the data from this test.

The coordinates of test sites which were occupied in Florida, Massachusetts, and Texas have been determined to 5-cm (2-in.) accuracy with VLBI. The GPS observations from these sites will be used for fiducial-point orbit correction, which can be visualized as a photographic snapshot of satellites from known locations, providing information that is used for instantaneous orbit determination. For the test, corrections for the ionosphere will be calculated from L1 and L2 GPS data, and atmosphere corrections will be calculated from water vapor radiometer data. The improved orbits will be used in place of the GPS broadcast orbits, in conjunction with ionospheric and atmospheric corrections, to explore possible improvements in GPS relative positioning accuracy.

During the past several years, the ability to determine vector baselines between GPS receivers has been demonstrated routinely at the level of 1 part per million [1 cm (0.4 in.) in all three coordinates of a 10-km (6-mi) baseline, for example]. It was expected that the spring 1985 test and others to follow would demonstrate 1-cm (0.4-in.) accuracy for baselines up to 100 km (60 mi) or more.

Crustal dynamics. The Earth's tectonic plates typically move relative to one another at rates of several centimeters per year. The plates often move erratically, and such behavior can be associated with earthquakes. If movements on the order of millimeters can be measured over distances of tens of kilometers, accumulated strain fields can be determined. The monitoring of strain fields of earthquake zones using GPS appears to be much easier and more accurate than alternate measuring techniques. Therefore, GPS offers a practical means of obtaining crucial strain data for use in earthquake prediction.

Future applications. The use of GPS promises revolutionary ease and accuracy not only for positioning measurements related to earthquake studies, but also for volcanics, subsidence, gravity, deflection of the vertical, and a host of other positioning and geodetic problems. Centimeter-level relative positioning for rapidly moving receivers is also possible. The carrier wavefronts of GPS signals originating from the satellites creates a dynamic phase space with a roughly 20-cm (8-in.) grid spacing that is predictable from the broadcast GPS message. If two GPS receivers starting at relative positions known to centimeter accuracy are able to predict the dynamic phase space and to count the wavefronts as the receivers move relative to one another, they can maintain the initial centimeter positioning accuracy. In the next decade, it is reasonable to expect that electronic circuitry for GPS receivers will be incorporated in very large-scale integrated circuits, and that positioning equipment will become pocket-sized and easily affordable, and will attain unprecedented accuracy. Navigation and tracking equipment will then be commonly used as a personal guide, or as a robotic navigator, for road, rail, water, air, and space transportation activities.

For background information *see* EARTHQUAKE; GEODESY; PLATE TECTONICS; SATELLITE NAVIGATION SYSTEMS in the McGraw-Hill Encyclopedia of Science and Technology.   [RANDOLPH WARE]

Bibliography: J. D. Bossler, C. C. Goad, and P. L. Bender, Using the Global Positioning System (GPS) for geodetic surveying, *Bull. Geod.*, 54:553–563, 1980; T. H. Dixon, M. P. Golombek, and C. L. Thornton, Constraints on Pacific Plate kinematics and dynamics with Global Positioning System measurements, *IEEE Trans Geosci. Remote Sens.*, GE-23: 491–501, 1985; J. J. Spilker, GPS signal structure and performance characteristics, *J. Inst. Navig.*, 25:121–146, 1978; R. H. Ware, C. Rocken, and J. B. Snider, Experimental verification of improved GPS-measured baseline repeatability using water vapor radiometer corrections, *IEEE Trans. Geosci. Remote Sens.*, GE-23:467–473, 1985.

Sea-level fluctuations

This article discusses some of the significant work that has been reported on the geological history of sea level. It presents a critical analysis of the Vail sea-level curve as well as isotopic studies relating changing ice volumes as a measure of sea-level change as far back as 100 million years ago.

Coastal onlap and sea-level change. As viewed from the continents, the level of the sea fluctuates on a wide range of time scales, from the approximately twice-daily tidal cycle to spans of hundreds of millions of years. The best stratigraphic record of sea-level change is found in the thick sedimentary successions that accumulate on subsiding continental margins produced by the rifting and separation of formerly contiguous continental masses. A few simple considerations show, however, that it is difficult to distinguish local changes of sea level (relative to a continent) from global or eustatic ones.

At any point on a continental margin, the tendency for sea level to rise or fall during a given interval of time is governed by the rate of change in eustatic sea level, and by the rate of subsidence minus the rate of sediment accumulation. The rate of subsidence generally increases seaward, but sediments are deposited preferentially in coastal areas and on the continental shelves. It is thus possible for sea level to rise relative to the outer part of a margin, but to fall with respect to the inner part, even when eustatic sea level is constant; and depending on the availability of sediment, seaward movement of one segment of a shoreline (regression) may be accompanied by landward movement of another segment (transgression).

Seismic stratigraphy. One approach to the investigation of sea-level fluctuations is the technique of seismic stratigraphy, which was developed for interpreting multichannel seismic reflection profiles (Fig. 1). Instead of attempting to gauge changes in bathymetry or the location of the shoreline, seismic stratigraphy makes use of regional stratigraphic discontinuities, known as unconformities or sequence boundaries. These are thought to form during times of relatively rapid sea-level fall, when patterns of sediment transport and deposition shift abruptly, and they are indicated in Fig. 1 by near-horizontal bold lines. In comparison with other measures of sea-level change, the development of such unconformities is insensitive to local variations in sediment supply, but the method provides information primarily about the timing of sea-level fluctuations, not about the magnitudes of rises and falls.

The gross geometry of unconformities and other stratal surfaces is recognizable in seismic sections because these surfaces are an important source of acoustic impedance contrasts. Acoustic impedance, or the product of rock density and the velocity of seismic waves traveling through the rock, is the

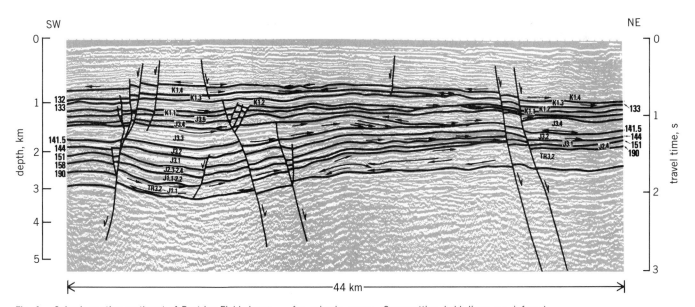

Fig. 1.   Seismic section northeast of Beatrice Field, Inner Moray Firth, North Sea, United Kingdom, showing interpretation of seismic sequences defined by the termination of seismic events (full arrows). Numbers with J and K prefixes identify sequences. Numbers on either side of section are estimated ages of the sequence boundaries in millions of years. Depth is given in kilometers and two-way travel time for seismic waves. Cross-cutting bold lines are inferred faults, with half arrows indicating apparent sense of displacement. 1 km = 0.6 mi. (*From P. R. Vail et al., Jurassic unconformities, chronostratigraphy, and sea-level changes from seismic stratigraphy and biostratigraphy, in J. S. Schlee, ed., Interregional Unconformities and Hydrocarbon Accumulation, AAPG Mem. 36, p. 130, 1984*)

property that determines whether or to what extent seismic energy is reflected at a given surface. Most reflection events seen on a seismic section are composites of reflections from individual interfaces, but the configuration of reflections mimics the configuration of stratal surfaces. For noise-free data, achievable vertical resolution is typically about one-eighth to one-fourth wavelength, or tens to hundreds of meters, depending on velocity and frequency, and it tends to deteriorate with depth. Although seismic sections resemble geological cross sections, not all seismic events possess primary stratigraphic significance. Examples of nonstratigraphic events are those produced by diagenetic boundaries and low-angle faults, together with features such as multiples, coherent noise, diffractions not migrated during data processing to their proper position, and energy returned from outside the plane of the section. An experienced interpreter generally can recognize spurious events, although they may mislead a novice.

Chronostratigraphy. The first step in seismic stratigraphic interpretation is to develop a chronostratigraphic framework by identifying unconformities on a grid of intersecting seismic lines. Chronostratigraphy refers to any procedure for achieving time correlation in layered sediments or rocks. Although unconformities usually represent depositional hiatuses of variable duration, most possess chronostratigraphic significance in that it is common for strata above an unconformity to be everywhere younger than strata below it. Exceptions are certain unconformities produced in the deep ocean adjacent to continents by the lateral migration of topographically intensified oceanic boundary currents, and some unconformities in tectonically active parts of the continents where the locus of tectonically driven subsidence migrates with time. Neither of these exceptions appears to be important in the shelf and nearshore regions of most continental margins produced by continental dispersal.

Unconformities are recognized in seismic sections by the oblique termination of seismic events, as shown by full arrows in Fig. 1. Some events terminate downward by onlap (updip) or by downlap (downdip), whereas others terminate upward by toplap or erosional truncation (Fig. 2). Toplap arises from sediment bypassing across the top of a prograding sedimentary wedge, whereas erosional truncation involves the removal of previously deposited sediment. Combinations of these stratigraphic relations are also possible, and seismic events locally parallel unconformities, especially where the latter pass laterally into correlative conformities, that is, boundaries for which there is effectively no depositional hiatus.

Dating unconformities. The next step in interpretation is to date the unconformities. This is done best by establishing the ages of strata overlying and underlying each unconformity where it becomes a conformity. Borehole or well data routinely are tied to seismic sections by means of synthetic seismograms derived from velocity and density logs, and relative ages are obtained from fossils (biostratigraphy). Where such logs are available, the major limitations to achievable age resolution are related to the lack of appropriately positioned boreholes, the use of cuttings rather than cores, errors or lack of resolution in biostratigraphy, and uncertainties about the numerical ages of biostratigraphic boundaries.

Sea-level changes. Dated unconformities correspond to times of rapidly falling relative sea level, and are therefore related to both subsidence history and eustacy. Some unconformities appear to be of global extent, and may be due to an overriding eustatic control. Other unconformities are less widespread, suggesting that rates of subsidence and sea-level change are generally comparable, a result supported by theoretical considerations of possible causes of eustatic fluctuations. The only known mechanism for changing sea level significantly faster than the rate of subsidence of a typical continental margin involves the storage of water on the continents in the form of glacial ice, but there is little

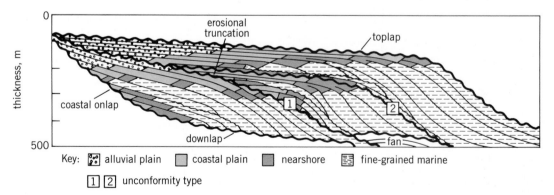

Fig. 2. Diagrammatic stratigraphic cross section showing sequence boundaries (wavy bold lines) defined by onlap, downlap, toplap, and erosional truncation (explained in text). Fine lines are geological time lines corresponding to stratal surfaces, and are parallel to events seen on a seismic section. Type 1 unconformities are present in submarine and subaerial environments, whereas type 2 unconformities are subaerial only. 1 m = 3.3 ft. (*After P. R. Vail et al., Jurassic unconformities, chronostratigraphy, and sea-level changes from seismic stratigraphy and biostratigraphy,* in *J. S. Schlee, ed., Interregional Unconformities and Hydrocarbon Accumulation, AAPG Mem. 36, p. 135, 1984*)

evidence for significant glacial ice during much of the past 200 million years, the interval for which most of the seismic stratigraphic data are available.

Eustatic curve. Estimates of the magnitudes of sea-level change have been attempted by using relative changes of coastal onlap, but there are difficulties with the method even if sea-level changes are regarded as relative ones. The published procedure for constructing charts of relative changes of coastal onlap is reproduced in Fig. 3, and a composite of global cycle chart for the Jurassic Period is shown in Fig. 4. The composite chart qualitatively allows for different rates of subsidence in different sedimentary basins, but is scaled from 1.0 to 0.0 arbitrary units because the rates of subsidence are uncertain. The coastal onlap curve originally was interpreted as a measure of relative sea-level change, and this led to considerable discussion as to why sea-level fluctuations should show a sawtooth pattern. It is clear, however, that the asymmetry is an artifact of the different methods used for calculating rises and falls, and the fact that onlap can occur in a broad range of alluvial to marine environments. In Fig. 3, the rise of 400 m (1300 ft) during cycle A includes an indeterminate component of subsidence, but the rapid fall of 450 m (1500 ft) between cycles A and B includes the differential subsidence (or tilting of sequence A) during cycles B to D (broken line in Fig. 3*a*). It could be argued in this specific case that because the stratal surfaces are parallel over the area of measurement in sequences B to D, no correction is needed for differential subsidence. However, this would imply that the uppermost position of onlap in sequence A was at a correspondingly higher elevation above sea level (located near the boundary between coastal and marine deposits). Lowest onlap in sequence B was close to sea level, and the downward shift in onlap would again overestimate the relative sea-level fall. Coastal aggradation in sequence A would be overestimated too, but to a lesser extent.

The derived short-term eustatic curve on the right side of Fig. 4 was constructed by smoothing the composite onlap curve and assuming that the unconformities are predominantly of eustatic origin. If this assumption is valid, the timing of eustatic falls is relatively well known, but the amplitudes of sea-level changes are not well established. The only constraint from seismic stratigraphy is the assumption that type 1 unconformities are probably associated with greater (or more rapid) sea-level falls than type 2 unconformities. A type 1 unconformity is one present in both submarine and subaerial environments. During formation, sediment tends to bypass the continental shelf and to accumulate preferentially on deep-sea fans (Fig. 2). Type 2 unconformities are subaerial only.

The long-term curve, which in Fig. 4 provides an envelope for short-term variations, is also an approximation. A more reliable estimate of long-term eustatic changes has been derived for the Creta-

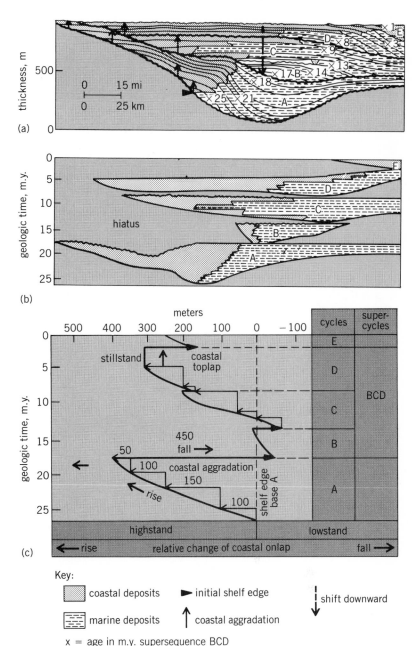

Fig. 3. Procedure for constructing regional chart of cycles of relative changes of coastal onlap, from estimates of coastal aggradation and downward shift in coastal onlap. (*a*) Stratigraphic cross section. (*b*) Chronostratigraphic chart. (*c*) Regional chart of cycles of relative changes of coastal onlap. The letters A to E are arbitrary labels for the five depositional cycles shown. A supercycle is a group of cycles during which there are only minor downward shifts in onlap. The numbers in *a* correspond to ages in millions of years (compare with *b* and *c*). 1 m = 3.3 ft. (*After P. R. Vail et al., Seismic stratigraphy and global changes of sea level, in C. E. Payton, ed., Seismic Stratigraphy: Applications to Hydrocarbon Exploration, AAPG Mem. 26, p. 78, 1977*)

ceous and Cenozoic from calculations of the bathymetry of the ocean basins through time, but insufficient data are available for the Jurassic. An additional limitation in applying oceanic data to long-term sea-level change on the continents is that the continents are not all characterized by the same hypsometry (the distribution of surface area as a function of elevation), and evidence exists that con-

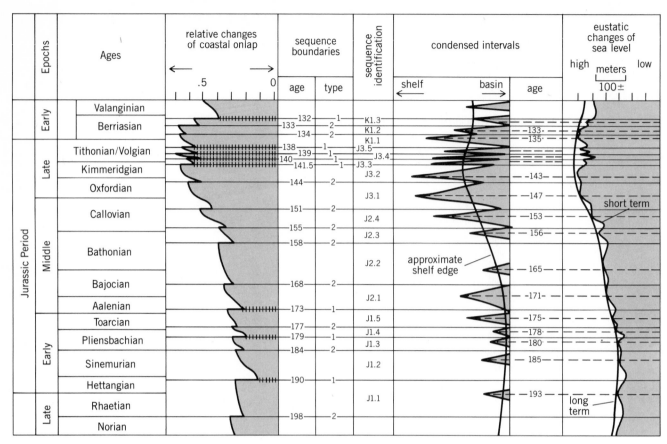

Key: ┼┼┼┼┼┼ times at which no coastal onlap occurs

Fig. 4.   Global cycle chart of the Jurassic supercycle showing age (in millions of years) and types of sequence boundaries (see Fig. 2), condensed intervals, relative changes of coastal onlap (see Fig. 3), and derived eustatic changes of sea level, assuming that changes in onlap are primarily of eustatic origin. Condensed intervals correspond with times of rapid sea-level rise, and sequence boundaries, with times of rapid sea-level fall. 1 m = 0.3 ft. (*After P. R. Vail et al., Jurassic unconformities, chronostratigraphy, and sea-level changes from seismic stratigraphy and biostratigraphy, in J. S. Schlee, ed., Interregional Unconformities and Hydrocarbon Accumulation, AAPG Mem. 36, p. 132, 1984*)

tinental hypsometry has changed through geological time. Thus a different long-term curve is needed for each continent.

[NICHOLAS CHRISTIE-BLICK]

Deductive stratigraphy. Studies of deep-sea sediments now place glacio-eustacy on an intellectually sound foundation which should allow prediction of eustatic sea-level fluctuation throughout perhaps 90% of the Phanerozoic. A chronostratigraphy based upon abiotic attributes of the world ocean is developing rapidly.

Deep-sea δ ^{18}O *record as an ice volume signal.* Whereas the vast majority of the surface on the Earth contains oxygen-16, a small portion contains the stable isotope, oxygen-18. Because water constructed with oxygen-16 possesses lighter molecules, these two isotopes of oxygen tend to become fractionated during the hydrologic cycle. During evaporation, oxygen-18 tends to be left behind; during precipitation, oxygen-18 tends to be the first molecule rained out of the clouds.

The significant geological consequence of this fractionation is that ice stored on the surface of the Earth tends to be largely depleted in oxygen-18. Ice stored on the continents is a semipermanent reservoir of isotopically light water. As the size of this reservoir varies, so sea level and the isotopic composition of the world ocean must vary. Calcium carbonate–secreting organisms (planktonic and benthic foraminifers are especially useful) capture this isotopic information in the oxygen isotopic composition of their shells. The measurement is made by mass spectrometric comparison of a sample to a standard (such as PDB calcite) and is reported as δ ^{18}O in parts per thousand relative to the standard. Measurement of the variation in isotopic composition of these shells down a deep-sea core actually represents a measurement of the variation in a continental ice volume and, thereby, fluctuation in glacio-eustatic sea level.

Numerous Quaternary deep-sea stratigraphic sections yield a similar pattern of oxygen isotopic variation downcore. Spectral analysis of these records yields the various frequencies long known to exist in the perturbations of the Earth's orbit. The principal orbital periodicities are as follows: precession, 19 thousand years (Ka) and 23 Ka; tilt, 41 Ka, and eccentricity, 95 Ka, 123 Ka, and 423 Ka. As soon

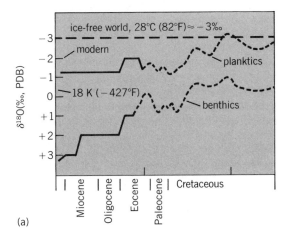

(a)

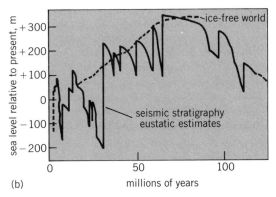

(b)

Fig. 5. Comparison of (a) the 100-million-year smoothed oxygen isotope glacio-eustatic record with (b) the seismic stratigraphy "relative sea level curve." The solid and broken-line curves in a are models developed by two separate research groups. The Y axes on the two curves are plotted to comparable scale [approximately 1‰ δ ^{18}O (calcite oxygen-18 content, relative to PDB standard) variation equals 91 m (300 ft) sea-level variation]. The oxygen isotope data suggest smaller events. Problems also exist concerning timing of events. These probably reflect major problems in inductive-mode data reduction in seismic stratigraphy. 1 m = 0.3 ft. (After R. K. Matthews, Oxygen isotope record of ice-volume history: 100 million years of glacio-eustatic sea-level fluctuation, in J. S. Schlee, ed., AAPG Mem. 36, pp 97-107, 1984)

as it became apparent to geologists that the Earth's ice volume had fluctuated in the past, some scientists, operating in deductive mode, predicted that these frequencies should be observed in the variation in global ice volume. Unfortunately, it took 50 to 100 years for observational technology to match the task set forward by the prediction.

An attempt to test this simple a priori model on δ ^{18}O data from Cretaceous and Tertiary deep-sea cores will be only partly successful. The vast majority of these isotopic data were gathered in inductive mode. The investigators did not have well-formed expectations as to what they should observe. More importantly, they did not sample with a test of the orbital frequencies hypothesis in mind. Whereas a test of the hypothesis would require a 5000-year sample interval, the vast majority of Cretaceous and Tertiary δ ^{18}O data on deep-sea cores were gathered

at approximately 200,000–500,000-year sample interval. Nevertheless, a few generalities are possible; these are depicted in Fig. 5a.

Perhaps the most encouraging piece of data in Fig. 5 is the observation of −3‰ planktonic δ ^{18}O values for mid-Cretaceous time. This is precisely the number which should be observed for an ice-free world with a tropical sea-surface temperature of 28°C or 82°F (today's value). The plate-tectonic reconstructions make the Cretaceous a most likely candidate for an ice-free world; it cannot be demonstrated unequivocally that a continent existed beneath a pole at this time. The fact that an ice-free mid-Cretaceous yields the expected δ ^{18}O value for a constant-temperature tropical surface ocean lends credence to this fundamental a priori assumption. Given constant tropical sea-surface temperature throughout the last 100 million years, δ ^{18}O values more positive than −3‰ reflect the existence of a semipermanent ice reservoir somewhere in the global system.

Another interesting feature of the Tertiary tropical planktonic δ ^{18}O record is the tendency toward isotopically heavier (more glacial) values from mid-Cretaceous through Oligocene time. The shift to more positive values occurs in two or more steps. Of these, the shift at or near the Eocene-Oligocene boundary is the best studied. When the isotopic data are compared to the seismic-stratigraphic "relative sea-level curve" (Fig. 5b), two problems are observed. First, the magnitude of the glacio-eustatic event is only about 50 m (160 ft), whereas seismic stratigraphy depicts a 400-m (1300-ft) sea-level fall in this same general time interval.

The second aspect of this event which is worthy of note is the discrepancy as to chronostratigraphic position. Biostratigraphy on deep-sea cores places the major isotopic event at or near Eocene-Oligocene boundary, whereas biostratigraphy of continental margins places the major seismic-stratigraphic event in the middle Oligocene. As a strong working hypothesis, this is perhaps a striking example of biostratigraphy as an inadequate chronostratigraphic framework within which to solve dating problems.

An abiotic chronostratigraphic framework. In late Quaternary studies, the δ ^{18}O ice volume signal itself has been used to establish an abiotic chronostratigraphic framework. This technology relies upon pattern recognition in the shape of a time series signal. Although this technology is in theory applicable throughout the Tertiary, a great deal of data will have to be generated to allow pattern recognition sufficient to establish a tight chronostratigraphic framework. Strontium isotopes appear to offer a much more straightforward opportunity to establish an abiotic chronostratigraphic framework for the last 50 to 150 million years.

As with the oxygen isotope technology, calcium carbonate precipitated from sea water samples the strontium isotopic composition of a well-mixed ocean. With regards to the last 150 million years or so, an elegantly simple reason exists to expect that

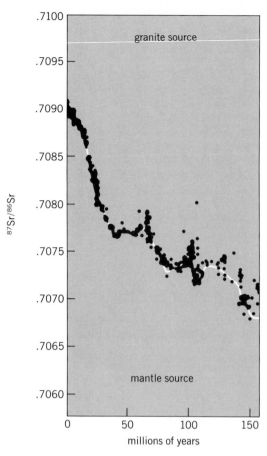

Fig. 6. Variation in strontium isotope composition of the shells of marine organisms through time. The well-mixed ocean averages the contribution of strontium isotopes from two sources. The fact that the relative importance of these two sources has varied with time affords a biotic chronostratigraphy which should be widely applicable. (*After W. H. Burke et al., Variation of seawater* $^{87}Sr/^{86}Sr$ *throughout Phanerozoic time, Geology, 10:516–519, 1982*)

the strontium isotopic composition of sea water should vary through time in a period. In the Cretaceous, far less high mountains existed than today, continents do not appear to have been significantly glaciated, and the rate of sea floor spreading appears to have been faster than today. All of these factors favor a strontium isotopic ratio for a Cretaceous sea water closer to mantle-derived values and an isotopic composition for modern sea water closer to granitic cratonic values. As indicated in Fig. 6, the simple model holds up beautifully for the last 40 million years and holds up reasonably well for the last 150 million years.

This technology holds promise for sorting out the various questions concerning Eocene-Oligocene or large middle-Oligocene relative coastal offlap. Analytic precision allows for chronostratigraphic precision on the order of a few hundred thousand years, whereas the biostratigraphic problems involved are on the order of several million years.

Perhaps more important, strontium isotopes promise to deliver a precise chronostratigraphic framework to high latitudes (such as Antarctica) for the first time. A better understanding of ice volume his-

tory on the Antarctic continent is clearly high priority to testing the glacio-eustatic interpretation of low-latitude planktonic δ ^{18}O stratigraphy.

Forward modeling. With a firm chronostratigraphic framework, a tectonic model for basin subsidence, and a good estimate of glacio-eustatic sea-level fluctuations, forward model problems which heretofore have only been dealt with inductively by the stratigrapher can be attacked deductively. Rather than studying the rocks in great detail and seeking to "explain the data," forward modeling can be used to characterize the interaction of basic subsidence with glacio-eustatic sea-level fluctuations. The model is then compared with the stratigraphic situation, and either more stratigraphic data are gathered to test the model prediction or the model is modified in light of the partial success of the model to match observations. And so the process goes on—modeling, observing, modeling again, and observing again—until model results and stratigraphic data converge.

The glacio-eustatic component to be modeled appears to be on the scale of tens of meters. These events are certainly within the error bars of any inductive approach to unraveling local tectonics from glacio-eustacy. Yet, glacio-eustatic events on the order of tens of meters can move shorelines by tens of kilometers. Similarly, subaerial exposure of bank margin carbonates has a profound effect on porosity and permeability of these rocks regardless of the amplitude of the glacio-eustatic signal. There should be good reasons to want to unravel the interrelationship between subsiding basin and the glacio-eustatic sea-level fluctuations. Forward modeling of these problems offers the most direct opportunity for success.

For background information *see* CONTINENTAL MARGINS; MARINE SEDIMENTS; SEA-LEVEL FLUCTUATIONS; SEA WATER; SEISMIC STRATIGRAPHY; STRATIGRAPHY in the McGraw-Hill Encyclopedia of Science and Technology.

[R. K. MATTHEWS]

Bibliography: W. H. Burke et al., Variation of seawater $^{87}Sr/^{86}Sr$ throughout Phanerozoic time, *Geology*, 10:516–519, 1982; J. Imbrie et al., The orbital theory of Pleistocene climate: Support from a revised chronology of the marine δ ^{18}O record, in A. L. Berger et al. (eds.), *Milankovitch and Climate, Part 1*, pp. 269–305, 1984; R. K. Matthews, Oxygen isotope record of ice-volume history: 100 million years of glacio-eustatic sea-level fluctuation, in J. S. Schlee (ed.), *Interregional Unconformities and Hydrocarbon Accumulation*, Amer. Ass. Petrol. Geol. Mem. 36, pp. 97–107, 1984; W. J. Morgan, Hotspot tracks and the early rifting of the Atlantic, *Tectonophysics*, 94:123–139, 1983; W. C. Pitman III, and X. Golovchenko, The effect of sealevel change on the shelfedge and slope of passive margins, in D. J. Stanley and G. T. Moore (eds.), *The Shelfbreak: Critical Interface on Continental Margins*, Soc. Econ. Paleontol. Mineralog. Spec. Publ. 33, pp. 41–58, 1983; P. R. Vail et al., Seismic stratigraphy and global changes of sea level, in C. E. Payton (ed.), *Seismic Stratigraphy: Applica-*

tions to Hydrocarbon Exploration, Amer. Ass. Petrol. Geol. Mem. 26, pp. 49–212, 1977; P. R. Vail, J. Hardenbol, and R. G. Todd, Jurassic unconformities, chronostratigraphy, and sea-level changes from seismic stratigraphy and biostratigraphy, in J. S. Schlee (ed.), *Interregional Unconformities and Hydrocarbon Accumulation*, Amer. Ass. Petrol. Geol. Mem. 36, pp. 129–144, 1984.

Ship design

The planing hull form is perhaps the oldest, simplest, and most ubiquitous high-speed marine craft in operation today (Fig. 1). Tens of thousands of these craft run on lakes, rivers, and coastal waters. Unfortunately, many planing craft designs have a well-deserved reputation for poor performance and have been properly stereotyped as underpowered craft which subject the structure and passengers to severe pounding, wetness, and discomfort when operating in a seaway. Fortunately, appropriate application of recently developed planing hull technology has resulted in the development of hull forms which have excellent performance in both calm water and waves.

Planing hull design. The planing hull form evolved to overcome the inherent hydrodynamic limitations associated with high-speed operation of the traditional displacement hull. It is useful to briefly compare both hull types in order to ascertain what changes were brought about by the demand for higher speeds and why the shape of the planing hull evolved as it has. This is best accomplished by reference to the speed-length ratio, $V_k/\sqrt{L}$, commonly used by naval architects to speed-characterize marine vehicles. In this notation, V_k is the boat speed in knots (1 knot = 0.5 m/s), and L is the waterline length of the boat in feet (1 ft = 0.3 m) at zero speed.

The translation of a hull through the water produces surface waves which travel at the same speed as the hull and have a length which is dependent upon the square of the boat speed. At $V_k/\sqrt{L} = 1.3$, the length of the generated wave is equal to the length of the boat, and at higher boat speeds ($V_k/\sqrt{L}$ greater than 1.3) the wave length is greater than the hull length (Fig. 2). Two detrimental effects are

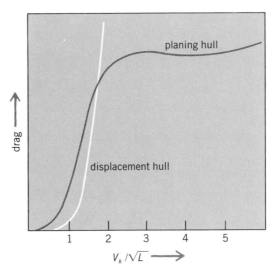

Fig 2. Plot of drag versus speed-length ratio ($V_k/\sqrt{L}$) for displacement hull and planing hull.

associated with the hull-generated waves: (1) there is an expenditure of ship power to create and maintain the traveling wave train, and (2) for that speed range where the generated waves are longer than the hull, the boat will trim to the slope of the wave it has generated—adding considerably to the water resistance of the hull and hence to the powering requirement.

In addition to the surface-wave formation associated with the ratio $V_k/\sqrt{L}$, there is a trade-off between dynamic and hydrostatic (buoyant) pressures which act on the hull bottom and support the weight of the boat. At low speeds corresponding to $V_k/\sqrt{L}$ less than, say, 0.60, the dynamic pressures can be neglected and the boat weight will be supported by buoyancy. At higher speeds dynamic pressures on the bottom increase as the square of the speed and can dominate the buoyant pressures.

For a displacement ship, the water flow along the convex curved bottom will develop negative or, more popularly, suction forces on the bottom. At nominal speeds these suction forces have only a small effect upon the trim, draft, and boat resistance. However, as the speed increases, the negative bottom pressures increase as the square of the speed, and result in large bow-up trim, large increases in draft, and enormous increases in resistance and required propulsion power. The displacement hull actually sinks deeper and deeper into the wave it is generating, and it is this hydrodynamic phenomenon which limits the speed of displacement hulls. In fact, the maximum economical speed of a well-designed displacement hull corresponds to a speed-length ratio, $V_k/\sqrt{L}$ of approximately 1.2. Thus a 25-ft-long (7.5-m) displacement craft can have a maximum speed of approximately 6 knots (3 m/s), a 100-ft-long (30-m) boat approximately 12 knots (6 m/s), and a 900-ft-long (270-m) ship approximately 36 knots (18 m/s).

The planing hull form is configured to develop positive hydrodynamic pressures on its bottom so

Fig. 1. Typical planing hull craft. *(Tiara/Slickcraft)*

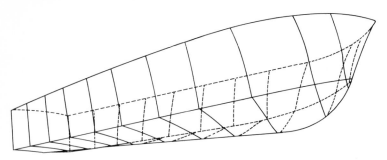

Fig. 3.   Details of hard-chine planing hull.

that its draft actually decreases with increasing speed, enabling it to ride higher on the wave it is generating, thus avoiding the enormous drag increases associated with displacement hulls when run at high speed. The sharp contrast between the drag of a typical displacement hull and that of a high-speed planing hull is particularly evident at speed-length ratios in excess of 1.2 (Fig. 2).

To attain positive dynamic pressures, the planing hull form eliminates convex curvature of the buttock lines so that the keel and chines are straight in the aft part. There is, of necessity, some convex curvature of buttock lines in the bow area—but this part is above the water at high speed. Whereas, in the displacement ship, all means are taken to reduce flow separation at the stern, and to preserve the smooth flow conducive to the recovery of pressure at the stern, in a planing boat the straight buttock lines are cut off clearly by the stern so as to induce early flow separation. The transverse section is typically a deadrise section with a sharp intersection of the bottom and sides to form a hard chine. Details of a high-speed planing hull are shown in Fig. 3.

Traditional configurations. Early planning hull designs were guided almost entirely by the requirement for high speed in calm water with minimal drag and consequently minimal power. The relationship between hull drag in smooth water and the primary planing variables at full planing speed shows that the minimum drag occurs at a boat trim of approximately 4° and a bottom deadrise angle of 0°. A review of the proportions of past planing hull designs indicates that the design features of a predominant number of boats had beamy forms (length-beam ratios between 3 and 4); had very low deadrise angles; and were loaded to cause the boat to run near 4° trim angle at maximum speed. If operation on a calm water surface were the only consideration, these hull features and proportions would be entirely satisfactory.

Unfortunately, high-speed planing craft must operate in waves where the hull bottom continuously impacts against the oncoming waves, resulting in large pitching and heaving motions of the hull, in heavy spray sheets which obliterate the driver's view and flood the decks; and in large impact forces which can damage hull structure and cause serious injury to passengers and crew. These actions in-

crease with boat speed and eventually become intolerable, thus thwarting the high-speed potential of those hulls which have been designed only for smooth-water operation. Accordingly, these traditional planing hull forms have obtained a well-deserved reputation for poor performance.

Modern configurations. The total performance of planing craft can be enhanced substantially by applying the following design principles.

1. *Increased hull length-beam ratio.* The designer should attempt to configure the planing bottom to be as long and as narrow as possible, consistent with the requirements of internal arrangements. Whereas traditional planing craft designs had length-beam ratios of approximately 3.5, the modern planing hull has a length-beam ratio in excess of 5.0. Such a modern hull has low powering requirements through its speed range and excellent performance in a seaway.

2. *Increased bottom deadrise.* The bottom pressures associated with hull-wave impact decrease substantially with increasing deadrise angle. Whereas traditional planing hulls have average deadrise angles of less that 10°, the modern planing hull has approximately 20° deadrise.

3. *Reduced running trim angle.* The seakeeping of planing hulls improves substantially with decreasing running trim angle. Whereas traditional planing hulls operated at trim angles approaching 4°, the modern hull operates at approximately 2° at high speed. The addition of trim control flaps at the transom provides a simple mechanism to adjust the boat trim to desired values.

4. *Double-chine hull.* It has been demonstrated that, all other conditions being equal, the narrower the beam, the smaller the hull-wave impact loads at high planing speeds. Figure 4 demonstrates the body plan for the modern double-chine planing hull, wherein the narrower inner beam width reduces the wave-hull impact loads and the wider beam width provides the necessary transverse stability at low speed. This hull form has proved very successful in recently constructed operational planing craft.

In summary, the continuing trend toward sustained high-speed operation in waves has stimulated the development of modern planing hull forms which

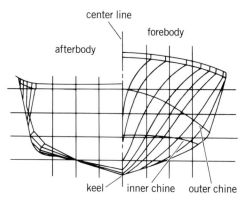

Fig. 4.   Body plan for modern double-chine planing hull.

can operate successfully in waves without compromising their calm-water performance.

For background information see SHIP DESIGN; SHIP POWERING AND STEERING in the McGraw-Hill Encyclopedia of Science and Technology.

[DANIEL SARITSKY]

Shock-wave chemistry

Study of chemistry in solids and liquids under conditions of high-pressure shock compression is one of the latest fields to emerge in chemical science. The shock-compression process involves extremes in pressure and high-speed motion. In the process, pressure is applied and released in durations of perhaps 1 microsecond. Under these conditions, matter is briefly forced into extreme states outside the range of more conventional experience. Understanding the chemical consequences of the shock process presents formidable challenges to both chemistry and shock-compression science. Application of the shock process to materials technology presents opportunities for chemical synthesis of new materials or modification of materials to achieve unusual properties. Probing the shock process by chemical techniques provides the means by which major new insight can be acquired into the state of shock-compressed matter.

Shock-compression process. A high-pressure shock-compression process is initiated in a solid or liquid when a sample is placed in intimate contact with the detonation of a high explosive or subjected to the impact of a high-velocity projectile. Exposure to intense pulsed radiation can also produce significant shock pulses. The compression and unloading pulse are inertial responses to the loading applied at a boundary of a sample. In solids, the shock pulses may have a complex shape with considerable structure in beginning, middle, and end phases. Unlike a static pressure process, the pressure and its temporal signature are explicitly controlled by density and wave speed properties of the sample under the particular conditions of the experiment. Wave speed is significantly influenced by a number of mechanical, structural, thermal, and chemical characteristics such that the usual experiment involves problems that are traditionally studied in a diverse range of scientific disciplines. Until recently, chemistry had been little considered. The self-controlled nature and short duration of the loading process presents special challenges and opportunities to experimental design and interpretation. As the compression occurs over a duration too short for significant heat conduction, there are shock-induced increases in temperature which also depend explicitly on inertial properties of the material.

Pressures and temperatures can be produced over a wide range. In solids, typical pressures achieved range three orders of magnitude from a few hundred megapascals to hundreds of gigapascals ($\sim 10^4$–10^7 lb/in.^{2}). Likewise, temperature increases vary from tens to thousands of degrees. Liquids, with their much larger compressions at corresponding pressures, are heated to substantially higher temperatures than solids. Solids, with their characteristic strengths, undergo extensive plastic deformation such that states with high densities of crystal defects are produced.

Unique materials observed in the natural shock-compression process of meteoritic impact (as seen, for example, in the Meteor Crater of northern Arizona) provide insight into opportunities for materials synthesis. It has been suggested that the life-forming process of large molecular synthesis from simple forms may have been the result of meteoritic impact.

Diamonds are now routinely produced for industrial processes from graphite by a shock process, and other industrial uses of shock compression in material processing are widespread.

Liquid substances. The simpler nature of the bonding of liquids compared to solids allows fundamental probing of the process of chemical changes with spectroscopic techniques. Emission spectroscopy experiments have been carried out which identify reaction products of benzene while it is under shock compression, and Raman spectroscopy measurements have been performed on benzene and water under shock compression. The shock-compression properties of a number of interesting liquids suspected of undergoing chemical changes have been studied. Detailed optical-band edge absorption measurements on shock-compressed carbon disulfide have been carried out. Static high-pressure spectroscopic techniques are being applied to liquids in a diamond cell for comparison to similar measurements under shock compression. The ability to apply these conventional spectroscopic tools to liquids under shock compression promises to provide detailed insight into chemical processes under these extreme conditions.

Shock synthesis of interesting polymers from the liquid state was shown to be possible many years ago. Polymers which are difficult or virtually impossible to polymerize under normal conditions are sometimes readily polymerized under shock compression. The fundamental spectroscopic studies of shock-induced reactions will provide a scientific basis for better control of synthesis with the shock process.

Solid substances. Chemistry in solids proceeds by quite different mechanisms than in liquids. The differences are a consequence of the strong atomic bonding in the solid state and the resulting low atomic mobility. As a result, solid-state reactivity and the speed of reactions and reaction products are controlled by the defects in the solid. Not only are the concentrations of defects critical to chemical reaction, but the particular types of defects (dislocations, point defects, and so forth) and their arrangements within the sample are controlling factors. The synthesis of new solid materials can have major impact on modern technology, and most of the field of shock-compression science is concerned with solid materials.

Solids have characteristic strengths under shock

compression which range from perhaps 100 MPa ($\sim 1.5 \times 10^4$ lb/in.^{2}) to 20 GPa ($\sim 3 \times 10^6$ lb/in.^{2}). When stresses greater than this are applied, the solid fails mechanically and flows due to plastic deformation. Under shock compression this plastic deformation occurs very rapidly, and a variety of defects are produced in large concentrations. Thus, study of chemical processes in shock-compressed solids involves detailed determination of shock-induced defects and their relation to solid-state reactivity. The influences of defects are then superposed on the effects of compression of the lattice and the increases in temperature.

Shock modification and shock activation of inorganic powders is being studied along with shock-induced chemical synthesis. It is found that catalytic activity can be increased by many orders of magnitude, that reaction temperatures and rates of solid-state reactions can be significantly affected, and that high-temperature dissolution of solids in liquids can be strongly accelerated. Structural transitions can also be strongly affected. Intermetallic aluminides have been synthesized which appear to have interesting properties, and ferrites with unusual magnetic properties can be modified or synthesized by shock compression. Shock-induced chemical synthesis in the solid state appears to be relatively easy to accomplish and subject to detailed control. The method appears to have considerable potential to affect materials technology.

For background information *see* CRYSTAL DEFECTS; SHOCK WAVE; SHOCK WAVES IN SOLIDS; SOLID-STATE CHEMISTRY in the McGraw-Hill Encyclopedia of Science and Technology.

[ROBERT A. GRAHAM]

Bibliography: G. A. Adadurov and V. I. Gol'danskii, Transformations of condensed substances under shock-wave compression in controlled thermodynamic conditions, *Russ. Chem. Rev.*, 50:948, 1981; R. A. Graham et al., Materials modification and synthesis under high pressure shock compression, *Annu. Rev. Materials Science*, publication pending; *Proceedings of the 15th International Symposium on Shock Waves and Shock Tubes*, Lawrence Berkeley Laboratory, July 29–August 1, 1985; Shock waves in condensed matter, *Proceedings of the American Physical Society Topical Conference on Shock Waves in Condensed Matter*, Spokane, Washington, July 22–25, 1985.

Skin

Artificial skin is a two-layer polymeric membrane that has been used experimentally with burn patients to induce partial regeneration of the two layers of skin, dermis and epidermis. The top layer of artificial skin comprises a silicone rubber membrane which protects the wound from infection and dehydration, and is removed 2 weeks following grafting. The bottom layer is a highly porous cross-linked network of collagen and chondroitin 6-sulfate which serves as a biodegradable matrix for the growth of new dermis. If the bottom layer has been seeded with a few epidermal cells, previously removed from the individual to be grafted, the artificial skin is capable of inducing growth of a new epidermis as well.

Skin structure. Skin is made up of two layers. The external layer, the epidermis, is being continuously turned over, sloughed off, and regenerated. It acts as a highly keratinized moisture seal that controls the water flux out of the body. The epidermis also controls the loss of body chemicals while preventing toxic substances and bacteria from entering. While the epidermis is made up almost entirely of cells and is mechanically rather weak, the inner layer of skin, the dermis, comprises a dense, tough network of collagen fibers and relatively few cells, and possesses superior mechanical strength.

The dermis provides a mechanically competent support for the epidermis while acting as the tough layer that protects internal organs against injury from external mechanical forces. If destroyed, as in a deep skin cut or in a fire, the dermis does not spontaneously regenerate, but instead scar tissue forms at the site of the injured tissue. Scar tissue is morphologically distinct from intact dermis.

Burn treatment. Fire can injure skin either superficially, destroying the epidermis alone (first-degree burn), seriously, destroying part of the dermis as well (second-degree burn), or completely, destroying the dermis through its entire thickness until muscles or bones exposed (third-degree burn). An excellent treatment for a third-degree burn consists in first removing all dead tissue and covering the exposed wound with an autograft. The latter is a layer of skin, consisting of the epidermis and about one-half the thickness of the dermis, that has been removed from an uninjured area of the patient's skin. Autografting causes additional trauma, but the grafts cover deep skin wounds very effectively, providing excellent protection from bacterial invasion and fluid loss while also minimizing contraction and scarring. Grafting deep wounds with transplant skin is also practiced widely, but these grafts must be removed after a few days before the onset of rejection by the patient's immunologic reaction. Synthetic polymeric materials have also been used to cover deep wounds, but they have to be removed after 10–20 days of grafting and replaced with autografts.

Artificial skin development. The development of artificial skin at the Massachusetts Institute of Technology (MIT) occurred in two phases. Stage 1 artificial skin, developed on the basis of studies with animals done in 1975, is capable of inducing regeneration of a new dermis (neodermis). Stage 2 artificial skin, developed on the basis of studies with animals in 1980, is capable of inducing regeneration not only of a neodermis but of a new epidermis (neoepidermis) as well.

Stage 1 skin. Stage 1 artificial skin is a highly porous, biodegradable membrane made of a cross-linked network of collagen and chondroitin 6-sulfate, a glycosaminoglycan (GAG). Since 1979, the biodegradable membrane, previously covered with a nonbiodegradable layer of silicone rubber, has been used to treat over 60 victims of serious burns at Boston's Massachusetts General Hospital. Both in animal experiments, which were initiated in 1970, and in studies with humans, it has been observed that Stage 1 membranes are capable of inducing synthesis of a neodermis whose microscopic structure is nearly identical to that of intact dermis. In particular, the morphology of collagen fibers, the extent of vascularization, the presence of specialized cells such as melanocytes and keratinocytes, and the presence of elastin fibers and of nerve tissue are all consistent with the morphology of intact dermis rather than that of scar.

Stage 2 skin. An improvement in the procedure for preparing the bilayer membrane led to development of Stage 2 artificial skin. In the new procedure, which was extensively worked out with guinea pigs, a dime-sized (about 0.4 in.^{2} or 2.4 cm^2) biopsy of the skin of the animal is removed and is treated with an enzyme solution until the least mature epidermal cells are separated out; these cells are immediately inoculated into a Stage 1 membrane by use of a centrifugation procedure. The stage 2 graft prepared in this manner is sutured onto a full-thickness skin wound. After 14 days a new epidermis (neoepidermis) spontaneously forms. The neoepidermis continues to mature until, by about 3–4 weeks, it has become indistinguishable from the intact epidermis in all apparent respects except for the absence of hair. In the meantime, formation of a mature neodermis proceeds under the neoepidermis, so that by the end of a month following grafting the neodermis possesses at least 50% of the mechanical strength of intact dermis. The maturation process continues for several weeks after grafting. Stage 2 grafts are remarkable in that they induce new synthesis of two morphogenetically distinct tissues in their correct final anatomical relationship.

Critical features of artificial skin. Critical features responsible for the performance of artificial skin include a number of physicochemical variables, such as the surface energy, the flexural rigidity, and the moisture flux rate through stage 1 membranes. These variables are responsible for the ability of the artificial skin to provide prompt closure of the deep wound. Almost immediately following grafting, however, enzymes present in the wound bed begin to break down the long-chain molecules of collagen into smaller fragments consisting of amino acids. The breakdown process is slowed down by the presence of GAG. This biodegradation process occurs simultaneously with another process, the migration of mesenchymal cells from the wound bed into the porous interior of the collagen-GAG (CG) membrane and the synthesis therein of a neodermis by these cells. The rate of biodegradation of the CG membrane must be critically adjusted in order for the graft to induce synthesis of a neodermis rather than a scar. Ongoing research at MIT also shows that it is necessary that the average pore diameter of the CG membrane should be adjusted to a

critical level. In addition, the seeding of the CG layer with epidermal cells prior to grafting is essential for synthesis of a neoepidermis.

Currently, Stage 2 membranes are being tested with patients to find out if the regrowth of an almost complete skin which occurs with animals can be duplicated in humans without a change in the procedure for preparing the artificial skin.

Extensive additional trials with patients in several clinical centers in the United States must be completed before artificial skin can become available for general use to treat burn patients.

For background information *see* INTEGUMENT; REGENERATION (BIOLOGY); SKIN in the McGraw-Hill Encyclopedia of Science and Technology.

[IOANNIS YANNAS]

Bibliography: J. F. Burke et al., Successful use of a physiologically acceptable artificial skin in the treatment of extensive burn injury, *Ann. Surg.*, 194:413, 1981; I. V. Yannas et al., Multilayer membrane useful as synthetic skin, U.S. Pat. 4,060,081, November 29, 1977; I. V. Yannas et al., Prompt, long-term functional replacement of skin, *Trans. Amer. Soc. Artif. Organs*, 27:19, 1981; I. V. Yannas et al., Wound tissue can utilize a polymeric template to synthesize a functional extension of skin, *Science*, 215:174, 1982.

Soil

Low-activity clay is a differentia used in *Soil Taxonomy* (which presents a system of soil classification in current use in the United States) to distinguish soils that have significant differences in the mineralogical composition of the clay fraction and thus in physical-chemical behavior and management. This article discusses low-activity clay soils in the United States, Central and South America, and Africa.

The term low-activity clay is used for soils with low cationic electrochemical potential or cation-exchange capacity (CEC) per unit weight of the clay fraction. Specifically, the term is used to define those clay-enriched subsurface horizons that meet the requirements of Kandic and have a total CEC of < 16 meq per 100 g clay (determined by ammonium acetate at pH 7), or an effective CEC of < 12 meq per 100 g clay (determined as the sum of exchangeable bases plus potassium chloride–extractable aluminum).

The management-related properties associated with soils having low-activity clay are uniquely different from analogs with high-activity clay. Examples of accessory physical and chemical properties of low-activity clay soils are: low effective cation-exchange capacity (CEC), low permanent charge relative to pH-dependent charge, moderately low specific surface area, low heat of wetting, low plasticity, low shrink-swell potential, high shear resistance (dependent more on internal friction than cohesion), low organic matter contents, low base saturation with reciprocally high exchangeable aluminum percentages, low plant nutrient reserves, high phosphorus and metallic trace-element retention, buffering of pH between 4.5 and 4.8 by exchangeable and interlayer aluminum, soil solution of low ionic strength, low retention of plant-available water, moderate to high permeability, dominance of geochemically stable 1:1 layer lattice aluminosilicates and iron and aluminum oxyhydroxide clays, and few weatherable minerals.

UNITED STATES

The extent of low-activity clay soils in the United States is over 54 million acres (22 million hectares). Most of the low-activity clay soils are Ultisols and Alfisols in the southeastern sector of the United States along the Southern Piedmont, Carolina and Georgia Sand Hills, and Southern Coastal Plain major land resource areas (Fig. 1).

Factors in soil formation. Factors involved in soil formation of low-activity clay soils include parent material, geological age, topography, climate, and biology.

Parent material. Low-activity clay soils have developed from many igneous, metamorphic, and sed-imentary bedrock types in the United States. Sedimentary deposits of marine and fluvial origin have been polycycled through erosional, deposition, and weathering environments. Many of the parent materials were initially acidic. The largest occurrence of these soils is on higher elevations of the Southern Coastal Plain, where subsoils and parent materials range from 18 to 35% in clay content (Fig. 1). This results in soils which are sufficiently permeable to favor deep leaching and weathering of parent materials that are initially rich in kaolinite and other low-activity clays.

The Western Coastal Plain has a smaller percentage of low-activity clay soils because the fluviatile deposits contain higher contents of bases, carbonates, and 2:1 layer lattice clay minerals. Further, loess additions appear to have been a major impediment to low-activity clay development because they provided a renewal vector for bases, weatherable minerals, 2:1 layer lattice clays, and silica. This slowed the advance of weathering, provided a geochemical environment that stabilized 2:1 layer lattice clays, and retarded formation of low-activity

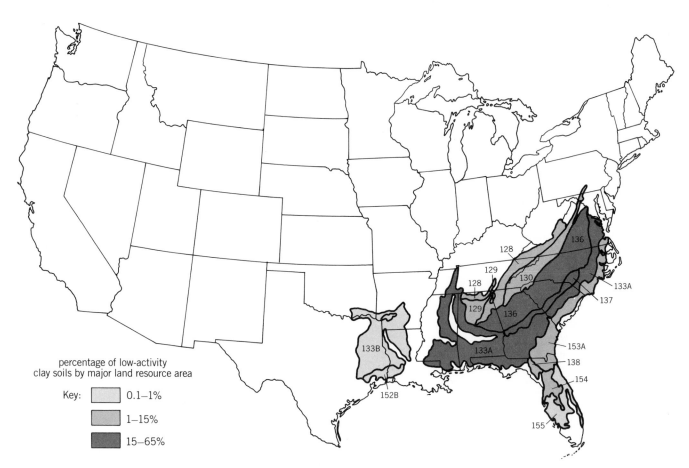

Fig. 1. Areal distribution of low-activity clay soils in mainland United States as a percentage of major land resource regions. Numbers identify major land resource areas: 128, Southern Appalachian Ridges and Valleys; 129, Sand Mountain; 130, Blue Ridge; 133A, Southern Coastal Plain; 133B, Western Coastal Plain; 136, Southern Piedmont; 137, Carolina and Georgia Sand Hills; 138, North Central Florida Ridge; 152B, Western Gulf Coast Flatwoods; 153A, Atlantic Coast Flatwoods; 154, South-Central Florida Ridge; 155, Southern Florida Flatwoods. (*After J. D. Nichols, L. P. Wilding, and D. E. Pettry, Low activity clay soils of the United States, in H. Eswaran, ed., Properties, Classification, and Management of Low-Activity Ultisols and Alfisols, Soil Management Support Services, USDA-Soil Conservation Service, 1985*)

percentage of low-activity
clay soils by major land resource area

Key:
0.1–1%
1–15%
15–65%

clays. Pyroclastic inputs to this region during Tertiary periods would have a similar impact on soils derived from exposed coastal plain sediments.

The parent materials for low-activity clay soils of the Southern Piedmont and Blue Ridge are weathered from granites, gneisses, and schists. These acid rocks develop soils high in kaolinite and other low-activity clays. The weathering zone extends below the base of the soil into several meters of saprolite (weathered rock) before hard bedrock is encountered.

The parent materials for low-activity clay soils in the Southern Appalachian Ridges and Valleys and the Sand Mountain major land resource areas are weathered from limestone, sandstone, siltstone, and shales. Many of the sandstone, siltstone, and shale deposits are composed of preweathered, kaolinite-rich materials.

Geological age. The low-activity clay soils are on landscapes that are considered Pleistocene or older. The ages of the geomorphic surfaces where low-activity clay soils predominate are not well dated. The youngest surfaces are probably older than 100,000 years. Low-activity clay soils do not occur on late Pleistocene river terraces or on coastal plain surfaces less than about 50,000 years in age. This is either because the composition of these terraces contained higher amounts of weatherable minerals and bases, or there was insufficient time for advanced soil weathering. Here, as in most other soil weathering relationships, there are parent material, climatic, and biological interactions with the age factor that confound chronology relationships.

Topography. The low-activity clay soils occur at elevations ranging from 23 ft (7m) to over 6600 ft (2000 m), but most of them occur between 100 and 330 ft (30 and 100 m). These soils are mainly on nearly level to gently sloping stable upland interfluves in the Southern Appalachian Ridges and Valleys, Sand Mountain, Blue Ridge, and Coastal Plain major land resource areas. They occur on narrow divides and steep-sided slopes in the Piedmont Plateau. Most of the soils are well drained, although some are moderately well drained or even poorly drained. In general, the internal drainage must be sufficient to move bases and products of weathering to lower depths in the soil or into the geologic substratum materials. When poorly drained, low-activity clay soils have formed in base-depleted sediments which are not readily base-recharged by ground or surface waters.

Steep slopes and lower topographic elements comprise fewer low-activity clay soils than more stable upland interfluves because: (1) truncation of soils on steep erosional surfaces exposes less weathered parent materials and rejuvinates the weathering cycle; (2) movement of eroded sediments from upper to lower slopes rejuvenates the weathering cycle in the transported sediments; and (3) neoformation of high-activity clay minerals and geochemical stability of weatherable 2:1 layer lattice clay minerals (micas, smectites, and vermiculites) are favored in lower topographic postions with upslope inputs of bases and soluble silica.

Climate. The mean annual temperature ranges from 55 to 73°F (13 to 23°C) with most of the low-activity clay soils in the United States occurring near 64°F (18°C). The mean annual precipitation ranges from 40 to 62 in. (1000 to 1600 mm), but the preponderance of low-activity clay soils occurs in areas with about 47 in. (1200 mm) of precipitation. In the Western Coastal Plains, the climatic factor does not favor low-activity clay soils because of lower total precipitation, lower effective precipitation (P-E index), and greater surface runoff due to short-duration, high-intensity storm events.

Thus, warm humid climates, where the low-activity clay soils predominate, favor elevated chemical reaction rates because of high temperatures, long periods of biological activity, and sufficient quantities of precipitation that soil weathering products can be removed rapidly and progressively from the soil system.

Biology. The low-activity clay soils developed under a climax forest vegetation primarily of pine and hardwoods. However, no direct tie can be made between areas of low-activity clay soils and vegetation (or other biological factors), because biology does not appear to be a major determinant in governing formation of low-activity clay soils. It seems reasonable, however, that areas subjected to vegetation with low base-cycling capabilities, such as pine trees, would favor more acid weathering conditions and a soil system favorable for formation of low-activity clays. Likewise, soils with active microbiological activity and acid organic litter generate decomposition products rich in soluble organic acids that complex and chelate metallic cations, especially Fe and Al. Such biological processes enhance weathering by serving as a source of acidity, as a sink for selected weathering products (potassium, iron, and aluminum), and as a carrier to readily translocate these constituents to lower soil horizons.

Mineralogical composition. Dominant mineralogical components in low-activity clay soils are kaolinite (sometimes halloysite); hydroxy aluminum interlayered 2:1 layer clay minerals (pedogenic chlorites); iron, aluminum, and manganese hydroxides, oxyhydroxides, and oxides; and other minerals highly resistant to weathering such as quartz, zircon, rutile, and anatase. The major iron minerals are goethite and hematite (sometimes ilmenite and maghemite from mafic-rich rocks), while the aluminous minerals are gibbsite and, to a lesser extent, boehemite. Most clay-size minerals found in low-activity clay soils are poorly crystallized, have smaller particle size and higher surface areas, and have more isomorphic structural substitution (that is, iron in kaolinite and aluminum in iron oxides) than reference mineral species.

The processes responsible for low-activity clay soils involve in-place transformation of primary weatherable minerals to geochemically stable secondary minerals or the geological accumulation of such preweathered pedogenic weathering products as parent materials from which low-activity clay soils are formed. Many of the low-activity clay soils in coastal plains regions of the United States are attributed to the latter, while those associated with igneous and metamorphic bedrocks are more likely related to the former.

According to pedogenic weathering concepts the geochemical changes leading to low-activity clay soils depend on specific environmental conditions, including porosity, weathering rates, mineralogical composition, intensity of rainfall, and effectiveness of leaching. The most important processes in the transformation of primary minerals leading to low-activity clay systems are removal of basic cations and desilication. Most of the secondary minerals are direct transformations. For example, feldspars alter directly through a solution phase into halloysite, kaolinite, or gibbsite. Iron oxyhydroxides are formed by oxidation of ferrous weathering products. In contrast, most pedogenic chlorites are likely transformed from micas, vermiculites, or smectites by hydroxy aluminum polymers crystallizing between the interlayer structures.   [LARRY P. WILDING]

AFRICA

The low-activity clay soils of Africa include all the soils in which the clay fraction has a low cation-

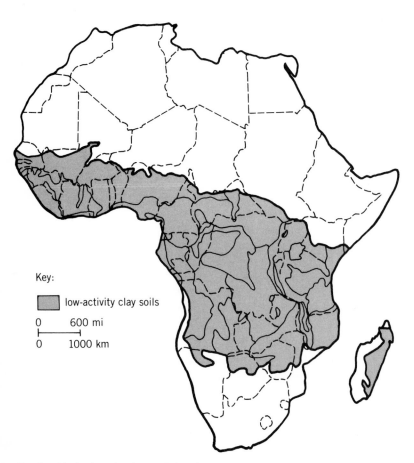

Key:

▨ low-activity clay soils

0   600 mi

0   1000 km

Fig. 2.   Distribution of soil associations in which low-activity clay soils occur in Africa. (*Compiled from the FAO-Unesco Soil Map of the World, 1977*)

exchange capacity, usually less than 24 meq/100 g clay. The CEC limit is higher than currently proposed to define Kandic horizon in *Soil Taxonomy*. They may have a horizon of clay accumulation or contain weatherable minerals in appreciable amounts. In terms of soil classifications, they belong to Oxisols, Ultisols, Alfisols, or Inceptisols. In Food and Agriculture Organization nomenclature they would be called Ferralsols, Acrisols, Luvisols, or Cambisols.

Figure 2 shows the broad soil regions of Africa. The area includes associations which contain soils in which the clay fraction has a low cation-exchange capacity.

The low-activity clay soils of Africa mostly formed in a thick mantle of detrital materials. Most sediments in which they developed are considered to have gone through more than one weathering cycle. In Africa they cover extensive areas both under savanna and rainforest vegetation. The length of the dry season may vary from zero to more than 6 months. They always received or presently receive intensive summer rainfall which causes strong leaching. Chemical decomposition and leaching are the main soil-forming processes responsible for the kind and the degree of weathering that the soil parent materials have undergone.

Stone-lines and ironstone are widespread features in the basement complex area of West Africa. Most sediments in which low-activity clay soils form were transported over variable distances. A great number of profiles show a characteristic layer sequence in which a topsoil overlies a stone-line which rests on rotten rock (saprolite) which grades into solid rock at great depth. Termites are often very active in mixing the soil materials above the stone-line.

The low-activity clays form mostly in well-drained soils under strong desilication and the residual concentration of kaolinitic minerals and sesquioxides; the removal of silica may have occurred under previous climatic conditions different from the present ones. An additional consequence to the strong weathering is the gradual decrease of the fine silt fraction (2–20 micrometers) to reach silt/clay ratios of less than 0.1 in the most advanced stages of weathering.

Most low-activity clay soils have strong colors in either yellow or red hues. These are the result of the presence of iron oxides which crystallized at higher temperatures than in temperate regions. Goethite and hematite are the main iron oxide minerals.

Recently studies have been made of goethite-containing samples from the eastern part of southern Africa, taken in nonhydromorphic highly weathered soils on old land surfaces derived mainly from sandstones and shales. All these goethites were highly aluminum-substituted (up to 32 mole % approximately) when overlying plinthite or ironstone; the aluminum substitution varied between 15 and 25 mole % when the subsoils were saprolites or bauxites. The degree of weathering and the acidity were considered conducive to high aluminum availability for its incorporation in the goethite structure. Hydromorphism caused a decrease in aluminum substitution, but it was not established whether there was a direct cause-and-effect relationship; there was a negative correlation between aluminum substitution and crystallinity. Organic matter seemed to lower the degree of crystallization.

Gibbsite is a common constituent in the clay fraction of soils developed on ultrabasic rocks, or when climatic conditions are such that leaching and desilication are extreme. Gibbsite found in young soils of Nigeria may have remained after other constituents had disappeared. The gibbsite could also have formed concurrently with halloysite in areas of the soil matrix where leaching is strongest. It may represent one of the initial products of weathering.

The major limitations to crop production in low-activity clay soils result from their low nutrient content. In the humid parts of Africa they have a low base status, and consequently a very low inherent fertility. Under virgin conditions, particularly under forest vegetation, they may have adequate nitrogen contents to support several crops after clearance, but they become rapidly deficient in nitrogen, due to fast mineralization of the organic matter. Leaching may also produce deficiencies in calcium, magnesium, and potassium after several years of cropping. Phosphorus, once the organic forms are depleted, becomes unavailable because of retention by iron oxides or because of precipitation with aluminum. In drier savanna climates the low-activity clay soils are exposed to severe erosion which may lead to deterioration of the soil resource.

[A. VAN WAMBEKE]

CENTRAL AND SOUTH AMERICA

Extensive regions of tropical America are occupied by low-activity clay soils. Their mineralogy reflects an advanced stage of weathering. The clay fraction is dominated by 1:1 clay minerals, especially kaolinite, iron oxides, and frequently gibbsite. Clay minerals with significant permanent charge such as smectites are usually absent. Quartz dominates the coarse fraction. These soils have been classified either as Ultisols and Alfisols or as Oxisols on the basis of the presence of argillic or oxic horizons, respectively. These Ultisols and Alfisols show remarkable differences in properties and management requirements from their counterparts in temperate regions.

Classification. This presents a problem in Central and South America. The Oxisol order was established to group those soils with characteristics associated with low-activity clays. However, these properties do not exclude the presence of an argillic horizon. In fact, in these tropical landscapes there is often a continuum of pedons with properties intermediate between modal Oxisols, without argillic horizons by definition, and Ultisols and Alfisols, with well-developed textural horizons. For the intergrades, classification by soil taxonomy is very difficult to establish, and the diagnostic criteria appears to be practically unoperational.

This problem is an important limitation of the present version of soil taxonomy. Different alternatives are being discussed, such as the introduction of a new diagnostic horizon, the kandic, or the definition of kandic great groups for low-activity clay Ultisols and Alfisols. Similar weaknesses are found also in other classification systems used in the tropics.

Distribution. A preliminary soil order map of tropical America is shown in Fig. 3. Oxisols constitute the most extensive soil order, covering 34% of the surface. Ultisols cover 22% and Alfisols 12%. Almost 60% of the acid, infertile soils of tropical America have a Udic moisture regime; another 30% are in ustic environments.

Low-activity clay soils have been recognized in almost every country in tropical America. They are fairly common in the higher rainfall areas of Central America, on gently sloping outwash plains of the Amazon and Orinoco basins, and on more dissected parts of the Brazilian and Guayanan shields. In udic environments on old landscapes the association Oxisols–low-activity clay Ultisols–Inceptisols is very frequent. In younger landscapes, a common association is low-activity clay Ultisols–low-activity clay Alfisols–Entisols.

In Brazil, Pale, Tropo, and Plinthic great groups of Udults and Ustults have been surveyed in the Transamazonic Highway region. These are the most extensive soils along the road. Pale, Rhodic, and Tropo great groups of Udalfs and Ustalfs are associated soils. In the Central Plateau, Paleustults, some Haplustults, and a few Plinthustults are found in dissected landscapes over acid gneisses. With more basic parent materials, Paleustalfs and Rhodustalfs are common.

In Peru, Paleudults seem to be the most extensive soils in the Amazonic plains. Plinthudults and Plinthic Paleudults are also developed from ancient alluvial sediments, in areas of inadequate drainage. Tropudults have been described throughout the hills and mountainsides with steep slopes.

In Colombia, highly weathered soils, sometimes with argillic horizons, have been identified in the Amazon region. It appears that in many places within the Amazon Basin, contrary to what has been so far believed, probably Ultisols and not Oxisols are predominant.

Soil management. Tropical America has significant potential for increase of agricultural production. However, several constraints must be overcome for its development. Many of these are soil-related constraints. In the humid tropics, population and soil distribution are closely related: people concentrate mainly on coastal and riverine alluvial plains or in volcanic areas. These are mostly high-base-status soils. Extensive areas of infertile low-activity clay soils and wet soils still remain available. In the subhumid zone, population is more dispersed but people still concentrate on the more fertile soils. Two complementary strategies for increasing food production emerge: to increase yield in existing farming systems on high-base-status soils, and to develop appropriate farming systems to expand the agricultural frontier in vast areas of acid, infertile soils.

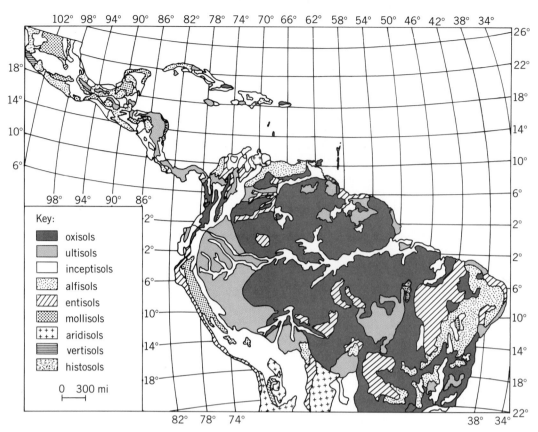

Fig. 3. Tentative soil order map for tropical America. (*After P. Sanchez and T. T. Cochrane, Soil constraints in relation to major farming systems of tropical America, in Priorities for Alleviating Soil-Related Constraints to Food Production in the Tropics, International Rice Research Institute, 1980*)

Solving soil-related constraints in low-activity clay soil areas is an important component of the second strategy. These constraints are mostly chemical: aluminum toxicity, phosphorus fixation, and nutrient deficiencies are among the most widespread.

Soil acidity and liming. The practice of liming low-activity clay soils has often given unsatisfactory results. Early recommendations were based on experience with soils of the temperate regions, and yield losses resulted. The approach of liming to raise the soil pH to near neutrality is not applicable to low-activity clay soils. The major exchangeable cation in these acid soils is aluminum. Crop production is drastically reduced when aluminum saturation is greater than 60% and tends to be optimum when it is zero. Aluminum essentially disappears from exchange positions as soil pH reaches values around 5.5. Lime rates sufficient to neutralize exchangeable aluminum are considerably lower than those required to raise the pH to near neutrality. In addition, liming to pH values of 6.5–7.0 may induce deficiencies of micronutrients (particularly boron, zinc, and manganese) and phosphorus fixation by calcium.

Soil acidity and varietal tolerance. Lime and fertilizers constitute one option to modify chemical status and improve production on low-activity clay soils. But the cost of these inputs on the farm will remain high and crop prices low, because of distance to markets and lack of transportation infrastructure, among other factors.

Another approach which requires minimal inputs is the breeding of species better adapted to this infertile, acid environment. There are forage grasses and legumes, tree crops, and starchy food crops which have been used by American tropical farmers for centuries. This genetic pool may be used for new breeding programs. Unfortunately, there are few cereal grains and food legumes in this group. Upland rice and cowpeas are the most promising crops.

Phosphorus fixation. Phosphorus fixation is another frequent constraint in these low-activity clay soils, particularly where surface horizons have medium or finer textures. It may also occur in sandy soils if the topsoil has been eroded. Phosphorus is fixed into slowly soluble forms by precipitation and sorption reactions with iron and aluminum compounds.

A high-input option to deal with this problem consists of a heavy initial fertilization (quenching the high phosphorus fixation capacity) followed by residual effects for several years. The high initial investment and uncertainty of long-term use of the land by the farmer are serious limitations to this approach for countries in the region.

A low-input option consists of applying phosphorus in bands to satisfy fixation capacity in a small soil volume. This practice may be combined with selection of cultivars tolerant to low available phosphorus levels and utilization of cheaper amendments to reduce phosphorus fixation and sources of phosphorus other than expensive superphosphates. This approach appears to be more appropriate for tropical America.

Land-use strategies. The traditional land-use system for most of the low-activity clay soil region of tropical America has been shifting cultivation. Although in its traditional form it may be considered an ecologically sound farming system, it does not offer a solution for supplying sufficient food to an increasing population. When pressure on land increases, the fallow period is shortened and the process of land degradation is triggered. Land-use strategies to increase food production in the region are conditioned by the biophysical environment and socioeconomic considerations.

Continuous crop production: a high-input option. A high-input technology is available to put some of these low-activity clay soils under continuous crop production. In the humid area, where topography is gentle, and wetness limitations are absent, three grain crops (rice–corn or peanuts–soybeans) may be produced annually with a sound land clearing system, mechanization, and appropriate lime and fertilizer inputs. Such a technology, originally developed at Yurimaguas in Peru, has been validated by local farmers. However, some economic and infrastructure considerations are needed. Investments are high, and market accessibility is needed to obtain inputs and sell the production. These are major constraints which render this strategy less attractive.

Continuous crop production: a low-input option. Greater immediate potential applicability among small farmers is foreseen in a low-input strategy. Research efforts are concentrated by the TROPSOILS Program in Peru in order to refine this approach. It is based on acid-tolerant cultivars to reduce liming, reduced tillage methods to allow residue management, and use of cheaper sources of fertilizers and better methods of application to reduce inputs. However, this option still presents some problems of stability as far as continuous cropping is concerned. Inclusion of a period of managed legume fallows may be a solution.

Legume-based, pasture production system. A great proportion of the beef of tropical America is produced on high-base-status soils where it should not compete with more intensive farming systems. A gradual displacement of beef production toward low-activity clay soil regions appears as an interesting alternative to increase overall food production.

Some of the main components of this strategy are: (1) land resource evaluation to select suitable soils; (2) appropriate land clearing systems; (3) persistent and compatible pasture grass and legume cultivars, with tolerance to high aluminum saturation, low available phosphorus, drought stress, and major disease and insect attacks; (4) nitrogen supply by bio-

logical fixation; (5) low-cost, low-reactivity rock phosphates, which became readily available at low pHs; and (6) low-cost pasture establishment methods. Preliminary results at CIAT in Colombia and other centers show that this approach is highly productive and profitable.

This technology is also very appropriate for sloping lands, where erosion hazard is high. Well-managed pastures provide a continuous protective cover against erosive agents.

Other farming systems. Secondary forest fallows improve soil conditions in shifting cultivation systems. The appropriate length of the fallow period is estimated to be around 12 years for humid tropical America. However, increases in population pressure often shorten this fallow period, and the ecological stability of the system is threatened. Leguminous fallows may be able to produce a similar effect on soil restoration in a shorter period of time. Kudzu (*Pueraria phaseolides*) is the most promising material.

In sloping lands, agroforestry is another technological alternative. Research on this topic continues to increase. Interesting results have been obtained with *Gmelina arborea* intercropping and peach palm (*Guilielma gasipaes*) and legume interrows. See SOIL CHEMISTRY.

For background information *see* SOIL; SOIL CHEMISTRY in the McGraw-Hill Encyclopedia of Science and Technology.

[FRANK G. CALHOUN]

Bibliography: B. L. Allen and D. S. Fanning, Composition and soil genesis, in L. P. Wilding, N. E. Smeck, and G. F. Hall (eds.), *Pedogenesis and Soil Taxonomy, 1. Concepts and Interactions,* Developments in Soil Science 11A, 1983; E. Bornemisza and A. Alvarado (eds.), *Soil Management in Tropical America,* Proceedings of a Seminar at CIAT, Colombia, February 10–14, 1974, North Carolina State University, Raleigh, 1975; FAO-Unesco, *Soil Map of the World,* vol. 6: *Africa,* Unesco, Paris, 1977; R. W. Fitzpatrick and U. Schwertmann, Al-substituted goethite: An indication of pedogenic or other weathering environments in South Africa, *Geoderma,* 27:335–347, 1982; J. C. Hughes, Crystallinity of kaolin minerals and their weathering sequence in some soils from Nigeria, Brazil and Colombia, *Geoderma,* 24:317–325, 1980; F. R. Moormann, Classification of Alfisols and Ultisols with low-activity clays, in H. Eswaran (ed.), *Properties, Classification and Management of Low Activity Ultisols and Alfisols,* Soil Management Support Services, USDA–Soil Conservation Service, Washington, D.C., 1985; J. D. Nichols, L. P. Wilding, and D. E. Pettry, Low activity clay soils of the United States, in H. Eswaran (ed.), *Properties, Classification and Management of Low-Activity Ultisols and Alfisols,* Soil Management Support Services, USDA–Soil Conservation Service, Washington, D.C., 1985; P. Sanchez, *Properties and Management of Soils in the Tropics,* 1976; P. Sanchez and T. T. Cochrane, Soil constraints in relation to major farming systems of tropical America, in *Soil-Related Constraints to Food Production in the Tropics*, Proceedings of a Symposium at International Rice Research Institute, Philippines, June 4–8 1979, IRRI, Los Banos, 1980; A. Van Wambeke et al., Oxisols, in L. P. Wilding, N. E. Smeck, and G. F. Hall (eds.), *Pedogenesis and Soil Taxonomy, II. The Soil Orders*, Developments in Soil Science 11B, 1983.

Soil chemistry

Low-activity soils recently have been an important area of soil chemistry research. This article discusses the effect of lime application on variable-charge soils and fertility management of low-activity soils.

Variable-charge soils. Variable-charge soils have often been described as having low-activity clays which are the products of high intensity or long duration of weathering. Low-activity clay is associated with low shrink-swell characteristics, low available water contents, and low cation-exchange capacities. All of these result from the low surface areas of the minerals in the clay fraction.

Soil minerals. As soils become more highly weathered, there is an accumulation of hydrous iron and aluminum oxides, kaolinite, and chloritized vermiculite which have low activity, while high-activity clay minerals such as smectites and micas do not form or else disappear.

Smectites and micas have a dominantly permanent charge arising from isomorphous substitution of an ion of lower valence in a position usually occupied by one of higher valence. Charges arising from this immutable feature of the lattice structure are termed permanent in that they are not influenced by conditions in the solution surrounding the particle. All but the oldest and most highly weathered soils are dominated by these types of minerals. On the other hand, low-activity clays and organic matter have few, if any, isomorphous substitutions and become charged as a result of the adsorption and desorption on their surface of potential-determining ions (usually hydrogen) from the soil solution. As a result, these charges are termed variable in that their magnitude and sign are determined by conditions (pH) in the surrounding solution, as shown by reactions (1). These charges constitute the cation-exchange capacity (CEC) when negative, and the anion-exchange capacity (AEC) when positive. The total charge is a product of surface-charge density, that is, the amount of charge per unit area, and the surface area. Thus low activity results from either low surface-charge density or surface area, and it is usually the former which determines activity.

In many highly weathered soils, there is a tendency for the soil pH to attain a value at which the variable-charge inorganic minerals have a net zero surface charge. The most weathered soils in the world would consist almost entirely of this type of material representing the classical cases of variable-charge soils. The majority of soils, however, are mixed systems, containing both permanent and variable-charge minerals in varying proportions, together with organic matter in surface horizons. Variation in negative and positive charge with pH for the systems discussed is shown in Fig. 1. One of the most important problems associated with dominantly variable-charge soils is acidity, manifested by a deficiency of basic exchangeable cations (Ca^{2+}, Mg^{2+}, K^+, Na^+) and a dominance of acidic exchangeable Al^{3+}. Thus acid variable-charge soils often need to be limed.

Limestone activity. Agricultural limestone (lime) usually consists of calcium carbonate ($CaCO_3$) or calcium magnesium carbonate ($CaCO_3 \cdot MgCO_3$), the former being referred to as calcitic limestone and the latter as dolomitic limestone. Although both materials are virtually insoluble in water, they are readily dissolved by the soil acid, behaving as bases, as shown in reactions (2) for acid permanent-charge sites and in reactions (3) for acid variable-charge sites (Exch = exchangeable).

The same amount of lime has different effects on permanent- and variable-charge components. In acid permanent-charge soils the introduced Ca^{2+} displaces the exchangeable Al^{3+}, precipitating $Al(OH)_3$, as shown in reactions (1), which means that the lime requirement to alleviate aluminum toxicity and replenish basic cation content may be calculated from the initial level of exchangeable aluminum.

The introduction of OH^- to variable sites results in their deprotonation, and therefore in the formation of more negative-charge sites as well as in a reduction in positive-charge sites. This shift to a more negatively charged system means that the cation-exchange capacity has been increased and therefore more Ca^{2+} is adsorbed. However, for an equivalent amount of lime, less adsorbed (exchangeable) calcium is produced in variable-charge systems. In acid soils containing organic matter, aluminum and iron are strongly adsorbed or complexed by the organic functional groups, rendering such potentially charged sites nonexchangeable. Lime precipitates these metals, freeing the functional groups,

$$\text{Surface-OHH}^+ \underset{-\,OH^-}{\overset{+\,H^+}{\longleftarrow}} \text{surface-OH} \overset{-\,H^+}{\underset{+\,OH^-}{\longrightarrow}} \text{surface-O}^- \qquad (1)$$

$$\begin{array}{ccc} \text{Positive charge} & \text{Neutral,} & \text{Negative} \\ & \text{pH increasing} \rightarrow & \text{charge} \end{array}$$

$$
\begin{aligned}
3CaCO_3 + 3H_2O &\rightarrow 3Ca^{2+} + 3HCO_3^- + 3OH^- \\
2\ \text{Exch}\ Al^{3+} + 6H_2O &\rightarrow 2Al(OH)_3\downarrow + 6H^+ \\
6H^+ + 3HCO_3^- + 3OH^- &\rightarrow 6H_2O + 3CO_2\uparrow
\end{aligned}
\qquad (2)
$$

$$2\ \text{Exch}\ Al^{3+} + 3CaCO_3 + 3H_2O \rightarrow 2Al(OH)_3\downarrow + 3\ \text{Exch}\ Ca^{2+} + 3CO_2\uparrow$$

$$
\begin{aligned}
3CaCO_3 + 3H_2O &\rightarrow 3Ca^{2+} + 3HCO_3^- + 3OH^- \\
3\ \text{Surface-OH}_2^+Cl^- + 3OH^- &\rightarrow 3\ \text{Surface-OH} + 3Cl^- + 3H_2O \\
3\ \text{Surface-OH} + 3HCO_3^- &\rightarrow 3\ \text{Surface-O}^- + 3H_2O + 3CO_2\uparrow
\end{aligned}
\qquad (3)
$$

$$
\begin{aligned}
3CaCO_3 + 3\ \text{Surface-OH}_2^+Cl^- &\rightarrow \\
3\ \text{Surface-O}^- + 3Ca^{2+} &+ 3Cl^- + 3CO_2\uparrow + 3H_2O
\end{aligned}
$$

which then behave as variable-charge sites. The reaction between liming products and permanent- and variable-charge surfaces is different but the end result is similar, namely, greater amounts of exchangeable basic cations on the particle surface which are the reservoir from which the soil solution is replenished.

The changes which take place when the pH of a variable-charge soil is raised is shown in Fig. 2. Exchangeable aluminum as well as manganese in solution, both of which are toxic to plants, are precipitated in insoluble forms while the level of phosphorus in the soil solution increases, reaching a maximum of about pH 5.5. Initially calcium in solution does not increase much, but above pH 6 there is a steady increase. On liming variable-charge mineral soils, the increased charge usually holds Ca^{2+} strongly (low amounts in solution initially), which means that other cations such as potassium and magnesium may be readily leached.

The acidity, and associated lack of basic cations, in low-activity-clay soils often extends deep into the soil profile. The adverse chemical environment prevents plant roots from extending into the subsoil to draw on lower-level moisture in times of drought. The application of lime to the soil surface normally does not alleviate subsoil acidity because of the mechanisms already described: the permanent charge in the upper soil horizons exchange the applied Ca^{2+} for Al^{3+} which reacts with OH^-, while the variable-charge sites deprotonate to neutralize OH^-, with the resulting increase in negative charge causing the adsorption of Ca^{2+}. Thus lime does not easily move from the surface soil. What is needed is a calcium source which will leach to the subsoil and displace exchangeable aluminum, and preferably raise subsoil pH. A promising material for the alleviation of subsoil acidity is gypsum ($CaSO_4 \cdot 2H_2O$). On some soils, gypsum has raised exchangeable calcium as well as pH of the subsoil, especially when applied in conjunction with lime. Actual mechanisms as to the role of calcium, aluminum, and the sulfate ion (SO_4^{2+}) in relieving subsoil acidity are not yet fully understood.

Exchange complex monitoring. The low cation-exchange capacity of low-activity-clay soils, combined with a usually high saturation by aluminum, means that another important plant nutrient, exchangeable potassium, is often in short supply. The addition of swamping amounts of calcium and magnesium could therefore produce severe potassium deficiency unless this element is also applied and the balance on the exchange complex monitored.

Appropriate laboratory methods must be used to characterize and estimate the base status of low-activity-clay soils. A more reliable estimate of cation-exchange capacity, and therefore of the base saturation, of soils containing a significant proportion of variable charge will be obtained if determinations are carried out under conditions in equilibrating solutions that approximate those in the field.

It is not completely clear whether a deficiency of

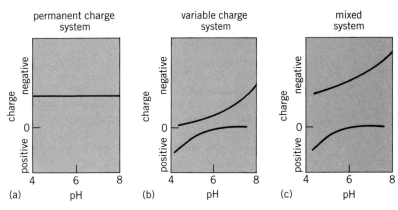

Fig. 1. Effect of pH on positive and negative charges. (*a*) Permanent-charge system. (*b*) Variable-charge system. (*c*) Mixed system.

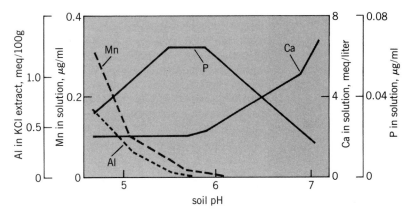

Fig. 2. Changes in the solubility or extractability of aluminum, manganese, phosphorus, and calcium in relation to the pH to which an Oxisol has been limed. (*After R. L. Fox, N. A. Saidy, and R. S. Yost, Illustrated Concepts of Tropical Agriculture, University of Hawaii, 1981*)

calcium or magnesium or a toxicity of aluminum is responsible for the poor results often achieved on low-activity-clay soils; much work in this area remains to be done. Fortunately, the addition of liming materials has the effect of overcoming both problems simultaneously, at least in the upper soil horizons. A greater understanding of basic soil processes, however, should result in the more efficient use of the various amendments mentioned above, consistent with long-term productivity and soil stability.

[M. E. SUMNER, G. P. GILLMAN]

Fertility management. Soils in the humid regions of the world generally contain low-activity clays. In their natural environment, these soils are acid and infertile. However, research in recent years has shown that these soils can be very productive with proper applications of lime and suitable fertilizers.

Lime. Soils with low-activity clays are inherently acid and contain low contents of available Ca^{2+} and Mg^{2+}. The principal exchangeable cation in these acid soils is exchangeable Al^{3+}. Examples of the

Table 1. Exchangeable Ca²⁺ and Al³⁺ and their percent saturation of the exchange complex of representative acid, low-activity clay soils*

Soil	pH	Exchangeable, $cmol(p+) kg^{-1}$		Saturation, %	
		Ca	Al	Ca	Al
Oxisol	4.5	0.45	1.15	26	70
Ultisol	4.2	0.26	1.83	9	64

*After E. J. Kamprath, Crop response to lime on soils in the tropics, in F. Adams (ed.), *Soil Acidity and Liming*, 2d ed., *Agronomy*, vol. 12, 1984.

exchangeable Ca^{2+} and Al^{3+} contents and saturation of cation exchange sites of acid soils are given in Table 1.

Without application of lime to these soils, crop plants suffer from Al^{3+} toxicities or Ca^{2+} deficiencies. However, when these soils were limed to neutrality (pH 7) as practiced with productive soils in temperate regions, poor results were often obtained. Raising the pH to 7 often resulted in other nutrients becoming less available and required large uneconomical amounts of lime because of the large pH-dependent charge associated with low-activity clays.

Liming practices which have proven to be successful on low-activity clay soils are based on reducing the exchangeable Al^{3+} saturation of the effective cation-exchange capacity (sum of the exchangeable cations) to zero. Lime rates which neutralize exchangeable Al^{3+} also supply adequate Ca^{2+} or Ca^{2+} and Mg^{2+} depending upon the lime source, and raise the pH sufficiently to eliminate any problems with H^+ ions. The amounts of lime required to neutralize exchangeable Al^{3+} are relatively low compared to rates required to raise the soil to pH 7.

The amount of exchangeable Al^{3+} can be used for determining the amount of lime to apply to acid soils. The amount of $CaCO_3$ required to neutralize a given amount of exchangeable Al^{3+} is given by Eq. (4). The quantity of exchangeable Al is multiplied

$$\text{Tons } CaCO_3 \text{ ha}^{-1} = 2 \text{ cmol } (\frac{1}{3} Al^{3+}) \text{ kg}^{-1} \quad (4)$$

by the factor 2 to take into account the fact that as the pH of the soil increases, the pH-dependent acidity ionizes and also reacts with the lime. Numerous studies have shown that lime rates based on Eq. (4) will reduce the exchangeable Al^{3+} to essentially zero.

Many acid soils have low amounts of exchangeable Mg^{2+}. Where exchangeable Mg^{2+} is less than 10% of the total exchangeable cations, dolomitic lime ($CaCO_3 \cdot MgCO_3$) should be used. If it is not available, then calcitic lime ($CaCO_3$) plus addition of a soluble magnesium fertilizer should be applied.

Phosphorus. Soils containing low-activity clays generally have very low amounts of native available phosphorus. Unless phosphorus fertilizers are added, crop yields are very low. When phosphorus fertilizers are added to the soils, the phosphorus reacts very quickly with the iron and aluminum ox-

ides to form iron and aluminum phosphates which are sparingly soluble. The amount of phosphorus fertilizer that has to be added to the soils to supply adequate levels of available phosphorus is a function of the soil texture (relative proportion of sand, silt, and clay). The iron and aluminum oxides are associated with the clay fraction, and as the clay content increases, the amount of iron and aluminum oxide also increases. Thus clay soils have a much higher capacity to absorb phosphorus than do sandy soils, and much higher rates of phosphorus fertilizer have to be added to clayey soils than sandy soils in order to increase the level of available phosphorus (Table 2).

There are several approaches to applying phosphorus fertilizers to low-activity-clay soils which are low in available phosphorus. One method is to apply a large broadcast application of phosphorus to bring the soil test level of available phosphorus to the recommended level. This is feasible with sandy soils which have a low capacity to absorb phosphorus. However, with soils which have a high phosphorus adsorption capacity, this approach may not be economical.

An alternative approach with soils having a high phosphorus adsorption capacity is to broadcast a moderate rate of phosphorus initially and then add a band application of phosphorus with each crop. [A band application is a localized application of fertilizer 4 to 6 in. (10 to 15 cm) either side of the plant at a depth of 4 to 6 in. in the soil.] Data on the response of corn to various rates and method of

Table 2. Relationship between soil texture and the amount of fertilizer phosphorus required to give the same level of available phosphorus in soils with low-activity clays*

Soil texture	P added, $mg kg^{-1}$	Available P, $mg kg^{-1}$
Sand	30	25
Fine sandy loam	60	25
Clay loam	170	25

*After J. R. Woodruff, *Relation of Phosphorus Availability to the Adsorption Maximum of Soils as Determined by the Langmuir Isotherm*, M.S. thesis, North Carolina State University, 1963.

Table 3. Response of corn to phosphorus fertilization of an Oxisol in Brazil*

P treatment†, $kg ha^{-1}$	Corn grain, tons ha^{-1}	
	Crop I	Crop IV
Broadcast 140	6.27	3.42
Broadcast 560	7.96	9.09
Broadcast 140 + 35 band	6.65	7.22
Band 35	4.56	6.03

*After R. S. Yost et al., Phosphorus response of corn on an Oxisol as influenced by rates and placement, *Soil Sci. Soc. Amer. J.*, 43:338–343, 1979.

†Broadcast phosphorus was applied only prior to crop I, while band applications were applied to all four crops. 1 kg ha⁻¹ = 0.89 lb/acre.

phosphorus application to a high phosphorus-adsorbing Oxisol in Brazil are given in Table 3.

The highest yields were obtained with the initial broadcast application of 560 kg P ha^{-1}, which continued to supply adequate phosphorus for the four crops. The initial broadcast application of 140 kg P ha^{-1} by itself was not sufficient to maintain a constant phosphorus supply for the four crops, but in conjunction with a band application of 35 kg P ha^{-1} to each crop a constant supply was maintained. With just a band application of 35 kg P ha^{-1} to each crop, yields were initially relatively low but were increased with each succeeding band application. Banding of the phosphorus fertilizers limits the amount of contact with the soil and thereby reduces rapid conversion of the phosphorus to less available forms.

Phosphorus soil tests will identify those soils which can supply adequate phosphorus for plant growth and those which are deficient. The amount of fertilizer phosphorus recommended will depend upon the soil test level, the soil texture, and the crop being grown.

Potassium. Soils with low-activity clays have been subjected to intensive weathering and leaching and generally contain only small amounts of total potassium. Because of this there is relatively little potassium released from nonexchangeable forms. The amount of potassium available for plant uptake in most instances is that which is exchangeable. When exchangeable potassium levels are less than 0.10 cmol(p^{+}) kg^{-1} (p^{+} is the charge on the proton), plants will often show potassium deficiency symptoms and will respond to potassium fertilization. With exchangeable potassium levels of >0.20 cmol(p^{+}) kg^{-1}, there is generally no response to potassium fertilization.

Since these soils have a low cation-exchange capacity, the amount of exchangeable potassium they can hold is somewhat limited. This is particularly true in areas of very high rainfall. On soils with low-activity clays, potassium fertilization practices should take into account the soil test level and the potassium requirements of the crops. Maintenance of exchangeable potassium levels above 0.20 cmol(p^{+}) kg^{-1} is not economical because of leaching of potassium in these soils.

Sulfur. The low-activity-clay soils contain relatively low amounts of total sulfur. Most of the sulfur is in the organic form, and release is not sufficient to supply enough available sulfate for sustained high yields. Application of sulfur-containing fertilizers is needed where intensive agriculture is practiced. Sulfate leaches readily from surface horizons which have been limed and have high levels of available phosphorus. The sulfate tends to accumulate in the B horizons of Oxisols and Ultisols and is available if accessible to plant roots. The need for sulfur fertilization depends upon the level of sulfate in the Ap horizon (the surface plow layer), the depth to the B horizon, and the amount of sulfate in the B horizon.

Micronutrients. The micronutrient content of soils is related to the parent material from which the soils have been formed. Low reserves of micronutrients are generally present in the highly leached, acid coarse-textured soils. Boron deficiencies are most likely to occur on the deep, sandy soils; zinc is deficient in many of the highly weathered soils; copper deficiencies are found in the poorly drained, high-organic-matter soils; and molybdenum deficiencies of legumes are common on acid, highly weathered soils.

The need for micronutrients should be based on soil tests, knowledge of the soil characteristics, and specific requirements of the crop.

Nitrogen. Low-activity-clay soils generally have low amounts of organic matter, and only small amounts of nitrogen are made available by mineralization of the organic nitrogen. Nitrogen fertilization must be based on crop requirements. Because nitrate is a highly mobile anion, only small amounts should be applied at planting, and then most of the nitrogen when the plant begins its rapid growth. *See* SOIL.

For background information *see* SOIL; SOIL CHEMISTRY in the McGraw-Hill Encyclopedia of Science and Technology.

[E. J. KAMPRATH]

Bibliography: F. Adams (ed.), *Soil Acidity and Liming*, 2d ed., Agronomy, vol. 12, 1984; C. S. Andrew and E. J. Kamprath, *Mineral Nutrition of Legumes in Tropical and Subtropical Soils*, CSIRO, Melbourne, Australia, 1978; R. L. Fox, N. A. Saidy, and R. S. Yost, *Illustrated Concepts in Tropical Agriculture*, University of Hawaii, 1981; F. E. Khasawaneh et al., *The Role of Phosphorus in Agriculture*, American Society of Agronomy, 1980; B. K. G. Theng, *Soils with Variable Charge*, N. Z. Soc. Soil Sci., Private Bag, Lower Hut, N. Z., 1980; G. Uehara and G. Gillman, *The Mineralogy, Chemistry and Physics of Tropical Soils with Variable Charge Clays*, 1980.

Soil potentials

Loss of agricultural land to other uses is a serious concern in the United States. The concept of soil potentials has been developed to provide a tool for resource conservationists, planners, and city and town officials. More clearly than any other available option, soil potential ratings define the potential production level of soils for specific types of agriculture. In addition, soil potential ratings are a first step in developing the Land Evaluation and Site Assessment (LESA) system, a program of the USDA Soil Conservation Service. The soil potential index range is used to determine relative groupings for this land evaluation.

The soil potential index (SPI), a relative numerical rating, is used to array soils according to their potential. The SPI is derived from indexes of soil production, costs of corrective measures, and costs for continuing limitations. The equation is SPI = $P - (CM + CL)$, where P is the index of production or yield, CM is the index of costs of corrective measures to overcome soil limitations, and *CL* is the index of costs resulting from continuing limitations.

Corn silage in New Hampshire. Soil potential classes were applied to indicate the potential of a soil for corn silage production in New Hampshire compared with soils elsewhere. The range in soil properties resulted in a fairly uniform progression from high to low values of soil potential index. Five rating classes were used: very high, high, medium, low, and very low. These rating classes were defined by a local committee for an individual soil survey area, adjusting nationwide definitions as proposed in the *National Soils Handbook* to suit the local situation.

The committee collected the necessary data and established a production standard. A list of the soils related to the production standard was developed; this served as a basis for the five soil potential rating classes. The production standard fixes the SPI of the best soil at 100 and arrays other soils in descending values.

Production standard. The production standard in New Hampshire was defined locally by agricultural experts. Actual production of each soil was compared to the standard by using a set of evaluation factors. The production standard in New Hampshire assumes that the site is intensively managed for corn silage production, that farming operations are timely, that lime and fertilizer are applied according to soil analysis, and that soil erosion is kept at or below the allowable soil loss value.

Soil evaluation factors. Six soil evaluation factors are used to determine soil potential ratings for corn silage in New Hampshire: slope, available water capacity to 40 in. (100 cm), depth to bedrock, rock fragments on the surface layer, water table level and soil permeability, and mean annual soil temperature. Each of these factors could have detrimental effects on the production of corn silage.

In developing the soil potential index, each soil evaluation factor was arrayed into two to six soil and site conditions. The degree of limitation and effects on use were assigned locally. The table shows an example of the Worksheet for Preparing Corrective

Measures for the rock fragment limitation. It is one of the six types of worksheets prepared by the local committee.

Corrective measures and continuing limitations. Since the costs of corrective measures are total initial costs and the costs resulting from continuing limitations are annual, economic analysis (amortization) was used to derive a common basis for corrective measures and continuing limitations. Amortization is sometimes called the partial payment or capital recovery factor. This factor will convert capital or initial cost to annual cost.

Sloping land and high inherent erodibility require protective measures. Protection can be provided through either no-till, a diversion system, stripcropping, or crop rotation. The local committee estimated the costs for these corrective measures with the basic intent that each measure would ensure that erosion did not exceed the allowable soil loss. Continuing limitations were identified as operation and maintenance costs and the costs resulting from reduced yield.

The costs for corrective measures for seasonal high water table and soil permeability were estimated by using the *New Hampshire Soil Conservation Service Drainage Guide*. Three soil permeability groupings were used to calculate costs. These costs reflect subsurface tile drainage systems that have closer spacing as the soil permeability becomes less. The continuing limitation costs were identified as maintenance costs for the tile drainage system and the costs resulting from reduced yield.

Rock fragments on the surface layer are a very common limitation in the upland glacial till soils in New Hampshire. The typical corrective measure is obstruction removal of fragments (see table). Even after this is done, there is a continuing limitation of equipment operation because rock fragments continue to work their way to the surface by frost action.

There are no feasible corrective measures for some soil limitations such as available water capacity. For these soils, costs were based on reduced yields for continuing limitations. Since irrigation is not a typical corrective measure in New Hampshire,

Worksheet for preparing corrective measures

Soil use: Corn silage			Area: Strafford County, New Hampshire						
Soil evaluation factor	Soil and site conditions	Degree of limitations	Effects on use	Typical corrective measures			Typical continuing limitations		
				Kind	Cost	Index	Kind	Cost	Index
Rock fragments in surface layer; percent cover	None	Slight	None	None	—	—	None	—	—
	.01%	Slight	None	None	—	—	Equipment operation	$10	1
	.01–3%	Severe	Equipment limitations	Obstruction removal	$127	16	Equipment operation	$10	1
	3 + %	Severe	Equipment limitations	No typical corrective measures	—	—	Surface boulders, stones, and cobbles	—	100

no corrective measures were assessed for the available-water-capacity soil limitation. Soils with low or very low available water capacity yield less than the standard; therefore, the reduced yield is a continuing limitation.

No corrective action is feasible to alter the depth to bedrock. The costs for reduced yield were used as a continuing limitation.

There are no corrective-measure costs for soil temperature. It was documented, however, that soils with a mesic soil-temperature regime [warm soils, with an annual soil temperature $> 47°F$ ($8°C$)] have a higher yield for corn silage production than soils with a frigid soil-temperature regime [colder soils, with an annual soil temperature $< 47°F$ ($8°C$)], if all the other soil evaluation factors and soil and site conditions are the same. Therefore, delayed planting and reduced yield were considered a continuing limitation for soils that had a frigid soil temperature.

Some soil limitations such as steep slopes and extremely stony surface cover had no realistic corrective measures. A cost index was assigned as a continuing limitation to those soil areas.

The local committee accepted the worst-case scenario to derive indexes whereby all of the costs for corrective measures and continuing limitations were summed for the soil map unit with the most soil limitations. This particular method has assured a workable spread in SPIs for corn silage in New Hampshire.

Development of ratings. Survey areas in New Hampshire commonly have more than 100 map units. The time needed to complete the Worksheets for Preparing Soil Potential Ratings can be quite lengthy. The table is just one of six Worksheets for Preparing Corrective Measures for corn silage. Each map unit and its soil properties must be cross-checked with these tables either manually or through some more efficient method to calculate the SPI. The worksheets themselves can be generated by a personal computer.

Assigning ratings. Soil potential rating classes are assigned from high to low by map units. All of the map units are arrayed from SPIs of 100 to 0. The class breaks relate to class definitions, and usually there are clusters of map units with similar SPIs to make these breaks.

The production standard, by definition, in New Hampshire has an SPI of 100 and is assigned a very high potential. On the other extreme, economic analysis indicates that map units with an SPI of less than 50 (about 12 tons or 11 metric tons) cannot maintain a viable corn silage production operation. Therefore, very low potential rating classes are assigned to these map units.

All of the assigned rating classes are compared with local knowledge. Occasionally adjustments are made, particularly for map units that border other rating classes.

[HENRY R. MOUNT; SIDNEY A. L. PILGRIM]

Range production in New Mexico. A soil potential index for the broad use of livestock grazing was developed for the soils of northern Catron County, New Mexico. Even with a broad-based use, a very specific definition of the performance standard was required. The results showed that the SPI has merit for broad area planning and land evaluation, but limited applicability for on-site planning because of variation of current range condition.

The study used a recently completed low-intensity soil survey in a dominantly rangeland area. The map units recognized were consociations, complexes, and associations. Elevation ranged from 6000 to 10,250 ft (1800 to 3120 m); and precipitation, from 10 to 20 in. (25 to 50 cm), increasing with elevation. No perennial streams exist within the area. The survey area consisted of 2,200,000 acres (890,000 hectares).

Range condition has deteriorated as a result of year-long continuous grazing on most of the area. Production and quality of forage have been lowered, and a high level of range management is necessary to achieve the potential production.

Performance standard. Livestock grazing implies many uses ranging from sheep to cattle use, summer use to winter use, and so on. The value of the range and potential vegetation varied for these different uses. The dominant local type of livestock operation was identified, and assumptions were made for range condition, use, management, and adequate conservation practices. The assumptions establish a correlation between the gross returns and the costs of obtaining the standard level of performance.

The performance standard was based on a specific set of conditions in regard to soil management and ranch management. The soil is managed for livestock grazing in a cow-calf-yearling operation; the range is in high good condition and properly grazed. Ranch management is of a high level. A deferred-rotation grazing program is included. Facilitating practices include cross-fencing, livestock water pipelines, troughs, storage tanks, wells, springs, brush management, and woodland selection-cut harvesting of pinyon-juniper stands. Pinyon-juniper woodland units have a 10 to 40% canopy cover. Revenues from wood products provide returns greater than the cost of thinning for forage production.

A production value of $213 per animal-unit-year (AUY) is based on several assumptions. Calves are retained and sold as long yearlings, and heifer replacements are selected from these yearlings. In droughty years, when forage is short, calves are sold when weaned. Conception rate is 90%, and calf crop percent is 80%. The death rate is 2%, and the cull rate is 18% on cows and 25% on bulls. Yearlings average 700 lb (320 kg) and bring $55 per hundredweight. Cull cows average 850 lb (380 kg) and bring $40 per hundredweight, and cull bulls average 1500 lb (675 kg) and bring $50 per hundredweight. These figures represent 418 lb (188 kg) of beef produced per AUY, or 581 lb (261 kg) per cow in the herd.

A performance standard of $4000 gross revenue per 640-acre (260-hectare) section was used, based on a suggested starting stocking rate of 18.4 AUY per section. This represents production on the best site of the area. The suggested initial stocking rate for range in excellent condition was used for base production figures. These stocking rates are tied to range and woodland sites that have been correlated to the soils.

The performance standard of $4000 was given an index value of 100. One index unit thus equals $40. Livestock water development and fencing were the essential practices needed to achieve the predefined management level. For soils with no limiting properties, the cost to develop the 40-section ranch to the defined standard was $705 per section per year.

Corrective measures and continuing limitations. The team identified soil properties that increased costs for installation or materials above those for the defined standard, and properties that cause continuing limitations on use or management. Those properties were rock fragments greater than 3 in. (7.5 cm) in diameter, restricted depth to bedrock or hardpan, slope, and flooding. Each limiting property was evaluated separately, and the increased cost per section per year above the standard was estimated. Index differences for continuing limitation were based on production, as continuing limitations cannot be overcome by corrective measures. Production yields were established in animal units per section for the standard. The yield for each soil was determined and was divided by the standard to get the index difference. Low yield is considered a continuing limitation. Miscellaneous land areas such as rock outcrop and Badlands, which have no production, pose a continuing limitation to management and use, and were assigned values of -20 on rock outcrop and -10 on Badlands.

Applications. The soil potential index for rangeland is one of many interpretive ratings for use of soil. The performance standard must be specifically defined. Soil potential ratings for livestock grazing are limited for use in on-site ranch planning, because the system does not take into account variables such as the current ecological conditions on a ranch. It does have merit in evaluating soils in terms of cost differences required for ranch development if corrective measures are applied. It has potential use in broad area planning, comparison of areas, and land evaluation.

[DAVID L. CARTER]

Southern states. The competition for land for urban development and agricultural production is becoming critical in the southern states. The land Evaluation and Site Assessment (LESA) system developed by the USDA Soil Conservation Service is designed to help local officials protect valuable farmland. The latest census of agriculture shows the southern states having 460,500,000 acres (186,400,000 hectares) of farmland. Important farmland is being lost at alarming rates near such metropolitan areas as San Antonio, Houston, Dallas, Oklahoma City, Memphis, and Atlanta, and across the state of Florida.

Agriculture is essential to the economy of the South and is important to the nation. According to *Agricultural Statistics* (1984), the southern states exceeded $38 billion in agricultural products. Although LESA is a tool to assist in protecting farmland, it is not a no-growth plan. It supports urban growth in a compact, efficient pattern by encouraging in-filling and discouraging leapfrogging patterns of urban development.

As in other regions, the LESA process for the southern states consists of two parts. The first part evaluates soil quality, determines the relative value of soils in the area, and classifies the soils by their suitability for agricultural use. Data for this evaluation come from soil survey information.

The second part of LESA considers other factors that influence decisions on agricultural land conversion such as land use adjacent to the site and availability of utility services. This information is developed by local planning officials with the assistance of local, state, and federal agricultural leaders.

LESA systems are being developed and tested across the South from Texas and Oklahoma to North Carolina and Florida. The most interest and activity is centered in Florida, where increasing numbers of new residents are seriously impacting the state's agricultural land.

Farmland evaluation. In implementing LESA, the first step is to identify the important farmland soils. Data from the soil survey interpretation records of each soil series are stored in a data bank at Ames, Iowa. The data are used to evaluate those properties that designate prime farmland, according to the *Federal Register*. These properties include: available water capacity, pH, salinity, sodicity, depth to water table, depth to rock, flooding frequency, erosion potential, permeability, and rock fragments. Some states have developed additional criteria to designate soils of statewide importance. Some areas have also designated certain soils to be of local importance.

In addition to the important farmland designation, LESA considers the capability classification and productivity of soils to arrive at 10 agricultural viability groups. Soils are arrayed by suitability for agriculture into these groups ranging from the best to the poorest.

Yields recorded on the soil interpretation records do not reflect costs of management or enduring practices; therefore, yields are adjusted to place all soils on a relative basis. The most successful method to accomplish this is soil potentials.

The final step in land evaluation is determining the relative value for each of the agricultural viability groups. The relative value is determined by dividing the soil potential index or adjusted yield of each group by the highest soil potential index or adjusted yield times 100. This will array the agricultural viability groups relative to each other. For example: assuming the highest adjusted yield of 60 bushels, the top group would have an index of 100 (60 divided by 60 times 100); or if the adjusted yield was 45, the group index would be 75 (45 divided by 60 times 100).

Site assessment. Following the development of Land Evaluations, which include soil potentials as the main driving force, Site Assessment evaluations are made. Site Assessments are made by local officials considering 12 criteria:

1. Land in nonurban use with 1 mi (1.6 km) radius.

2. Land in nonurban use on perimeter.

3. How much of site has been farmed in 5 of the last 10 years.

4. If the site is protected for farming by local zoning.

5. How close the site is to urban area.

6. How close the site is to utility lines.

7. If the farm unit is an average size or larger.

8. How much of remaining farm becomes unfarmable.

9. If the site has adequate farm support services.

10. If the site has on-farm investment.

11. If the conversion jeopardizes remaining farms.

12. If the use is so incompatible with agricultural uses as to cause other conversions.

Advantages. Soil potentials are a valuable tool in the southern states where the cost of protecting the soil may be reflected in the yields but not in the net profit. Arraying soils by performance or yield alone often results in misaligned groupings. Two soils may have identical yields and appear to be of equal value. However, if one of these soils required considerable investment to overcome limitations or to protect the soil from excess erosion, the effective yield or relative value would be different. The concept of soil potentials is used to compensate for these costs and thus arrive at a realistic relative value.

Procedures developed to rate soil potentials supplement other soil interpretations currently available to users of soil surveys. An important difference is that potentials emphasize feasibility of use rather than the negative connotation of limitations.

For background information *see* SOIL; SOIL CONSERVATION in the McGraw-Hill Encyclopedia of Science and Technology.

[DEWAYNE WILLIAMS]

Bibliography: L. J. Bartelli et al. (eds.), *Soil Surveys and Land Using Planning*, 1966; D. E. McCormack, Soil potentials: A positive approach to urban planning, *J. Soil Water Conserv.*, 29(6):258–262, 1974; M. J. Rogoff et al., Computer-assisted ratings of soil potentials for urban land uses, *J. Soil Water Conserv.*, 35(5):237–241, 1980; Soil Conservation Service, USDA, Durham, *New Hampshire Drainage Guide*, 1984; *Soil Potential Ratings for Corn Silage and Grass-Legume Hay*, Cheshire County Conservation District, New Hampshire, February 1984; *Soil Potential Ratings for Corn Silage and Grass-Legume Hay*, Sullivan County Conservation District, New Hampshire, July 1983; Soil Survey Staff, Soil Conservation Service, USDA, *National Soils Handbook*, 1983; L. E. Wright et al., LESA: Agricultural land evaluation and site assessment, *J. Soil Water Conserv.*, 38(2):82–86, 1983.

Solution mining

Recent developments in solution mining have been highlighted by advanced computer modeling, testing procedures, and environmental safeguards. While continually depressed markets and lengthy environmental permitting processes have kept mining companies from major commitments to new operations, solution mining will probably capture a larger role in mining when markets improve.

As mining companies wrestle with diminishing ore grades, solution mining emerges as an alternative to conventional mining for metals such as uranium, copper, gold, and silver as well as for evaporites such as salt and soda ash. Although solution mining has many variations, in-place leaching is the most revolutionary. Based on injecting solutions into an ore deposit to dissolve the desired minerals and on recovering the mineral-laden solution from adjacent vertical wells (Fig. 1), the method is frequently

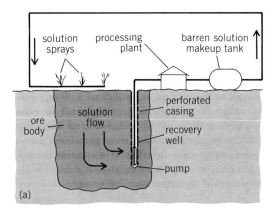

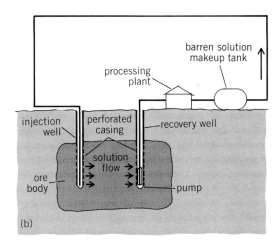

Fig. 1. In-place solution mining. (*a*) Surface solution application for exposed ore body. (*b*) Injection via vertical well for buried ore body. Solution recovery is accomplished through vertical wells, processing plant, and solution regeneration or makeup tank. (*After P. G. Chamberlain and M. G. Pojar, The status of gold and silver leaching operations in the United States, in W. J. Schlitt et al., eds., Gold and Silver: Leaching, Recovery, and Economics, Proceedings from 110th AIME Meeting, Chicago, February 22–26, 1981*)

identified with true solution mining. In spite of economical, environmental, and safety benefits, several problems prevent solution mining from realizing its full potential. Recent improvements, however, have helped boost mineral recovery, protect groundwater supplies, restore aquifers, and expand the potential target minerals for solution mining. Rapid advances in computer modeling, tailored for solution mining, have contributed to much of the progress.

Uranium. Fed by early successful operations, many recent developments have focused on uranium production. Computer models have been developed that predict solution movement and chemical reactions during leaching. In addition, there are available many groundwater models that can be adapted to the modeling of leaching solution movement in formations. Data can now be plugged into minicomputers to determine the best well configuration, pumping schedules, and reagent concentrations. The influence of permeability, mineralogy, and deposit configuration on uranium recovery and reagent consumption can be accurately determined in advance of actual operations.

The strong interest in developing environmentally compatible leaching solutions has spawned considerable research in replacements for the prevailing ammonium carbonate–bicarbonate solutions. The ammonia ions are difficult to flush out of a formation after solution mining. One promising strategy is to first flush the formation with inexpensive potassium chloride and then to leach it with potassium carbonate–bicarbonate, thus eliminating ammonia. Such new leaching solutions and the complex mineralogy in uranium deposits have also created the need for fast, inexpensive field tests to verify solution applicability. A recently developed three-well test (Fig. 2) has cut on-site testing time from 6 months to 3 weeks at half the cost of previous tests, which required a five-spot configuration. In the three-well test, small-diameter sampling wells are drilled to

form a right angle with the injection-production well, shown in the center of Fig. 2. Water samples are then drawn from the sampling wells by nitrogen pressure in a sampling tube, and no pump is required.

To stop leaching solutions from escaping from wells above or below leaching zones, wells can now be pressurized to detect leaks in their casing. Recent research also reveals that sophisticated geophysical monitoring instruments may occupy a significant role in detecting solutions escaping from a leaching zone.

Production declines from individual wells have plagued commercial operations. Recent experiments show that microorganisms can accumulate in well screens, thereby clogging the screen openings. The experiments also show that adding hydrogen peroxide to solutions kills the microorganisms, thus minimizing productivity losses. Techniques are also now available for restoring good flow from wells by using high-pressure water jets lowered into the well to cut new openings through the well casing into the surrounding formation.

Copper. Although commercial in-place solution mining operations for copper preceded those for uranium, copper market doldrums and the unexpectedly low recovery from early operations have discouraged use of this method. Currently, only depleted underground mines are leached via vertical wells drilled from the surface. In contrast to the leaching of porous sandstones for uranium, copper deposits usually must be blasted to allow for ready passage of leaching solutions. Early operators found that the flow of solutions through the blasted rock was different from the flow through granular rock. Also, the leaching solutions did not interact predictably with minerals in the deposits. However, evolving computer models plus extensive field testing now assure better advance prediction of solution movement and solubility rates; engineers can plan specific well spacings, locations, and pumping characteristics.

A new concept that involves boosting solution injection pressures to achieve good flow rate while eliminating costly blasting is gaining acceptance. This expands the range of conditions under which solution mining operations become profitable. Recently a large copper deposit has been tested preparatory to leaching a deposit that cannot be economically mined otherwise.

Other metals. In-place solution mining for gold and silver is receiving attention because of the many successful heap leaching operations, which involve ore that has been mined and piled on the surface for leaching. The U.S. Bureau of Mines has studied possible cost advantages and has evaluated the techniques that have been developed for copper leaching. Unfortunately, the toxicity of the main solvent for gold and silver, a dilute cyanide solution, presently precludes its injection into ore deposits. Consequently, mining companies and research groups are experimenting with less toxic solutions before

Fig. 2. A three-well test that can be used to determine the amenability of an ore body to leachants. (*From J. R. Pederson, ed.,* Bureau of Mines Research in 1982, *1983*)

seriously considering commercial operations. A mixture of ferric sulfate and thiourea (the most promising candidate) is too expensive for present use. The Bureau of Mines recently experimented with water at one site to ascertain the probable solution movement patterns in typical gold deposits, under the assumption that the chemistry problems could be resolved eventually.

Manganese also may be solution-mined in the future. In a recent spin-off of World War II technology, laboratory tests and cost studies favored leaching with aqueous sulfurous acid (SO_2). Application of the solutions could be similar to that for copper deposits. Either sulfur or pyrite would be roasted on site to generate SO_2. However, low-cost foreign supplies of manganese preclude present commercial development in the United States. Recent laboratory studies have demonstrated that nickel-cobalt laterite deposits could also be leached. These embryonic concepts will require years of further testing to achieve commercial production.

Evaporites (salt and soda ash). Salt has been commerically solution-mined from wells for decades. During the 1980s, several companies tested similar concepts to produce sodium carbonate (soda ash) from trona deposits, primarily in the Green River Basin of southwestern Wyoming. These methods have not advanced beyond experimentation in the United States because of a dwindling market and lengthy environmental permitting processes. Two American companies have entered into foreign development ventures in Turkey and Botswana; successful tests there would probably spur reintroduction of the method for domestic trona deposits in the late 1980s. A promising technology developed for trona deposits in the Green River Basin of Wyoming features pumping a solvent such as dilute sodium hydroxide into injection wells under pressure and pumping the sodium carbonate–bearing solution from adjacent wells. This method also involves a regeneration process for the sodium hydroxide solution.

Nahcolite and dawsonite, two other sodium carbonate minerals, could be similarly leached from deposits in northwestern Colorado. At least one experimental well constructed in these deposits yielded good results.

Solution mining operations continue for buried trona at the Searles Lake and Owens Lake deposits in California to recover the sodium carbonate as well as potash and borax. At the Searles Lake deposit, a unique salt-gradient solar pond warms the solutions to increase solubility. The solar ponds collect solar radiation and store its thermal energy. The heat then transfers to solutions circulated through the pond with trivial energy costs (Fig. 3).

The recent developments in solution mining for salt have focused on properly designing the cavities left after solution mining. Structurally stable cavities provide excellent underground storage. Such cavities often store natural gas. They are also being considered for gaseous or liquid hazardous chemical

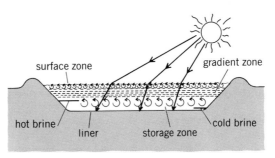

Fig. 3.   Salt-gradient solar pond for storing solar heat to warm the brine leaching solutions and thus increase their efficiency. The Sun's thermal energy is stored in the lower saline layer of the pond, where it transfers to the brine solutions piped through the pond. (*After J. Giulianelli, S. Carpenter, and W. Dowler, Feasibility of solution mining for sodium carbonate at Searles Lake, California, in W. J. Schlitt and J. B. Hiskey, eds., Interfacing Technologies in Solution Mining: Proceedings of the 2d SME International Solution Mining Symposium, Denver, November 18–20, 1981, 1982*)

waste disposal sites. Computer models stemming from sophisticated rock mechanics studies now enable engineers to design cavities for their structural integrity over a long time.

For background information *see* Mining; Rock Mechanics in the McGraw-Hill Encyclopedia of Science and Technology.   [PETER G. CHAMBERLAIN]

Bibliography: J. K. Ahlness and M. G. Pojar, *In Situ Copper Leaching in the United States: Case Histories of Operations*, Bur. Mines Inform. Circ. 8961, 1983; P. G. Chamberlain and M. G. Pojar, *Gold and Silver Leaching Practices in the United States*, Bur. Mine Inform. Circ. 8969, 1984; W. J. Schlitt and J. B. Hiskey (eds.), *Interfacing Technologies in Solution Mining: Proceedings of the 2d SME International Solution Mining Symposium, Denver, November 18–20, 1981*, Society of Mining Engineers of AIME, 1982.

Sonar

When the United States declared the Exclusive Economic Zone (EEZ) in 1983, it claimed sovereign rights and jurisdiction over an area that extends 200 nautical miles (365 km) seaward from its shores, in effect increasing the nation's area by an amount equivalent to about two-thirds that of the 48 contiguous states and Alaska combined. This vast submerged region is mostly unknown geologically. The small percentage of the region that is relatively well known has significant economic potential, including oil and gas, polymetallic sulfides, cobalt-rich manganese crusts, sand and gravel, phosphates, gold, and other commodities.

Side-scan sonar is a tool that can provide a reconnaissance plan view of relatively wide expanses of the sea floor. Side-scan sonar was originally developed for military applications but it has since found ready use in marine research. The system is composed of a ship-towed instrument package, streamlined somewhat like a fish, which sends out a beam of sound that images (insonifies) the sea floor, and

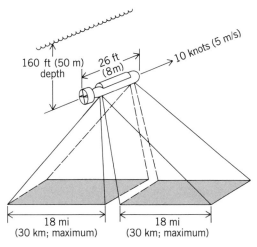

Towing configuration of the GLORIA side-scan sonar fish, showing the maximum area that can be insonified.

then records the backscattered acoustic energy. The result is analogous to that obtained by using light: hard objects, like rock surfaces, give strong reflections of sound energy and so appear "bright" on the sonar records, and soft materials, such as deep-sea ooze, adsorb sound energy and so appear "dark."

Although side-scan sonar has been available to marine geologists for many years, its development has been limited. Conventional side-scan sonars can insonify swaths that are only about 0.6-mi (1-km) wide. These instruments must be towed close to the sea floor; consequently, they are generally restricted to water depths of several hundred feet because of the long electrically conductive tow cables required, and to ship speeds of less than about 3 knots (1.5 m/s). Recently however, two systems, SeaBeam and SeaMARC, developed in the United States, are able to obtain mesoscale coverage of the sea floor with swath widths of about 1.8 and 6 mi (3 and 10 km), respectively, and a third system, GLORIA (Geological Long-Range Inclined Asdic), developed in the United Kingdom by the Institute of Oceanographic Sciences (IOS), can obtain accurate large-scale reconnaissance views of the sea floor. Because of its unique design, GLORIA can insonify a maximum 36-mi-wide (60-km) swath (18 mi or 30 km to either side of the ship's track; see illus.). The GLORIA instrument is towed only 160 ft (50 m) below the sea surface and thus can survey at relatively fast ship speeds (10 knots or 5 m/s), allowing coverage of an area of as much as 10,700 mi^2 (27,700 km^2) per day. It can operate in water as shallow as 500 ft (150 m) and possesses no maximum water-depth limit.

The program. The first efforts to map the EEZ systematically was a program of the U.S. Geological Survey (USGS) called EEZ-SCAN, a regional reconnaissance survey of the Pacific EEZ adjacent to the west coast. The objective of EEZ-SCAN was to provide a foundation for subsequent, more detailed studies focused on areas identified as being of special interest. The program was a cooperative scientific study by the USGS and IOS that takes advantage of the unique capabilities of the GLORIA side-scan sonar system. Concurrent with the use of GLORIA, seismic-reflection profiling and high-resolution subbottom and bathymetric profiling and measurements of the Earth's magnetic field were run. The survey, which began in April and ended in August 1984, completely insonified the area extending from the edge of the continental shelf seaward to the 200-mi (365-km) limit, from the Mexican border to the Canadian border.

Gloria system. The GLORIA system is composed of a side-scan sonar fish that is 26 ft (8 m) long and weighs 2.2 tons (2 metric tons) in air. The sonar array is composed of two rows of 30 transducers each, on either side of the fish. The transducers send out a burst of energy with a frequency of 6.5 kHz and a 100-Hz bandwidth. The incoming "reflections" from the port and starboard sides are recorded separately in a digital format. The system is able to identify features as small as 8 in. (20 cm) high, features separated by as little as 150 ft (45 m) in the direction perpendicular to the ship's track, and features on the order of 330 ft (100 m) long in the direction parallel to the ship's track.

Processing GLORIA data. The GLORIA data were recorded in 12-bit format, then compressed into 8-bit (256 digital numbers) format, a band sufficiently wide to fully record the dynamic response of the sea floor. Once the data are recorded in this digital format, processing to generate spatially rectified and enhanced images is relatively easy on a modern digital computer. Several different types of corrections were necessary before the images could be fully interpreted.

Most of the required corrections involved geometrical corrections. The side-scan data were recorded in a form that relates the time it takes sound to travel in a slightly curved path through the water from the GLORIA fish to a given point on the sea floor and back again to the fish. This elapsed time, when converted to a position on the sea floor, is called the slant range and it is not the true position. A geometric correction must be made that calculates the true position of each pixel (data point) on the Earth's surface.

The next correction required was to take into consideration the attenuation of sound in sea water, a function that decreases sonar energy by approximately $1/R^3$, where R is the distance downrange from the source. The consequence of this phenomenon is that an unprocessed image gets very dark and is difficult to interpret in the far range. The correction is made by means of calculating coefficients normalized to range which inverts the attenuation so that the returned images have uniform intensities.

Another set of calculations was required to compensate for the changes in the ship's speed along its track. The GLORIA system transmitted a pulse of sound every 30 s, but the ship traveled a variable distance in each 30-s interval, depending on sea state, wind, and currents. Consequently, each line of recorded data had to be registered to its correct geographic position on the Earth's surface.

The last type of processing that completes the routine processing involves high- and low-pass spatial filtering to eliminate noise and some artifacts from the data.

Preliminary results. The photomosaic constructed from the results of EEZ-SCAN provides a new and informative perspective of sediment transport patterns and oceanic crust history. For example, during the project more than 100 previously uncharted seamounts were identified as were outcrops of oceanic basement in the Baja Seamount Province west of southern California. The trends in basement in this region seem to indicate that either a failed propagating ridge or a failed leaky transform fault strikes northeast-southwest across a large part of the province. The inundation of the seamount province by sediment transported from the north is evident on the mosaic. Arguello, Monterey, Delgada, Astoria, and Cascadia canyons and channels can be traced either from the upper slope to the limit of the EEZ or to depositional lobes on their respective fan systems. Several of the channels meander, but none bifurcate into distributary systems as suggested by existing models for submarine fan deposition. Monterey and Delgada fans show bedforms developed on extensive areas of their surfaces, but the origin of these bedforms is unknown. A puzzling area of previously uncharted sediment drift is seen just west of Astoria Fan.

Juan de Fuca and Gorda Ridge systems and the intervening Pioneer, Mendocino, and Blanco fracture zones clearly caused a marked effect on the sedimentation patterns. The asymmetric spreading of Gorda Ridge is seen by the fanning of linear ridges away from the axial valley, and the previously postulated propagating ridge structures and pseudofaults are clearly delineated.

The convergent margin of northern California, Oregon, and Washington is clearly contrasted to the strike-slip margin of central California and the borderland of southern California. Compressive anticlines at the base of the slope characterize the compressive margins, whereas a relatively steep and gullied slope and rise characterizes the passive margin.

For background information *see* MARINE GEOLOGY; MARINE SEDIMENTS; SONAR in the McGraw-Hill Encyclopedia of Science and Technology.

[JAMES V. GARDNER]

Sound-reproducing systems

Since the discovery of the phonograph, all sound production, storage, and reproduction systems have handled a replica of a sound waveform. Each of these elements will eventually be replaced with digital counterparts, which differ in that the signal which is stored and modified is a mathematical representation of the sound.

Digital representation of audio signals. All digital audio systems contain converters to translate the audio signal into a digital form and back to audio. There are several methods for conversion, but pulse-code modulation (PCM) is by far the most common method used. In PCM conversion the audio signal, which is a continuous function of amplitude versus time, is rapidly sampled at regular intervals and measured for amplitude. The numbers representing these amplitudes are stored or signal-processed until it is desired to listen, at which point a complementary conversion is performed to replicate the audio waveform. The ultimate accuracy of the system is determined by both how frequently the signal is sampled and how many gradations are used to measure amplitude.

There are several advantages to this method. Any distortion, noise, or changes in transmission speed may contaminate the digital signal, but as long as the binary values can be recovered the signal will retain complete integrity. The digital signal is easily analyzed or processed through conventional digital computers, using mathematical techniques that are readily available, and can be stored in the same equipment used for data storage in digital computers.

Error correction. As with any digital storage system, there must also be a means for the detection and correction of any errors that may occur with damage to the storage medium. Because of the immense quantities of data generated, digital audio systems are typically designed for maximum storage density. Thus the probability of an error appearing in the data is increased proportionately.

In order to counteract this error probability, extra data words that serve to check parity or add redundancy are inserted with the data. Since these new data words are themselves subject to the same error sources as the original ones, absolute data integrity cannot be ensured. There is, however, a mitigating factor to be considered in the case of digitized audio signals. Unlike computer software, where a single incorrect bit can render a file invalid, a single error in digital audio will result in only a brief moment of increased distortion. It cannot be assumed that a random error will always make a small change on a given data value. For this reason it is still necessary to detect the erroneous data element, but it is less important to find out exactly what the correct value should have been. Once the erroneous element is identified, it can be replaced by an entirely new word that has a value which is an average of the previous good data point and the subsequent one. This form of error concealment is known as interpolation.

Since each storage medium is subject to unique sources of error, a different method of error correction is designed for each. A standard is selected that will have a high probability of detecting an error and at least a fair chance of correcting the error, while adding as little as possible to the amount of data stored. Typical methods can result in as few as 0.005% of the total errors going undetected. The total amount of errors, however, is dependent on the condition of the medium. An error rate of about one per second is considered to be around the maximum allowable for quality reproduction.

A program containing a large amount of errors can be copied while employing error correction. The resultant version will have the damaged data restored. As long as the errors were not so severe as to cause some form of error concealment, the new copy will be as good as the original recording.

Storage. Any storage system used by computers can be used in digital audio. Since a typical digital audio system develops over 40 kilobytes of 16-bit data words each second, it has only recently become affordable to store audio signals in digital form. The most cost-effective means of data storage is magnetic tape, whose main disadvantage for the computer field is the difficulty of random access. This is less of a problem in audio because the signals are inherently sequential. Many digital recorders use magnetic tape in a conventional fixed-head configuration. Others convert the signal into a video picture of black-and-white pixels (Fig. 1) permitting the use of a standard helical scan video recorder for storage.

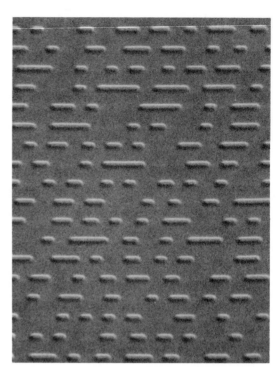

Fig. 3.   Microscopic pits in a compact disk, seen here as bumps on the underside of the disk. (*Courtesy of Margaret Goreau*)

Fig. 1.   Display of digital audio data on a video screen.

Fig. 2.   Compact disk.

A dramatic advance in the mass dissemination of recorded material is the compact disk (CD), a play-only medium that is manufactured in a fashion analogous to phonograph record pressing (Fig. 2). This 120-mm (4.75-in.) disk can contain over 70 min of stereo program in the form of microscopic pits 0.6 micrometer wide and 1.6 micrometers deep (Fig. 3). These pits rank among the smallest features manfactured, being comparable in size to the circuit element detail in advanced very large-scale integrated circuit (VLSI) semiconductor devices. The disk is read with a laser and photodiode that detect scattering of the light from the recorded side of the disk. The signal is then checked for errors and sent to a digital-to-analog converter for playback.

As a digital storage medium, the compact disk can hold up to 500 megabytes of read-only data. Applications are being developed to use the compact disk as a means to store complex software or large data bases. This represents a unique reversal of roles where the digital audio industry is developing technology for the computer industry. Efforts have been undertaken to develop a read/write erasable version of the compact disk.

Signal processing. While in digital form, the audio signal can be manipulated for analysis, as well as for musical or sonic effect. In fact, any process that can be described through mathematical algorithms can be employed on a digitized signal. The earliest application of digital audio was for time delay and the simulation of reverberation. Processing capabilities now include filtering, attenuation, summing, and modulation. It is also possible to change pitch or tempo independently of each other. Extensive processing such as that required by music recording studios is presently limited by the amount of computer power available at a reasonable cost, but the situation is expected to improve at a rapid rate.

Digital synthesis of sound. Probably the most dramatic application resulting from the innovation of digital audio is in the production or synthesis of music. Ever since the art of music began with stones and logs as instruments, the composer has continually been able to improve the subtleties of musical expressions through technological improvements in the instruments themselves. Each step in the evolution of musical instruments has brought improvements in both sound quality and playability. With the use of computers geared toward the production or modification of digitized sound, the composer is free to produce any possible sound. This includes those recorded from actual acoustic sources such as sounds from all other musical instruments. In addition, the input device may be similar to that of keyboard, wind, stringed, or percussion instruments. Composers can thus have virtually limitless possibilities in terms of the sounds available, the mechanics of composition, and the means of expression in performance.

For background information *see* COMPUTER STORAGE TECHNOLOGY; DISK RECORDING; MAGNETIC RECORDING in the McGraw-Hill Encyclopedia of Science and Technology.

[JOHN MONFORTE]

Bibliography: J. Monforte, The digital reproduction of sound, *Sci. Amer.*, 251(6):78–84, December 1984; K. C. Pohlmann, *Principles of Digital Audio*, 1985.

Space flight

Numerous flights of the United States space shuttle, with short intervals of time between liftoffs, dominated 1985. A fleet of three reusable shuttle vehicles permitted teams of astronauts, including representatives from the U.S. Congress, Mexico, Saudi Arabia, and European countries, to access space. Despite a series of nagging launch postponements and engine shutdowns, the ability to deploy satellites, repair spacecraft, conduct complicated research tasks, all from Earth-orbiting shuttle vehicles, was repeatedly demonstrated during the year.

Augmenting the shuttle program by launching vehicles from both the east and west coasts of the United States was delayed beyond an earlier projected March 1986 date as U.S. Air Force and National Aeronautics and Space Administration (NASA) planners agreed to forestall the first space shuttle launch from Vandenberg Air Force Base, California.

During 1985, two dedicated missions for the Department of Defense (DOD) were flown as part of the shuttle program; shuttle mission 51-C took place on

January 24 and shuttle mission 51-J on October 3. The two military flights placed a total of three classified payloads into high Earth orbit.

Economic competition in developing space services was heightened throughout the year as the uncrewed French Ariane launcher scored a series of successful flights, placing communications satellites into orbit. Ariane launch rates are to soon increase as an additional launching pad becomes operational. Ariane did, however, suffer one failure, tossing its satellite cargo into the ocean. The Soviet Union announced the creation of its first civilian space agency, Glavkosmos. As the main administration for the creation and use of space technology for the economy and for scientific research, the agency is unique in the military structure of Soviet space operations. Glavkosmos will concentrate, in its early years, on remote sensing from space to survey crops and other resources.

In many ways, 1985 could have been seen as a lull before the storm. A variety of space firsts were to be accomplished in 1986, primarily centered upon the launching of 15 payload-carrying shuttles. Unfortunately, the tragic explosion of shuttle *Challenger* after a January 28, 1986, liftoff from the Kennedy Space Center, and the death of its seven crew members, caused a major reevaluation of the shuttle program.

Significant space launches in 1985 are listed in the table.

UNITED STATES SPACE ACTIVITY

In 1985, 17 United States launches carrying a total of 34 payloads were placed in orbit. Launches were grouped as follows: science, 1; commnications, 16; crewed, 9; electronic intelligence, 1; navigation, 3; air-traffic control test, 1; and miscellaneous, 3.

Space Transportation System (STS) missions. A record-setting nine space shuttle missions were completed during 1985, including the maiden voyage of *Atlantis*, the latest addition to the shuttle fleet. Joining sister ships *Columbia*, *Challenger*, and *Discovery*, *Atlantis* brought to four the number of piloted space vehicles capable of orbital flight.

Shuttle astronauts expanded their abilities to deploy spacecraft, rescue and repair satellites, routinely carry out spacewalks, use the shuttle as an orbiting scientific platform, and construct assemblies in space larger than the confines of the shuttle's cargo bay. In carrying out these important tasks, the foundation for even greater activities in orbit has been set.

Shuttle mission 51-D. After nearly 2 months of delays, and just missing a sixth postponement by less than a minute due to poor weather, shuttle *Discovery* rocketed toward space at 8:59 A.M. EST on April 12, 1985. The space plane circled Earth at altitudes of approximately 185 to 286 mi (298 to 460 km) and at a 28.5° orbit. On board were Karol Bobko as commander, Donald Williams as pilot,

Significant space launches in 1985

Payload	Date	Payload country or organization	Purpose and comments
Discovery, mission 51-C	1/24/85	U.S.	First dedicated shuttle mission for the Department of Defense
Discovery, mission 51-D	4/12/85	U.S.	*Telesat* satellite deployed, with crew failing to salvage *LEASAT* spacecraft
Challenger, mission 51-B	5/6/85	U.S.	*Spacelab 3* mission yields bounty of scientific results in materials science, astronomy, and life sciences
Soyuz T-13	6/6/85	U.S.S.R.	Two Soviet cosmonauts head for docking with *Salyut 7* space station
VEGA 1	6/10/85	U.S.S.R.	Soviet interplanetary craft deploys French balloon and surface lander in atmosphere of Venus
VEGA 2	6/14/85	U.S.S.R.	Soviet interplanetary craft ejects French balloon and surface lander in atmosphere of Venus
Discovery, mission 51-G	6/24/85	U.S.	Three communications satellites ejected; international crew included French and Arab astronauts
Challenger, mission 51-F	7/29/85	U.S.	*Spacelab 2* mission achieved research goals in solar and atmospheric physics, high-energy astrophysics, and life sciences
Discovery, mission 51-I	8/27/85	U.S.	In-space repair of *LEASAT* communications satellite; deployment of three satellites
International Cometary Explorer (*ICE*)	9/11/85	U.S.	*ICE* becomes first spacecraft to probe a comet, Comet Giacobini-Zinner
Atlantis, mission 51-J	10/3/85	U.S.	Second dedicated Department of Defense mission
Challenger, mission 61-A	10/30/85	U.S.	West German *Spacelab D-1* mission concentrated on materials processing and life sciences; first eight-person crew in space

Rhea Seddon, David Griggs, and Jeffrey Hoffman as mission specialists, Charles Walker as payload specialist, and Senator "Jake" Garn as payload specialist/congressional observer, the first public official to fly aboard a space shuttle.

Deployed from the shuttle during the nearly 7-day flight were two communications satellites: a Canadian *Telesat 1* (*Anik C-1*) and a *LEASAT 3* (*Syncom IV-3*) to be leased by the Navy. Also on board the vehicle were two student experiments; an American Flight Echocardiograph (AFE); a pharmaceutical production unit called the Continuous Flow Electrophoresis System; a Protein Crystal Growth Experiment; and a series of toys to be used and filmed by the astronauts to demonstrate basic laws of physics to students.

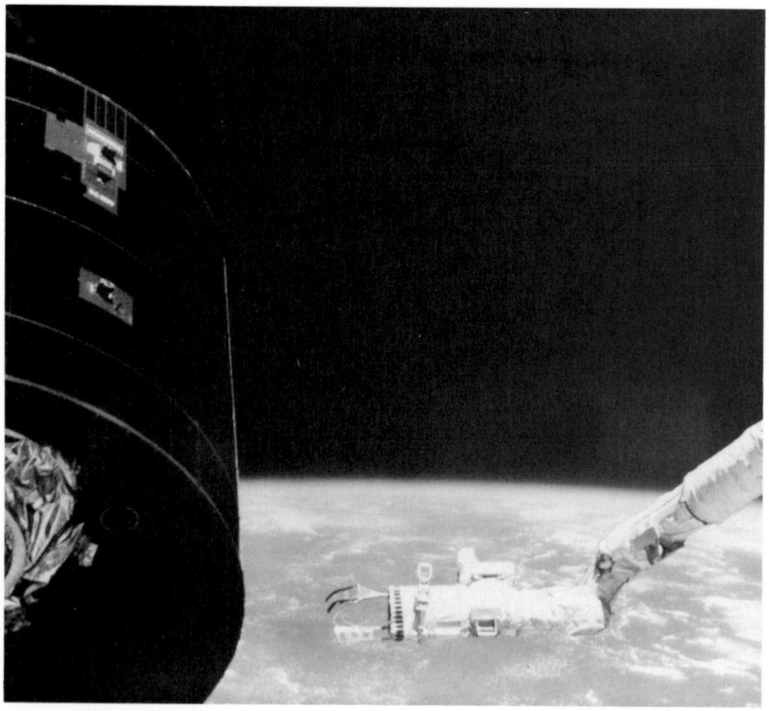

Crew members successfully ejected Canada's *Anik* satellite early in the shuttle mission. Deployment of the *LEASAT* was marred when a failed timing mechanism on the satellite prevented the satellite from boosting itself into geosynchronous altitude. Shuttle mission controllers on the ground pieced together the first "real-time" space repair in the shuttle era. By extending the shuttle flight and working around the clock, ground teams concocted a satellite rescue plan. *Discovery* astronauts cannibalized parts from the shuttle's cabin and jury-rigged "flyswatters" for attachment to the space plane's robot arm during a 3-h unrehearsed spacewalk by Hoffman and Griggs. By using the modified robot arm, several attempts were made to hit a lever visible on the outside of *LEASAT* (Fig. 1). Three hard contacts with the lever were completed, but failed to move the switch, and the 15,200-lb (6840-kg) communications satellite was left in a useless orbit.

On April 19, *Discovery* touched down at the Kennedy Space Center. A faster-than-normal touchdown speed, caused by a 10 mi/h (5 m/s) tailwind, led to the failure of two shuttle brakes and a blown tire. A later inspection of the spacecraft revealed over a hundred tiles were missing. In addition, a hole the width of a dinner plate was found in the orbiter's

Fig. 1. Unsuccessful attempt to activate malfunctioning *LEASAT 3* communications satellite, using the space shuttle's robot arm with flyswatter-like devices attached. (*NASA*)

left outboard elevon, apparently caused during *Discovery*'s reentry.

Shuttle mission 51-B. The European-built *Spacelab* left aboard space shuttle *Challenger* on May 6, 1985, at 12:02 P.M. EDT. Due to schedule slippage in readying *Spacelab 2*. *Challenger* was manifested to loft *Spacelab 3* on this flight. Placed into an orbit of 219 mi (352 km) at 57° to the Equator, *Spacelab 3* (Fig. 2) housed multiple scientific investigations in five research disciplines: materials science, life sciences, fluid mechanics, atmospheric physics, and astronomy. Twelve experiments were developed by United States scientists, two by French scientists, and one by Indian scientists.

The mission 51-B crew comprised Robert Overmyer as commander, Frederick Gregory as pilot, Don Lind, William Thornton, and Norman Thagard as mission specialists, with Lodewijk van den Berg and Taylor Wang as payload specialists. The mission was the first where a principal investigator, Taylor Wang—the chief scientist in charge of a particular project—flew with the experiment in space.

All but one of the experiments or board returned information. Among *Spacelab 3*'s accomplishments were the growth of crystals, produced without the gravity-induced convection of Earth; experimental data on the behavior of a free-floating fluid suspended by sound waves, which may lead to containerless materials processing; information on convective flows in rotating spherical bodies; chemical and physical analysis of atmospheric composition; video images of the South Pole auroral oval (Fig. 3); data on high-energy particles streaming toward Earth; and detailed biological and physiological measurements on 24 rats and 2 squirrel monkeys.

With a week of experimentation behind them, the seven-person crew and all 26 animals landed at a desert site of the Edwards Air Force Base in California. The switch to the desert location from the shuttle runway in Florida was made due to the landing problems encountered during the previous shuttle flight, mission 51-D.

Shuttle mission 51-G. The eighteenth flight of the shuttle program saw *Discovery* lift off on June 24, 1985, at 7:33 A.M. EDT. This mission was distinctively international as NASA flew the first French and Arab astronauts and a cargo that included American, Mexican, and Arabian domestic communications satellites.

Members of the flight were shuttle veteran Daniel Brandenstein as commander, pilot John Creighton, mission specialists Shannon Lucid, Steven Nagel, and John Fabian, and payload specialists Patrick Baudry from France and Sultan Salman Al-Saud from Saudi Arabia. *Discovery* roared into a 190 by 207 mi (304 by 331 km) orbit, and achieved an orbital inclination of 28.45°.

Crew members deployed a record three satellites. The *Morelos-A* satellite was the first of two domestic communications satellites to be orbited on behalf of the Secretariat of Communications and Transportation in Mexico. The *Arabsat 1B* was placed into

Fig. 2. Astronauts working in *Spacelab 3* module. (*NASA*)

space for the Arab Satellite Communications Organization, while a *Telstar 3-D* domestic communications satellite was orbited for the American Telephone and Telegraph (AT&T) Company. All three communications spacecraft were spring-ejected from *Discovery*, complete with a small attached rocket located on the bottom of each satellite. Using this Payload Assist Module booster and a built-in engine, the trio of spacecraft achieved their designated orbit at 22,300 mi (35,680 km) above the Earth. *See* COMMUNICATIONS SATELLITE.

During the flight, a *Spartan 1* payload was deployed and then retrieved. The 2223-lb (1000-kg) rectangular structure included 300 lb (135 kg) of experiments and was the first in a series of shuttle-launched, short-duration "free-flyers" designed to extend the capabilities of sounding-rocket-class ex-

Fig. 3. Auroral phenomenon photographed from space shuttle *Challenger* on mission 51-B. The high orbital inclination of this mission permitted over 500,000 video images to be taken of the South Pole auroral oval. (*NASA*)

periments. Deposited in space by the astronaut-guided robot arm, *Spartan* performed, among other duties, medium-resolution mapping of the x-ray emission from extended sources and regions. After gathering data for 2 days, *Spartan* was retrieved by the robot arm and repositioned within *Discovery*'s cargo bay. Although early depletion of *Spartan 1*'s attitude-control gas limited the amount of science data gathered, the collection was equivalent to the combined observation time of 144 instrument-laden sounding rockets.

A French Echocardiograph Experiment and French Postural Experiment were conducted by payload specialist Baudry to obtain on-orbit data regarding the response to weightlessness of the human body's cardiovascular and sensorimotor systems. Sultan Al-Saud photographed expanses of Saudi Arabia with a hand-held camera. The photographs are being studied by Saudi scientists to review geological features, sand dune morphology, hydrogeological features, turbidity in the Red Sea, urban areas, and forestry. Al-Saud also mixed various concentrations of oils with water and photographed the separations under fluorescent light.

In support of the Strategic Defense Initiative Organization, a High-Precision Tracking Experiment was mounted in the shuttle's middeck side hatch window. The 8-in.-diameter (20-cm) retroreflector was targeted from the Earth by a ground Laser Beam Director projected from a test facility on the island of Maui, Hawaii. A first attempt failed while a second succeeded, proving the ability of ground-based lasers to precisely track objects in low Earth orbit.

Deemed one of the most trouble-free missions of the shuttle program, *Discovery* landed 7 days after takeoff at Edwards Air Force Base.

Shuttle mission 51-A. On July 29, 1985, shuttle *Challenger* took to the skies at 5:00 P.M. EDT, carrying in its payload bay the equipment which constituted *Spacelab 2*. A dozen instruments were aboard to conduct research in seven scientific disciplines: solar physics, atmospheric physics, plasma physics, high-energy astrophysics, infrared astronomy, technology research, and life sciences. Much of the equipment was loaded on three unpressurized pallets and one special structure in *Challenger*'s cargo bay.

Minutes into the flight, two temperature sensors on the shuttle's center main engine turbopump failed, causing automatic shutdown of that engine. Flight controllers spent anxious minutes calculating that the shuttle could reach low Earth orbit by relying only on the vehicle's two remaining engines. An abort-to-orbit flight plan allowed the mission to continue, but at a greatly reduced altitude; an orbit of 195 mi (314 km) was achieved, far short of the desired 380-mi (610-km) orbit. Due to the abort, several experiments were compromised. Inclination to the Equator was 49.5°.

Astronaut Charles Fullerton was mission commander, with Roy Bridges, Jr., as pilot and Karl Henize, Anthony England, and F. Story Musgrave

as mission specialists, and Loren Acton and John-David Bartoe as payload specialists.

Throughout the 8-day mission, 13 experiment teams, 11 from the United States and 2 from the United Kingdom, directed investigations from the shuttle. A European-built Instrument Pointing System (IPS) caused some frustation for the first half of the mission when it could not acquire and track the Sun. Attached to the IPS were an array of three solar instruments and one atmospheric device. By using the IPS, sensors were accurately fixed on targets as the shuttle moved through space with a relative accuracy of 2 arc-seconds.

Telescopes pointed at the Sun's disk permitted crew observation of sunspots, prominences, filaments, and granules. The first observations ever made of a spicule—a high-velocity jet of gas moving into the Sun's corona—were recorded by a solar ultraviolet high-resolution instrument. A Solar Optical Universal Polarimeter inexplicably started up late in the mission after scientists had given up hope of activating the instrument. Its abbreviated operation, however, allowed measurement of the strength, structure, and evolution of magnetic fields in the solar atmosphere.

Detectors within a small helium-cooled infrared telescope, intended to provide an all-sky survey of infrared sources, were surprisingly saturated by a strong background source some 500 to 1000 times more intense than the natural background from dust in the solar system. Scientists believe that the detector saturation was caused by the outgassing of materials located within the cargo bay of the shuttle. This possibility, if verified, could decrease the use of a shuttle as the site for future infrared research. A 2-ton (1800-kg) egg-shaped cosmic-ray detector operated well throughout the flight and recorded about 2.4×10^7 particle events.

Although the 8-day mission was plagued with multiple instrument failures, astronaut and ground teamwork led to 85% of the flight objectives being met. The role of astronauts to promote on-orbit research, while doubling as trouble-shooting repair personnel, was demonstrated throughout the mission. *Challenger* landed at Edwards Air Force Base on August 6.

Shuttle mission 51-I. Perhaps the most daring of all shuttle missions to date began at 6:58 A.M. EDT on August 27, 1985, when shuttle *Discovery* was launched into a 219-mi (352-km) orbit inclined 28.5° to the Equator to salvage the dormant *LEASAT 3/Syncom IV-F3* communications satellite, lost to space after deployment by mission 51-D astronauts. The crew consisted of commander Joe Engle, pilot Richard Covey, and mission specialists James D. van Hoften, John Lounge, and William Fisher.

Three communications satellites were ejected during the early phases of the flight: the *ASC 1* for the American Satellite Company to provide voice, data, facsimile, and video conferencing communications services to more than 450 businesses and government agencies; the *AUSSAT 1* satellite for the Australian government to provide domestic telecommunication services; and a *LEASAT 4/Syncom IV-F4* satellite, the fourth spacecraft in the LEASAT system which is being leased by the Department of Defense. All three satellites were ejected as planned, and subsequently achieved separate 22,300-mi (35,680-km) orbits. The *LEASAT 4*, however, later experienced a failure in its primary communications circuit, rendering the satellite inoperative.

Special maneuvers edged the *Discovery* to within 35 ft (11 m) of *LEASAT 3*. Garbed in extravehicular-activity (EVA) space suits, astronauts Fisher and van Hoften positioned themselves in the shuttle's cargo bay. Riding on the end of the robot arm, van Hoften, his feet restrained, was maneuvered to the slowly spinning satellite by astronaut Lounge who remained in the shuttle's cabin. Stopping the satellite's spin by his hand, van Hoften attached a grappling bar to *LEASAT*. Astronaut van Hoften and his prize catch were lowered to the edge of the shuttle's payload bay. Working for more than 7 h—the longest EVA in the history of the United States space program—Fisher and van Hoften "hot-wired" *LEASAT*, electrically bypassing the failed timer which prevented the satellite from igniting a booster engine that would blast it to a correct orbital slot. A second spacewalking stint by the astronaut duo was carried out a day later. Perched on the end of the robot arm, van Hoften used muscle power to set the repaired *LEASAT* spinning into open space. Minutes later, an antenna locked against the side of *LEASAT* righted itself, a first indication that the repair-in-space operations had succeeded.

Discovery landed at Edwards Air Force Base after 7 days of flight. The results of van Hoften and Fisher's work were not realized until late October. After weeks of bathing the *LEASAT 3* in sunlight, to allow propellant to thaw out, a radio signal from the ground successfully commanded the astronaut-"jump-started" communications satellite to a geosynchronous orbit.

Shuttle mission 61-A. The Deutschland Spacelab Mission D-1 was the first of a series of dedicated West German missions on the space shuttle. Seen as a milestone in space cooperation between the United States and Europe, shuttle *Challenger* was fitted to carry a *Spacelab* module chartered for $65 million by West Germany. *Spacelab D-1* was managed by the Federal German Aerospace Research Establishment (DFVLR) for the German Federal Ministry of Research and Technology. *Challenger* was launched at 12:00 noon EST on October 30, 1985, into a 200-mi (320-km) circular orbit inclined 57° to the Equator.

Challenger carried an eight-member crew—the largest ever flown in space—commanded by Henry Hartsfield. The other crew members were pilot Steven Nagel; mission specialists James Buchli, Guion Bluford, and Bonnie Dunbar; and European payload specialists Reinhard Furrer (DFVLR—German), Ernst Messerschmid (DFVLR—German), and Wubbo Ockels (European Space Agency—Dutch).

Working in two 12-h shifts, astronauts operated experiments around the clock within the pressurized confines of the *Spacelab* module.

A unique aspect of the mission was its management, not from the NASA Houston, Texas, control center, but from a German Space Operations Center situated in Oberpfaffenhofen near Munich. A team of 160 flight specialists worked with the orbiting astronauts to fulfill scientific objectives.

A total of 76 scientific experiments was carried on the *Spacelab*. Among the experimental facilities were melting furnaces, apparatus for observing fluid physics phenomena, and chambers to provide specific environmental conditions for living test objects. In the life sciences, a small botanical garden was tended during the mission. Frog eggs were fertilized to yield data on the influence of microgravity on organs necessary for balance. Also among the life science experiments was a Vestibular Sled, which exposed astronauts to defined accelerations in an attempt to study the function of the inner ear.

A majority of experiments tested the attributes of microgravity in relationship to industrial processes. In one experiment, an isothermal heating oven melted specimens of metals such as molybdenum and nickel to a temperatue of 2900°F (1600°C) to create a hoped-for alloy of unique strength.

Challenger made a flawless touchdown at Edwards Air Force Base on November 6. In doing so, a new steering system for the shuttle fleet was cleared for future use, designed to ease the strain on shuttle main landing gear.

Space science flights. On September 11, 1985, the 1050-lb (473-kg) *International Cometary Explorer (ICE)* became the first spacecraft to encounter a comet. The spacecraft had been launched in 1978 as the *International Sun-Earth Explorer* but was redirected to accomplish other duties, including an interception with Comet Giacobini-Zinner. Nudged out of its orbit in 1982, *ICE* was sent near the Moon on multiple swing-bys in order to retarget the probe for its comet rendezvous.

Approaching the celestial object after years of orbital gymnastics, the probe plowed through the tail of the comet at a speed of approximately 45,000 mi/h (20 km/s). *ICE* sliced through the 14,000-mi-wide (22,500-km) tail of the comet, 4800 mi (7700 km) behind the cometary nucleus, and emerged from the tail approximately 20 min later.

Prior to the intercept, concern was raised that *ICE* would be destroyed by fast-moving dust particles and other material in the comet's tail. The milestone in space science produced a number of surprising results. Most importantly, the spacecraft survived hundreds of high-speed dust impacts with no apparent harm. Although some data transmitted back to Earth by the *ICE* spacecraft confirmed the traditional portrait of a comet (water vapor ions were measured which suggest the "dirty snowball" model is correct), other data received were unexpected.

Among these was the detection of electrical wave (plasma) disturbances and high-speed, molecular species coming from the comet long before the rendezvous. Scientists had theorized that first detection might occur just a few hours before the spacecraft crossed the comet's tail. A broad, bow-shock-like phenomenon, unlike the bow shocks previously detected around the terrestrial planets, was observed. A transition region around the head of the comet was revealed, an area in which the solar wind has been heated, compressed, and slowed. Cometary ions were directly sampled for the first time, and trapped magnetic field lines, draped around the head of the comet, were detected. Many other plasma and wave phenomena were registered, including some more intense than any observed by the *ICE* payload during 7 years in interplanetary space. In contrast to the hot electrons on the outskirts of the comet, the object's tail was found to consist of a dense, narrow structure of cool plasma.

After its encounter with Comet Giacobini-Zinner, *ICE* moved onward and passed between the Sun and Comet Halley on October 31, 1985. The interplanetary spacecraft was also scheduled to pass within 19.5×10^6 mi (31.4×10^6 km) of Comet Halley in late March 1986. The 1986 "upstream" pass of Comet Halley by *ICE* was considered important because it would allow the probe to transmit data on the solar-wind state upstream from Halley while Earth-based telescopes observed the effect of the solar wind on Halley's tail.

ICE will return to the vicinity of Earth in July 2012. Plans are being discussed to retrieve the spacecraft, perhaps for eventual museum display.

AMPTE project. The Active Magnetospheric Particle Tracer Explorers (AMPTE) completed their work during March 1985. AMPTE was a joint undertaking by the United States, West Germany, and the United Kingdom. AMPTE probe experiments made use of three satellites to determine how the solar wind interacts with the Earth's magnetosphere. Over the course of the project's life, which began in 1984, concentrations of barium and lithium were deposited by the satellites from both outside and inside the boundary of the Earth's magnetosphere. As the released chemicals were energized by solar wind, artificial comets were produced, observable by scientists stationed on Earth. Preliminary results indicated that less than 1% of the solar wind gains access to the magnetosphere under the conditions in which the releases took place.

Voyager at Uranus. Launched by the United States in August 1977, *Voyager 2* continued to make its way toward the planet Uranus for an encounter on January 24, 1986. Early images broadcast to Earth from the interplanetary probe during 1985 suggested that a surprising ring system surrounds the globe. Meanwhile, scientists analyzing *Voyager 1* and *2* data, relayed from the spacecraft in 1980 and 1981 on their way past Saturn, discovered a new satellite orbiting that planet.

SOVIET SPACE ACTIVITY

In 1985, 98 Soviet launches carrying a total of 119 payloads were placed in orbit. Launches were

grouped, in general, as follows: science, 2; communications, 14; weather, 3; crewed or crew-related, 5; reconnaissance, 34; electronic intelligence, 9; natural resources/oceanographic, 8; early warning, 7; radar calibration, 3; store and data-dump communications, 23; and navigation and geodetic, 11.

The year saw the Soviet Union continue its crewed program, expanding the abilities of cosmonauts to serve as space repair personnel in order to rehabilitate an aging *Salyut 7* space station. In addition, the year was marked by achievements in uncrewed space science missions, primarily at the planet Venus.

Space station activity. Cosmonauts Vladimir Dzhanibekov and Viktor Savinykh rode their *Soyuz T-13* into space, blasting off from the Baikonur cosmodrome in Kazakhstan on June 6, 1985. They manually docked with the *Salyut 7* space station 2 days later. Unique to the early phase of the mission was a more than 48-h transit time between cosmonaut liftoff and space station docking. Taking a route normally utilized by uncrewed *Progress* resupply vessels, the *Soyuz T-13* crew apparently made use of the longer transit time to test new spacecraft controls and a computer system. The flight came as a surprise to Western space watchers as Soviet officials had earlier proclaimed that the then-4-year-old *Salyut 7* had been deactivated. The *Salyut* had been empty since October 2, 1984, after another set of cosmonauts had established a record-setting endurance milestone of 237 days in space.

Once on board, however, commander Dzhanibekov and flight engineer Savinykh were faced with a severely limited station, drifting out of control and hampered by a crippled electrical system. Working in freezing temperatures and dangerous conditions, the cosmonauts managed to return the station to livable conditions. It took 10 days to restore heat, power, drinkable water, and radio communications. A *Progress 24* transport vehicle, launched on June 21, brought the crew additional supplies and equipment. A docking port on one end of the *Salyut* was freed by discarding the *Progress* supply ship on July 15.

In early August the cosmonauts ventured outside the near-derelict station in new full-pressure space suits of a semirigid type and fitted additional solar panels onto *Salyut 7*'s solar array. A new type of spacecraft was docked to the *Salyut 7*, 2 days after its liftoff on July 19. Designated *Cosmos 1669*, the craft is apparently similar in outside design to that of the *Progress* spacecraft but is modified to hold different cargo and work independently of the space station. The *Cosmos 1669* module was eventually decoupled from the *Salyut* and reentered the Earth's atmosphere on August 30. Its full range of stated abilities was not demonstrated.

A new crew of cosmonauts was launched in a *Soyuz T-14*, arriving at the space station on September 18. Mission commander for the flight was Vladimir Vasyutin, joined by flight engineer Georgiy Grechko and research cosmonaut Alexander Volkov. The three-person crew joined the long-duration duo—Dzhanibekov and Savinykh—who had by then exceeded 100 days in orbit. A steady work load of remote sensing, astronomical observations, growing plants, and technology demonstrations was maintained by the crew.

On September 26, a milestone in piloted space flight took place with the first rotation of crews in orbit, a step closer to the goal of permanently staffing space stations. Returning to Earth, cosmonauts Dzhanibekov and Grechko left Vasyutin, Volkov, and long-term space flier Savinykh on board the *Salyut 7*. The overlapping of crews brought closer the day of continuous operations of space habitats.

Linking to the *Salyut 7* on October 2, a *Cosmos 1686* module lengthened the total space station complex and provided added living volume for the trio of cosmonauts. This new attachment, replete with its own maneuvering engine, was also used to orient the *Salyut*.

In a surprise development, the three *Salyut 7* crew members had to make an emergency return to Earth on November 21 aboard *Soyuz T-14* when cosmonaut Vasyutin required medical attention for an undisclosed ailment. Placed in a hospital, Vasyutin was released several weeks later. Prior to their forced return, Vasyutin and Volkov had spent 65 days in space, and Savinykh had experienced some 168 days of long-term microgravity.

At the close of 1985, the *Salyut 7/Cosmos 1686* complex floated empty in a 186-mi (300-km) circular orbit, inclined 51.6° to the Equator. Soviet space scientists announed that a new version of *Salyut* would be orbited in 1986, equipped with more docking ports and advanced capabilities.

Venus probes. Soviet *VEGA* probes arrived at Venus in June, deploying both a French-built balloon probe and surface lander. *VEGA 1* released its balloon and lander on the night of June 10–11, 1985. With an earthwide antenna network tuned in, 2 days of radio data from the French balloon sonde was received. The lander part of the mission dropped through the thick Venusian atmosphere, coming to rest near the Venus equator. Unfortunately, the *VEGA 1* lander deployed its surface sampler arm 10 to 15 min too early, upward of 10 mi (16 km) above the terrain of Venus. Despite the malfunction, over 20 min of data was relayed from the Venusian surface before the harsh environment silenced the lander.

On the night of June 14–15, 1985, *VEGA 2*'s lander, which had also deployed a balloon during its descent, came to rest 1000 mi (1600 km) to the south and slightly east of the *VEGA 1* lander touchdown site. This time, the landing craft did obtain a surface sample, analyzing the material and surroundings for 36 min.

Floating free in the atmosphere of Venus at 34 mi (55 km) altitude, the 10-ft-diameter (3-m), helium-filled balloons were caught in wind speeds much higher than predicted. Experiment designers had expected 20–30 mi/h (10–15 m/s) turbulence, not the 150 mi/h (70 m/s) velocity indicated by balloon

instruments. Although a nephelometer—designed to measure the size and density of cloud partices—on one of the balloons failed to operate, a total of over 90 h of useful data was sent earthward from the balloons.

With their deployments complete at Venus, the *VEGA 1* and *2* spacecraft continued onward to a rendezvous with Comet Halley in March 1986. The dual missions by the Soviet Union signal a new complexity in spacecraft operations.

ASIAN SPACE ACTIVITY

Japan achieved its first space missions beyond Earth orbit, rocketing two probes toward Comet Halley during 1985. *Sakigake* (*MS-T5*) was launched on January 7, 1985, from the Kagoshima Space Center by the Japanese Institute of Space and Astronautical Science (ISAS). The *MS-T5* probe was essentially a demonstration model designed to prove that a follow-on Japanese probe can survive the rigors of deep-space travel. *Sakigake* carried three space plasma experiments and was targeted to slip through Comet Halley's tail region, at a range of over 600,000 mi (965,000 km), in the early days of March 1986. From that distance, the Japanese probe was designed to perform important solar-wind ion-flux studies, plasma-wave analysis, and magnetic-field studies.

A second Japanese spacecraft, *Planet-A*, also named *Suisei*, was rocketed to an immediate ascent from Earth on August 18, 1985. Housed within the *Planet-A* payload was an ultraviolet camera to image the abundance and distribution of hydrogen in Comet Halley's coma. Also on board was a solar-wind experiment with a spherical electrostatic energy analyzer to measure the energy and direction of electrons and ions surrounding the comet. *Planet-A* was targeted to encounter Comet Halley in March 1986 at a distance somewhere between 60,000 and 120,000 mi (96,000 and 193,000 km) on the Sun side of the comet.

China orbited its seventeenth satellite on October 21, 1985. The spacecraft was maneuvered into a 106 by 240 mi (171 by 386 km) orbit, inclined 62.9° to the Equator. Seventeen days later, a hemispherical capsule weighing over 4000 lb (1850 kg) returned to Earth, ejected from the satellite. The capsule apparently carried the results of the satellite's photographic reconnaissance of the Earth, believed to be used for both military and civilian applications.

The official *China Daily* newspaper reported in 1985 that the country plans to enlarge its space efforts, offering international services to other countries. Plans were revealed to commercialize China's CZ-3 booster rocket in an effort to launch non-Chinese satellites.

EUROPEAN SPACE ACTIVITY

On July 2, 1985, the European Space Agency (ESA) launched the *Giotto* space probe to encounter Comet Halley in March 1986. Placed into geostationary transfer orbit by an Ariane 1 launcher from

Kourou, French Guiana, *Giotto* began its 8-month trek toward the comet. *Giotto* carried a scientific payload of 10 experiments, including spectrometers, multicolor cameras, a light-sensing photopolarimeter, charged particle detectors, dust impact detectors, and a magnetometer to sense Comet Halley's electrified plasma tail.

The *Giotto* encounter period with Halley was scheduled to last 4 h, with the spacecraft voyaging within 310 mi (500 km) of the object. *Giotto* carried a circular aluminum shield to protect it from the Halley dust cloud on its approach. The close-in images of the cometary nucleus provided by *Giotto*'s cameras were expected to reveal surface details down to 65 ft (20 m) across. Many predicted that the "comet kamikaze" mission of *Giotto* would not survive the beating from cometary dust particles.

European space activity also included several launches of the Ariane booster. On February 8, 1985, an Ariane 3 rocket boosted into orbit an *ARABSAT F1* satellite for the Arab Satellite Communications Organization (ARABSAT). On the same Ariane, a *Brasilsat 1* communications satellite was lofted spaceward for the Empresa Brasileira de Telecomunicações (EMBRATEL), the central telecommunications authority in Brazil.

An American *GSTAR 1* satellite for the GTE Spacenet firm was placed into orbit by an Ariane 3 booster on May 7, 1985. Also on board the launcher was a French communications satellite, a *Telecom IB*, orbited on behalf of the French Telecommunications Ministry.

The Ariane suffered a launch failure on September 12, 1985, its third in 15 launches. Ground teams blew up the booster less than 10 min into its flight when it veered off course. Lost in the explosion were two communications satellites, a United States *Spacenet F3* belonging to GTE and a European *ECS 3* spacecraft owned by EUTELSAT, a telecommunications consortium. The failure was made more embarrassing as French president François Mitterrand observed the destruction of the Ariane.

For background information *see* COMET; COMMUNICATIONS SATELLITE; MAGNETOSPHERE; SPACE BIOLOGY; SPACE FLIGHT; SPACE PROBE; SPACE PROCESSING; SPACE SHUTTLE; VENUS in the McGraw-Hill Encyclopedia of Science and Technology.

[LEONARD DAVID]

Bibliography: Aviat. Week Space Technol., issues from November 19, 1984, through November 18, 1985; *NASA Activities*, December 1984 through November 1985; *Science News*, December 1, 1984, through November 16, 1985; *Space World*, December 1984 through November 1985.

Spin glass

Just as a glass is a solid in which the spatial distribution of the atoms is amorphous rather than crystalline, so in its most particular definition spin glass refers to a magnetic state in which microscopic magnetic moments (spins) order in direction in a nonperiodic fashion. Additionally, the state is characterized by very slow equilibration after perturbation and significant history dependence. This type of order is found at low temperatures in magnetic alloys having competing interactions between the spins. Attempts to understand the origin and behavior of the spin glass state in such systems has, however, led to the appreciation of several new concepts and to the recognition that the fundamental ingredients for these concepts are much more widespread in occurrence than spin glasses. Thus, lessons learned from the study of spin glasses have had important impact in statistical mechanics, complex global optimization, models of biological memory, and possible new computational procedures, as well as in other branches of solid-state physics.

Properties. Among the characteristic properties of spin glasses are: (1) the onset of an apparent nonperiodic orientational freezing of microscopic magnetic moments at some characteristic temperature; (2) an associated change of slope with temperature of the very low field susceptibility (ratio of induced magnetization to applied magnetic field); (3) an extreme sensitivity to magnetic field in rounding out this singularity; (4) a distinct dependence of the susceptibility in the low-temperature spin-glass region on the measurement procedure, the magnetization obtained by cooling in the measuring field being higher than that obtained if the cooling is performed before applying the field; (5) a similar preparation-dependence of magnetization remaining after applying a field; (6) a very slow temporal decay of this remanent magnetization; (7) a relatively sharp temperature of onset of this "irreversibility," even in fields sufficient to cause significant rounding of the susceptibility, the onset temperature reducing with increasing field; and (8) in real experimental examples, the cusp in the susceptibility being unmatched by a corresponding readily observable singularity in the heat capacity. The first two of these properties indicate an amorphous antiferromagnetic order, but it is properties 4–7 which suggest that this new state has the fundamentally new features which have led to the appreciation of a relevance beyond magnetism.

Disorder and frustration. There appear to be two key ingredients for spin-glass behavior to be possible. One is quenched spatial randomness of the location of the magnetic atoms or of the strength or sign of their interactions, effectively frozen over the characteristic times of magnetic ordering. The second is known as frustration and refers to the fact that individual magnetic moments receive competing ordering instructions via different routes. This arises because the interaction between pairs of atomic moments varies with their separation, sometimes being ferromagnetic (tending to line up the moments), sometimes antiferromagnetic (tending to make them point in opposite directions). The combination of frustration and disorder leads to different competitions even for pairs of moments the same distance apart, because of their different environments. The details of this disorderly frustration are thought to be of secondary relevance, having only quantitative rather than qualitative consequence, and much of the progress in theoretical understanding has come from models with spins on all sites but with random (ferromagnetic and antiferromagnetic) interactions.

Theoretical study and computer simulation. The starting point of the theory of conventional periodic magnetism is an approximation known as mean field theory. For conventional pure systems this is very straightforward, and recent interest has been in its modification due to thermal fluctuations, particularly in the study of critical phenomena. Spin glasses are, however, already highly nontrivial even in mean field theory, and much of the current conceptualization has grown out of its study. This study has led to the development of sophisticated and novel mathematical procedures which themselves have had wider application.

Complementary to theoretical and experimental studies, an important role has been played by computer simulation in which experiments are effectively performed on a large (or special-purpose) computer. In such experiments thermal processes are mimicked by allowing the system to make random spin flips, with a probability determined by the Boltzmann factor $\exp(-\delta E/k_B T)$, where δE is the energy change such a spin flip would engender, T is the temperature, and k_B is Boltzmann's constant. This procedure is known as Monte Carlo simulation. As well as being able to perform the experiments on exactly the same models as studied theoretically, the computer experiments offer the invaluable opportunity to examine properties for which no real experimental probe exists, such as microscopic correlations between replicas with identical quenched controlling disorder but with spin structures evolving independently.

Fractal energy landscapes. If a schematic graph is constructed of an appropriate measure of the energy of a spin-glass system as a function of the orientations of the spins, a "landscape" is found with many hills and valleys with statistical self-similarity on any scale; effectively, this is what is known as fractal behavior. A schematic example (Fig. 1) has many local minima, the walls of any large valley containing secondary valleys whose walls have tertiary valleys, and so on. By contrast, a conventional pure ferromagnet may be construed (Fig. 2) as having a single valley with smoothly rising walls (if the representation is chosen appropriately). These qualitative figures have two interpretations: (1) energy as a function of a complete microscopic description (in which case the abscissa is a schematic one-dimensional representation of the multidimensional space of spin orientations; and (2) free energy as a function of a macroscopic characterization of the state (for example, the magnetization in the case of the ferromagnet), the free energy being a measure of both internal energy and thermal disorder whose minimum gives the favored macrostate for a system maintained at constant temperature. The first of these descriptions is temperature-independent, while in the latter case the landscape changes as the

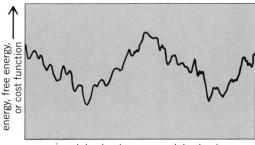

microstate structure, macrostate structure, or complete parameter characterization

Fig. 1   Schematic plot of energy or free energy for a spin glass, or cost function for an NP-hard optimization problem. For a spin glass the ordinate is energy and the abscissa is a characterization of the microstate structure, or the ordinate is free energy and the abscissa is macrostate structure. For a general optimization problem the ordinate is the cost function and the abscissa is a complete parameter characterization.

temperature is varied. *See* FRACTALS.

These figures can be used to explain many aspects of the differences in the behaviors of spin glasses and conventional magnets. Consider a system wishing to achieve its global (free) energy minimum but prepared in a nonminimal state. For a conventional system the state parameters can be changed gradually so that the energy runs downhill to the correct minimum, but the spin glass would become stuck in a higher secondary valley if only downhill energy changes were permitted. In fact, thermal excitations allow uphill energy moves too, determined by the Boltzmann probability discussed above, but the response is much slower, especially at low temperatures. Second, a change in external parameters, for example the application of a magnetic field, modifies the landscape, both locally and globally. The fractal system can respond quickly to a change in local properties but will be slow to reach a new global optimum, in contrast to the conventional system for which these minima are identical.

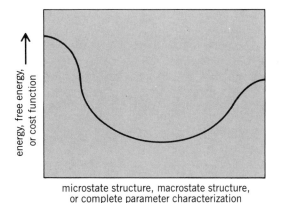

microstate structure, macrostate structure, or complete parameter characterization

Fig. 2   Schematic plot of energy or free energy for a conventional ferromagnet, or cost function for a polynomial optimization problem. The ordinate and abscissa are as in Fig. 1.

A related property concerns phase changes from paramagnet to ordered phase. A pure ferromagnet orders when a free energy valley at finite magnetization arises as the temperature is lowered past the transition temperature. As the temperature is lowered, the optimal magnetization increases but the single valley evolves continuously with no further transition. By contrast, the ordering of a spin glass corresponds to the many-valley situation discussed above, but with the actual valley structure varying as the temperature is reduced, leading to an effectively continuous sequence of transitions.

Application to global optimization. Many practical organizational problems involve global optimization under various different demands. Mathematically they can be posed as the minimization of some cost function which quantifies the weight to be attached to the demands. If the demands involve incompatibility (optimizing one precludes optimizing another), then the minimization problem is often NP (nondeterministic-polynomial) hard and its solution on a deterministic computer cannot be guaranteed in a time (number of operations) polynomial in the number of variables. Such problems are notoriously difficult, but also quite common.

The energy of a spin glass is a cost function whose minimization is normally NP-hard; yet thermodynamic experience suggests that if slowly cooled any system will achieve its ground state (or a close approximation). It has been argued that NP-completeness implies the same sort of fractal cost-function landscape as characterizes a spin glass energy, and the application of a computer-simulational analog of slow cooling has been suggested as a useful technique for more general application to NP-hard optimization.

Other applications. A system with many locally minimal "energy" states has been proposed as a model for memory. The neuronal firing states are the analog of the spin states above, the learning process is modeled by modifying the controlling interactions in response to experience, and the search for those minima corresponds to the recall process. The addition of an analog of temperature permits a measure of "noise." A similar system has been built into a prototype parallel computing device, known as the Boltzmann machine.

Irreversibility and metastable freezing effects also have been observed in several other solid-state contexts, presumably again with their origin in random frustration.

For background information *see* FERROMAGNETISM; MONTE CARLO METHOD; OPTIMIZATION; PHASE TRANSITIONS; SPIN GLASS in the McGraw-Hill Encyclopedia of Science and Technology.

[DAVID SHERRINGTON]

Bibliography: J. J. Hopfield, Neural networks and physical systems with emergent collective computational abilities, *Proc. Nat. Acad. Sci. USA*, 79:2554, 1982; S. Kirkpatrick, C. D. Gelatt Jr., and M. P. Vecchi, Optimization by simulated annealing, *Science*, 220:671–680, 1983; B. Mandelbrot, *The Fractal Geometry of Nature*, 1982; J. L. van Hemmen and I. Morgenstern (eds.), *Proceedings of the Heidelberg Colloquium on Spin Glasses*, 1983.

Squid

Cephalopods are the most advanced invertebrates, and squid are in many respects the most advanced cephalopods. While octopuses are probably more intelligent, there are no larger invertebrates than squid, and only flying insects are faster. Some estimates suggest that there may be almost as many squid in the sea as fishes.

Although they are relatives of such sedentary and slow-moving creatures as snails and oysters, squid have turned their molluscan body plan into a powerful jet propulsion system which allows them to compete directly with some of the fastest animals in the sea. Recent studies indicate that this combination of ancestry and competition has produced unique adaptations to an enormous energy requirement that affects not only anatomy, biochemistry, and physiology, but even the life histories in the group.

Swimming efficiency. Fishes and cetaceans swim by undulating (wiggling) through the water, which allows them to move forward by pushing large volumes of water backward at low speeds. In contrast, squid draw small volumes of water into their mantle cavities (Fig. 1) and pump the water out at high speed. This gives them the same forward momentum as fishes according to Newton's second law, but the propulsion system is much less efficient because the energy required to accelerate the water to high speed goes up as the square of speed and can never be recovered. This gives squid a low Froude efficiency, which means that they use more energy than fishes to move as fast or as far.

Metabolic rate. In order to escape from or capture fishes, squid must expend more energy, and as a result they have the highest metabolic rates so far recorded for poikilotherms of their size at the studied temperatures. Their maximum metabolic rates during high-speed swimming are two to three times

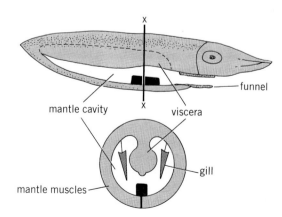

Fig. 1. Features of the jet propulsion system of squid. The dimensions and placement of a differential pressure telemetering device are shown. (*After E. R. Truman, Locomotion of Soft Bodied Animals, American Elsevier, 1975*)

Comparison of swimming and metabolic performance for a well-studied fish and three species of squid

Characteristics*	Sockeye salmon, *Oncorhynchus nerka*	Arrow squid, *Illex illecebrosus*	Long-fin squid, *Loligo pealei*	Market squid, *Loligo opalescens*
Temperature, °C	15	15	22	14
Total length, m	0.37	0.42	0.24	0.20
Mass, g	500	400	100	40
Critical speed, m s^{-1}†	1.35	0.76	0.60	0.36
Active metabolism, ml O_2 kg^{-1} h^{-1}‡	480	1047	950	862
Standard metabolism, ml O_2 kg^{-1} h^{-1}§	40	313	320	254
Rest metabolism, ml O_2 kg^{-1} h^{-1}	—	202	300	239
Scope for activity, ml O_2 kg^{-1} h^{-1}	440	734	630	608
Gross cost of transport, J kg^{-1} m^{-1}	1.9	7.6	9.6	12.6
Net cost of transport, J kg^{-1} m^{-1}	1.7	5.4	6.1	7.7
Oxygen debt, ml kg^{-1}	329	167	—	73

*°F = (°C × 1.8) + 32; 1 m = 3.3 ft; 1 g = 0.035 oz.
†Highest sustainable speed.
‡Oxygen consumption at critical speed.
§Extrapolated value at zero speed.

the comparable values for fishes, and at rest their rates are nearly five times those of fishes (see table). It seems that, like a racing car engine, the squid's metabolic system is designed for high performance, and when the system slows down its fuel economy becomes worse. This demand for fuel has given squid large appetites and requires a fast, efficient digestive system. The system seems to work well since squid given unlimited food eat more than twice as much as fishes, and not only keep up with their energy demands but also manage to grow much faster. The arrow squid, *Illex illecebrosus*, common in the Western North Atlantic, grows to over 1 lb (0.5 kg) in less than 9 months; at comparable temperatures fishes might take 2–4 years to reach this size.

Life history. The evidence is not complete, but it appears that such fast growth is a characteristic of nearly all squid. There is no good evidence that any squid lives more than 2 years, and some tropical and subtropical species reach maturity and die in less than 6 months. All observations suggest that squid are semelparous, that is, they breed only once in their lifetime. They store nutrients in body tissues and then in one big surge convert a high proportion of these nutrients to reproductive products and never recover. Giant squid (*Architeuthis*) weighing up to 1320 lb (600 kg) may live longer, but this is not certain. The jumbo squid, *Dosidicus gigas*, from the rich Humboldt Current off South America, can weigh over 110 lb (50 kg), and there is reasonable evidence that it lives less than 2 years.

One of the most interesting questions for teuthologists (those who study teuthids, or squid) is why nature would produce such large, complex animals only to have them die after a year or two. Although cephalopods as a class have been around since the Early Cambrian (575 million years ago), the shell-less coleoid cephalopods (which includes all living forms except *Nautilus*) are actually a recent innovation and have been competing directly with pelagic fishes for only about 50 million years. One might conclude that squid play the "short-but-merry life"

game more efficiently than fishes, since most fishes live longer and are iteroparous, that is, they reproduce repetitively. However, it is not clear why squid have failed to exploit this fish strategy. Perhaps there just has not been enough evolutionary time, but some aspect of their physiology seems to constrain them either from living a long time or from cyclic reproduction. It may be that there is no mechanism to allow only one egg to develop or that the hormone system initiating maturation cannot be turned off, but there also appear to be more fundamental problems associated with energy storage and the cephalopod's high-energy life-style.

Lipid storage. One problem is a limited ability to digest, synthesize, and store lipids. Lipids are the most concentrated form of nutritional energy, packing five to ten times as much energy into a given volume of tissue as protein or carbohydrate. Fishes can have lipid stores in muscle equal to 30% of their total body weight which, combined with their relatively low metabolic rates and transportation costs, allows them make long migrations or overwinter without feeding. Cephalopods store lipid only in their digestive glands, and these stores rarely amount to more than 5–10% of their body weight.

This appears to have a dramatic effect on reproductive patterns. Octopuses stay in one place and probably spend their entire lives collecting enough lipid to produce one batch of eggs. (It is essential that the hatchlings have energy reserves to keep them alive until they find food.) Young squid, in contrast, may travel thousands of kilometers (1 km = 0.6 mi) to find rich sources of food, often taking advantage of current systems to help them on their way. As adults, they feed on fishes, and manage to collect large lipid reserves, but may have to burn some of it to fuel a spawning migration that ensures their offspring are in the right place to feed on microscopic food and use the same currents. Because of their high cost of transport, even adult squid probably cannot fuel migrations from reserves. The large variations in the size of squid stocks from year to year and the success of the spawning migration may reflect the availability of food along the route. Squid are obligate schoolers, and in some species larger animals appear to practice "social cannibalism" on smaller members of their school when food supplies are short. This not only would ensure that some squid survive to spawn, but also would select the most successful squid for breeding stock. In any case, the remaining lipid stores are all used in gamete production, and the postspawning adults are left with no reserves in an area where little food suitable for them is available. The apparent consequence is that squid are unable to recover sufficiently to survive and compete with fishes having much greater reserves. Thus they are constrained to semelparous life histories.

Energy costs. Since energy costs seem to weigh so heavily in determining annual reproductive success, and therefore the size of commercial squid fisheries, considerable interest in exactly how squid swim has developed. Squid will readily swim in tunnel respirometers (aquatic "treadmills" which recirculate water at various speeds), allowing the oxygen they consume in swimming to be measured. This gives a very good estimate of the energy consumed, because squid have a limited capacity for anaerobic metabolism. Some fishes can actually swim for hours by breaking glucose down to lactic acid while accumulating an oxygen debt, but squid can only make a few jets without oxygen, by producing octopine, an alternate end product made in small amounts.

Some results of respirometry experiments with both squid and fishes are shown in the table. Fishes can swim much faster than squid, but squid compensate for their lack of speed with quick acceleration in almost any direction from their funnels which swivel, similar to the jets on a STOL aircraft. Squid are most streamlined at the tail where the fins used for control are located, and they normally swim backward (head last). They can reverse almost instantly, however, and swim head first at about half of their usual maximum speed. Even though squid swim at less than half the speed of comparable fishes, it takes more energy. Figure 2 suggests that

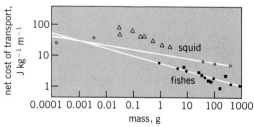

Key:
- sockeye salmon ∘ squid
- other fishes △ crustaceans

Fig. 2. Comparison of the effect of size on the cost of transport (energy cost per unit weight of locomotion at the most efficient speed) for fish, squid, and crustacean. The slopes of the regression lines show that the economies of scale are greater for fishes than for squid.

on a long migration things may be even worse; at the most efficient speeds the cost of transport (the energy needed to move a given weight of animal a given distance) is almost 10 times higher for a jet-propelled squid than for an undulatory fish. Squid's costs are, in turn, about one-tenth those of their chief prey, the paddling crustaceans.

Despite energy problems, there are many records of squid traveling several thousand kilometers. One tagged *Illex illecebrosus* traveled at least 1248 mi (2000 km) from northern Newfoundland to Maryland in 100 days, thus averaging about 10 in. (0.25 m) per second. By swimming steadily at this speed, it would have taken 2000 kilocalories (8400 kilojoules) to fuel the trip, equivalent to the energy in six squid. However, if the squid traveled at 2.5 ft (0.75 m) per second, a more efficient speed, it could have rested on the bottom two-thirds of the time and saved 400 kcal (1700 kJ). At lower bottom temperatures it might save a further 100 kcal (420 kJ) reducing the total cost to only four squid equivalents. The negatively buoyant squid may make further savings by climbing slowly to the surface and then gliding back down. Estimating how much this would save is difficult, but daily vertical migrations by squid to considerable depths are well known.

There are not enough data on the natural swimming patterns of squid to accurately calculate the cost of such a trip, but evidence is rapidly accumulating, since squid swimming is easy to study. Because nearly all the work squid do is focused on pressurizing the water inside their mantles, this work can be measured directly by measuring the pressure. By comparing the work done in swimming to the oxygen consumed, the efficiency of swimming can be measured in a way never possible with undulating fishes. Electronic minaturization even allows construction of pressure-measuring devices which fit inside the squid's mantle (Fig. 1) and telemeter the information ultrasonically as the squid swims. These will monitor both the behavior patterns and energy consumption of migrating squid directly in nature.

For background information *see* CEPHALOPODA; ENERGY METABOLISM; SQUID in the McGraw-Hill Encyclopedia of Science and Technology.

[RON O'DOR]

Bibliography: P. R. Boyle, *Cephalopod Life Cycles*, vols. 1 and 2, 1983, 1986; J. M. Gosline and M. E. DeMont, Jet-propelled swimming in squids, *Sci. Amer.*, 252(1):96–103; R. K. O'Dor, Respiratory metabolism and swimming performance of the squid, *Loligo opalescens*, *Can. J. Fish. Aquat. Sci.*, 39:580–587, 1982; E. R. Truman, *The Locomotion of Soft Bodied Animals*, 1975.

Star

Recent x-ray and ultraviolet observations indicate that phenomena seen on the Sun also occur on many types of stars, but often with vastly greater energies and scales. Such solarlike phenomena include hot outer layers called coronae and chromospheres, rapid enhancements of radiation called flares, and the acceleration of relativistic particles leading to x-ray and microwave radiation. Not all types of stars exhibit these phenomena, but most stars similar to or cooler than the Sun probably do, and cool stars that are young, are rapidly rotating, or are components of close binary systems exhibit the most energetic phenomena. There is now persuasive evidence that turbulent magnetic fields are responsible, directly or indirectly, for most of the x-ray and ultraviolet emission.

Most stars in the Galaxy would appear to the eye as yellow or red because their emitted energy is produced predominantly by gas at the relatively cool temperatures of 2500–7000 K (4000–12,100°F). The emitting layer, called the photosphere, is the deepest layer from which light can escape from a star. Such stars are commonly called cool stars because of their low photospheric temperatures, although astronomers classify them into various subgroups according to their photospheric temperatures and emitted spectra. These subgroups are called the F stars (6000–7000 K or 10,300–12,100°F), G stars (5300–5900 K or 9100–10,200°F), K stars (3900–5200 K or 6600–8900°F), and M stars (2500–3800 K or 4000–6400°F). The Sun is a G-type star.

One of the surprises of recent space research has been the discovery that cool stars are often very bright x-ray and ultraviolet emitters. This was unexpected because gas at the cool temperatures of the photospheres of such stars emits very little energy in the untraviolet and none at all in x-rays. Hot ionized gas is necessary to emit ultraviolet and x-radiation by such processes as electron excitation of ions or recombination of ions with electrons. High temperatures are needed to ionize the gas and to provide electrons with enough kinetic energy to excite the ions into their upper energy levels. A tenuous gas will emit mostly ultraviolet radiation in emission lines when it is roughly 10^5 K (1.8×10^5°F), and will emit primarily in the x-ray region when it is hotter than about 10^6 K (1.8×10^6°F). Thus, ultraviolet and x-radiation indicates the presence of hot gas in the layers above the photospheres of cool stars that is not apparent in their optical radiation.

Since the terrestrial atmosphere absorbs all ultraviolet and x-radiation, telescopes in space are necessary for stellar observations. Most ultraviolet observations of cool stars have been obtained with the *International Ultraviolet Explorer* (*IUE*) satellite, while the second *High Energy Astronomical Observatory* (known as *HEAO-2* or the *Einstein Observatory*) has provided the bulk of x-ray observations upon which the study of stellar coronae is now based.

Solarlike phenomena. Above the solar photosphere lies warm gas at 5000–20,000 K (9000–36,000°F) that is called the chromosphere because of the bright red emission in the hydrogen Balmer alpha line (656.3 nanometers) observed above the solar limb during total eclipses. Stellar chromo-

spheres are readily identified by bright emission in the cores of resonance lines formed at chromospheric temperatures. The presence of these emission lines in ultraviolet and optical spectra indicates that essentially all cool stars have chromospheres, although the strength of these emission features varies greatly from star to star.

Above the solar chromosphere lies a thin transition region with gas at 20,000–1,000,000 K (36,000–1,800,000°F) and an extended corona with gas at several millions of degrees. Ultraviolet and x-ray observations have revealed a pattern in which only certain types of cool stars have hot gas indicative of atmospheric regions analogous to the solar transition region and corona. Apparently, all cool

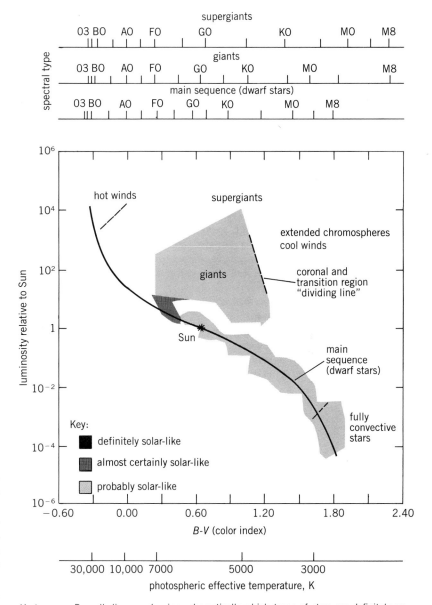

Hertzsprung-Russell diagram showing schematically which types of stars are definitely or probably solarlike on the basis of direct or indirect indicators (such as x-ray and ultraviolet emission from coronae and chromospheres) of strong, turbulent magnetic fields. Also identified are regions where massive stellar winds occur and where hot gas is apparently absent. °F = (K × 1.8) − 460. (*After J. L. Linsky, Nonradiative activity across the H-R diagram: Which types of stars are solar-like?, Solar Phys., vol. 100, 1985*)

stars with surface gravities similar to the Sun have such hot gas, although the amount of hot gas is highly variable from star to star. Such stars are called cool dwarfs and are located along the main sequence in a plot of stellar luminosity versus spectral type (see illus.). As stars exhaust their core hydrogen nuclear fuel, they evolve from the main sequence to become luminous giants and extremely luminous supergiants. Observations with the *International Ultraviolet Explorer* and *Einstein* spacecraft show that giants of spectral type F, G, and the beginning of K have hot gas indicating solarlike transition regions and coronae, but the cooler giants and most supergiants do not. The absence of hot gas in such stars has not yet been explained adequately, but these stars have chromospheres that extend out several radii and are expanding at rates indicating the loss of 10^{-10} to 10^{-6} solar mass per year.

Stellar age and rotational velocity. Two interesting correlations have emerged with respect to stellar parameters. The first is an increase in the surface flux (radiated energy per unit area of the stellar surface) in ultraviolet emission lines and x-rays with increasing stellar rotational rate. There is some dispute, however, as to the functional form of this correlation; in particular, whether the dependence is on the second power of the equatorial rotational velocity or is a function of the rotational period.

The second correlation is with age. The very youngest cool stars, called T Tauri stars, are evolving from their nebular environment toward the less luminous main sequence. Although their ultraviolet and x-ray spectra are often heavily obscured by local gas and dust, their spectra indicate surface fluxes 10^3 to 10^4 times larger than average solar values and thus heating rates in the chromospheric and coronal layers 10^3 to 10^4 times those typical of the Sun. The age of stars after they evolve to the main sequence can be determined by various techniques, such as their lithium abundance, and clusters of stars with common ages have been studied extensively by the *International Ultraviolet Explorer* and *Einstein*. Three such clusters are the Pleiades (age about 10^8 years), Ursa Major (1.6×10^8 years), and the Hyades (6×10^8 years). Ultraviolet and x-ray surface fluxes for stars in these clusters are typically 100 times that of the average Sun and decline exponentially with age, decreasing by a factor of e (2.7) in about 1.2×10^9 years for x-rays and 2.6×10^9 years for chromospheric lines.

As stars age, they lose angular momentum through mass loss and slowdown. One commonly held theory is that the observed decreases in x-ray and ultraviolet emission from cool stars with age and with decreasing rotational velocity are two aspects of the same phenomenon. As mentioned above, magnetic fields appear necessary for heating the hot outer layers of cool stars. These fields in turn are thought to be generated by dynamo processes occurring deep in the convective zones of cool stars by the interaction of convection and rotation. Thus, as stars age they rotate slower, the dynamo generation

of field decreases, the heating rates decrease in the outer layers, and the ultraviolet and x-ray surface fluxes decrease. Recent observations, however, indicate that this picture is too simplistic. X-ray flux measurements of the Pleiades, for example, show no correlation with rotational velocity as predicted by this theory, indicating that another age-dependent parameter, perhaps the depth of the convective zone, may play an important role.

Activity in close binary systems. Most stars are members of binary systems, and the proximity of a companion can alter the evolution and shape of a star. When binary stars are closer than about 10 stellar radii (corresponding to a 20-day orbital period for solar-mass stars), then tidal forces can speed up the rotation of a cool star until the rotational and orbital periods are synchronous, as for the Moon in the Earth-Moon system. Depending on whether or not the stars are in contact or have streams of matter flowing from one star to another, these systems are called RS Canum Venaticorum (noncontact and no streaming gas), Algol (noncontact but with gas streams), or W Ursae Majoris (contact) systems.

The *International Ultraviolet Explorer* and *Einstein* have found these systems to be unusually strong ultraviolet and x-ray emitters, even though their component stars have often evolved somewhat from the main sequence and are thus old. The strong emission is generally believed to result from the efficient dynamo generation of magnetic fields in stars with deep convective zones and tidally induced rapid rotation. Stars in RS CVn-type systems are also noted for large periodic changes in their visible light that are interpreted as being due to large, dark starspots that appear and then disappear as the star rotates. By analogy with sunspots, these spots are dark because of strong magnetic fields. Several observations now show that the ultraviolet and x-ray emission is largest when the starspots are on the visible hemisphere, providing further evidence that the heating mechanisms are magnetic in character.

Flares. The most energetic phenomena on the Sun are flares, which are observed as rapid enhancements of x-ray, ultraviolet, and hydrogen Balmer alpha light rising on time scales of seconds and decaying on time scales of minutes. These flares are often accompanied by strong radio emission, acceleration of particles to relativistic energies, and the ejection of matter from the Sun. They are thought to be caused by the rapid annihilation of magnetic fields in the corona. Flares have been detected by the *International Ultraviolet Explorer* and *Einstein*, as well as earlier satellites, on a number of M dwarf stars, on T Tauri stars, and in RS CVn-type systems, but with energies as large as 10^5 that of large solar flares.

Despite the vastly greater energies for flares on the M dwarf stars, they have many properties in common with solar flares, including similar rise and decay time scales, peak coronal temperature before peak soft x-ray flux, and similar flare loop lengths.

The larger radiated energies in the M dwarf flares appear to result from higher densities in the flaring loops. Flares in RS CVn-type systems have much longer rise and decay times and may be produced when magnetic loops of the two stars in the system interact. Little is known about flares on T Tauri stars, except that they are very energetic. Strong and often polarized radio emission during flares on M dwarf stars and RS CVn-type systems indicates the presence of relativistic electrons in magnetic fields.

For background information *see* BINARY STAR; SATELLITE ASTRONOMY; STAR; STELLAR EVOLUTION; SUN; ULTRAVIOLET ASTRONOMY; X-RAY ASTRONOMY in the McGraw-Hill Encyclopedia of Science and Technology.			[JEFFREY L. LINSKY]

Bibliography: J. P. Caillault and D. J. Helfand, The Einstein soft x-ray survey of the Pleiades, *Astrophys. J.*, 289:279–299, 1985; J. L. Linsky, Nonradiative activity across the H-R diagram: Which types of stars are solar-like?, *Solar Phys.*, vol. 100, 1985; R. Rosner, L. Golub, and G. S. Vaiana, On stellar x-ray emission, *Annu. Rev. Astron. Astaphys.*, 23:413–452, 1985; T. Simon, G. Herbig, and A. M. Boesgaard, The evolution of chromospheric activity and the spin-down of solar-type stars, *Astrophys. J.*, 293:551–574, 1985.

Statistics

Agricultural scientists have developed new statistical methods for agronomic applications. This article discusses the statistical significance of technical agricultural data and the evaluation of statistics from the perspective of field research.

Significance level. Choosing an appropriate significance level is a key step in the statistical analysis and interpretation of data from experiments. Recent studies indicate that assessment of risks associated with various kinds of errors can be helpful, as can the determination of their relative seriousness. A criterion for management of these risks can then be used to choose the significance level which minimizes the consequences of the errors which are committed.

Types of errors. Testing hypotheses is an essential part of the scientific method. The scientist sets up a specific hypothesis and seeks evidence which forms the basis for either rejecting or not rejecting the hypothesis. Usually the evidence is obtained by collecting and statistically analyzing data from a well-planned experiment which has been designed with the objective of testing the specific hypothesis.

Because the experimental material is a sample from the population about which the scientist wishes to draw conclusions, the results of the experiment are not infallible. The scientist would like either correctly to reject the hypothesis if it is really false, or correctly not to reject the hypothesis if it is really true. However, the experimental results are always affected by uncertainty caused by uncontrollable random variation inherent in the experimental material and by other nontreatment-related variation

introduced during conduct of the experiment. Thus the scientist will sometimes reach an incorrect conclusion; this can occur in one of three ways which are called type I, type II, and type III errors.

For example, a scientist may want to know if the yield performances of two cultivars differ under a certain set of environmental conditions. The hypothesis to be tested is that the true yield difference, δ, is equal to zero. A type I error is committed if δ is actually zero but the scientist concludes that it is not and thereby rejects a true hypothesis. A type II error is committed if δ is not zero but the scientist concludes that it is and thereby fails to reject a false hypothesis. A type III error is committed if δ is not zero but the scientist either concludes that δ is positive when it actually is negative or concludes that δ is negative when it actually is positive. Type III errors are sometimes called reverse decisions because the scientist rejects the hypothesis but reaches a conclusion which is the opposite of the truth.

It is impossible for a scientist never to commit one or more of these kinds of errors even if the scientist never uses statistics. An important goal in statistical analysis should be to minimize the consequences of the errors that are committed.

Frequencies of errors. The scientist would like the frequencies of the three types of errors to be small. Here the frequencies will be expressed as comparisonwise error rates. The type I error rate, α, is called the significance level. It is selected by the scientist and can be fixed at any desired value between 0 and 1. Traditional and common choices are $\alpha = 0.01$ and 0.05. The choice of α has profound effects on the type II and type III error rates and should be based on the seriousness of type I and type III errors relative to the seriousness of type II errors.

The type III error rate, γ, is generally quite small; it can never be greater than $\alpha/2$ even for a very small magnitude of δ. Fortunately the value of γ decreases rapidly toward zero as the magnitude of δ increases, but the magnitude of γ does increase as α increases.

The type II error rate, β, is also affected by the choice of α and by the magnitude of δ. For many combinations of α and δ, the error of deciding there is no difference when there really is a difference is more probable than the error of deciding that there is a difference when there really is not. That is, the type II error rate is often greater than the type I error rate.

The type II error rate can be reduced either by increasing the value of α or by increasing the magnitude of δ relative to its standard error. The traditional recommendation of statisticians has been to improve experimental precision by the use of more replications, refinement of experimental procedures, selection of more homogeneous experimental material, use of blocks in appropriate experimental designs to control variation, and use of covariates to help measure uncontrollable variation.

If the scientist provides reasonable estimates of δ

and the experimental error variance, then statisticians can calculate the number of replications required to achieve specific target values of α and β. This approach has great merit, especially in areas of basic research where the goal is establishment of "scientific truth" and where it is very important to have low probabilities for all three types of errors to ensure a high level of certainty with regard to the conclusions.

Risk assessment–risk management approach. In cases of applied research and in the extension of applied research results to the nonacademic community, resources often are not available to permit the traditional approach. Another approach has been suggested that involves two phases to deal with uncertainties: risk assessment and risk management.

Risk assessment deals with determination of the seriousness of errors and quantification of the risks associated with them. This phase includes four steps: (1) determine the relative seriousness of the three types of errors, (2) decide upon a reasonable relative loss curve, (3) incorporate knowledge of the error rates (α, β, and γ) with the relative loss curve to obtain a risk function, and (4) compute the weighted average risk for the experiment through the use of a reasonable frequency distribution for the magnitudes of δ.

After an adequate assessment of the risks, some criterion for risk management can be applied to control the risks in an optimal, or at least acceptable, manner. The suggested risk management criterion is to choose as the optimal significance level that value of α which minimizes the consequences of errors by minimizing the weighted average risk for the entire experiment.

This approach has similarities to the development of the Bayesian k-ratio t test (sometimes called the Bayes least significant difference) by R. A. Waller and D. B. Duncan, which involves the concept of the seriousness of making wrong decisions (type I and type III errors) relative to the failure to detect real differences (type II errors).

The problem of selecting an optimal significance level for corn hybrid performance trials has been studied with this approach. Relative seriousness of errors was evaluated from the viewpoints of farmers, seed companies, and personnel of institutions conducting the trials. A linear loss curve and a linear relative risk function were developed. Weighted average linear risks were computed based on empirical evidence that frequencies of occurrences of differences between hybrids were uniformly distributed. The optimal significance level was determined to be in the range α = 0.20 to 0.40.

Assessment and management of risk should be adaptable to many agricultural production and management practices. Researchers, extension personnel, industrial agriculturists, and farmers all need to assess the relative seriousness of errors from their own perspectives. For some practices the relative seriousness may be quite small, whereas for others

it may be very large. The determination of relative seriousness should consider the monetary costs associated with each type of error. But other, less tangible costs need to be considered as well. Researchers and extension personnel must consider influences upon their reputations as unbiased sources of reliable information. There are costs to society in general which must be considered in addition to costs to an individual. Erosion control and other soil conservation practices, for example, have implications for both present and future members of society at large as well as for present and future farmers.

Use of the risk assessment–risk management approach should not be a substitute for other fundamental principles of designing experiments. But when used in conjunction with other sound statistical practices, it can help reduce the consequences of the three types of errors.

[SAMUEL G. CARMER]

Field research. Agricultural researchers conduct field experiments to evaluate various theories or hypotheses. These experiments are conducted under conditions in which the environment is not fully controlled and in which the experimental sites vary. Thus, scientists must always question whether the experimental observations represent true biological differences or random variation. The proper use of statistical analysis is essential as an aid in answering this question and in making data easier to interpret.

Field researchers are interested in assessing the relative merits of several experimental treatments. In some cases one or more of the treatments may give nearly the same experimental result. However, judgment must often be made as to which treatment is superior or inferior based on observations that include both true treatment effects and an element of random chance due to variable field, biological, and environmental conditions. Thus, the investigator is never certain whether the observed differences among treatments are due to treatment effects or due to chance. Procedures have been developed whereby the probability can be calculated that an observed difference among treatments could occur due to chance. When these calculations are made (statistical analysis) and an observed difference is greater than would be expected due to chance, the treatment difference is said to be statistically significant.

Statistical significance. Statements of significance imply an associated probability statement indicating that the differences are of such magnitude that they would occur due to chance with only a certain frequency. Commonly used probability levels are 10, 5, or 1%. Statisticians refer to this probability as the probability of a type I error, that is, the probability that a difference will be judged to be real when in fact it is due to chance. Errors in making judgments about treatment differences can also arise by concluding that a difference is due to chance when in fact it is real. The failure to declare a true difference significant is a type II error. Experimenters can increase the chances of declaring true differences significant by accepting a higher probability of a type I error.

The presence of statistical significance among treatments does not necessarily imply biological or economic importance. Conversely, differences that are of a magnitude to be important biologically may not be significant statistically. One goal in designing field experiments is to keep the size of a chance difference small so as to reduce the occurrence of situations in which biologically important differences are not statistically significant. Random variation, number of replications, and choice of experimental design determine the size of the experimental error.

Experimental design. Even though they must balance design needs with available resources, investigators should always give high priority to minimizing experimental errors. Treatments in agricultural experiments are often arranged so that all experimental units (plots of ground, animals, and so forth) making up a replication are grouped together. This practice is referred to as blocking. Wise blocking is important. In situations where variation within a block is large or where block by treatment interactions is likely, it may be advantageous to repeat one or more treatments within a block. The repeats can be used to estimate within blocks experimental error and to statistically test block by treatment interactions. A more accurate estimate of experimental error can result. Not only can the repeated treatments be used to estimate experimental error, but they can also provide information on treatment effects if the analysis of variance takes into consideration unequal subclass numbers.

Covariants can be effective in reducing error. Most appropriate in field studies is a covariant that accounts for inherent productivity of experimental units on a unit-by-unit basis. A blank or uniformity trial can provide this information. A method that has proven useful involves determining a calibration value for each field plot from the performance of the plot's neighbor. A plot in the midst of others producing poorly can also be expected to produce poorly. In using this approach, an expected value is calculated for each plot and used as a covariate. An advantage of the approach is that a preliminary period or blank trial is not required. Application of this technique has been reported to have reduced experimental errors by 36% when compared to the errors from a randomized complete block design without covariate analysis. The approach is especially useful for agronomic studies on variable sites.

It is always possible to increase the likelihood of declaring a difference statistically significant by increasing the number of observations in each treatment mean. Investigators should construct critical statistical tests prior to establishing a field experiment. If properly done, this will help evaluate the design, help assess the chances of detecting differences, and point out tests in which the number of observations is unreasonably small. With factorial treatment sets, adding a level of a factor or even adding an additional factor may be as helpful as adding a replication. Additional treatments may also broaden the inference space and provide more information on important interactions.

Assumptions. Tests of statistical significance have traditionally been made by using a type I error probability of 5 or 1%. Some researchers have recently adopted a 10% criterion. Type II errors (failing to declare a true difference stastistically significant) may be more important in some experiments. Modern statistical packages routinely calculate the probability of a type I error. The probability of a type II error is not commonly calculated and depends on the magnitude of a type I error. Investigators must decide which is more serious and take steps to minimize the more serious error. Both type I and type II errors cannot be minimized simultaneously. In cases where the consequences of declaring a false positive are not as serious as failing to detect a true difference, investigators should consider accepting as statistically significant those differences that could occur due to chance as frequently as 15 or 20% of the time. This would decrease the frequency of type II errors.

For background information *see* AGRONOMY; ESTIMATION THEORY; STATISTICS in the McGraw-Hill Encyclopedia of Science and Technology.

[V. L. LECHTENBERG]

Bibliography: V. L. Anderson and R. A. McClean, *Design of Experiments*, 1974; R. J. Buker and D. D. Alvey, Productivity covariance: An improved field plot analysis, *Agron. Abstr.*, p. 570, 1979; R. J. Buker, D. D. Alvey, and W. E. Nyquist, Covariance techniques for using a moving average to reduce errors, *Agron. Abstr.*, p. 4, 1972; S. G. Carmer, Optimal significance levels for the application of the least significant difference in crop performance trials, *Crop. Sci.*, 16:95–99, 1976; S. G. Carmer and W. M. Walker, Pairwise multiple comparisons of treatment means in agronomic research, *J. Agron. Educ.*, 14:19–26, 1985; S. C. Pearce, *Field Experimentation with Fruit Trees and Other Perennial Plants*, Tech. Comm. 23, Comm. Bur. Hort. Plantation Crops, Earl Malling, Maidstone, Kent, 1953; R. A. Waller and D. B. Duncan, A Bayes rule for the symmetric multiple comparisons problem, *J. Amer. Stat. Ass.*, 64:1484–1503, 1969 (*Corrigenda*, 67:253–255, 1972).

Stellar evolution

The existence of objects having masses between the lowest-mass stars and the planets has been largely conjecture until recently. Theories of stellar evolution predict that the lowest-mass objects capable of burning hydrogen in their interiors have masses of the order of 0.08 times the mass of the Sun ($M_\odot$). Large numbers of these objects with ages of the order of 10^9 to 10^{10} years have been observed in the vicinity of the Sun. Direct measurements of masses of stars in binary systems tend to confirm this theoretical lower mass limit. By comparison, the planet

Jupiter has a mass of about 0.001 $M_\odot$ and cannot produce energy through nuclear reactions. Jupiter is seen principally through its reflected sunlight. Satellite and far-infrared observations have shown, however, that it does have some internal energy source because it radiates about twice as much energy as it receives from the Sun. It is thought that this energy source derives from the gravitational energy converted to heat during past rapid and present slow contraction phases.

Existence of very low-mass objects. The interesting questions concern objects having masses less than 0.08 $M_\odot$: whether there are many such objects or few, and whether there is a continuum ranging from the lowest-mass stars to the giant planets. A number of important astronomical problems rest upon answers to these questions. For example, the observations of stellar motions and densities as a function of height above the galactic plane in the neighborhood of the Sun may require the existence of 50% more mass than is observed in the form of stars and the interstellar medium. It is not known whether this could indicate the existence of a substantial amount of dark matter in the solar neighborhood, perhaps in the form of stars of mass less than 0.08 $M_\odot$. The numbers of these very low-mass stars affects the initial mass function, the frequency distribution of masses at the time of formation of the Galaxy. The initial mass function in turn affects theories of the formation, structure, and evolution of galaxies and star clusters. Finally, it is important to the understanding of the formation of the solar system and its place in the Galaxy to know whether any other very low-mass objects exist. *See* COSMOLOGY.

Detection methods. There are two methods of finding very low-mass objects: direct detection and inference by indirect means. The direct techniques include photometric surveys in the optical and infrared. The principal problem in detecting such objects directly is that they must be intrinsically very faint and difficult to observe with present detectors. These brown dwarfs may burn deuterium at some time during their evolution and become observable, but this period is relatively short, lasting about 10^7 to 10^8 years. Thus, it is improbable that an object could actually be observed at this stage of its evolution. Any such object detected by direct means would radiate energy mostly through contraction, analogous to Jupiter.

The indirect techniques include searches for proper motion or radial velocity variations in binary star systems containing very low-mass objects. Although the photometric surveys have not been successful, proper motion studies have revealed a number of stars which apparently have substellar companions. Historically notable among these is Barnard's star, which has long been suspected to have a companion of about Jupiter's mass. There is some controversy about the existence of this companion, and there is as yet no direct confirmation. For other stars the evidence is more convincing.

Application of speckle interferometry. The recently developed technique of speckle interferometry has been successfully applied to these problems. This method is essentially a direct-detection technique which allows the observer to overcome the effects of turbulence in the Earth's atmosphere and thereby measure very small angular separations between neighboring objects. It is well suited to finding companions to other stars.

T Tauri companions. An infrared companion has been found 0.6 arc-second south of T Tauri, a star that is the prototype of a class of pre–main sequence stars of about 1 solar mass. Because of the relatively low luminosity and temperature of the companion, it is thought to be a very low-mass star. There is some speculation that it may be a giant protoplanet. Recently, another object even closer to T Tauri has been found, which may also be of this class. Searches around stars which are known astrometric binary (or multiple) systems on the basis of their proper motion perturbations have revealed companions in several cases. It seems unlikely, however, that any of these companions is in fact a true brown dwarf or giant planet.

VB 8 companion. The most spectacular success has been the observation of an object near one of Van Biesbroeck's stars, VB 8. Astrometrists have shown that VB 8 undergoes small fluctuations in its proper motion that could be attributed to a companion with a mass of 0.045 $M_\odot$. In 1984, this companion was directly observed in the infrared, using speckle interferometry techniques. Measurements in two colors have revealed that the companion has a surface temperature of about 1360 K (2000°F) and a luminosity 3×10^{-5} that of the Sun, both far lower than corresponding quantities for any known star. Comparison of these data with theoretical calculations indicate that the object may indeed have a mass as low as 0.03–0.04 $M_\odot$. This is the lowest mass object yet found outside the solar system. While there are current arguments about whether the object is a brown dwarf or a giant planet, clearly it is an object of substellar mass. This method of discovery and other similar ones now under way at other observatories can be expected to yield more exciting results in the next few years.

For background information *see* BINARY STAR; SPECKLE; STAR; STELLAR EVOLUTION in the McGraw-Hill Encyclopedia of Science and Technology.

[H. M. DYCK]

Bibliography: H. M. Dyck, T. Simon, and B. Zuckerman, Discovery of an infrared companion to T Tauri, *Astrophys. J. Lett.*, 255:L103–L106, 1982; R. S. Harrington, V. V. Kallarakal, and C. C. Dahn, Astrometry of the low-luminosity stars VB8 and VB10, *Astron. J.*, 88:1038–1039, 1983; D. W. McCarthy, R. G. Probst, and F. J. Low, Infrared detection of a close, cool companion to Van Biesbroeck 8, *Astrophys. J. Lett.*, 290:L9–L13, 1985; A. G. D. Phillip and A. R. Upgren (eds.), *Nearby Stars and the Stellar Luminosity Function*, 1983.

Stratigraphy

Recent studies in the field of stratigraphy have involved chemical signatures frozen in the rock record, and the trangressive-regressive depositional sequences within the Carboniferous.

Chemostratigraphy. Chemostratigraphy, the use of elemental abundances in stratified rocks as a supplement to other procedures in stratigraphy, makes use of new techniques in rock analysis that provide abundance of 50 or more elements from a specimen relatively quickly. Chemical analysis of stratified rocks is not new; in the past, the primary focus of chemical analysis of sedimentary rocks was on certain suites of rocks (black shales, for example) or rocks from a selected stratigraphic unit. Few elements were analyzed. Now abundances of many elements may be obtained relatively rapidly, and elemental abundance in individual units of stratified rock may be used to recognize and trace a rock unit from place to place as well as to suggest environmental conditions during deposition. Chemical data may be used as well to establish correlations among facies and, in conjunction with biostratigraphic data, to recognize boundaries between units in the geologic time scale.

Use of neutron activation analysis allows rapid determination of the abundance of a relatively large number of elements. A system for automated neutron activation analysis was described in 1982. In this system, uranium is measured by delayed neutron counting, and the other reported elemental concentrations are determined by conventional reduction of gamma-ray spectra of the radioactive isotopes. The automated system is calibrated against a rock standard provided by analyses from the U.S. Geological Survey, the U.S. Bureau of Standards, and the Canadian Geological Survey. Stability of the system is checked periodically against the standard.

Chemical signatures. The suite of element abundance information obtained from a rock unit may be examined for a chemical signature by which the unit may be recognized and traced. That signature may be a relatively great abundance of one or several elements. Changes seen in elemental abundances in a stratigraphic sequence of layered rocks may suggest change in the depositional environment not obvious from nonchemical examination of the rocks. Trace-metal abundances may reveal unsuspected but potential resources for certain metals.

The chemical signature of a rock is a function of both its depositional environment and its postdepositional history. Because postdepositional processes can result in change in elemental concentrations, a signature that is most useful in stratigraphy probably will be developed in those rocks in which the elements are essentially "frozen" in amounts and states that were present during deposition. Accordingly, a search has been carried out for those rocks most likely to possess elemental concentrations frozen in them since deposition. The search has indicated that black shales deposited under anoxic, marine waters possess such characteristics. Four aspects of black shales appear to make them most useful in chemostratigraphy: (1) Rates of deposition commonly are small; thus these rocks are likely to be reflective of chemical conditions in the overlying water. (2) Lack of bioturbation limits chemical interactions to diffusion and mineral equilibria and kinetics. (3) Poor porosity and permeability during and particularly after diagenesis inhibit later migration of ions in or out of the sediment. (4) Sulfides and other minerals formed in anoxic conditions are thermodynamically and kinetically "stable" with respect to oxic-minerals in all but surface environments.

Furthermore, in any study in which a relatively complete sequence of strata is essential, as in the case of geologic time unit boundaries, marine anoxic conditions tend to develop in waters over the slopes and outer parts of the shelves. Dark, organic-rich mudstones tend to accumulate under these waters.

Applications. The Ordovician age Dictyonema Shale in Balto-Scania (southern Norway, southern Sweden, Estonia, and the adjacent part of Denmark) is recognized primarily by its unique fossil content, the graptolite *Dictyonema flabelliforme*. In the absence of that graptolite, the unit is similar to many other black shale units. In those areas where the characteristic graptolite is not present and stratigraphic position in a recognizable sequence of rock units may not be determined, the Dictyonema Shale is essentially unrecognizable. The unit displays a distinctive chemical signature, however, by which it may be recognized and traced. That signature has been identified by analysis of many *D. flabelliforme*–bearing samples from the unit. The chemical signature includes relatively great quantities of antimony, arsenic, molybdenum, uranium, and vanadium. In 1981 it was demonstrated that the signature is so distinctive and persistent that the unit may be traced into terrain where it has been metamorphosed. The chemical signature has been used as well to trace the unit through an area of limited rock exposure in southern Scandinavia and into Estonia. The chemical signature has been identified only in the Balto-Scanian Dictyonema Shale (Fig. 1). Age-equivalent shales, most of which bear *D. flabelliforme*, in Great Britain and North and South America contain elemental concentrations different from those in the Balto-Scanian Dictyonema Shale.

Elemental concentrations in a shale or clay layer at the biostratigraphically determined Cretaceous-Tertiary boundary give that layer a distinct chemical signature: high concentrations of iridium and osmium. The signature has been found in rock sequences that formed under many different environments both on land and in the seas of the time. The Cretaceous-Tertiary boundary is identified on fossil faunal or floral change. The chemical signature is used to supplement biostratigraphic data, to identify the boundary in rock sequences in which fossil data

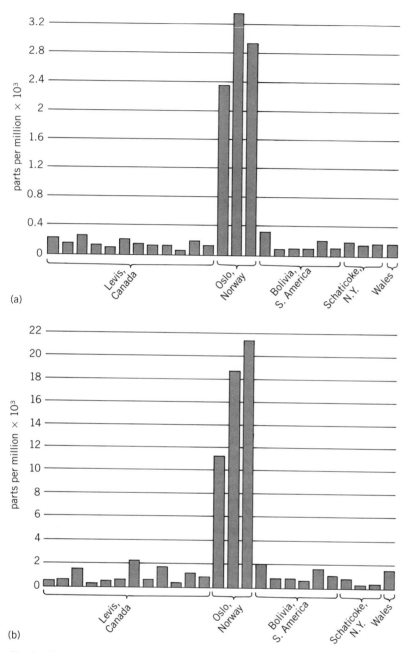

Fig. 1. *Dictyonema* zones; elemental abundances in samples of *D. flabelliforme*—bearing shales and coeval rocks in North and South America and Europe. The abundances indicate one component of the chemical signature of the Balto-Scanian Dictyonema Shale. (*a*) Antimony-122. (*b*) Vanadium-52.

under waters reflective of different degrees of anoxic conditions in ancient oceans will bear elemental concentrations suggestive of the specific anoxic conditions close to the water-sediment interface. Change in anoxic conditions or a change from anoxic to oxic conditions in ocean bottom waters will be reflected in change in elemental abundances. Such a change may be widespread, and the chemical data suggestive of change may be used as a supplement to biostratigraphic data in establishing contemporaneities of boundaries between units in the time scale. Potentially, such a change could be recognized widely and be helpful in establishing the contemporaneity of a time unit boundary in rock sequences formed in several different environments. For example, a major change in anoxic conditions was recognized in strata in the Ordovician-Silurian boundary interval at Dob's Linn near Moffat, Scotland. That change most likely seems to be related to Late Ordovician glaciation. During that glaciation, deep ocean waters would have become ventilated more fully than they had been during the long nonglacial time that preceded the glaciation.

The change reflective of glaciation seen in marine strata may be helpful in establishing the contemporaneity of those strata with strata formed in certain nearshore marine environments in which the effects of glaciation may be recognized. For example, glacial-eustatic sea-level lowering will be reflected in shallow marine sequences.

[W. B. N. BERRY]

Carboniferous transgressive-regressive deposition. More than 40 transgressive-regressive depositional sequences are present in Carboniferous shallow marine successions on the world's stable cratonic shelves. Eight more occur in lower Permian successions. These sequences were synchronous events resulting from eustatic sea-level changes and were time-equivalent based on currently available fossil knowledge. Most of these depositional sequences contain fossils of rapidly evolving late Paleozoic tropical, subtropical, and temperate shelf faunas, which provide firm age correlations. These transgressive-regressive sequences averaged about 2 million years and ranged from 1.2 to 4.0 million years in duration.

Abrupt stratigraphic changes of species and genera in faunal assemblages, which commonly are so useful for correlation, in most cases occur across unconformities that introduce new marine transgressive events and new depositional sequences. The

are ambiguous or lacking, and to trace the boundary with precision into rock suites formed under many different environments.

Based upon trace-metal abundances recognized in analysis of many black shales and mudstones, it was suggested that four zones which are reflective of different degrees of anoxic conditions may be present above the water-sediment interface in oceanic settings where anoxic waters are present at depth. Modern examples include deep waters in fiords off British Columbia, those in the Cariaco Trench, and those in the Black Sea. Sediment that accumulated

Fig. 2 Correlation of transgressive-regressive deposits on shallow marine shelves in parts of Europe and North America. Tn = Tournaisian; V = Viséan; Fm = formation; Ls = limestone; Gp = group; A, B, C, and D = unnamed parts of European Middle Carboniferous; E = Eumorphoceras (European zone, the lower part of this interval is based on ranges of two species of this genus of goniatite cephalopod; intervals are conventionally referred to as E_1 and E_2); ? = questionable boundaries or questionable presence of beds.

Stage	N.W. EUROPE Mesothem	MOSCOW BASIN AND SOUTHERN URALS	MISSISSIPPI VALLEY AND ILLINOIS BASIN	MIDCONTINENT REGION (Kansas and Oklahoma)	NORTH CENTRAL TEXAS	WEST TEXAS (GLASS MOUNTAINS)
LOWER PERMIAN — SAXONIAN	260 Ma	ARTINSKIAN	LOWER PERMIAN — not present	LOWER PERMIAN — red beds and evaporites	nonmarine deposits — WICHITA-ALBANY GROUP	LOWER PERMIAN — LEONARDIAN
AUTUNIAN / SAXONIAN	270 Ma — nonmarine deposits	ASSELIAN SAKMARIAN		GEARYAN / CIMMARONIAN	LOWER PERMIAN — CISCO GROUP	WOLFCAMPIAN
	290 Ma	GZHELIAN				
UPPER CARBONIFEROUS — STEPHANIAN	nonmarine deposits	KASIMOVIAN	UPPER PENNSYLVANIAN — VIRGILIAN (Matton Fm.) / MISSOURIAN (Bond Fm.)	UPPER PENNSYLVANIAN — VIRGILIAN / MISSOURIAN	UPPER PENNSYLVANIAN — VIRGILIAN / MISSOURIAN — CANYON GROUP	UPPER PENNSYLVANIAN — VIRGILIAN (Gaptank Fm.) / MISSOURIAN
CANTABRIAN	296/310 Ma? (D?, C)	MOSCOVIAN — Myachkovian, Podolian	LOWER AND MIDDLE PENNSYLVANIAN — DESMOINESIAN (Modesto Fm., Carbondale Fm., Spoon Fm.)	MIDDLE PENNSYLVANIAN — DESMOINESIAN	MIDDLE PENNSYLVANIAN — DESMOINESIAN — STRAWN GROUP	S.E. ARIZONA — MIDDLE PENNSYLVANIAN — DESMOINESIAN — LIMESTONE / HOROQUILLA
MIDDLE CARBONIFEROUS — WESTPHALIAN	deposition / hiatus (B, A)	Kashirian / Vereyan	ATOKAN (Abbott Fm.)	ATOKAN	Kickapoo Creek Gp / Bend Group — ATOKAN	ATOKAN
NAMURIAN	315 Ma	BASHKIRIAN	MORROWAN (Caseyville Fm.)	LOWER PENNSYLVANIAN — MORROWAN	LOWER PENNSYLVANIAN — MORROWAN — Not deposited on higher parts of Eastern Shelf or Bend Arch	LOWER PENNSYLVANIAN — MORROWAN — Black Prince Ls.
	320 Ma — ?, ?	SERPUKHOVIAN				
LOWER CARBONIFEROUS — VISÉAN	E2, E1, V3c, V3b, V3a, V2b, V1b V2a, V1a	VISÉAN — OKIAN, MALINOVIAN, YASNOPOLYAN	MISSISSIPPIAN — CHESTERIAN / MERAMECIAN (evaporites)	MISSISSIPPIAN — Tectonically active shelf margin transgressive regressive subdivision not attempted	MISSISSIPPIAN — Barnett Fm. / Chapell Ls.	MISSISSIPPIAN — CHESTERIAN / MERAMECIAN (Paradise Fm.)
TOURNAISIAN	hiatus / deposition — Tn3, Tn2, Tn1b	TOURNAISIAN — CHERNY-SHINIAN, LIKHVINIAN	OSAGEAN (bone bed) / KINDERHOOKIAN			OSAGEAN — Escabrosa Gp (evaporites) / KINDERHOOKIAN
	360 Ma — **DEVONIAN** basin ←→ shelf					

hiatuses between these transgressive-regressive sequences may represent as much (if not more) time as is actually recorded in the stratigraphic record. The presence within late Paleozoic shelf strata of these numerous, worldwide synchronous unconformities of considerable duration suggests that the fossil record is very incomplete and is mostly related to times of sea-level highs. Such a fossil record could easily be misinterpreted as a punctuated evolution showing an extremely irregular mutation rate. When late Paleozoic shelf strata are viewed as a set of transgressive-regressive sequences within the framework of regional depositional histories, more precise correlations are possible than if based on faunal assemblages by themselves.

These transgressive-regressive sequences are stratigraphic units characterized by a relatively conformable succession of genetically related strata which are bounded at their top and base by unconformities or by correlative conformities. Two terms that share similar application are mesothem and synthem, but these have not gained wide acceptance in North American literature. A mesothem is a middle-scale depositional succession that is larger than a cyclothem, is smaller than a sequence, and represents deposition during a major transgressive-regressive event. A synthem is a stratigraphic unit bounded above and below by unconformities with no restraints on size.

Depositional sequences for the Carboniferous and lower Permian are shown in Fig. 2. These include shelf areas containing tropical, subtropical, and warm-temperate marine faunas which are used as regional standard sections.

Lower Carboniferous. In the British Isles, marine transgressions and regressions having time intervals in the range of 1 to 3.5 million years have been recognized as widespread and synchronous in Dinantian, Namurian, and Westphalian rocks. These were attributed to eustatic sea-level fluctuations.

Portions of the Chesterian and Morrowan successions on the southern Ozark shelf have a number of the Namurian transgressions and regressions which have been correlated with individual mesothems in the British succession based on goniatite cephalopod faunas. Thus, these transgressions and regressions are worldwide, and the time duration of a single transgression and regression can be resolved with current paleontological knowledge.

The correlation of lower Carboniferous strata on the southern part of the Russian Platform with those of northwestern Europe is well established. The subdivisions of the Tournaisian and Viséan can be readily correlated in most parts of the world, including the western North American Cordillera, northern Canada, Australia, and southern, southeastern, and eastern Asia. The correlation within the lower Carboniferous is particularly good worldwide, except for the type area of the North American Mississippian Subsystem where the faunal assemblage was strongly provincial.

In the Viséan, evaporites on shelves were widespread at two times, initially during the early Viséan (Chadian) and again in the early late Viséan (Asbian). These times of evaporite deposition are remarkably consistent and are geographically widespread on many shelves.

Middle Carboniferous. On most shelves, the lower-middle Carboniferous boundary is within a long hiatus, and the Chokierian and Alportian stages are missing. This was a time when relative sea level remained generally low, and only occasional and brief transgressions reached on to more than the edges of the shelves. These successions tend to be thin and mostly clastic, suggesting much colder temperatures worldwide than in the Viséan.

The Westphalian of northwestern Europe is mainly continental and interrupted by thin tongues of brackish-marine bands which demonstrate 10 transgressions and regressions. A similar number of depositional sequences occur in normal marine deposits on the Russian Platform and on the southern shelves of North America. These marine deposits can be readily correlated, even though the foraminiferal assemblages contain a large percentage of endemic species and genera.

Upper Carboniferous. Sediments near the middle-upper Carboniferous boundary show many features in common with those near the early to middle Carboniferous boundary. Major hiatuses occurred on most shelves, clastics are common and carbonates rare, and the faunas underwent major changes because of evolution and extinction. At this time, extinctions included many of the remaining genera which had survived from the lower Carboniferous. New lineages in the upper Carboniferous became the stocks from which evolved the diverse, warmer-water, Permain faunas. In the middle part of the upper Carboniferous, at least two successive times of deep stream erosion cut across these shelf sediments.

Permian. Correlation of lower Permian strata is well established through the Irenian horizon of the Artinskian. The upper Permian Kazanian section and many other shelf sections pass into facies that may be brackish or supersaline at one or other times, and many shelves became entirely nonmarine. Also, upper Permian strata are difficult to correlate because of strongly developed provincial and latitudinal faunas. Most upper Permian Tethyan marine faunas are known only from active tectonic belts where transgressive-regressive sequences, as identified on stable shelves, are not easily recognizable.

Interpretation. These transgressions and regressions are comparable to the megacyclothems of Kansas and are of considerably longer duration than most Illinois basin-type cyclothems. A middle and upper Carboniferous transgressive-regressional depositional sequence may show four or five cyclothems or partial cyclothems within it, particularly in areas rich in clastic sediment sources. In general, the depositional sequences are easily indentified in areas possessing cyclothems because, being high on the shelf, the regressive phases are expressed by weathered surfaces and deeply incised river valleys cut into the underlying beds.

Many types of evidence support interpretations of repeated worldwide transgressions and regressions of shorelines onto stable shelves during Carboniferous and Permian time. Stratigraphic evidence shows the broad extent of these shoreline displacements across large areas of stable cratonic shelves. Faunal correlations indicate that these produced synchronous deposits on different shelves around the world. As a consequence, these features are correlated with eustatic changes in sea level.

The causes for these eustatic sea-level changes are not known; however, they were most likely the result of a combination of several different mechanisms. For example, the volume of ocean basins may change as a result of heating or cooling of oceanic crust, of increase or decrease in ocean trench activity at plate boundaries, and of orogenic activity along continental margins. Also, as demonstrated by the changes in sea level during the Pleistocene, climatic changes causing accumulation of ice sheets may result in eustatic changes of at least 330 ft (100 m) or more. It has been established that the middle and late Carboniferous was both a time of rapid plate motions and rapid sea-floor spreading and, in parts of Gondwana and Angara, a time of widespread glaciation.

During the Carboniferous, pieces of continental crust were assembled to form Pangaea and resulted in the closing of an equatorial east-west tropical seaway between western North America and eastern Europe. This strongly influenced middle Carboniferous ocean water circulation and the distribution of shallow, warm-water, marine faunas. In addition, sea level and surface-water temperature generally were less near the lower/middle and middle/upper Carboniferous boundaries which contributed to the higher-than-average rates of extinctions. Warm-adapted, shallow marine-shelf faunas on opposite coasts of Pangaea could disperse to the other major tropical shelf area only across shelves that were in cooler water. Eustatic changes of sea level, along with climatic fluctuations, provided variable filter routes along these shallow marine shelves. Many middle and late Carboniferous and early Permian faunal lineages evolved independently in the two tropical areas and had rare, and commonly short-lived, dispersals into other areas. Within the late Paleozoic, extinctions and apparent ecological displacements, which are recorded for many species of shallow-water marine faunas, are closely associated with these transgressive-regressive sequences.

For background information see ACTIVATION ANALYSIS; ANOXIC ZONES; CARBONIFEROUS; FACIES (GEOLOGY); STRATIGRAPHY in the McGraw-Hill Encyclopedia of Science and Technology.

[C. A. ROSS; J. R. P. ROSS]

Bibliography: D. G. Gee, The *Dictyonema*-bearing phyllites at Nordaunevoll, eastern Trondelag, Norway, *Norsk Geol. Tidsskrift.*, 61:93–95, 1981; M. Q. Hunt et al., Chemostratigraphy: Supplement to biostratigraphy, *Geol. Soc. Amer. Abstr. Programs*, 16(6):546, 1984; M. M. Minor et al., An automated activation analysis system, *J. Radioanal. Chem.*, 70:459–471, 1982; W. H. C. Ramsbottom, Eustacy, sea level, and local tectonism, with examples from the British Carboniferous, *Yorkshire Geol. Soc. Proc.*, 43:473–482, 1981; C. A. Ross and J. R. P. Ross, Carboniferous and Early Permian biogeography, *Geology*, 13:27–30, 1985; C. A. Ross and J. R. P. Ross, Late Paleozoic depositional sequences are synchronous and worldwide, *Geology*, 13:194–197, 1985; W. B. Saunders, W. H. C. Ramsbottom, and W. L. Manger, Mesothemic cyclicity in the mid-Carboniferous of the Ozark shelf region?, *Geology*, 7:293–296, 1979; P. Wilde et al., Anoxic facies in the Lower Paleozoic ocean, *Geol. Soc. Amer. Abstr. Programs*, 16(6):694, 1984.

Structural geology

Fission track dating using the minerals apatite and zircon has become a common technique recently to determine the ages and rates of uplift for mountain ranges. This technique is a relatively new and valuable method for quantifying vertical movements within the crust of the Earth along major fault systems. It is a vertical equivalent to paleomagnetic studies that have been in use since the mid-1960s to quantify horizontal movements of the Earth's crust. The use of fission track dating to determine uplift ages is an indirect approach that is relatively easy to implement.

Fission tracks. Apatite [$Ca_5(PO_4)_3(F,Cl,OH)$] and zircon ($ZrSiO_4$) minerals invariably contain uranium as an impurity within their crystal structures. These two minerals are used most frequently for fission track dating because of availability, chemical stability, and low blocking temperature (the temperature at which fission tracks are annealed). They are common accessory minerals in intermediate and silicic igneous rocks, as well as present in minor amounts in some clastic sedimentary rocks. Fission tracks are submicroscopic trails that represent radiation damage caused by the fission, or splitting apart, of radioactive uranium within the crystal lattice of a mineral. Above the blocking temperatures, 220 ± 18°F (105 ± 10°C) for apatite and above 347 ± 36°F (175 ± 20°C) for zircon, the fission tracks are annealed and thus disappear. Below the blocking temperature, any uranium that fissions leaves a permanent track in the mineral.

Uranium decays at a known rate, resulting in a predictable amount of fission per unit time for a given amount of uranium. Therefore, after determining the amount of uranium in a sample, the time that has passed since the mineral cooled below its blocking temperature can be calculated by counting the fission tracks. Consequently, fission track ages are actually ages of cooling, not directly ages of uplift.

Uplift rates. In order to use fission track ages as evidence for uplift, the temperature distribution within the Earth must be considered. On a regional scale. temperature increases with depth in the Earth; this geothermal gradient normally averages between 60 and 90°F per mile (20 and 30°C per kilometer). Hence, with only rare exceptions, rocks at depths shallower than approximately 2.5 mi (4 km) are cooler than the apatite blocking temperature. Therefore, the fission track ages of apatite collected from rocks in mountain ranges can be interpreted as the age of uplift from below to above a depth of approximately 2.5 mi (4 km). Similarly, zircon fission track ages represent uplift from below to above approximately 4 mi (7 km) of depth. During uplift, overlying material is removed by erosional processes resulting in rocks being exposed at the surface that were once buried by 2.5–4 mi (4–7 km) or more of material.

Uplift rates for mountains can be calculated in two possible ways. First, from a single rock sample containing both apatite and zircon it is possible to determine when the sample was uplifted above 4 mi (7 km) from the zircon fission track age, and to determine when it was subsequently uplifted above 2.5 mi (4 km) from the apatite fission track age. The difference in depths (4 mi − 2.5 mi = 1.5 mi; 7 km − 4 km = 3 km) divided by the difference in time between the zircon and apatite ages yields an uplift rate. Alternatively, by obtaining one kind of mineral, either apatite or zircon, from two samples, one near the base of a mountain and one near the top, it is possible to determine an independent evaluation of uplift rate. This approach entails dividing the vertical distance between the two samples by the difference in their fission track ages for the same type of mineral. These two methods provide valid differences in uplift rates because they may record uplift rates at different times in the past.

Applications. The use of fission track dating in the earth sciences has seen a rapid increase over the past few years. Fission track dating of apatite has been used to calculate an average uplift rate for the Wasatch Mountains in northern Utah of 0.2 in./yr (0.4 mm/yr) over the past 5 million years. The Wasatch Fault, which bounds the north-south-trending Wasatch Mountains on the west, is an active normal fault, with the Wasatch Mountains on the east being uplifted relative to the Great Salt Lake Valley on the west. Movement on the fault has in the past caused large earthquakes (magnitude 6 and greater) with accompanying potential of mass destruction. Estimates of the amount of movement on the fault gained from fission track uplift rates can be used to estimate the recurrence rate of large earthquakes. By using a rough estimate of 13 ft (4 m) of fault movement during a large earthquake and the uplift rate of 0.2 in./yr (0.4 mm/yr), an estimated recurrence rate of 1000 years for large earthquakes on the Wasatch Fault can be determined. Because the Wasatch Fault, similar to most large faults in the world, is made up of many segments, the recurrence rate for the whole fault is probably more accurately placed at 200 years. Knowledge of this type is very useful in determining building codes and emergency procedures planning for local governments.

Another study using fission track dating of apatite was completed recently in southern Alaska. In this study, fission track ages were used to time the horizontal movement of large blocks of the Earth's crust. This can be accomplished because horizontal movements resulting in compressional forces are accompanied by thrust faulting and uplift. Much of southern Alaska consists of continental crustal blocks, most on the order of 60 mi (100 km) wide and several hundred miles long. These blocks have been scraped off the western coast of the United States and Canada by northward-moving oceanic crust which underlies the Pacific Ocean. The portion of California west of the San Andreas Fault, including the Baja peninsula, San Diego, Los Angeles, and San Francisco, is a present-day example of a block which is being moved northward by the motion of the Pacific plate in a similar fashion to a cardboard box on a conveyor belt. The dense (density = 3.2 g/cm^3) Pacific oceanic plate is being subducted under the less dense (density = 2.7 g/cm^3) Alaskan continental crust along the Aleutian Trench just offshore the southern coast of Alaska. When continental blocks that are riding on top of the Pacific plate reach the Aleutian Trench, they are scraped off and accreted to the southern coast of Alaska. During accretion, uplift occurs that can be dated by using the fission track method. Uplift ages calculated for the Chugach Mountains show accretion at about 68 million years ago. Knowledge of these accretionary events aids in understanding the development of Cook Inlet, an oil-producing basin.

Fission track dating in general is a new field of study and is still in a period of improvement and refinement. Its use in determining uplift ages for a particular mountain range is tied to the thermal history of that area and is therefore subjected to additional uncertainties. The work that has been done so far is very encouraging, and additional field studies will certainly improve the credibility and understanding of fission track dating. Ongoing laboratory experiments will also improve routine dating techniques and increase the accuracy of age determination. *See* FAULT AND FAULT STRUCTURES.

For background information *see* FAULT AND FAULT STRUCTURES; FISSION TRACK DATING; PLATE TECTONICS; STRUCTURAL GEOLOGY in the McGraw-Hill Encyclopedia of Science and Technology.

[BRONSON W. HAWLEY]

Bibliography: B. W. Hawley, R. L. Bruhn, and S. H. Evans, Jr, Vertical tectonics in a forearc region, southern Alaska, using fission track dating of apatite grains and flexural beam modeling, *Geol. Soc. Amer. Abstr. Programs*, 16:(6):533, 1984; C. W. Naeser et al., Fission-track ages of apatite in the Wasatch Mountains, Utah: An uplift study, *Geol. Soc. Amer. Mem.*, no. 157, pp. 29–36, 1983; R. R. Parrish, Cenozoic thermal evolution and tectonics of the Coast Mountains of British Columbia, 1. Fission track dating, apparent uplift rates, and patterns of uplift, *Tectonics*, 2(6):601–631, 1983.

Sulfide phase equilibria

One recent area of investigation in the study of sulfide phase equilibria involves the representation of four-component (quaternary) sulfide systems. An understanding of these systems is based on the Gibbs' phase rule, which states that under equilibrium conditions in a closed system the numbers of degrees of freedom (F) is related to the number of components (C) and to the number of phases (P), as expressed by the relationship, $F = C - P + 2$. Here the number 2 represents one variable for temperature and one for pressure. Other variables are possible, but usually not considered. If in such a system $F = 0$ (no degrees of freedom), so-called invariancy exists. Univariancy occurs when $F = 1$ and divariancy when $F = 2$.

System representations. In considering the thermodynamics of ore or rock systems, only temperature, pressure, and composition variables are treated. Usually, possible additional variable parameters such as electric or gravitational fields are not considered; each of these would add one extra degree of freedom to the systems.

The study of phase relations in a four-component system is based on one-, two-, and three-component systems. A one-component, or unary, system, in which there is no compositional variation, can be completely represented by a two-dimensional pressure-temperature (P-T) diagram. A two-component, or binary, system requires three independent parameters—one for pressure, one for temperature, and one for composition—for complete diagrammatic presentation. The number of compositional variables in a system with n components is $n - 1$, since n components add up to 100% of the composition and the concentration of the nth component is determined as 100% minus the percentage sum of the other components. The construction of the three-dimensional block diagram required for presentation of a complete binary system is time-consuming and not very practical since it cannot readily be published. In order to stay within the two dimensions of the printed page, such a system can be projected unto a plane through an artist's eye, in which case distortions and inaccuracies are unavoidable, or it can be projected onto a plane from three-dimensional space along the pressure (P), temperature (T), or composition (X) axis. This produces temperature-composition (T-X), pressure-composition (P-X) or pressure-temperature (P-T) projections respectively. The T-X projection is the most commonly found one in the literature, but P-T projections certainly also are widely used. The T-X diagrams represent projections, and the pressure can vary, sometimes dramatically, with relatively small changes in temperature or composition. This is demonstrated by the Fe-S system, in which at 450°C (842°F) the pressure varies from about 10^{-30} atm (10^{-25} pascal) over pure iron to a little more than 1 atm (10^5 pascals) over pure sulfur.

Complete diagrammatic presentation of a three-component, or ternary, system requires four independent parameters—one for pressure, one for temperature, and two for composition. Such a three-component system, therefore, is contained within the boundaries of a four-dimensional prism, which cannot be visualized since the human eye can see at the maximum three dimensions. In order to illustrate such a system, a series of three-dimensional diagrams can be drawn, each at a given constant pressure or temperature or constant concentration of any one component. The most common way to do this is to retain the compositional variables in an equilateral triangle base and then vary pressure or temperature along an axis situated normal to this base. In the resulting trigonal prisms, for instance, the axis normal to the triangular base may be used as the temperature variable and the phase relations drawn at a selected constant pressure. Thus, for each desired pressure there will be one trigonal prism. Viewed in sequence, this series of prisms will make it possible for variations in this phase relations in the ternary system to be visualized at changing compositions, temperatures, and selected pressures.

The phase relations in four-dimensional space are commonly projected along the pressure axis onto a three-dimensional trigonal prism which contains the two compositional independent variables in its equilateral triangular base and in which the axis normal to this base represents the temperature variable. In such prisms, pressure varies considerably even with small changes in temperature or composition. For further simplification in illustration, it is common practice to specify the temperature, or temperatures, at which it is desired to visualize the phase relations in such three-component systems. These phase relations at a constant temperature will appear in the triangle which results when a plane parallel to the triangular composition base intersects the trigonal prism. Such diagrams are constant-temperature intersections of a trigonal prism which in turn represents a projection from four-dimensional space. These triangles represent isothermal (constant-temperature) conditions. They have little meaning unless the temperature is specified. In addition, pressure changes, sometimes dramatically, in such diagrams with even small changes in composition.

Complete diagrammatic presentation of a four-component system requires three-composition parameters and one additional parameter each for temperature and pressure. The four components are plotted, one at each of the corners of an equilateral tetrahedron. In this tetrahedron each corner represents an element, each edge represents a binary system, and each equilateral surface triangle represents a ternary system. All possible compositional combinations and mixtures of the four components will lie somewhere on the surface or in the interior of this tetrahedron. In the illustration the elements iron, lead, zinc, and sulfur are plotted at the corners of such a tetrahedron. The minerals formed from these elements are of common occurrence in a large number of ore deposits. Three dimensions are required for the compositional parameters in this quaternary

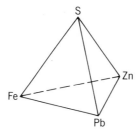

Elements Fe, Pb, Zn, and S are plotted at the corners of an equilateral tetrahedron. Three dimensions are required for expression of the three independent composition variables. Complete illustration of the phase relations in this system at variable temperature and pressure requires five-dimensional space. Phase relations in this system can be illustrated by a three-dimensional tetrahedron, as shown, if the pressure variable is removed by projection along the pressure axis and if the temperature is constant. In such isothermal diagrams the pressure is not constant, but varies considerably even with small changes in composition.

system. If it is desired to vary temperature and pressure as well, the full presentation of such systems would be possible only in five-dimensional space. The human eye cannot visualize the phase relations displayed in a five-dimensional prism, so the representation must be simplified through projections or by keeping certain variables constant. From five-dimensional space, phase relations may be projected along the pressure axis onto a four-dimensional prism which permits variations of all composition variables as well as of temperature. In this prism, pressure is not constant. It may vary considerably with even small composition or temperature changes. The phase relations displayed in this four-dimensional prism cannot be visualized. However, if the variation in temperature is removed by keeping it constant, isothermal phase relations can be illustrated by use of a tetrahedron. A separate tetrahedron must be used for each temperature at which it is desired to show the phase relations in a quaternary system. In these isothermal three-dimensional illustrations, pressure varies widely even with small changes in composition. Since illustrations in publications are limited to two-dimensional space, it is sometimes convenient to project phase relations in such an isothermal tetrahedron onto one of its boundary triangles. In the illustration, for instance, the phase relations might be projected at some selected temperature along the sulfur axis onto the Fe-Pb-Zn triangle. In other instances, the phase relations are illustrated under isothermal conditions in a quaternary system when the concentration of one component is constant. As an example, if the sulfur concentration is specified as in the illustration, a triangle which is produced when a plane parallel to the Fe-Pb-Zn base intersects the tetrahedron at the desired sulfur concentration and temperature will provide the required representation.

Ore systems. The compositions of most of the sulfide minerals occurring in numerous ores plot within the boundaries of four-component systems. For instance, the minerals in some common iron-lead-zinc

sulfide ores plot within the Fe-Pb-Zn-S system. Such minerals are pyrrhotite ($Fe_{1-x}S$), pyrite (FeS_2), galena (PbS), and sphalerite [ZnS, or more correctly (Zn,Fe)S since Fe replaces some Zn in this mineral's crystal structure]. Other examples are provided by the minerals in copper-iron-nickel sulfide ores of the Sudbury type deposits. Such occurrences contain a high percentage (often 70% or more) of pyrrhotite, usually 5–15% pentlandite ([Fe,Ni]$_9$S$_8$), 5–15% chalcopyrite (CuFeS$_2$), and 1–5% pyrite. These mineral compositions all lie within the boundaries of the Cu-Fe-Ni-S systems. Studies of the phase relations in such systems over large pressure and temperature ranges provide information concerning the variation in the compositions of minerals, the stability fields of these minerals alone or in assemblages, solid solutions among coexisting phases, and reaction rates. When this knowledge is applied to the pertinent ore minerals in nature, valuable data may result, bearing on the pressures and temperatures that prevailed when the ores were deposited, on the chemistry of the ore-carrying solutions, on the mechanisms of ore formation, and on the postdepositional history of the ores involving varying degrees of reequilibration of ore assemblages during their periods of cooling and during metamorphic events. Such information may be useful in the derivation of theoretical and practical knowledge not only of ore formation but also of how to extract metals from their ores, and might even assist in the search for new ore deposits.

For background information *see* PHASE RULE; SULFIDE PHASE EQUILIBRIA in the McGraw-Hill Encyclopedia of Science and Technology.

[GUNNAR KULLERUD]

Surface spectroscopy

Surface spectroscopy is used to investigate, characterize, and understand the physical and chemical properties of the outermost atomic layers of materials. Spectroscopic methods for probing surfaces and interfaces address fundamental and applied questions in catalysis, corrosion oxidation, electrochemistry, materials science, and semiconductor technology. The two major classes of methods used are shown in Fig. 1.

Numerous experimental probes of surfaces exist which utilize the scattering (Fig. 1a) of neutral species (atoms or molecules), photons, or charged particles (electrons, ions or positrons). The most widely practiced surface spectroscopies use electron scattering or electron emission, stimulated for example, by high-energy photons as in photo-electron emission (photoemission for short). The use of electron spectroscopies for surfaces arises since electrons strongly interact with most solids and cannot penetrate deeply or escape from any but the topmost atomic layers. Solid particles such as thermal-energy helium-atoms do not even penetrate most solids, and their scattering reflects properties slightly outside the surface. These electron or particle-based spectroscopies must be performed under ultrahigh vacuum conditions (pressures of 10^{-13} atmospheres

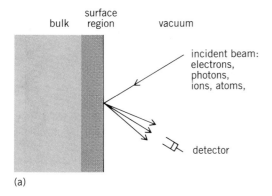

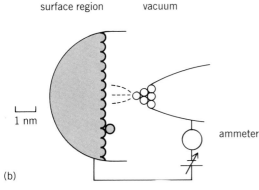

Fig. 1. Two major groups of surface spectroscopic methods used to study surfaces. (*a*) Scattering methods are macroscopic probes. (*b*) Tunneling method has atomic scale lateral resolution.

or 10^{-10} Pa). This limits the type of surfaces and the conditions or environment under which these surfaces can be analyzed. Energy lost by electrons, atoms or photons can be analyzed to reveal the fundamental excitations of the surface region. The ability to detect such excitations and precisely resolve these energies is limited by the technologies used in the measurement process. It has been improvements in these technologies that have permitted many of the surface spectroscopies used today.

Surface electronic structure. Studies of surface electronic structure can be categorized by the type of energy level examined.

Core states. The energy levels of the most strongly bound electrons in the atoms (core states) are studied by x-ray photoelectron spectroscopy (also known as electron spectroscopy for chemical analysis, ESCA) and provide a unique signature of these atoms. Foreign atoms and their concentration at the surface can be readily distinguished with this method. The continuously tunable radiation provided by synchrotron radiation facilities has revolutionized this field. The ability to vary photon energy makes it possible to adjust the outgoing photoexcited electron energy and permits "tuning" to select minimal electron escape depths so as to optimize surface sensitivity. Changing the electron collection angle relative to the sample also allows depth information to be achieved. Small but discernible differ-

ences in the energies of these core states can be used to distinguish different chemical states, such as oxidation states, or species at the surface (Fig. 2).

The excitation strength of these deep-lying (core) electrons can also be modified by the surrounding atoms in the solid. Varying the photon energy varies the absorption of x-rays by, say, a foreign atom, since the outgoing photoelectrons are back-reflected from the surrounding atoms and interfere. Analysis of this interference structure, called extended x-ray adsorption fine structure (EXAFS), provides local surface structural information about the bond distances and bonding directions for this atom. The intense, strong source of x-rays from synchrotrons has also made it possible to see the weak x-ray diffraction features arising from atoms on the surface.

Valence electron states. Surface spectroscopy using the more weakly bound valence electrons has also progressed rapidly. These electrons are more sensitive to details of geometry and bonding at the surface. On a clean surface the relation of the energy of these electrons to the momentum they are allowed to take in the (usually crystalline) material is fundamental to details of surface bonding. These energy-momentum relations can be directly probed by angle-resolving the photon-excited electrons, and are usually compared to detailed theoretical calculations. Such comparisons performed on bulklike states have provided a rigorous test of electronic structure theory. Surface electronic states depend on the geometric structure of the surface atoms and can be used as a test for proposed surface structures. Other measurements, for example, of the polarization dependence of the photoemitted electrons, provide information useful to identify the symmetry and general character of these electron states, without performing extensive calculations.

For molecules bound to the surface these higher-lying valence electron states reflect the molecular

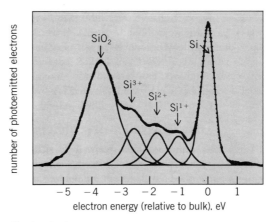

Fig. 2. Surface core level spectra of the silicon (Si)$2p_{3/2}$ core level for a 1.1-nm thick oxide grown on silicon. The levels of pure unoxided silicon and fully oxidized silicon dioxide (SiO_2) are shown. Several other intermediate or mixed oxidation states are observed and occur only at the Si/SiO_2 interface. (*After G. Hollinger and F. J. Himpsel, Probing the transition layer at the SiO_2-Si interface using core level photoemission, Appl. Phys. Lett., 44:93–95, 1984*)

structure and can be used as a fingerprint of its chemical nature. Surface reaction processes and products have been identified from such spectra.

Another class of electron spectroscopy involves detecting electrons emitted from the surface when, for example, an excited metastable atom (usually helium) deexcites by kicking electrons from the surface. Here, the valence electron states that are probed by the deexcitation process are usually spatially located outside the surface.

Unoccupied states. While many of these methods have been developed and used to probe electrons occupying energy levels of atoms in the surface region, interest exists in the energy levels that would arise if additional electrons were to be placed at the surface. These may be energy levels to which an electron can be excited or which become filled during a chemical reaction. One method developed to probe these states is inverse photoemission. Here an electron incident on the surface drops down into one of these unoccupied electron states by photon emission. The spectrum of the emitted photons and the incident electron energy defines the energies of the unoccupied or available energy states. Unfortunately, the emission of photons per incident electron is very inefficient (1 out of every 10^8) due to the existence of numerous nonradiative deexcitation processes.

Another method used to indirectly probe the unoccupied states is to probe the energy needed to excite an electron to an unoccupied energy state. Here, a monoenergetic incident electron beam is reflected from the sample and energy-analyzed. Fundamental excitations are detected from the electrons that have lost a well-defined amount of energy. These excitations need not only be between filled and unoccupied states but can be collective excita-

tions, such as plasmons and phonons. One advantage of this method, called electron energy loss spectroscopy (EELS), is that by using sufficiently high resolution, vibration spectroscopy of atoms or molecules at the surface can be achieved in the same experiment (Fig. 3).

Surface vibrational spectroscopy. The energy levels that characterize vibrational motions at surfaces include the collective phonon modes of surface atoms as well as local vibrations of foreign atoms and molecules bound to surfaces. These energies are much smaller than typical electronic transition energies and require significantly higher resolution.

The most widely practiced surface vibrational spectroscopy is high-resolution electron energy loss spectroscopy (HREELS). Here electron scattering is performed by using a monoenergetic beam and electron energy analyzers to achieve a resolution of 5–10 meV. As with electron loss spectroscopy, the experimental conditions can be selected to selectively excite vibrations perpendicular or parallel to the surface. Variation of the incident electron beam energy to selectively trap impinging electrons in certain electronic states near the surface also makes possible selective probing of specific vibrational modes and even observation of high-lying vibrational overtones.

Another new area of surface vibrational spectroscopy is the study of the vibrations (or phonons) of the clean surface itself. As in angle-dependent photoemission spectroscopy, angle-dependent vibrational measurements of the surface phonons with electron loss or inelastic atom scattering makes possible the measurement of the energy and momentum of the surface phonon states. Such measurements when compared to theoretical calculations provide a better understanding of the atomic forces and geometry of atoms at surfaces.

The largest application of surface vibrational spectroscopy is to the study of chemical reactions of molecular species on surfaces. This includes determining the nature of bonding and the bond directions as well as identifying possible products formed after chemical reaction with the surface. Various novel chemical species have been identified on surfaces.

While electron scattering–based vibrational spectroscopy has been the most widely utilized approach, it has limited resolution and must be performed in vacuum. Optical surface vibrational spectroscopies are being developed to provide higher resolution to study more complex systems or to be applied in nonvacuum environments such as in electrochemical cells or gas-phase reactors. For infrared-based methods these advantages are at the expense of a limited spectral range. Recent applications of Fourier-transformed infrared spectroscopy to surfaces have been successful and hold much promise. Raman (vibrational) spectroscopy, which uses visible light, has also been possible on surfaces. Other nonlinear optical effects arising at surfaces and interfaces are also being developed.

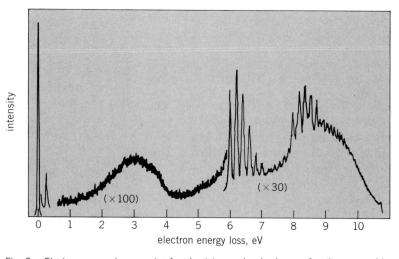

Fig. 3. Electron energy loss spectra for about two molecular layers of carbon monoxide (CO) adsorbed at low temperatures on an evaporated silver (Ag) film. Electronic transitions of the adsorbed molecule, including vibrational subbands, occur at approximately 6 and 8 eV; molecule-metal charge-transfer excitations from a broad band occur around 3 eV; and carbon monoxide molecular vibrations are observed at 0.2 eV. The incident beam energy is 10.8 eV. (*After D. Schmeisser, J. E. Demuth and Ph. Avouris, Metal molecular charge transfer excitations on silver films, Chem. Phys. Lett., 87:324–326, 1981*)

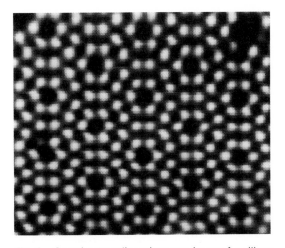

Fig. 4. Scanning tunneling microscopy image of a silicon (111) crystal surface. The lightest tone of the image is only 0.2 nm above the darkest tone. The distance between the dark holes is 2.7 nm. The white spots correspond to individual silicon atoms. An imperfection in this crystal surface is seen in the upper right-hand corner.

Scanning tunneling microscopy. In this new method (Fig. 1b), a fine probe tip is scanned over a surface (in vacuum) at a distance of 0.5–1 nanometer and monitors the tunneling current, which depends exponentially on this distance. By raster-scanning the tip over the surface and examining the tunnel current, or the position of the tip required to maintain a constant tunnel current, images of the surface with atomic-scale lateral resolution are produced. Variations in electronic states along the surface also modify the tunneling probability and affect the tunnel current. Thus these images contain both geometric and electronic information (Fig. 4). While complicated to perform due to the small distances involved and the mechanical vibrations possible between the tip and surfaces, this method has provided insight to many surface structure problems on both metal and semiconductor surfaces. The electronic contribution to these images depends on the bias voltage placed between the tip and surface, and can be utilized to obtain spectroscopic information on an atomic scale.

For background information *see* ELECTRON SPECTROSCOPY; LATTICE VIBRATIONS; PHOTOEMISSION; SPECTROSCOPY; X-RAY CRYSTALLOGRAPHY in the McGraw-Hill Encyclopedia of Science and Technology. [JOSEPH E. DEMUTH]

Bibliography: F. R. Aussenegg, A. Leitner, and M. E. Lippitsch, *Surface Studies with Lasers*, 1983; G. Binnig and H. Rohrer, The scanning tunnelling microscope, *Sci. Amer.*, 253(2):50–56, August 1985; J. E. Demuth and Ph. Avouris, Surface spectroscopy, *Phys. Today*, **36**(11):62–68, November 1983; H. Ibach (ed.), *Electron Spectroscopy for Chemical Analysis*, Topics in Current Physics 4, 1977; H. Ibach and D. L. Mills, *Electron Energy Loss Spectroscopy and Surface Vibrations*, 1982; R. Vanselow and R. Howe (eds.), *Chemistry and Physics at Solid Surfaces*, vol. 6, 1986.

SWATH ship

The performance capability and design of SWATH ships have been under study since the early 1970s, but only now are actual ship designs being introduced. The acronym SWATH stands for small-waterplane-area twin-hull. Concepts of this type have also been described as semisubmerged platforms (SSP) and trisected (TRISEC) ships, but all have the same basic concept. Most of the buoyant volume that supports the vessel's weight is placed well below the free surface, and most of the usable volume is placed above the surface of the water. These two volumes are connected by a third component, the struts which pierce the free surface of the water (Fig. 1) and connect the lower hulls to the cross-structure. The widley spaced twin-hull configuration is used to provide the necessary roll stability. The purpose of this configuration is to reduce the waterplane area near the sea surface, since the waveinduced exciting forces are primarily proportional to this waterplane area. Reducing this exciting force reduces the motions caused by the waves. This is the key to and the primary attribute of the SWATH concept: a SWATH ship can have better seakeeping characteristics than a similarly sized conventional ship because of the small waterplane area built into the design. *See* MARINE ENGINEERING.

Development. The origin of the SWATH concept can be traced back to 1943, when a SWATH aircraft carrier was proposed to the British Admiralty (Fig. 1g and b). The concept received little attention until 1973, when the SSP *Kaimalino* was placed in service as a 220 metric-ton (240-short-ton) range support boat in Hawaii for the Naval Ocean Systems Center (NOSC). More recently a series of SWATH designs have been built and put into service in Japan. These ships include a 350-metric-ton (385-short-ton) passenger ferry, the *Seagull*; two 240-metric-ton (265-short-ton) coastal hydrographic survey vessels, the *Kotosaki* and the *Ohtori*; and a 3500-metric-ton (3850-short-ton) deep-ocean support ship, the *Kaiyo*. In the United States, SWATH ships are limited to several small fishing vessels. Although large United States SWATH designs have so far remained on the drawing board, several Navy designs for auxiliary ships such as the ocean-survey (T-AGX) and the ocean-surveillance (T-AGOS) designs may be built before 1990. All of the current planned applications of the SWATH ship require the stable platforms that the concept can provide: stable ferries, so that passengers and crew experience less discomfort and seasickness; and stable platforms, so that equipment handling is easier and less dangerous, and helicopter pilots find it easier to land on the deck.

Difficulties. Although SWATH ship designs use the same technology (structure, propulsion plants, machinery, and so forth) as conventional monohull ships, the designer must be aware of several unique features that are inherent in the SWATH design. First, since the SWATH design has a small water-

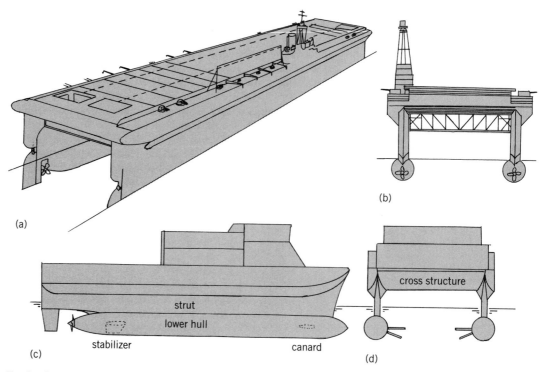

Fig. 1. Development of the SWATH concept: (*a*) perspective view and (*b*) transverse view of Creed and Lewis seadrome (1943), and (*c*) profile view and (*d*) transverse view of modern ships.

plane area, the change in draft due to the change in displacement is always more for a SWATH than for an equivalently sized monohull (low tons per inch immersion), and SWATH ships are less stable in pitch than a monohull (low pitch stiffness). A SWATH ship is more stable in roll than an equivalent monohull. The very large beam is dictated by the need for adequate damaged stability and for good seakeeping qualities. The low tons per inch immersion and low pitch stiffness make the SWATH a poor choice for ship designs which have large variable payloads such as most cargo-carrying vessels, because the cargo will cause large changes in draft and trim.

Second, a SWATH ship design usually has more surface area than conventional monohulls. There is more wetted surface exposed to the water because the SWATH configuration is not an efficient method of enclosing the volume of the ship. This extra wetted area means higher frictional drag in calm water. This high frictional drag is somewhat offset by better drag performance in actual seaways when compared to the monohull, and better speed in higher seas.

Seakeeping performance. The primary benefit of the SWATH ship is its admirable seakeeping performance. As stated above, the small waterplane area of the design reduces the wave-induced exciting forces acting to cause ship motions. This small waterplane area coupled with the separation of the two struts leads to SWATH ships with greater transverse stability (roll) and less longitudinal stability (pitch) than a similarly sized monohull. If the SWATH and monohull are of similar weight, the SWATH responds most to longer waves. To adjust this response, a SWATH designer must adjust the pitch stiffness, roll stiffness, and waterplane area (heave stiffness) so that the SWATH design's peak motion response occurs in longer waves which do not occur as often as shorter waves. In these longer waves, the SWATH will contour over the crests instead of pushing through the waves and thereby will reduce deck wetness and slamming. For a given ship design, ship speed, wave size, and ship size there is an optimum combination of pitch stiffness, roll stiffness, and waterplane area that provides minimum motions. An example of the excellent seakeeping performance of the SSP *Kaimalino* is shown in Fig. 2, where its motions are compared with those of two U.S. Coast Guard ships. The *Mellon* is a 3100-metric-ton (3400-short-ton) ship and should therefore have better seakeeping capabilities. The *Kaimalino* is better than both ships at all headings. (Significant values are shown in Fig. 2; these are the averages of the one/third highest excursions. Single amplitude is the average of the significant peak and trough values, and double amplitude is the sum of those two values.)

Canards and stabilizers. One motions problem for the SWATH occurs at higher speeds. As the speed of the ship increases, a nonuniform pressure field develops over the lower hulls due to the proximity of the free surface and its mirror-image effect on the flow field. The forces associated with this so-called Munk moment are proportional to speed squared and can become quite large. Since the pitch stability is somewhat small, an unappended

SWATH can take on large pitch angles at high speed. To counter this problem and to improve the seakeeping characteristics of the SWATH concept, most designs have horizontal fins placed on the inboard side of each hull, both forward (canards) and aft (stabilizers). These fins are usually adjustable so that at high speeds, the Munk moment can be countered and the pitch angle kept near zero. An additional benefit can be obtained from the canards and stabilizers by deflecting these fins as each wave passes the ship when moving through a seaway. These control surface deflections generate forces that counteract the wave-induced exciting forces and therefore reduce ship motions. Since the waterplane and therefore the exciting forces are small, the forces from modest-sized fins are quite effective at reducing ship motions. Figure 3 shows the results of full-scale seakeeping tests on the *Seagull* with and

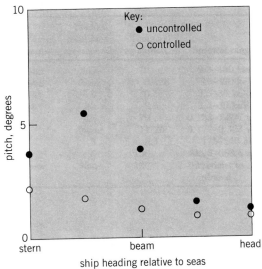

Fig. 3. Measured single-amplitude pitch motions of the passenger ferry *Seagull* in sea state 5 at 24 knots (12 m/s), at various ship headings relative to seas, with and without control by horizontal fins.

without this "automatic motions control" system.

Propulsion. In other areas of performance and design the SWATH ship is quite similar to conventional ships. The methods of propulsion are much the same as for other ships. A typical SWATH has two propellers, one on each hull, powered by perhaps an electric motor in each hull. The motor's power is supplied by diesel or gas turbine generator sets located in the cross-structure. By using electric motors, the ship designer can reduce the need for air-intake and exhaust ducting into the lower hulls. On small vessels (typically less than 300 metric tons or 330 short tons) the power can be transmitted to the lower hulls by either chain drives or bevel gear z-drives. For ship-sized SWATH designs the electric drive is the easiest solution to the power transmission problem.

Structural design. The structural technology which goes into a SWATH ship design is no different from the design of conventional ships. Because the SWATH configuration is less efficient at enclosing volume, the structural weight per volume is larger for a SWATH design. A designer must be careful to look at two areas of the structural design; the bottom of the cross-structure can be prone to slamming onto the water surface in very high seas, which can cause large local structural loads; and the designer must be careful to eliminate stress concentrations where the various sections of the hull are connected.

For background information *see* NAVAL SHIP; SHIP DESIGN; SHIP POWERING AND STEERING in the McGraw-Hill Encyclopedia of Science and Technology.

[ROBERT THOMAS WATERS]

Bibliography: J. L. Gore (ed.), SWATH ships, *Nav. Eng. J.*, 97(2):83–112, special edition, Feb-

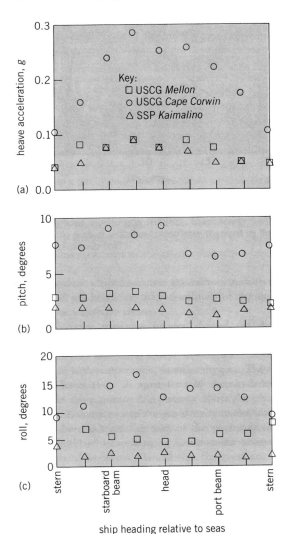

Fig. 2. Comparison of the motions of the SWATH ship SSP *Kaimalino* (220 metric tons or 240 short tons) with the U.S. Coast Guard ships *Mellon* (3100 metric tons or 3400 short tons) and *Cape Corwin* (100 metric tons or 110 short tons) at various ship headings relative to seas. (a) Significant double-amplitude heave. 1 g = 32.2 ft/s² = 9.8 m/s². (b) Significant double-amplitude pitch. (c) Significant double-amplitude roll.

ruary 1985; T. Inui (ed.), *Proceedings of the 13th Office of Naval Research (ONR) Symposium on Naval Hydrodynamics*, Shipbuilding Research Association of Japan, 1981; J. P. Sikora, A. Dinsenbacher, and J. E. Beach, A method for estimating lifetime loads and fatigue lives for SWATH and conventional monohull ships, *Nav. Eng. J.*, 95(3):63–85, May 1983.

Television

Recent developments in television technology include; (1) research on a high-definition television system which would have about four times as many resolvable picture elements as present television; (2) the establishment of low-power television service in the United States; and (3) the appearance of satellite master-antenna television systems, which combine traditional systems with reception from satellites.

HIGH-DEFINITION TELEVISION (HDTV)

Television systems now in use are based on scanning the scene with 500–600 lines of picture elements at a rate of 25–30 complete pictures per second. A complete picture is displayed in two interlaced fields, each containing half the number of lines. The scanning standards were adopted just before and after World War II on the basis of the existing technology and have remained essentially unchanged, constrained by compatibility with the large number of television receivers.

Recent attention has been given to a higher-definition television system that would substantially improve the quality of the viewed image, making it comparable to or better than that presented in the motion picture theater. The distinguishing characteristics of such a high-definition television (HDTV) display are a large screen (of the order of 10 ft^2 or 1 m^2 or greater), a larger width-to-height ratio (5:3 or greater), and higher resolution (about twice the number of scan lines and more than twice the number of resolvable picture elements (pixels) along each line as compared with present television).

Japanese system. An HDTV system with these characteristics has been developed in Japan. The wide-screen image is scanned with 1125 lines, 2:1 interlace, at a 60-Hz field rate. The width-to-height (aspect) ratio is 5:3, close to the value used in the cinema. It has been proposed that this system be adopted as a single worldwide HDTV standard, not to be encumbered by the degrading effects of the standards conversions that accompany the exchange of television programs between the 50- and 60-Hz scanning standards in use.

Because of the four-times-larger number of resolvable pixels and the increased image width, the channel bandwidth required for the transmission of the Japanese system is about five times that required for television at present (about 30 MHz). The proposed system is incompatible with existing television receivers and would also require a new generation of television studio equipment, distribution channels, and transmitting equipment.

Evolutionary compatible approach. Primarily because of the economic impact of these factors, several counterproposals have been offered, suggesting ways by which an HDTV service could be introduced in a more evolutionary manner, maintaining a degree of compatibility at each stage of the evolution. However, the evolutionary compatible approach could not possibly yield a single worldwide HDTV standard, because of the constraint of compatibility, especially with the different field rates in use. Thus future generations of television broadcasters and producers would be handicapped by an even larger plethora of different release media in international program exchange.

The evolutionary compatible approach to high-definition television can be described roughly in three stages: (1) improvements in existing television systems made possible by new technology that permit a significantly enhanced picture quality within the constraints of the present scanning standards, (2) enhanced or extended-definition television (EDTV) systems that require larger channel bandwidths (for example, twice that of current television) but provide compatible enhancements in resolution or aspect ratio; and (3) compatible HDTV that would permit existing television receivers to display the image at normal television resolution while new HDTV sets would provide the full high-definition picture quality.

Improved television. The perceived resolution of interlaced television displays is lower than that possible from the channel bandwidth because of masking of higher-resolution components by the interfering line-scanning structure. The visibility of the scan lines and flickering of horizontal edges and other regions of high detail actually impair the perceived sharpness of the image. These so-called artifacts of the displayed picture are a direct result of the interlace process. With advances in solid-state circuit technology, it will be possible in the early 1990s to store an entire field of pixels on a single chip of silicon; thus displays can be converted to scan the entire image "progressively" from top to bottom in the time of one field. Further improvements will be incorporated into the television studios with digital signal processing technology that will permit better-quality standard-definition signals to be derived from higher-definition cameras.

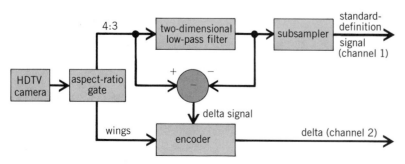

Fig. 1.   Two-channel compatible high-definition television (HDTV) system.

Extended-definition television. The existence of new distribution media, such as the direct broadcasting satellite and improved home videocassettes and disks, will provide channel capacity about double that of present television, permitting an enhanced television service with picture quality intermediate between improved television and high-definition television. Alternate methods for encoding the color information into the video signal to avoid the cross-color and dot-crawl artifacts of today's composite television standards will be possible. Increased horizontal resolution and wider image aspect ratio can also be accommodated by the wider bandwidth. Although these enhanced systems are not strictly compatible with the improved television systems, they are based on the same scanning parameters and can be simply converted in the direct broadcast satellite receiver or videocassette player to be viewed on a standard-definition television set.

Compatible HDTV. Proposals for compatible HDTV transmission or distribution systems generally incorporate a two-channel concept. One of the channels carries the image in a standard-definition format, and the other constitutes a difference or delta channel containing the higher-resolution components. The second channel can also include the wings or side-panel information of the wide-screen image.

In this system (Fig. 1), high-definition camera scans the scene with, say, 1050 lines at a 60-Hz field rate or 1250 lines at 50 Hz, corresponding to twice the number of lines (525 or 625) of the standard with which compatibility is desired. An aspect ratio gate selects the center 4:3 aspect ratio from the wide-screen signal. A two-dimensional digital filter reduces both the horizontal and vertical resolution of this gated signal by a factor of two, and a subsampler selects alternate horizontal samples and alternate vertical lines for encoding into the standard-definition signal format. Prior to the subsampling, the filtered signal is subtracted from the high-definition signal to derive the difference signal which contains no low-resolution energy. This delta signal can be encoded into the second channel that also incorporates (for example, by multiplexing) the side-panel high-definition signals. In some proposed configurations, the first channel contains a compatible EDTV signal of twice standard bandwidth; thus the filter reduces only the vertical resolution. Other alternatives that reduce the temporal update rate of the high-resolution delta signal have been proposed.

Prospects for a world standard. The widespread desire to avoid multiple incompatible HDTV production and distribution systems is, unfortunately, in conflict with an equally strong desire to use the HDTV production system as the source for EDTV and standard-definition television programs. If such programs are produced in high-definition television at a field rate different from that of the intended distribution system, the motion artifacts produced by the standards conversion process defeat much of

the enhancement desired. Future improvements in the effectiveness of standards conversion equipment could possibly remove this dilemma, but the urgency of reaching worldwide agreement on a common standard is outpacing the rate of progress of the technology.   [KERNS H. POWERS]

LOW-POWER TELEVISION (LPTV) BROADCAST STATIONS

Broadcast stations authorized to operate in the low-power television (LPTV) service are facilities limited in transmitter output so as to provide reception in only a local area. Although low-power stations providing localized television service had been used for many years in Canada and countries outside North America, rules establishing the LPTV service were not adopted in the United States until 1982.

LPTV stations providing programming from a local source, or received from a centralized originator and distributed via satellite to the local station for rebroadcast on a conventional television channel, were a natural outgrowth from the translator service. As originally conceived and authorized, translators picked up television signals from distant stations, amplified those signals, and rebroadcast them on a new channel. They served to provide reception in areas too distant from originating stations for acceptable service utilizing conventional home receiving equipment and in areas where terrain blockage prevented satisfactory service directly from the originating station.

Power. The power permitted for use by an LPTV station is based solely on transmitter output. On VHF channels 2 through 13, the permitted power is 10 W. For UHF channels 14 through 69, the permitted transmitter power output level is 1000 W. As for full-service television stations, the power specified is that emitted at the peak of the synchronizing pulse. Average power for the video channel during normal programming is approximately 40% of the peak power. Power for the frequency-modulated aural channel is normally 10% of the peak visual power.

The Federal Communications Commission (FCC) places no limit on effective radiated power or on antenna height. Effective radiated power is the product of the transmitter power output and the gain of the transmitting antenna reduced by the loss in the transmission line connecting transmitter to antenna.

Service range. Typical service ranges for LPTV stations are given in Table 1. Omnidirectional operation has been assumed with antenna gains and transmission line losses typical for the service. In some instances, the avoidance of interference to other stations may require reduced power, resulting in a diminution of service-area radius. Depending upon the location of the transmitting facility with respect to the particular market to be served, a directional antenna may be used, increasing radiation in some directions and reducing radiation in other directions. In such instances, the maximum extent of the service area may be greater than indicated in

Table 1. Typical service ranges of low-power television stations

Channel no. and frequency	Effective radiated power, kW	Height above average terrain		Radius of service area	
		ft	m	mi	km
2–6 (54–88 MHz)	0.030	200	61	3.3	5.3
	0.025	500	152	5.0	8.0
	0.020	1000	305	6.6	10.6
7–13 (174–216 MHz)	0.080	200	61	3.6	5.8
	0.070	500	152	5.1	8.2
	0.060	1000	305	6.8	10.9
14–69 (470–806 MHz)	18	200	61	8.2	13.2
	16	500	152	12.5	20.1
	13	1000	305	16.3	26.2

Table 1 for those directions favored, but reduced in other directions.

Terrain considerations and siting. Considerations of terrain are of great importance for all television transmissions, but particularly for the LPTV service. Whereas the diffracted signal behind an obstructing hill or building may still be satisfactory for reception in the case of a full-power television station, the limited power in the LPTV service cannot tolerate diffraction losses if satisfactory service is to achieved. Dense foliage, such as that provided by a grove of trees between the point of transmission and the point of reception, may also produce excessive loss, particularly in the UHF portion of the spectrum.

These considerations dictate the need for careful siting of the LPTV transmitting facility with respect to the market to be served. To the maximum extent possible, line of sight must be provided to the majority of receiving antennas within the target area.

Station assignments. Full-service television stations are assigned on the basis of a preplanned table reserving particular channels for specific communities. That Table of Allotments is coordinated with the needs of neighboring countries. In contrast, LPTV (and translator) station assignments are made only with regard to the avoidance of interference, either caused or received. Furthermore, LPTV stations are accorded a "secondary" status with respect to full-service stations. A new, full-service station assigned in accordance with the Table of Allotments, or changes in the Table of Allotments, need not consider impact upon existing or proposed LPTV operations. Some threat to the very existence of an LPTV station is, therefore, inherent, but experience with the longer-established translator service subject to the same restrictions shows that the loss of an assignment because of the introduction of a full-service station is a rare occurrence.

Protection from LPTV interference. Full-service stations are afforded the maximum protection from LPTV facilities. They are protected to their Grade B contours. Rules of the FCC specify the Grade B contour signal levels as 47 dB above a microvolt per meter (dBu) for channels 2 through 6, 56 dBu for channels 7 through 13, and 64 dBu for channels 14 through 69. The signal contours specified for the

protection of LPTV stations from other similar operations are 62 dBu for stations on channels 2 through 6, 68 dBu for stations on channels 7 through 13, and 74 dBu for stations on channels 14 through 69.

Criteria for the avoidance of interference to full-service stations from LPTV stations are as follows.

1. An LPTV station may not be located within the Grade B contour of a cochannel or first-adjacent-channel TV broadcast station.

2. The protection ratio for cochannel operations depends upon whether or not the LPTV station uses an offset carrier. For offset operation, the visual carriers must be displaced either 10 kHz or 20 kHz from each other and the frequency stability of the visual transmitter must be such as to maintain the frequency within 1 kHz of the licensed frequency. For offset operation, the interfering 10% of the time field strength must be at least 28 dB lower than the desired median field strength. For nonoffset operation, a 45-dB ratio applies.

3. In the UHF band, an LPTV station may not be located within the Grade B contour of a television broadcast station if operation is proposed on a channel either 14 or 15 channels above the channel in use by the television broadcast station. The purpose of this restriction is to avoid image interference to the visual and aural carriers.

4. For the avoidance of intermediate-frequency interference, a UHF LPTV station must be at least 62 mi (100 km) from the site of a television broadcast station operating on the seventh channel above the LPTV channel.

5. For the avoidance of intermodulation interference, a UHF LPTV station may not be located closer than 20 mi (32 km) from the transmitter site of a television broadcast station operating on the second, third, fourth, or fifth channel above or below the channel of the LPTV station.

6. For first-adjacent-channel protection, the LPTV signal strength may be 6 dB greater than that of a VHF television broadcast station operating one channel above the LPTV station channel. The applicable ratio is 12 dB if the protected VHF TV broadcast station is one channel below the LPTV station channel. For UHF, the applicable ratio of undesired-to-desired television field strength for first

adjacent channels is 15 dB.

The mutual protection of LPTV stations calls for less rigor than that applicable to protection of full-service television stations. All stations are protected from cochannel and first-adjacent-channel interference. The UHF LPTV stations are protected also from image and intermediate-frequency interference.

[JULES COHEN]

SATELLITE MASTER-ANTENNA TELEVISION (SMATV)

Satellite master-antenna television (SMATV) is the union of an old technology with a relatively new one. It combines a traditional master-antenna television distribution system carrying off-air broadcast television channels received by standard VHF and UHF antennas with additional cable television channels received by satellite. Each of the component technologies will be reviewed individually.

Community antenna television (CATV). The first cable systems were constructed in mountainous areas where, due to the nearly line-of-sight propagation characteristics of television transmissions, antennas located in valleys surrounded by mountains received little if any television signals. In many cases, an antenna was erected on a mountaintop facing the broadcast station and a cable was run down the side of the mountain, which was then connected to television receivers throughout the valley through a distribution network.

This distribution network consists of three major components: coaxial cable, actives, and passives.

Since the energy levels of television signals become attenuated by their passage through the cable, active amplifiers must be inserted along the transmission path to boost the signals. Passive electronic circuits are inserted along the cable to equalize signal levels or split the signals into two separate paths. Some advanced CATV systems have two-way capabilities and, in addition to delivering multiple television channels to the home, may also transmit information from the home back to the cable system office.

Master-antenna television (MATV). Master-antenna television systems resemble community-antenna television systems in that they distribute signals from a common antenna or origination point to a large number of television receivers. The FCC defines an MATV system as a distribution network for television signals serving apartments or dwellings under common ownership, and further states that the cables used to distribute the signal must be contained on private property and cannot cross public right-of-ways. Typically, MATV systems are found in apartment buildings or complexes where a single, well-positioned antenna provides better reception and is less unsightly than a multitude of individual antennas. The headend, or origination point, of an MATV system can be as simple as a single broadcast television receive antenna in combination with a single broadband amplifier, or sophisticated enough to include separate antennas and signal processors for each television channel. The distribution portion of the system is generally composed of a

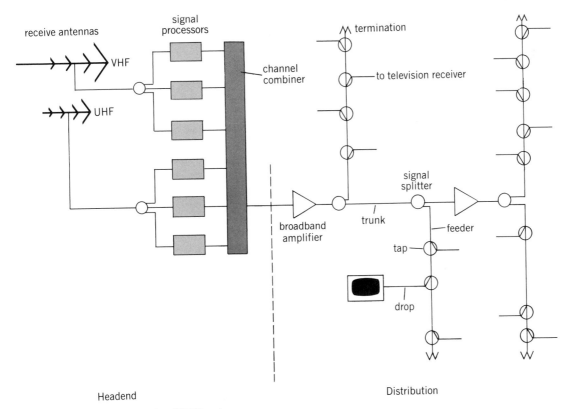

Fig. 2. Master-antenna television (MATV) system.

main cable or trunk that runs through the center of the complex and connects to feeder cables which bring the television service around each floor or level. A portion of the signal is then removed from the feeder cables by the installation of passive taps, which deliver the proper amount of signal to each television set via drop cables (Fig. 2).

Satellite reception. In 1975, one cable programmer placed its pay television service on a satellite, enabling cable systems to downlink the signal and distribute it to their subscribers. Prior to that time, pay programmers had to make hundreds of copies of each program and ship them across the country for cable systems to air individually in their own studios. Because of its cost effectiveness, many other cable programmers moved to the satellite as a means of program distribution, soon creating a demand for additional satellites to service cable systems.

Originally, a satellite television receive-only antenna (TVRO) or dish necessary to receive these signals needed to be a minimum of 15 ft (4.5 m) in diameter and cost over $10,000. A complete TVRO consists of a parabolic reflector or dish, a feed horn which collects the signals reflected by the dish, a low-noise amplifier for preamplification, and a tunable satellite receiver. As dish technology advanced, the cost of purchasing a TVRO began to decrease. Concurrently, more powerful satellites designed for the distribution of cable television signals were being constructed and launched. This further brought dish prices down because the more powerful

the satellite, the smaller the dish which is required to receive its signals.

At first, a cable system with several thousand subscribers was necessary to support the purchase of a TVRO. As dish prices continued to drop, it became cost-effective for systems with only a few hundred customers to justify implementing this technology. Since MATV systems typically service 50 to 200 apartments or units, it eventually became feasible to consider adding a satellite dish to these systems as well. Operators of MATVs installed TVROs on their properties and added these signals to their existing MATV systems (Fig. 3). For many operators, however, there are both technical and legal obstacles to overcome.

Technical obstacles. The satellite dish must be placed at a location where it can view the satellites in their geostationary arc 22,300 mi (35,900 km) above the Equator without any physical blockages. Even the leaves on a tree will attenuate the signal if they interfere with the line of sight of the dish. The location of the antenna must also be free from terrestrial interference, because some microwave transmitters also operate at the same frequencies as the satellite and could interfere with proper reception.

Broadcast stations are usually carried on the MATV system at the same channel numbers as they would be received over the air; however, this does not have to be the case as signal processors can be designed to convert stations to any television channel. If channels in the VHF band are already occu-

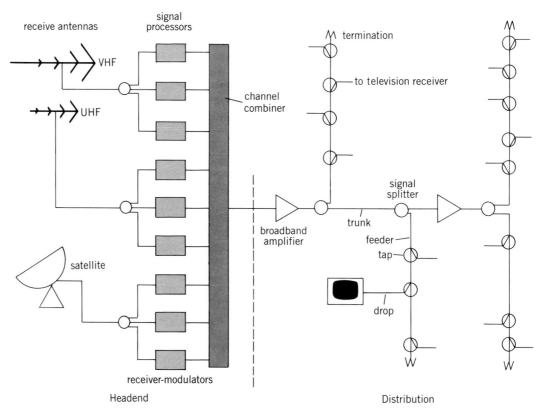

Fig. 3. Satellite master-antenna television (SMATV) system.

Table 2. CATV frequency allocation table

Band	Channels	Frequency, MHz
Low VHF	2–6	54–88
Midband	A–I	120–174
High VHF	7–13	174–216
Superband	J–W	216–300

pied, satellite-delivered signals are usually placed in the midband or superband of the electromagnetic spectrum (Table 2). These bands are utilized by the cable industry to carry their additional television channels which can be received by a standard television set through the addition of a cable converter box. The FCC allows cable to carry channels at these frequencies even though other users such as police, taxi, and aircraft two-way communications are also transmitting there, because the cable itself is a closed system and signals should not leak out to interfere with over-the-air users, and vice versa. Most MATV systems were not originally designed with the same leakage integrity against interference as CATV systems. As a result, if other channels on the MATV system are added in these shared bands, the entire distribution portion of the system must often be rewired to protect against signal leakage and its resulting interference.

Legal obstacles. In addition to the FCC rules governing the operation of a MATV system, there are restrictions governing the reception and distribution of satellite signals. Two types of channels are available from cable television satellites: advertiser-supported services and pay services. The advertiser-supported channels derive their revenues through selling commercial time, and allow reception of their programming by the general public. Pay services derive their revenues solely from a monthly fee charged to each viewing household or subscriber. These services require a contract between the MATV operator and programmer for the distribution of their signals. Some pay services scramble their satellite signals in order to prevent reception by unauthorized dishes. In these cases, a remotely energized descrambler is issued to users with which there is a properly executed distribution agreement.

For background information *see* CLOSED-CIRCUIT TELEVISION; COLOR TELEVISION; COMMUNICATIONS SATELLITE; TELEVISION; TELEVISION SCANNING; TELEVISION STANDARDS in the McGraw-Hill Encyclopedia of Science and Technology.

[WILLIAM W. RIKER]

Bibliography: S. Bassett, Low-budget high-altitude LPTV pays off for Colorado ski towns, *Private Cable*, 4(6):20–24, June 1985; G. Beakly (ed.), *Satellite Communication Symposium*, Scientific Atlanta (Georgia), 1981; *CCIR Recommendations and Reports*, vol. 11, 1986; Federal Communications Commission, *Notice of Proposed Rule Making*, BC Docket no. 78–253, RM-1932, September 9, 1980; Federal Communications Commission, *Report and Order*, BC Docket No. 78–253, March 4, 1983; T. Fujio, A study of high definition TV system in the future, *IEEE Trans. Broadcast*, BC-24(4):92–100, December 1978; W. Grant, *Cable Television*, 1983; R. J. Shiben, N. W. Cornell and S. J. Lukasik, *Report and Recommendations in the Low Power Television Inquiry*, Federal Communications Commission Broadcast Bureau, September 9, 1980; K. Simons, *Technical Handbook for CATV Systems*, 1968; D. E. Troxel et al., Two-channel picture coding system-1: Real-time implementation, *IEEE Trans. Commun.*, COM-29(12):1841–1848, December 1981; B. Wendland, *SMPTE Television Technology in the 80's*, 1981.

Temperature measurement

Within the next few years the International Practical Temperature Scale of 1968 (IPTS-68) will probably be superseded by a new version of the scale that is simpler, is closer to thermodynamic temperatures, and extends to much lower temperatures than IPTS-68. Before the new scale can be adopted, however, there remain a number of problems to be resolved, and these make it unlikely that the scale can be introduced before 1989. In this article the shortcomings of the existing scale will be described, together with an outline of the various difficulties at present holding up the introduction of its successor. First, however, it is worth recalling the reasons why it is necessary to have an International Temperature Scale.

Reasons for an International Scale. The IPTS-68 exists today for very nearly the same reasons as brought into being the first International Temperature Scale in 1927. These were: to provide a common basis for scientific and industrial temperature measurements, and to ensure that the basis chosen would allow full advantage to be taken of the high reproducibility available from certin secondary thermometers—at that time the platinum resistance thermometer, the platinum/platinum-rhodium thermocouple, and the optical pyrometer. The reproducibility of these standard secondary thermometers was considerably better than that of the various primary (that is, thermodynamic-temperature-measuring) thermometers. It followed that the International Scale could be so defined that temperatures measured on it were more reproducible than were the best measurements of thermodynamic temperature.

It is still the case that primary thermometers, such as the constant-volume gas thermometer or total radiation thermometer, are less reproducible than the best secondary thermometers such as the platinum resistance thermometer. This advantage in reproducibility of the secondary thermometers over the primary thermometers is now much less than it was in 1927. The best gas or total radiation thermometers now have reproducibilities in the room temperature range of 1 or 2 millikelvins, which is only about a factor of 10 worse than that of the best platinum resistance thermometer. To achieve such a reproducibility with one of these primary thermometers is a very complex and time-consuming task, although this in itself is not a sufficient reason for requiring an International Scale; that remains the superior reproducibility of the secondary thermometers.

Shortcomings of IPTS-68. It was implicit in the first International Temperature Scale that temperatures measured on it would be as close as possible to thermodynamic temperatures. The question of just how close the next scale, if it is to be acceptable, should be to thermodynamic temperatures must be answered by considering the prospective needs of science and industry. If an International Temperature Scale departs significantly from thermodynamic temperature T, it leads to values of temperature, T_{68} in the case of IPTS-68, which are consistent neither with the definition of the kelvin nor with the consequent values of the Boltzmann constant k or the molar gas constant R. Under these conditions, if the scale is used in the measurement of quantities that are functions of T, the results will not be consistent with physical theory. Such discrepancies would only be significant, of course, in the case of very precise measurements, given the magnitudes of $T - T_{68}$ known to exist at present. However, measurements of quantities which depend on $d\phi/dT$ (the rate of change of ϕ with respect to T), where ϕ is some physical parameter, will be significantly in error if the measurements are made in temperature ranges where $T - T_{68}$ changes rapidly with T or where $d(T - T_{68})/dT$ (the rate of change of $T - T_{68}$ with respect to T) is discontinuous. It will then be impossible to distinguish between a real physical effect (for example, a specific heat anomaly) and a spurious one that is due solely to one of these two scale errors.

It is now known that IPTS-68 departs significantly from thermodynamic temperatures. The difference $T - T_{68}$ is most evident at room temperature and above, where it amounts to -10 mK at 40°C, -27 mK at 100°C, -80 mK at 460°C, and $+400$ mK (the maximum difference) at 800°C. Differences also exist below 0°C and reach a maximum of about 10 mK at temperatures near 100 K. At 630°C, $d(T - T_{68})/dT$ is discontinuous to the extent of about 0.2%.

Since about 1975, sufficient reliable experimental evidence has been obtained for these differences to be sufficiently well known over practically the whole range of IPTS-68, and this is no longer the principal impediment to drawing up a new scale.

Extension to lower temperatures. IPTS-68 extends down only to 13.81 K (the triple point of equilibrium hydrogen). Any new scale must extend to much lower temperatures, at least down to 0.5 K, the lower limit of the 1976 Provisional Scale (EPT-76). This scale was introduced as a provisional measure to meet a pressing need for an internationally agreed-upon Scale in the range below 13.81 K, pending a new International Practical Temperature Scale.

Although details of how the new scale will be defined below about 30 K have yet to be finalized, the broad lines have been agreed upon and this would now no longer be an impediment to drawing up the new scale. From room temperature down to at least the triple point of neon (approximately 24 K), the scale will be defined, as at present, in terms of a platinum resistance thermometer and a specified interpolation function or functions (although these functions are sure to be much simpler than that of IPTS-68). From the triple point of neon down to the boiling point of helium (approximately 4.2 K), the scale will probably be defined by a constant-volume gas thermometer, used not as a primary thermometer but as an interpolating device calibrated at 4.2 and 24 K. It is likely that provision will also be included to allow the scale to be established by a plantinum resistance thermometer as an alternative to the gas thermometer in the range from 13.81 up to 24 K. Below the boiling point of helium the new ^{3}He and ^{4}He vapor-pressure scales will be incorporated into the new scale to extend it down to at least 0.5 K and perhaps down to 0.3 K.

There are no plans at present to extend the scale into the very low-temperature range below 0.3 K and reaching down to a few millikelvins or below. Temperatures in this range are usually measured in any case by thermometers that can in many respects be considered primary thermometers; magnetic, nuclear orientation, or noise thermometers, for example.

Impediment to new scale. The main difficulty holding up progress toward a new International Practical Temperature Scale is the replacement of the platinum platinum/10% rhodium thermocouple by the high-temperature platinum resistance thermometer for the range 630 to 1064°C. There is as yet no design of platinum resistance thermometer which has proved satisfactory for use up to 1064°C (the freezing point of gold). Although a number of designs have been tried they are all extremely difficult to manufacture, all are very expensive and suffer from various defects in performance related either to electrical leakage or drift due to contamination. If no successful substitute can be found, the new scale will have to continue to be defined in this range by the platinum platinum/10% rhodium thermocouple. This thermocouple, despite its relative cheapness and ease of manufacture, suffers from the major weakness of having very poor reproducibility—only about 0.2 K. However, work aimed at producing a satisfactory design of a high-temperature platinum resistance thermometer is in progress in a number of national standards laboratories.

For background information *see* GAS THERMOMETRY; LOW-TEMPERATURE THERMOMETRY; TEMPERATURE MEASUREMENT; THERMOCOUPLE; THERMOMETER in the McGraw-Hill Encyclopedia of Science and Technology.   [T. J. QUINN]

Bibliography: Bureau International des Poids et Mesures, *Reports of the Consultative Committee on Thermometry*; H. Preston, T. J. Quinn and R. P. Hudson, *Metrologia*, 21:75–79, 1985; T. J. Quinn, *Temperature*, 1983.

Theoretical ecology

Environmental legislation and common discourse often contain references to the health or integrity of an ecosystem, attributes without rigorous definitions. It is now possible, however, to define and measure a related property called ecosystem ascendency, which in a single index quantifies both the magnitude of system activity and the degree of organization inherent in the pattern of interactions among taxa. The increase of ecosystem ascendency describes the process of community growth and development, and the limits to this increase are reflected in various components of the measure. Ascendency is an admixture of ideas from thermodynamics, economic input-output analysis, information theory, and cybernetics. That the index rests upon such firm theoretical foundations contributes to the power of ascendency to unify several hypotheses of development pertaining both to ecosystems and to other cybernetic ensembles.

Growth and development are processes. It is only natural, therefore, that their mathematical definitions be related to the magnitudes and the juxtaposition of the constituent ecological processes. The most concrete measure of any process is the amount of material or energy transfer which accompanies the change; for purposes of description this intensity may be assumed to reflect and integrate all the more detailed aspects of the process. For example, the measured feeding rate of a herbivorous copepod in the ocean characterizes and embodies the densities and relative distributions of the predator and its prey, the feeding behavior of the herbivore, the morphologies and genetic makeups of copepod and phytoplankton, and so forth.

Derivation of ascendency. In a community of n species there are at most n^2 bilateral exchanges of a given medium, and these flows may be regarded as a network connecting the compartments. The network of energy flowing among the five "taxa" of the Cone Spring ecosystem is depicted in Fig. 1. The size of the network (that is, the exchanges as distinct from the contents of the nodes) is taken to be simply the aggregation of all the existing flows. Mathematically, if T_{ij} represents the magnitude of the flow from species i to species j, then the total amount of flow associated with the ecosystem becomes Eq. (1), where T_{oj} represents the inputs to

$$T = \sum_{i=0}^{n} \sum_{j=1}^{n+2} T_{ij} \qquad (1)$$

species j from outside the system (for example, primary production, or advection of medium into ecosystem), $T_{i,n+1}$ the output of useful medium from taxon i (such as by advection, or harvest by humans), and $T_{i,n+2}$ the amount of medium dissipated by i. In input-output analysis, T is called the total system throughput, and in economics a partial sum of T for the network of a national economy is the gross national production. There is strong and familiar precedent for reckoning community size in terms of total flow activity.

Although the mathematical description of network organization is moderately complicated, the underlying idea is nonetheless straightforward. A network is said to be organized or well articulated when an event (output) at any given compartment engenders other events (inputs) at only a limited number of other nodes (taxa). That is, if one knows that a quantum of medium has left a particular taxon, then one knows with relative confidence which species will receive the flow. Such articulation has been equated with the average mutual information of the flow structure as defined by Eq. (2), where K is a scalar constant. For example, knowledge that a

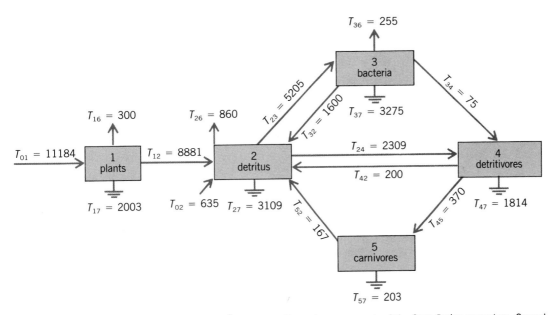

Fig. 1. Schematic of the flows of energy (kcal/m^2·y) between the major components of the Cone Spring ecosystem. Ground symbols represent dissipations.

$$A = K \sum_{i=1}^{n} \sum_{j=1}^{n} (T_{ji}/T) \log \left(T_{ji}T / \left[\sum_{k=1}^{n+2} T_{jk} \right] \left[\sum_{m=0}^{n} T_{mi} \right] \right) \quad (2)$$

quantum of flow has left a particular compartment in the hypothetical network in Fig. 2a gives one no idea as to which node the medium will enter next. The configuration is totally inarticulated, and A for the system is identically zero. By contrast, the network in Fig. 2b has the same total flow as 2a; however, knowing where the output exits tells one unequivocally where it is going. The value for A in Fig. 2b (K log 4) is the maximum possible for a network of four compartments. Real four-component networks of flow would possess values for A intermediate to these theoretical extremes.

The usual convention in information theory is to set $K = 1$, but the scalar constant may be used to give physical dimensions to the system being studied. By setting $K = T$ the factor of organization is scaled by an index of size, and the product becomes the network ascendency.

Because growth may be regarded as an increment in size, and development as an augmentation of organization, growth and development become two aspects of a unitary process—the increase of ascendency. Ecosystems may be said to grow and develop insofar as their associated networks of material or

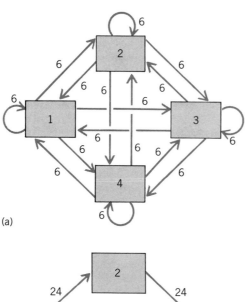

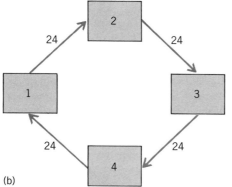

Fig. 2. Hypothetical ecosystem networks. (a) Network of inarticulated flows. (b) Network of maximally articulated flows.

energy flows increase in ascendency.

Community reconfiguration. No real system can grow and develop without bounds, and the limits on increasing A are readily apparent. To visualize the constraints, it helps to imagine an ecosystem recovering from a major environmental shock which has obliterated all but a few of the original species. Initially the surviving taxa have ample resources opened to them by the sudden demise of their competitors. As Alfred Lotka observed, the advantage under conditions of abundance goes to those species best able to increase their throughputs. Consequently, initial recovery occurs largely as increasing total activity T. Available resources are finite, however, and medium cannot be recycled indefinitely because of obligatory losses at each transfer. Inevitably, the rise in T begins to slow, and the subsequent increase in ascendency occurs mostly through the reconfiguration of the flows.

To follow the effects of community reconfiguration upon A, it is helpful to expand A into four intrinsically nonnegative terms, as in Eq. (3). The mathe-

$$A = C - (E + S + R) \quad (3)$$

matical forms of C, E, S, and R derive from information theory, and all bear some resemblance to that for A given above. Only their qualitative characteristics will be described below.

The quantity C exceeds the value of A for all real configurations; it represents the capacity of the existing system for development. During the intermediate stages of ecological succession, ascendency increases mostly as a result of increasing capacity. A larger number of species and a more equitable distribution of activity among the taxa both serve to augment C. In the example of the recovering ecosystem, new species are continually arriving from less disturbed communities, and the available resources become spread over a more diverse ensemble of living things. This trend toward greater diversity cannot continue forever, because an increasing n and a finite T guarantee that some compartments eventually will share in only a minuscule fraction of the overall activity. Such insubstantial species are always subjected to the chance of extinction by the small perturbations which are a part of any real environment.

During the final stage of development, A increases primarily because of decreases in any of the last three terms, which collectively are called the system overhead. The first overhead term, E, is generated by the exports of usable medium, the $T_{i,n+1}$, whence a diminishing E mirrors the tendency for mature systems to become more self-contained. In principle, nothing prohibits E from shrinking to zero; however, most real systems are themselves components of hierarchically larger systems. Should the outputs of the smaller system be cybernetically linked to its own inputs in the next hierarchical order, then reduction of E beyond a certain amount could imperil the system's own inputs. The most poignant example of this limitation comes from international economics, where a cutback in oil exports by producer states depressed the global economy and eventually worked hardships upon the oil producers.

The dissipation of medium (the $T_{i,n+2}$) gives rise to S, which by the second law of thermodynamics is guaranteed to remain positive in real networks. Ilya Prigogine was the first to formalize the tendency for systems developing near their limits to minimize their dissipation. However, decreasing dissipation as a way of increasing A is effective only during the latter stages of development. During initial growth, A is increased more rapidly by inflating T and C at the relatively minor concomitant expense of increasing S.

The last term, R, reflects the redundancy of pathways connecting any two species. In the absence of disturbances, competition should diminish flows along the less efficient pathways, resulting in the relatively less diverse and more streamlined configurations often observed in mature ecosystems. But researchers have pointed out a connection between redundant pathways and ecosystem reliability. When some of the several pathways connecting two taxa are differentially perturbed, the less impacted routes may compensate for the disturbance. All real systems are subjected to some degree of environmental disturbance; hence R may decrease only to that level which, on the average, is sufficient to compensate for the vicissitudes of the given environment.

No real course of system evolution can be cleanly divided into the three phases as discussed; any one of the ways of increasing A may be occurring at any time. What is apparent at all times, however, is that the flow (and the occasional ebb) of the process called growth and development may be charted in strict quantitative fashion. The phenomenon central to the temporal behavior of ecosystems and connecting them with the actions of economic communities, ontogenesis, or social and political systems is now subject to scrutiny by numbers—a significant step toward a powerful and comprehensive theory of living systems.

For background information *see* ECOLOGICAL SUCCESSION; ECOSYSTEM; INFORMATION THEORY (BIOLOGY) in the McGraw-Hill Encyclopedia of Science and Technology.

[ROBERT E. ULANOWICZ]

Bibliography: H. Hirata and R. E. Ulanowicz, Information theoretical analysis of ecological networks, *Int. J. Syst. Sci.*, 15:261–270, 1984; R. E. Ulanowicz, *Growth and Development: A Phenomenological Perspective*, monograph, in review; R. E. Ulanowicz, An hypothesis on the development of natural communities, *J. Theor. Biol.*, 85:223–245, 1980.

Thermoregulation

It has long been known that day-flying butterflies regulate their body temperature by taking advantage of sunshine for basking. Night-flying Lepidoptera (usually moths) do not have such a readily available heat source. However, they also require often quite

specific and high body temperatures in order to generate sufficient work output from their flight muscles so that they can stay airborne. This article examines comparative aspects of moth thermoregulation in the contexts of body mass and habitat.

Sphinx moths. Most sphingids are fast-flying, narrow-winged, nocturnal moths that feed on nectar while they hover. The majority of species are tropical, and they range in weight from about 200 mg to over 6 g. In most species, flight is not possible until the temperature of the thorax (which contains the flight muscles) is near 104° F (40°C), and thoracic temperature (T_{th}) is regulated [maintained independent of ambient temperature (T_a)]. The T_a at which sphinx moths fly ranges from near 50°C (10°C) to 95°F (35°C). Resting moths have body temperature similar to that of their environment, and before starting to fly they warm up by a process like shivering, during which the upstroke and downstroke muscles of the wings are contracted nearly simultaneously, rather than alternately as in flight. During flight no heat is produced specifically for thermoregulation. Rather, all of the heat that then elevates body temperature is a by-product of the flight metabolism.

The internally generated heat during flight does not automatically result in optimum muscle temperatures. Because of their large mass, vigorous flight metabolism, and flight at high ambient temperature, these moths more commonly produce excess heat during flight. But the moths stabilize thoracic tem-

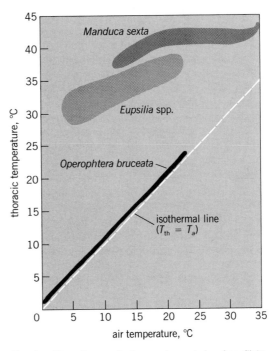

Fig. 1. Thoracic muscle temperatures during free flight of a sphinx moth, *Manduca sexta*; a winter noctuid moth, *Eupsilia* spp.; and a winter geometrid moth, *Operophtera bruceata*. The thoracic temperatures indicated also reflect the range of ambient temperature over which the moths were able to remain in continuous free flight. °F = (°C × 1.8) + 32. (*After B. Heinrich and T. P. Mommsen, Flight of winter moths near 0°C, Science, 228:177–179, 1985*)

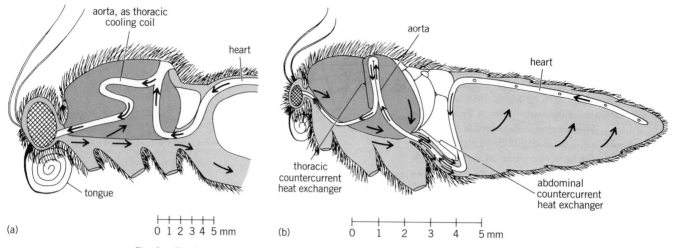

Fig. 2.   Circulatory system of two moths. *(a)* Sphinx moth, showing the typical loop in the descending portion of the aorta, which dissipates heat to head and abdomen. *(b)* Sag- ittal section of a winter cuculinid moth, in which the abdominal countercurrent heat exchanger retards heat flow to the head. Arrows indicate direction of blood flow.

perature (Fig. 1) by a physiological control of heat loss that involves the circulatory system.

As in other insects, the sphinx moths' circulatory system is an open one (the blood bathes the tissues and there are no capillaries), although blood is pumped by a pulsatile vessel, the heart. The heart is a muscular tube extending along the middorsal line of the abdomen into the thorax. In the thorax the pulsatile vessel, now called the aorta, continues on into the head. Blood enters the heart through ostia in the abdomen and is pumped through the thorax into the head and then, following pressure differences, percolates back into the abdomen. In sphinx moths the aorta makes a loop through the flight musculature (Fig. 2a). This loop, which receives the cool blood from the abdomen, functions as a cooling coil as it picks up heat from the working thoracic muscles during flight. The heat from the blood is then dissipated both from the head and the abdomen. As a result of this cooling mechanism sphinx moths are able to maintain a stable thoracic temperature while flying over a large range of air temperatures.

Winter owlet moths. Most owlets (Noctuidae) are small moths (100–300 mg) that fly in the summer on warm nights. However, in one subfamily of north temperate owlets (Cuculiinae) commonly called winter moths, the adults emerge in the fall and fly in the winter at air temperatures as low as 32°F (0°C). These moths also need to maintain an elevated thoracic temperature in order to fly. In the laboratory they have been observed to warm up by shivering from ambient temperature as low as 27.6°F (−3°C) and to maintain thoracic temperature some 54°F (30°C) above air temperature.

Because of their small size (and hence rapid convective cooling), and the prevailing low temperature of their environment, the cuculinid moths face the opposite thermal problem that sphinx moths face. Temperatures in the winter seldom exceed 32°F (0°C), and activity of these moths is constrained not

by excess heat generated by metabolism of flight, but by low temperatures.

The remarkable ability of these moths to generate sufficiently high thoracic temperature to maintain enough power output for flight is due to several anatomical features that function in the conservation of heat within the thorax. The thorax is covered with a dense layer of insulating "fur" (technically pile). This fur consists of modified scales (that color Lepidoptera wings), elongated to near 0.08 in. (2 mm.) In addition, a series of air sacs (Fig. 2b) in the thorax (where they function also as ears) and the abdomen (where they function also as bellows in ventilation) act to insulate the thorax from conductive heat loss to the abdomen. Abdominal temperature remains within 0.7°F (0.4°C) of ambient temperature, while thoracic temperature may increase over 54°F (30°C) above ambient temperature. During flight, abdominal temperature also remains unusually low, rising no more than 3.6°F (2°C) above air temperature.

How is a temperature gradient of over 54°F (30°C) maintained between thorax and abdomen despite blood flow between the two body parts? Part of the answer may lie in the circulatory anatomy (Fig. 2b). The only functional connection for exchange of materials between the thorax and the abdomen is a relatively long (0.1 in. or 3 mm), narrow passage in the anterior ventrum of the abdomen that accommodates the esophagus, the ventral nerve cord, and the aorta. All of the blood from the cool abdomen necessarily flows through the aorta in this passage, and all of the blood heated in the thorax returning to the abdomen must flow over and around this vessel. Since the aorta with its cool blood is bathed in the host returning blood, countercurrent heat exchange is inevitable as long as there is bloodflow. This heat exchange brings heat back into the thorax that would otherwise be lost to the abdomen.

A second countercurrent heat exchanger is located in the thorax. Within the thorax the aorta ascends between the thoracic muscles to the scutellum, and then descends again, but without making a large loop or "cooling coil" as in sphinx moths. Instead, the descending portion of the aorta is pressed closely against the ascending vessel before proceeding anteriorly into the head. Heat from blood in the descending loop should be recovered by the ascending portion of the loop, and this countercurrent heat exchange should retain heat in the thorax that would otherwise be lost through the head. Not being able to use head and abdomen as heat radiators, these moths are unable to fly at an air temperature greater than 71.6°F (22°C) [note *Eupsilia* spp. in Fig. 1] because they then overheat from their own flight metabolism.

Geometrid winter moths. A second group of winter moths (of the Geometridae) also emerges in the fall. Of these, *Operophtera bruceata* and *Alsophila pometaria* fly in late November in the eastern United States until the first heavy snowfalls stop their activity. The females are wingless, and the males (which do not feed) fly both at night and in the daytime at temperatures as low as 27.6°F (−3°)C.

The winter geometrids are small moths (males weight near 10 mg) that are uninsulated. They neither shiver nor bask, and thoracic temperature during flight is within 1.8°F (1°C) of air temperature (Fig. 1).

In many ectothermic animals the catabolic enzymes are specialized to operate at the body temperatures the animal normally encounters. In the moths, however, catalytic efficiencies and overall activities of citrate synthase and pyruvate kinase per gram of thorax are nearly identical in the winter noctuids that fly with thoracic temperature near 95°F (35°C), in a sphinx moth *(Manduca sexta)* that flies with a thoracic temperature near 104°F (40°C), and in the winter geometrids that fly with a thoracic temperature near 32°F (0°C). Part of what allows these insects to fly with a muscle temperature below the freezing point of water lies in their ability to fly with a very low energy expenditure.

Geometrids, and other nocturnal nonfeeding Lepidoptera, generally have large wings relative to their body mass (low wing-loading), and this characteristic confers a low cost of flight. The wing-loading of winter noctuids is 40–50 mg/cm², but in the winter geometrids wing-loading is 10–14 times lower than this. It is also lower than that in other geometrids of both tropical and temperature regions. Low wing-loading that decreases the cost of transport appears also to allow these geometrids to fly with a low muscle temperature. When thoracic temperature is 32°F (0°C), for example, the moths support flight with an energy cost only 4 cal per gram of thorax per minute, which is similar to the energy output of shivering of the winter noctuids at a thoracic temperature of 50°F (10°C). Because of their low wing-loading, the geometrids are able to support flight at one-tenth the metabolic rate required by the noctuid moths. It appears that with moths, the morphology affecting thermoregulation (including the circulatory system as well as wing morphology) can vary radically, while the adaptations at the enzyme level for temperature are apparently highly conservative.

For background information *see* CIRCULATION; INSECT PHYSIOLOGY; LEPIDOPTERA; THERMOREGULATION in the McGraw-Hill Encyclopedia of Science and Technology.

[BERND HEINRICH]

Bibliography: B. Heinrich, Nervous control of the heart during thoracic temperature regulation in a sphinx moth, *Science*, 169:606–607, 1970; B. Heinrich, Temperature regulation in the sphinx moth, *Manduca sexta*, *J. Exp. Biol.*, 54:141–166, 1971; B. Heinrich and T. P. Mommsen, Flight of winter moths near 0°C., *Science*, 228:177–179, 1985.

Transformer

The incentive to develop new types of transformers originates in pressures to lower manufacturing costs, as well as to conserve energy with transformers having lower losses. Transformer manufacturing costs have been reduced by the adoption of foil (sheet) windings for low-voltage coils and simplified constructional methods. Tank costs for liquid-cooled transformers have been significantly reduced by the use of corrugated finned tanks. The development of improved core materials has assisted in the production of transformers having lower losses. There is also an increased need for specialized fire-resistant transformers that do not present environmental hazards. This need is being met by improved synthetic liquid coolants and resin-encapsulated dry-type transformers.

Winding developments. The introduction of foil windings for low-voltage coils has provided solutions to many of the problems associated with distribution and small power transformers. Foil windings consist of thin sheets of conductors which occupy the full axial length of the coil; the conductor material may be either copper or aluminum. Comparison between a conventional spiral strip-wound low-voltage coil and a foil winding shows that the helical form of the spiral coil is wasteful of winding space. The magnetic imbalance of the primary and secondary windings, due to the helical effect, produces axial forces on spiral coils during short-circuit conditions. With conventional designs these forces are constrained by substantial end frames and coil support blocks. The adoption of foil windings for low-voltage coils results in the elimination of axial forces, or they are made to be compressive on the high-voltage coils. This enables a much simplified winding support method to be adopted, as compressive forces on high-voltage coils may be easily withstood.

The use of resin-coated insulation materials adds further strength to this winding method by the heat-bonding of adjacent winding turns and layers. High-voltage coils are wound with each layer occupying the full width of the low-voltage foil; by this means coils may be produced in the minimum possible time, with the simplest form of winding construction. Further material economies and reduction in overall transformer dimensions are obtained by the adoption of windings having rectangular shape, in preference to the round form formerly used. Cores are also of rectangular form, employing one width of lamination; this further simplifies the manufacturing process.

Tank manufacture. The heat generated by the core and windings of a liquid-cooled distribution transformer is transferred to the cooling surface by conduction and convection. To dissipate the heat, it is usual to increase tank cooling on ratings above approximately 50 kVA. Various methods have been used to achieve this, including several designs of tube and pressed steel panel radiators. With these designs of cooling system the radiators are attached to the tank body by bolted flanges or welding.

The corrugated-tank principle is to form cooling fins from the tank walls, thus providing adequate cooling by using the minimum oil content in the narrow hollow fins (Fig. 1). This method of tank construction is gaining in use. The introduction of automatic folding and welding machinery has enabled achievement of high production rates with associated cost reductions. The overall dimensions of a corrugated-tank transformer are usually smaller than the equivalent conventional design. The corrugated design of the tank is suitable for constructions that are free-breathing, with or without conservator, gas-cushion–sealed, or completely oil-filled and sealed.

In any tank or oil-preservation system it is desirable to limit the transfer of moisture and atmospheric pollutants from the gas in contact with the oil to the transformer insulation system. Moisture affects paper insulation by reducing dielectric

strength and increasing the deterioration rate of mechanical properties. These damaging processes may be minimized by fitting a conservator with a dehydrating breather, which limits the interchange of atmospheric gas with the oil. This system has the disadvantage of requiring regular replacement or drying of the desiccant material; in remote installations this can be difficult to achieve. A recent improvement has been the introduction of completely filled sealed transformers of corrugated-tank construction. The changes in volume of the insulating liquid are resiliently absorbed by the bellows action of the corrugated-tank sides. This design of sealed transformer offers favorable aging properties and requires a minimum of maintenance. The fully sealed design is ideally suited to pole-mounted distribution transformers for rural distribution schemes.

Improved magnetic materials. One of the constraints the transformer engineer has to take into account is the flux density at which a core may be operated. Increasing flux density results in greater unit loss in the core steel; the performance curves of electrical steels have a nonlinear form. Increasing the core flux density results in decreasing transformer volume, but iron loss increases and ultimately the steel becomes magnetically saturated. The electrical performance of conventional silicon steel is being improved by several means, and new methods of steel production can produce materials with significantly improved performance.

The material types compared in Fig. 2 illustrate the differences in performance level obtained with several currently obtainable materials. The 28M4 grade of material is widely used for distribution transformer cores. By reducing the material thickness, eddy-current losses can be reduced and overall loss improvement made. The 28M4 grade is 0.28 mm nominal thickness, whereas the RGO23 is 0.23 mm thick. Thinner materials are commercially available down to 0.15 mm; thinner plates imply increased core building times, and the overall economics of using this material have to be considered. The most promising method of loss reduction in conventionally produced steels is illustrated by the material designated ZDKH 0.23. By scribing the surface of high-permeability core steel with a laser beam, it is possible to induce tensile stress in the material, which has the effect of refining the magnetic domain spacing. The eddy-current loss of transformer laminations is proportional to the magnetic domain wall spacing; hence, refining the domain spacing has the effect of reducing losses.

Innovative methods of core-steel production offer the potential for materials with vastly improved properties, as illustrated by the curve for 2605S-C material in Fig. 2. This material, commercially known as Metglass or amorphous metal, is produced by ejecting molten metal alloy onto a cooled rotating wheel which produces a very high rate of cooling. This technique, known as planar flow casting, produces a thin ribbon of steel, almost devoid of grain structure, which exhibits very low specific loss.

Fig. 1. A 315-kVA ground-mounted corrugated-tank distribution transformer.

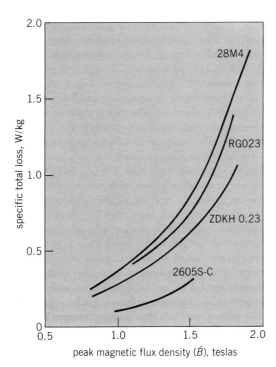

Fig. 2. Comparative performance of magnetic materials used in transformer cores. Specific total loss in the core steel is shown as a function of peak magnetic flux density.

Cores formed from this material have been built into distribution transformers with encouraging results, and many are in service for evaluation. However, amorphous core steels have low saturation induction densities and are fragile and, at present, expensive. *See* MAGNETIC MATERIALS.

Low-flammability transformers. For the majority of transformer installations the oil-cooled type is the natural choice. However, there are many instances where mineral oils can present an unacceptable fire risk, and so an alternative coolant must be selected. There are several options. The first choice to be made is between liquid- or air-cooled types, both being further divided by choices of insulation system.

Liquid-cooled types. The environmental hazards presented by polychlorinated biphenyl materials are well known, and considerable development work has been undertaken to produce an acceptable alternative. There are four main groups of high-fire-point liquids: silicone, heavy hydrocarbons, halocarbons, and complex esters. Some of the materials, such as the halocarbon group, are totally nonflammable, but the environmental acceptability of certain halogen-containing materials is in dispute. Most of the other available liquids are flammable, having fire points in excess of 570°F (300°C), which is considerably in excess of the 300°F (150°C) of mineral oil. However, such material properties as heat of combustion and energy content must also be considered. Certain of these materials may be used as direct substitutes for mineral oils and polychlorinated biphenyls, which is advantageous during initial manufacture as

well as retrofilling activities. There are instances where extremely hostile environmental conditions preclude the use of dry-type transformers, and a choice must be made from the available liquid-cooled types.

Air-cooled types. Where installation circumstances permit, the optimum choice of a low-flammability transformer is considered by many authorities to be the dry type. In this instance the choice lies between the conventional class C (or class H) design and the cast-resin type. Class H and C dry-type transformers feature windings of disk (pancake) or layer form, having temperature rises limited to 225°F (125°C) and 270°F (150°C), respectively. The insulation system consists of Nomex or high-temperature-enamel–covered conductors which are impregnated in suitable varnishes, silicone, or polyester, after winding. The coils are substantially unprotected from surface contaminants and can be vulnerable to the effects of moisture and atmospheric pollutants, requiring care in installation and maintenance. Recent development work on this type of transformer includes the production of designs suitable for operation under conditions of seismic shock. This feature is required for certain transformers installed in nuclear power stations which may be at risk from seismic disturbances.

Expertise gained in the production, handling, and casting of resins has enabled the development of the cast-resin distribution transformer (Fig. 3). The vulnerability of conventional dry-type transformers may

Fig. 3. A 1600/2000-kVA resin-encapsulated transformer, cooled by natural air convection with an increase in rating given by forced-air (fan) cooling.

be overcome by encapsulating windings with suitable resins. Epoxy resins, combined with mineral fillers, have proved eminently suitable for this application. Transformers of this type are hard to ignite and do not continue to burn when the source of heat is removed.

Development of cast-resin transformers has concentrated on improvement of resins to reduce internal thermal stresses and to reduce manufacturing costs. Ratings and impulse levels have been raised with transformers of up to 10 MVA currently in service.

For background information *see* EDDY CURRENT; ELECTRICAL INSULATION; HEAT TRANSFER; MAGNETIC MATERIALS; TRANSFORMER in the McGraw-Hill Encyclopedia of Science and Technology.

[K. FREWIN]

Bibliography: D. J. Bailey and L. A. Lowdermilk, *Amorphous Steel Core Distribution Transformers,* General Electric, 1983; K. Frewin, Cast resin transformers increase in popularity, *Electr. Power,* pp. 133–136, February 1985; D. G. Say and K. Frewin, Amorphous steel core distribution transformers, *Electr. Times,* p. 9, October 24, 1980.

Transuranium elements

Currently 17 elements are known beyond uranium, the last of the elements found in macroscopic amounts in nature. These 17 artificial transuranium elements appear in two series in the periodic system, the actinide series from element 89 through 103 and the transactinide series from element 104 through 109. The form of the periodic table shown in Fig. 1, incorporating an actinide *5f* electron series paralleling in chemical properties the *4f* electron lanthanide series, was proposed by G. T. Sea-

borg in 1945. Since its introduction, the actinide concept has furnished the framework for all chemical research in the transuranium area. But within this framework, many of the most important discoveries in recent years concern the differences in chemical behavior between the heavy actinides and their lanthanide homologs rather than their similarities.

The detailed study of the heavy actinides has become possible only since the early 1970s as milligram quantities of the relatively long-lived isotopes curium-248 (^{248}Cm), berkelium-249 (^{249}Bk), and californium-249 (^{249}Cf) as well as the relatively short-lived einsteinium-253 (^{253}Es) became available from a production program centered at the Department of Energy's High Flux Isotope Reactor (HFIR) located in Tennessee. These isotopes have made it possible to investigate many basic properties of these elements, including the structures of the metals and their compounds by x-ray and electron diffraction, thermodynamic functions (such as the heats of sublimation, heats of solution, and complexation constants), oxidation-reduction reactions, and absorption and emission spectra. Furthermore, these isotopes, along with ^{254}Es, serve as targets at heavy-ion accelerators for the production of even heavier elements. For the elements fermium (Fm; element 100) and beyond, the quantities available for experimentation are measured in atoms rather than in weighable amounts, and the chemistry of lawrencium (Lr; element 103) and rutherfordium (Rf; element 104) has, indeed, even been studied in some instances with only one atom per experiment. In this case a number of experiments are repeated to build up statistics. It is clear under such circumstances that considerable ingenuity is required to develop original procedures, and the results should be confirmed by using several different approaches where possible. The most reliable chemical methods for one-atom-at-a-time chemistry have been found to be solvent extraction, ion exchange, and gas chromatography. These methods have in common that they put the one atom or a few atoms through a large number of reactions per experiment so that the statistical equivalent is achieved of the more usual case of having a large number of atoms go through one reaction per experiment.

Structural chemistry and ionic radii. The availability of milligram quantities of long-lived isotopes of curium, berkelium, and californium, together with improvements in data collecting and handling techniques, has allowed the precise determination of the structures of a number of compounds of these elements through x-ray studies, some done by single-crystal methods, with many more done by powder methods. Only limited results have been obtained on einsteinium, because of the short half-life of einsteinium-253 and its consequent intense alpha activity. Elements beyond einsteinium cannot be produced, at this time, in quantities large enough for structural studies.

The ionic radii of the trivalent actinides have

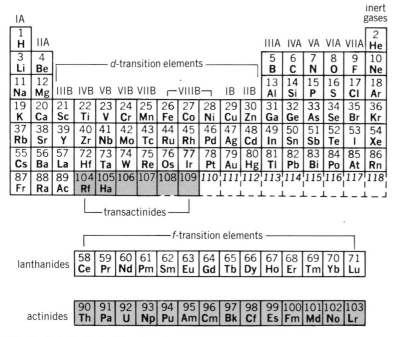

Fig. 1. Periodic chart of the elements.

been measured for coordination number 6 in picometers as uranium (U) = 104.1, neptunium (Np) = 101.7, plutonium (Pu) = 99.7, americium (Am) = 98.2, curium = 97.0, berkelium = 94.9, californium = 93.4, and einsteinium = 92.5. When these are compared to the corresponding lanthanide values, they are found to be close to 3 pm larger in the 5f element than in the 4f homolog. This empirical correlation allows the prediction that lawrencium, for example, will be found to have an ionic radius of 88 pm compared to the value of 84.8 for lutetium (Lu).

The organometallic tricyclopentadienide compounds of the actinides have been synthesized through californium. In a particularly elegant study, single crystals of $Bk(C_5H_5)_3$ and $Cf(C_5H_5)_3$ were grown for a precise determination of their structures. They were found to be closely similar to the lighter lanthanides and actinides. This implies that the metal-ligand bonding is of the same nature.

Hydrides of the actinides through californium have been prepared and their structures determined for comparison to the lanthanides. The dihydride of californium is particularly interesting because it shows a tendency toward divalency in californium that does not appear in the preceding actinides or in the homologous lanthanides.

Examples are known of binary compounds of the actinides with nitrogen (N), phosphorus (P), arsenic (As), and antimony (Sb) from thorium (Th) to californium. These compounds have been shown to have the sodium chloride (NaCl) crystal structure. Although the nature of the bonding in these compounds is only partially understood, it is generally thought to be semimetallic.

The metals from americium through californium have been subjected to exceedingly high pressures that cause the localized 5f orbitals to delocalize from their atomic sites and enter into the metallic bonding. Under these very high pressures, the structures of the metals change from their usual simple arrangement to the complicated α-uranium structure. The α-uranium structure is known to be characteristic of itinerant 5f electron behavior. Since the mechanism of the transition of electrons from a localized atomic state to a delocalized metallic state is a fundamental question for understanding bonding in metals, these high-pressure studies as a function of atomic number progressing up the 5f electron series offer a unique opportunity to the theorist.

Thermodynamic properties. The availability of milligram quantities of the actinides through einsteinium has made it possible to measure the heats (enthalpies) of sublimation of these metals. These enthalpies are important because they enter into the Born-Haber cycle. Although the ionization potentials and heats of hydration (or lattice energies) are needed for a true analysis, the heats of sublimation of the metals by themselves fortunately offer a useful guide to trends in valency (oxidation states) in the actinides compared to the lanthanides (Fig. 2). In the lighter end of the series through americium, the

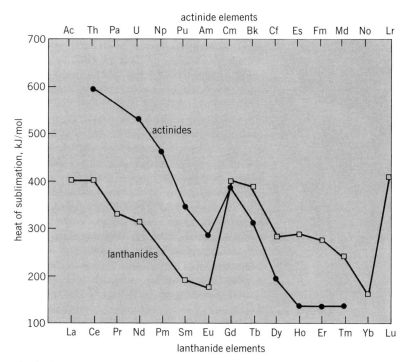

Fig. 2. Heats of sublimation of actinide and lanthanide metals.

heats of sublimation of the actinides are higher than the corresponding lanthanides. In the middle of the series, at curium and gadolinium (Gd), they are equal, and in the heavy end of the series they are lower. The higher heats of sublimation correspond to the stabilization of the higher valencies in thorium, protactinium (Pa), uranium, neptunium, and plutonium over the corresponding lanthanides. The higher enthalpy of americium is the result of its being a trivalent metal whereas europium (Eu) metal is divalent. Also europium is readily reduced to the 2+ ion in aqueous solution whereas this is impossible for americium. After curium the actinides trend toward having lower heats of sublimation than the lanthanides rather than higher as before. In fact, the value for californium is similar to samarium (Sm) rather than to its homolog dysprosium (Dy). This corresponds to the behavior of californium being rather similar to samarium and quite different from dysprosium. In particular, californium and samarium are both easily reducible to the divalent state whereas dysprosium is not.

The trend toward divalency begun at californium culminates in nobelium (No) being most stable in the divalent state. In fact, aqueous chemical studies of No^{2+} have shown that it behaves in its chemistry with common ligands such as acetate, oxalate, and citrate similarly to calcium (Ca) and strontium (Sr) in the alkaline-earth series. Also, the ionic radius of nobelium was measured to be 105 pm, which compares to 99 pm for calcium and 113 pm for strontium. Such behavior is in marked contrast to that of the lanthanides. A solvent extraction study has shown, however, that lawrencium is most stable in the 3+ state as expected for the last member of the actinide series.

The heats of sublimation of fermium and mendelevium (Md) were deduced from gas chromatography measurements of metallic atoms of these elements, produced a few atoms at a time using a cyclotron. The gas chromatography column was calibrated with macroscopically available elements to show that the heats of adsorption of metallic atoms in it correlated with their heats of sublimation. Such work illustrates the ingenuity that must go into the chemical studies of the actinides and transactinides from curium to hahnium (Ha) since they are available only in quantities ranging from milligrams to single atoms.

Reduction potentials. Similar ingenuity has gone into measuring the formal reduction potentials of the heavy actinides. These measurements again reflect the trend toward divalency shown by the heats of sublimation. Radiocoulometry and radiopolarography methods that will operate on the basis of a few hundred atoms per experiment have been developed. Of particular interest is the value for the No^{2+}/No^0 couple determined by radiocoulometry as -2.5 V versus the standard hydrogen electrode. This can be compared to the values of -2.9 V for Ca^{2+}/Ca^0 and Sr^{2+}/Sr^0, and -2.4 V for magnesium (Mg). The values of the $3 \rightarrow 2$ reduction potentials in volts versus the standard hydrogen electrode have been determined as: curium $= -3.7$, berkelium $= -2.80$, californium $= -1.60$, einsteinium $= -1.55$, fermium $= -1.15$, mendelevium $= -0.15$, and nobelium $= +1.45$.

Trend toward divalency. A prime result of the recent experimental studies has been the discovery that there is a trend toward divalency in the last half of the actinides that does not exist in the lanthanides. This trend reflects a larger separation of the f and d orbital energies in the actinides than in the lanthanides in the heavier end of the series. The opposite is true in the light end. Differing relativistic effects on the electronic configurations of the two series are an important influence on their differing chemical properties. The development of relativistic atomic and molecular calculational codes is now allowing a beginning to be made in understanding the comparative chemistry of the two series within the framework of the Dirac theory.

For background information *see* ACTINIDE ELEMENTS; CHEMICAL BONDING; CHEMICAL STRUCTURES; ELECTRON CONFIGURATION; ORGANOMETALLIC COMPOUND; OXIDATION-REDUCTION; RELATIVISTIC QUANTUM THEORY; TRANSURANIUM ELEMENTS in the McGraw-Hill Encyclopedia of Science and Technology.

[O. L. KELLER]

Bibliography: A. J. Freeman and C. Keller (eds.), *Handbook on the Physics and Chemistry of the Actinides*, vol. 3, 1985; E. K. Hulet, Chemistry of the elements einsteinium through element 105, *Radiochim. Acta*, 32:7–23, 1983; J. J. Katz, G. T. Seaborg, and L. R. Morss (eds.), *The Chemistry of the Actinides*, 2d ed., 1985; O. L. Keller, Jr., Chemistry of the heavy actinides and light transactinides, *Radiochim. Acta*, 37:169–180, 1984.

Tremolite

The mineral tremolite, $Ca_2Mg_5Si_8O_{22}(OH)_2$, is one of a very large and chemically complex group of minerals known as amphiboles. Tremolite occurs almost exclusively in metamorphic rocks, particularly metamorphosed dolomites and serpentinites, suggesting that it only forms (is stable) at temperatures that are typical of metamorphism (200–800°C or 390–1470°F). Experimental studies within the last several decades have confirmed this hypothesis and, in addition, have shown that tremolite is stable to a maximum pressure of about 23–27 kilobars (2300–2700 megapascals). Furthermore, recent experimental evidence indicates that tremolite changes to a slightly calcium-depleted and magnesium-enriched composition at its highest temperature of stability. Such small but noticeable changes directly affect thermodynamic data derived from experimental studies involving tremolite and may prove useful as an indicator of the pressure and temperature conditions at which tremolite formed in nature.

Maximum thermal stability. The maximum thermal stability of tremolite is governed by dehydration reaction (1), where quartz (SiO_2) can exist in either

$$\text{Tremolite} \rightleftarrows 3 \text{ enstatite} + 2 \text{ diopside}$$
$$+ \; \alpha\text{- or } \beta\text{-quartz} + H_2O \quad (1)$$

the α or β form, depending on the particular pressure and temperature being considered, and enstatite and diopside have the compositions $MgSiO_3$ and $CaMgSi_2O_6$, respectively. Reaction (1) has been experimentally studied by synthesizing each of the minerals individually, mixing them together in the stoichiometric proportions of reaction (1), sealing this mixture along with some extra water (to act as a catalyst and hydrostatic pressure medium) into an inert capsule, and subjecting the capsule to high confining pressures and temperatures. The capsule is opened at the conclusion of the experiment and the contents examined with the microscope or with x-ray diffraction techniques to determine if tremolite has grown at the expense of the right-hand assemblage of reaction (1) or vice versa.

Most of the experimental data on reaction (1) have been obtained at (water) pressures of 0.2–2.0 kilobars (20–200 MPa). These data are represented by the solid portion of dehydration boundary (1) in the illustration. The remainder of this boundary is broken to indicate that there is moderate uncertainty in its location beyond the range of 0.2–2.0 kilobars (20–200 MPa); however, the entire dehydration boundary has been calculated from thermodynamic data that are consistent with the experimental data in the range 0.2–2.0 kilobars (20–200 MPa), so that its location is probably accurate to ±30°C (54°F).

Several features of the dehydration boundary are notable. First, there is a strong increase (nearly 400°C or 720°F) in the upper thermal stability of tremolite as the pressure increases from atmospheric pressure (0.001 kilobar or 0.1 MPa) to about 3 kilobars (300 MPa). This behavior is commensurate with the experimentally determined dehydration boundaries of other amphiboles. Second, there is a change in the slope of the boundary at about 7.5 kilobars (750 MPa), such that the upper thermal stability of tremolite decreases with increasing pressure above 7.5 kilobars (750 MPa). This curious reversal in the influence of pressure is attributable to the greater compressibility of water, relative to most crystalline substances, at high pressures. That is, the net volume of the products of reaction (1) becomes smaller than that of tremolite above 7.5 kilobars (750 MPa), and the slope of reaction (1) switches from positive to negative according to the Clausius-Clapeyron thermodynamic relation. Third, the calculated dehydration temperature of tremolite at "bench top" or atmospheric pressure is shown to be about 480°C (896°F), which is considerably lower than the 700–800°C (1290–1470°F) that is usually observed experimentally. The source of this discrepancy is not known, but may be related to the sluggish diffusion of H_2O out of the crystalline framework of tremolite when heated in air without the catalytic effect of high water pressures.

Maximum pressure of stability. The maximum pressure to which tremolite is stable is governed by reaction (2); this reaction has not been extensively

$$\text{Tremolite} \rightleftarrows 2 \text{ diopside} + \text{talc} \qquad (2)$$

investigated because it is very difficult to induce a perceptible growth of tremolite at the expense of diopside plus talc [$Mg_3Si_4O_{10}(OH_2)$], or vice versa, even at the highest temperatures and pressures for reaction (2) where reaction rates are accelerated. Some experimental results on reaction (2) have been reported, and these are represented by the solid portion of reaction boundary (2) shown in the illustration. The broken portion of the line is an extrapolation to lower temperatures to suggest that reaction (2) may also define the lower temperature of stability of tremolite.

Compositional variations. It is apparent that most, if not all, experimental studies involving synthetic tremolite have dealt with a tremolite that does not have the ideal composition $Ca_2Mg_5Si_8O_{22}(OH)_2$. Essentially all attempts to synthesize tremolite at the conditions most favorable for its synthesis (700–850°C or 1290–1560°F, 2–12 kilobars or 200–1200 MPa) have yielded a mixture of minerals consisting of about 88% (by weight) tremolitic amphibole, 11% diopside, and 1% quartz (the latter being dissolved in the water). From the aspect of mass conservation, the tremolitic amphibole in this mixture must be depleted in calcium, which is used to form the diopside, and enriched in magnesium. In other words, the amphibole that is obtained synthetically has the calculated composition $(Ca_{1.8}Mg_{0.2})Mg_5Si_8O_{22}(OH)_2$. This has been confirmed by direct chemical analyses of individual, synthetic crystals. This observation creates an interesting problem: pure tremolite

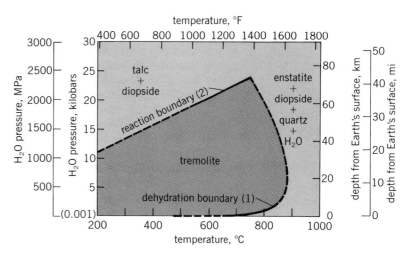

Water pressure and temperature region (shaded) where tremolite is stable. Depth scale on the right assumes a constant rock density of 3.3 g/cm^3 and that the water pressure equals the total pressure.

can be observed in nature, yet cannot be made synthetically. In order to solve this problem, it is necessary to determine whether or not the mixture tremolitic amphibole, diopside, and quartz is a stable assemblage. Experimental investigations dealing with the synthesis of tremolite over a range of chemical compositions suggest that equilibrium is attained in the range of 700–850°C (1290–1560°F) and 2–12 kilobars (200–1200 MPa) and that magnesium-enriched tremolite is stable. Pure tremolite, therefore, must form either at very low temperatures (perhaps less than 500°C or 930°F) or under special geological conditions that are not readily reproduced in experimental studies.

Natural occurrence. The shaded region in the illustration represents the maximum range of pressures and temperatures where tremolite is stable. Unfortunately, this region may not accurately represent the stability limit of tremolite in the presence of certain minerals or in the presence of chemical components other than H_2O, CaO, MgO, and SiO_2. For example, if tremolite comes in contact with the mineral forsterite (Mg_2SiO_4), then the upper thermal stability of tremolite will be governed by reaction (3). The boundary for reaction (3) lies about 50–

$$\text{Tremolite} + \text{forsterite} \rightleftarrows$$
$$5 \text{ enstatite} + 2 \text{ diopside} + H_2O \quad (3)$$

100°C (90–180°F) lower than the boundary for reaction (1). In general, the pressure and temperature region where tremolite is stable is reduced whenever tremolite comes in contact with a mineral with which it can react. The stability range of tremolite is also affected, though in a less predictable manner, by the substitution of other chemical components (such as FeO or Al_2O_3) for the components in tremolite. This so-called solid solution may increase or decrease the stability of tremolite by many hundreds of degrees. It is important, therefore, to consider carefully both the composition and coexisting mineral assemblage of a particular natural tremolite before attempting to determine the conditions under which it was formed.

Perhaps the most interesting outcome of recent experimental studies on tremolite is the recognition that tremolite has a variable chemical composition that can be expressed by the ratio $Mg/(Mg + Ca)$. This is of importance for two reasons. First, anyone seeking to derive thermodynamic data from experimental data involving tremolite should be aware that tremolite has a variable composition, and should make appropriate allowances in the calculations. Second, it may be possible to calibrate experimentally the $Mg/(Mg + Ca)$ ratio of tremolite as a function of temperature (and possibly pressure), and thus, it may be possible to determine the condition of formation of tremolitic amphibole from its composition. A schematic graph of the magnesium content of tremolitic amphiboles as a function of temperature based on empirical observations of natural amphiboles has been proposed but has not been calibrated experimentally.

For background information *see* AMPHIBOLE; CHEMICAL THERMODYNAMICS; HIGH-PRESSURE MINERAL SYNTHESIS; SOLID SOLUTION; TREMOLITE in the McGraw-Hill Encyclopedia of Science and Technology.

[DAVID M. JENKINS]

Bibliography: F. R. Boyd, Hydrothermal investigations of amphiboles, in P. H. Abelson (ed.), *Researches in Geochemistry*, pp. 377–396, 1959; B. W. Evans, Amphiboles in metamorphosed ultramafic rocks, in D. R. Veblen and P. H. Ribbe (eds.), Amphiboles: Petrology and Experimental Phase Relations, *Mineral. Soc. Amer. Rev. Mineral.*, 9B:98–113, 1982; M. C. Gilbert and R. K. Popp, End-member relations, in D. R. Veblen and P. H. Ribbe (eds.), Amphiboles: Petrology and Experimental Phase Relations, *Mineral. Soc. Amer. Rev. Mineral.*, 9B:231–268, 1982; H. A. Yin and H. J. Greenwood, Displacement of equilibria of OH-tremolite and F-tremolite solid solution, I. Determination of the P-T curve of OH-tremolite, *Trans. Amer. Geophys. Union*, 64(18):347, 1983.

Varve

Marine sediments from the ocean floors generally are bioturbated extensively and mixed. Samples taken from these sediments represent hundreds to thousands of years of deposition, depending on sediment accumulation rates, and do not permit study of decadal or annual change. Occasionally, however, burrowing organisms are excluded from the ocean floor by the presence of a stable bottomwater layer depleted in dissolved oxygen (less than 0.1 ml O_2 per liter); thus, any sediment particles coming either from the continents or from the ocean surface layer will be laid down undisturbed, and will provide a continuous record through time of input variability.

Marine varved sediments. When the fluxes from the ocean or from the continents are offset seasonally, these sediment sections will display an alternating sequence of color-coded thin bands that can

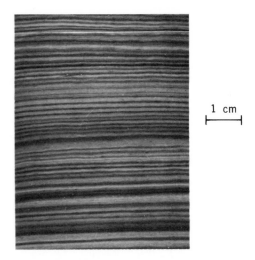

Fig. 1. Marine varves from the Gulf of California. Scale equals 5-cm, positive print of an original x-radiograph. DSDP Leg 64, Site 480, around 26 ft (8 m) below the sediment surface. (*A. Soutar, DSDP-IPOD*)

be used to date them visually, by counting laminae pairs, once the seasonal source of individual laminae pairs are determined. Furthermore, the seasonal laminae exhibit quantitative and qualitative imprints of both continental climate (that controls sediment input from the continents through runoff or eolian transport) and oceanic climate (that controls sedimentation and composition of skeletal remains of phytoplankton and zooplankton). The seasonal couplets of sediment laminae are termed marine varves by analogy to glaciolacustrine varves which were defined originally from Quaternary deposits of glacial lakes in Scandinavia, where an alternating sequence of light-coarse and dark-fine laminae were deposited, due to spring melting and summer production; here, each couplet represents one year of sediment deposition.

Recent varved marine sediments (Fig. 1) have been found primarily along continental margins where, due to strong eastern boundary currents coupled with trade wind systems, strong coastal upwelling is produced seasonally. The great amount of settling organic matter depletes, through oxidation, the amount of dissolved oxygen in the water column, and is responsible for producing near-anoxic bottomwaters. If these conditions last over long periods of time, they will prohibit the development of a sediment infauna that normally would mix sediment pulses through bioturbation. These conditions prevail off southwest Africa (Walvis Bay), Somalia, southwest India, the coast of Peru, and in the Gulf of California in water depths between 1310 and 2600 ft (400 and 800 m). They are found also in restricted basinal environments where, due to sluggish circulations, almost no exchange of bottomwaters occurs or is cut off by shallow sills (for example, Norwegian fjords, Saanich Inlet off British Columbia, borderland basins along California, the Orca Basin in the Gulf of Mexico, and the Black Sea).

The regional-geographical distribution of varved

sediments might be much more common than presently known. This is due to the fact that their retrieval requires usage of special sampling-coring equipment, such as the Soutar box corer, that permits undisturbed recovery of primary sediment

structures in very watery facies (water content of up to 96%).

Fossil varved marine sediments are found in a large variety of deposits of various ages that originally resembled one of the environments described

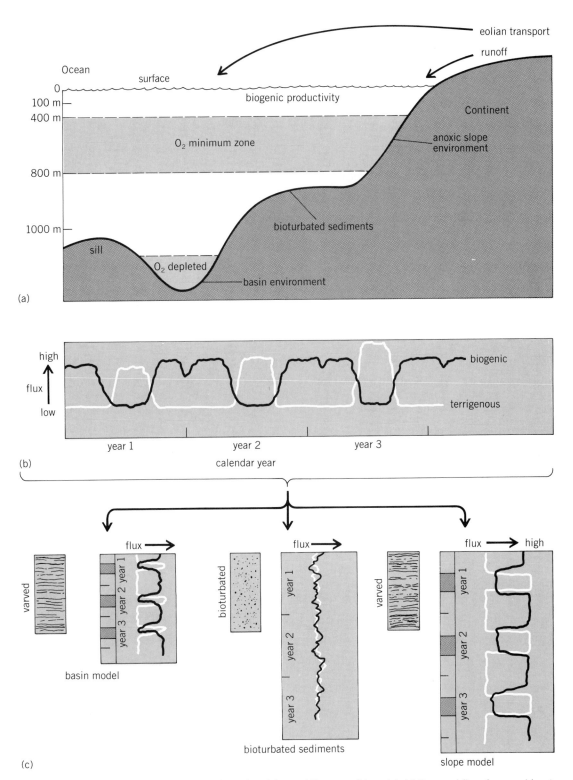

Fig. 2. Schematic model of varved sediment formation. (*a*) Occurrences of varved sediments in an anoxic slope and basin environment. Note bioturbated sediments at intermediate depth. (*b*) Seasonal production of sedimentary particles from land (terrigenous) and from within the photic zone of the ocean (biogenic). (*c*) Preserved flux of seasonal input in the varved sediment facies in the slope and basin environment. Note total mix of seasonal fluxes under oxygenated bottom conditions due to bioturbation. 1 m = 3.3 ft.

above. The most pronounced sequences are found within the siliceous facies of the Monterey Formation in California of Tertiary age.

Depositional environments. Two major types of depositional environments are responsible for formation and preservation of varves (Fig. 2): (1) the basin system where, due to restricted bottomwater circulation, renewal of oxygen-depleted water is limited; and (2) the continental slope system where, due to the presence of a stable midwater (1300–2600 ft or 400–800 m water depth) oxygen minimum zone that intersects continental margins, an infauna is excluded. In both systems, the primary controlling factor is the increased production of organic marine matter in the ocean surface layer that, by oxidation during settling, will consume oxygen.

The other controlling factor is the seasonally offset flux of various sedimentary components. In case this seasonal offset does not exist, for instance, when synchronous or continuous fluxes prevail, no varves will be formed.

Dating. Individual varve thicknesses range from fractions of millimeters to millimeters, pending on the amount of material deposited during each season (Fig. 1). Due to the calendar nature of varved sediments, they can be simply dated by counting laminae couplets downcore, either visually or by using density differences displayed on x-radiographs.

One of the major problems encountered is the determination of the exact age of surface layers, which are lost frequently due to coring disturbances. Short-lived radioisotope dating, using excess ^{210}Pb and ^{137}Cs, are used to determine true ages of surface sediments; in addition, sediment variable fluctuation is compared with measured instrumental data, such as sea-level change, surface water temperature change, and runoff data. Individual laminae consist of either large amounts of terrigenous clay- or silt-sized particles formed after major runoff from the continents occurred, due to increased precipitation, or large amounts of marine biogenic skeletal particles formed during increased productivity in the surface waters, due to enhanced upwelling phenomena. The skeletal particles are either remains of phytoplankton (diatoms, coccolithophorids) or zooplankton (foraminifera, radiolaria).

Time series. Long, continuous sediment sections with seasonal resolution contain an invaluable record of climatic change that goes far beyond the instrumental and historical record. Once additional time series become available (so far only the Santa Barbara Basin and Gulf of California varves have been studied), they will provide climate modelers with the necessary high-resolution time series to be used in their models, similar to tree-ring records, except that they contain a much more diverse record of various parameters that include both oceanic and continental signals. Oceanic signals can be interpreted by determining the assemblage variability and relating its change to the presence or absence of surface water masses; this is a source of information about surface water temperature and change in primary productivity, and details major changes in current circulation. The variable amounts, composition, and grain-size distribution of minerals permit inferences about increased or decreased runoff, change in source area, and increased or decreased eolian transport. Similar studies can be carried out to determine short-range changes in fossil varved sections, and they will permit determination of high-resolution time series for a large variety of geographic and age sections.

Interpretation. With the routine use of the hydraulic piston corer in the Deep Sea Drilling Program, more undisturbed marine sections have become available. One such drill hole recovered 500 ft (152 m) of partially varved sediments from the Gulf of California that represent an almost complete record from the Present through much of the Holocene, spanning roughly a 10,000-year time interval. Very recent sections in the Gulf of California were used to establish criteria for the true varve nature of these sediments. By subsampling the most recent 100-year record and determining microfossil assem-

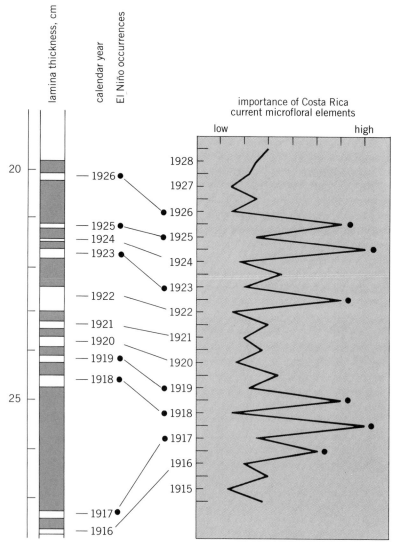

Fig. 3. Time series of ocean climate variability from 1915 to 1928 from a seasonal record from the Gulf of California and its correlation to the Southern Ocean Oscillation (El Niño). Shaded areas represent the portions of the core being plotted.

blage change (Fig. 3), a striking similarity was found between the increase of floral microfossils characteristic of the coastal Costa Rica Current during El Niño occurrences and an increase in primary productivity in the Central Gulf. Recent phytoplankton productivity measurements, taken during the most recent El Niño of 1983, showed similar trends that contrast with what has been found off Peru, where primary productivity is severely depressed during El Niño events. Frequently, El Niño occurrences in the Gulf are associated with an increased runoff. By studying other portions of this record, it will be possible to identify major climatic changes that include both local and overregional changes over a 10,000-year time interval. Similar studies can be carried out in the fossil record by using a floating time scale in a broad variety of geographic areas and over various time periods.

For background information *see* ANOXIC ZONES; OCEAN CURRENTS; PHYTOPLANKTON; SEA WATER; SEA-WATER FERTILITY; VARVE in the McGraw-Hill Encyclopedia of Science and Technology.

[HANS SCHRADER]

Bibliography: S. E. Calvert, Origin of diatom-rich varved sediments from the Gulf of California, *J. Geol.*, 76:546–565, 1966; H. Schrader and T. R. Baumgartner, Decadel variations of upwelling in the central Gulf of California, in J. Thiede and E. Suess (eds.), *Coastal Upwelling: Its Sediment Record*, pt. B: *Sedimentary Records of Ancient Coastal Upwelling*, pp. 247–276, 1983. A. Soutar, S. R. Johnson, and T. R. Baumgartner, In seach of modern depositional analogs to the Monterey Formation, in R. E. Garrison and K. Pisciotto (eds.), *The Monterey Formation and Related Siliceous Rocks of California*, Soc. Econ. Paleontol. Mineral. Pac. Sect., Spec. Publ., pp. 123–147, 1981.

Venus

Venus is similar to the Earth in size, density, and solar energy falling on its surface. It has been called the Earth's sister planet. However, Venus is shrouded by a thick, unbroken layer of sulfuric acid clouds, and its surface cannot be seen from Earth or from the many spacecraft sent by the United States and the Soviet Union to fly by the planet. An important question is whether Venus has an active geology like that of the Earth, and whether it has volcanoes, earthquakes, and plate tectonics. New evidence from many different sources indicates that it does, and that Venus is still recovering from an immense volcanic eruption that occurred in 1978.

Exploration. Since the beginning of the space age, Venus has been a prime target for spacecraft exploration: almost two dozen space missions have probed it since the first interplanetary mission, *Mariner 2*, was launched in 1962. Since the late 1970s a burst of new information from the United States and Soviet space probes has vastly increased knowledge of the planet, and shown it more similar to Earth (at least in its geology) than previously expected. The United States *Pioneer Venus* mission arrived at Venus in 1978 along with the Soviet Venus

missions *Venera 11* and *12*. *Pioneer Venus* included an orbiter (still in operation), three widely spread atmospheric probes, and instruments to measure directly the upper atmosphere before the probe carrier burned up. *Venera 11* and *12* made measurements as they flew by the planet, and dropped two landers. *Venera 13* and *14* (flyby and landers in 1982), *Venera 15* and *16* (radar mapping orbiters in 1983), and *Vega 1* and *2* (flyby, landers, and balloons in 1985) all provided new information about Venus.

An exciting discovery based on this new information is the likelihood that volcanoes are currently active on Venus and that a massive volcanic eruption occurred there around 1978. The evidence for this conclusion comes from geologic studies of the surface by radar, observations of lightning in the atmosphere, pictures and chemical analysis of surface rocks, and remote sensing of the Venus clouds.

Surface geology. The most basic evidence for volcanic activity on Venus comes from studies of the surface showing volcanic landforms. Because of the thick clouds surrounding the planet, these studies have been carried out by radar. When Venus is closest to the Earth, its surface can be studied by ground-based radars, particularly the giant radar telescope at Arecibo, Puerto Rico. Only a small part of the Venus surface can be seen at these times, but radar investigations show a massive uplifted region named Beta Regio. As early as 1977, it was proposed that this region contained a massive volcano 435 mi (700 km) across, larger than the Hawaii volcano chain but smaller than Olympus Mons (the largest volcano on Mars). In 1978 the *Pioneer Venus Orbiter* carried a radar which mapped almost the entire surface of Venus at a resolution of 40–60 mi (60–100 km). Several investigators associated with that study interpreted the geology of the Beta region. Their analysis suggested at least two giant volcanoes, perhaps associated with a broad uplift and rift, such as occurs on the Earth in East Africa. Another area of the Venus surface, named Atla, also showed topographic features which resemble volcanoes. Although the radar studies of Venus showed landforms due to volcanoes and rifting, nothing resembling the Earth system of midocean ridges, coastal mountains, and a global system of interlocking plates has been seen. On the Earth, this system of plate tectonics explains much of the observed geologic activity.

The Soviet spacecraft *Venera 15* and *16*, sent to orbit Venus in 1983, carried radars with 50 times better resolution than that of the *Pioneer Venus Orbiter* (they were able to detect features as small as 0.6 to 1.2 mi or 1 or 2 km wide). Unfortunately, these spacecraft carried only enough attitude control fuel to map about one-third of the planet: this did not include the Beta or Alta region. However, these high-resolution radar data showed numerous volcanic features, ranging from circular basins several hundred kilometers across created by the collapse of giant volcanoes, to fields of small conical structures resembling terrestrial volcanic cinder cones. Every closer look at the Venus surface shows more evidence of geologic activity. The next mission to look through the Venus clouds with radar is the Venus Radar Mapper (VRM), to be launched by the United States in 1988. It will resolve features 10 times smaller than those visible to *Venera 15* and *16*, and is planned to study more than 90% of the surface.

Lightning. In 1978, the *Venera 11* and *12* landers detected electromagnetic signals that appear to arise from lightning near the surface, along with acoustic signals that may be caused by thunder. Over a longer period, extending from 1978 to 1983, the electric field detector on the *Pioneer Venus Orbiter* detected radio bursts which could be emitted by lightning. The sources of these radio bursts are clustered around the Beta and Alta regions, the same areas having volcanic topography. A plausible interpretation is that lightning strokes arise in hot plumes of volcanic gases, a common phenomenon on Earth.

Surface rocks. Pictures and chemical analyses of Venus rocks have been obtained from measurements that were transmitted by the various successful *Venera* landers. The most recent of these are from the *Venera 13* and *14* landers, whose pictures show a flat plain of broken platelike rocks. This panorama suggests a volcanic outflow plain, composed of several overlying thin volcanic lava flows that have been subsequently eroded and segmented. This interpretation is not unique, however.

Strong evidence for recent volcanic activity comes from the Soviet chemical measurements of the surface. The amounts of calcium and sulfur in the surface are not in chemical equilibrium with the amount of sulfur gases (mostly sulfur dioxide) measured in the atmosphere. To be in balance, sulfur in the crust and the atmosphere should combine with the calcium in the rocks to form calcium sulfate. The fact that the surface has not achieved chemical equilibrium with the atmosphere shows that not enough time has passed for this chemical reaction to proceed to equilibrium. Either the rocks or the sulfur gases in the atmosphere must be geologically re-

cent additions. Volcanic activity is a natural mechanism for both exposing new rock and releasing gases into the atmosphere.

Observations of clouds. The evidence described above only indicates that volcanoes have occurred on Venus in the geologically recent past, probably in the last million years. Direct evidence that volcanoes are erupting now comes from recent observations of the Venus clouds made by the *Pioneer Venus Orbiter*. On Earth, major eruptions (like that of El Chichon in 1982) inject haze particles and sulfur gases into the stratosphere, where they may stay for months or years. On Venus, a similar phenomenon has been observed. In 1978, the amount of sulfur dioxide gas above the Venus clouds was 10–100 times more than upper limits set by observations of the previous decade. Further, a haze of small particles covered the entire planet, and especially the poles, making them unusually bright. Both these occurrences were noted by the *Pioneer Venus* investigators, but the connection with volcanoes was not made.

Because the *Pioneer Venus Orbiter* has continued to orbit the planet since 1978, it has been able to monitor these constituents above the clouds. A remarkable discovery was made: both the sulfuric acid and the haze have been gradually disappearing, until only about 10% of the 1978 amounts remain (see illus.). The amounts of these two constituents are now approaching more normal, pre-1978 values. Old Earth-based data show that a similar phenomenon may have occurred in the late 1950s. An obvious explanation of these recurrent events is that occasionally major volcanic eruptions inject sulfur dioxide gas above the clouds. The haze is particles of sulfuric acid created by the action of sunlight on the sulfur dioxide (what is called acid rain on the Earth). These haze particles gradually fall out of the atmosphere, leaving the more normal situation between eruptions. Calculations show that a volcanic explosion sufficient to create these effects must have been larger than the Krakatoa eruption of 1882.

A volcanic eruption has still not been seen on Venus because the clouds continue to block the view. Further, no single piece of evidence is entirely definitive in proving the existence of current volcanic activity. Nonetheless, the evidence from geology, lightning, surface chemistry, and remote observations of sulfur dioxide taken together is very persuasive. All this seems to point to the realization that Venus, like the Earth (but unlike the Moon and Mars), is still geologically active. Heat, molten lava, and volcanic gases continue to escape from the Venus interior, sometimes in an explosion so large that its effects can be seen above the planet's clouds.

For background information *see* PLANETARY PHYSICS; PLATE TECTONICS; VENUS; VOLCANOLOGY in the McGraw-Hill Encyclopedia of Science and Technology.

[L. W. ESPOSITO]

Bibliography: L. W. Esposito, Sulfur dioxide: Ep-

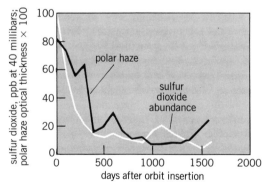

Consistent decline of both sulfur dioxide (SO₂) and polar haze since *Pioneer Venus Orbiter* insertion on December 4, 1978. Data on polar haze optical thickness are from the *Pioneer Venus Orbiter* cloud photopolarimeter (CPP). Data on sulfur dioxide abundance in parts per billion, inferred at an atmospheric pressure level of 40 millibars or 4 kilopascals (approximately 44 mi or 70 km altitude), are from the *Pioneer Venus Orbiter* ultraviolet spectrometer (UVS).

isodic injection shows evidence for active Venus volcanism, *Science*, 223:1072–1074, 1984; D. M. Hunten et al. (eds.), *Venus*, 1983; R. G. Prinn, The volcanoes and clouds of Venus, *Sci. Amer.*, 252(3):46–53, March 1985.

Weather forecasting and prediction

Technological advances over the past four decades have dramatically altered the way that weather forecasts are produced. Forecasts that were once based on subjective interpretation of hand-drawn weather maps are now produced by large computers fed with global data sets and lengthy equations. While forecasting is far from being an exact science, considerable progress has been made, but at a pace perhaps too slow to be perceived by the general public. In fact, by the middle 1970s, 48-h forecasts of winds and pressures aloft were as accurate as 24-h forecasts were in the 1950s. Progress has been slower since then, but as the 1990s approach, meteorologists are cautiously optimistic that a more rapid increase in accuracy is possible as significant improvements are anticipated in many of the components that go into making a forecast.

Mathematical models. Weather forecasts are produced by mathematical models of the atmosphere that are solved numerically by high-speed computers. Although many different types of models exist, there are basic features that are common to most of them. The atmosphere may be considered as a stack of thin layers that envelop the Earth. Most weather is generated in the lowest layer, the troposphere, which is 6–12 mi (10–20 km) thick. In three-dimensional models of the atmosphere, the troposphere is subdivided into thousands of smaller volumes, roughly 120 × 120 mi (200 × 200 km) in the horizontal direction, 0.6 mi (1 km) in the vertical. The size of these boxes is governed by two considerations. First, observations of the atmosphere above the surface are made 120–240 mi (200–400 km) apart. Second, the amount of data needed to describe atmospheric conditions in all these boxes strains even the enormous capabilities of today's supercomputers. For example, the number of 120 × 120 × 0.6 mi (200 × 200 × 1 km) boxes needed to describe the atmosphere to a depth of 9 mi (15 km) is over 191,250. Since 10–20 numbers are required to describe the atmosphere in each box, 2–4 million values must be stored to begin a global forecast. Better model resolution (for example, 30–60 mi or 50–100 km spacing) is possible with present computers only if smaller regional models are used. Even these models strain present computer power because the finer resolution requires shorter time intervals between forecast steps; that is, a weather feature traversing smaller boxes has to be looked at more often to describe it properly. Thus computer forecasts are stepped forward only 5–10 min at a time. Since over 600 billion calculations may be needed for a 24-h forecast, the grid spacing must be chosen large enough so that the prediction is completed in time to disseminate to the local forecaster.

Once the dimensions of the model are set, forecasts are made by solving equations governing the local change of wind velocity, temperature, pressure, and moisture. The physical principles on which these equations are based are well known: Newton's second law of motion (accelerations are due to the sum of the forces acting on air parcels); first law of thermodynamics (temperature changes are due to work done on a parcel plus external sources or sinks of heat); and conservation of mass and moisture. These laws are formulated to be appropriate for atmospheric motion, and a numerical procedure is developed to solve them on digital computers. One of the persistent problems in atmospheric modeling is the difficulty in formulating the physical processes that affect the atmosphere on scales less than 60–120 mi (100–200 km), such as cumulus convection and local topographic features, without adding more unknown variables to the model.

Computer-produced forecasts for the United States are made by regional models twice a day by the National Meteorological Center (NMC) for use by the National Weather Service (NWS) forecasters. In addition, global forecasts are produced by spectral models, in which each meteorological field is decomposed into 40 wave components of different wavelength, phase, and amplitude. Predictions are made for each component, and the waves are summed to provide the forecast fields. These models are stepped forward (integrated) for up to 10 days to provide medium-range (3–10 days) guidance. Long-range forecasts (greater than a month) are not produced by deterministic forecast models but by empirical methods based on statistical relationships between past and present climatic conditions. At the other end of the scale, extreme weather events such as severe thunderstorms, tornadoes, and flash floods cannot be forecast explicitly due to their small dimensions, but conditions likely for their occurrence can be assessed from computer forecasts of the larger-scale flow.

NMC forecasts. Typical NMC forecasts are produced in the following sequence: (1) Wind, pressure, temperature, and moisture observations at and above the surface over the entire globe are made, transmitted over the Global Telecommunication System (which connects all United Nations countries), organized, and checked for errors. The data come primarily from rawinsondes (radio-tracked balloon-borne instruments), satellites, commercial aircraft, and surface stations, ships, and buoys. This process begins at 0000 and 1200 GMT and takes 1–2 h. (2) The data are then interpolated to the centers of the grid volumes (objective analysis) and adjusted somewhat to remove unwanted noise (initialization). (3) The terms in the forecast equations are calculated by using the initial data, yielding local tendencies of wind, pressure, temperature, and moisture. Once the tendencies are evaluated (integrated) for each box, the resulting new values are used to calculate the terms in the equations again. This process is repeated until the forecast is completed. The predictions are usually available 2–4 h after the observation time. (4) The forecasts are transmitted to the local National Weather Service (NWS) offices in numerical and graphical form. Since the model predictions are for wind and pressure patterns above the surface, meteorologists interpret the model output in terms of local weather conditions. Statistical relationships between model output and quantities not predicted by the model (for example, maximum and minimum temperatures, and precipitation probabilities) are also evaluated to aid the forecaster.

Forecast quality. The skill of the computer forecasts is closely monitored by verifying the predicted fields against the observed fields. Figure 1 provides a general picture of present forecast quality compared to what is theoretically possible. The quantity being verified is the height of the 500-millibar-pressure (50-kilopascal) surface, which fluctuates about an average height near 3.5 mi (5.5 km) above the surface. The curved, shaded region represents current skill level (as measured by the root-mean-square error) and shows the forecast error getting worse with forecast length. The horizonal broken line is the error if climatological values are used as the forecast for each day. The solid curve labeled "persistence" represents the error if the initial conditions are used for all the forecasts. Since the latter two "forecasts" can be made without a model, a good model should improve on persistence in the short term and remain better than climatology over the long term. The figure shows that current models remain better than climatology for 5–7 days but that there is much room for improvement toward what is theoretically possible (lower curve). Figure 1 applies to upper air patterns which may be forecast with some skill up to a week, but the skill of precipitation forecasts, for example, is much less.

The NWS has been evaluating its forecast skill

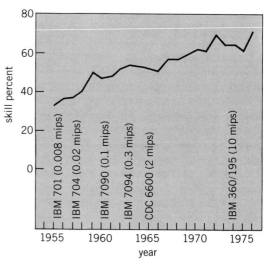

Fig. 2.   Record of annually averaged skill for National Meteorological Center (NMC) 36-h 500-millibar-height (50-kPa) forecasts over North America. 100% skill represents a nearly perfect forecast. The computers used at the NMC to make the forecasts are identified along the horizontal scale along with their approximate capacities in millions of instructions per second (mips). (*After F. G. Shuman, Numerical weather prediction, Bull. Amer. Meteorol. Soc., 59:9–17, 1978*)

since the 1950s, and steady, if uneven, progress has been made. Figure 2 represents another measure of skill for 36-h forecasts of the 500-millibar (50-kPa) heights for the 1955–1976 period. Most of the upswings in the curve are due to improved models which were used as greater computer power became available. Since 1976 the rate of improvement has been less, but may begin to increase again as a result of new forecast models that began running in 1985, made possible by a supercomputer.

Improvement in forecasting the weather elements that directly affect people has been slower. Forecasts of whether it will rain or not have improved, but little progress has been made in predicting the actual amounts of rain, especially in summer when the horizontal extent of the heavy precipitation area is smaller. Local temperature forecasts have improved very little over the past decade, but large temperature errors are much rarer. A significant improvement has occurred in recent years in issuing timely watches and warnings for severe weather, but most of the credit is due to improved sensors such as radar and satellites and faster communication rather than to the numerical models.

Sources of error. As has been noted above, weather prediction is not an exact science; analysis of sources of uncertainty remains incomplete. Sources of errors that cause forecasts to deteriorate can be identified in three categories: errors in the initial state, errors in the numerical models, and the inherent limit to atmospheric predictability.

Initial state. These errors include observational errors and errors introduced by the analysis and initialization stages. However, the most serious problem is the absence of data in two respects. First,

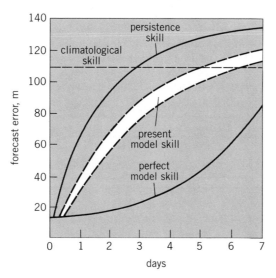

Fig. 1.   Comparison of actual and theoretical forecast errors as a function of time for numerical weather prediction. The quantity being verified is the height of the 500-millibar-pressure (50-kPa) surface. 1 m = 3.3 ft.

land-based observations are not taken over 80% of the globe (due to oceanic, arctic, desert, and mountainous regions), and second, even over the well-observed areas the data density is insufficient to resolve the details of important small-scale (240 mi or 400 km or less) phenomena. For example, when a squall line 60 mi (100 km) wide is traversing the central United States during an observing period, not enough information about it can be obtained from a 240-mi-spaced (200-km) observing network to allow successful prediction for even a few hours. In short, forecasts cannot be made in the absence of proper observations.

Numerical models. There are several possible sources of error. The mathematical approximations needed to solve the model equations introduce a significant error, primarily due to the limited horizontal and vertical resolution of the models (truncation error). The equations themselves may not be capable of describing atmospheric motion exactly. The model formulation of physical processes such as solar heating and convection may be crude. The lower boundary (surface) values of soil moisture, ocean temperature, terrain roughness, and so on, are not always known or included. Errors due to lateral boundary conditions arise in regional models.

Inherent limit to atmospheric predictability. As Fig. 1 suggests, even a perfect model could not produce a good forecast longer than 2 weeks. This arises from the unavoidable errors that exist in the observations which grow with time during the forecast period. In fact, even with perfect observations (no measurement error), the model and true atmosphere differ initially because of the energy in the weather features not sampled by the observing systems. For example, if a thunderstorm exists in nature but is not included in the initial state of the forecast model (which cannot resolve such a feature), the energy transfer from that storm to the larger scales will not be simulated in the model as it occurs in nature. Thus the model atmosphere will differ from the actual one, and this difference will grow with time, rendering a perfect model useless after 2 weeks.

Future prospects. Although the inevitability of the 2–3-week limit to numerical forecasts may appear frustrating, state-of-the-art forecasting is far from this limit (Fig. 1). In fact, meteorologists are in a position to make a significant advance over the next decade. This will be accomplished by: (1) advances in observing technology, allowing more frequent, dense, complete, and accurate observations; (2) advances in understanding of small-scale phenomena, leading to more accurate model formulation of physical processes; and (3) increased computer power, enabling modelers to take advantage of the advances that have been made in meteorology.

Changes in observing systems will be the most dramatic. Satellite-based observations provide global coverage now, but their accuracy is inferior to rawinsondes and they suffer from poor vertical resolution. New microwave and laser-based remote

sounding techniques from satellite-borne instruments should greatly alleviate these problems. The rawinsonde network may also be augmented by vertically looking Doppler-based wind profilers, to provide continuous upper-air wind observations. Improved temperature and moisture sounding technology is also being developed. For smaller phenomena, a new network of Doppler radars (NEXRAD) will provide high-resolution (0.6 mi or 1 km) information on storm wind structure and precipitation rates. High-speed satellite-based communication systems will ensure that the data reach to (and are relayed from) the modeling centers more quickly.

Advances in knowledge of atmospheric behavior are more difficult to anticipate. Certainly the wealth of new data from the above-mentioned observing systems will increase meteorologists' understanding of the structure and evolution of small-scale weather events. With additional knowledge, modelers can test new formulations of the important physical processes and, in a slow but accumulating manner, increase the reality of the numerical models.

The largest improvement to the forecasts will probably come from increasing the resolution of the models. The new data sets will allow analyses of 60-mi-scale (100-km) phenomena or less. Then, if the simulation of the physical processes has also improved, the model grid resolution can be lowered to 15–30 mi (25–50 km). The data density alone does not justify this, but if the larger-scale flow is analyzed and forecast well, the higher resolution will allow small-scale phenomena to develop in the model from internal instabilities and external forcing. Only a very large increase in computing power will make this possible, and this appears to be forthcoming. Today's supercomputers can operate at 200–400 millions of instructions per second (mips). At this rate, current models can complete their forecasts in 20–60 min. The new models discussed above will do 10–50 times more calculations, but computers are being built that will operate at 10,000 mips by late 1986 and 1 million mips by the middle 1990s. Sophisticated color graphics and interactive image processing techniques will allow both the forecasters and the public to see clearly the current and forecast weather patterns. The challenge to meteorologists is to harness these prodigious tools to make a significant impact on future weather forecasts.

For background information *see* COMPUTER; DATA COMMUNICATIONS; METEOROLOGICAL INSTRUMENTATION; METEOROLOGICAL SATELLITES; WEATHER FORECASTING AND PREDICTION in the McGraw-Hill Encyclopedia of Science and Technology.

[FREDERICK H. CARR]

Bibliography: K. A. Browning (ed.), *Nowcasting*, 1982; G. J. Haltiner and R. T. Williams, *Numerical Prediction and Dynamic Meteorology*, 1980; R. A. Kerr, Pity the poor weatherman, *Science*, 228:704–706, 1985; F. G. Shuman, Numerical weather prediction, *Bull. Amer. Meteorol. Soc.*, 59:5–17, 1978.

McGRAW-HILL YEARBOOK OF SCIENCE AND TECHNOLOGY

List of Contributors

List of Contributors

A

Abeles, Dr. Benjamin. Senior Research Associate, Exxon Research and Engineering Company, Corporate Research Science Laboratories, Annandale, New Jersey. AMORPHOUS SOLID.

Abramson, Prof. Nils. Royal Institute of Technology, Department of Production Engineering, Industrial Metrology, Stockholm, Sweden. HOLOGRAPHY.

Adelman, Dr. William J., Jr. Laboratory of Biophysics, NINCDS, National Institutes of Health, Department of Health and Human Services at the Marine Biological Laboratory, Woods Hole, Massachusetts. BIOPOTENTIALS AND IONIC CURRENTS.

Agabian, Dr. Nina. Acting Scientific Director, Naval Biosciences Laboratory, University of California, Berkeley, Naval Supply Center. MEDICAL PARASITOLOGY—coauthored.

Ahmadjian, Dr. Vernon. Department of Biology, Clark University, Worcester, Massachusetts. LICHENS—in part.

Andrews, Dr. Robert S. Senior Operations Officer, Deep Observation and Sampling of the Earth's Continental Crust, Inc., Washington, D.C. BORING AND DRILLING (MINERAL).

Angel, Dr. Roger. University of Arizona, Steward Observatory. OPTICAL TELESCOPE.

Appelquist, Prof. Thomas. Professor of Physics, Yale University. ADDITIONAL DIMENSIONS OF SPACE-TIME—feature.

B

Baird, Dr. Curtis J. Assistant Professor of Anatomy, Loma Linda University, School of Medicine. ENDOCRINE MECHANISMS.

Baltay, Prof. Charles. Columbia University in the City of New York, Department of Physics, Nevis Laboratories, Irvington, New York. PARTICLE DETECTOR—in part.

Barbacci, Prof. Mario R. Associate Director, Project Engineering, Software Engineering Institute, Carnegie-Mellon University, Pittsburgh. COMPUTER SOFTWARE.

Barber, Prof. Stanley A. Professor of Agronomy, Purdue University. PLANT MINERAL NUTRITION—in part.

Bardsley, Prof. J. N. Physics Department, University of Pittsburgh. POSITRONIUM.

Bartholomew, Dr. M. J. Division of Geology and Mineral Resources, Montana Bureau of Mines and Geology, Butte. MASSIF.

Berry, Dr. William B. N. Museum of Paleontology, University of California, Berkeley. STRATIGRAPHY—in part.

Betzig, Dr. Eric. School of Applied and Engineering Physics, Cornell University. OPTICAL MICROSCOPE—coauthored.

Bickford, Dr. M. E. Department of Geology, University of Kansas. BASEMENT ROCK.

Bodey, Dr. Gerald P. Professor of Medicine, and Chief, Section of Infectious Diseases, Department of Internal Medicine, University of Texas System Cancer Center, M.D. Anderson Hospital and Tumor Institute, Texas Medical Center, Houston. PSEUDOMONAS AERUGINOSA.

Boyd, Dr. Douglas P. Department of Radiology, University of California, San Francisco. COMPUTERIZED TOMOGRAPHY.

Brassell, Dr. Simon C. Royal Society Research Fellow, University of Bristol, School of Chemistry, Bristol, England. ORGANIC GEOCHEMISTRY.

Brau, Dr. Charles A. Los Alamos National Laboratory, Los Alamos, New Mexico. LASER—in part.

Bray, Dr. Elizabeth A. Department of Botany and Plant Sciences, University of California, Riverside. ABSCISIC ACID.

Bruck, Dr. David K. Department of Biology, University of California, Los Angeles. PLANT TISSUE SYSTEMS.

Buhyoff, Dr. Gregory J. Professor of Forest Biometrics, School of Forestry and Wildlife Resources, College of Agriculture and Life Sciences, Virginia Polytechnic Institute and State University. FOREST MANAGEMENT AND ORGANIZATION—in part.

C

Calhoun, Dr. Frank G. Texas A&M University, Department of Soil and Crop Sciences, Texas Agricultural Experiment Station, College of Agriculture. SOIL—in part.

Calundann, Gordon W. Celanese Research Company, Celanese Corporation, Summit, New Jersey. LIQUID CRYSTALS—coauthored.

Carmer, Prof. Samuel G. Professor of Biometry, University of Illinois at Urbana-Champaign, College of Agriculture, Department of Agronomy. STATISTICS—in part.

Carr, Dr. Frederick H. Associate Professor, University of Oklahoma, School of Meteorology, Norman. WEATHER FORECASTING AND PREDICTION.

Carter, Dr. David L. Area Soil Scientist, USDA–Soil Conservation Service, Las Cruces, New Mexico. SOIL POTENTIALS—in part.

Castleman, Prof. A. Welford, Jr. Professor of Chemistry, Pennsylvania State University, Dover Laboratory. ATOM CLUSTERS.

Chakoumakos, Dr. Bryan C. University of New Mexico, Department of Geology. PYROCHLORE.

Chamberlain, Peter G. U.S. Department of the Interior, Bureau of Mines, Washington, D.C. SOLUTION MINING.

Checkai, Dr. Ronald T. Research Associate, U.S. Plant, Soil and Nutrition Laboratory, Ithaca, New York. HYDROPONICS.

Christie-Blick, Dr. Nicholas. Department of Geological Sciences and Lamont-Doherty Geological Observatory of Columbia University, Palisades, New York. SEA-LEVEL FLUCTUATIONS—in part.

Chung, Dr. Tai-Shung. Celanese Research Company, Celanese Corporation, Summit, New Jersey. LIQUID CRYSTALS—coauthored.

Cizewski, Dr. Jolie A. Yale University, A. W. Wright Nuclear Structure Laboratory. NUCLEAR STRUCTURE.

Cohen, Jules. President, Jules Cohen & Associates, P.C., Consulting Electronics Engineers, Washington, D.C. TELEVISION—in part.

Coleman, Dr. David C. Research Professor, Department of Entomology, University of Georgia. UNDERGROUND BIOLOGY—feature, coauthored.

Cooney, Timothy M. President, Forest Resources Systems Institute, Florence, Alabama. FOREST MANAGEMENT AND ORGANIZATION—in part.

Cooper, Dr. Neil R. Department of Immunology, Scripps Clinic and Research Foundation, Research Institute of Scripps Clinic, La Jolla, California. CELLULAR IMMUNOLOGY—in part.

Corney, John V. Sir Alexander Gibb & Partners, Consulting Engineers, Earley, Reading, England. HYDRO-ELECTRICITY.

Crasemann, Prof. Bernd. Director, Chemical Physics Institute, University of Oregon. ATOMIC STRUCTURE AND SPECTRA.

Creutz, Dr. Michael. Brookhaven National Laboratory, Department of Physics, Associated Universities, Inc., Upton, New York. ELEMENTARY PARTICLE—in part.

Croat, Dr. John J. Assistant Engineering Manager, Delco Remy, General Motors Corporation, Anderson, Indiana. MAGNETIC MATERIALS—in part.

Crossey, Dr. Laura J. Department of Geology/Geophysics, University of Wyoming. DIAGENESIS.

Crowley, Bucky. Vice President, Lincoln Technology, Inc., Needham, Massachusetts. PRINTING.

Croxdale, Dr. Judith G. Associate Professor of Botany, University of Wisconsin—Madison, Department of Botany. APICAL MERISTEM.

Cruickshank, Dr. Michael J. U.S. Department of the Interior, Geological Survey, Office of Energy and Marine Geology, Reston, Virginia. MARINE MINING.

Curran, Patrick. Allied Corporation, Metglas Products Department, Parsippany, New Jersey. METALLIC GLASSES—coauthored.

D

David, Dr. Leonard W. Director of Research, National Commission on Space, Washington, D.C. SPACE FLIGHT.

Dawson, Dr. Gaynor W. Vice President, ICF Northwest, ICF Technology Inc., Richland, Washington. LAND RECLAMATION.

DeCristofaro, Dr. Nicholas. Allied Corporation, Metglas Products Department, Parsippany, New Jersey. METALLIC GLASSES—coauthored.

Demuth, Dr. Joseph E. International Business Machines Corporation, Thomas J. Watson Research Center, Yorktown Heights, New York. SURFACE SPECTROSCOPY.

Dilcher, Prof. David. Professor of Paleobotany, Department of Biology and Department of Geology, Indiana University. PALEOBOTANY—in part.

Duffy, Dr. Frank Hopkins. Associate Professor of Neurology, Harvard Medical School, Director of Developmental Neurophysiology, Children's Hospital, Boston. ELECTROENCEPHALOGRAPHY.

Dugan, Patrick R. Dean, Ohio State University, College of Biological Sciences. MICROBIAL GEOCHEMISTRY—in part.

Dyck, Dr. H. Melvin. University of Hawaii at Manoa, Institute for Astronomy, Honolulu. STELLAR EVOLUTION.

Dyer, Dr. Betsey Dexter. Assistant Professor of Biology, Wheaton College, Norton, Massachusetts. CELL (BIOLOGY).

E

Erb, Dr. Karl A. Oak Ridge National Laboratory, operated by Martin Marietta Energy Systems, Inc., Oak Ridge, Tennessee. NUCLEAR PHYSICS—in part.

Esposito, Dr. L. W. University of Colorado, Boulder, Laboratory for Atmospheric and Space Physics. VENUS.

F

Fermanian, Dr. Thomas W. Extension Turfgrass Specialist, University of Illinois at Urbana-Champaign, College of Agriculture, Cooperative Extension Service. AGRONOMY.

Forrest, Prof. S. R. Department of Electrophysics, University of Southern California, Los Angeles. OPTICAL DETECTORS.

Forward, Dr. Richard B., Jr. Associate Professor of Zoology, Duke University Marine Laboratory, Pivers Island, Beaufort, North Carolina. CHAETOGNATHA—coauthored.

Frewin, K. Engineering Manager, GEC Distribution Transformers Limited, member of the GEC Power Engineering Group, Broadstairs, Kent, England. TRANSFORMER.

Friedmann, Dr. Paul G. Senior Consultant, Allied Corporation, Corporate Technology, Morristown, New Jersey. FOOD MANUFACTURING—coauthored.

G

Gallagher, Dr. T. J. Department of Electrical Engineering, University College, Dublin, Ireland. FLUIDS.

Gardner, Dr. James V. U.S. Department of the Interior, Geological Survey, Branch of Pacific Marine Geology, Menlo Park, California. SONAR.

Gillman, Dr. G. P. University of Georgia College of Agriculture, Agronomy Division, Department of Agronomy. SOIL CHEMISTRY—in part.

Goetz, Dr. Alexander F. H. Jet Propulsion Laboratory, California Institute of Technology, and Department of Geological Sciences, University of Colorado, Boulder. REMOTE SENSING—in part.

Graham, Dr. Robert A. Distinguished Member of the Technical Staff, Division 1131, Sandia National Laboratories, Albuquerque, New Mexico. SHOCK-WAVE CHEMISTRY.

Grantz, Dr. David A. U.S. Department of Agriculture, Agricultural Research Service, Sugarcane Research Laboratory, Experiment Station HSPA, Aiea, Hawaii. PHYSIOLOGICAL ECOLOGY (PLANT).

Greiner, Prof. Dr. Walter. Institut für Theoretische Physik der Universität Frankfurt/M., Frankfurt, West Germany. NUCLEAR PHYSICS—in part.

H

Haim, Prof. Albert. Professor of Chemistry, State University of New York at Stony Brook. CHEMICAL DYNAMICS.

Hale, Dr. Mason E. Department of Botany, National Museum of Natural History, Smithsonian Institution. LICHENS—in part.

Harootunian, Alec. School of Applied and Engineering Physics, Cornell University. OPTICAL MICROSCOPE—coauthored.

Harris, Prof. Richard W. Department of Environmental Horticulture, University of California, Davis. ARBORICULTURE.

Hartley, Dr. Daniel L. Director, Combustion and Applied Research, Sandia National Laboratories, Livermore, California. COMBUSTION CHEMISTRY AND PHYSICS—feature.

Hatcher, Prof. Robert D., Jr. University of South Carolina, Department of Geology, Tectonics Program, Columbia. CONTINENTAL DRILLING—feature.

Hawley, Dr. Bronson W. University of Utah, Department of Geology and Geophysics, College of Mines and Mineral Industries. STRUCTURAL GEOLOGY.

Hay, Dr. William W. Director, Museum, University of Colorado, Boulder. PALEOCLIMATOLOGY.

Hayes, William C. "Electrical World," McGraw-Hill Publications Company, New York. ELECTRICAL UTILITY INDUSTRY.

Heffner, Dr. Robert H. P-10 Group Leader, Los Alamos National Laboratory, Physics Division, Los Alamos, New Mexico. MUON SPIN RELAXATION.

Heinrich, Prof. Bernd. Professor of Zoology, University of Vermont. THERMOREGULATION.

Hench, Prof. Larry L. Professor of Materials Science and Engineering, University of Florida, Gainesville. GLASS AND GLASS PRODUCTS.

Hille, Dr. Merrill B. Department of Zoology, University of Washington, Seattle. DEVELOPMENTAL BIOLOGY.

Hunt, Dr. John M. Woods Hole Oceanographic Institution, Woods Hole, Massachusetts. PETROLEUM.

I

Illing, Dr. Rainer M. E. National Center of Atmospheric Research, High Altitude Observatory, Boulder, Colorado. SATELLITE ASTRONOMY.

Isaacson, Dr. Michael I. School of Applied and Engineering Physics, Cornell University. OPTICAL MICROSCOPE—coauthored.

J

Jansky, Dr. Donald M. President, Jansky Telecommunications, Inc., Washington, D.C. COMMUNICATIONS SATELLITE—in part.

Jeffery, Prof. William R. Marine Biological Laboratory, Woods Hole, Massachusetts. CELL DIFFERENTIATION.

Jenkins, Dr. David M. Assistant Professor of Geology, University Center at Binghamton, State University of New York, Department of Geological Sciences and Environmental Studies. TREMOLITE.

Jones, Prof. T. B. University of Leicester, Department of Physics, Leicester, England. IONOSPHERE.

K

Kalfayan, Dr. Laura. University of North Carolina at Chapel Hill, School of Medicine, Department of Biochemistry and Nutrition. GENE.

Kamprath, Prof. Eugene J. North Carolina State University, School of Agriculture and Life Sciences, Academic Affairs, Extension and Research, Department of Soil Science. SOIL CHEMISTRY—in part.

Kapetanakos, Dr. Christos A. Head, Adv. Beam Tech. Br., Department of the Navy, Naval Research Laboratory, Washington, D.C. PARTICLE ACCELERATOR.

Katz, Dr. Joseph. Jet Propulsion Laboratory, California Institute of Technology, Pasadena. LASER—in part.

Keck, William J. COMSAT General Corporation, El Segundo, California. COMMUNICATIONS SATELLITE—in part.

Keller, Dr. O. L., Jr. Director, Transuranium Research Laboratory, Oak Ridge National Laboratory, operated by Martin Marietta Energy Systems, Inc., Oak Ridge, Tennessee. TRANSURANIUM ELEMENTS.

Kelly, Dr. Michael T. Professor of Pathology, Vancouver General Hospital, University of British Columbia. AUTOMATED SYSTEMS IN MICROBIOLOGY—feature.

King, Dr. T. A. Atomic, Molecular and Polymer Physics Group, Schuster Laboratory, University of Manchester, England. LASER PHOTOBIOLOGY—in part.

Kozlowski, Dr. T. T. Director of the Biotron and WARF Senior Distinguished Research Professor of Forestry, University of Wisconsin—Madison. PLANT-WATER RELATIONS.

Krauss, Dr. Lawrence M. Department of Physics, J. W. Gibbs Laboratory, Yale University. PARTICLE DETECTOR—in part.

Krishnan, Prof. Mahadevan. Yale University, Applied Physics, Becton Center. ISOTOPE (STABLE) SEPARATION.

Kronenberg, Dr. Andreas K. U.S. Department of the Interior, Geological Survey, Menlo Park, California. QUARTZ.

Kullerud, Prof. Gunnar. Professor of Geosciences, Purdue University. SULFIDE PHASE EQUILIBRIA.

Kuzmanoff, Dr. Konrad. Stanford University, Department of Biological Sciences. PLANT PHYSIOLOGY—in part.

L

LaBarbera, Dr. Michael. Associate Professor, University of Chicago, Department of Anatomy. ECOLOGICAL INTERACTIONS.

Lechtenberg, Dr. V. L. Associate Director, Agricultural Research, Purdue University. STATISTICS—in part.

Leupold, Mathias J. Francis Bitter National Magnet Laboratory, Massachusetts Institute of Technology. MAGNETIC FIELD—coauthored.

Levine, Dr. Joel S. Senior Research Scientist, Atmospheric Sciences Division, NASA Langley Research Center, Hampton, Virginia. PRECAMBRIAN.

Lewis, Prof. Aaron. School of Applied and Engineering Physics, Cornell University. OPTICAL MICROSCOPE—coauthored.

Linsky, Dr. Jeffrey L. Joint Institute for Laboratory Astrophysics, University of Colorado, National Bureau of Standards, Boulder. STAR.

Liou, Prof. J. G. Stanford University, School of Earth Sciences, Department of Geology. FACIES (GEOLOGY).

Lippard, Prof. Stephen J. Professor of Chemistry, Massachusetts Institute of Technology. CHEMOTHERAPY.

Liu, Prof. C. H. University of Illinois at Urbana-Champaign, Department of Electrical and Computer Engineering. METEOROLOGICAL RADAR.

Liu, Dr. C. T. Oak Ridge National Laboratory, Oak Ridge, Tennessee. ALLOY—coauthored.

London, Dr. David. Assistant Professor, School of Geology and Geophysics, University of Oklahoma. PEGMATITE.

M

Ma, Prof. Yi Hua. Professor and Head, Chemical Engineering, Worcester Polytechnic Institute, Worcester, Massachusetts. FOOD ENGINEERING.

Maas, Eugene V. Research Leader, U.S. Department of Agriculture, Agricultural Research Service, Pacific Basin Area, U.S. Salinity Laboratory, Riverside, California. PLANT MINERAL NUTRITION—in part.

Macdonald, Dr. Eoin H. Consulting Mining Engineer, Seaforth, N.S.W., Australia. PLACER MINING.

McGinnis, Dr. William J. Department of Molecular Biophysics and Biochemistry, Yale University. HOMEO BOX.

McMurry, Dr. Thomas J. Postdoctoral Fellow, University of California, Berkeley, Department of Chemistry. CHELATION—coauthored.

Madaras, Dr. Ronald J. Lawrence Berkeley Laboratory, University of California, Berkeley. PARTICLE DETECTOR—in part.

Mak, Dr. Tak W. Senior Staff Scientist, Ontario Cancer Institute, and Professor, Department of Medical Biophysics, University of Toronto. IMMUNOGENETICS.

Mason, Dr. John O. Medical Research Council, Laboratory of Molecular Biology, Cambridge, England. MONOCLONAL ANTIBODIES—coauthored.

Matthews, Prof. Robley K. Brown University, Department of Geological Sciences. SEA-LEVEL FLUCTUATIONS—in part.

Metcalf, Dr. Harold. Department of Physics, State University of New York at Stony Brook. ATOMIC BEAMS.

Mokhoff, Dr. Nicolas. Senior Editor, "Computer Design," PennWell Publications, Mineola, New York. COMPUTER NETWORKING.

Molinari, Dr. Robert L. U.S. Department of Commerce, National Oceanic and Atmospheric Administration, Atlantic Oceanographic and Meteorological Laboratory, Miami. OCEAN-ATMOSPHERE RELATIONS.

Monforte, Dr. John. School of Music, Music Engineering, University of Miami, Coral Gables, Florida. SOUND-REPRODUCING SYSTEMS.

Moore, Dr. Gordon P. Assistant Professor of Biological Sciences, University of Michigan, and Group Leader of Molecular Biology, New Technology Research, E. I. Du Pont and Company, Inc., North Bellerica, Massachusetts. GENETICS.

Moore, Dr. Gregory F. Associate Professor, University of Tulsa, College of Engineering and Applied Sciences, Department of Geosciences. MARINE SEDIMENTS—coauthored.

Morgan, Dr. M. Granger. Head, Department of Engineering and Public Policy, and Professor, Engineering and Public Policy/Electrical and Computer Engineering, Carnegie-Mellon University. ELECTROMAGNETIC FIELD.

Morris, Dr. Mark R. Department of Astronomy, University of California, Los Angeles. GALAXY.

Morrison, Dr. Larry E. Standard Oil Company (Indiana), Standard Oil Research Center, Naperville, Illinois. IMMUNODIAGNOSTICS.

Mount, Henry R. Assistant State Soil Scientist, USDA, Soil Conservation Service, Durham, New Hampshire. SOIL POTENTIALS—in part.

N

Nei, Prof. Masatoshi. Professor of Population Genetics, University of Texas, Health Science Center at Houston, Graduate School of Biomedical Sciences, Genetics Centers, Houston. HUMAN GENETICS.

O

O'Dor, Ron K. Department of Biology, Dalhousie University, Halifax, Nova Scotia. SQUID.

Odum, Prof. Howard T. Graduate Research Professor, Department of Environmental Engineering Sciences, University of Florida, Gainesville. HUMAN ECOLOGY.

Ormrod, Prof. Douglas P. University of Guelph, Ontario Agricultural College, Department of Horticultural Science. AIR POLLUTION.

Overshott, Prof. K. J. Professor of Magnetics, Wolfson Centre for Magnetics Technology, Department of Electrical and Electronic Engineering, University College, Cardiff, Wales. MAGNETIC MATERIALS—in part.

Owen, Dr. Robert M. Oceanography Program, Department of Atmospheric and Oceanic Science, University of Michigan. CLIMATIC CHANGE—coauthored.

P

Parker, Dr. Charlotte. Associate Professor of Microbiology, University of Missouri—Columbia, School of Medicine, Department of Microbiology. BORDETELLA PERTUSSIS.

Paylor, Earnest D., II. Jet Propulsion Laboratory, Pasadena, California. REMOTE SENSING—in part.

Peirson, Dr. David R. Associate Professor, Wilfrid Laurier University, Department of Biology, Waterloo, Ontario. PLANT PHYSIOLOGY—in part.

Phillips, Dr. William D. Physicist, Electricity Division, Center for Basic Standards, U.S. Department of Commerce, National Bureau of Standards. ATOMIC BEAMS—coauthored.

Pilgrim, Dr. Sidney A. L. State Soil Scientist, U.S. Department of Agriculture, Soil Conservation Service, Durham, New Hampshire. SOIL POTENTIALS—in part.

Post, Prof. Daniel. Virginia Polytechnic Institute and State University, College of Engineering, Engineering Science and Mechanics. INTERFEROMETRY.

Powers, Dr. Kerns H. Staff Vice President, Communications Research, RCA Laboratories, David Sarnoff Research Center, Princeton. TELEVISION—in part.

Prasad, Ray. Senior Staff Engineer, Intel Corporation, Hillsboro, Oregon. CIRCUIT (ELECTRONICS).

Price, Prof. T. Douglas. Professor and Chairman, Department of Anthropology, University of Wisconsin—Madison. ARCHEOLOGY.

Prisbrey, Prof. Keith A. Professor of Metallurgical Engineering, University of Idaho, College of Mines and Earth Resources, Department of Metallurgical and Mining Engineering. GOLD METALLURGY.

Prusik, Dr. Thaddeus. Allied Corporation, Corporate Technology, Morristown, New Jersey. FOOD MANUFACTURING—coauthored.

Q

Quinn, Dr. T. J. Le sous-directeur, Bureau International des Poids et Mesures, Sévres, France. TEMPERATURE MEASUREMENT.

R

Raymond, Prof. Kenneth N. Professor of Chemistry, University of California, Berkeley. CHELATION—coauthored.

Rea, Dr. David K. Associate Professor, University of Michigan, Department of Atmospheric and Oceanic Science, Oceanography Program. CLIMATIC CHANGE—coauthored.

Reay, Dr. David A. International Research & Development Co., Ltd., Fossway, Newcastle-upon-Tyne, England. HEAT PIPES.

Redmond, Dr. D. Eugene, Jr. Yale University, Neurobehavioral Laboratory, School of Medicine. ANXIETY STATES—coauthored.

Revsbech, Dr. Niels Peter. Department of Ecology and Genetics, University of Aarhus, Denmark. MICROBIAL ECOLOGY.

Riker, William W. Executive Vice President, Society of Cable Television Engineers, Inc., West Chester, Pennsylvania. TELEVISION—in part.

Riopel, Dr. James L. Department of Biology, University of Virginia. PARASITIC PLANTS.

Rogers, John D. Los Alamos National Laboratory, Los Alamos, New Mexico. ENERGY STORAGE.

Rose, Dr. Elise. Research Associate, University of Wisconsin—Madison, Department of Horticulture. ETHYLENE.

Rose, Dr. Peter H. President, Semiconductor Equipment Operations, Eaton Corporation, Ion Beam Systems Division, Beverly, Massachusetts. TAILORED MATERIALS—feature.

Ross, Dr. Charles A. Chevron U.S.A. Inc., Southern Region, Exploration, Land & Production, Houston. STRATIGRAPHY—in part.

Ross, Dr. J. R. P. Chevron U.S.A. Inc., Southern Region, Exploration, Land & Production, Houston. STRATIGRAPHY—in part.

Ross, Dr. Nancy L. Department of Earth and Space Sciences, State University of New York, Stony Brook. HIGH-PRESSURE PHASE TRANSITIONS.

Rothwell, Prof. Gar W. Ohio University, Department of Botany. PALEOBOTANY—in part.

Ruddle, Dr. Nancy H. Associate Professor, Yale University, Department of Epidemiology and Public Health, School of Medicine. CANCER (MEDICINE).

Russell, Dr. Scott D. Assistant Professor of Botany, and Director, Electron Microscopy Laboratory, University of Oklahoma, Department of Botany and Microbiology. FERTILIZATION.

S

St. John, Dr. Theodore V. Laboratory of Biomedical and Environmental Biology, University of California, Los Angeles. UNDERGROUND BIOLOGY—feature, coauthored.

Sander, Prof. Leonard. Professor of Physics, University of Michigan, Harrison M. Randall Laboratory of Physics. FRACTALS.

Savitsky, Dr. Daniel. Director, Davidson Laboratory, Stevens Institute of Technology, Hoboken, New Jersey. SHIP DESIGN.

Schrader, Prof. Hans. National Science Foundation, Submarine Geology and Geophysics, Division of Ocean Sciences, Washington, D.C. VARVE.

Schwartz, Dr. Bernard. IBM Corporation, Thomas J. Watson Research Center, Yorktown Heights, New York. ELECTRONICS.

Schwarz, Prof. John H. California Institute of Technology, Charles E. Lauritsen Laboratory of High Energy Physics, Pasadena. ELEMENTARY PARTICLE—in part.

Shapira, Dr. Dan. Oak Ridge National Laboratory, operated by Martin Marietta Energy Systems, Inc., Oak Ridge, Tennessee. NUCLEAR PHYSICS—in part.

Sherrington, Prof. David. Imperial College of Science and Technology, Department of Physics, Blackett Laboratory, London, England, and Statistical Physics Group, Schlumberger-Doll Research, Ridgefield, Connecticut. SPIN GLASS.

Shipley, Dr. Thomas H. University of Texas, Institute for Geophysics, Austin. MARINE SEDIMENTS—coauthored.

Sinclair, Prof. Robert. Department of Materials Science and Engineering, Stanford University. ELECTRON MICROSCOPE.

Slemmons, Prof. David B. Professor of Geology, Department of Geological Sciences, Mackay School of Mines, University of Nevada, Reno, and Consulting Geologist. FAULT AND FAULT STRUCTURES.

Slovic, Dr. Paul. Research Associate, Decision Research, Perceptronics, Eugene, Oregon. INFORMATION PROCESSING (PSYCHOLOGY).

Spear, Dr. Frank S. Associate Professor of Geology, Rensselaer Polytechnic Institute, Department of Geology. METAMORPHISM.

Stamatoff, Dr. James B. Celanese Research Company, Celanese Corporation, Summit, New Jersey. LIQUID CRYSTALS—coauthored.

Steinhardt, Prof. Paul J. University of Pennsylvania, Department of Physics. QUASICRYSTALS.

Stephenson, Dr. Michael D. Business Manager, Electronics, Chubb Fire Security Limited, Electronics Division, Bognor Regis, West Sussex, England. FIRE DETECTION.

Stiegler, Dr. J. O. Division Director, Metals and Ceramics, Oak Ridge National Laboratory, operated by Martin Marietta Energy Systems, Inc., Oak Ridge, Tennessee. ALLOY.

Strayer, Dr. Michael R. Physics Division, Oak Ridge National Laboratory, Oak Ridge, Tennessee. NUCLEAR PHYSICS—in part.

Strom, Dr. Stephen E. University of Massachusetts at Amherst, Department of Physics and Astronomy, Astronomy Program. INFRARED ASTRONOMY.

Stuart, Dr. Kenneth. Issaquah Health Research Institute, Issaquah, Washington. MEDICAL PARASITOLOGY—coauthored.

Sumner, Prof. Malcolm E. University of Georgia College of Agriculture, Agronomy Division, Department of Agronomy. SOIL CHEMISTRY—in part.

Sweatt, Dr. Andrew J. Wake Forest University, Bowman Gray School of Medicine, Department of Anatomy, Winston-Salem, North Carolina. CHAETOGNATHA.

T

Tachibana, Dr. Kazuo. Suntory Institute for Bioorganic Research, Osaka, Japan. CHEMICAL ECOLOGY.

Tang, Dr. Man-Chung. President, DRC Consultants, Inc., Flushing, New York. BRIDGE—in part.

Taylor, Dr. A. C. Department of Zoology, University of Glasgow, Scotland. RESPIRATORY SYSTEM (INVERTEBRATE).

Taylor, Dr. Jane R. Yale University, Neurobehavioral Laboratory, School of Medicine. ANXIETY STATES—coauthored.

Terhorst, Dr. Cox P. Dana-Farber Cancer Institute, Boston. CELLULAR IMMUNOLOGY—in part.

Terzian, Prof. Yervant. Chairman, Department of Astronomy, Cornell University. COSMOLOGY.

Todd, Dr. Stanley G. Manville Exploration, Billings, Montana. ORE AND MINERAL DEPOSITS.

Troesch, Dr. Armin W. Professor and Director, Ship Hydrodynamics Laboratory, University of Michigan, Department of Naval Architecture and Marine Engineering. MARINE ENGINEERING.

Trost, Dr. Barry M. Vilas Research and Helfaer Professor of Chemistry, Department of Chemistry, University of Wisconsin—Madison. ORGANIC CHEMICAL SYNTHESIS.

Twin, Prof. P. J. Daresbury Laboratory, Science and Engineering Research Council, Warrington, England. PARTICLE DETECTOR—in part.

U

Ulanowicz, Prof. Robert E. University of Maryland, Center for Environmental and Estuarine Studies, Chesapeake Biological Laboratory, Solomons, Maryland. THEORETICAL ECOLOGY.

V

van Loon, Dr. L. C. Department of Plant Physiology of the Agricultural University, Wageningen, Netherlands. PLANT PATHOLOGY.

Van Thiel, Dr. David H. Professor of Medicine, Chief of Gastroenterology, University of Pittsburgh, School of Medicine. ALCOHOLISM.

Van Wambeke, Prof. Armand. Professor of Soil Science, New York State College of Agriculture and Life Sciences, Cornell University, Department of Agronomy. SOIL—in part.

Vincent, Dr. G. Michael. Professor of Medicine, University of Utah, and Chairman, Department of Medicine, LDS Hospital, Salt Lake City. LASER PHOTOBIOLOGY—in part.

W

Wallich, Paul. Associate Editor, "IEEE Spectrum," New York. OFFICE AUTOMATION—feature.

Ware, Dr. Randolph. University of Colorado at Boulder, National Oceanic and Atmospheric Administration, Cooperative Institute for Research in Environmental Sciences. SATELLITE NAVIGATION SYSTEMS.

Waters, Dr. Robert Thomas. U.S. Navy, Washington, D.C. SWATH SHIP.

Weldon, Dr. William F. Technical Director, University of Texas, Austin, Bureau of Engineering Research, Center for Electromechanics. ELECTRICAL POWER ENGINEERING.

Wibbens, Russell P. Executive Vice President, American Institute of Timber Construction, Englewood, Colorado. BRIDGE—in part.

Wilchek, Dr. Meir. Department of Biophysics, Weizmann Institute of Science, Rehovot, Israel. CHROMATOGRAPHY.

Wilding, Prof. Larry P. Texas A&M University, Department of Soil and Crop Sciences, College of Agriculture, Texas Agricultural Experiment Station. SOIL—in part.

Willey, Dr. Ruth L. University of Illinois at Chicago, Department of Biological Sciences. EUGLENIDA.

Williams, Dr. DeWayne. Soil Scientist, USDA, Soil Conservation Service, South National Technical Center, Fort Worth, Texas. SOIL POTENTIALS—in part.

Williams, Dr. Gareth. Medical Research Council, Laboratory of Molecular Biology, Cambridge, England. Monoclonal antibodies—coauthored.

Wilson, Dr. John T. Research Microbiologist, Subsurface Processes Branch, U.S. Environmental Protection Agency, Robert S. Kerr Environmental Research Laboratory, Ada, Oklahoma. Microbial geochemistry—in part.

Wolff, Prof. Peter A. Professor of Physics, and Director, Francis Bitter National Magnet Laboratory, Massachusetts Institute of Technology. Magnetic field—coauthored.

Wood, Dr. William I. Senior Scientist, Genentech, Inc., South San Francisco. Genetic engineering.

Yannas, Prof. Ioannis. Professor of Polymer Science and Engineering, Massachusetts Institute of Technology, Department of Mechanical Engineering. Skin.

Yule, Dr. A. B. University College of North Wales, Department of Marine Biology, Marine Science Laboratories, Gwynedd, Anglesey. Barnacle.

McGRAW-HILL YEARBOOK OF SCIENCE AND TECHNOLOGY

Index

Index

Asterisks indicate page references to article titles.